DISCARDED

D0078873

The Atmospheric Environment

The Atmospheric Environment
Effects of Human Activity

Michael B. McElroy

Princeton University Press
Princeton and Oxford

Published by Princeton University Press, 41 William Street, Princeton,
New Jersey 08540

In the United Kingdom: Princeton University Press, 3 Market Place,
Woodstock, Oxfordshire OX20 1SY

Library of Congress Cataloging-in-Publication Data

McElroy, Michael B.
The atmospheric environment : effects of human activity / Michael B. McElroy.
p. cm.
Includes bibliographical references and index.
ISBN 0-691-00691-1 (acid-free paper)
1. Atmosphere. 2. Nature--Effect of human beings on. I. Title.

QC861.3 .M34 2002
551.51--dc21 2001032106

British Library Cataloging-in-Publication Data is available

This book has been composed in 10/12 Berkeley

Printed on acid-free paper. ∞

www.pup.princeton.edu

Printed in the United States of America

10 9 8 7 6 5 4 3 2 1

For
Stephen, Brenda, Lisa, Zoe, and Abby,
and the future they represent

Brief Contents

Detailed Contents

Chapter 4: Properties of Water

Chapter 5: Atmospheric History and Composition

Chapter 6: Energy Budget of the Atmosphere

Chapter 7: Vertical Structure of the Atmosphere

Chapter 8: Horizontal Motion of the Atmosphere

Chapter 9: Dynamics of the Ocean

Chapter 10: Climate through the Ages

Chapter 11: The Carbon Cycle

Chapter 16: Stratospheric Ozone (IV): Influence of Dynamics

Chapter 17: The Chemistry of the Troposphere

Chapter 18: The Chemistry of Precipitation

Chapter 19: Prospects for Climate Change

Preface

Atmospheric science has enjoyed a remarkable period of growth over the past quarter-century. New analytical techniques, and the opportunity to make measurements not only at the surface but also from balloons, aircraft, rockets, and orbiting satellites, have provided a data base allowing us to define the composition of the atmosphere to a precision and accuracy scarcely imaginable 25 years ago. Measurements of key reactions in the laboratory, combined with carefully crafted field studies, have provided a framework for the interpretation of these data, leading to unprecedented new insights into the nature of the atmosphere. We have come to see the atmosphere, in both its physical and chemical dimensions, as an indispensable component of the support system for life on Earth. Belatedly, we have come to realize that we have the capacity to change the composition of the atmosphere, not just locally but globally, with uncertain but potentially serious consequences not only for ourselves but for much of the rest of life on our planet. For too long, we have used the atmosphere as a dump for waste. We have an obligation now to assess the consequences of our actions and adjust our practices accordingly.

Who would have thought, when an inventive chemist at General Motors more than 50 years ago found a means to combine carbon, chlorine, and fluorine atoms to create an entirely new class of chemical compounds, that the products of his ingenuity would have the potential to alter the flux of biologically harmful ultraviolet radiation reaching Earth's surface? Should we have anticipated that release of these otherwise benign compounds into the atmosphere could punch a hole in Earth's protective ozone shield that would first appear, of all places, over Antarctica? When we launched the industrial revolution fueled by coal 250 years ago, should we have forecast that accumulation in the atmosphere of carbon dioxide, the end-product of the combustion, would have the potential to alter global climate? Should we have anticipated that nitrogen- and sulfur oxides, by-products of combustion, could contribute to the phenomenon of acid rain? Should we have foreseen that nitrogen oxides in combination with hydrocarbons of both natural and anthropogenic origin could lead to the production of ozone concentrations in surface air high enough to pose a problem not only for human health but also for vegetation? Should we have anticipated that disposal of human and animal waste and the application of nitrogen-based fertilizers to agricultural crops could contribute to a buildup of nitrous oxide in the atmosphere or that domesticated ruminants and rice cultivation could be responsible for an increase in the abundance of methane? The fact is we failed to foresee any of these consequences. We are in the business now of playing catch-up, fine-tuning our actions to minimize potential future damage. The message from the past is that it is easier to innovate technologically than it is to fully assess in advance the consequences of our technology.

The issues addressed in this text are all at some level interrelated. We cannot expect to understand the stratosphere without understanding the processes responsible for changes in the abundances of carbon dioxide, nitrous oxide, and methane. A change in climate has implications for the dynamics of the stratosphere and ultimately for the abundance and distribution of ozone in the stratosphere. Warming of the lower atmosphere can trigger cooling of the stratosphere, exacerbating conditions responsible for the loss of ozone in polar regions. In turn, changes in the stratosphere can alter the chemistry of the lower atmosphere, with implications for both the composition of the atmosphere and the climate.

In a real sense, the atmosphere should be considered an extension of the ocean and biosphere; on longer time scales, it is inextricably linked not just to the ocean and biosphere but also to the solid planet. Mining of fossil fuels should be viewed as contributing to an acceleration of the global forces responsible for apportionment of carbon between sediments, on the one hand, and the atmosphere, ocean, and biosphere, on the other. Acid rain should be seen as a response of the biota to a disturbance of the biogeochemical cycles involved in the mobilization and distribution of sulfur and nitrogen. By placing the present environmental disturbances in a larger geologic perspective, the hope is that we may develop a more informed appraisal of their significance.

This book was written with multiple constituencies in mind. My hope is that it can provide a useful introduction to environmental issues for a general science audience. Over the past 25 years I have used some of the material to teach a course on the atmosphere directed at nonscientists as part of the core curriculum at Harvard University. Instituted in 1979, the core is designed to introduce undergraduates to "approaches to knowledge in seven areas considered indispensable to the contemporary student: foreign cultures, historical study, literature and the arts, moral reasoning, quantitative reasoning, science and social analysis." It is assumed that the clientele for a science core course should have no special training in either science or mathematics other than that available in the curriculum of a typical American high school. Our objective was to develop an account of how the atmosphere works as a physical and chemical system and to introduce relevant scientific concepts without resorting to the use of mathematics beyond the level of algebra. The material should be presented not as a body of facts but in a fashion communicating the quantitative dimensions of the subject. Over the course of a semester, we covered most of the material in Chapters 1–8 and selected parts of Chapters 11–18.

I have also used the material to teach a course directed at students concentrating (majoring) in Environmental Science and Public Policy (ESPP) at Harvard: the audience in this case is scientifically more literate. Students in ESPP are required to take college-level courses in biology, chemistry, and mathematics (in addition to economics and government). The curriculum is essentially equivalent to the program followed by students aiming to pursue postgraduate education in medicine (premeds). For the ESPP audience, we used much of the same material as for the core, downplaying the elementary material in Chapters 2–4 while adding portions of Chapters 9, 10, 19, and 20. We were able to proceed at a faster pace. Even for this audience, though, we found it useful to review parts of the material in Chapters 2 and 3 (notably the discussions of angular momentum, circular motion, and the distinction between equilibrium and kinetic descriptions of chemically reacting systems).

I attempted to provide an up-to-date account of contemporary studies of climate (Chapter 10) and stratospheric and tropospheric chemistry (Chapters 13–18) with references to contemporary literature. My hope is that the material in this part of the text can serve a useful function for beginning graduate students. For the most part, I have relied on professional peer-reviewed literature, but in some cases, notably in parts of the climate chapter, I have attempted to provide a synthesis that might have been first aired more appropriately for a professional audience in a peer-reviewed journal. I did so because I thought it important to offer at least some conjecture of how the disparate new sources of climate data might fit together and because I was anxious not to further postpone the long, lonely labor involved in bringing this project to a timely end. The material in question is explicitly identified in the text.

For my early introduction to the challenges and excitement of atmospheric science, I am indebted to the late D. R. Bates and to J. W. Chamberlain, A. Dalgarno, T. M. Donahue, D. M. Hunten, and R. M. Goody. I am grateful for their friendship, tutelage and unselfish support at formative stages of my career. My thanks to W. S. Broecker, from whom I learned much of what I know about the ocean and paleoclimate. At Harvard, I have been privileged over the past 30 years to work with a large number of talented graduate students, research associates, and faculty colleagues. I acknowledge my debt to J. G. Anderson, J. W. Elkins, B. F. Farrell, P. F. Hoffman, D. J. Jacob, D. B. A. Jones, F. Knox-Ennever, W. M. Hao, P. F. Hoffman, J. A. Logan, B. D. Marino, K. Minschwaner, M. J. Prather, R. J. Salawitch, D. Schrag, H. R. Schneider, C. M. Spivakovsky, N. D. Sze, J. S. Wang, S. C. Wofsy, J. H. Yatteau and Y. L. Yung, with a special thanks to Cecilia McCormack, who, for so many years, has kept our research group at Harvard organized, solvent, and cheerful. I am grateful to R. J. Salawitch, my friend, companion, and critic on many late evenings, who generously provided many of the illustrations used in the stratospheric portion of the book, to J. W. Elkins, responsible for many of the Figures used in the chapter on composition, and to R. A. Reck and J. S. Wang, who read earlier drafts of the manuscript and made many valuable suggestions for its improvement. I could not have brought the project to completion without the help of Andrew Abban, Ryan Lim, Arnico Panday, and Steve Won. A special thanks to Arnico and Ryan who brought order to chaos at critical stages and made numerous valuable suggestions to improve the overall presentation of the material. Finally, my thanks to the government agencies that sponsored my work over the years at Harvard: the Department of Energy, the National Aeronautics and Space Administration, and the National Science Foundation.

Introduction

This book concerns the complex interdependent system that regulates the environment for life on Earth. Our objective is to provide a comprehensive account of the science underlying a number of environmental issues that have captured the attention of the public worldwide over the past several decades. These issues include regional air pollution, acid rain, depletion of stratospheric ozone (O_3), destruction of tropical forests, and, most perplexing of all, the problem of climate change—the possibility of a change in climate induced by increasing concentrations of greenhouse gases such as carbon dioxide (CO_2), methane (CH_4), nitrous oxide (N_2O), and industrial chlorofluorocarbons (CFCs).

Our emphasis is on the atmosphere. The issues we address range in scale from global to local. They involve, to a large extent, problems associated with the use of fossil fuels: coal, oil, and natural gas.

Our modern industrial economy depends on the availability of these fuels. They provide the electricity we use to light our homes and workplaces, the energy consumed to heat and cool the environments in which we spend a large part of our lives, the fuel we use to drive our cars, trucks, trains, ships, and airplanes, the fertilizers and mechanization that permit us to produce supplies of food sufficient to satisfy the needs of an ever increasing population, and the goods and services we take for granted as essential ingredients of modern life.

The industrial revolution was fueled initially by coal, replacing diminishing reserves of wood. The invention of the heat pump, the internal combustion engine, and the means to generate and distribute electricity changed the way we lived and spawned a myriad of technologies designed to improve the quality of our lives. But there were costs, often unanticipated. Even during the early stages of the industrial revolution, air quality problems associated with fossil fuel use were apparent in England and in the United States. Later, we were forced to confront directly the problem of dirty, smoke-laden air when thousands of people died, and even more became seriously ill, in a series of air pollution disasters in Donora, Pennsylvania, and in London, England, in the late 1940s and early 1950s. It was relatively easy to deal with the problem of sooty, dirty air. The solution was to burn cleaner fuels, to remove particles from smokestacks, or to build higher stacks and send the problem somewhere else. But the smogs of Donora and London were merely harbingers of more serious problems to come: acid rain, photochemical smog, and now, the most serious challenge of all, the threat of global climate change.

There is a troubling pattern to the history of the environmental problems caused by our use of fossil fuels and the responses we have selected to deal with these problems. The problems have tended to surface *after* we introduced the technologies that ultimately were the source of the problems. When we installed high smokestacks to disperse the emissions from coal- and oil-fired power plants, we were unaware of the phenomenon of acid rain. When we began our love affair with the automobile, we did not suspect that the interaction of sunlight with hydrocarbons and oxides of nitrogen could create an environment for production of ozone at levels harmful to public health and to the photosynthetic capacities of plants. The capital investments in roads, cities, and industries depending on fossil fuels were so large that it was easier to look for piecemeal solutions and to avoid potentially more effective and more comprehensive strategies for dealing with the underlying cause of the problems. It was easier to search for technological fixes to specific problems rather than seek alternate, more environmentally acceptable solutions to our energy needs. The challenge defined by the threat to the global climate system posed by emissions of large and growing concentrations of carbon dioxide makes it all the more urgent that we seek a comprehensive solution. If we fail to meet this challenge, we run the risk that the global life support system we take for granted may fail and our civilization may suffer irreparable damage.

It is clear that the composition of the atmosphere is changing today at a rate unprecedented in the recent history of Earth. The abundance of CO_2 has increased during the past century by almost 25% and is expected to rise a further 50% over the next several decades. The level of CH_4 has grown by more than a factor of two. Concentrations of CO_2 and CH_4 are higher today than they have been at any other time in at least the past 450,000 years. Production of N_2O on a global basis exceeds consumption by almost 40%. Concentrations of CFCs have been increasing at a rapid rate since the atmosphere was first exposed to these compounds approximately 50 years ago. These changes are human-induced (anthropogenic), a by-product of our way of life. In an extremely short span of time, human beings have changed the composition of Earth's atmosphere.

1.1 Problems of Global Scale: The Ozone Layer and Climate Change

Ozone, present mainly in the stratosphere, provides a shield protecting the surface of Earth from otherwise lethal doses of ultraviolet solar radiation. The ozone layer is especially effective at blocking the solar radiation that would be the most damaging to DNA. As a result, small changes in O_3 can result in significant increases in the incidence of human skin cancer and can have similar effects on other living organisms. Epidemiological studies suggest that the fractional increase in the number of new cases of skin cancer is approximately two to three times larger than the fractional decrease in O_3. Decreases of several percent in O_3 have been observed over large regions of Earth in recent years. Particularly dramatic reductions, as much as 60%, are detected over Antarctica during spring, with bothersome signs that the phenomenon observed in the Antarctic may be spreading to the Arctic.

The issues posed by rising levels of CFCs, which are involved both in the destruction of stratospheric O_3 and in climate change, are in some respects the simplest to deal with from the public-policy standpoint. These gases are produced exclusively by chemical industry. They decompose in the stratosphere, where their constituent chlorine atoms are involved in relatively well-understood reactions contributing to accelerated loss of O_3. CFCs survive in the atmosphere for long periods of time; they can last in the atmosphere for centuries. Consequently, effects on O_3 are persistent. We know, thanks to a well-organized research program involving both government and industry, that the reductions in O_3, especially those observed in polar regions, are due largely to effects of industrial CFCs and related compounds containing bromine.

Governments responded to the threat to stratospheric O_3 with a comprehensive strategy designed to minimize the impact of industrial chemicals on stratospheric O_3. The strategy involved reducing emissions of the more hazardous gases and replacing them with compounds with a less serious or less persistent impact on the environment. This approach was outlined in a remarkable document known as the Montreal Protocol, developed at a meeting of governments in Montreal in September 1987. The large reduction in O_3 over Antarctica and the threat of a similar phenomenon over the Arctic accelerated demands for action to protect the integrity of Earth's protective ozone shield, while increasing the level of public awareness of the fragility of the global environment.

Unlike the relatively straightforward case of CFCs and stratospheric ozone depletion, the issue of climate change is inherently complex. It is more difficult to deal with the problems posed by rising levels of CO_2, CH_4, and N_2O, which are present as natural components of the atmosphere. Increases in the concentrations of these gases are expected to contribute to global warming. Our understanding of the processes affecting the natural distribution of these gases is still incomplete, and debate over the potential effects of greenhouse gases is fierce. We know that burning of fossil fuels (coal, oil, and natural gas) provides a large source of CO_2 to the atmosphere, more than six billion tons of carbon per year at present (more than 20 billion tons of CO_2). Additional CO_2 is produced by deforestation and development, which are thought to contribute as much as two billion tons of carbon per year. These activities represent the major anthropogenic sources of CO_2. On the other side of the ledger, CO_2 can be taken up by trees, stored in soils, and incorporated in the ocean. These influences are responsible for the major sinks of CO_2. A complete understanding of atmospheric CO_2 must account for these disparate sources and sinks. A policy designed to reduce the rate of increase of CO_2 in the atmosphere can take multiple directions. It could seek to reduce the rate at which the gas is introduced into the air by burning of fossil fuels or by deforestation; this would involve a plan focused on strategies to reduce the sources of CO_2. Alternatively, it could focus on sinks of CO_2 by seeking to enhance the rate at which the gas is taken up by vegetation and soils, encouraging reforestation and soil conservation, or devising means to accelerate uptake of carbon by the ocean. Crafting of a responsive policy to reduce CO_2 requires a comprehensive understanding of the atmosphere, biosphere, soil, and ocean.

While impressive progress has been made over the past several years to develop the requisite understanding of the global carbon cycle, it is clear that our knowledge of this complex topic is still rudimentary. Our understanding of the budgets of CH_4 and N_2O is even more primitive. Moreover, there are significant uncertainties in our ability to predict the response of climate to an increase in greenhouse gases. Under the circumstances, one can appreciate the reluctance of governments to take action on the threat of climate change. There are two extreme responses: one, promoted by several key players, argues for a vigorous research program, maintaining that the danger posed by global warming lies in the future, whereas the risk of economic hardships associated with premature action is imminent; the other calls for a more immediate response, emphasizing the potentially devastating effects on the biosphere of a possible abrupt shift in global climate. The dilemma is nicely defined in an anecdote related by the former British prime minister, John Major, in a speech delivered at the *Sunday Times* International Conference on the Environment in London on 8 July 1991:

> There is a Victorian picture which many of you may know. It is called: "the last day in the old home". The ancestral home has been sold up to pay for the father's debt. The removal men are in the hall. The feckless father is toasting the future while his wife weeps. His small son and heir stands at his knees.

Major concluded, "I do not want to be in the position of that father, squandering my children's global home and inheritance for my own immediate wants." Elsewhere, he noted that "the threat of global warming is real; the spread of deserts, changed weather patterns with potentially more storms and hurricanes, perhaps more flooding of low-lying areas, and possibly even the disappearance of some island states." More recently, Vice President Gore announced to world leaders at the Kyoto Conference, "Whether we recognize it or not, we are now engaged in an epic battle to right the balance of our Earth, and the tide of this battle will turn when the majority of people in the world become sufficiently aroused by shared sense of urgent danger to join an all-out effort."

Religious leaders also attest to the gravity of the crisis. Pope John Paul II, in his message delivered on 1 January 1990 in celebration of the World Day of Peace, noted that

> the gradual depletion of the ozone layer and the related greenhouse effect has now reached crisis proportions as a consequence of industrial growth, massive urban concentrations and vastly increased energy needs. Industrial waste, the burning of fossil fuels, unrestricted deforestation, the use of certain types of herbicides, coolants and propellants: all of these are known to harm the atmosphere and the environment. The resulting meteorological and atmospheric changes range from damage to health to the possible submersion of low-lying lands.

Elsewhere in the same speech he noted that

> theology, philosophy and science all speak of a harmonious universe, of a cosmos endowed with its own integrity, its own internal, dynamic nature. This order must be respected. The human race is called to explore this order, to examine it with due care and to make use of it while safeguarding its integrity.

In the final analysis, the problems of global environmental change are linked with the rapid increases in human population and with aspirations of peoples around the world for improved standards of living. The relative affluence of the developed world is due in no small measure to its profligate use of energy. Industrial society has an insatiable appetite for fossil fuels. Five percent of the world's population, those who live in North America, account for almost 25% of the global emission of CO_2, and the performance of Europe and Japan is only marginally better. Can developing societies such as China and India be expected to forgo the advantages already reaped by the industrialized world due to its exploitation of fossil fuels? Are global environmental issues of sufficient urgency to affect strategies for industrial development? Combustion of fossil fuels has a deleterious impact on not only the global but also the local and regional levels. Is it possible, and perhaps even reasonable, that local and regional issues should have a larger influence than global concerns in the formulation of policy for the developing world?

1.2 Problems of Regional and Local Scale

There are three important problems on the local and regional scales related to the use of fossil fuels. Emission of particulate material associated with the combustion of coal results in a variety of respiratory problems. It was responsible for the devastation wrought by the killer smog that visited a number of major industrial cities during the 1940s and 1950s, notably in London in 1952. That particular problem was dealt with by restricting use of coal to major power plants and by installing devices to remove particulate material before it was released to the air. A second problem of local air pollution concerns the production of elevated concentrations of O_3 near the ground, and it has proven to be a more difficult problem. High levels of O_3 cause damage to plants and are responsible for a variety of pulmonary problems in human beings. The combination of nitric oxide (NO), hydrocarbons, and sunlight provides an ideal medium for the synthesis of O_3. Nitric oxide is emitted by automobiles and by power plants, while hydrocarbons are produced not only by automobiles and a variety of industrial sources but also by vegetation. It has proven exceptionally difficult to control emissions of these compounds, due to the multiplicity of potential sources.

The third problem on the local and regional scales is also associated with emission of NO and is compounded by the release of sulfur dioxide (SO_2), formed as a product of the combustion of coal and oil. Nitric oxide is converted in the atmosphere to nitric acid (HNO_3), while SO_2 is transformed to sulfuric acid (H_2SO_4). These compounds are transported over distances as large as thousands of kilometers before they are removed either by precipitation, as rain and snow, or, directly from the gas phase, by contact with vegetation. Returning to Earth, they may be responsible for acidification of soils, rivers, and lakes. Acidification of soils can lead to release of toxic elements such as aluminum, resulting in serious damage to plants and various forms of aquatic life.

It is clear that significant costs are associated with the adverse environmental impacts of fossil fuels on the local and regional levels. Is it possible that the expense associated with abatement of these problems could be sufficient to justify an alternate pattern of development? Is it possible that developing economies might judge the benefits of fossil fuels insufficient to justify their costs? The answer to these questions increasingly will depend on the availability and cost of alternatives in addressing the fundamental needs of developing societies and the provision of services taken for granted in modern industrial economies. We might expect to look to the first world for creative solutions, for imaginative new sources of energy, and for environmentally compatible strategies for development. Yet the first world is a product of the industrial revolution and has little incentive to abandon its addiction to fossil fuels. The problem is that energy supplied by fossil fuels is cheap, particularly when the costs of associated environmental damage are ignored. Creative solutions for the environmental challenges of the developing world may require more realistic pricing of fossil fuels, creating incentives for the introduction of environmentally benign alternatives.

1.3 Roadmap of the Book

Our discussion of the atmosphere is preceded by a review of basic physical and chemical principles in Chapters 2 and 3. Physical processes of special relevance for an understanding of climate provide the focus for the following nine chapters. Properties of water, in many respects the most important constituent of the atmosphere, are treated in Chapter 4. Current ideas on the origin and evolution of the atmosphere and the chemical, physical, and biological processes affecting its composition are discussed in Chapter 5. The energetics of the atmosphere are discussed in Chapter 6, which also introduces a simple model for the greenhouse effect implicated in the issue of global warming. The physical processes responsible for the variation of pressure, density, and temperature with altitude are treated in Chapter 7. Chapter 8 provides an introduction to the dynamics of the atmosphere, outlining ideas concerning the nature of global circulation and the factors responsible for the variation of climate with latitude and season. The structure and dynamics of the ocean and its role in climate are treated in Chapter 9. Perspectives on the range of climates that visited Earth in the past are presented in Chapter 10.

Chapter 11 begins the more chemistry-oriented component of the text with a focus on the carbon cycle and the processes that affect the abundance of carbon dioxide. The cycles of nitrogen, phosphorus, and sulfur, essential nutrients for life, are treated in Chapter 12. In combination, Chapters 11 and 12 serve to emphasize the significance of biologically mediated processes in regulating the composition of the atmosphere, with a particular emphasis on CO_2 and N_2O. They provide a framework with which to assess the impact of human activity on the biosphere, emphasizing the importance of interactions coupling the atmosphere inextricably to the ocean, the solid planet, and life.

Of the following six chapters, photochemical processes—the response of the atmosphere to incident sunlight—provide the focus for five. The chemistry of stratospheric O_3 is treated in a series of three chapters beginning with Chapter 13. The relatively simple chemistry of a pure oxygen atmosphere is introduced in Chapter 13. The influence of free radicals—reactive compounds formed from hydrogen, nitrogen, chlorine, and bromine—is the subject of Chapter 14. The role of reactions taking place on particle surfaces, key to understanding the large-scale losses of ozone observed in recent years at high latitudes, is treated in Chapter 15. An integrated approach to the stratosphere, accounting both for chemistry and dynamics, is presented in Chapter 16. The photochemistry of the lowest region of the atmosphere, the troposphere, is the subject of Chapter 17. Chapter 18 completes this sequence with a discussion of the chemistry of precipitation and the issue of acid rain.

The final chapters of the text are concerned specifically with the issue of climate change. Chapter 19 describes the challenge involved in efforts to model the impact of projected future emissions of greenhouse gases on climate. The potential significance of the environmental change induced by these emissions is illustrated using results from a coupled ocean-atmosphere climate model developed by the Hadley Centre at the U.K. Meteorological Office. We conclude in Chapter 20 with a summary of ongoing efforts by the international community to confront the threat to the climate system defined by the contemporary buildup of greenhouse gases.

Basic Physical Concepts

2

Science is concerned with measurement and prediction of the behavior of natural systems. In this chapter, we introduce a number of basic physical quantities:

1. A set of consistent units (the *cgs* system) for measurement of **length**, **time**, and **mass**
2. The **vector**, exemplified particularly by the concepts of velocity and acceleration
3. The notion of **force**, **work**, and **energy** (both potential and kinetic)
4. **Temperature**, as a measure of the kinetic energy of molecules in a gas
5. **Pressure**, as a measure of force per unit area
6. The relation between pressure, temperature, and density expressed in the **perfect gas law**
7. The concept of **angular momentum**

The objective is for readers to develop familiarity with these ideas and to become comfortable carrying out simple calculations making use of these concepts.

2.1 Length, Time, and Mass

Basic to the concept of measurement is the need for a system of units. The fundamental units used by physicists are based on ideas we use in everyday life. We could not survive without a means to measure **distance** and **time.** In what follows we will use the centimeter (cm) as our unit of distance and the second (sec) as our unit of time.

There is a third quantity as essential as length and time: **mass.** Mass is a measure of the amount of matter in an object. This definition is more subtle than those for length and time. We are familiar with the notion of weight; weight is the force that mass exerts due to gravity. When we say that an object is heavy or light, we are describing its weight. One way to find the weight of an object is to place it on a scale and measure the scale's displacement due to the pull of gravity on the object. If the same measurement were carried out on the moon, where the force of gravity is less than that of Earth, we would obtain a different reading. Scientists need a more absolute measure, one that does not vary from place to place. We refer to this quantity as mass, and in the *cgs* system it is measured in grams.

Mass is an intrinsic property of an object. For a chemist, the reference standard is the mass of a hydrogen atom, 1.67×10^{-24} grams. If we know what an object is made of (its atomic composition), we can determine its mass. A hydrogen atom has a mass of 1.67×10^{-24} grams whether the atom is on Earth, on Mars, or in the interior of the Sun. A cubic centimeter of liquid water contains 3.3×10^{22} molecules of H_2O. The mass of an oxygen atom is 16 times that of a hydrogen atom. Each H_2O molecule, composed of two hydrogen atoms and one oxygen atom, has a mass of 3.0×10^{-23} grams. The cubic centimeter of liquid H_2O has a mass of 1 gram. How did we get those numbers? Through simple addition and multiplication:

$$\text{mass of one molecule of } H_2O = 2\,(\text{mass of H atom}) + (\text{mass of O atom})$$

$$\text{mass of 1 cm}^3 \text{ of } H_2O = (\text{mass of one } H_2O \text{ molecule})\,(3.3 \times 10^{22}\ H_2O \text{ molecules})$$

We refer to the mass contained in a cubic centimeter, our standard unit of volume, as the **mass density**, denoted by ρ. Liquid H_2O has a mass density of 1 gram per cubic centimeter, written as 1 g cm^{-3}. In practice, our scales are calibrated to allow for the fact that they are employed normally to weigh objects at sea level on Earth. It is good to remember that the scale records weight and that there is an important distinction between weight and mass. Only the latter is an intrinsic property of an object.

2.2 Speed, Velocity, and Acceleration

Given the fundamental units for length, mass, and time, taken for present purposes as centimeters, grams, and seconds (we refer to the corresponding system of units as the *cgs* system), we can now introduce other quantities, some familiar, others less so. **Speed** measures distance traveled per unit of time. We can calculate the average speed of an object by recording the time, t, required to move a distance d. The average speed, v, is given by $v = d/t$, and, with our choice of units for length and time, v has units of cm per sec, written as cm sec^{-1}. Speed is an example of a **scalar quantity.** It is completely specified by a single number, with appropriate units.

There are other quantities that cannot be specified solely by a number, quantities that require information on direction. Imagine an aircraft taking off from Boston. Suppose that, an hour later, the plane has traveled a distance of 300 miles. If you wanted to locate the plane, you would need more information. If all you knew was its distance from Boston, you might draw a circle of radius 300 miles centered on Boston; the plane could be anywhere on the circumference of this circle. To locate it precisely, you would need to know the direction in which it had traveled. If you knew that the plane had flown due north, that would be enough to pin it down exactly (setting aside the question of its altitude). Quantities that require information on direction as well as magnitude are known as **vectors.**

An important quantity related to speed is **velocity.** Velocity is a vector quantity. The magnitude of the velocity vector is given by the speed. The direction of motion is given by the direction of the velocity vector, $\underline{v}$. (It is customary to indicate that a specified quantity is a vector by underlining the symbol denoting it. You may have seen different conventions to indicate vectors in other texts, such as boldface or inclusion of an arrow above the symbol.)

Acceleration measures the rate of change of velocity, the change in velocity per unit of time. It has dimensions, with our choice of units, of cm per sec per sec, written as cm sec^{-2}. Like velocity, acceleration is a vector quantity: it measures the rate of change of speed along the direction of motion. A positive value for acceleration indicates that the object's speed is increasing along the direction of motion; a negative number indicates that the speed is decreasing—the object is slowing down.

To change the velocity of an object, that is, to impart an acceleration, we must apply an external agent, a **force.** Like velocity and acceleration, force is a vector quantity.

2.3 Force

Newton's law of motion (named for Sir Isaac Newton, 1642–1727) defines the relation between force, mass, and acceleration:

$$\underline{F} = m\underline{a}, \tag{2.1}$$

where $\underline{F}$ defines the magnitude and direction of the force, m is the mass of the object, and $\underline{a}$ is the associated acceleration. More precisely, Newton's law of motion states that force is equal to the rate of change of linear momentum, where **momentum** is a vector quantity given by the product of mass and velocity, $m\underline{v}$. Thus, keep in mind that equation (2.1) is appropriate only if mass is constant. A rocket, for example, accelerates by firing off mass and thus shedding associated momentum; the mass of the rocket plus its fuel decreases with time as the rocket accelerates, and the acceleration is directly related to the decrease in mass.

Football coaches are familiar with Newton's first law; they know that a 180-pound (8.2×10^4 g) running back can generate as much momentum as a 270-pound (1.2×10^5 g) lineman, so long as he can move 50% faster. Consider the dimensions of the quantities on the right-hand side of equation (2.1): if m is expressed in g and $\underline{a}$ in cm sec^{-2}, it follows that the product, $m\underline{a}$, and thus the force, should have dimensions of g cm sec^{-2}. Force is such a centrally important quantity that the unit of force merits a separate name. We refer to the unit of force in the *cgs* system as the **dyne (dyn).** A force of 1 dyn would cause the speed of a 1 g object to increase by 1 cm sec^{-1} in 1 second. You may be more familiar with the *mks* (**m**eters, **k**ilograms, **s**econds) system of units, in which the unit of force is known as the **newton (N).** This is simply force measured in units of meters, kilograms, and seconds. Table 2.1 summarizes the basic elements of the *cgs* and *mks* systems and the factors involved in converting from one system to the other.

Perhaps the most important force for present purposes is the force of gravity. We find empirically that objects of masses m_1 and m_2 are attracted to each other with a force F proportional to their masses. We find further that the force drops off as the square of the distance separating the objects

Table 2.1 Basic elements of the cgs and mks systems of units

	cgs *unit*	mks *unit*	*Conversion*
Length	centimeter	meter	1 m = 100 cm
Time	second	second	1 sec = 1 sec
Mass	gram	kilogram	1 kg = 1000g
Velocity	cm sec^{-1}	m sec^{-1}	1 m sec^{-1} = 100 cm sec^{-1}
Acceleration	cm sec^{-2}	m sec^{-2}	1 m sec^{-2} = 100 cm sec^{-2}
Momentum	g cm sec^{-1}	kg m sec^{-1}	1 kg m sec^{-1} = 10^5 g cm sec^{-1}
Force	dyne = g cm sec^{-2}	newton = kg m sec^{-2}	1 N = 10^5 dyn
Energy	erg = dyn cm = g cm^2 sec^{-2}	joule = N m = kg m^2 sec^{-2}	1 J = 10^7 erg
Work	erg	joule	1 J = 10^7 erg
Power	erg sec^{-1}	watt = J sec^{-1} = kg m^2 sec^{-3}	1W = 10^7 erg sec^{-1}

increases (i.e., it is inversely proportional to the square of the separation). This behavior may be described mathematically according to the expression

$$F = \frac{Gm_1m_2}{r^2}, \tag{2.2}$$

where G is a constant and r measures the separation of the objects. Here F denotes the magnitude of the force F. The relation expressed by (2.2) may be considered a law of nature; it was formulated by Sir Isaac Newton, the scientist responsible for much of our understanding of classical physics. The value of G, 6.67×10^{-8} dyn cm^2 g^{-2}, is determined experimentally. We refer to G as the gravitational constant. An object of mass m is attracted toward the center of Earth with a force of magnitude equal to

$$F = \frac{GmM}{R^2}, \tag{2.3}$$

where M is the mass of Earth (5.969×10^{27}g) and R is the separation of the object from the center of mass of Earth. Such a force would cause the object to accelerate toward the mass center of Earth. The magnitude of the acceleration, given by g, may be obtained using Newton's law of motion (be careful to note that g as used here and in subsequent equations refers to the acceleration of gravity, not to grams):

$$mg = \frac{GmM}{R^2}, \tag{2.4}$$

Solving for g, we find

$$g = \frac{GM}{R^2}, \tag{2.5}$$

Example 2.1: Given that the radius of Earth is 6372 kilometers (km), determine the gravitational acceleration (g) at sea level (altitude = 0 km) and at an altitude of 100 km (1 km is equivalent to 100,000 cm, written in scientific notation as 10^5 cm).

Answer: At sea level, R (radius of Earth) = 6372 km = 6.372×10^8 cm. Using this value in equation (2.5), we get

$$g = \frac{(6.67 \times 10^{-8}\text{dyn cm}^2\ \text{g}^{-2})(5.969 \times 10^{27}\text{g})}{(6.372 \times 10^8\text{cm})^2}$$
$$= 980.6 \text{ cm sec}^{-2}$$

At 100 km altitude, $R = 6472$ km $= 6.472 \times 10^8$ cm. Thus

$$g = \frac{(6.67 \times 10^{-8}\text{dyn cm}^2\ \text{g}^{-2})(5.969 \times 10^{27}\text{g})}{(6.472 \times 10^8\text{cm})^2}$$
$$= 950.5 \text{ cm sec}^{-2}$$ ■

Since the change of g corresponding to a 100 km change in altitude is small, and since most of the discussion in this book refers to the region of the atmosphere below 100 km altitude, we shall adopt everywhere, without loss of significant accuracy, the value of g calculated for the surface of Earth. The force of gravity acting on an object of mass m, directed toward the center of Earth, has magnitude mg.

2.4 Vectors

To describe the motion of a particle we need a coordinate system, a frame of reference. Suppose we are interested primarily in the motion of a particle in the horizontal plane. Given an origin, or reference point, two numbers are required to specify the particle's location. One number serves to describe the location of the particle north or south of the reference point; the other defines its location to the east or west. Positive numbers may be used to indicate distances to the north or east; negative numbers specify south or west. In Figure 2.1a, imagine that the origin of the coordinate system defines a particular location, Boston for example. Location A may be reached by traveling 100 km north, then 50 km east, or alternatively, using a boat or plane for at least the initial part of the journey, we could travel first 50 km east, then 100 km north. Location A may be represented by either the number pair (50, 100) or the vector $50\underline{i} + 100\underline{j}$. The first number in the number pair gives the displacement to the east; the second defines position to the north. The vectors $\underline{i}$ and $\underline{j}$ are known as unit vectors; they define unit displacements in the east and north directions, respectively. The quantity $50\underline{i} + 100\underline{j}$ is shorthand for "Move 50 kilometers east, then 100 kilometers north." Position B in Figure 2.1a may be specified by either (–50, –100) or $-50\underline{i} - 100\underline{j}$.

We can generalize the example above to include a third dimension, height. Vertical displacement can be indicated by adding a third number to the pair representing the two-dimensional vectors above. Similarly, vertical displacement

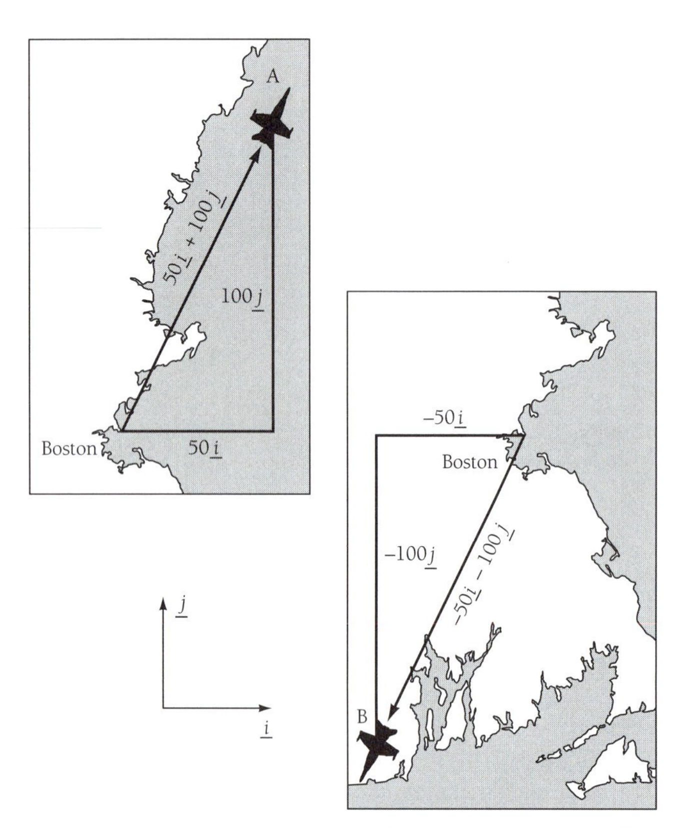

Figure 2.1a Vector displacement. Top: The position of a plane 50 km east and 100 km north of Boston is represented by the vector ($50\underline{i} + 100\underline{j}$). Bottom: A position 50 km west and 100 km south of Boston is represented by ($-50\underline{i} - 100\underline{j}$).

can be indicated as a multiple of a unit vector $\underline{k}$. If point A in Figure 2.1b indicates an aircraft at an altitude of 1 km above Earth, it can be represented by either (50, 100, 1) or $50\underline{i} + 100\underline{j} + 1\underline{k}$. An aircraft at the same altitude above point B would be defined by (–50, –100, 1) or $-50\underline{i} - 100\underline{j} + 1\underline{k}$.

Vector quantities such as velocity, acceleration, and force may be indicated in a manner similar to that used to describe the vector position of a particle. Suppose an airplane is moving at a constant speed v in a northeasterly direction, as indicated in Figure 2.2. The magnitude of the velocity, the speed, can be represented by the length of a directed arrow, as indicated. The direction of motion is given by the orientation of the arrow. The coordinate system is ruled off in units of speed (cm sec^{-1}) for the situation summarized in Figure 2.2. The airplane moving in the northeasterly direction from the origin, O, has a component of velocity in both the easterly and the northerly direction (it is moving simultaneously to the east and north). The angle θ denotes the inclination of the direction of motion with respect to the east direction. Thus, the component of velocity in the east direction is obtained by multiplying the actual speed by the **cosine** of θ ($\cos\theta$). Similarly, the magnitude of the velocity in the north direction is determined by multiplying the actual speed by the cosine of the angle with respect to the north axis, which is equivalent to the **sine** of θ ($\sin\theta$).

Example 2.2: Suppose a toy airplane is moving at a speed of 10 cm sec^{-1} in a direction inclined 30° from the east axis (see Figure 2.2). It starts at the origin, with coordinates (0, 0). What will be its coordinates in 10 seconds?

Answer: In general, a vector may be resolved into its components in three mutually perpendicular directions—east, north, and vertical, for example—using the cosine rule. Here we consider only the east and north directions.

First we calculate the speed the particle moves in each direction, and then the distance it travels in each direction after 10 seconds.

$$
\begin{aligned}
\text{East:}\quad \text{speed} &= (10\ \text{cm sec}^{-1})(\cos 30°) \\
&= 8.66\ \text{cm sec}^{-1} \\
\text{distance} &= \text{speed} \times \text{time} \\
&= (8.66\ \text{cm sec}^{-1})(10\ \text{sec}) \\
&= 86.6\ \text{cm} \\
\text{North:}\quad \text{speed} &= (10\ \text{cm sec}^{-1})(\cos 60°) \\
&= 5.0\ \text{cm sec}^{-1} \\
\text{distance} &= (5.0\ \text{cm sec}^{-1})(10\ \text{sec}) \\
&= 50.0\ \text{cm}
\end{aligned}
$$

The new coordinates are (86.6, 50.0). ■

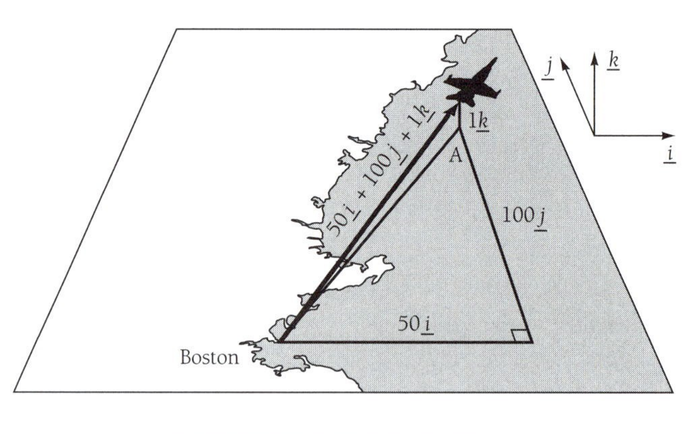

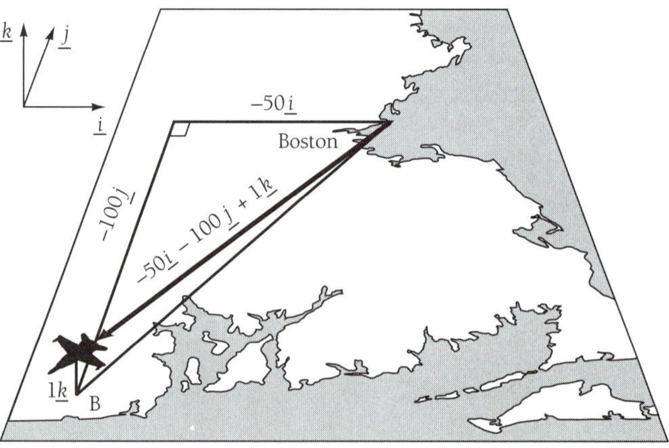

Figure 2.1b Representing a plane's position in the three dimensions with vectors. Top: The plane flying at an altitude of 1 km is 50 km east and 100 km north of Boston Logan ($50\underline{i} + 100\underline{j} + 1\underline{k}$). Bottom: It is 50 km west and 100 km south of Boston Logan ($-50\underline{i} + 100\underline{j} + 1\underline{k}$).

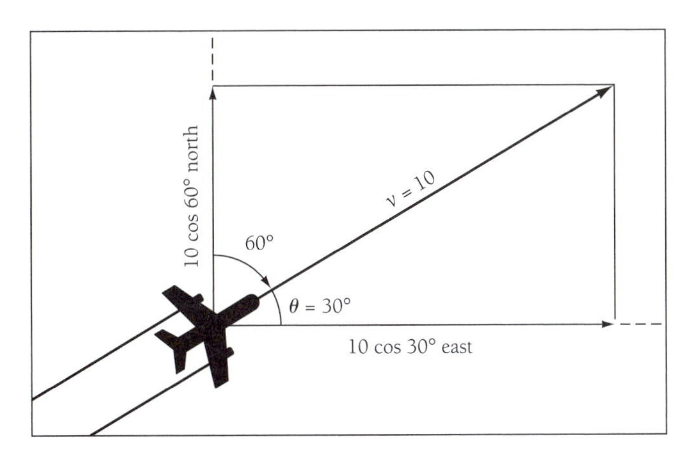

Figure 2.2 Figure illustrating components of motion. Components in the easterly and northerly directions are obtained by multiplying by cos 30° and cos 60°, respectively. Direction of motion is indicated by the diagonal arrow. The speed is 10 units per second.

Equation (2.1) can be used to specify separate relations between force, mass, and acceleration in the east-west, north-south, and vertical directions. This economy of notation justifies introduction of the rather abstract concept of a vector and the even more puzzling notion of an equation relating the magnitudes and directions of different vectors.

2.5 Work and Energy

A force moving an object a distance d along the direction of the force is said to do **work** on the object. The work done is measured by Fd, where F is the magnitude of the force. If the velocity of the object when the force is first applied is given by v_1, and if the velocity is v_2 after the object has been displaced a distance d, we can use Newton's law of motion to show that

$$\text{work} = Fd = \frac{1}{2}m(v_2)^2 - \frac{1}{2}m(v_1)^2 \tag{2.6}$$

To calculate the change in speed of the particle if its motion is opposed by the force, we use the same equation (2.6) with F changed to –F.

The quantity $\frac{1}{2}mv^2$ (appearing with v_1 and v_2 in the equation) is known as the kinetic energy of the particle. Expressed in words, equation (2.6) states that the work done on the particle by the applied force is equal to the increase in kinetic energy. If the force opposes the motion, the kinetic energy of the particle decreases by an amount equal to the work done against the applied force.

Suppose a ball is thrown in the air, leaving the hand with speed v_1. We know from experience that the ball will rise to some height, then return to the ground. In rising, its motion will be opposed by the force of gravity, mg. If we denote the velocity of the ball after it is released by v_2 and the altitude of the ball above the point of its release by z, equation (2.6) gives us

$$-mgz = \frac{1}{2}m(v_2)^2 - \frac{1}{2}m(v_1)^2 \tag{2.7}$$

At the top of the trajectory, the velocity of the ball will be zero; that is, $v_2 = 0$. Substituting $v_2 = 0$, we can rearrange equation (2.7) to find the maximum height (z_{max}) that the ball reaches:

$$z_{max} = \frac{1}{2}\frac{(v_1)^2}{g} \tag{2.8}$$

Equation (2.8) defines the height of the ball at the top of its trajectory.

Suppose the ball is launched from height z_1 with velocity v_1 as before. Suppose it has speed v_2 at height z_2. Equation (2.6) allows us to write

$$-mg(z_2 - z_1) = \frac{1}{2}m(v_2)^2 - \frac{1}{2}m(v_1)^2 \tag{2.9}$$

Equation (2.9) can be rearranged to give

$$\frac{1}{2}m(v_1)^2 + mgz_1 = \frac{1}{2}m(v_2)^2 + mgz_2 \tag{2.10}$$

We refer to mgz_1 and mgz_2 as the **potential energies** of the ball at height z_1 and z_2, respectively. As the ball rises, mgz_2 increases while the kinetic energy decreases proportionally. We can imagine that kinetic energy is converted to potential energy. At the top of the trajectory, the conversion of kinetic energy to potential energy is complete. As the ball falls back to the ground, kinetic rises up and potential energy decreases accordingly. Kinetic energy is used up in working against the force of gravity on the upward flight. Kinetic energy is gained due to work done by gravity on the downward leg. In the *cgs* system, energy is expressed in units of **ergs**, a compact representation for the fundamental dimensions g cm^2 sec^{-2}. A particle of mass 1 g moving at a speed of 100 cm sec^{-1} has a kinetic energy of 5000 erg $(\frac{1}{2}mv^2 = \frac{1}{2}(100)^2)$; similarly, if a particle of mass 1 g is displaced vertically a distance of 5 cm, its potential energy will change by a comparable amount, 4900 erg ($mgz = 1 \times 980 \times 5$).

2.6 Temperature

Atoms and molecules are in constant motion. The average speed of the typical constituent of the atmosphere is about 2×10^4 cm sec^{-1}. **Temperature**, denoted by T, provides a measure of the kinetic energy of the average molecule or atom in a gas, through the relation

$$\frac{1}{2}mv^2 = \frac{3}{2}kT, \tag{2.11}$$

where k is the Boltzmann constant, named in honor of the Austrian physicist Ludwig Boltzmann (1844–1906). The value of k depends on the units used to define temperature. Using (2.11) as a definition of temperature, and assuming that k is a positive number, we must select a system of units in which temperature is always positive; otherwise we would face the impossibility of trying to rationalize a negative value for kinetic energy.

Physicists favor a scale for measurement of temperature named in honor of the nineteenth-century Irish physicist Lord Kelvin (1824–1907). All temperatures are represented by positive numbers in the **Kelvin system.** Negative temperatures are impossible in the Kelvin system because a temperature of 0 K corresponds to the extreme condition where all molecular motion ceases—the kinetic energy falls to zero. As temperature increases, molecules pick up speed; that is, their average kinetic energy increases. Using the *cgs* system of units for length, mass, and time, with temperature expressed in K, we find that the Boltzmann constant, k, has the value 1.38×10^{-16} ergs K^{-1}.

We are more familiar in everyday life with measurements of temperature expressed in either the Celsius or Fahrenheit scales. The Celsius scale is based on the properties of water; the zero point corresponds to the temperature at which water freezes, while the 100° C mark represents the temperature at which water boils. The Celsius scale is used in most parts of the world, with the notable exception of the United States. In the United States, temperatures are usually expressed in Fahrenheit. The Fahrenheit scale, named in honor of the German physicist who invented the mercury thermometer (Gabriel Fahrenheit, 1686–1736), locates the freezing point of water at 32° F, with the boiling point at 212° F. The zero point on the Fahrenheit scale was selected to represent record cold temperatures experienced during German winters; the 100° point was set to approximate the temperature of the human body.

A visitor from the United States, tuning into the summer weather forecast in England or Canada, might be alarmed to hear the announcer cheerfully predict a noontime temperature of 25°. The visitor would be reassured, however, by the knowledge that a temperature of 25° on the Celsius scale is equivalent to a comfortable 77° on the Fahrenheit scale. We can use the following formulas to shift from one system of units to another:

$$\begin{aligned} T(\mathrm{K}) &= T(^\circ\mathrm{C}) + 273.15 \\ T(^\circ\mathrm{C}) &= \tfrac{5}{9}[T(^\circ\mathrm{F}) - 32], \end{aligned} \tag{2.12}$$

where K, °C, and °F indicate the Kelvin, Celsius, and Fahrenheit scales, respectively.

It is important to remember that the physical constants quoted above, such as k, are given for a system of units where temperature is represented by the Kelvin scale. You should probably carry out all calculations using temperatures expressed in K; in this way you can avoid absurd results, negative values of kinetic energy, for example. This will be important later on for problem solving. If you wish, after a calculation is complete you can convert to units with which you are more comfortable. Life would have been simpler if Kelvin's derivation of his absolute, rigorously consistent scale for temperature had preceded Fahrenheit's invention of the thermometer!

2.7 Escape Velocity

We introduced earlier the concept of a force and emphasized specifically the importance of gravitational force. It is the force of gravity that binds the atmosphere to Earth; we owe the very existence of the atmosphere to the presence of gravity. We can show that any nonaccelerating object is permanently bound to Earth's surface so long as its vertical speed is less than about 10 km sec^{-1}. An object leaving the surface with a vertical speed of less than 10 km sec^{-1} will have its entire store of kinetic energy converted to potential energy, its vertical speed thus slowing to zero, and will subsequently plunge back to the surface.

The speed below which objects are gravitationally bound to Earth is known as the **escape velocity** (v_{esc}). It may be calculated by determining the speed for which the initial kinetic energy of a particle is just enough to allow it to escape the gravitational field of Earth; this would occur when a particle reaches the high point of its trajectory at an infinite distance from Earth. This may be obtained using the formula

$$\frac{1}{2}m(v_{esc})^2 = \frac{GMm}{R}, \tag{2.13}$$

where R defines the distance from the center of Earth, G is the gravitational constant introduced in equation (2.2), M is the mass of Earth, and m is the mass of the particle. Equation (2.13) may be derived using an argument similar to that employed above to estimate the height reached by an object fired vertically from the surface of Earth. It differs from the earlier treatment in that we must take into account the decrease in the strength of gravitational force with increasing distance from the center of the planet. We must do this because we are now taking into account the motion of a particle at a large distance from Earth. The right-hand side of equation (2.13) defines the work done against the gravitational field in moving to infinity a particle of mass m located at an initial distance R from the center of the planet. Solving (2.13) for v_{esc}, we find that

$$v_{esc} = \left(\frac{2GM}{R}\right)^{1/2} \tag{2.14}$$

Notice that the escape speed is independent of the mass of the particle. Speeds needed to escape a number of objects in the solar system, calculated using equation (2.14), are summarized in Table 2.2.

For comparison, we list in Table 2.3 average speeds calculated using equation (2.11) for hydrogen and nitrogen atoms at a temperature of 300 K. The fraction of the available number of atoms or molecules of mass m at temperature

Table 2.2 Escape speeds for selected objects in the solar system

Body	*Escape speed (cm sec^{-1})*
Mercury	4.3×10^5
Venus	1.0×10^6
Earth	1.1×10^6
Moon	2.3×10^5
Mars	5.0×10^5
Ceres (asteroid)	5.9×10^4
Jupiter	6.0×10^6
Saturn	3.5×10^6
Uranus	2.2×10^6
Neptune	2.5×10^6

Table 2.3 Average speeds for selected gases at 300 K

Gas	*Speed (cm sec^{-1})*
H	2.7×10^5
He	1.4×10^5
N	7.3×10^4
O	6.8×10^4
H_2O	6.4×10^4
N_2	5.2×10^4
O_2	4.8×10^4

T with speed v in the range v to $v + \Delta v$, where Δv denotes a small increment in speed, is given by

$$f(v)\Delta v = 4\pi\left(\frac{m}{2\pi kT}\right)^{3/2}\left(\exp\left(\frac{-mv^2}{2kT}\right)\right)v^2\Delta v \quad (2.15)$$

The derivation of the complex formula (2.15) will not be given here: f is known as the velocity distribution function, and it depends on v. Figure 2.3 illustrates this dependence for two different values of T for hydrogen and oxygen atoms.

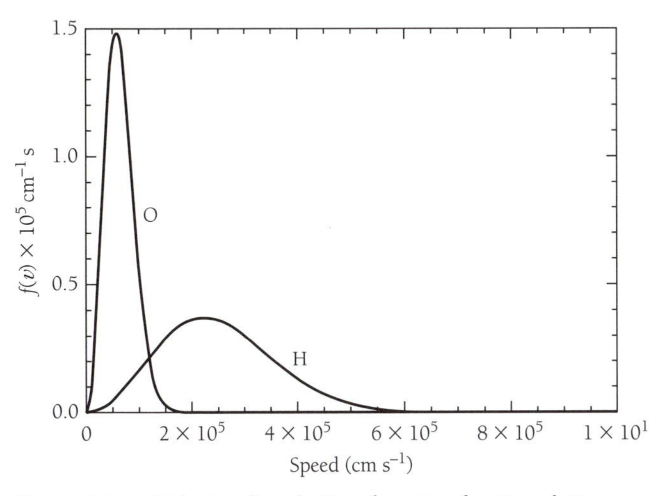

Figure 2.3a Velocity distribution function for H and O atoms, for T = 300 K.

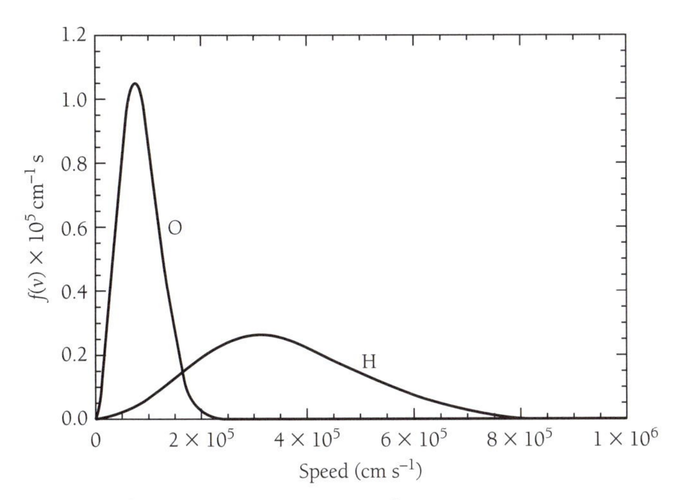

Figure 2.3b Same as Figure 2.3a, for T = 600 K.

A trivial fraction of the most abundant gases in Earth's atmosphere, N_2 and O_2, have sufficient speed to escape Earth's gravitational field. For Jupiter, the temperature at its cloud tops is about 140 K. Even hydrogen is stable on Jupiter, a fact that accounts for the extremely high abundance of H_2 in the planet's atmosphere. On the other hand, it is relatively easy for gases to escape the weak gravitational field of the moon, and thus the moon has little atmosphere. Tables 2.2 and 2.3 provide a theoretical support for this observed fact.

2.8 Pressure

The distribution of the atmosphere with respect to altitude is determined largely, as we shall see, by the influence of the gravitational force and a second force, equally important, associated with **pressure.** Pressure measures the force applied to a unit area of surface. In the *cgs* system, pressure has units of dyn cm^{-2}.

Consider an athletic woman weighing 132 pounds (mass equal to 6×10^4 g). She is pulled toward Earth with a force of 6×10^7 dyn. Imagine that her feet have a surface area of 300 cm^2, corresponding to a shoe size of about 8. The feet impose a pressure on the ground equal to about 2×10^5 dyn cm^{-2}. If she tried to stand on a thick blanket of powdered snow, on a ski slope for example, she would probably sink up to her ankles: the snow would be unable to accommodate the pressure exerted by her feet. The following example calculates the change in pressure that would arise if the woman were to stand on a pair of skis.

Example 2.3: Suppose the aforementioned woman dons a pair of skis with a surface area of 3×10^3 cm^2. Using the same data, determine the pressure exerted on the snow.

Answer:

$$\text{Pressure} = \frac{\text{force}}{\text{area}} = \frac{6 \times 10^7\text{dyn}}{3 \times 10^3\text{cm}^2} = 2 \times 10^4\text{dyn cm}^{-2}$$ ∎

The diminished pressure could be taken up by the snow with minimal compaction, and our athlete would be ready to float on the surface of the snow, assuming that her sense of balance and other skills were up to the challenge.

On the other hand, suppose she decided to turn her attention to the ice rink. Donning a pair of skates with relatively sharp blades, her weight is now distributed over a scant 2 cm^2. The pressure exerted on the ice is an impressive 3×10^7 dyn cm^{-2}, equivalent to 440 pounds per square inch, almost 30 times the pressure exerted by the atmosphere. A pressure of this magnitude is sufficient to melt the ice under the blade of the skate. (Water has the remarkable and unusual property of having a higher density in the liquid phase than in the solid phase. This property causes solid water to melt under pressure. It also allows ice to float on the top of a lake during winter, insulating the liquid below and thus allowing aquatic life to survive even under the most extreme winter conditions. The properties of water are considered further in Chapter 5.) The woman is now set to glide on the cushion of liquid under her skates. Skating, with all its grace, would be impossible were it not for the high pressure generated by the blade and the thermodynamic properties of water.

Pressure at any position in a fluid is equivalent to the weight of all the material above an imaginary horizontal surface of area 1 cm^2. Consider the function of a **barometer**, a device used to measure pressure, as illustrated in Figure 2.4. We can set up the barometer fairly easily; all we need is a test tube more than 76 cm long and open at one end, a supply of mercury, and an open container. We fill the test tube with mercury; this eliminates all the air from the inside of the tube. Then we flip it over and quickly insert the open end in the bath of mercury, as indicated. The mercury in the open container senses the pressure of the atmosphere at its surface. This pressure is transmitted readily through the mercury. The mercury moves up or down the tube until the pressure at the point where the tube intersects the surface of the container is equal to the pressure of the atmosphere. At this point all forces are balanced and the mercury comes to rest. The height of the column of mercury is proportional to the weight of mercury in the tube. The tube can be calibrated to provide a simple measure of atmospheric pressure. A column of mercury 76 cm high has a weight equivalent to the weight of the atmosphere above a surface of the same cross-sectional area at sea level. An increase in atmospheric pressure, perhaps associated with a change in weather, forces the mercury to rise in the barometer; conversely, a fall in atmospheric pressure causes a drop in the level of the mercury.

A barometer similar to that in Figure 2.4 could be constructed using water, rather than mercury, as the diagnostic fluid. The problem with a water barometer is that the column of fluid required to support the pressure of the atmosphere would have to be about 10 m high, almost 33 feet. The relatively compact nature of the mercury barometer is a consequence of the high mass density of mercury, about 13 times that of water (mass density is defined as mass per unit volume). The advantage of the water barometer is that it can be used to measure extremely small changes in pressure; a minute change in pressure induces a readily measurable fluctuation in the height of the water.

Pressures in the atmosphere are normally expressed not in the *cgs* units of dyn cm^{-2} but in units known as **millibars.** A pressure of 1 millibar (mb) is equivalent to 10^3 dyn cm^{-2}. The pressure of the atmosphere at sea level is approximately 1013 mb. The pressure of the atmosphere fluctuates with changing weather patterns. It typically varies from about 980 to about 1050 mb at midlatitudes at sea level. As discussed below, fair weather is normally associated with high pressure, unsettled weather with low pressure. The highest recorded sea level pressure, 1084 mb, is reported to have been measured in December 1968 at Agata, Siberia; the lowest recorded

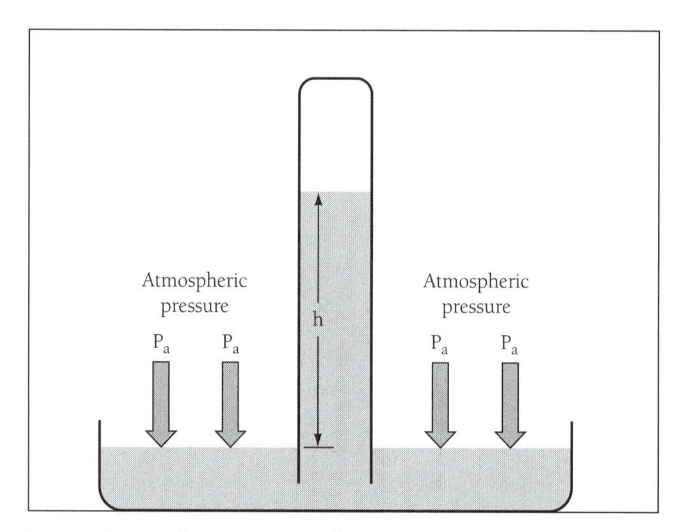

Figure 2.4 Schematic view of a mercury barometer. Remember that $\rho gh = P_a$. The height (h) of the mercury in the column is proportional to the pressure P_a with which air presses down on the mercury outside of the column.

pressure was 870 mb during Typhoon Tip in October 1979. Imagine the force experienced by the walls of an airtight house during Typhoon Tip. The pressure differential across the walls would have been as large as 100 mb (the change in pressure inside would lag behind the change outside). A differential pressure of this magnitude corresponds to a force of 10^5 dyn cm^{-2}, comparable to the force delivered by a powerful bomb.

2.9 The Perfect Gas Law

The pressure, density, and temperature of a gas are not independent quantities. They are related according to an equation known as the **equation of state.** For a gas of relatively low density, a condition satisfied by the atmosphere, the equation of state assumes the simple form

$$p = nkT, \quad (2.16a)$$

where n defines the **number density** of molecules, the number of molecules per unit volume (in the *cgs* system, n is expressed in molecules per cm^3, abbreviated as cm^{-3}). Expressed in the form of (2.16a), the equation of state is known as the **perfect gas law.**

Chemists favor a different expression of the perfect gas law. They prefer to measure pressure in units of atmospheres (atm), density in units of **moles (mol)** per liter (l) (a mole is equivalent to 6.02×10^{23} molecules, where 6.02×10^{23} is known as **Avogadro's number**, discussed further in Section 3.3). It is no longer appropriate in this case to use k as the constant in the equation relating p, n, and T. The alternate, chemists' expression for the perfect gas law is

$$P = NRT, \quad (2.16b)$$

where P is given in atm, N in mol l^{-1}, and T in K. R, known as the **universal gas constant,** has the numerical value 0.08206, with units of atm l mol^{-1} K^{-1}.

We can rationalize the form of the perfect gas law by considering the momentum content of a gas confined in a container with elastic walls. Molecules are buzzing in all directions. From time to time they strike the walls. We assume that the kinetic energy of the molecules is conserved on collision with the walls; this is why we describe the walls as "elastic." The molecules must change direction as they encounter the surface; that is, their velocity vectors must change. A change in velocity corresponds to either an acceleration or a deceleration. In either event, by Newton's law of motion, there is an associated force. The force exerted on the wall, per unit area, is equivalent to the pressure of the fluid.

Consider a coordinate system centered on one of the walls of the container as indicated in Figure 2.5. Let the direction of motion of the molecules be oriented perpendicular to the wall. Focus attention on molecules moving toward the wall along the $\underline{i}$ direction. Assume that the speed of these molecules, v (in units of cm sec^{-1}), is equal in magnitude to the mean thermal speed of the gas, given in terms of temperature by equation (2.11). Consider a volume with unit cross-sectional area oriented perpendicular to the wall. A molecule originating at a distance v cm from the wall has sufficient speed to reach the wall in one second (this follows from our definition of speed, distance traveled per unit time). Let f denote the fraction of the molecules moving toward the wall along the $\underline{i}$ axis at any given time. (If you find $\underline{i}$-$\underline{j}$-$\underline{k}$ coordinates confusing, think of them as akin to x-y-z coordinates.) Molecules moving toward the wall along the $\underline{i}$ axis with speed v and capable of reaching a square centimeter of the wall's surface area in one second must be contained initially in the volume indicated in Figure 2.5. The number of molecules contained in this volume is given by multiplying the number density, n, by the magnitude of the volume, V. The number of molecules striking the surface along the $\underline{i}$ direction in one second is fnv. Each of the molecules striking the surface imparts a momentum of $2mv$ to the wall. The molecules reverse direction, maintaining constant speed and kinetic energy. The momentum transferred to the surface per second is obtained by multiplying the number of molecules impacting per second by the momentum delivered per molecule, $fnv \times 2mv$. This quantity is numerically equal to the pressure imparted to the surface:

$$p = 2fnmv^2 \tag{2.17}$$

or, substituting for mv^2 using (2.11),

$$p = 6fnkT \tag{2.18}$$

There is no preferred direction of motion for molecules in a gas; we are as likely to find molecules moving in one direction as in another. To compute f, it seems reasonable to assume that on average we might find one third of the molecules moving along each of the vector directions $\underline{i}$, $\underline{j}$, and $\underline{k}$. Of the molecules moving along the $\underline{i}$ direction, we may assume that half are moving toward the wall and half moving away. In this case, f equals $\frac{1}{6}$ and (2.18) is equivalent to (2.16).

The derivation of the perfect gas law given here is necessarily approximate. A more rigorous derivation could be

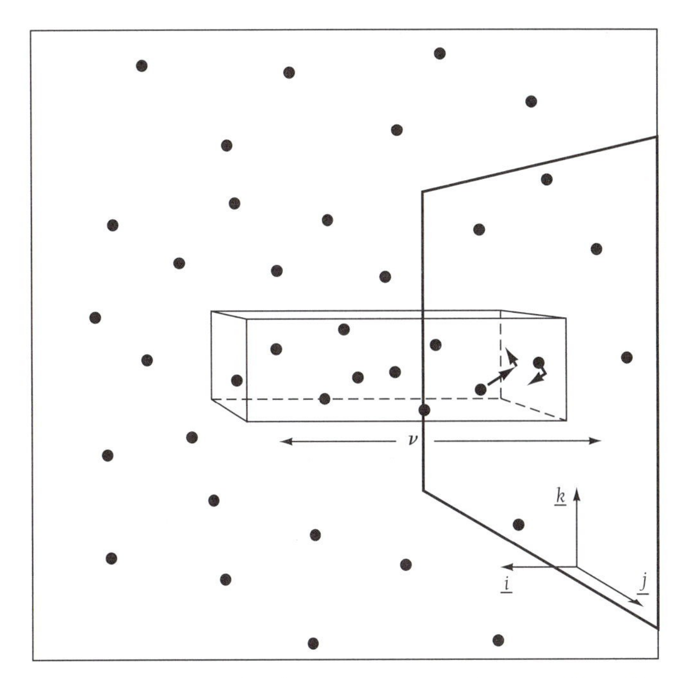

Figure 2.5 Gas molecules exert a pressure force on the wall to the right. Only molecules located at a distance of less than v(cm) can reach the wall during the interval of 1 sec.

offered using the kinetic theory of gases, but this would require a level of mathematical complexity beyond the scope of the present analysis. Our derivation captures, however, the essential physics of the perfect gas law; all we need is common sense, a picture of a gas, and Newton's law of motion. We can readily accept that the rate of transfer of momentum from the gas to the walls of the container should be proportional to the number of molecules available to strike the walls. The less obvious dependence on temperature is explained by the analysis presented above.

2.10 Angular Momentum

Consider a particle of mass m moving in a circle at constant tangential speed. The orientation of the velocity vector is constantly changing in this situation; the particle is turning so as to remain at a constant distance from the center of the circle. A change in velocity with time corresponds to an acceleration; by Newton's law, this acceleration must be associated with a force. In a counterintuitive manner, the particle's force is directed *toward* the center of the circle. This force ensures that the particle remains in circular motion; otherwise it would tend to fly off the circumference of the circle. As an example of circular motion, consider a ball attached to a string. As illustrated in Figure 2.6, a man sets the ball in circular motion by twirling the string around his head. The force that maintains the ball at a constant distance from the the man's hand is the tension in the string, a force directed toward the center of the circle. The movement of Earth around the Sun provides another example of nearly circular motion. The operative force in this case is the force of gravity. Suppose that we were to haul in the string attached to the

Figure 2.6 Circular motion. Velocity is tangent to the circular path, while the force is directed inwards, perpendicular to the path.

twirling ball. We can show, using Newton's law of motion, that the magnitude of the quantity mvr is conserved, where v defines the speed of the object in circular motion. If the radius of the circle, r, were reduced by a factor of two, the speed of the ball would double. The quantity mvr is known as the **angular momentum** of the ball.

Angular momentum is a vector. The magnitude of the angular momentum vector is defined as the product of mvr and the sine of the angle between the position and velocity vectors of the particle, that is, $mvr \sin\theta$; we can readily show that this reduces to mvr for the case of circular motion. The angular momentum vector, in vector notation, is expressed as $m\underline{r} \times \underline{v}$ (The symbol × here means to take the cross product of two vectors; it should not be confused with the multiplication sign.) The cross product of two vectors, $\underline{A} \times \underline{B}$, yields a third vector oriented perpendicular to both $\underline{A}$ and $\underline{B}$. Its magnitude is equal to the product of the magnitudes of $\underline{A}$ and $\underline{B}$ times the sine of the angle between $\underline{A}$ and $\underline{B}$. The vectors $\underline{v}$ and $\underline{r}$ lie in the plane defined by the circular motion of the particle. With our definition, the angular momentum vector must be oriented perpendicular to this plane but can point either up or down. To resolve the ambiguity, we use the following rule of thumb (literally) to determine the direction of $\underline{A} \times \underline{B}$: choose a direction for the resulting vector, place your thumb in this orientation, line up your fingers with the direction of $\underline{A}$, and see if the natural curl of your fingers directs you toward the direction of $\underline{B}$. If it doesn't, then $\underline{A} \times \underline{B}$ is oriented in the direction opposite to that of your initial guess. The procedure is illustrated in Figure 2.7. Practice it until you are comfortable with your ability to determine the direction of an arbitrary vector product $\underline{A} \times \underline{B}$. If you fail to master the **right-hand rule**, you can always fall back on a cookbook approach. If the object's motion is in a counterclockwise sense as you look at it, the angular momentum vector is directed toward you—this is a simple recipe for determining the orientation of an angular momentum vector.

Newton's law of motion, equation (2.1), allows us to calculate the rate of change over time of linear momentum

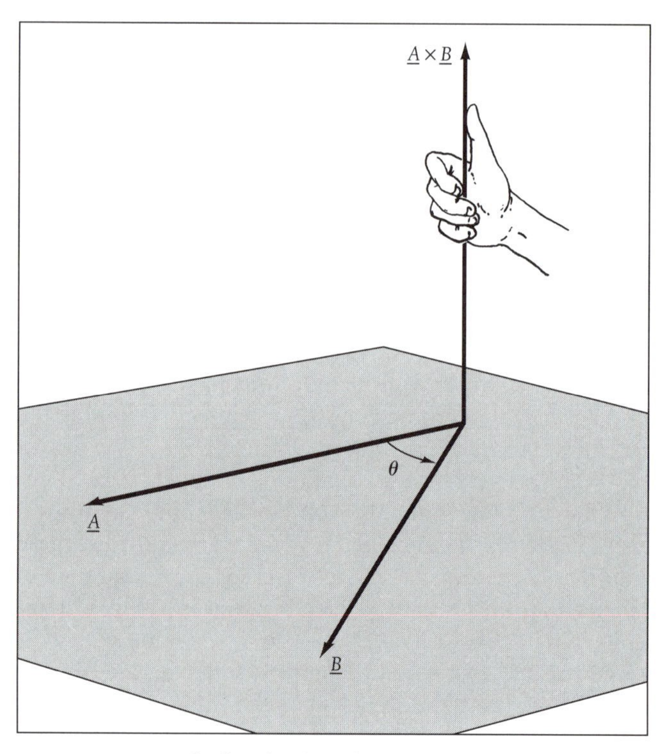

Figure 2.7 "Right-hand rule" of cross product.

arising as a consequence of an imposed force $\underline{F}$. For a particle of fixed mass m, the law simply defines the rate of change of its velocity v. Expressed in the language of differential calculus, equation (2.1) may be written in the form

$$\underline{F} = m\frac{d\underline{v}}{dt} \tag{2.19}$$

where

$$\frac{d\underline{v}}{dt} = \lim_{\Delta t \to 0} \frac{\underline{v}(t+\Delta t) - \underline{v}(t)}{\Delta t} \tag{2.20}$$

Here, the notation $\lim_{\Delta t \to 0}$ indicates that the subsequent expression, $(\underline{v}(t+\Delta t) - \underline{v}(t))/\Delta t$, should be evaluated for vanishingly small values of Δt. Put another way, (2.20) provides a means to calculate the velocity at time $t + \Delta t$, $\underline{v}(t + \Delta t)$, given information on the velocity at time t, $\underline{v}(t)$, and the magnitude and direction of the imposed force $\underline{F}$:

$$\underline{v}(t+\Delta t) = \underline{v}(t) + \Delta t\frac{\underline{F}}{m} \tag{2.21}$$

If $\underline{v}$ is changing rapidly, we need to be careful to employ this relation only for very small values of Δt. Recall that, strictly speaking, the relation is only valid for the limiting case where Δt tends to zero. If a force is such that you expect the corresponding velocity to change significantly only on a time scale of about an hour, you can be confident that predictions made for $\underline{v}$ on a time scale of minutes will be reasonably reliable.

To derive an expression for the rate of change of angular momentum over time, we first write the angular momentum vector in the form

$$\underline{L} = m\underline{r} \times \underline{v} \tag{2.22}$$

Differentiating both sides of this equation with respect to time, we find that

$$\frac{d\underline{L}}{dt} = \left(m\frac{d\underline{r}}{dt}\times\underline{v}\right)+\left(m\underline{r}\times\frac{d\underline{v}}{dt}\right) \tag{2.23}$$

But

$$\frac{d\underline{r}}{dt} = \underline{v} \tag{2.24}$$

Thus

$$\frac{d\underline{L}}{dt} = (m\underline{v}\times\underline{v})+\left(m\underline{r}\times\frac{d\underline{v}}{dt}\right) \tag{2.25}$$

But, by definition, the cross product of a vector and itself is zero (the angle between $\underline{v}$ and itself is zero, and the sine of this angle is also equal to zero). Hence, the first term on the right-hand side of (2.25) vanishes. Using Newton's law of motion as given by (2.19), (2.25) may be rewritten in the form

$$\frac{d\underline{L}}{dt} = \underline{r}\times\underline{F} \tag{2.26}$$

The vector

$$\underline{\tau} = \underline{r}\times\underline{F} \tag{2.27}$$

is referred to as the **torque** exerted by the force $\underline{F}$.

Torque is a vector quantity oriented perpendicular to the plane defined by the vectors $\underline{r}$ and $\underline{F}$. In the *cgs* system, torque has units of grams cm^2 sec^{-2}, or dyn cm. Angular momentum has units of grams cm^2 sec^{-1}, or erg sec.

Consider the situation depicted in Figure 2.6. In pulling the string, the man applies a force along the radial direction. The magnitude of the associated torque is thus equal to zero ($\underline{r}$ and $\underline{F}$ are oriented in the same direction). It follows that

$$\frac{d\underline{L}}{dt} = 0, \tag{2.28}$$

hence $\underline{L}$ must be constant. For the motion of a particle of mass m and velocity $\underline{v}$ in a circle of radius R, the magnitude of the angular momentum vector is given by

$$L = mRv, \tag{2.29}$$

where L and v denote the scalar magnitudes of $\underline{L}$ and $\underline{v}$, respectively. The conservation of L implies that if we reel in the string in Figure 2.6, the speed of the particle must increase to compensate for the decrease in R, such that

$$mRv = \text{constant} \tag{2.30}$$

A decrease in R by a factor of two requires a corresponding increase in v. To increase or decrease the angular momentum of a particle, we need to apply a torque; that is, we need to apply a force with a component either in the direction, or opposed to the direction, of motion of the particle.

The trick used by a skater to spin faster by drawing in her arms is a consequence of the conservation of angular momentum. Likewise, if an air parcel is forced to change latitude, it will tend to spin up or spin down, depending on whether the change in latitude is positive (increasing) or negative (decreasing). If the parcel moves to a higher latitude, the effect is akin to the string being pulled in: the parcel is now closer to the axis of rotation of Earth, and the effective radius of the circle it traverses as it moves west to east is reduced. Beginning from an initial condition where it was stationary, it will speed up. As a result of the increase in speed (in moving from a lower latitude), the air parcel appears to acquire an additional component of motion to the east. Particles appear to turn to the right with respect to their north-south (meridionial) motion in the Northern Hemisphere. The apparent deflection is oriented to the left in the Southern Hemisphere, a fact we can readily appreciate by considering the increase in the apparent easterly velocity of a particle moving south or in the westerly velocity of a particle moving north. We shall return to this topic later in Chapter 8 in connection with our discussion of the Coriolis effect.

2.11 More on Circular Motion

Consider a particle of mass m moving at constant speed v along the circumference of a circle of radius R, as illustrated in Figure 2.8. The velocity at time t is given by $\underline{v}(t)$. At a later time $(t + \Delta t)$, when the particle has rotated through an angle $\Delta\phi$, the velocity is given by $\underline{v}\,(t + \Delta t)$. The velocity vector at any point P on the orbit is oriented in the direction of the local tangent to the circle. It is clear from the diagram that over the time interval Δt the velocity vector must turn through an angle $\Delta\phi$. (The relative orientations of $\underline{v}\,(t + \Delta t)$ and $\underline{v}\,(t)$ are illustrated in Figure 2.9.) It is also clear that the direction of $\underline{v}$ must change with time; that is, $\underline{v}$ must change. Since a change in $\underline{v}$ with respect to t corresponds to an acceleration, it is apparent that the particle in circular motion must be subject to an acceleration. This acceleration (a vector) is given by

$$\underline{a} = \frac{d\underline{v}}{dt} = \lim_{\Delta t\to 0}\frac{\underline{v}(t+\Delta t)-\underline{v}(t)}{\Delta t} \tag{2.31}$$

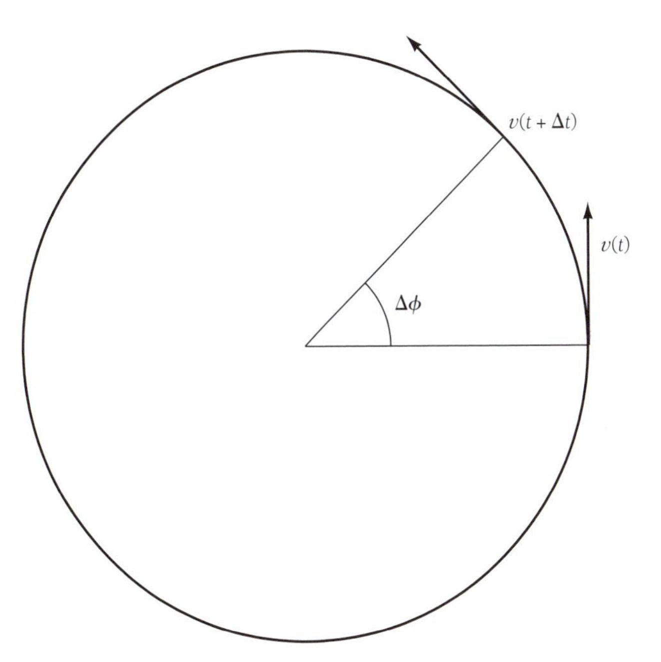

Figure 2.8 Velocity of a particle in circular motion is given by $v(t)$ at time t and by $v(t + \Delta t)$ at time $t + \Delta t$.

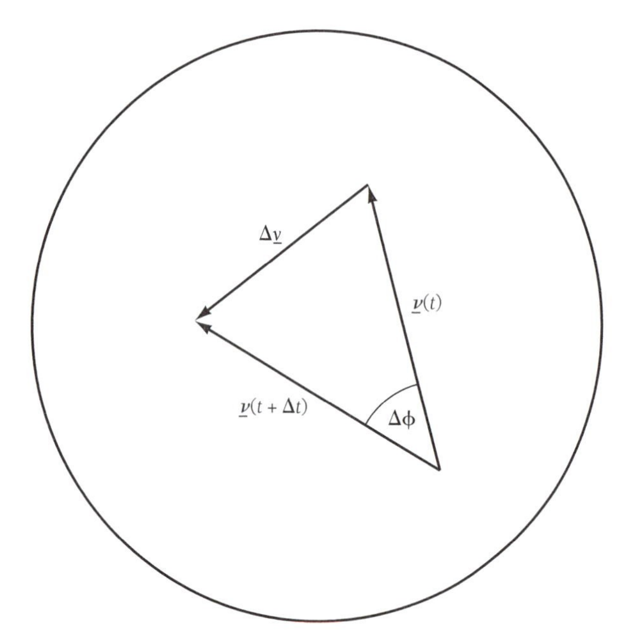

Figure 2.9 The velocity vectors for the particle in Figure 2.8. During the time interval Δt, the velocity changes by Δv.

The picture in Figure 2.9 indicates that $\Delta \underline{v}$ must be oriented toward the center of the circle. (If you have trouble seeing this, think of what must happen as $\underline{v}(t+\Delta t)$ collapses back to $\underline{v}(t)$: the difference $\Delta \underline{v}$ must line up perpendicularly to $\underline{v}(t)$.) The magnitude of $\Delta \underline{v}$, Δv, is given by

$$\Delta v = v\Delta\phi, \tag{2.32}$$

where v denotes the magnitude of $\underline{v}$. Thus, the magnitude of the acceleration is given by

$$a = \lim_{\Delta t \to 0} v\frac{\Delta\phi}{\Delta t} = v\frac{d\phi}{dt} \tag{2.33}$$

The quantity $d\phi/dt$ is known as the **angular velocity**, or **angular frequency**, associated with circular motion and is defined by

$$\Omega = \frac{d\phi}{dt} \tag{2.34}$$

The angular velocity is expressed in radians per second (remember that 2π radians corresponds to 360 degrees of rotation around a circle). It follows that the time required for the particle to progress once around the circle is given by

$$T = \frac{2\pi}{\Omega} \tag{2.35}$$

Over a time T the particle moves a distance $2\pi R$. Its rotational speed is thus given by

$$v = \frac{2\pi R}{\left(\frac{2\pi}{\Omega}\right)} = R\Omega, \tag{2.36}$$

and the magnitude of the acceleration (directed toward the center of the circle) is given by

$$a = v\Omega = \frac{v^2}{R} = \Omega^2 R \tag{2.37}$$

Here, we expressed the angular velocity in terms of v by rewriting (2.36) in the form

$$\Omega = \frac{v}{R} \tag{2.38}$$

Equation (2.37) defines what is known as the **centripetal** (center-seeking) acceleration. For the situation depicted in Figure 2.6, centripetal acceleration is imparted by the center-directed tension in the string. If the string is released, the tension will go to zero; the particle is no longer forced to describe circular motion and it will fly off in a straight line (at constant speed) in the direction defined by the tangent at the point where the string is released.

It is important to remember that Newton's law of motion applies only if we carry out our measurements from the perspective of what is known as an **inertial coordinate system.** It is customary to think of the fixed stars as providing a basis for defining a standard inertial system. If the coordinate system we choose is accelerating relative to the standard reference system, Newton's law of motion in its conventional form ($\underline{F}=m\underline{a}$) is no longer valid. Suppose we wish to use a coordinate system defined relative to fixed positions on Earth. We must be careful to remember that this coordinate system is rotating; it is accelerating relative to an inertial reference. A particular position on Earth's surface, except for the poles, viewed from an inertial perspective, exhibits centripetal acceleration. Viewed from the vantage point of a coordinate system rotating with Earth, the net acceleration of a fixed point is zero. But we know that a real force corresponding to the centripetal acceleration acts to keep the particle in circular motion. We resolve the dilemma by introducing a false force to cancel the centripetal force when we choose to use the perspective of the noninertial rotating frame. This force must be equal in magnitude but opposite in direction to the centripetal force in order to keep the particle at rest in the rotating frame. We refer to this force as the **centrifugal** force. It is important to remember that this is not a real force in the Newtonian context. It must be included, though, if we try to use Newton's law of motion in the context of a coordinate system rotating with Earth. As indicated in Figure 2.10, the centrifugal force is directed away from Earth perpendicular to the axis of rotation. For a particle of unit mass moving with speed v in a zonal (west to east) direction at latitude λ, the centrifugal force has magnitude

$$F_{centr} = \frac{v^2}{R\cos\lambda}, \tag{2.39}$$

as may be inferred from the expression for the centrifugal acceleration given by (2.37). For a particle at rest on the surface of latitude λ, using (2.36) for v, the centrifugal force has magnitude

$$F_{centr}^{rest}(\lambda) = \Omega^2 R\cos\lambda \tag{2.40}$$

The centrifugal force is largest at the equator, where it acts in a direction precisely opposite to the force of gravity.

Example 2.4: Consider a satellite in a circular orbit in Earth's equatorial plane. Calculate the magnitude of the radius of the orbit for which the satellite will remain directly overhead of a fixed position on Earth.

Answer: For circular motion in an orbit of radius R, the force of gravity drawing the satellite toward the center of mass of Earth

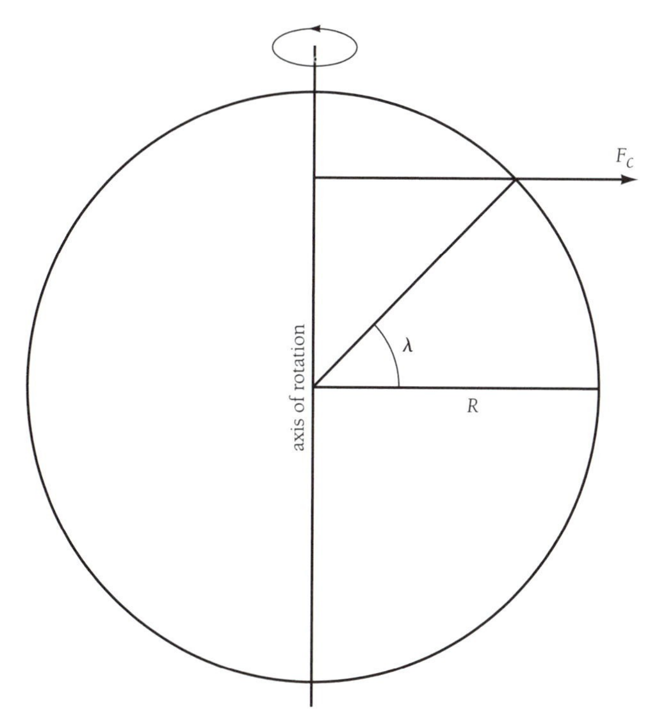

Figure 2.10 Direction of the centrifugal force at latitude λ.

must be exactly equal but opposite in sign to the centrifugal force pushing it away. Suppose that the angular velocity of the satellite is given by Ω. The magnitude of the centrifugal force, given by equation (2.37), is $m\Omega^2 R$, where m is the particle mass. The gravitational force at radial distance R, given by (2.4), is GmM/R^2, where M is the mass of Earth. Equating the magnitudes of these forces,

$$m\Omega^2 R = \frac{GmM}{R^2}$$

Thus

$$R^3 = \frac{GM}{\Omega^2}$$

Assume that Ω is appropriate for solid-body rotation of Earth. Then

$$\Omega = \frac{2\pi}{8.64 \times 10^4} \text{ sec}^{-1}$$

where the denominator defines the length of the day, measured in seconds. Thus,

$$\Omega = 7.3 \times 10^{-5} \text{ sec}^{-1}$$

Taking $G = 6.67 \times 10^{-8}$ dyn cm^2 g^{-2} with $M = 5.98 \times 10^{27}$g, it follows that

$$R^3 = \frac{(6.67 \times 10^{-8})(5.98 \times 10^{27})}{(7.3 \times 10^{-5})^2} \text{cm}^3$$
$$= 7.49 \times 10^8 \text{cm}^3$$

It follows that

$$R = 4.2 \times 10^9 \text{cm},$$

which corresponds to a planetocentric distance equal to about 6.6 times the radius of Earth. ■

2.12 Summary

We introduced in this chapter a systematic way of thinking about length, time, and derivative quantities such as velocity and acceleration. We discussed the concept of mass and described the relationship between force, mass, and acceleration defined by Newton's law of motion. We introduced kinetic energy, potential energy, the concept of work, and temperature as a measure of the kinetic energy of molecules in a fluid. We discussed pressure, density, and temperature, together with composition, as the key quantities needed to define the physical state of a gas such as the atmosphere, quantities related according to the perfect gas law.

We found that the concept of linear momentum was useful in describing the linear motion of a particle. By Newton's law of motion, force is equal to the rate of change of linear momentum. For a particle in circular motion, it is more instructive to focus on the concept of angular momentum. We showed that Newton's law of motion can be recast in a form more applicable to circular motion. In doing so, we introduced the concept of torque, a vector quantity that serves a role analogous to that of force in Newton's law of motion but that is more appropriate for circular motion. We showed that Newton's law of motion, applied to circular motion, requires that the rate of change of angular momentum be given by the torque. In the absence of torque, angular momentum is conserved.

We noted that Newton's law of motion is appropriate only for the description of motion in an inertial frame of reference. If we wish to describe motion from the vantage point of an observer fixed with respect to the surface of Earth, we must remember that Earth is rotating, that our coordinate system is accelerating with respect to an inertial reference system. We can account for this complication while retaining the basic form of Newton's law of motion by introducing a new, apparent force known as the centrifugal force (corresponding to centrifugal acceleration). Using the concept of centrifugal force allows us to apply Newton's law of motion in the description of motion as viewed from the perspective of an observer in a rotating reference system.

Basic Chemical Concepts

3

The physicist is interested in quantities such as mass, velocity, force, acceleration, energy, and angular momentum, as discussed in Chapter 2. The chemist focuses on a detailed examination of the properties of matter. Here we ask, What is an atom? How is an atom assembled from its basic constituents, known as protons, neutrons, and electrons? How do atoms link to form complex structures such as molecules? We discuss the concepts of oxidation and reduction, the significance of acids and bases, and the notion of chemical equilibrium. We conclude with an introduction to photochemistry. We will apply the material developed here to both the atmosphere and ocean in more detail later. The discussion of acid-base chemistry will be particularly relevant for the treatment of carbon chemistry developed in Chapter 11 and for the treatment of acid rain as presented in Chapter 18.

Section 3.1 introduces the properties of atoms and the processes by which they link to form molecules. The nature of chemical bonding is discussed in Section 3.2. Properties of acids and bases are described in Section 3.3, which includes an introduction to the concept of chemical equilibrium. A molecular level treatment of energy is presented in Section 3.4, highlighting the distinction between energy associated with electronic motion and that with vibrational and rotational motion of the nuclei. Reaction energetics is discussed in Section 3.5. Summary remarks are presented in Section 3.6.

3.1 Atoms and Molecules

Earth and its atmosphere are composed of vast numbers of minute particles known as **atoms.** We may think of atoms as the building blocks of matter. Often, atoms are joined together in more complex units known as **molecules.** Atoms are composed of nuclei and electrons. Nuclei are positively charged, while electrons carry a negative charge. The laws of electromagnetism dictate that oppositely charged particles attract each other, while particles of the same charge repel. The attraction of opposites constrains the electrons of an atom to remain in relatively close proximity to the nucleus and is responsible for the existence of atoms and molecules as composite entities.

The simplest atom, hydrogen, has a single electron. The hydrogen electron describes an almost spherical orbit around the nucleus, maintaining an average separation of about 5×10^{-9} cm. Atomic physicists, accustomed to very small length scales, find it convenient to use a unit for distance other than the centimeter. They adopt as the standard of length a distance of 10^{-8} cm, referred to as one **angstrom** (Å), honoring the Swedish physicist Anders Ångström (1814–1874). The average distance separating the electron and the nucleus of a hydrogen atom is about 0.5 Å.

The nucleus itself is divisible. The nucleus of an atom is composed of protons and neutrons. A proton has an electric charge equal and opposite to the charge of an electron. Neutrons in contrast are electrically neutral. Protons and neutrons have approximately the same mass. By standards we are accustomed to in everyday life, protons and neutrons are tiny: they individually weigh 1.67×10^{-24} g. However, they are giant in comparison with electrons, the mass of an electron is only 9.1×10^{-28} g. Protons and neutrons are bound together in the nucleus of an atom by a force known as the **nuclear force.** Nuclear forces are much stronger than the gravitational and electrostatic forces we have encountered so far. This allows protons to remain in close proximity to each other, despite the mutual repulsion resulting from their electromagnetic interaction.

The properties of an atom are determined by the number of protons in its nucleus, a quantity known as the **atomic number.** Each **element** (hydrogen or oxygen, for example) is associated with a specific number of protons in its nucleus. In its electrically neutral (normal) form, the number of electrons orbiting the nucleus of an element is exactly equal to the number of protons in the nucleus. An individual element may occur with differing numbers of neutrons. Atoms formed in this manner are known as **isotopes** of the element and have similar chemical properties. The total number of protons plus neutrons defines the **mass number**. The nucleus of the oxygen atom, for example, contains eight protons. There are three distinct isotopes of oxygen, containing eight, nine, and ten neutrons, respectively. The most abundant isotope of oxygen has eight neutrons, corresponding to a mass number of 16. The second most abundant has ten neutrons, hence a mass number of 18, while the isotope with nine neutrons, with a mass number of 17, is relatively rare.

Water (H_2O), composed of molecular aggregates each containing two atoms of hydrogen (H) joined to one atom of oxygen (O), may contain oxygen of all three isotopic types. The lighter molecules, formed from oxygen with a mass number of 16, evaporate more readily from the liquid phase than their heavier counterparts. This has an interesting consequence; water vapor in the atmosphere and in precipitation is isotopically light compared to water in the ocean, from which it is ultimately derived. In comparison with water in the ocean, water vapor in the atmosphere contains a relatively larger abundance of oxygen atoms with a mass number of 16 relative to atoms with mass numbers of 17 and 18. This physical effect has been exploited by geochemists to open a powerful window to the study of the past climatic states of our planet, as we shall see in Chapter 10.

During ice ages, vast quantities of water evaporated from the ocean, precipitated from the atmosphere, and then were stored in ice sheets covering large parts of the American and European continents—the quantity of water removed from the ocean was sufficient to lower the sea level by more than a hundred meters. As expected, water in ice sheets was isotopically light, while water remaining in the ocean was correspondingly heavy. The isotopic composition of oxygen in ocean water is recorded in the composition of shells of organisms that grow in the sea. Large numbers of shells formed by organisms that lived in the past are preserved in the sediments of the ocean today. By extracting a core of sediment from the seafloor, we can recover the shells of these ancient organisms, and, by analyzing their isotopic composition, we can look back in time, drawing quantitative conclusions regarding the mass of water removed from the ocean and stored in continental ice sheets as a function of time. In this fashion, scientists have been able to reconstruct the composition of the ocean and to identify the changing patterns of climate—alternating ice ages and interglacials—characterizing the state of our planet for the past several million years.

In similar fashion, water samples removed from the ancient ocean can be recovered by drilling through present-day vestigial ice sheets in Greenland and Antarctica. Isotopic measurements of this water can be used to infer the temperature of the atmosphere from which the water precipitated, providing an invaluable record of high-latitude temperatures. Such ice also contains air bubbles trapped by accumulating snow; analysis of this air provides a beautiful and indispensable record of atmospheric composition dating back 450,000 years. This record includes indisputable evidence for the global significance of human activity over the past several centuries. Carbon dioxide (CO_2) accounted for about 280 molecules per million molecules, or **parts per million (ppm)**, of the atmosphere since the end of the last ice age, about 15,000 years ago, until the early part of the nineteenth century. The concentration of CO_2 has steadily risen to about 360 ppm at present, reflecting additions of CO_2 to the atmosphere due to destruction of forests and burning of fossil fuels. We shall return to this topic in Chapter 11, when we discuss the function of the global carbon cycle.

We can think of an atom as a miniature solar system. In this analogy, the nucleus, at the center, is the Sun. Electrons play the role of the planets, orbiting the parent nucleus. The universe contains many atoms, many suns. According to the laws of quantum mechanics, electrons are required to occupy constrained and well-defined orbits, arranged in what quantum physicists call **shells.** The first shell, closest to the nucleus, is known as the **K shell** and can accommodate two electrons. The second shell, somewhat more removed, can hold up to eight electrons and is known as the **L shell.** Hydrogen, with a single proton, has a single electron; in its most stable, or lowest, energy state, the electron in hydrogen occupies the K shell. Helium (He), with two protons, has two electrons: this completes the capacity of the K shell. Lithium (Li), with three protons, has three electrons: two occupy the K shell, while the third falls into the more extended, less tightly bound L shell. The universe of atoms can be constructed by adding protons to nuclei, and electrons to shells, until their carrying capacity is exhausted. Adding electrons to the L shell, after lithium, we sequentially form beryllium (Be), boron (B), carbon (C), nitrogen (N), oxygen (O), fluorine (F), and neon (Ne). Neon, with ten protons and ten electrons, exhausts the capacity of both the K and L shells.

The universe of atoms is summarized in Plate 1 (see color insert) and Table 3.1 with a diagram known as the **periodic table**, attributed to the Russian physicist Dmitri Mendeleev (1834–1907), who first devised this arrangement of the elements. The chemical properties of atoms, their ability to form composite entities by bonding to other atoms, are primarily determined by the number of electrons in the outermost shell of electrons. The periodic table is organized so that elements in a given column have similar arrangements of their **valence** (outermost) electrons. Thus, atoms in the same column of the periodic table have similar chemical properties, and it is this feature that makes the table so extraordinarily useful to chemists.

As we shall see, atoms go to great lengths to either give up or borrow electrons from other atoms in order to develop a fully populated outer shell of electrons. Atoms like helium or neon have it made from the start, since their outer shells are already fully populated. As a consequence, they lead a rather lonely existence; we refer to atoms with complete shells as inert or **noble gases**, an appropriate title given their obviously aloof and superior status. The family of noble gases, including argon (Ar), krypton (Kr), and xenon (Xe), in addition to helium and neon, appears in the rightmost column of the periodic table.

On an atomic basis, the most abundant constituent of the atmosphere is nitrogen, followed by oxygen and hydrogen. These elements do not normally appear singly as atoms. Rather, they are present as aggregates of atoms, as molecules. Nitrogen appears as molecular nitrogen, N_2. Oxygen is present mostly as O_2, while water vapor, H_2O, is the predominant form of hydrogen.

3.2 Chemical Bonds

Before we can begin to appreciate atmospheric chemistry, we need to understand the nature of the forces binding atoms together in molecules. Why is it that some molecules, such as N_2, are exceptionally stable while others, such as O_2, are relatively reactive? When we draw in a breath of air, we indiscriminately inhale N_2 and O_2 and everything else in the air. Molecular nitrogen passes through our lungs and is exhaled (returned to the atmosphere) without changing its chemical state. Molecular oxygen, on the other hand, combines with carbon and is released as carbon dioxide, CO_2. The comparatively inert character of N_2 is related ultimately to the strength of the forces binding the atoms together in the associated molecular compound. It is important that we understand the forces allowing particular groups of atoms to exist as stable molecular entities.

The simplest interatomic force is that responsible for bonding a sodium (Na) atom to chlorine (Cl) in sodium chloride, NaCl. Sodium belongs to the family of elements known as the **alkali metals** (a name derived from the Arabic word *alquilim*, meaning the ashes of the plant saltwort, used in making soda ash). The alkali metals occupy the first column of the periodic table and are distinguished by having a single electron in their outermost populated shell. Chlorine is a member of the class of elements known as the **halogens.** The halogens, missing a single electron to complete their outermost orbital shell, occupy the column penultimate to the right of the periodic table. It is relatively easy to remove the valence electron from an alkali metal; given a choice, the element would prefer to lose its lonely outermost electron and assume the characteristics of a noble gas, with its remaining orbital shells fully occupied. In a similar fashion, chlorine is eager to gain an electron to complete its outermost shell. Aspirations of both elements are satisfied in the molecule NaCl. We can think of this molecule as a combination of a sodium atom that has given up an electron and a chlorine atom that received it. The sodium atom, missing an electron, has a net positive electric charge, while the chlorine atom, with its extra electron, has a compensatory charge of the opposite sign. The atoms are held in place in the molecule by the electrostatic force associated with these oppositely signed charges. We refer to bonding of this type as **ionic.** The configuration of NaCl is illustrated schematically in Figure 3.1.

The simplest electrically neutral molecule is that formed by the combination of two atoms of hydrogen, H_2. The electrons in H_2 are shared. If the two protons were to coalesce, forming a single nucleus, the molecule would resemble a helium atom. In practice, though, the positive charges keep the protons apart. The electrons are shared between the two protons. They provide on the average a concentration of negative charge between the protons, as schematically illustrated in Figure 3.2; the resulting electrostatic forces account for the stability of the composite entity. When electrons are shared, bonding is said to be **covalent.**

The protons in H_2 maintain a separation of about 0.74 Å, which may be compared with the average displacement of about 2.8 Å separating the nuclei in NaCl. The relatively large size of the NaCl molecule may be attributed to the size of its component atoms. Stripped of one electron, a sodium atom (denoted by Na^+) has a radius of about 1 Å, while the chlorine atom with its additional electron (denoted by Cl^-) has a radius of about 1.8 Å. If we attempted to squeeze the atoms closer, their electron shells would overlap, giving rise to a strong opposite repulsive force.

We may attribute the great stability of the nitrogen molecule to the fact that the component atoms share three pairs of electrons. Each atom is able, at least part of the time, to complete its outer shell by borrowing electrons from its neighbor. It requires more than twice as much energy to fragment N_2 as to break apart H_2. The oxygen molecule, in which two pairs of electrons are shared, presents an intermediate case. We refer to the bond connecting the atoms in an N_2 molecule as a **triple bond**; that joining the atoms in O_2 is a **double bond,** while that linking the atoms in H_2 is a **single bond.** Sometimes, to emphasize the nature of the bonds, chemists prefer to indicate nitrogen, oxygen, and hydrogen molecules with the symbols N≡N, O=O, and H–H, rather than the more economical notation, N_2, O_2, and H_2. We shall generally opt for economy in what follows.

Table 3.1 Chemical symbols of the elements

Symbol	Element	Symbol	Element	Symbol	Element
Ac	Actinium	Ge	Germanium	Po	Polonium
Ag	Silver	H	Hydrogen	Pr	Praseodymium
Al	Aluminum	Ha	Hahnium	Pt	Platinum
Am	Americium	He	Helium	Pu	Plutonium
Ar	Argon	Hf	Hafnium	Ra	Radium
As	Arsenic	Hg	Mercury	Rb	Rubidium
At	Astatine	Ho	Holmium	Re	Rhenium
Au	Gold	Hs	Hassium	Rf	Rutherfordium
B	Boron	I	Iodine	Rh	Rhodium
Ba	Barium	In	Indium	Rn	Radon
Be	Beryllium	Ir	Iridium	Ru	Ruthenium
Bh	Bohrium	K	Potassium	S	Sulfur
Bi	Bismuth	Kr	Krypton	Sb	Antimony
Bk	Berkelium	La	Lanthanum	Sc	Scandium
Br	Bromine	Li	Lithium	Se	Selenium
C	Carbon	Lr	Lawrencium	Sg	Seaborgium
Ca	Calcium	Lu	Lutetium	Si	Silicon
Cd	Cadmium	Md	Mendelevium	Sm	Samarium
Ce	Cerium	Mg	Magnesium	Sn	Tin
Cf	Californium	Mn	Manganese	Sr	Strontium
Cl	Chlorine	Mo	Molybdenum	Ta	Tantalum
Cm	Curium	Mt	Meitnerium	Tb	Terbium
Co	Cobalt	N	Nitrogen	Tc	Technetium
Cr	Chromium	Na	Sodium	Te	Tellurium
Cs	Cesium	Nb	Niobium	Th	Thorium
Cu	Copper	Nd	Neodymium	Ti	Titanium
Db	Dubnium	Ne	Neon	Tl	Thallium
Dy	Dysprosium	Ni	Nickel	Tm	Thulium
Er	Erbium	No	Nobelium	U	Uranium
Es	Einsteinium	Np	Neptunian	V	Vanadium
Eu	Europium	O	Oxygen	W	Tungsten
F	Fluorine	Os	Osmium	Xe	Xenon
Fe	Iron	P	Phosphorus	Y	Yttrium
Fm	Fermium	Pa	Protactinium	Yb	Ytterbium
Fr	Francium	Pb	Lead	Zn	Zinc
Ga	Gallium	Pd	Palladium	Zr	Zirconium
Gd	Gadolinium	Pm	Promethium		

Covalent bonding allows the synthesis of structures more complex than the **diatomic** (two-atom) species discussed above. The principles are similar. In the water molecule, hydrogen atoms are linked to the oxygen atom by single bonds. The oxygen atom completes its outer shell by borrowing two electrons, one from each of the hydrogen atoms. At the same time, the oxygen atom contributes two electrons to the partnership, one to each hydrogen atom, allowing the shells of the hydrogen atoms also to be filled.

Carbon, located near the middle of the second row of the periodic table, is a "switch-hitter": it can be stabilized by either giving up or receiving electrons. The central atom in carbon dioxide (CO_2), carbon, is linked to each of the oxygen atoms by a double bond. The carbon atom in CO_2 surrenders two electrons to each oxygen atom, completing the L shell of the oxygen atoms while simultaneously adjusting itself to an electron configuration analogous to helium. The carbon atom in CO_2 is said to be **oxidized**; it has given up electrons. The carbon atom in CH_4 forms single bonds with each of the four hydrogen atoms. It is on the receiving side of electron transfer in this case, completing its L shell by borrowing four electrons, one from each of the hydrogen atoms; the carbon atom in CH_4 is said to be **reduced.**

The oxidation (or reduction) state of an element in a compound is represented by a number obtained according to the following rule. We assign an oxidation number of -2 to each oxygen atom in the compound (assuming that oxygen, gaining two electrons, carries a charge of -2), while to each hydrogen atom we associate a number of $+1$ (assuming that hydrogen, surrendering one electron, acquires a charge $+1$).

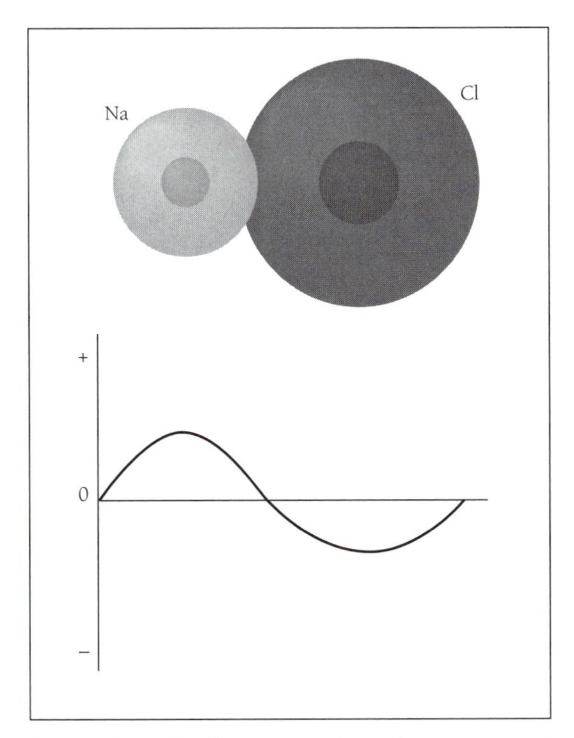

Figure 3.1 Configuration of NaCl. Lower graph shows the distribution of charge.

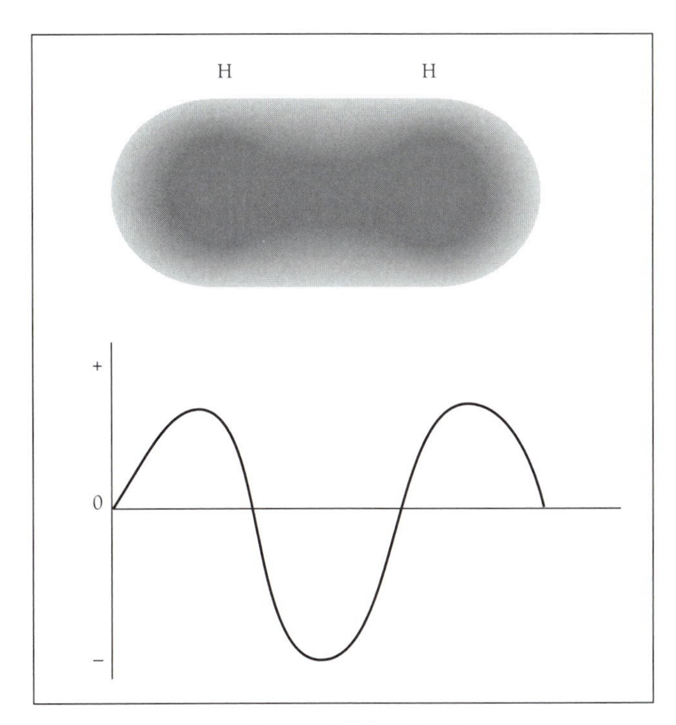

Figure 3.2 Same as Figure 3.1, but for H_2.

The oxidation number of the element is determined by subtracting the sum of the numbers assigned to each oxygen and hydrogen atom from the electrical charge of the compound. Thus, the sum of the oxidation numbers of each constituent equals the residual charge of the compound.

Example 3.1: Determine the oxidation number of the specified elements in the following compounds:

a) carbon in carbon dioxide (CO_2)
b) carbon in methane (CH_4)
c) nitrogen in ammonium ion (NH_4^+)
d) nitrogen in nitrate ion (NO_3^-)

Answers:

a) We assign an oxidation number of −2 for each oxygen.
$C + 2(O) =$ electrical charge of compound
$C + 2(-2) = 0$
$C = +4$

b) We assign an oxidation number of +1 for each hydrogen.
$C + 4(H) = 0$
$C + 4(+1) = 0$
$C = -4$

c) $N + 4(+1) = +1; N = -3$
d) $N + 3(-2) = -1, N = +5$ ■

Elements with positive oxidation numbers, such as the carbon atom in CO_2 or the nitrogen atom in NO_3^-, are said to be oxidized (they gave up electrons). Elements with negative oxidation numbers, such as carbon in CH_4 or nitrogen in NH_4^+, are said to be reduced (they were on the receiving side of electron transfer). The oxidation number of elements in compounds with identical nuclei (referred to as **homonuclear**), such as H_2, N_2, and O_2, where the rule stated above is inapplicable, is assigned a value of zero—electrons are shared equally, and there is no net transfer of charge.

In addition to stable molecules such as N_2, O_2, H_2O, and CO_2, the atmosphere also includes an important group known as **free radicals.** A free radical is a compound characterized by an odd number of electrons. Free radicals are exceptionally reactive; they go to great lengths to pair off their lonely electrons. Among the more important radicals in the atmosphere are NO, NO_2, OH, HO_2, Cl, ClO, Br, and BrO. As we shall see, much of atmospheric chemistry is concerned with complex chains of reactions involving radicals. These chains play a dominant role in the synthesis of O_3 in urban environments; also, they are pivotally involved in removal of O_3 from the stratosphere. We return to a more detailed discussion of radical chemistry in Chapters 13, 14, 15, and 17.

3.3 Acids and Bases

The atmosphere contains a suite of soluble gases known as **acids** (from the Latin word *acidus,* meaning sour). An acid's characteristic property is the hydrogen content of its associated compound that, when placed in water, **dissociates** (comes apart), and is released as a positively charged proton. The dissociation process for sulfuric and nitric acid (H_2SO_4 and HNO_3), two of the more important acidic components of the air, may be summarized by reaction equations as follows:

$$H_2SO_4 \rightarrow H^+ + HSO_4^- \tag{3.1}$$

$$HNO_3 \rightarrow H^+ + NO_3^- \tag{3.2}$$

The arrows in (3.1) and (3.2) indicate transformation of the reactants, appearing on the left, to the products, appearing on the right. The protons formed in (3.1) and (3.2) attach immediately to one or more water molecules. The proton in (3.1) and (3.2) should be thought of not as an isolated H^+ but rather as a larger molecular unit such as H_3O^+ (formed by attaching the proton to a water molecule) in the liquid phase, or as an even more complex aggregate such as $H_9O_4^+$ (produced by bonding H^+ to as many as four neighboring water molecules). We can sidestep this complication by denoting the proton released in (3.1) and (3.2) by H^+ (aq), explicitly distinguishing between the relatively complex form of H^+ in aquatic media as compared with the relatively simple configuration of the element in the gas phase.

Protons, being small and mobile, are exceptionally reactive. They can fit in almost anywhere, displacing a diversity of elements in a variety of compounds. Add a metal such as zinc to a strong acid and you can watch it dissolve before your eye. Marble statues preserved from antiquity crumble under the attack of **acid rain.** Aluminum, stable for thousands of years in soils, is leached to groundwater by acid rain and transported to streams and lakes. Fish die. It is astonishing to think that all of this damage is effected by simple protons.

Complementing the acids is a family of compounds known as **bases,** which have the ability to absorb protons in solution. Ammonia gas, NH_3, provides a simple example of a base; added to solution, NH_3 absorbs a proton and is converted to NH_4^+ by the reaction

$$NH_3 + H^+ \text{ (aq)} \rightarrow NH_4^+ \tag{3.3}$$

Water can serve as both an acid and a base. Its role as an acid is illustrated by the reaction

$$H_2O \rightarrow H^+ + OH^-, \tag{3.4}$$

while its function as a base is exemplified by

$$H^+ + H_2O \rightarrow H_3O^+ \tag{3.5}$$

The composite reaction is summarized by the equation

$$H_2O + H_2O \leftrightarrow H_3O^+ + OH^- \tag{3.6}$$

The reaction indicated in (3.6) can proceed either to the right or to the left. In equilibrium, the number of reactions per unit time producing H_3O^+ and OH^- is exactly equal to the number of compensatory reactions involved in removal of these species; otherwise, water would decompose completely, transforming irreversibly to H_3O^+ and OH^-.

Thermodynamic equilibrium corresponds to a condition where all possible reactions, including those involving the radiation field (light), are reversible. Consider a reaction of substance A with substance B forming substances C and D, represented by

$$A + B \leftrightarrow C + D \tag{3.7}$$

In thermodynamic equilibrium, the concentrations of A, B, C, and D may be shown to be related according to the equation

$$\frac{[C][D]}{[A][B]} = K, \tag{3.8}$$

where [X] denotes the concentration of species X and K is a quantity known as the **equilibrium constant.** The value of K depends on temperature and pressure, on the nature of the compounds A, B, C, and D, and on the choice of units used to specify concentration. Concentrations in liquid media are most frequently specified in units of moles per liter (moles l^{-1}). A **liter** is defined as a volume of 1×10^3 cm^3, equivalent to a cube measuring 10 cm on each side.

A **mole** of a chemical substance X corresponds to a number of individual units (atoms, molecules, etc.) of X equal to 6.02×10^{23}. The number 6.02×10^{23} is known as **Avogadro's number,** named after the Italian physicist Amedeo Avogadro (1776–1856). The utility of the molar concept arises as a consequence of the relation between Avogadro's number and the mass number of a compound. A mole of hydrogen atoms (with a mass number of 1) has a mass of 1 g; a mole of oxygen atoms (with a mass number of 16) has a mass of 16 g, while a mole of oxygen molecules (each molecule containing two atoms of O corresponding to a mass number of 32) has a mass of 32 g. A liter of water has a mass of 1×10^3 g, corresponding to 55.6 moles of H_2O molecules [(1000 g)/(18 g mol^{-1})]. It is important to remember that a mole is a measure of the number of units of a substance; it is not a measure of mass, although the latter can be readily derived if we know the mass number of the constituent material.

Returning to reaction (3.6), in equilibrium we can write

$$\frac{[H_3O^+][OH^-]}{[H_2O]} = K \tag{3.9}$$

The concentration of H_2O is not significantly altered by chemical reactions in solution. It is convenient, and usual, to set $[H_2O]$ equal to 1 in applying concepts of thermodynamic equilibrium to the liquid phase; this requires a simple redefinition of units for the equilibrium constant K. With $[H_3O^+]$ and $[OH^-]$ expressed in units of mol l^{-1}, K has the value 10^{-14} mol^2 l^{-2}.

The concentration of positive charge must exactly equal the concentration of negative charge in any finite-sized volume; any imbalance would be immediately removed by the strong associated electrostatic force field. Thus, for pure water,

$$[H_3O^+] = [OH^-] \tag{3.10}$$

Substituting for $[OH^-]$ in the revised equation (3.9), we find

$$[H_3O^+]^2 = 10^{-14}\text{mol}^2\ l^{-2} \tag{3.11}$$

It follows that the concentrations of H_3O^+ (equivalent to H^+(aq)) and OH^- in pure liquid water are both equal to 10^{-7} mol l^{-1}.

More complex mixtures contain charged species in addition to H^+(aq) and OH^-. These species must be considered in writing an equation for the charge balance analogous to (3.10). In general, the concentration of H^+(aq) can differ significantly from $[OH^-]$. Concentrations of H^+(aq) and OH^- are constrained by (3.9), however, to satisfy the relation

$$[H^+(aq)][OH^-] = 10^{-14} mol^2\ l^{-2} \quad (3.12)$$

The concentration of $H^+(aq)$ is expressed conventionally in terms of a quantity known as **pH**. The pH of a solution is defined by the relation

$$pH = -\log_{10}([H^+(aq)]), \quad (3.13)$$

where $[H^+(aq)]$ is measured in units of mol l^{-1}. The pH of pure water is 7.0. Solutions with a pH of less than 7.0 are said to be **acidic**; solutions with a pH higher than 7.0 are defined as **basic**. Values of pH for some common solutions are given in Table 3.2.

The strength of an acid is measured by the degree of dissociation of the compound in solution. For a representative acid, HA, strength can be calculated using the equilibrium constant, *K*, for the corresponding dissociation reaction:

$$H_2O + HA \leftrightarrow H_3O^+ + A^- \quad (3.14)$$

The concentrations of HA, $H^+(aq)$, and A^- satisfy the equilibrium equation

$$\frac{[H^+(aq)][A^-]}{[HA]} = K \quad (3.15)$$

The concentration of A^- is equal to that of HA when $[H^+(aq)]$ equals *K*. Defining a quantity pK analogous to pH according to

$$pK = -\log_{10}(K), \quad (3.16)$$

where *K* is measured in units of mol l^{-1}, we see that dissociation of HA is essentially complete when the pH of the solution is greater than its pK.

Values of pK for selected acids are given in Table 3.3. Compounds characterized by large values of *K* and small values of pK dissociate readily and are classified as **strong acids.** Somewhat arbitrarily, we associate the family of strong acids with values of *K* larger than 10^{-6} (values of pK less than 6). With this convention, hydrochloric (HCl), nitric (HNO_3), and sulfuric (H_2SO_4) acids are classified as strong; carbonic acid (H_2CO_3), formed by adding CO_2 to water, is defined as a **weak acid**.

As we shall see, the presence of CO_2 in the atmosphere ensures that rain should be naturally acidic; under pristine conditions, we expect a pH for rain in equilibrium with the current level of atmospheric CO_2 of about 5.6. The pH of rain falling in eastern parts of North America today is typically less than 4.0, often as low as 3.0. The surplus acidity is due mainly to sulfuric and nitric acid formed from SO_2 and NO_2 introduced to the atmosphere as by-products of fossil fuel combustion. The pH of rain falling in the western United States is often as high as 7.0; acids in this instance are neutralized, in part by NH_3 produced mainly from cattle feedlots, in part by carbonate dust blown into the air from naturally occurring alkaline soils of the West. We return to this topic later in connection with the discussion of the nitrogen cycle in Chapter 12 and the discussion of acid rain in Chapter 18.

Table 3.2 pH *for selected solutions*

Solution	*pH*
lime juice	1.8
vinegar	3.0
carbonated water	5.6
pure water	7.0
sea water	8.4
stomach antacid	10.0
household ammonia	11.0
typical drain cleaner	13.0

Table 3.3 pK *for selected solutions*

Acid	*Reaction*	*pK*
hydrochloric	$HCl \leftrightarrow H^+ (aq) + Cl^-$	very small
hydrofluoric	$HF \leftrightarrow H^+ (aq) + F^-$	3.2
acetic	$CH_3COOH \leftrightarrow H^+ (aq) + C_2H_3O_2^-$	4.7
carbonic	$H_2CO_3 \leftrightarrow H^+ (aq) + HCO_3^-$	6.4
water	$H_2O \leftrightarrow H^+ (aq) + OH^-$	14.0

3.4 Energy on the Molecular Level

An atom or molecule can respond to an input of energy in a number of different ways. If we smash an atom with a fast moving electron or proton, or with another atom or molecule, the atom can speed up, gaining kinetic energy. If the energy of the collision partner is large enough, in addition to changing the kinetic energy of the target, we can induce a change in the orbital motion of the electrons in the atomic target. This could also be effected by absorption of light. We could, for example, force the electron in the normal, ground state of the hydrogen atom to flip from the K to the L shell by delivering an amount of energy equivalent to about 1.6×10^{-11} ergs. To cause such an internal adjustment in a mole of hydrogen atoms (with a mass of 1 g) would require the expenditure of about 10^{13} ergs of energy, equivalent to 2.3×10^5 **calories** (1 calorie is equal to 4.19×10^7 ergs). To place this in context, one calorie would suffice to raise the temperature of a gram of liquid water by 1° C. Obviously, large quantities of energy are required to alter the electronic structure of an atom or molecule.

The smallest particles in nature obey a set of rules distinct from, though consistent with, those discussed earlier in connection with the larger macroscopic world. Properties of microscopic particles are prescribed by the laws of **quantum mechanics.** According to quantum mechanics, the motion of electrons in an atom or molecule is restricted to certain well-defined energy states or levels, associated, for example, with the orbital shells described above. A molecule can store energy internally, not only in orbital motion of its bound electrons but also through motion of its component nuclei. There are two important modes of nuclear motion: the molecule as a whole can **rotate** about its center of mass; in addition, the nuclei can **vibrate**, like particles linked by a spring. The different modes of nuclear motion are illustrated for a diatomic molecule in Figure 3.3. The particular state of a molecule is specified by a set of (quantum) numbers separately defining the condition of its electrons and nuclei. The state of an atom is completely defined by specifying the shells occupied by its electrons.

As noted above for hydrogen, relatively large quantities of energy are required to change the electronic state (orbital configuration) of an atomic system. If the transition is to be effected by absorption of light, we need radiation of relatively short wavelength (ultraviolet, for example, in the case of hydrogen). Much smaller quantities of energy (longer wavelengths of light) suffice to alter the vibrational state of a molecule. It is even easier to induce a change in rotation. As we shall see in Chapter 6, the atmosphere is bathed in two separate radiation fields, one originating in the sun, consisting mainly of relatively short wavelengths in the visible and ultraviolet portions of the spectrum, the other arising largely at the surface, comprising for the most part longer wavelengths in the infrared portions of the spectrum. Absorption of sunlight by atmospheric gases is primarily associated with changes in the electronic structure of molecules. Absorption of surface radiation is largely due to changes in the vibrational and rotational states of molecules.

Molecular oxygen, for example, absorbs sunlight at wavelengths of less than 2400 Å. The electronic configuration of the molecule is altered. In its new state the molecule is unstable; its nuclei fly apart spontaneously, resulting in the production of two separate atoms of oxygen. The process is summarized as follows:

$$h\nu + O_2 \rightarrow O + O, \tag{3.17}$$

where $h\nu$ denotes a photon of light. Reaction (3.17) exemplifies a process known as **photodissociation**; it defines the first important step in the synthesis of stratospheric O_3.

Not all molecules are able to alter their vibrational and rotational states by absorption of light. The capacity to do so is confined to molecules with a particular configuration of electric charge, a distribution associated with what is known as an **electric dipole.** At large distances, a dipole looks like a barbell, with clusters of positive and negative charge on either end of the bar (molecule), as illustrated in Figure 3.4. This condition is satisfied, for example, by the water molecule. Carbon dioxide is a linear molecule in its ground (lowest energy or most stable) state. The molecule can bend, however; in this configuration it also has the capacity to interact with radiation. The separate vibrational modes of CO_2 are depicted in Figure 3.5, while the vibrational modes of H_2O are illustrated in Figure 3.6. Water and CO_2, together with O_3, are the most important absorbers of infrared radiation in the atmosphere. Together, they are responsible for

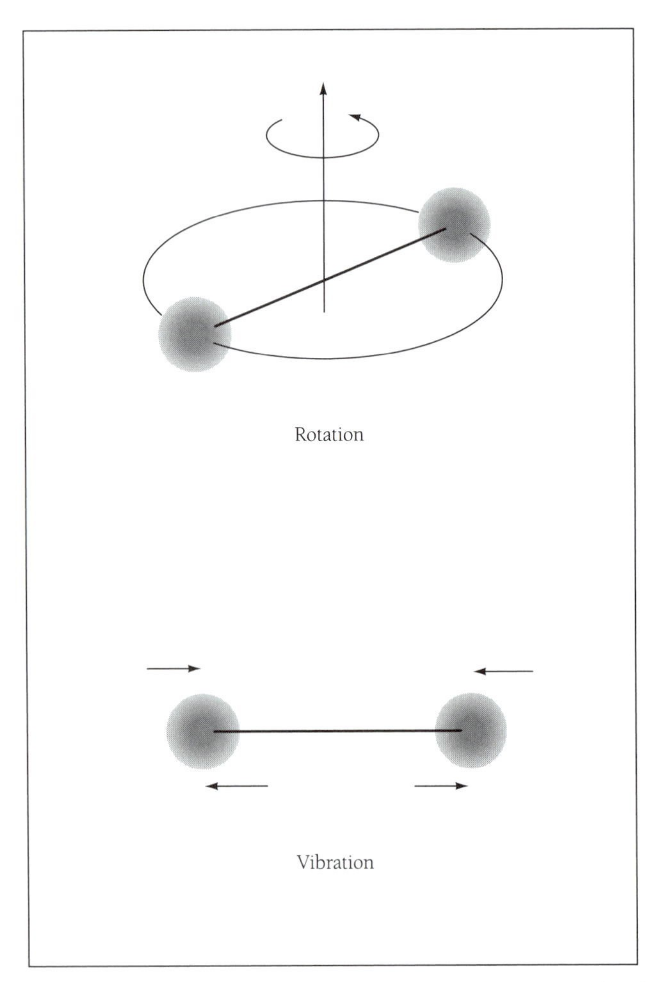

Figure 3.3 Modes of nuclear motion for a diatomic molecule.

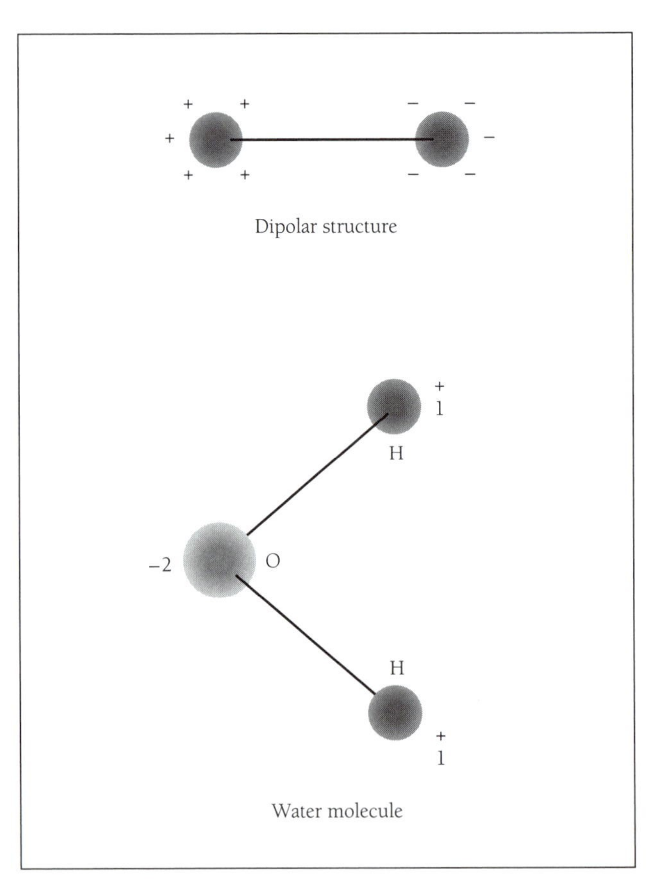

Figure 3.4 Charge distribution for an electric dipole (upper diagram) and an indication of the approximate dipolar configuration of the water molecule (lower diagram).

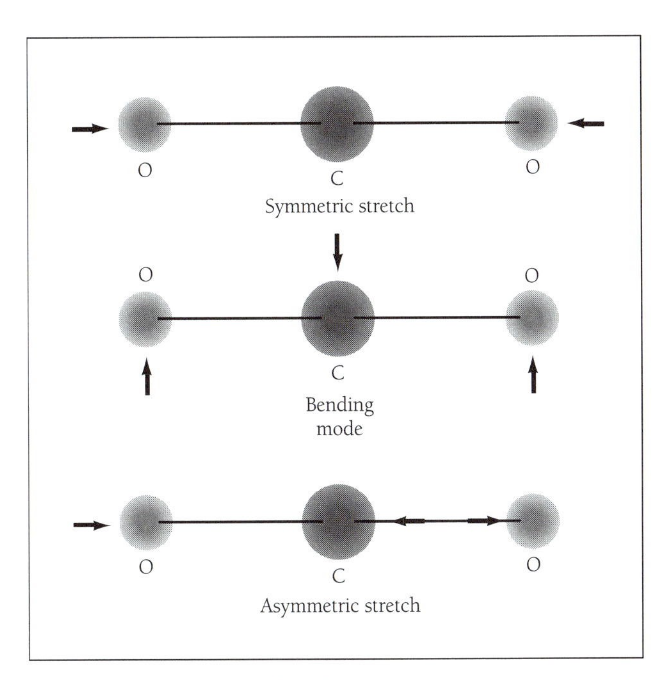

Figure 3.5 Vibrational modes of CO_2. Source: UPL 1994.

raising the temperature of Earth by almost 40° C. It is interesting to note that the major constituents of the atmosphere, the homonuclear diatomic molecules N_2 and O_2, are essentially transparent to light, not only in the infrared but also for the most part in the visible portion of the light spectrum. In the absence of H_2O, CO_2, and O_3, Earth would be freezing cold, and life as we know it would be impossible.

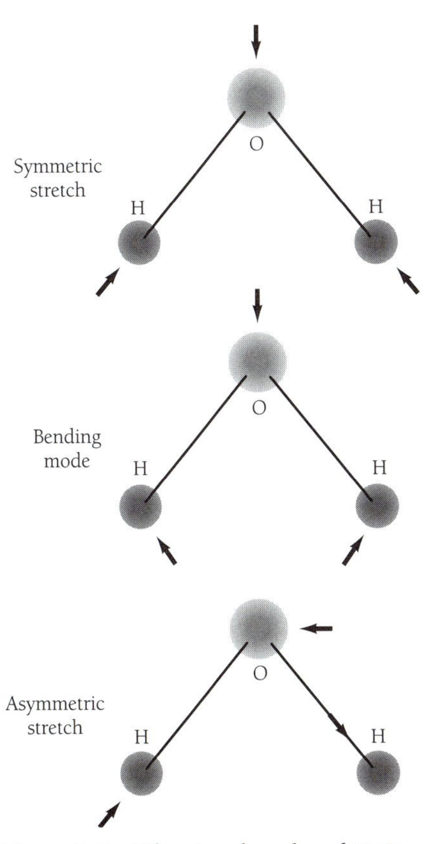

Figure 3.6 Vibrational modes of H_2O.

3.5 The Energetics of Reactions

The first law of thermodynamics states that if heat is added to a system, it may be used either to increase internal energy or to do work. Expressed in mathematical form (see Section 7.3), the first law of thermodynamics is given by the relation

$$\Delta Q = \Delta E + P\Delta V, \tag{3.18}$$

where ΔQ measures the quantity of heat added to the system, ΔE defines the changes in internal energy, P and V denote the pressure and volume of the system, respectively, and $P\Delta V$ provides a measure of the work expended in changing the volume by an amount ΔV.

Suppose that, as a result of the addition of heat, the system is transformed from an initial state A to a final state B. With obvious notation, the change in internal energy is given by

$$\Delta E = E_B - E_A \tag{3.19}$$

If the change in the state of the system takes place at a constant pressure P, the work expended in the transition from A to B may be written in the form

$$P\Delta V = P(V_B - V_A) = PV_B - PV_A \tag{3.20}$$

It follows that

$$\Delta Q = (E_B - E_A) + (PV_B - PV_A) \tag{3.21}$$

or

$$\Delta Q = (E_B + PV_B) - (E_A + PV_A) \tag{3.22}$$

Defining a quantity H by the relation

$$H = E + PV, \tag{3.23}$$

the first law of thermodynamics may be rewritten as

$$\Delta Q = (H_B - H_A) \tag{3.24}$$

or

$$\Delta Q = \Delta H \tag{3.25}$$

The quantity H defined in equation (3.23) is known as the **enthalpy** of the system.

Note that, if a system evolves at constant pressure from state A to state B, the quantity of heat expended in the transition ΔQ is given simply by the difference between the final and initial values of enthalpy (ΔH). If ΔH is positive (if the enthalpy of state B is higher than that of state A), heat is expended in the transition from A to B (ΔQ is positive). On the other hand, if ΔH is negative (H_B is less than H_A), heat will be released (ΔQ will be negative).

Note also that the quantity of heat evolved (or expended) in the transition from A to B is independent of the path followed in effecting the transition. For example, if the transition from A to B were to proceed through an intermediate state C, the net change in heat would still be expressed as a difference between the enthalpies of the final (B) and initial (A) states. The heat evolved (or expended) depends on the *difference* between the enthalpies of the states involved in the transition.

It is useful to introduce the concept of **standard states** of the elements, meaning states for which enthalpies may be conveniently set equal to zero. Enthalpies for compounds assembled from specific combinations of elements may then be expressed with reference to zero points established for the appropriate standard states.

Standard states for elements are selected on the basis of the form in which we expect to find the elements under typical laboratory conditions for temperatures of interest. In what follows, we focus attention on reactions taking place at room temperature (298 K). Standard states for a number of key elements are indicated in Table 3.4.

When the standard state is identified as a gas, it is assumed that the pressure is low enough so that the gas may be considered from a thermodynamic point of view as a perfect gas. If the standard state is identified as either a liquid or a solid, the state is assumed to be present at a pressure of 1 atm.

Heats of formation for a number of compounds of interest in the atmosphere are given in Table 3.5. Heats of formation are expressed in terms of the change in enthalpy involved in assembling a compound from the appropriate mix of its component elements, assuming that the elements are initially supplied in their standard states. Changes in enthalpies listed here are expressed in units of **kilocalories per mole** (kcal mol^{-1}), that is, the values of ΔH given here define the heat (in kcal) expended (positive values of ΔH) in forming a mole of a specific compound. Consider CH_4 for example. To form a molecule of CH_4 from the elements in standard states, we begin with 1 mole of C as graphite and 2 moles of H as H_2. The change in enthalpy involved in forming 1 mole of CH_4 at a temperature of 298 K is equal to -17.88 kcal mol^{-1}; that is, 17.88 kcal of heat are released in forming a mole of CH_4 at 298 K.

The following examples illustrate how the data in Table 3.5 may be used to estimate the changes in enthalpy associated with specific reactions. If the change in enthalpy is negative, heat is released in the reaction and the reaction is said to be **exothermic.** If the change in enthalpy is positive, heat must be supplied and the reaction is said to be **endothermic.** If the value of ΔH is large and positive, it is unlikely that the reaction can proceed under atmospheric conditions.

Table 3.4 Standard states for selected elements at 298 K

Element	*Standard state*	*ΔH (298 K)*
H	H_2, gas	0
C	C(graphite), solid	0
N	N_2, gas	0
O	O_2, gas	0
F	F_2, gas	0
S	S(rhombic), solid	0
Cl	Cl_2, gas	0
Br	Br_2, liquid	0

Example 3.2: Estimate the change in enthalpy associated with the reaction $N + NO \rightarrow N_2 + O$ at a temperature of 298 K.

Answer: Using the data in Table 3.5, the changes in enthalpy associated with formation of N, NO, N_2, and O are given by

$\Delta H(N) = +113.0$ kcal mol^{-1}

$\Delta H(NO) = +21.57$ kcal mol^{-1}

$\Delta H(N_2) = 0$ kcal mol^{-1}

$\Delta H(O) = +59.57$ kcal mol^{-1}

The change in enthalpy associated with formation of the reactants N and NO is given by $\Delta H(\text{reactants}) = +113.0 + 21.57 = 134.57$ kcal mol^{-1}.

The change in enthalpy associated with the products is given by $\Delta H(\text{products}) = 0 + 59.57 = 59.57$ kcal mol^{-1}.

The change in enthalpy associated with the reaction is given by

$$\begin{aligned}\Delta H(\text{reaction}) &= \Delta H(\text{products}) - \Delta H(\text{reactants})\\ &= 59.57 - 134.57\\ &= -75.0 \text{ kcal mol}^{-1}\end{aligned}$$

The reaction is strongly exothermic. ■

Example 3.3: Estimate the change in enthalpy associated with the reaction $N + O_2 \rightarrow NO + O$ at a temperature of 298 K.

Answer: The changes in enthalpies associated with formation of N, O_2, NO, and O are given by

$\Delta H(N) = +113.0$ kcal mol^{-1}

$\Delta H(O_2) = 0$ kcal mol^{-1}

$\Delta H(NO) = +21.57$ kcal mol^{-1}

$\Delta H(O) = +59.57$ kcal mol^{-1}

(All values are expressed in units of kcal mol^{-1}.)

Thus,

$\Delta H(\text{reactants}) = +113.0$ kcal mol^{-1}

$\Delta H(\text{products}) = +21.57 + 59.57 = +81.14$ kcal mol^{-1}

$\Delta H(\text{reaction}) = 81.14 - 113.0 = -31.86$ kcal mol^{-1}

This reaction is also strongly exothermic but less so than the reaction in Example 3.2. ■

Example 3.4: Estimate the change in enthalpy associated with the reaction $ClO + BrO \rightarrow Cl + Br + O_2$.

Answer: The changes in enthalpies associated with formation of ClO, BrO, Cl, Br, and O_2 are given by

$\Delta H(ClO) = +24.4$ kcal mol^{-1}

$\Delta H(BrO) = +26.0$ kcal mol^{-1}

$\Delta H(Cl) = +28.9$ kcal mol^{-1}

Table 3.5 Changes in enthalpy associated with formation of 1 mole of listed species at a temperature of 298 K

Species	ΔH_f (298) (kcal/mol)	Species	ΔH_f (298) (kcal/mol)	Species	ΔH_f (298) (kcal/mol)	Species	ΔH_f (298) (kcal/mol)
H	52.1	C_2H_5	28.4 ± 0.5	CH_3CF_3	−179 ± 2	BrNO	19.7
H_2	0.00	C_2H_6	−20.0	CF_2CF_3	−213 ± 2	BrONO	25 ± 7
O	59.57	CH_2CN	57 ± 2	CHF_2CF_3	−264 ± 2	$BrNO_2$	17 ± 2
$O(^1D)$	104.9	CH_3CN	15.6	Cl	28.9	$BrONO_2$	<11
O_2	0.00	CH_2CO	−11 ± 3	Cl_2	0.00	BrCl	3.5
$O_2(^1\Delta)$	22.5	CH_3CO	−5.8	HCl	−22.06	CH_2Br	40±2
$O_2(^1\Sigma)$	37.5	CH_3CHO	−39.7	ClO	24.4	$CHBr_3$	6±2
O_3	34.1	C_2H_5O	−4.1	ClOO	23.3 ± 1	$CHBr_2$	45 ± 2
HO	9.3	CH_2CH_2OH	−10 ± 3	OClO	22.6 ± 1	CBr_3	48 ± 2
HO_2	2.8 ± 0.5	C_2H_5OH	−56.2	$ClOO_2$	>16.7	CH_2Br_2	−2.6 ± 2
H_2O	−57.81	CH_3CO_2	−49.6	ClO_3	52 ± 4	CH_3Br	−8.5
H_2O_2	−32.60	$C_2H_5O_2$	−6 ± 2	Cl_2O	19.5	CH_3CH_2Br	−14.8
N	113.00	CH_3COO_2	−41 ± 5	Cl_2O_2	31 ± 3	CH_2CH_2Br	32 ± 2
N_2	0.00	CH_3OOCH_3	−30.0	Cl_2O_3	37 ± 3	CH_3CHBr	30 ± 2
NH	85.3	C_3H_5	39.4	HOCl	−18 ± 3	I	25.52
NH_2	45.3	C_3H_6	4.8	ClNO	12.4	I_2	14.92
NH_3	−10.98	n-C_3H_7	22.6 ± 2	$ClNO_2$	3.0	HI	6.3
NO	21.57	i-C_3H_7	19 ± 2	ClONO	1.3	CH_3I	3.5
NO_2	7.9	C_3H_8	−24.8	$ClONO_2$	5.5	CH_2I	52 ± 2
NO_3	17.6 ± 1	C_2H_5CHO	−44.8	FCl	−12.1	IO	<28
N_2O	19.61	CH_3COCH_3	−51.9	CCl_2	57 ± 5	INO	29.0
N_2O_3	19.8	$CH_3COO_2NO_2$	−62 ± 5	CCl_3	17 ± 1	INO_2	14.4
N_2O_4	2.2	F	19.0 ± 0.1	CCl_3O_2	2.7 ± 1	S	66.22
N_2O_5	2.7 ± 1	F_2	0.00	CCl_4	−22.9	S_2	30.72
HNO	23.8	HF	−65.14 ± 0.2	$CHCl_3$	−24.6	HS	34.2 ± 1
HONO	−19.0	HOF	−23.4 ± 1	$CHCl_2$	23 ± 2	H_2S	−4.9
HNO_3	−32.3	FO	26 ± 5	CH_2Cl	29 ± 2	SO	1.3
HO_2NO_2	−12.5 ± 2	F_2O	5.9 ± .4	CH_2Cl_2	−22.8	SO_2	−70.96
C	170.9	FO_2	6 ± 1	CH_3Cl	−19.6	SO_3	−94.6
CH	142.0	F_2O_2	5 ± 2	ClCO	−5 ± 1	HSO	−1 ± 5
CH_2	93 ± 1	FONO	−15 ± 7	$COCl_2$	−52.6	HSO_3	−92 ± 2
CH_3	35 ± 0.2	FNO	−16 ± 2	CHFCl	−15 ± 2	H_2SO_4	−176
CH_4	−17.88	FNO_2	−26 ± 2	CH_2FCl	−63 ± 2	CS	67 ± 2
CN	104 ± 3	$FONO_2$	2.5 ± 7	CFCl	7 ± 6	CS_2	28.0
HCN	32.3	CF	61 ± 2	$CFCl_2$	−22 ± 2	CS_2OH	26.4
CH_3NH_2	−5.5	CF_2	−44 ± 2	$CFCl_3$	−68.1	CH_3S	29.8 ± 1
NCO	38 ± 3	CF_3	−112 ± 1	CF_2Cl_2	−117.9	CH_3SOO	18 ± 2
CO	−26.42	CF4	−223.0	CF_3Cl	−169.2	CH_3SO_2	−57
CO_2	−94.07	CHF_3	−166.8	$CHFCl_2$	−68.1	CH_3SH	−5.5
HCO	10 ± 1	CHF_2	−58 ± 2	CHF_2Cl	−115.6	CH_2SCH_3	32.7 ± 1
CH_2O	−26.0	CH_2F_2	−107.2	CF_2Cl	−67 ± 3	CH_3SCH_3	−8.9
COOH	−53 ± 2	CH_2F	−8 ± 2	COFCl	−102 ± 2	CH_3SSCH_3	−5.8
HCOOH	−90.5	CH_3F	−56 ± 1	CH_3CF_2Cl	−127 ± 2	OCS	−34
CH_3O	4 ± 1	FCO	−41 ± 15	CH_2CF_2Cl	−75 ± 2		
CH_3O_2	4 ± 2	COF_2	−153 ± 2	C_2Cl_4	−3.0		
CH_2OH	−3.6 ± 1	CF_3O	−157 ± 2	C_2HCl_3	−1.9		
CH_3OH	−48.2	CF_3O_2	−148 ± 5	CH_2CCl_3	17 ± 2		
CH_3OOH	−31.3	CF_3OH	−214 ± 5	CH_3CCl_3	−34.0		
CH_3ONO	−15.6	CF_3OOCF_3	−360	CH_3CH_2Cl	−26.8		
CH_3ONO_2	−28.6	CF_3OOH	−184 ± 4	CH_2CH_2Cl	22 ± 2		
$CH_3O_2NO_2$	−10.6 ± 2	CFOF	−183 ± 3	CH_3CHCl	17.6 ± 1		
C_2H	133 ± 2	CH_3CH_2F	−63 ± 2	Br	26.7		
C_2H_2	54.35	CH_3CHF	−17 ± 2	Br_2	7.39		
C_2H_2OH	30 ± 3	CH_2CF_3	−124 ± 2	HBr	−8.67		
C_2H_3	72 ± 3	CH_3CHF_2	−120 ± 1	HOBr	−14 ± 6		
C_2H_4	12.45	CH_3CF_2	−71 ± 2	BrO	26 ± 5		

JPL. "Chemical Kinetics and Photochemical Data for Use in Stratospheric Modeling," *(Pasadena, Cal: Jet Propulsion Lab, 1994): 194*

$\Delta H(Br) = +26.7$ kcal mol^{-1}

$\Delta H(O_2) = 0$ kcal mol^{-1}

Thus,

$\Delta H(\text{reactants}) = +24.4 + 26.0 = +50.4$ kcal mol^{-1}

$\Delta H(\text{products}) = +28.9 + 26.7 = +55.6$ kcal mol^{-1}

$\Delta H(\text{reaction}) = +55.6 - 50.4 = 5.2$ kcal mol^{-1}

This reaction is endothermic, but only slightly so. ■

Example 3.5: Estimate the change in enthalpy associated with the reactions (1) $OH + HCl \rightarrow H_2O + Cl$ and (2) $OH + HF \rightarrow H_2O + F$.

Answer: Changes in enthalpies involved in formation of the various species implicated in reactions (1) and (2) are

$\Delta H(OH) = +9.3$ kcal mol^{-1}

$\Delta H(H_2O) = -57.81$ kcal mol^{-1}

$\Delta H(Cl) = +28.9$ kcal mol^{-1}

$\Delta H(F) = +19.0$ kcal mol^{-1}

$\Delta H(HCl) = -22.06$ kcal mol^{-1}

$\Delta H(HF) = -65.14$ kcal mol^{-1}

The change in enthalpy involved in reaction (1) is evaluated as follows:

$\Delta H(\text{reactants}) = +9.3 - 22.06 = -12.76$ kcal mol^{-1}

$\Delta H(\text{products}) = -57.81 + 28.9 = -28.91$ kcal mol^{-1}

$\Delta H(\text{reaction 1}) = -28.91 - (-12.76) = -16.15$ kcal mol^{-1}

The corresponding quantities for reaction (2) are

$\Delta H(\text{reactants}) = +9.3 - 65.14 = -55.84$ kcal mol^{-1}

$\Delta H(\text{products}) = -57.81 + 19.0 = -38.81$ kcal mol^{-1}

$\Delta H(\text{reaction 2}) = -38.81 - (-55.84) = +17.03$ kcal mol^{-1}

Reaction (1) is exothermic, while reaction (2) is endothermic. The difference is significant in terms of the relative impacts of Cl and F on stratospheric O_3; chlorine and fluorine atoms are converted in the stratosphere to HCl and HF respectively by reactions with CH_4. The chlorine atom bound up in HCl can be liberated, however, by reaction (1). The fluorine atom in HF, on the other hand, is stable. Fluorine tends to accumulate in the stratosphere as unreactive HF. Chlorine persists in significant abundance in the form of the radicals Cl and ClO implicated in removal of O_3. ■

3.6 Summary

We have seen that atoms are assembled from specific combinations of electrons, protons, and neutrons. The chemical properties of an atom are mainly determined by the number of valence electrons, indicated by the column that the atom occupies in the periodic table. The larger the number of vacancies in the outer shell, the greater the reactivity of a particular atom. When atoms react, they form more stable compounds known as molecules. We introduced the notion of the chemical bond, the bonding of atoms associated either with the exchange (ionic) or sharing (covalent) of electrons in the outer orbital shells of atoms. Electrons are usually exchanged or shared to create more stable structures. Highest stability is achieved when the exchange or sharing of electrons results in completion of the outer shells of atoms composing molecular aggregates. In the most important example, the N_2 molecule includes three pairs of shared electrons, accounting for the exceptional stability of the molecule representing the predominant constituent of the atmosphere. Radicals include an odd number of electrons; unpaired electrons account for the exceptional reactivity of radicals. We noted that compounds are termed acids or bases, depending on their tendency to give up or receive protons in a solution. The pH of a solution provides a measure of the concentration of H^+. We introduced the concept of chemical equilibrium, the condition where rates for reactions proceeding in one direction precisely balance rates for reactions in the opposite direction. Concentrations of species in equilibrium adjust to provide this balance. The chemistry of species in solution often approximates this ideal state of equilibrium. Equilibrium provides a less satisfactory approximation, however, for gases in the atmosphere, particularly for those affected by absorption of solar radiation. Ozone is formed as a consequence of the absorption of ultraviolet sunlight by molecular oxygen: reactions involving radicals (as we shall see in Chapters 13 through 17) play an important role in regulating its abundance. We offered a brief introduction to the chemical processes responsible for the interaction of radiation with molecules, specifically noting the importance of vibrational and rotational motion in the context of the transfer of heat between the surface and the atmosphere and between different regions of the atmosphere and the atmosphere and space (the greenhouse effect). (We return to this matter in more detail in Chapters 6 and 7.) Finally, we introduced the concept of enthalpy and outlined the procedures to be followed in evaluating whether a specific reaction is exothermic or endothermic.

Properties of Water

4

Water, as we will see in Chapter 5, is the most important volatile gas, setting Earth apart from its nearest planetary neighbors, Mars and Venus. Water is present on Earth in three phases: ice, liquid, and gas. Liquid water is essential for life as we know it. In many environments, such as deserts and polar ice caps, its absence is the limiting factor in the growth of organisms. More than 70% of Earth's surface is covered by liquid water, and gaseous water is the third most abundant constituent of the atmosphere after N_2 and O_2. Water plays a critical role in the climate, both as the most important greenhouse gas and, in the form of clouds, as the screen limiting transmission of sunlight to the surface. It is the most important atmospheric cleansing agent, removing pollution from the air. Changing phase, water can liberate and consume impressive quantities of energy. In floods, it can cause vast property damage, exacting an enormous toll in loss of human life. Falling to the surface as snow, it can bring a bustling city to its knees. It can change the contours of a coast. Water is the most important agent in the weathering of surface rocks, breaking down even the toughest materials by entering cracks and changing volume on freezing. The significance of water is difficult to overestimate. It is essential that we fully understand the properties of this powerful compound.

4.1 The Molecular Perspective

We previously discussed the differences between solids, liquids, and gases in general terms. Molecules in the solid and liquid phases are more tightly bonded to their neighbors than in the gas phase. Despite the denser packing and more stringent constraints imposed by frequent collisions, molecules in the liquid phase are in a state of constant motion, as is the case for their analogues in a gas. This was demonstrated more than 150 years ago by a botanist named Robert Brown. He observed that tiny particles, less than 10^{-4} cm in diameter, suspended in an otherwise stagnant liquid, began to immediately move, assuming a complicated, apparently random pattern as they diffused through the fluid. He correctly attributed this motion to the transfer of momentum to particles (due to collisions with molecules of the fluid) and showed that the motion of the particles grew more intense as the temperature of the liquid was allowed to increase. The average kinetic energy of the molecules in a liquid is determined by temperature, just as it is for molecules in a gas.

Molecules in a liquid are trapped in what physicists refer to as **potential wells.** Imagine a complex depression in an otherwise flat surface and suppose that a ball bearing is placed at some depth in the depression (or well) and released. Assume that the surface is frictionless. The ball bearing will begin to describe a complicated path up, down, and around the depression. Its motion will be confined to depths at or below the level at which it was released. If additional energy is imparted to the ball bearing, it will rise to higher levels in the depression. If the supplied energy is large enough, the ball bearing may exit the depression and run off across the flat surface. In a liquid the analogue to the depression is the force field imposed on a molecule by its molecular neighbors. From time to time, a molecule can acquire excess energy in a favorable collision at the expense of its partners. Under the right circumstances, it can escape the confines of the potential well in the liquid and move into the gas phase, just as the ball bearing escapes the geometric depression. The process by which molecules are transferred from the liquid to the gas phase is known as **evaporation.** The reverse process, by which molecules are converted from the gas to liquid phase, is referred to as **condensation.**

If we were to sample the kinetic energies of molecules in a liquid, we would find that they exhibit a range of values. The average energy would be given by $\frac{3}{2}kT$. Some of the molecules would have energies significantly larger than $\frac{3}{2}kT$, while energies of others would be much less. The fraction of molecules with energy greater than E can be described by a relation known as the **Boltzmann distribution** and can be shown to be proportional to $\exp(-E/kT)$. To escape the liquid, molecules must have energy greater than a threshold value. It follows that the rate at which molecules leave a liquid sensitively depends on temperature. Liquids stable at low temperature can be converted to gas with the addition of energy (heat) and an associated increase in temperature. Evaporation involves a loss of a liquid's more energetic molecules. Unless additional energy is supplied, a liquid's temperature will drop as a consequence of evaporation. Experimentally, it has been found that 10,519 calories (1 calorie equals 4.18×10^7 erg) of heat are required to evaporate one mole of liquid water at constant pressure. A similar quantity of energy is released when one mole of H_2O vapor is converted to liquid. Evaporation of H_2O plays an important role in cooling the ocean; subsequent condensation and precipitation make major (positive) contributions to the energy budget of the atmosphere.

The equilibrium value for the vapor pressure over a liquid is determined by temperature and by the chemical properties of the liquid. Consider a liquid placed in an evacuated container, as illustrated in Figure 4.1a. Initially the gas pressure is zero. Higher energy molecules immediately begin to evaporate from the surface of the liquid. The rate of evaporation is regulated solely by the liquid's temperature. Gas pressure builds up in the container's head space. With the increase in gas pressure, significant numbers of molecules in the gas phase begin to impact the surface of the liquid, which captures some of these molecules. Over time, an equilibrium is established (Figure 4.1b). The number of molecules leaving the liquid eventually equals the number that return; the rate of evaporation exactly balances the rate of condensation.

Now suppose that the container is fitted with a moveable piston. If the piston is depressed, as illustrated in Figure 4.1c, the molecules in the vapor phase are confined

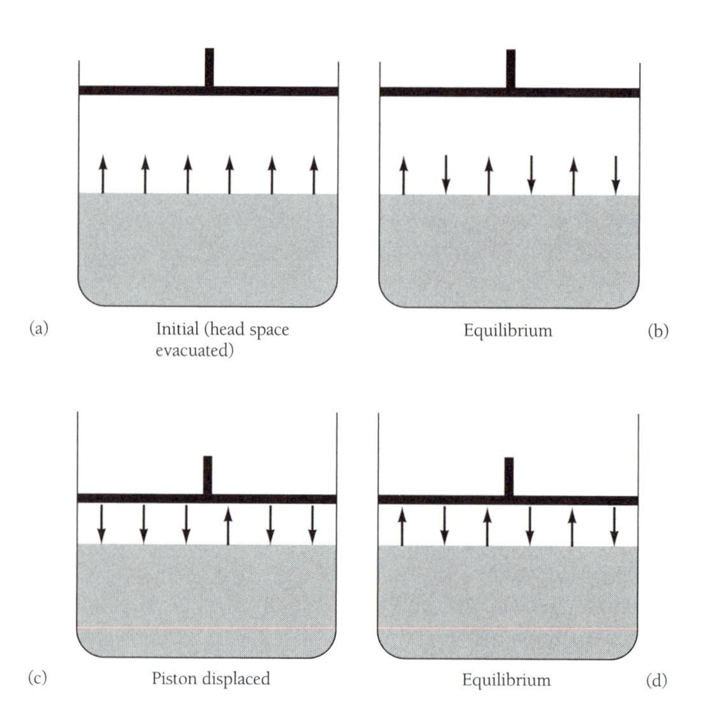

Figure 4.1 Pressure of vapor over a liquid indicating the approach to equilibrium. Case a illustrates the condition in which the headspace is initially evacuated. Case b shows the condition where evaporation is balanced by a return of vapor to the liquid. Cases c and d illustrate conditions where the headspace is restricted by downward displacement of the piston.

to a smaller space, and the vapor pressure increases. If temperature remains constant, the rate of evaporation remains fixed at the value that applied before the piston moved. The rate of condensation rises, due to the increase in gas pressure, reflecting the larger number of molecules hitting the surface. The total number of molecules in the gas phase decreases until the number of molecules hitting the surface declines to the value that applied before the piston moved; that is, until the vapor pressure is restored to its original value (Figure 4.1d). If the temperature of the liquid increases, the rate of evaporation rises and a higher concentration of vapor is required to restore equilibrium: the vapor pressure in equilibrium increases as the temperature rises.

Similar considerations apply to equilibrium between vapors and solids. We refer to the transfer of molecules from solid to vapor as **sublimation.** The reverse process is known as **deposition.** Sublimation of a mole of H_2O vapor from ice requires the addition of 11,955 cal of heat, slightly more than was needed to evaporate a comparable quantity of vapor from the liquid. Conversion of a mole of H_2O as ice to a mole of H_2O as liquid requires a heat source of 1436 cal, equal to the difference between the heats (per mole) for sublimation and evaporation (11,955−10,519 cal mol^{-1}). Like climbing a hill, energy is expended in moving from ice to liquid to vapor. Keep in mind that melting ice at room temperature seems effortless, but in reality it involves the expenditure of significant quantities of energy. Energy is returned as water transforms from vapor back to liquid and ice. Water can evolve directly from ice to gas, or the reverse. The cost or benefit in terms of energy is the same as if the transition had proceeded through the intermediate liquid. This is necessary, otherwise the imbalance in energy could be used to drive a perpetual motion machine, a device with the capacity to do work without expending energy. As in life, there are no free lunches in thermodynamics.

Frequent examples in the everyday world remind us of the significance of the vast quantities of energy involved in changing the phase of water. Some of us have experienced the fit of shivering that occurs on emerging from a swimming pool on a hot, dry summer day, even when the temperature of the water in the pool is less than that of the air. Water on the skin evaporates rapidly, resulting in an appreciable cooling, leading to a readily sensible chilling of the skin. On the other hand, a hot humid day may find people unable to cool down. The perspiration readily evaporating on dry days to cool our skin does not evaporate so easily on humid days; evaporation is partially offset by condensation. Have you ever stopped to think why it is that a glass of water with ice stays cool longer than one without? The answer is that a large portion of the heat conducted to the liquid is expended in melting the ice, rather than in raising the temperature of the liquid.

Water possesses a property rare among most substances: its solid phase (ice) is less dense than its liquid phase. For this reason, ice floats in a glass of water. This unusual property results from the special arrangement of H_2O molecules in ice allowing for a maximum number of hydrogen-bonded interactions between neighboring molecules, with large hexagonal holes that contribute vacant space to the ice structure, resulting in a decrease in its overall density (see Figure 4.2). An important consequence of this property of ice is that, as a lake freezes in winter, the resulting ice remains at the surface. If the entire top of the lake freezes, the water below is insulated, allowing most of the lake to remain liquid and relatively warm. If ice were denser than liquid, lakes would entirely freeze in winter, as newly formed ice would fall to the bottom, and many forms of aquatic life would be doomed to extinction.

It is easier for a molecule of water to move from the liquid to the vapor (gas) phase than for it to transfer from ice to vapor; lower energy molecules are able to effect the transition in the former case. As a consequence, the pressure of water vapor in equilibrium with ice is a more sensitive function of temperature than is the corresponding quantity for a liquid; the equilibrium vapor pressure increases more rapidly with rising temperature over ice than it does over liquid.

4.2 Phase Diagrams

The conditions for equilibrium between the three phases of a substance can be represented simultaneously on a graph known as a **phase diagram.** The phase diagram for water is illustrated by Figure 4.3a. Equilibrium between phases exists only for temperatures and pressures indicated by the lines separating the areas labeled "ice," "liquid," and "vapor." The curves for the three phases intersect for pure water at a temperature of 273.0098 K (0.0098°C) at a pressure of 6.1 mb. For this unique configuration, known as the **triple point,** all three phases (ice, liquid, and vapor) are simultaneously in equilibrium. Ice is unstable at temperatures higher than the triple point; it melts. Liquid is unstable at temperatures below the triple point; it freezes.

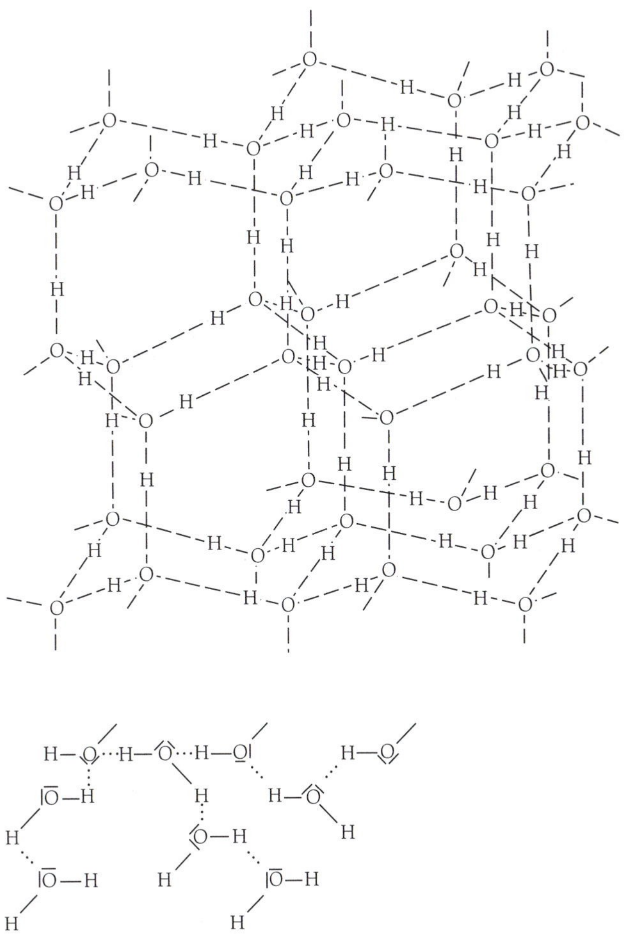

Figure 4.2 Molecular structure of solid water (top) and liquid water (bottom).

The melting-point temperature (the solid line separating the liquid and solid phases) decreases slightly as the value of the total pressure applied to the ice increases; meaning, it is affected by the presence of other gases in addition to H_2O. It is equal to 273 K (0°C) at a pressure of 1 atm (Figure 4.3b): it is this property of water that was initially used to set the zero point for the Celsius scale of temperature. We can understand the shift in the melting point by recalling that the density of water ice is less than the density of its liquid: application of pressure to an ice crystal causes it to more easily revert to the more dense liquid form. The pressure effect on melting is relatively modest under atmospheric conditions; the melting point line in Figure 4.3a is nearly vertical.

The dashed line in Figure 4.4 indicates vapor pressures for H_2O over ice and liquid under conditions where the

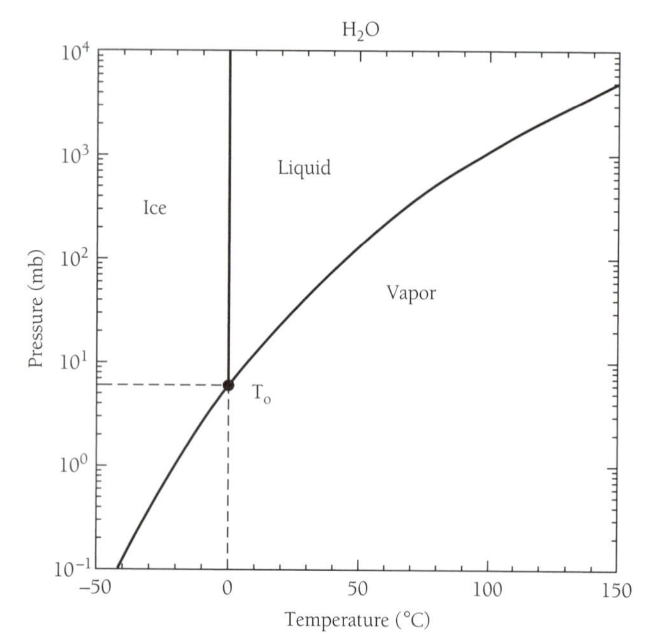

Figure 4.3a Phase diagram for H_2O. T_0 is the triple point. Vapor pressures of H_2O are given in units of mb on the vertical axis.

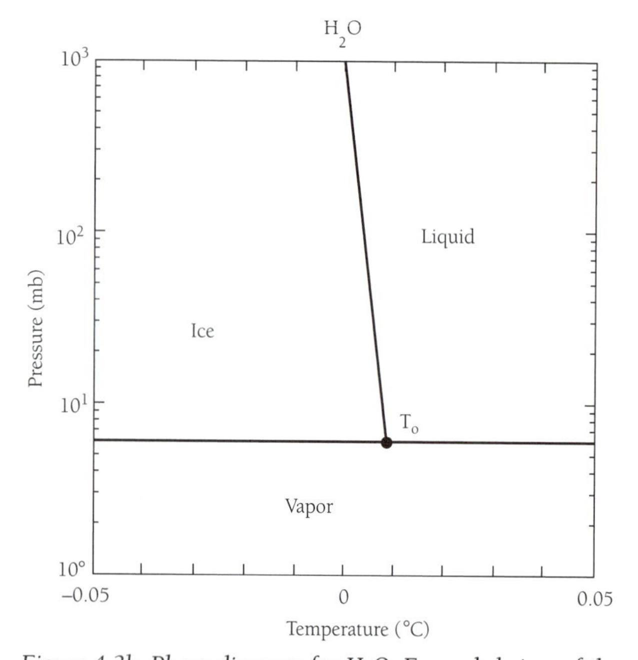

Figure 4.3b Phase diagram for H_2O. Expanded view of the region in Figure 4.3a around the triple point, illustrating the decrease in melting-point temperature with rising pressure.

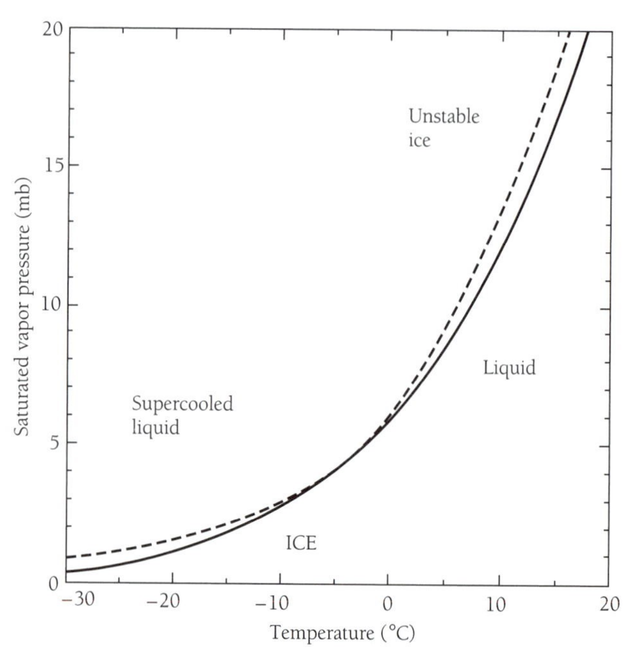

Figure 4.4 Equilibrium vapor pressure of solid and liquid H_2O. Vapor pressures in equilibrium with unstable phases are shown by dashed lines. The dashed lines refer to supercooled liquid at temperatures below 0°C., to the vapor pressure over unstable ice at temperatures above 0°C.

condensed phases are unstable. Thus it indicates the vapor pressures applicable for liquid H_2O below 0°C or for ice above 0°C. Liquid water at a temperature below the freezing point is said to be **supercooled.** Persistence of water as a liquid below the freezing point reflects the relatively complex transition that must occur as water is transformed from the more or less random organization of molecules in a liquid to the highly ordered arrangement characterizing the crystalline structure of a solid. Supercooled water can persist in the atmosphere to temperatures as low as −40°C. Freezing is normally initated on the surface of particles: it is easier for the ordered structure of ice to form in this relatively stable environment than in the disordered interior of a liquid. A particular surface's efficiency in promoting the growth of ice from a supercooled liquid depends on the solid's structure, on the compatibility of the solid's surface-molecular arrangement with the structure of ice. Introduction of a few ice particles can result in the rapid freezing of a supercooled liquid (ice provides the ideal substrate for further growth of ice), and this process is thought to play an important role in producing particles large enough to precipitate from the atmosphere (as discussed later in this chapter).

Figure 4.4 includes an extension of the ice vapor's pressure curve to temperatures above the melting point. The solid is much less stable than the liquid in this regime, as indicated by the dashed line at temperatures above 0°C. The **second law of thermodynamics** states that processes in nature occur spontaneously so as to increase the disorder of a system, to increase a quantity known as **entropy.** The relative disorder of the liquid compared to the solid ensures the rapid demise of ice at temperatures above 0.0098°C.

Imagine an apparatus consisting of two test tubes sharing a common head space as depicted in Figure 4.5 a and b. Suppose that a sample of liquid water is inserted into one of the test tubes, a slab of water ice into the other. Imagine that the head space was initially evacuated and that the system was isolated from the external atmosphere. The unit is surrounded by a bath at fixed temperature, so that the temperature of all parts of the apparatus is held constant.

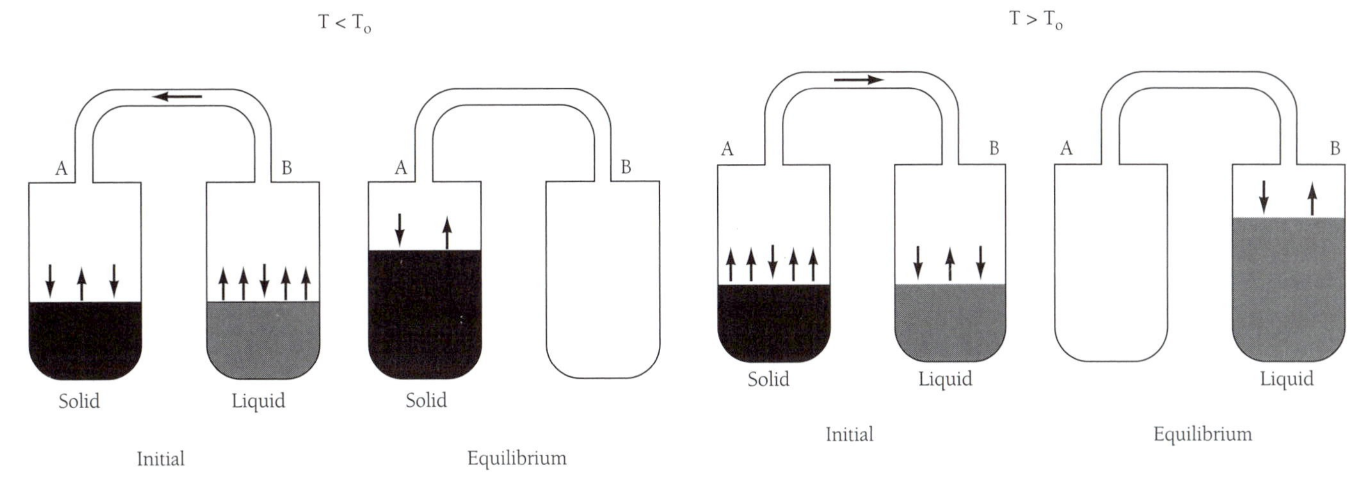

Figure 4.5a Equalibrium between solid and liquid H_2O for T less than T_0 (the triple point temperature).

Figure 4.5b Equalibrium between solid and liquid H_2O for T greater than T_0 (the triple point temperature).

Consider the situation illustrated in Figure 4.5a, where the temperature is set equal to a value less than that at the triple point for water, T_0. The vapor pressure over the ice in test tube A is initially lower than that over the liquid in test tube B. In this case, H_2O vapor will move from test tube B to A, driven by the pressure difference in the head space. The excess vapor arriving at B will deposit, forming additional ice. The pressure differential will be maintained as evaporation and deposition proceed, until all of the H_2O in the liquid phase is transformed to ice. At this point the vapor pressure is everywhere constant and equal to the equilibrium vapor pressure over ice at the given temperature.

A similar sequence ensues, with transfer of H_2O in the opposite direction, if the temperature is set equal to a value greater than the triple point (Figure 4.5b). The vapor pressure is now higher over the ice than over the liquid. Transfer of H_2O proceeds through the vapor phase from ice to liquid and continues until the ice is eliminated, or until it melts. As before, the vapor pressure is finally uniform, in this case equal to the equilibrium value for the vapor pressure over the liquid at the given temperature.

The equilibrium vapor pressure over ice is exactly equal to that over liquid at the triple point temperature, T_0. In contrast to the situation at other temperatures, there is no net force, no pressure differential, in the head space at temperature T_0 to drive vapor between the condensed phases. For this unique configuration, the three phases of water are simultaneously stable.

According to the vapor pressure diagram in Figure 4.3, vapor in equilibrium with liquid water reaches a pressure of 1013 mb (1 atm) at a temperature of 373 K (100°C). Vapor is present under this circumstance in gas bubbles throughout the liquid. The pressure of the liquid, assuming that its depth is small, is essentially equal to the pressure of the overlying atmosphere, 1 atm at sea level. With a further increase in temperature, the pressure of the gas bubbles exceeds the pressure of the surrounding liquid. The incipient bubbles grow in size. The mean density of air in the bubbles is less than the density of the liquid; the bubbles rise and burst at the surface, transferring vapor from the liquid to the gas phase. Eventually, all of the liquid is eliminated, replaced by an identical quantity of water in the gas phase. The formation, growth, and upward transfer of bubbles, with the associated loss of liquid, defines the phenomenon of **boiling**. If we try to boil a pot of water at an elevated altitude, in the mile-high city of Denver for example, the critical vapor pressure is reached at a lower temperature; the water boils sooner. Heat is used to evaporate liquid rather than to increase its temperature, and it is not possible to reach a temperature of 100°C before all of the H_2O has been converted to vapor. The enterprising cook must be aware of this effect of altitude and take appropriate corrective measures.

The phase diagram for CO_2 is illustrated in Figure 4.6. Since solid CO_2 is more dense than liquid CO_2, the melting point temperature rises as the pressure on the solid is increased. The partial pressure of CO_2 in Earth's atmosphere is about 0.35 mb; CO_2 vapor with this pressure would be in equilibrium with CO_2 ice at a temperature of about 183 K (−90°C). At higher temperatures, solid CO_2 sublimes directly to gaseous CO_2. Frozen CO_2, known as **dry ice**, is commercially available. Exposed to the atmosphere, it disappears without a visible trace. Dropped in a container of water, gaseous CO_2 bubbles out of the liquid, resembling boiling water. If enough dry ice is added to liquid water, the water freezes: heat in the water is used up by the CO_2 ice as it sublimes.

The phase diagrams of H_2O and CO_2 play an important role in determining the composition of the atmospheres of Mars, Earth, and Venus. Carbon dioxide is the major constituent of the Martian atmosphere (pressure and temperature at the surface are about 5 mb and 220 K, respectively). The white caps observed on the poles of Mars are composed mainly of solid CO_2, and it is thought that the pressure of the Martian atmosphere is controlled by an equilibrium with the CO_2 ice caps. The vapor pressure of H_2O is exceedingly small, limited by the low values of surface temperature; water on Mars should be present primarily as ice. The temperature of Earth allows water to exist in all three phases, whereas CO_2 in the atmosphere is present solely as

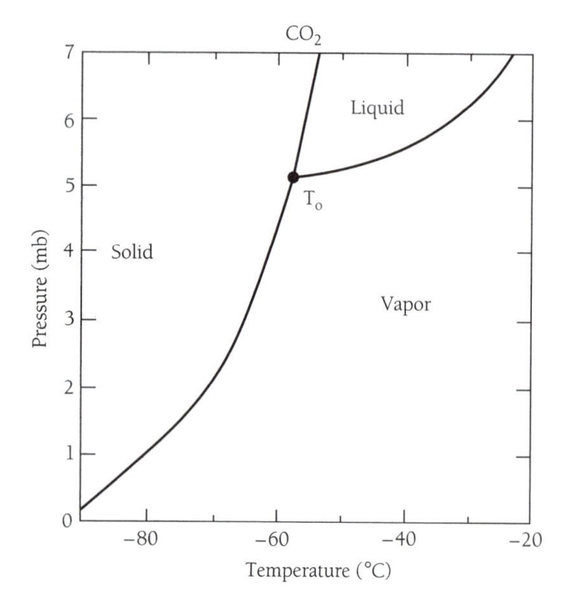

Figure 4.6 Phase diagram for CO_2. Vapor pressures of CO_2 are indicated on the vertical axis.

a gas. The dominant component of the atmosphere of Venus—pressure and temperature at the surface being equal to 80 atm (80,000 mb) and 700 K, respectively—is CO_2, which is believed to be in chemical equilibrium with carbonate rocks on the surface. We expect that the bulk of Venus's H_2O should be present in the atmosphere as vapor: temperatures are too high to support either liquid or solid water on the surface.

4.3 Measuring the Abundance of H_2O in the Atmosphere

Meteorologists employ a variety of conventions to define the water content of air. **Specific humidity** measures the mass of H_2O vapor expressed as a fraction of the total mass of air contained in unit volume. The **mass mixing ratio** of H_2O is defined by the mass of H_2O expressed as a ratio with respect to the mass of air in unit volume, *excluding* H_2O. The **mixing ratio by volume** of H_2O is defined by the ratio of the number of H_2O molecules with respect to the total number of air molecules in unit volume, *excluding* H_2O. **Relative humidity** measures the vapor pressure of H_2O as a ratio with respect to the saturation vapor pressure; that is, with respect to the maximum amount of vapor the air can hold, given its temperature and constraints imposed by the equilibrium vapor pressure curve, as presented in Figure 4.3. In many respects, mass or volume mixing ratios provide the most useful scientific measures of the abundance of H_2O. The quantity most commonly quoted on television and radio, however, is relative humidity. Relative humidity, as we shall see, provides the best measure of personal comfort or discomfort on a hot summer day. If pressure and temperature are known, any one of the measures of water noted above can be used to evaluate the others.

The following example illustrates the different conventions used to define the abundance of water vapor in air.

Example 4.1: Suppose that the temperature is 30°C (86° F) and that the pressure of the atmosphere, excluding the contribution from water vapor, is equal to 1000 mb (10^6 dyn cm^{-2}). If the atmosphere was saturated with respect to liquid at 30°C, the vapor pressure of H_2O would equal 42.43 mb. Imagine that the mixing ratio by volume of H_2O is 2×10^{-2}. For this amount of water vapor, calculate the number density (number of molecules cm^{-3}), partial pressure (contribution of H_2O to total pressure), mass density, specific humidity, mass mixing ratio, and relative humidity.

Answer:

Expressing temperature in K and pressure in dyn cm^{-2}, the perfect gas law, equation (2.17a), can be used to evaluate the dry air **number density**, n:

$$\begin{aligned} n &= \frac{p}{kT} \\ &= \frac{10^6 \text{dyn cm}^{-2}}{1.38 \times 10^{-16} \text{erg K}^{-1} \times 303 \text{ K}} \\ &= 2.39 \times 10^{19} \text{cm}^{-3} \end{aligned} \tag{4.1}$$

It follows, with our assumptions, that the number density of H_2O, denoted by n^*, is given by

$$\begin{aligned} n^* &= 2 \times 10^{-2} \times 2.39 \times 10^{19} \text{cm}^{-3} \\ &= 4.78 \times 10^{17} \text{cm}^{-3} \end{aligned} \tag{4.2}$$

The **partial pressure** of H_2O, expressed by p^*, may be obtained using the perfect gas law, multiplying n^* by kT

$$\begin{aligned} p^* &= 4.78 \times 10^{17} \text{cm}^{-3} \times 1.38 \times 10^{-16} \text{erg K}^{-1} \times 303 \text{ K} \\ &= 2.00 \times 10^4 \text{dyn cm}^{-2} \\ &= 20 \text{mb} \end{aligned} \tag{4.3}$$

The mass density of dry air, ρ, is given by the product of the number density (n), the mean molecular mass of the atmosphere in atomic units (28.97), and the mass corresponding to an atomic unit (1.66×10^{-24} g).

$$\begin{aligned} \rho &= 2.39 \times 10^{19} \text{cm}^{-3} \times 28.97 \text{ au} \times 1.66 \times 10^{-24} \text{g au}^{-1} \\ &= 1.15 \times 10^{-3} \text{g cm}^{-3} \end{aligned} \tag{4.4}$$

Similarly, the mass density of H_2O, ρ^*, is defined by

$$\begin{aligned} \rho^* &= 4.8 \times 1017 \text{cm}^{-3} \times 18 \text{ au} \times 1.66 \times 10^{-24} \text{g au}^{-1} \\ &= 1.43 \times 10^{-5} \text{g cm}^{-3}, \end{aligned} \tag{4.5}$$

where 18 is the mass of H_2O in atomic units.

For the situation envisaged above, the **specific humidity** (SH) is given by

$$\begin{aligned} \text{SH} &= \frac{1.4 \times 10^{-5} \text{g cm}^{-3}}{1.15 \times 10^{-3} \text{g cm}^{-3} + 1.4 \times 10^{-5} \text{g cm}^{-3}} \\ &= 1.23 \times 10^{-2} \end{aligned} \tag{4.6}$$

The mass mixing ratio (MMR) is

$$\begin{aligned} \text{MMR} &= \frac{1.4 \times 10^{-5}\text{g cm}^{-3}}{1.15 \times 10^{-3}\text{g cm}^{-3}} \\ &= 1.22 \times 10^{-2}, \end{aligned} \tag{4.7}$$

and the relative humidity (RH) is

$$\begin{aligned} \text{RH} &= \frac{20 \text{ mb}}{42.43 \text{ mb}} \times 100\% \\ &= 47.1\% \end{aligned} \tag{4.8}$$

Relative humidity is expressed conventionally in units of percent; the factor of 100% in (4.8) is required to convert the ratio of pressures to this scale. ■

Relative humidity can change either as a consequence of a change in the water content of the air (an increase or decrease in specific humidity, or equivalently the H_2O mass mixing ratio) or as a result of a change in temperature. A decrease in temperature, with specific humidity held constant, would lead to an increase in relative humidity. Conversely, fixing the abundance of water, an increase in temperature, would result in a decrease in relative humidity. In the example above, the relative humidity would rise to 86% if the temperature dropped to 20°C; assuming in both cases that the abundance of H_2O remained constant, it would decrease to 27% if the temperature rose to 40°C.

The level of personal discomfort in summer depends on both temperature and relative humidity. If the relative humidity is low, the body can cool by evaporation. The rate of evaporation depends on the difference between the actual and the equilibrium vapor pressures; cooling by evaporation is inefficient when the relative humidity is high. The National Weather Service has developed an index that attempts to take into account effects of both temperature and relative humidity on human discomfort, the so-called heat index. It identifies an effective temperature, as sensed by a human subject, for a range of actual temperatures and relative humidities, as summarized in Figure 4.7. An air temperature of 110°F at low relative humidity is equivalent, with this index, to an effective temperature of about 100°F. The level of discomfort would be similar if the temperature were about 25°F cooler, 85°F, but if the relative humidity approached 100%. Serious health effects are associated with extended exposure to effective temperatures above about 100°F. The proud residents of Tucson, Arizona, may claim that the desert heat is really not that bad as the temperature climbs to 110°F, boasting that "it's a dry heat." To be sure, it is just as uncomfortable for the Texan in Houston at 85°F with a relative humidity near 100%, but in either case the discomfort is serious.

There are two other measures of water abundance in common use, the **dew point** and the **frost point.** As noted above, the relative humidity of an air mass with a fixed abundance of water increases as the temperature drops. Eventually the relative humidity rises to 100%. If air at this point is saturated with respect to liquid, we refer to the temperature at which saturation occurs as the dew point. If saturation reflects an equilibrium of vapor with ice, the temperature is identified as the frost point.

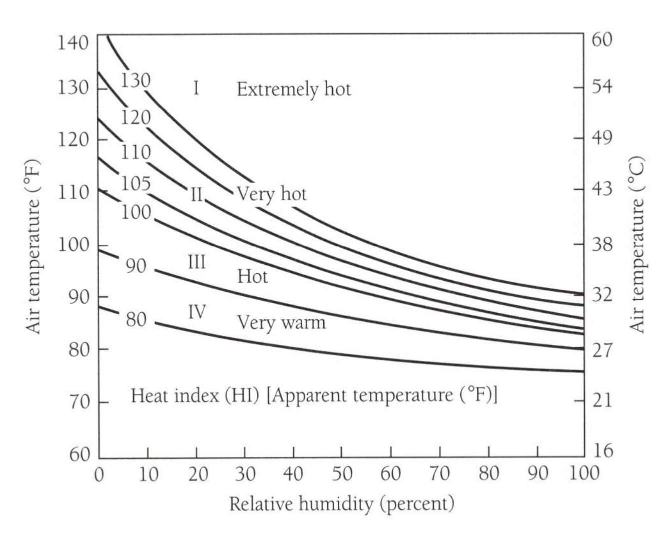

Category	Apparent temperature (°F)	Heat syndrome
I	130° or higher	Heatstroke or sunstroke *imminent*.
II	105°–130°	Sunstroke, heat cramps, or heat exhaustion *likely*, heatstroke *possible* with prolonged exposure and physical activity.
III	90°–105°	Sunstroke, heat cramps, and heat exhaustion *possible* with prolonged exposure and physical activity.
IV	80°–90°	Fatigue *possible* with prolonged exposure and physical activity.

Figure 4.7 Human discomfort index. The apparent temperature expressed in °F is given by the intersection of the values for air temperature and the relative humidity. As indicated in this Figure, apparent temperatures range from 80°F to 130°F. Source: Aherns 1994.

Example 4.2: Suppose that the vapor pressure of H_2O is 10.2 mb with the temperature at 16°C (60°F). Using Figure 4.8a, estimate the relative humidity and the dew point.

Answer: We start off by locating the vertical line on Figure 4.8a corresponding to a temperature of 16°C. It intersects with the saturation vapor pressure curve at a partial pressure of 17.6 mb. To find the relative humidity, we divide the actual vapor pressure by the saturation vapor pressure, and multiply by 100%:

$$\text{RH} = \frac{10.2 \text{ mb}}{17.6 \text{ mb}} \times 100\% = 58\%$$

To calculate the dew point, we locate the point where the horizontal line corresponding to a vapor pressure of 10.2 mb intersects the saturation vapor pressure curve. From that point we look straight down to find a temperature of 7°C (45°F). That is the temperature at which 10.2 mb of vapor pressure saturates the air; that is, the dew point.

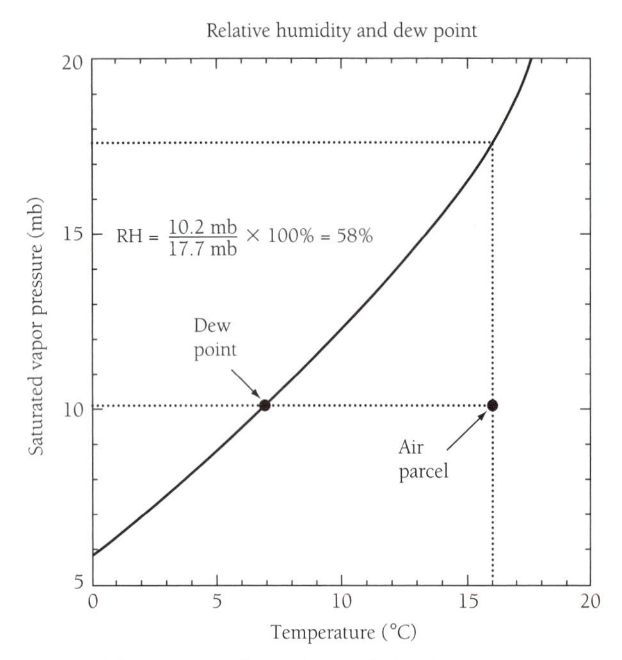

Figure 4.8a Relative humidity and the dew point.

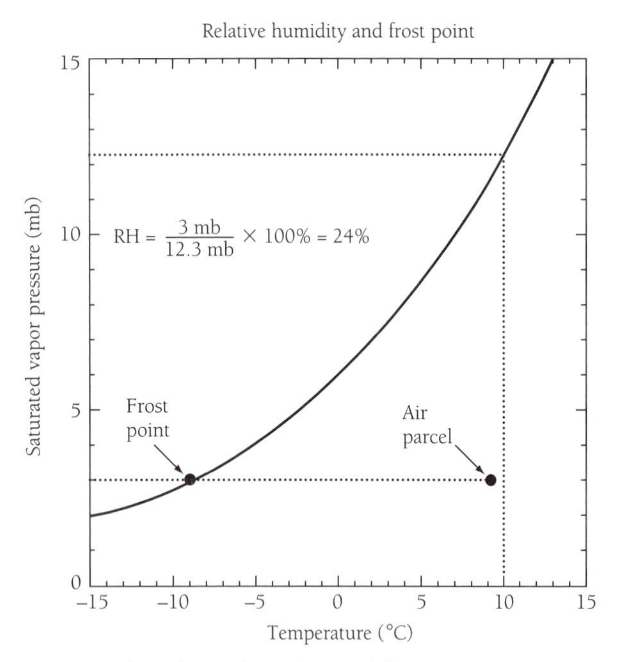

Figure 4.8b Relative humidity and frost point.

If the nighttime temperature drops below 45°F, we might expect the night to be foggy and most probably we would find dew on exposed surfaces in the morning. ■

Example 4.3: For air with a temperature of 10°C (50°F) and a vapor pressure of H_2O equal to 3.0 mb, calculate the relative humidity and the frost point.

Answer: We use the same method as in Example 4.2. However, to get a closer view of the saturation vapor pressure curve at a lower temperature, we use Figure 4.8b. We calculate a relative humidity of:

$$RH = \frac{3 \text{ mb}}{12.5 \text{ mb}} \times 100\% = 24\%$$

The temperature at which 3 mb of water vapor saturates the air is less than the freezing point of water. We calculate a frost point of −9°C (16°F). ■

Clouds provide the most dramatic visible example of condensation present in the atmosphere. (Fog is nothing more than a cloud reaching the ground.) Clouds are usually associated with an upward motion of air. As the atmosphere rises, it cools, and the relative humidity increases. We return to this topic in Chapter 7, within the context of discussing dry and moist adiabats and the processes involved in formation of clouds and precipitation.

4.4 Summary

This chapter highlighted the importance of water and the transformations that take place between its separate phases—ice, liquid, and gas. We pointed out that large quantities of heat (energy) are either expended or released in conjunction with transitions from ice to liquid, ice to gas, and liquid to gas. We introduced the concept of the equilibrium vapor pressure and the phase diagram, pointing out that vapor pressure over either ice or liquid may be expected to increase with increasing temperature. Ice is unstable at temperatures above 0°C. All three phases can coexist at the triple point, at the temperature of 0.0098°C. We discussed the variety of approaches available to specify the abundance of H_2O vapor in the atmosphere: specific humidity, mass mixing ratio, relative humidity, and either dew point or frost point. We emphasized the relevance of these concepts to the treatment of clouds and precipitation, setting the stage for a more focused discussion in Chapter 7.

Atmospheric History and Composition

5

While there are scores of detectable chemical species in the atmosphere, the number of elements represented by these compounds is surprisingly limited. Nitrogen, oxygen, carbon, hydrogen, and the noble gas argon (Ar) account for all but a few parts in one hundred thousand of the atmosphere. If we add the noble gases neon, helium, krypton, and xenon, we can accommodate all but one part in one billion. As we shall see, the atmosphere's chemical complexity is largely a consequence of the influence of life; to explain the comparative simplicity of the atmosphere's elemental composition, we need to examine the processes that led to the origin of the atmosphere in the first place.

We begin in Section 5.1 with a description of the processes that led to the formation of Earth from the primitive solar nebula. The importance of life for the early evolution of the atmosphere is discussed in Section 5.2. Perspectives on the recent changes in atmospheric composition are developed in Section 5.3. A particular objective in Section 5.3 is to provide a context within which to assess the significance of contemporary changes in atmospheric composition arising as results of various forms of human activity. Summary remarks are presented in Section 5.4.

5.1 Perspectives on the Origin and Early Evolution of Earth as a Planet

Our story begins almost five billion years ago with the formation of the solar system from a spinning nebula of gas and dust. Initially, the nebula was hot. As time elapsed, the nebula cooled, mainly by loss of heat through radiation (emission of light) from its edges. Gases near the center of the nebula collected to form the Sun. Materials condensing elsewhere eventually assembled to form the diverse suite of planets, satellites, rings, meteorites, and comets characterizing our solar system today. The cooling sequence resulted in segregation of condensable materials with time and position in the nebula. Elements condensing early at relatively high temperature, such as iron, furnished the rocky building blocks of the inner planets, Mercury, Venus, Earth, and Mars. Ices, such as water, methane, and ammonia, forming at lower temperatures, provided the major ingredients for the outer planets, Jupiter, Saturn, Uranus, and Neptune. The inner planets acquired, in addition to their rocky materials, a small quantity of low temperature ices, presumably formed during the late stages of evolution of the nebula.

As Earth formed, it began to heat up, largely as a consequence of energy released by the decay of radioactive elements such as uranium, thorium, and potassium in its rocky substrate (with a contribution from heat released in conjunction with gravitational accretion). Heavier elements, such as iron, settled to the core. More volatile elements, such as hydrogen, carbon, nitrogen, and the noble gases, were vented from primitive volcanoes or released from late infalling icy material, similar to that contained today in cometary material, to form the early atmosphere and ocean. Hydrogen was released mainly as water with a small quantity of H_2, carbon as CO_2 with a little CO, and nitrogen as N_2. Most of the hydrogen vented from the interior, or captured from the outside over geologic time, is present today in the ocean as water. Carbon is preserved mainly in sediments as calcium carbonate, $CaCO_3$, and in the organic remains of organisms that once lived at the surface: nitrogen persists in the atmosphere as N_2.

The atmospheres of Mars and Venus, we think, had an origin similar to that of Earth. Mars, further from the Sun, is much colder than Earth; its surface temperature is about 220 K (−53°C). Venus, nearer to the Sun, is sizzling hot, with a surface temperature of 750 K (477°C). Carbon dioxide is the most abundant constituent of both planets' atmospheres. The surface pressure on Mars is about 200 times less than that on Earth; the surface pressure on Venus is about 80 times larger. Both atmospheres contain significant quantities of N_2, at levels of a few percent relative to CO_2. Water is present in both atmospheres but in trace amounts relative to CO_2. It is thought that the differences between Mars, Earth, and Venus may be largely attributed to effects of temperature.

Suppose that the quantity and composition of gas vented into the primitive atmosphere of Mars was similar to that released by Earth. At the cold temperature of Mars, CO_2 and H_2O would precipitate from the atmosphere, forming ice on the surface. Nitrogen would accumulate in the atmosphere as N_2. Some of the nitrogen would be subsequently lost by escape into space, a consequence of the relatively low strength of the planet's gravitational field. The Martian atmosphere today contains amounts of CO_2 and H_2O as large as they could possibly be, given the planet's current temperature. Contemporary polar caps store large quantities of CO_2 and H_2O. Everywhere we look in the rocks strewn across the surface, there are indications of H_2O in chemically bound form. Present evidence is consistent with the view that the total quantity of CO_2 and H_2O on Mars today may be comparable to that on Earth, but only a small fraction of either compound, limited by temperature, is present in the atmosphere. It appears that conditions were very different in the past. Photographs from Mariner and Viking spacecrafts reveal a surface on Mars visited at least once in the past, and perhaps many times, by raging floods. Climate change may be as common an historical feature for Mars as it has been for Earth.

In the atmosphere of Venus the quantity of carbon as CO_2 is roughly comparable to that in the sediments of Earth. The abundance of N_2 in both atmospheres is similar. Water is the anomaly. If Venus had evolved a quantity of water similar to Earth, we would expect it to have an atmosphere dominated by steam (H_2O vapor) with a surface pressure of about 1000 atm. However, observations of the atmosphere of Venus indicate that this is not the case. There are two possible explanations for the lack of water in the atmosphere of Venus. It is most likely, in our opinion, that Venus evolved with much less water than Earth, a consequence of warmer temperatures (and less icy volatiles) in the region of the nebula where the planet formed. Carbon dioxide would have been the dominant constituent of Venus's early atmosphere, as it is today, and the abundance of H_2O would have been comparable to N_2. Subsequent escape of hydrogen and oxygen would have resulted in a steady depletion of H_2O, accounting for the large abundance of deuterium (D) relative to hydrogen observed in the contemporary atmosphere of Venus. The light form of hydrogen, H, would escape more readily than the heavier form, represented by D. Deuterium represents approximately two parts in ten thousand of the hydrogen present on Earth; the relative abundance in Venus's atmosphere is 100 times larger. According to a second view, Venus formed with an abundance of H_2O similar to Earth but lost its steamy early atmosphere by rapid escape in the early years of its history. Enrichment of deuterium with respect to hydrogen occurred slowly during subsequent evolution, similar to the situation envisaged for the low-water scenario. Present information is insufficient to permit a definitive choice between these competing possibilities.

It is likely that the composition of Earth's atmosphere was initially dominated by CO_2, as is the case today for Mars and Venus. The second most abundant constituent would have been N_2, followed by H_2O and Ar. Water, the most abundant volatile released from the interior, or captured from the outside, would have precipitated rapidly to form the primitive ocean. Significant quantities of carbon would have dissolved in the ocean forming carbonate-rich sedimentary rocks, lowering the level of atmospheric CO_2. This, presumably, defined the early environment for life on Earth.

5.2 The Early Evolution of Life on Earth and Its Atmospheric Consequences

It is clear that life evolved early on Earth; the oldest surviving rocks contain fossils dating back almost four billion years. The precise steps that led to the evolution of self-replicating organisms are unclear, but it is apparent from the geologic record that they proceeded rapidly. Some believe that the action was in the atmosphere, that lightning and ultraviolet radiation, interacting with atmospheric gases, resulted in a synthesis of complex organic molecules providing the initial building blocks for life. A second possibility is that life began in the ocean, in the vicinity of hot springs distributed along zones of sea floor spreading, for example, or in earlier analogues to these environments, in regions of the ocean where fresh material emerges from the interior of Earth, replacing matter withdrawn at subduction zones (regions where surface materials converge and descend into interior regions of the planet).

Deep-sea vents support a remarkable biological system at present. Bacteria, drawing energy from the oxidation of sulfide contained in hot (300°C) spring water, form the bottom of a relatively brief food chain, supporting a dense population of worms and clams living in close proximity to the vents, almost 3 km below the sea surface. Water emanating from the vents contains a variety of trace metals and other elements essential for life. Nitrate (NO_3^-) or nitrite (NO_2^-), formed from acids produced in the early atmosphere, could have provided the oxidant for the synthesis of the first self-replicating organisms. It would be interesting to design an experiment in the laboratory attempting to simulate conditions at the vents in the prebiotic environment, to study effects of the addition of various oxidants to a sulfide, metal-rich solution. Identification of the chemical compounds formed in this environment could provide valuable clues to the origin of life, opening a window to one of the great unsolved mysteries of nature.

Oxygen was a minor constituent of the early atmosphere. Indeed, it is thought that the primitive organisms populating our planet in its early years would have found O_2 toxic, even in relatively modest concentrations. **Photosynthesis** is the process by which solar energy is used to transform CO_2 and H_2O to organic matter. The bulk or net reaction may be written in the form

$$CO_2 + H_2O + h\nu \rightarrow CH_2O + O_2, \qquad (5.1)$$

where CH_2O is a chemical short-hand for the organic matter formed as a consequence of photosynthesis. It is thought that the first organisms to evolve with the capacity to tap solar energy through photosynthesis were able to dispose of the oxygen released in (5.1) by binding it to a convenient reduced compound, such as ferrous iron. Eventually, these organisms would have been limited in their life cycle by the availability of a suitable sink for their O_2 waste product. Over time, an organism evolved with an enzyme allowing it to withstand the presence of free O_2.

This oxygen-adapted organism would have had a wonderful advantage; it could grow using readily available substrates, such as CO_2 and H_2O, and it had at its disposal an almost infinite reservoir, either the atmosphere or the ocean, in which to dump waste. The organism and its eventual progeny flourished. Oxygen accumulated in the atmosphere. Organisms sensitive to O_2 were either eliminated or banished to isolated niches, such as swamps, where peculiar local physical conditions provided a measure of protection from O_2. The development and successful propagation of the O_2-mediating enzyme led to the first great instance of global pollution. It opened up new niches for life, paving the way for the evolution of organisms, such as ourselves, with the capacity to exploit the presence of free oxygen for energy extraction, while at the same time eliminating vulnerable species. The sequence of events that resulted in the profusion of O_2 unfolded almost three billion years ago and led to the development of an atmosphere roughly comparable to the one present today.

With the exception of the noble gases, the atmosphere may be considered in many respects as an extension of life. This is particularly true for the more abundant elements carbon, oxygen, nitrogen, and sulfur, which are repetitively and continuously exchanged in various chemical forms among the atmosphere, biosphere, soils, oceans, and sediments.

A carbon atom present as CO_2 in the atmosphere today was part of the structural material of a plant or animal, either on land or in the sea, only a few years ago, and will be again soon. Methane, the second most abundant carbon compound in the atmosphere, is produced by burning organic matter and by microorganisms functioning in anaerobic (oxygen-deprived) environments. Carbon monoxide, the third most abundant carbon compound, is a product of combustion and is also formed photochemically in the atmosphere as an intermediate in the oxidation of organic gases.

Accumulation of O_2 in the atmosphere is associated with burial in sediments of organic carbon formed by photosynthesis, the reaction described in equation (5.1). Oxygen produced by photosynthesis is primarily consumed by the reverse reaction,

$$CH_2O + O_2 \rightarrow CO_2 + H_2O, \qquad (5.2)$$

resulting in no net production of O_2 (the energy evolved is not released as a photon but is communicated to the environment as heat or kinetic energy). A fraction of the organic carbon produced by photosynthesis, however, may be removed from the near surface region and incorporated in sediments. The companion O_2 is stranded in the atmosphere until organic carbon is returned to the surface, by the uplift of sediments for example, and is made available once again to combine with O_2 by the reaction in equation (5.2). The lifetime of organic carbon in sediments is long, hundreds of millions of years, as we shall see in Chapter 11. This allows for slow accumulation of large quantities of organic carbon in sediments, with complementary growth of O_2 in the atmosphere. Even if life on Earth were to terminate, O_2 would persist in the atmosphere for millions of years, until sedimentary uplift supplied sufficient quantities of organic carbon to deplete the reservoir of O_2 accumulated over long periods in the past.

Nitrogen in biologic tissue is ultimately derived from N_2, the most abundant constituent of the atmosphere. It is returned to the air mainly as N_2 and N_2O after it has been used by living organisms, with small quantities of NO (nitric oxide) formed as products of microbial activity in soils and in the ocean. As we shall see, N_2O decomposes in the stratosphere, primarily forming N_2, with a small but significant yield of NO. Nitric oxide derived from N_2O in the stratosphere plays an important role in the chemistry of atmospheric O_3. Microbial processes in soils and combustion are responsible for additional release of NO to the troposphere, where NO is transformed photochemically to NO_2, HNO_3, and other oxides of nitrogen. Nitrogen oxides, in combination with CO and hydrocarbons, play a pivotal role in the synthesis of O_3 in urban smog. They also contribute, with sulfate, to the phenomenon of acid rain.

Various forms of reduced sulfur, notably hydrogen sulfide (H_2S) and dimethyl sulfide [$(CH_3)_2S$], are released to the atmosphere as a consequence of biological activity. Reduced sulfur is photochemically converted in the atmosphere to sulfur dioxide (SO_2) and ultimately to sulfuric acid

(H_2SO_4). Sulfur dioxide is also emitted directly from volcanoes and as a by-product of combustion and the smelting of metal ores. Other contributions to the atmospheric sulfur budget include carbonyl sulfide (COS) and carbon disulfide (CS_2). Sulfur from volcanoes is vented in part directly to the stratosphere where it is oxidized to sulfate, forming a dispersed persistent layer of small particles. These particles, scattering short-wavelength sunlight at high altitude, are visible for months following a major volcanic eruption as a spectacular purple haze in twilight.

Sulfate aerosols are also thought to play a role as condensation centers for the production of clouds in the troposphere. Some scientists believe that selected organisms in the sea can exercise a direct, immediate influence on climate by releasing $(CH_3)_2S$, which, when converted to sulfate, can contribute to an enhanced local condensation of H_2O, with a consequent increase in cloudiness. The idea that the biota can affect the climate directly is scarcely new as a generality but has found imaginative public expression recently in the **Gaia theory** proposed by the British scientist James Lovelock (Gaia is the Greek goddess of the earth). According to this theory, the global atmosphere-ocean-biosphere-soil system functions as a single organism. As the parts of a human body combine to regulate the health of the whole, so also the components of Gaia cooperate to ensure the integrity of life as a global phenomenon. To follow the Gaia metaphor further, our immediate challenge is to gauge the health of the global life-support system. There are alarming signs that the organism is sick. It exhibits a variety of complex symptoms, such as acidification of rain, depletion of stratospheric ozone, and rapid changes in the abundance of the gases CO_2, CH_4, and N_2O. The question is whether we can diagnose the ill and take remedial action before the sickness is allowed to spread.

5.3 Perspectives on the Changes in Composition Associated with Human Activity

Table 5.1 presents a summary of the contemporary composition of the atmosphere. As indicated in the table, the atmosphere contains a bewildering array of chemical species. Some, such as N_2, O_2, Ar, Ne, He, N_2O, and the industrial chlorofluorocarbons, are distributed more or less uniformly, at least in the lower atmosphere; that is to say, their relative abundances are essentially constant and independent of time and spatial location. Others, such as CO, O_3, H_2O, SO_2, and the oxides of nitrogen (NO_2, NO, and HNO_3), are variable both in time and space. The abundance of CO_2 varies seasonally in the lower atmosphere in response to a combination of photosynthesis, respiration, and decay (see Chapter 11).

In general, the variability of a particular species depends on its lifetime. Variability for short-lived species is highest in the vicinity of sources or sinks. For example, fossil fuel combustion is responsible for an important source of CO. As a consequence, the abundance of CO is anomalously high in or near cities, or in the vicinity of major highways. Ozone is a photochemical product. Its abundance is greatest in the stratosphere, where the flux of ultraviolet solar radiation is most intense (see Chapters 13–16). High concentrations of O_3 are also observed in industrial regions of the lower atmosphere, where the compound is formed as a result of photochemical reactions fueled by inputs of hydrocarbons, carbon monoxide, and oxides of nitrogen (see Chapter 17). The abundance of H_2O is strongly correlated with temperature. As we would expect based on the preceding discussion of

Table 5.1 Atmospheric composition

Gas	*Volume mixing ratio*
Nitrogen (N_2)	0.78
Oxygen (O_2)	0.21
Water (H_2O)	0.04 to 0.005 (variable)
Argon (Ar)	0.0093
Carbon Dioxide (CO_2)	350×10^{-6}
Neon (Ne)	18.2×10^{-6}
Ozone (O_3)	2×10^{-8} to 10^{-5} (variable, increasing with altitude into the stratosphere)
Helium (He)	5.2×10^{-6}
Methane (CH_4)	1.6×10^{-6}
Krypton (Kr)	1.1×10^{-6}
Hydrogen (H_2)	0.5×10^{-6}
Nitrous Oxide (N_2O)	0.32×10^{-6}
Carbon Monoxide (CO)	0.1×10^{-6}
Chlorofluorocarbons (CCl_3F, CCl_2F_2)	3.0×10^{-9}
Carbonyl Sulfide (COS)	0.4×10^{-9}
Methyl Bromide (CH_3Br)	0.1×10^{-9}

the thermodynamics of water (Chapter 4), the mixing ratio of H_2O vapor is highest near the surface and is particularly elevated in the tropics. The mixing ratio of H_2O varies over a wide range of values (orders of magnitude) as a function of latitude and altitude. The change is especially large at high latitudes during winter. For many purposes, variations in the abundance of H_2O may be most usefully expressed in terms of changes in relative humidity rather than mixing ratio; the range of values observed for the former is notably more limited. In contrast to CO, O_3, and H_2O, molecular nitrogen is exceptionally stable. In this case, the lifetime is measured in units of tens of millions of years (see Chapter 12). As a result, the mixing ratio of N_2 is essentially constant from pole to pole, from the surface to altitudes as high as 100 km.

This text is primarily concerned with changes in atmospheric composition taking place today as a result of various forms of human activity. It is important to establish a context for these changes. How do the changes observed in the contemporary environment compare with past variations, changes that took place before the human presence assumed the dominant role it exercises today? For the most part, direct measurements of atmospheric composition are available only for the immediate past; with few exceptions, the direct instrumental record is confined to the last 50 years. Analyses of air trapped in ancient ice provide a means to extend the instrumental record back in time. This opportunity, developed only recently, has revolutionized our understanding of atmospheric composition. It clearly attests to the unique importance of the modern human influence. It demonstrates, moreover, that changes in atmospheric composition are not simply confined to the modern era. Significant variations, associated most notably with major shifts in climate, are observed on a range of time scales. The ice-core record currently available provides a perspective extending back to 450,000 years before the present (B.P.) in the case of CO_2 (time horizons for other gases are more limited as yet). We may anticipate extensions of the compositional record as new cores are drilled and as increasingly more precise analytical methods are applied to the analysis of air trapped in both new and currently available cores.

Figure 5.1 presents a summary of changes observed for CO_2 over the past millennium. Data obtained from analyses of ice cores are complemented for the most recent period by results from direct measurements of air sampled at the South Pole. Prior to about A.D. 1600, the concentration of CO_2 averaged about 280 ppm. There is an indication of a small dip in concentration roughly coincident with the peak in the Little Ice Age (see Chapter 10). The rapid increase in CO_2 over the past two centuries is attributed to a combination of emissions due to the combustion of fossil fuels and the release of carbon from soils and vegetation associated with conversion of land to agriculture (see Chapter 11).

The record in Figure 5.2 covers four major glacial epochs lasting in each case about 100,000 years (see Chapter 10). The changes in CO_2 depicted in Figure 5.2 are thought to arise mainly as the result of a redistribution of carbon between the biosphere, soil, and atmosphere, on the one hand, and the ocean, on the other. Under glacial conditions, carbon is sequestered in the deep ocean, a consequence primarily of climate-related changes in ocean circulation with associated changes in ocean biology and chemistry (see Chapters 10 and 11). The variations in CO_2 depicted in Figure 5.2 are correlated closely with changes in climate, as indicated by a comparison of the gas record with changes observed in the isotopic composition of the ice from which the gas was extracted. The time resolution of the data is not sufficient, however, to resolve the question of whether the changes in CO_2 precede or lag behind the changes in climate. A careful examination of the pattern of change associated with the termination of the last major ice age suggests

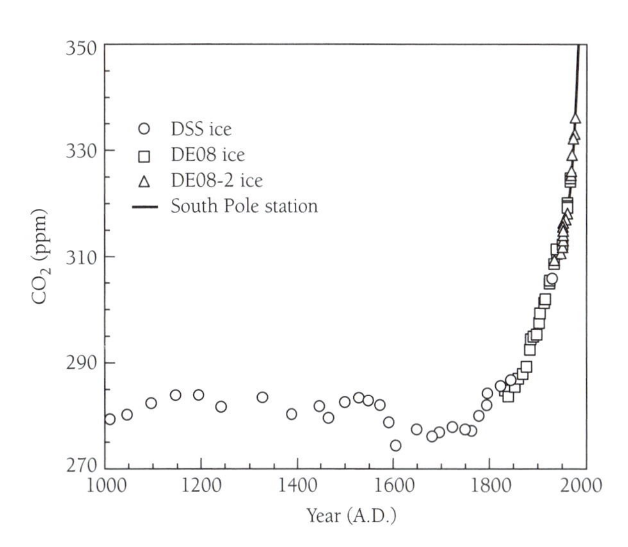

Figure 5.1 Variation of CO_2 (ppm) over the past 1000 years. Source: Etheridge et al. 1996.

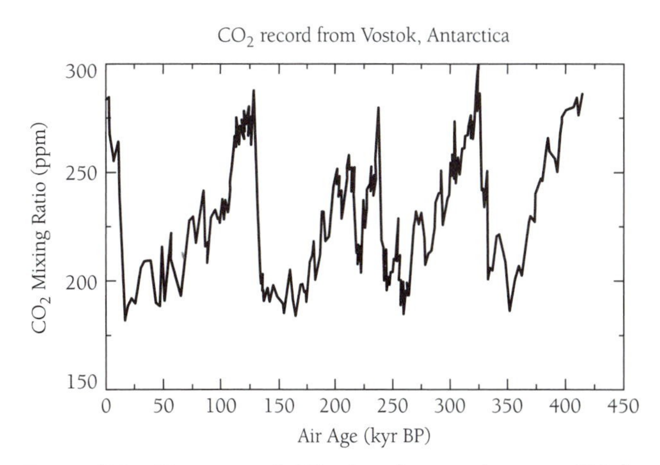

Figure 5.2 Variation of CO_2 (ppm) over the past 450 kyr. Source: Barnola et al. 1999.

that climatic variations occur first and that changes in CO_2 develop later, apparently in response to the change in climate.

Given the close association of changes in CO_2 with climatic changes observed over the recent geologic past, one might be tempted to exploit this association to obtain an empirical estimate for climatic sensitivity to changes in CO_2. It is doubtful, however, that such an approach would be justified. It is clear that a variety of factors—changes in the seasonal pattern of sunlight incident at different latitudes and changes in ocean circulation, for example—are implicated in the variations exhibited by the climate system. Absent a model to isolate effects of all of the separate influences, it would be misleading to attribute a specific effect to any particular cause. It is *likely* that changes in CO_2 were implicated in the large climatic swings observed over the recent geologic past. The complexity of the climate system would appear to preclude a more specific *quantitative* conclusion.

The change in the abundance of CH_4 over the past millennium is illustrated in Figure 5.3. Changes over the past several decades are presented in Figure 5.4, while a longer-term perspective, extending back through the last ice age, is displayed in Plate 2 (see color insert). The abundance of CH_4 was relatively constant at about 650 ppb from A.D. 1000 up to the early part of the eighteenth century. The trend over the past several centuries is characterized by a rapid increase beginning about A.D. 1750, climbing to a level close to 1800 ppb today. The increase in CH_4 over this period is proportionally larger than the increase in CO_2: a factor of about 2.5 for CH_4, as compared to about 30% for CO_2. As indicated in Figure 5.4, the rate of increase of CH_4 slowed significantly in the early 1990s. The budget of CH_4 and the factors responsible for the modern rise in concentration are discussed in Chapter 17. The slow-down in growth recently observed should not, we suggest, be interpreted as a guarantee that the contemporary trend of relatively modest growth will continue in the future.

The concentration of CH_4 was lower during the last ice age (from 110 kyr B.P. to about 20 kyr B.P.; see Plate 2) as compared with concentrations observed over the more recent interglacial period (the past 20 kyr): about 400 ppb as compared to about 650 ppb. Significant variability is observed during the glacial period, much larger than is seen for CO_2. Changes are particularly rapid when they involve a transition from low to high concentrations of CH_4. Increases approaching a factor of two are observed over time scales as brief as centuries (similar in magnitude and rapidity to the change observed over the past few hundred years). The fluctuations in CH_4 observed under glacial conditions are attributed primarily to changes in tropical climate (see Chapter 10), with warm, moist conditions favoring enhanced emission of CH_4. The rapidity of the transition from low to high concentrations of CH_4 depicted in Plate 2 indicates that changes in tropical climate can rapidly occur. Observations of CH_4 provide an important means to link changes in the tropical climate to climatic changes inferred for other regions.[1] The evidence (see Chapter 10) suggests that the fluctuations in tropical climate implied by the data for CH_4 are not merely local; they are associated with large-scale, rapid, and essentially synchronous rearrangements of the global climate system.

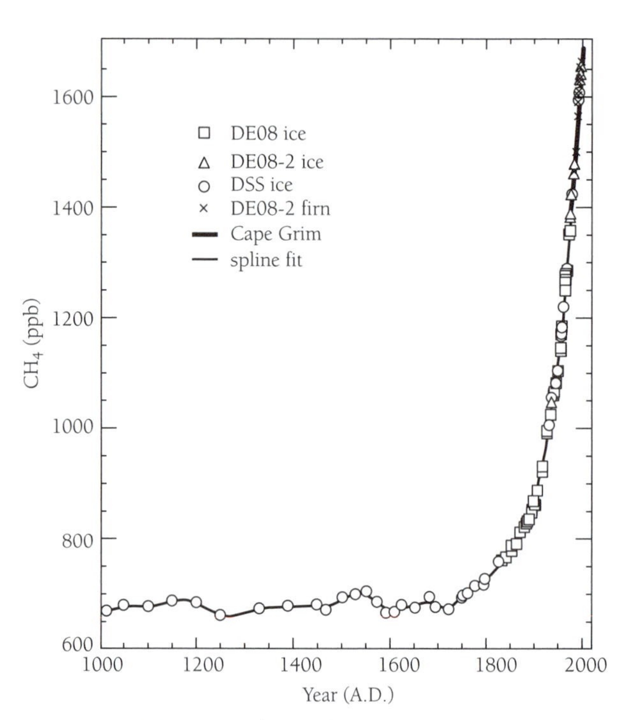

Figure 5.3 Variation of CH_4 (ppb) over the past 1000 years. Source: Etheridge et al. 1998.

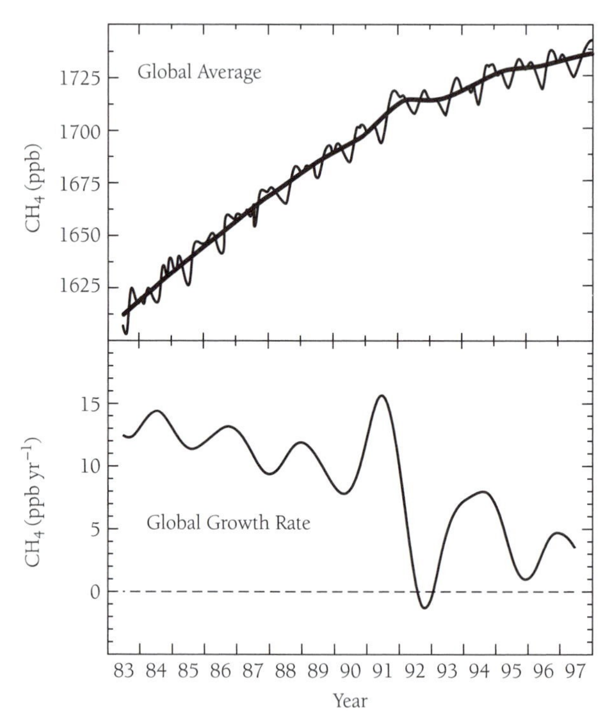

Figure 5.4 Variation of CH_4 (ppb) since 1983 (upper panel) and estimates for the growth rate (ppb yr^{-1}) (lower panel). Source: Dlugokencky et al. 1998.

A fourth important contributor to the global greenhouse effect (together with H_2O, CO_2, and CH_4) is N_2O. In addition to its role as a greenhouse gas, N_2O plays an important role in the chemistry of the stratosphere. Reaction of an electronically excited state of atomic oxygen ($O(^1D)$) with N_2O represents the dominant source of stratospheric NO. As discussed in Chapters 10–16, reactions involving NO play a critical role in the catalytic chemical cycles responsible for removal of O_3.

Changes in N_2O over the past 45 kyr are indicated in Figure 5.5. They suggest that, at the peak of the last ice age (20 kyr B.P.), the abundance of N_2O may have been as low as about 190 ppb. Associated with the glacial to interglacial transition, the concentration of N_2O rose from about 190 ppb to about 270 ppb. As illustrated in Figure 5.6, the concentration of N_2O ranged from about 286 to about 278 ppb between A.D. 1200 and A.D. 1600, with a suggestion of a decline during the Little Ice Age. The abundance of N_2O has risen steadily over the past several centuries, as illustrated in Figures 5.7 and 5.8. A more detailed perspective on the changes observed over the past several decades is presented in Figure 5.8.

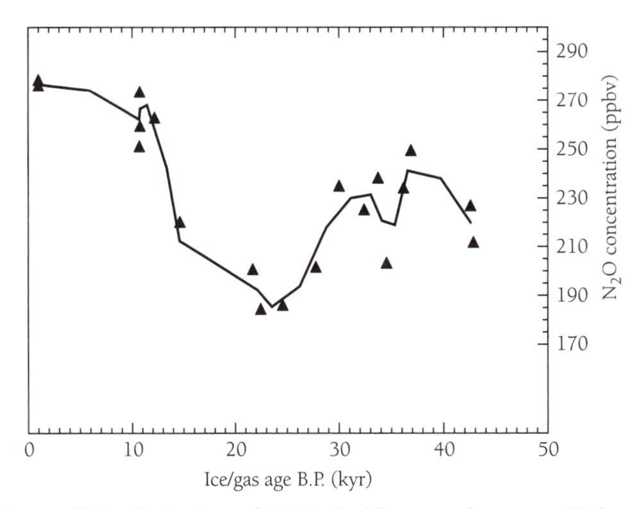

Figure 5.5 Variation of N_2O (ppb) over the past 45 kyr. Source: Lewenberger and Siegenthaler 1992.

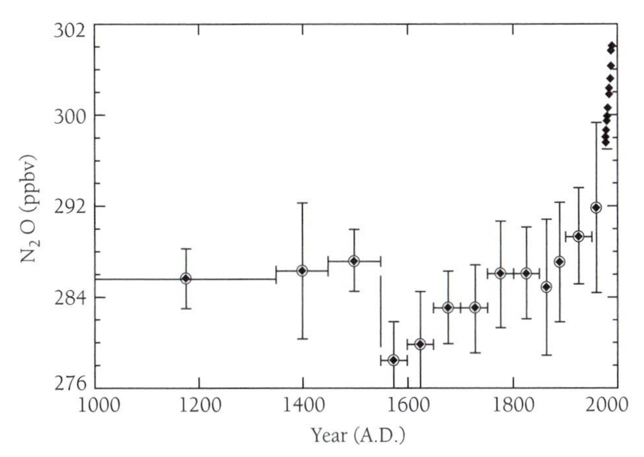

Figure 5.6 Variation of N_2O (ppb) over the 1000 years. Source: Khalil and Rasmussen, 1992.

Factors influencing the N_2O budget are discussed in Chapter 12. We suggest that microbial activity, stimulated by intensive application of nitrogen-based fertilizers to agricultural systems and by supplies of ever increasing quantities of organic matter associated with disposal of human and animal waste, is primarily responsible for the recent rise in N_2O.

Another greenhouse gas drawing attention is SF_6 (sulfur hexafluoride). The lifetime of this species in the atmosphere is

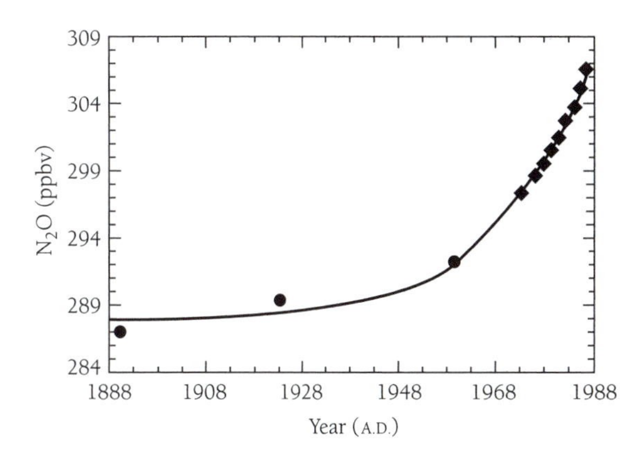

Figure 5.7 Variation of N_2O (ppb) over the past 100 years. Source: Khalil and Rasmussen, 1992.

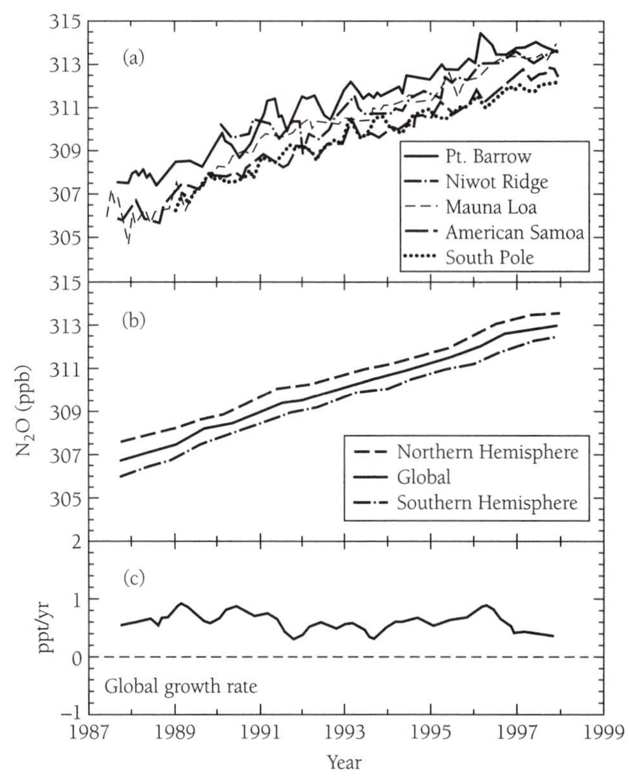

Figure 5.8 Variation of N_2O (ppb) since 1987. Data included in upper panel (a) summarize in situ measurements from sites included in the NOAA/CMDL network of stations as indicated. The middle panel (b) presents average values of concentrations for the Northern Hemisphere, Southern Hemisphere and globe. The lower panel (c) indicates the global growth rate (ppb yr^{-1}). Source: Elkins 1999.

exceptionally long, about 3200 yr. It is estimated that the potential of SF_6 to alter the climate is about 20,000 times greater on a per molecule basis than that of CO_2 (Intergovernmental Panel of Climate Change, or IPCC, 1996). In recognition of these unusual properties, SF_6 was included in the suite of compounds regulated under the Kyoto Protocol. The atmospheric concentration of the gas is presently increasing at a rapid rate, as illustrated in Figure 5.9. The results in Figure 5.9 reflect measurements of air in flask samples taken from a variety of locations around the world. As indicated, the concentration of SF_6 has risen from less than 1 ppt in 1978 to more than 4 ppt today. Industrial production for 1996 was estimated at 7600 tons; emissions of SF_6 into the atmosphere in the same year accounted for a source equivalent to about 80% of production (World Meteorological Organization, 1999).

The chlorofluorocarbons (CFCs), chlorocarbons, hydrochlorofluorocarbons (HCFCs), and bromocarbons represent classes of compounds, largely industrial in origin, which have been implicated in the accelerated loss of stratospheric O_3 observed in recent years (see Chapters 13–16). Most abundant of the CFCs are CCl_3F (CFC-11) and CCl_2F_2 (CFC-12). These compounds were extensively used in the seventies and eighties as fluids for air conditioning systems and as refrigerants and foaming agents. Together with a number of other industrial compounds, such as CCl_4 (carbon tetrachloride) and a class of brominated species known as the **Halons**, these gases were effectively regulated under the Montreal Protocol (1987) as amended and extended in London (1990), Copenhagen (1992), Vienna (1995), and most recently in Montreal (1997). Production of CFC-11 and CFC-12 was capped by developed countries beginning in 1989, with production being terminated in 1996. Production by developing countries was frozen as of 1999, with a total ban scheduled to enter into effect in 2010. The success of the Montreal process is clearly attested to by the data presented in Figures 5.10–5.14.

Growth rates for CFC-11 and CFC-12 have decreased significantly since 1989. The concentration of CFC-11 reached a maximum in 1994 and has now begun, as regulations have entered into effect, to exhibit an anticipated long-term decrease (Figure 5.10). The concentration of carbon tetrachloride has been declining since 1991 (Figure 5.12), while the concentration of methyl chloroform attained its maximum value in 1992 (Figure 5.13). Growth in CFC-12 has slowed significantly in recent years (Figure 5.11), with similar decreases in growth rates observed for the Halons (Figure 5.14). In contrast, concentrations of HCFCs, introduced as substitutes for CFCs in the nineties, have begun to exhibit a rapid increase (Figure 5.15). Production of HCFCs is scheduled to end, however, in developed countries in 2030, with a comparable ban set to go into effect for developing countries in 2040.

5.4 Summary

As we have seen, the atmosphere initially developed as an inevitable consequence of processes involved in the formation of Earth from the parent solar nebula. Its early composition was influenced by not only exchange of materials with

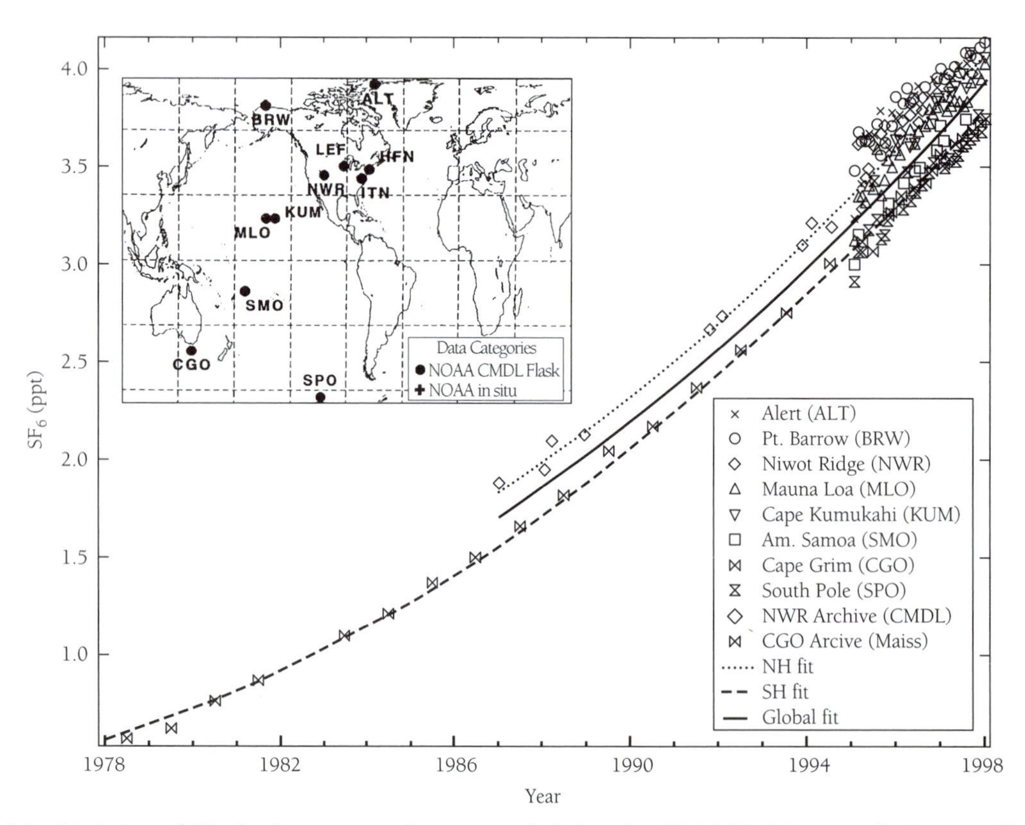

Figure 5.9 Variation of SF_6 (ppt) as measured at sites included in the NOAA/CMDL network. Source: Elkins 1999.

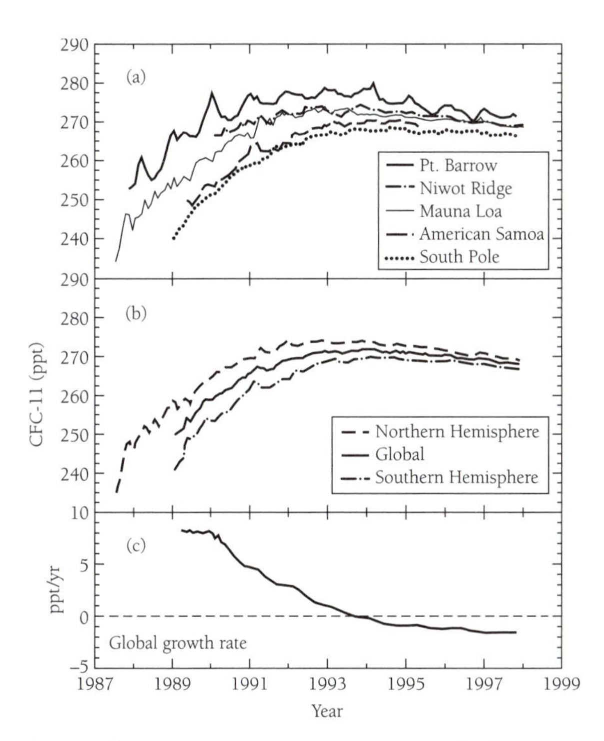

Figure 5.10 Variation of tropospheric CFC-11 (ppt) since 1987. Data for individual NOAA/CMDL stations are presented in the upper panel (a) as indicated. Composite results for the Northern Hemisphere, Southern Hemisphere and globe are presented in the middle panel (b). Global growth rate (ppt yr^{-1}) is shown in the lower panel (c). Source: Elkins 1999.

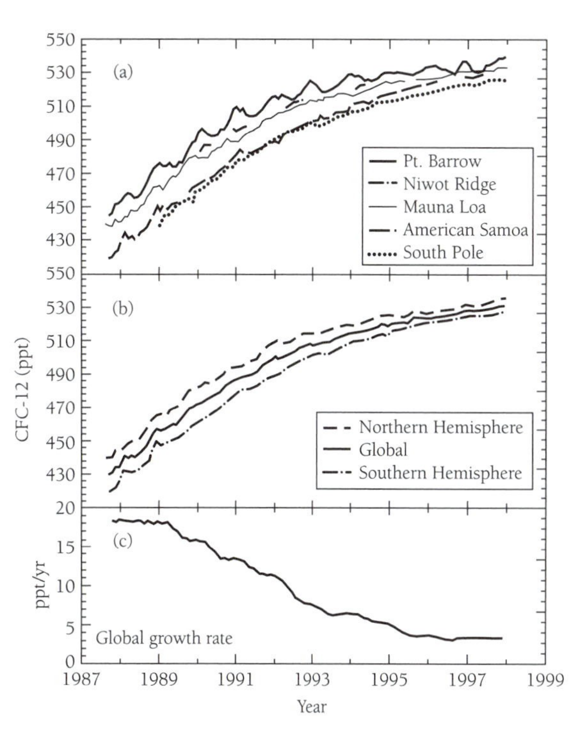

Figure 5.11 Variation of CFC-12 (ppt) since 1987. Data for individual NOAA/CMDL stations are presented in the upper panel (a) as indicated. Composite results for the Northern Hemisphere, Southern Hemisphere and globe are presented in the middle panel (b). Global growth rate (ppt yr^{-1}) is shown in the lower panel (c). Source: Elkins 1999.

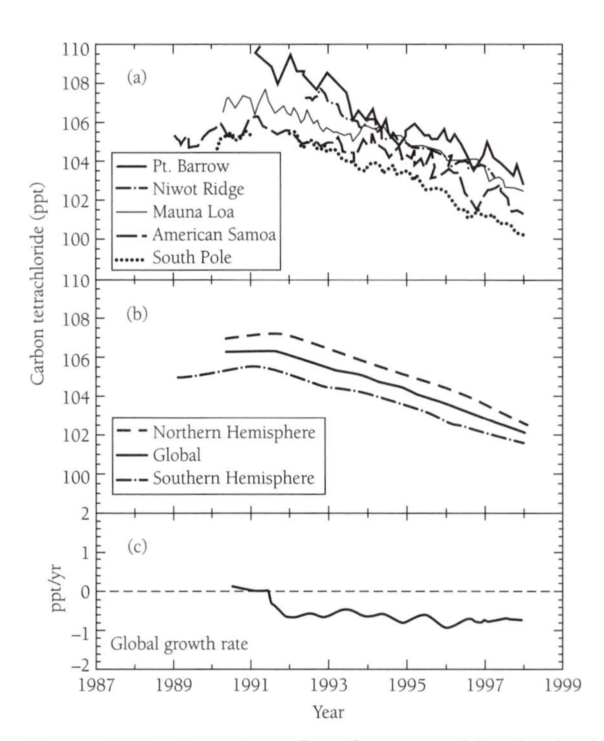

Figure 5.12 Variation of carbon tetrachloride (ppt) since 1989. Data for individual NOAA/CMDL stations are presented in the upper panel (a) as indicated. Composite results for the Northern Hemisphere, Southern Hemisphere and globe are presented in the middle panel (b). Growth rate (ppt yr^{-1}) is shown in the lower panel (c). Source: Elkins 1999.

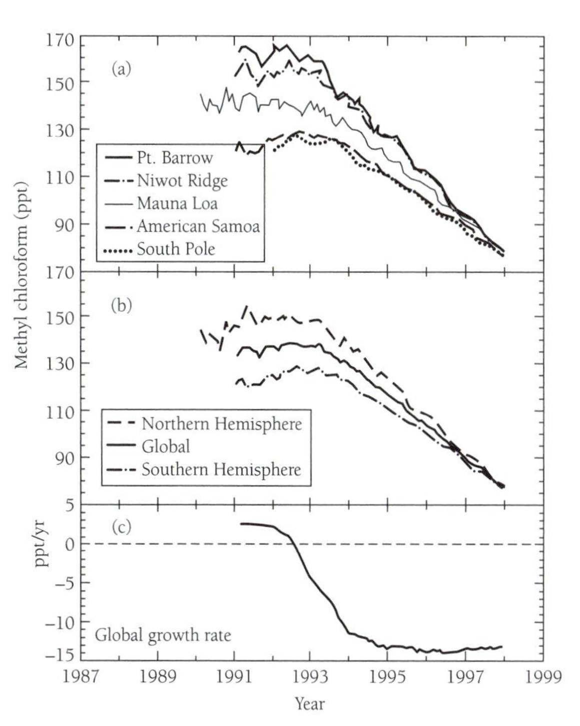

Figure 5.13 Variation of CH_3CCl_3 (ppt) since 1990. Data for individual stations are presented in the upper panel (a) as indicated. Composite results for the Northern Hemisphere, Southern Hemisphere and globe are presented in the middle panel (b). Growth rate (ppt yr^{-1}) is shown in the lower panel (c). Source: Elkins 1999.

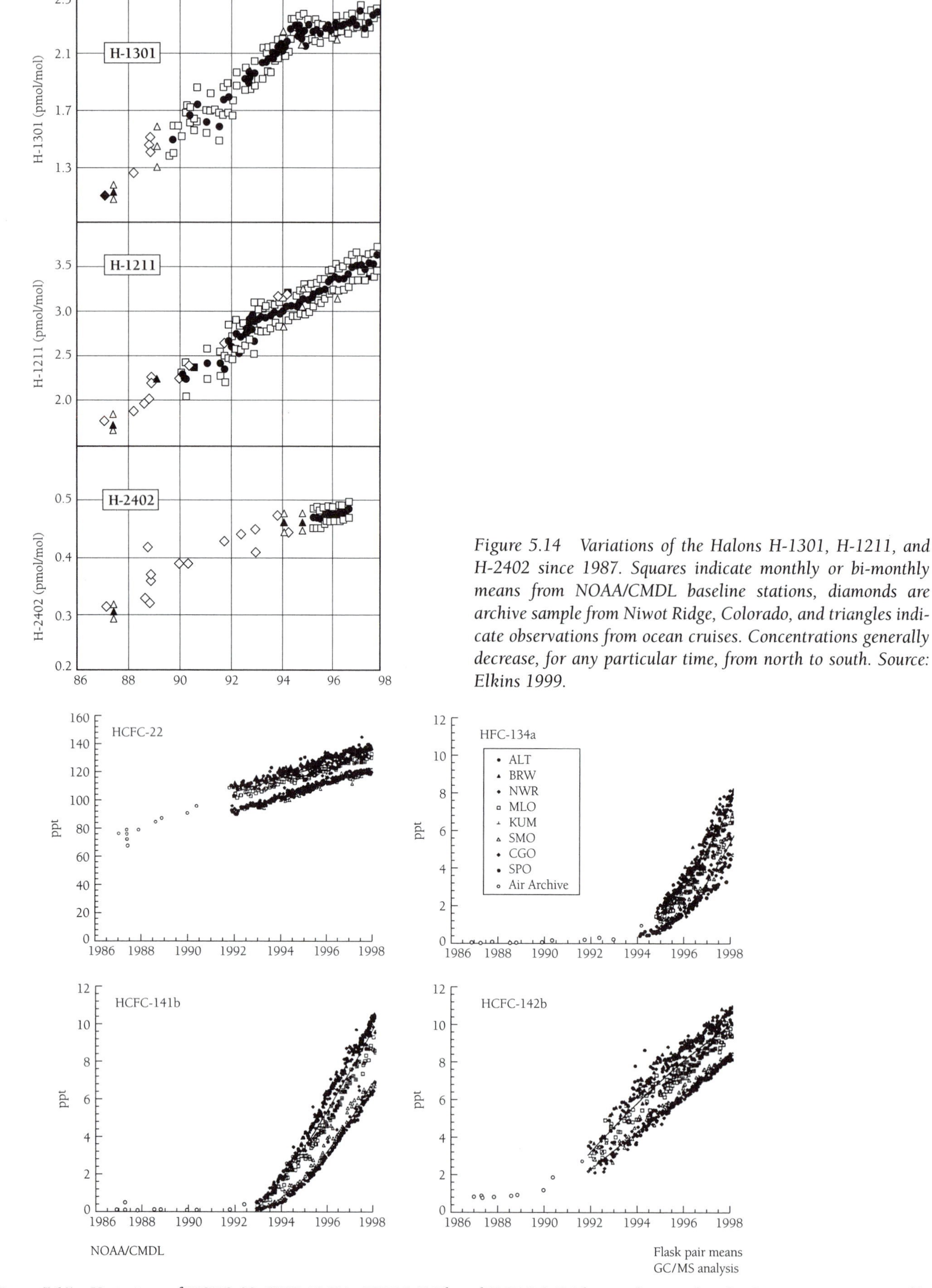

Figure 5.14 Variations of the Halons H-1301, H-1211, and H-2402 since 1987. Squares indicate monthly or bi-monthly means from NOAA/CMDL baseline stations, diamonds are archive sample from Niwot Ridge, Colorado, and triangles indicate observations from ocean cruises. Concentrations generally decrease, for any particular time, from north to south. Source: Elkins 1999.

Figure 5.15 Variations of HCFC-22, HFC-1349/a, HFC-1416/b and HCFC-1426/b over the past decade. Concentrations generally decrease, for any particular time, from north to south. Source: Elkins 1999.

the ocean and solid planet but also by the interaction with sunlight. Life developed early and had a major influence on the subsequent composition of the atmosphere. Eventually, life forms evolved with the capacity to tolerate large concentrations of free oxygen. Oxygen accumulated as a major constituent of the atmosphere, accompanied by photochemical reactions resulting in production of O_3. Ozone provided a protective shield screening organisms at the surface from otherwise lethal exposure to ultraviolet radiation. With the passage of time, life forms spread from the ocean to land. Plants and animals appeared, and the geochemical cycles responsible for the distribution of life-essential elements, such as carbon, nitrogen, phosphorus, and sulfur, were altered accordingly. Finally, man evolved and began to exert his influence on nature. Harnessing vast stores of energy (represented by organic matter in fossil reservoirs)and by clearing and converting land to agriculture and animal husbandry, he developed the capacity to change the composition of the atmosphere on a global scale simply by dint of sheer numbers. Our species is responsible for a release of chemical species into the atmosphere for which there are no natural analogues. The question is whether we can develop the wisdom to anticipate the consequences of our actions and respond accordingly.

Energy Budget of the Atmosphere

6

The temperature of Earth's atmosphere-surface-ocean system is ultimately controlled by the absorption and spatial redistribution (movement from one location to another) of energy supplied by the Sun. Solar energy is delivered to Earth mainly in the form of light: photons with wavelengths primarily in the visible portion of the spectrum, readily detectable with the human eye. Earth emits into space a quantity of energy almost exactly equal to that which it receives from the Sun. Terrestrial energy is emitted primarily at long wavelengths (the infrared), invisible to the naked eye but readily detectable with modern optical instrumentation. The near balance between energy absorbed and emitted is essential; otherwise Earth would either rapidly warm up or quickly cool off.

We begin in Section 6.1 with an introduction to the electromagnetic spectrum. The origin and nature of the radiation field emitted by the Sun is discussed in Section 6.2. Radiation emitted by Earth is treated in Section 6.3. An account of the global energy balance—a description of the processes by which solar radiation absorbed by Earth is converted to heat and eventually radiated back into space in the infrared—is presented in Section 6.4. A simple model for the greenhouse effect is given in Section 6.5. Summary remarks are presented in Section 6.6.

6.1 Electromagnetic Radiation

The intensity of light, or radiation, emitted by an object as a function of wavelength is known as its **spectrum.** The spectrum of a thermodynamically ideal object, known as a **blackbody**, is determined solely by its temperature. The concept of an ideal blackbody arises from consideration of a system of matter and radiation confined to a container of fixed temperature with perfectly absorbing walls, where energy is continually exchanged between matter and radiation. The system is in the ultimate steady state: at all wavelengths production and loss of radiation are in precise balance, and there is no net loss or gain of energy by the body. (It is assumed that the body emits and absorbs radiation at all wavelengths.) Under such conditions, the intensity of the radiation field is everywhere uniform; its wavelength dependence is defined in terms of a function named in honor of the German physicist Max Planck (1858–1947), who, along with Albert Einstein, is responsible for much of our understanding of the nature of electromagnetic radiation.

Intensities of radiation emitted by blackbodies of different temperature are presented in Figure 6.1. The energy emitted by a blackbody of temperature T (K) is a maximum at a wavelength λ (cm) given by

$$\lambda_{\max} = \frac{0.29\,\mathrm{cm\,K}}{T} \tag{6.1}$$

This is known as **Wien's displacement law**, or Wien's law. The total radiant energy crossing a unit area of a blackbody's surface in unit time, in a direction perpendicular to the surface, is equal to σT^4, where σ is the **Stefan-Boltzmann constant**, equal to 5.67×10^{-5} erg cm^{-2} K^{-4} sec^{-1}.

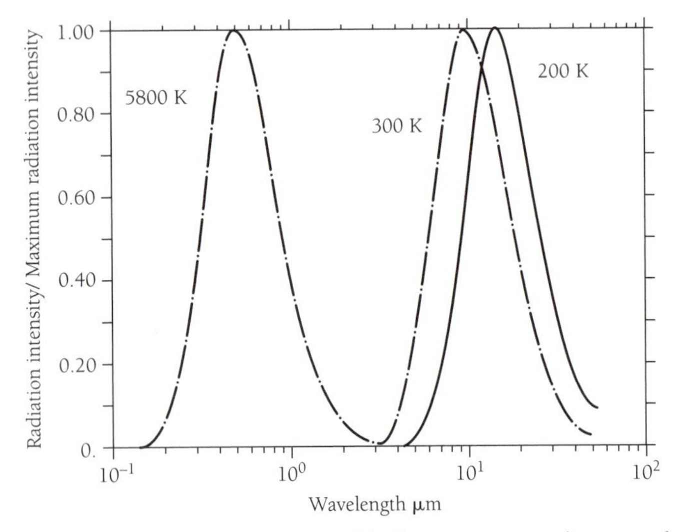

Figure 6.1 Intensity of the blackbody emission as a function of wavelength for temperatures of 200K, 300K and 5800K.

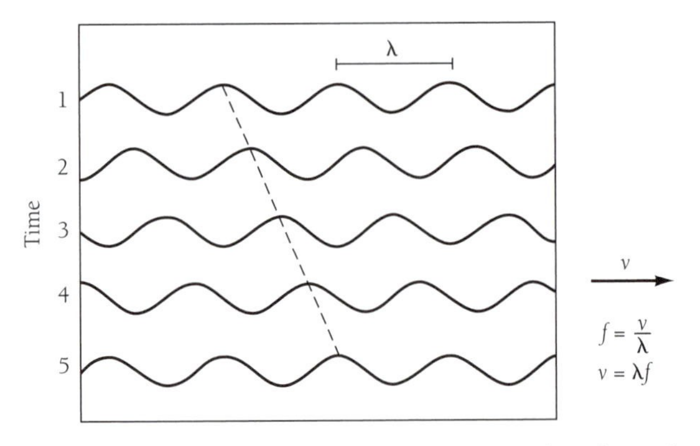

Figure 6.2 The relationship between the wavelength, λ, frequency, f, and the velocity of a wave, v.

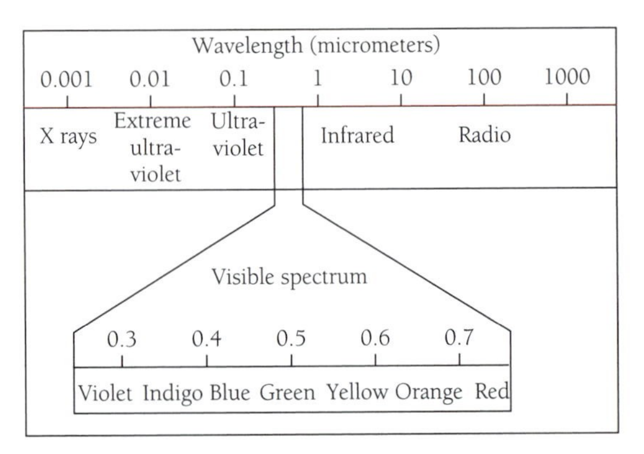

Figure 6.3 The electromagnetic spectrum.

Light, or radiation, propagates through space as an electromagnetic wave. The intrinsic electric and magnetic fields composing the wave oscillate in space and time with a wavelength characteristic of the color of the light. **Wavelength** (λ) defines the distance from the crest of one wave to the next, as illustrated in Figure 6.2. The **frequency** of the wave (*f*) is given by the number of wave cycles that pass a particular reference point in unit time. The **velocity** of the wave (v) is equal to the speed at which individual components of the wave propagate through space. Wavelength, frequency, and velocity are related according to the expression $v = \lambda f$, as indicated in Figure 6.2. Light travels at a fixed velocity, independent of wavelength or frequency. We denote the velocity of light by *c*; experimentally, it is found to have a magnitude of 3×10^{10} cm sec^{-1}.

The energy of a light particle or photon, E_f, is given in terms of the frequency of the radiation field by a relation first derived by Einstein:

$$E_f = hf, \tag{6.2}$$

or equivalently,

$$E_f = \frac{hc}{\lambda}, \tag{6.3}$$

where *h* is **Planck's constant**, 6.62×10^{-27} erg sec, and *c* is the velocity of light. Photons of short wavelength have relatively high energy. Absorption of radiation at short wavelengths can result in a major rearrangement of the structure of the absorbing atoms or molecules. In contrast, long-wavelength photons have lower energy. As discussed in Chapter 3, absorption of long-wavelength radiation may lead to changes in the rotational or vibrational state of molecules but is generally ineffective in altering their structure.

The different regions of the electromagnetic spectrum are summarized in Figure 6.3. We refer to light with wavelength less than 100 Å as **X rays.** Radiation between about 100 and 1250 Å is termed **extreme ultraviolet**. Wavelengths in the solar spectrum associated with X-rays and extreme ultraviolet radiation are absorbed by the atmosphere above 80 km with relatively high efficiency. Absorption of this energetic radiation is responsible for ejection of electrons from atoms and molecules in the upper atmosphere (photoionization), producing the partially ionized region of the atmosphere known as the **ionosphere.** Radiation in the **ultraviolet** portion of the spectrum between 1250 and 3000 Å is absorbed over a range of altitudes in the middle atmosphere, as illustrated in Figure 6.4, contributing to dissociation of O_2 and production of O at high altitudes and to production of O_3 in the stratosphere. Radiation between 3000 and 7000 Å constitutes the **visible** portion of the spectrum. Sunlight in this spectral interval penetrates to the surface, with transmission through the atmosphere limited mainly by particulate matter and clouds. The visible portion of the spectrum is subdivided, from short to long wavelengths, into violet, indigo, blue, green, yellow, orange, and red spectral regions, as indicated in Figure 6.3. Radiation greater than 7000 Å, extending to about 100 μm (10^{-2} cm), is referred to as **infrared.** Transmission of radiation in this spectral interval is limited mainly by clouds and by trace constituents of the atmosphere such as H_2O, CO_2, O_3, CH_4, CO, N_2O, SF_6, and the industrial chlorofluorocarbons. Energy radiated by the surface in the infrared region is trapped, partially at least, by the atmosphere and is reradiated back to the surface. Atmospheric trapping of infrared radiation is primarily responsible for what is known as the **greenhouse effect**, as discussed below. The **microwave** por-

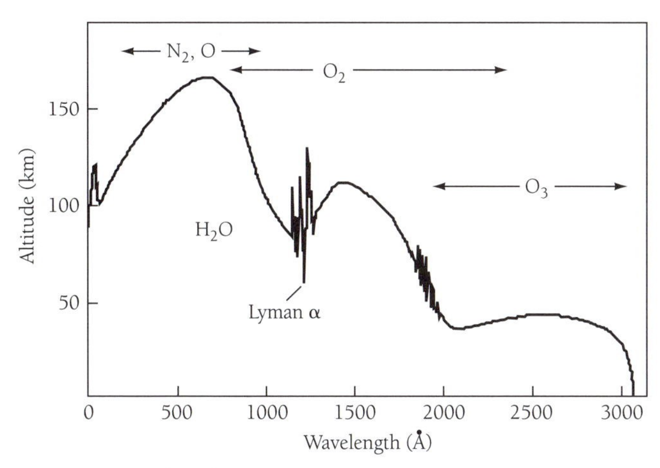

Figure 6.4 Altitude to which ultraviolet sunlight penetrates Earth's atmosphere. The indicated height corresponds to the level of maximum energy deposition at each wavelength. Source: Chamberlain and Hunten 1987.

tion of the spectrum extends from about 100 μm to 1000 μm; radiation at longer wavelengths falls into what is known as the **radio region** of the spectrum.

6.2 Solar Radiation

Radiation in the visible portion of the solar spectrum originates from a region of the solar atmosphere known as the **photosphere.** A schematic representation of the different segments of the solar atmosphere is presented in Figure 6.5. The temperature of the photosphere ranges from about 4000 to 6000 K. The spectrum of the Sun at visible wavelengths approximates that of a blackbody at a temperature of 5800 K, as illustrated in Figure 6.6. Radiation emitted by the Sun at shorter wavelengths is more intense than we would expect for a blackbody of this temperature, reflecting the fact that short-wave radiation from the Sun emanates mainly from high-temperature, high-altitude regions of the solar atmosphere. Solar atmospheric temperature increases with altitudes above the photosphere in the **chromosphere** and **corona**, reaching values as high as 10^6 K in the outer regions of the corona, as illustrated in Figure 6.7. Extreme ultraviolet and X-radiation are emitted from the upper chromosphere and corona. The intensity of solar radiation at these wavelengths is orders of magnitude larger than would be the case if the temperature of the Sun's outer atmosphere were the same as that of the photosphere.

The Sun's radiant energy is supplied by nuclear reactions in the solar core. The high temperature of the solar atmosphere's outer region is maintained by the transfer of energy in the form of hydromagnetic waves originating deep within the Sun. These waves propagate vertically, growing in amplitude as they move toward regions of lower density in the outer solar atmosphere. Eventually, they break like ocean waves on a beach. Their energy is deposited in the solar atmosphere's outer layers, raising the temperature of the corona and upper chromosphere to values high enough to fuel the escape of gas from the Sun. The transfer of energy from deeper regions of the solar atmosphere to the

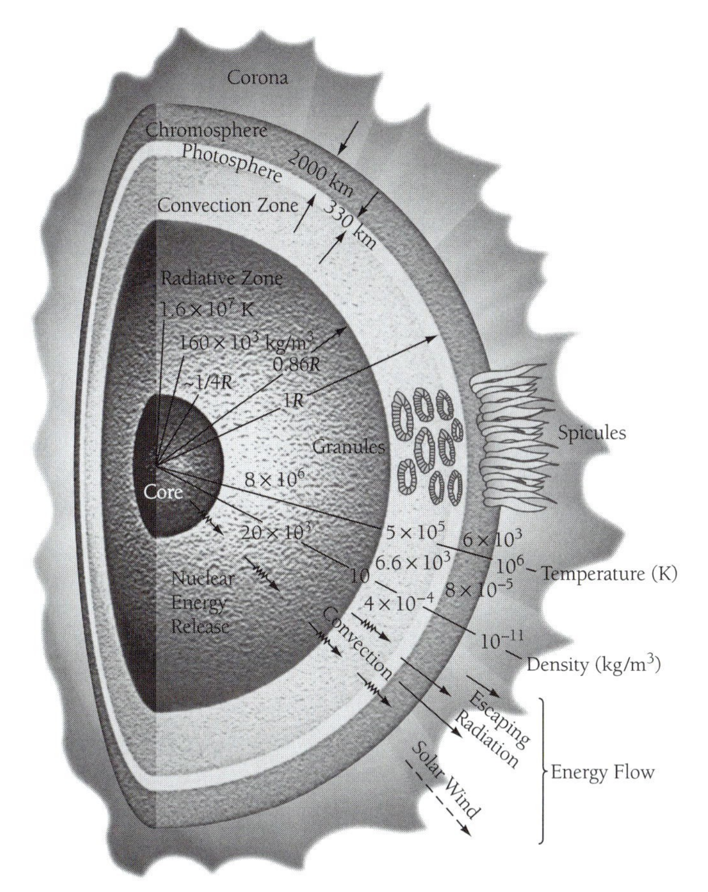

Figure 6.5 Schematic view of the Sun. Source: Pasachoff 1979.

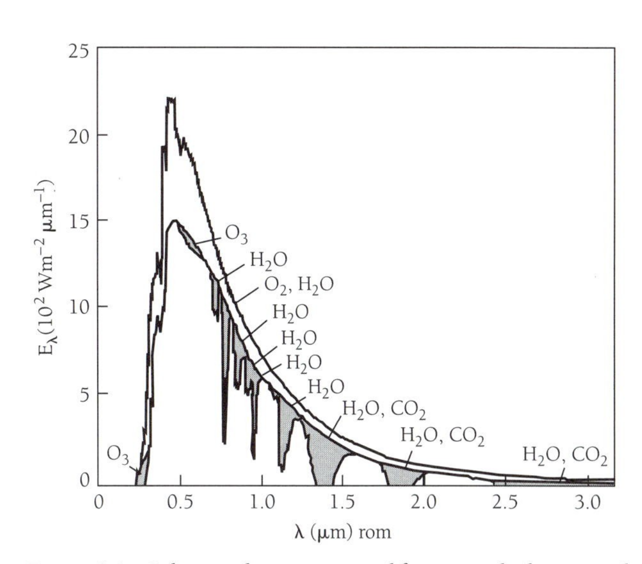

Figure 6.6 Solar irradiance as viewed from outside the atmosphere of Earth (upper curve) indicating the influence of absorption by O_2, O_3, H_2O, and CO_2 (lower curve). Source: Peixoto and Oort 1992.

outer layers of the chromosphere and corona is variable, with a characteristic period of about 11 years, as indicated in Figure 6.8. This results in changes in the output of radiant energy from the Sun, particularly in the extreme ultraviolet and X-ray regions of the spectrum and in variations in the strength of the **solar wind**, the stream of ionized plasma emitted by the corona.

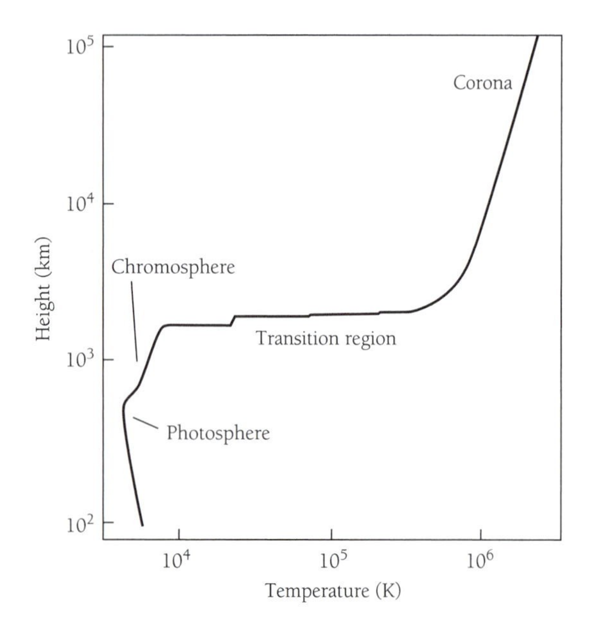

Figure 6.7 The temperature structure of the Sun. Source: Zeilik and Smith 1987.

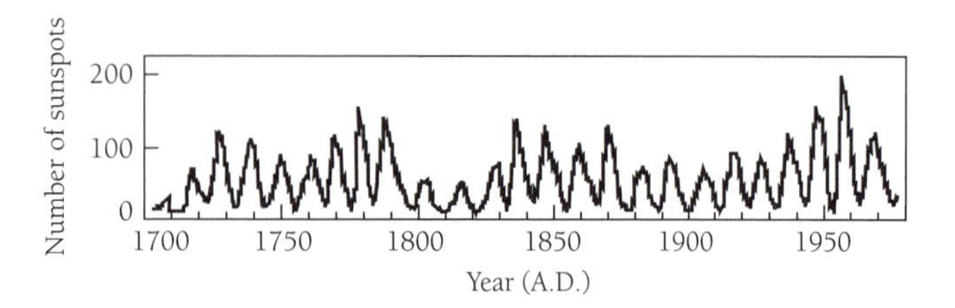

Figure 6.8 Sunspot numbers as a function of time. Source: Eddy 1977.

Variations in the flux of extreme ultraviolet- and X-radiation reaching Earth lead to detectable changes in the density and structure of the ionosphere, affecting the propagation of radio waves around Earth. Changes in the strength of the solar wind modulate the flux of energetic particles reaching Earth from the cosmos, so-called **cosmic radiation**, influencing the atmospheric production rate of an unstable isotope of carbon containing eight neutrons, ^{14}C.[1] Variations in the flux of solar plasma at Earth cause electric currents to flow in outer regions of Earth's magnetic envelope, the **magnetosphere**, inducing changes in the strength and direction of Earth's magnetic field that are readily detectable at the surface. Changes in the output of solar plasma are also responsible for spectacular optical displays at a high latitude in the atmosphere, the aurora phenomenon. The aurora's luminosity is associated with an excitation of atoms and molecules in the upper atmosphere caused by collisions with electrons energized as a consequence of the complex interaction between the solar wind and the outer regions of Earth's magnetic field. All these phenomena vary with solar activity.

A record of solar activity dating back to the early eighteenth century is presented in Figure 6.8. It shows the characteristic 11-year cycle and offers clear evidence that solar activity, measured by the number of spots observed on the visible disc of the Sun, is exceptionally variable. There are indications of an extended period of unusually low solar activity in the seventeenth century, an interval termed the **Maunder minimum** by solar physicists. There is an intriguing association of this interval of low solar activity with the Little Ice Age, an unusually cold period extending from about 1400 to 1900. Despite clear evidence that the Sun is variable and that this variability is sufficient to lead to significant changes in the strength and orientation of Earth's magnetic field, the frequency of auroral activity, and the rate of ionization associated with absorption of cosmic rays, there is little indication that the observed changes in solar activity are sufficient to account for the large changes observed for Earth's past climates. The Sun's total energy output appears to be remarkably constant. Satellite measurements, summarized in Figure 6.9, suggest that changes are limited to fractions of a percent of the average value of the total output of solar energy, too small to exert a measurable direct impact on climate.

6.3 Terrestrial Radiation

Viewed from space, Earth's spectrum looks like the back of a double-humped camel, as illustrated in Figure 6.10. The short-wavelength hump is contributed by solar radiation reflected by clouds, the atmosphere, and the surface on the day side of the planet. The long-wavelength structure is due to thermal emission from the atmosphere and surface. The shape of the short-wavelength component resembles that of the solar spectrum greater than about 3000 Å; it approximates the emission of a blackbody at a temperature of about 6000 K, truncated at shorter wavelengths due to absorption of sunlight in the atmosphere by O_3. The long-wavelength component resembles that of a blackbody with an average temperature of about 255 K. The intensity, and spectral distribution, of radiation in the long-wavelength portion of the spectrum varies from place to place and time to time, reflecting spatial and temporal variability in the temperature of the atmosphere and surface.

The importance of the role played by H_2O, CO_2, O_3, CH_4, and N_2O in atmospheric absorption and emission of infrared radiation is illustrated in Figure 6.11a-b. Long-wavelength radiation is transmitted directly from the surface to space only in certain well-defined regions of the spectrum, called **windows**, for example between 8 and 9 μm, or between 10 and 14 μm. The intensity of radiation emitted into space in window regions can be interpreted to obtain a measure of surface temperature. Radiation at other wavelengths can be analyzed to infer temperatures as functions of altitude in the atmosphere, assuming that the height distribution of the absorbing gas is either known or may be derived from the data. Spectra recorded by the Nimbus 4 satel-

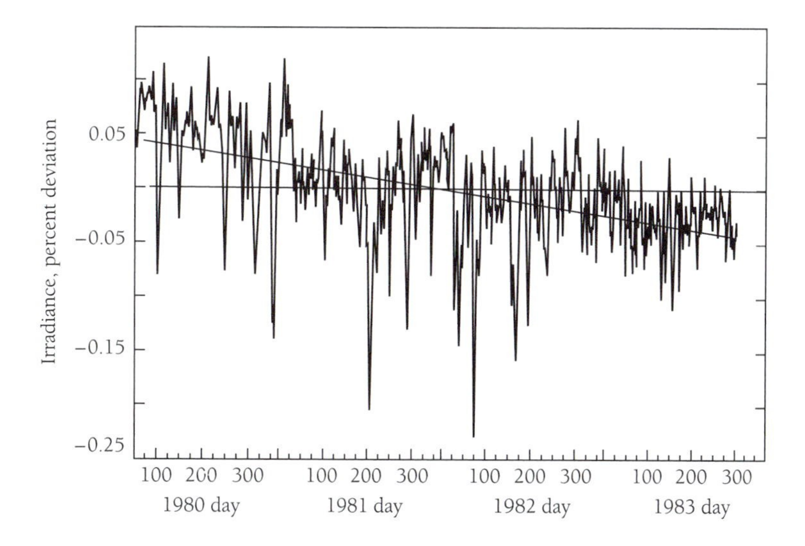

Figure 6.9 Percent variation in emission of solar energy from 1980 to 1983. Numbers on the horizontal axis indicate days measured from January 1 in individual years. The mean value for this period was 1367.6 watt m^{-2}. Source: Wilson 1984.

lite over the Sahara, the Mediterranean, and Antarctica are presented in Figure 6.12. For ease of interpretation, the figure includes estimates of radiation intensity as a function of the wavelength expected for emission by a blackbody at various temperatures.

Good agreement between the spectrum of the Sahara and the blackbody reference curve for a temperature of 320 K over the spectral interval 10–12 μm suggests that emission in this spectral interval provides a reliable measure of surface temperature. Emission in the vibrational-rotational band of CO_2 centered at 15 μm reflects the variation of temperature with altitude and covers a range of temperatures from 320 to 220 K. The low values of the effective blackbody temperature in the 15 μm band of CO_2 indicates the importance of absorption in this band in limiting direct transmission of radiation from the surface into space. It attests to the importance of CO_2 as a trap for infrared radiation emitted by the surface, i.e., as a **greenhouse gas**; radiation absorbed by the atmosphere in this spectral interval is partially radiated back to Earth, where it contributes to surface heating and to a significant increase in ground temperature. The CO_2 absorption band is important also for the Mediterranean. It is less significant for Antarctica, where surface temperatures are much colder, probably near 180 K; the relatively high temperature near the center of the 15 μm band of CO_2 over Antarctica reflects the fact that temperatures in the emitting region of the atmosphere for this wavelength interval are about 40 K higher than temperatures near the ground; in this case emission most probably originates from the stratosphere.

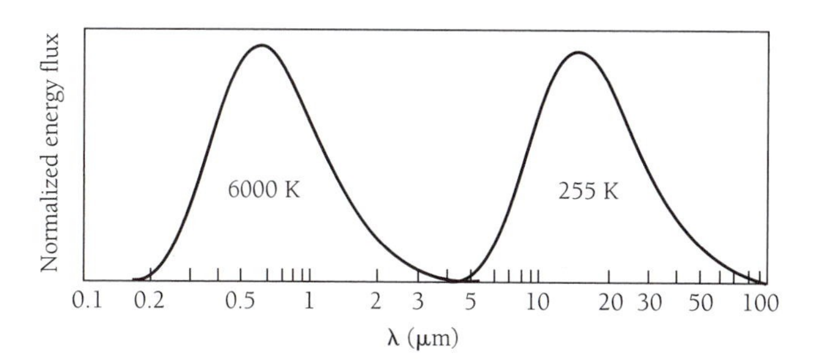

Figure 6.10 Schematic representation of the radiant energy of Earth, as a function of wavelength (λ), as viewed in space. Source: Peixoto and Oort 1992.

6.4 Global Energy Balance

An illustrative budget for the energetics of Earth's global surface-atmosphere system is presented in Figure 6.13. The left side of Figure 6.13 summarizes the fate of incident solar energy. It begins with an input of 100 units of solar energy at the top of the atmosphere. Of this, roughly 20 units are reflected into space by clouds. Four units are reflected into space from the surface, and 6 units are scattered back to space by the air. Approximately 16 units of solar energy are absorbed by the atmosphere, mainly by aerosols (small liquid or solid particles suspended in air), O_3, CO_2, and H_2O. An additional 4 units are absorbed by clouds. The balance, 50 units, is absorbed at the surface. The fraction of the total incoming solar energy reflected into space [(20 + 10)/100 = 0.30] defines what is known as the **albedo** of Earth.

Of the 50 units of solar energy absorbed at the surface, a relatively small fraction, 12% or 6 units, is emitted directly into space as illustrated on the right side of Figure 6.13, The bulk of the energy absorbed at the surface is transferred to the atmosphere, through a combination of infrared radiation (14 units), latent heat (24 units), and sensible heat (6 units). The flux of infrared radiation emitted by the surface, as indicated on the right side of Figure 6.13, reflects a difference between two relatively large terms of approximately equal

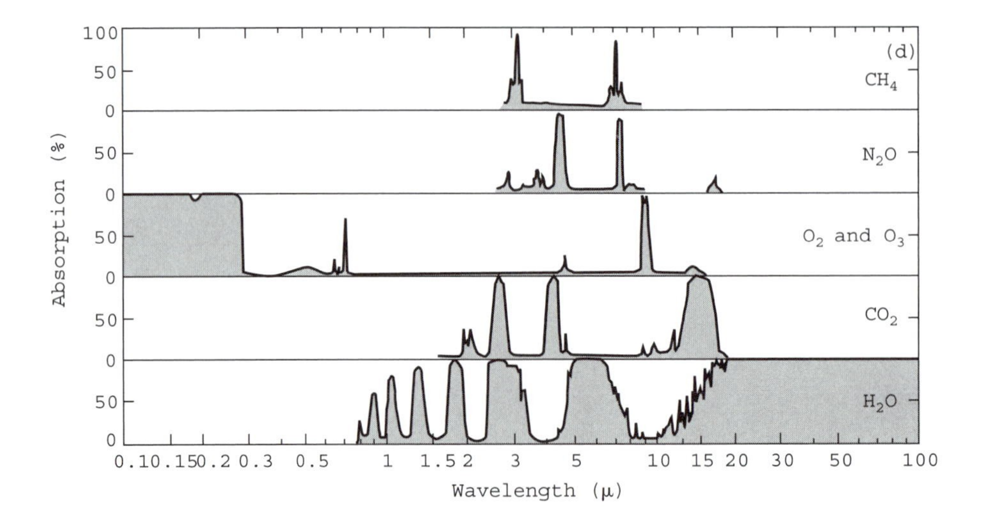

Figure 6.11a Absorption of radiation between the surface of Earth and the top of the atmosphere, as a function of wavelength, for several radiatively active gases. Source: Peixoto and Oort 1992.

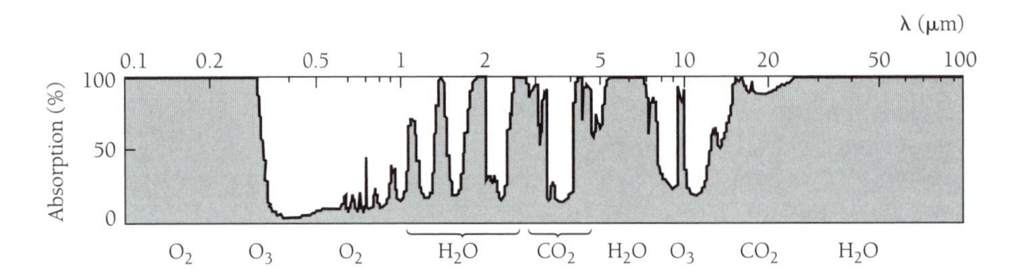

Figure 6.11b Absorption of radiation between the surface of Earth and the top of the atmosphere, as a function of wavelength. The dominant absorbing gas in each spectral region is indicated on the bottom of the Figure. Source: Peixoto and Oort 1992.

magnitude. It represents the *net* flux of infrared radiation emitted by the surface, thus accounting for a balance between radiative energy emitted by the surface and radiative energy absorbed from the atmosphere. Earth's surface is warmer on average than its atmosphere. This circumstance accounts for the fact that exchange of infrared radiation between the surface and the atmosphere is responsible for a net cooling of the surface. In Figure 6.13 the term "latent heat" denotes the influence of energy transferred from the surface to the atmosphere in the form of H_2O vapor. The surface cools as a result of evaporation. Energy removed from the surface by evaporation is deposited in the atmosphere when water vapor condenses to form precipitation. In Figure 6.13 the term "sensible heat" refers to energy transferred from the (warm) surface to the (cool) atmosphere by conduction.

Of the 70 total units of solar energy absorbed by the atmosphere-surface system, 64 units are returned into space in the form of infrared radiation from the atmosphere (the balance is accounted for by the emission of infrared radiation from the surface directly into space, as noted above). High-altitude clouds and infrared absorbing gases such as H_2O and CO_2 are primarily responsible for the emission of infrared radiation from the atmosphere. Radiation is emitted into space from upper levels of the troposphere, a region characterized by an average temperature of about 255 K. In the absence of infrared absorptive species in the atmosphere, energy absorbed from the Sun would necessarily be balanced by emission of infrared radiation into space from the surface. As indicated below, the flux of energy absorbed by Earth from the Sun would be balanced by the emission of infrared energy from a blackbody at a temperature of 255 K. In the absence of infrared absorptive species in the atmosphere, the surface temperature would be a freezing cold 255 K. The atmosphere serves as an insulating blanket maintaining the surface at a relatively balmy average temperature of about 290 K. The role of the atmosphere in regulating the transfer of infrared radiation from Earth into space is known as the **greenhouse effect**.

It should be emphasized that there is no particular merit to the specific budget presented in Figure 6.13. It was developed merely to provide a general picture of the approximate magnitude of the various contributions to the energetics of

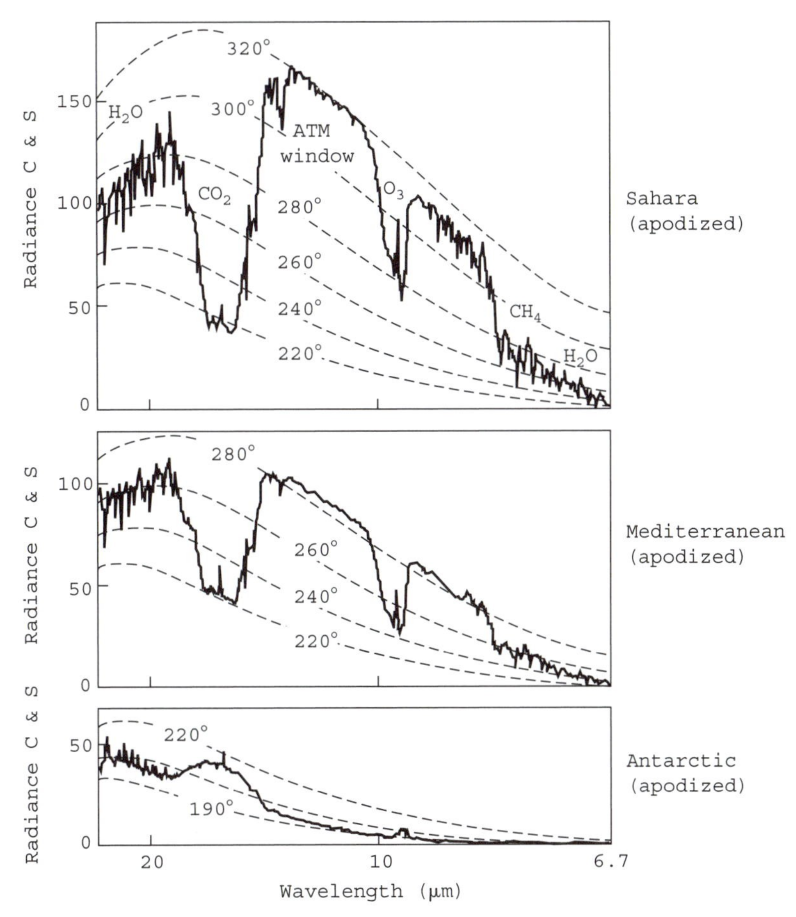

Figure 6.12 Infrared emission spectra of Earth recorded by instrumentation aboard the Nimbus 4 satellite. The dashed curves represent blackbody irradiances at several temperatures. Source: Hanel et al. 1971.

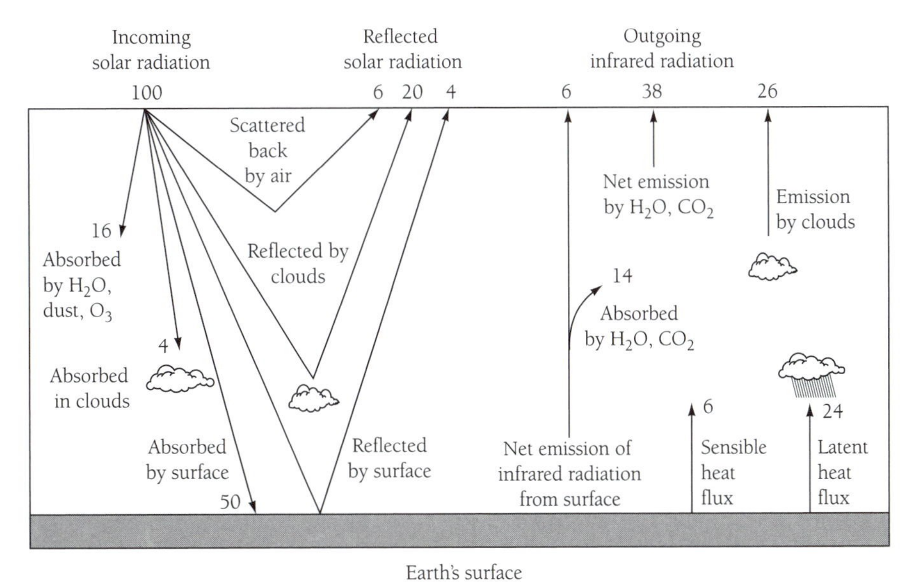

Figure 6.13 The radiation budget of Earth's surface-atmosphere system, for solar (visible radiation), on the left, and planetary (infrared) radiation, on the right. Source: Peixoto and Oort 1992.

the atmosphere-surface system. The specifics of the global budget may vary from time to time. It is possible, for example, for the ocean to either store or release heat, such that the global, annually averaged heat budget could depart temporarily in any given year from a steady state. In one year rather than another, clouds could play a greater or lesser role in reflecting solar radiation back to space. The significance of the budget in Figure 6.13 is largely pedagogical; it illustrates in a general way the relative importance of the processes involved in energy exchange between the surface, atmosphere, and space. We turn now to a more quantitative, less descriptive treatment of the problem.

We begin by calculating the temperature of the surface of a planet devoid of atmosphere, assuming that the albedo of this hypothetical planet is equivalent to that of the real Earth, with its complex surface of continents and oceans and its atmosphere of reflecting and absorbing gases, clouds, and aerosols. We assume that the quantity of solar energy absorbed by our hypothetical Earth is exactly equal to the amount of infrared energy radiated into space. The flux of solar energy, crossing a unit area in unit time, perpendicular to the direction of propagation of sunlight, and at a distance from the Sun equivalent to the mean orbital radius of Earth, can be measured with instrumentation on a satellite that is orbiting Earth (it is necessary to reach above Earth to eliminate effects of absorption by the atmosphere). This quantity is known as the **solar constant**, F.

Example 6.1: The Sun emits radiation at a rate of 3.9×10^{26} W. Calculate the solar constant, the flux of energy crossing a unit area normal to the solar beam at the orbital position of Earth. Express the answer in units of W m^{-2}.

Answer: The total amount of solar energy crossing the surface of an imaginary sphere centered on the Sun should be a constant independent of the radius of the sphere (there is no significant sink for energy in the effectively empty circumsolar medium). If the flux (W m^{-2}) crossing a unit area of the surface of a sphere of radius R is given by F, it follows that

$$F\left(4\pi R^2\right) = 3.9 \times 10^{26}$$

Thus

$$F = \frac{3.9 \times 10^{26}}{4\pi R^2}$$

Substituting for R the average separation between the Sun and Earth, we find that

$$F_{\text{Earth}} = 1379 \text{ W m}^{-2}$$

Values for solar radiation flux at other planetary radii can be readily obtained by scaling the value of F_{Earth} as follows:

$$F_R = F_{\text{Earth}} \left(\frac{R_{\text{Earth}}}{R}\right)^2$$

where R_{Earth} and R define the average seperations from the Sun of the Earth and the planet in question. ■

Example 6.2: Express the solar constant in units of (a) W cm^{-2}, (b) erg sec^{1} m^{-2}, and (c) erg sec^{-1} cm^{-2}.

Answer:

(a)
$$1 \text{ m}^2 = 10^4 \text{ cm}^2$$
$$1379 \text{ W m}^{-2} \times (1 \text{ m}^2 / 10^4 \text{ cm}^2) = 0.1379 \text{ W cm}^{-2}$$

(b)
$$1 \text{ W} = 10^7 \text{ erg sec}^{-1}$$
$$1379 \text{ W m}^{-2} \times (10^7 \text{ erg sec}^{-1} / 1 \text{ W}) = 1.379 \times 10^{10} \text{ erg sec}^{-1} \text{ m}^{-2}$$

(c)
$$1 \text{ m}^2 = 10^4 \text{ cm}^2$$
$$1.379 \times 10^{10} \text{ erg sec}^{-1} \text{ m}^{-2} \times (1 \text{ m}^2 / 10^4 \text{ cm}^2) = 1.379 \times 10^6 \text{ erg sec}^{-1} \text{ cm}^{-2}$$ ■

Example 6.3: For comparison, compute the energy flux crossing a 1 cm^2 area on the floor beneath a 100 W lightbulb on the ceiling 3 meters overhead.

$$Flux = \frac{100 \text{ W}}{4\pi R^2}$$
$$= \frac{100 \text{ W}}{4\pi (3\text{m})^2}$$
$$= 0.884 \text{ W m}^{-2}$$
$$= 8.84 \times 10^6 \text{ erg sec}^{-1}\text{m}^{-2}$$
$$= 884 \text{ erg sec}^{-1}\text{cm}^{-2}$$ ■

The total solar energy intercepted by Earth in unit time can be obtained by multiplying the solar constant by the effective target area of Earth. The appropriate geometry is illustrated in Figure 6.14. If the total energy intercepted by Earth is represented by E, and the solar constant is given by F, E is defined by the equation

$$E = F \times \pi R^2, \tag{6.4}$$

where R is the radius of Earth. The albedo of Earth (A) denotes the fraction of the incident solar energy reflected back to space; the fraction of incident solar energy absorbed by the surface is given by $(1 - A)$. Consequently, the energy absorbed by Earth is obtained by multiplying E by $(1 - A)$:

$$E_{\text{abs}} = (1 - A) \times F \times \pi R^2 \tag{6.5}$$

We assume that Earth radiates as a blackbody at temperature T. The energy emitted by unit area in unit time is given then by σT^4, where σ, again, denotes the Stefan-Boltzmann constant. The total energy emitted by the surface, E_{emit}, is obtained by multiplying this quantity by the area of the surface, $4\pi R^2$:

$$E_{\text{emit}} = \sigma T^4 \times 4\pi R^2 \tag{6.6}$$

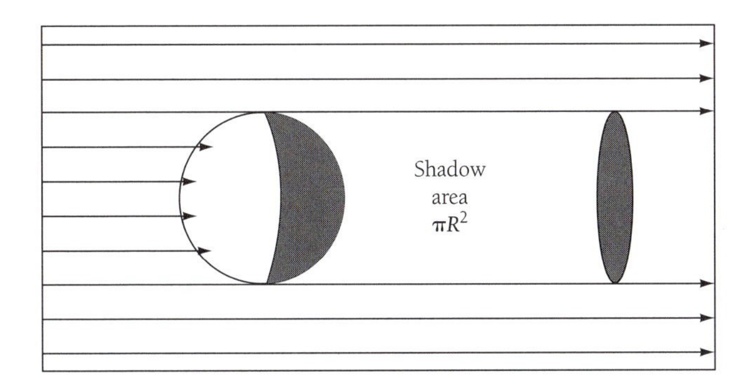

Figure 6.14 The shadow formed by an illuminated sphere has a circular shape with a radius R equal to the radius of the sphere.

Equating E_{abs} and E_{emit}, canceling the common factor πR^2, and rearranging the terms yields an equation for T:

$$T = \left(\frac{F(1 - A)}{4\sigma}\right)^{1/4} \tag{6.7}$$

We refer to the temperature obtained in this fashion as the **effective temperature** of a planet.

Example 6.4: Using a value for σ of 5.67×10^{-8} W m^{-2} K^{-4}, and magnitudes of F and A appropriate for Earth, calculate the effective temperature for Earth.

Answer:

$$T = \left(\frac{F(1-A)}{4\sigma}\right)^{1/4}$$

$$= \left(\frac{1379 \text{ W m}^{-2} (1 - 0.30)}{4(5.67 \times 10^{-8} \text{ W m}^{-2} \text{ K}^{-4})}\right)^{1/4}$$

$$= 255.4 \text{ K}$$ ■

The average temperature of Earth, without an atmosphere and with the same albedo as today, would be 20°C below the freezing point of water. An ocean, if it existed, would be frozen, at least at its surface. This would result in an even colder temperature, since the albedo of such a planet would be presumably larger than that of Earth today: $(1 - A)$ would be reduced, and the value of T would be less than calculated here. As we shall see, the presence of an atmosphere results in the temperature of Earth's surface being considerably higher than the effective temperature.

6.5 Layer Model for the Atmosphere

We now develop a model, adopted from Goody and Walker (1972), that can account, at least approximately, for the influence of the atmosphere. Suppose that the surface at a temperature of T_g, is enveloped by an atmospheric layer characterized by a uniform temperature T_a. The model is illustrated in Figure 6.15. Suppose that the atmosphere is transparent to visible radiation; incident sunlight penetrates with 100% efficiency to the surface, where it is reflected with an albedo A. Assume that the atmosphere is opaque to infrared radiation; infrared radiation emitted by the surface is absorbed by the atmosphere with 100% efficiency. As before, the energy absorbed by the surface is given by equation (6.5). In addition to the energy absorbed in the visible portion of the spectrum, the surface receives energy at infrared wavelengths from the atmosphere, which is assumed to emit light as though it were a blackbody at its ambient temperature, T_a. Balancing sources (the left-hand side of the equation) with losses (the right-hand side) of energy at the surface, we have

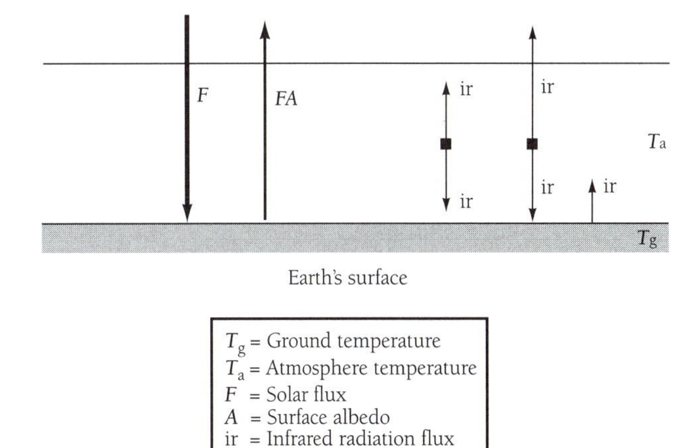

Figure 6.15 Schematic view of the layer model.

$$F \times \pi R^2 \times (1 - A) + \sigma(T_a)^4 \times 4\pi R^2 = \sigma(T_g)^4 \times 4\pi R^2 \tag{6.8}$$

Energy balance for the combined atmosphere-surface system requires that the infrared energy radiated into space by the atmosphere must be equal to the quantity of energy absorbed by the surface:

$$F \times \pi R^2 \times (1 - A) = \sigma(T_a)^4 \times 4\pi R^2 \tag{6.9}$$

We have implicitly assumed here that the thickness of the atmosphere is very small compared to the radius of Earth (thus we use the same value for R in equations 6.8 and 6.9). Equation (6.9) is identical to equation (6.7), with T in (6.7) replaced by T_a. Emission into space originates exclusively in the atmosphere; the temperature of the atmosphere is consequently identical to the effective temperature discussed above. Combining equations (6.8) and (6.9) [using (6.9) to substitute for the solar absorption term, $F\pi R^2$ $(1 - A)$ in (6.8), and canceling the common factors], we find

$$\sigma(T_g)^4 \times 4\pi R^2 = 2\sigma(T_a)^4 \times 4\pi R^2$$

$$T_g = (2)^{1/4} T_a \tag{6.10}$$

Example 6.5: Compute T_g, given T_a equal to 255.4 K, as before.

Answer:

$$T_g = (2)^{1/4} T_a$$

$$= (2)^{1/4} (255.4 \text{ K})$$

$$= 303.7 \text{ K}$$ ■

This provides a simple example of the operation of the greenhouse effect. Irradiation of the surface by a single atmospheric layer is sufficient to raise the temperature of the surface by 48 K, compared with the temperature that would apply in the absence of an atmosphere.

The layer model can be easily generalized to allow for an arbitrary number of atmospheric layers. Suppose that the atmosphere includes N layers, sequentially numbered from the top down. The temperature of the first (highest) layer, T_1, is equal to the effective temperature of the planet, T_{eff}. The temperature of the second highest layer, T_2, is given by

$$(T_2)^4 = 2(T_{\mathrm{eff}})^4 \tag{6.11}$$

The temperature of the ith layer, T_i, is given by

$$(T_i)^4 = i(T_{\mathrm{eff}})^4, \tag{6.12}$$

and the ground temperature, for N layers, is given by

$$(T_g)^4 = (N+1)(T_{\mathrm{eff}})^4, \tag{6.13}$$

which implies

$$T_g = (N+1)^{1/4} T_{\mathrm{eff}} \tag{6.14}$$

The layer model can be used to obtain an additional important insight into the variation of atmospheric temperature with height. Consider a thin skin of gas at the outer fringe of the atmosphere, as indicated in Figure 6.16. Denote the temperature of the skin by T_s. Gas at high altitude gains energy by absorbing infrared radiation emerging from the atmosphere below. It finds a radiation source in the infrared equivalent to that of a blackbody at a temperature equal to the temperature of the highest altitude layer of the atmosphere: radiation with a temperature equal to the effective temperature of the atmosphere. Since the density of the skin layer is small, it absorbs only a fraction, ϵ, of the energy from below. Likewise, the upward radiation from the skin is $\epsilon\sigma(T_s)^4$ with a similar value for the flux downward. The energy balance for the skin is expressed thus by the equation

$$\epsilon \times \sigma\left(T_{\mathrm{eff}}\right)^4 \times 4\pi R^2 = 2 \times \epsilon \times \sigma\left(T_s\right)^4 \times 4\pi R^2, \tag{6.15}$$

where T_{eff} defines the effective temperature of Earth (T_a for the one-layer model above). The factor of two on the right-hand side of equation (6.15) accounts for the fact that the skin can dispose of energy by emitting radiation both up and down. Solving this equation for T_s gives

$$T_s = \frac{T_{\mathrm{eff}}}{(2)^{1/4}} \tag{6.16}$$

Using the value obtained above for T_{eff}, equation (6.16) implies a value for T_s of 213 K. We refer to T_s as the **skin temperature** of the planet. In the layer model, temperature decreases steadily with altitude, eventually falling to the height-independent skin temperature.

We turn attention now to a discussion of the assumptions implicit in the layer model. To what extent, and to what accuracy, might we expect the layer model to provide a useful representation of conditions in the real atmosphere? The model depicts an atmosphere divided into distinct strata,

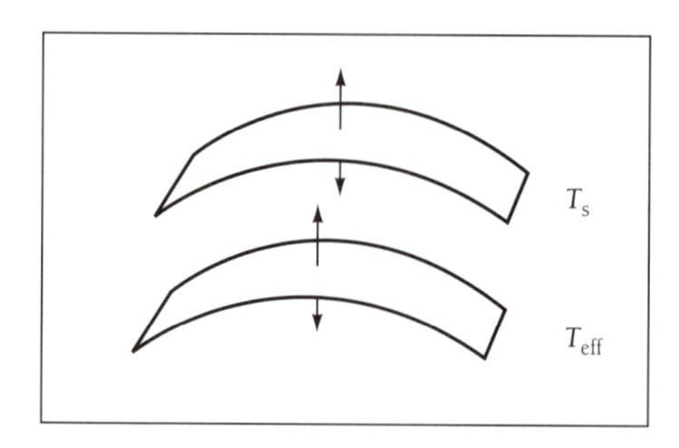

Figure 6.16 Skin temperature energy balance.

each of which is assumed to be characterized by a constant temperature. Infrared radiation emitted by a given layer is captured with 100% efficiency by its immediate neighbors, either above or below. If radiation is emitted by the topmost layer in the upward direction, it is assumed to escape into space. If it is emitted by the bottommost layer in the downward direction, it is absorbed by the surface. In the real atmosphere, energy transfer from level to level by infrared radiation proceeds by a more probabilistic, or stochastic, process. Emission at a particular level results in heating of the atmosphere over a range of neighboring levels, in contrast to the localized heating depicted in the layer model. It is also necessary to allow for the dependence on wavelength of the infrared radiation emitted and absorbed by the atmosphere. This reflects in part the abundance and height distribution of specific gases such as H_2O, CO_2, and O_3, in part the efficiency with which these molecules absorb or emit radiation of a specific wavelength. The spectrum of infrared radiation emitted by the atmosphere differs significantly from that of an ideal blackbody. Radiation emitted into space at different wavelengths originates from different levels of the atmosphere, as was clearly indicated in Figure 6.12.

Realistic simulation of atmospheric temperature structure also requires a treatment of the absorption of solar radiation by the atmosphere. The layer model presented above assumes that solar energy is absorbed solely at the surface. As seen earlier, a fraction of the solar energy captured by Earth is absorbed by the atmosphere. Though small in absolute terms compared to the total energy absorbed by the planet, as indicated in Figure 6.13, energy absorbed by the atmosphere has a disproportionately large effect, not only on atmospheric temperature but also on surface temperature. Energy absorbed at upper levels of the atmosphere can escape relatively easily into space by emission in the infrared. The total energy emitted into space must balance the total energy absorbed from the Sun. Heating of the upper layers of the atmosphere is thus compensated by cooling at lower levels; energy otherwise available to heat the surface is able to escape directly into space.

We can allow for the influence of solar atmospheric heating in the layer model as follows. Consider first the case where all of the solar energy absorbed by the planet is absorbed in the uppermost layer. Assume that the energy absorbed in this layer is the same as the total energy absorbed by the planet in the example above. It follows, as before, that the temperature of the top layer must equal the effective temperature of the planet, 253 K. As before, energy radiated into space at infrared wavelengths by this layer must equal the total energy absorbed from the Sun. Consideration of energy balance for the underlying layers indicates that the atmospheric temperature must be everywhere constant in this example, equal to the effective temperature discussed earlier, regardless of the number of layers included in the model. The temperature of the surface is the same as that of the atmosphere.

Now consider the case where solar energy is absorbed exclusively in the lowest layer of the atmosphere. It may be shown, irrespective of the number of strata in the model, that the atmospheric temperature profile is identical to that obtained earlier when solar energy was assumed to be deposited at the surface. Surface temperature is reduced when heat is deposited in the atmosphere rather than at the surface. In the present instance, it is equal to the temperature of the lowest atmospheric layer,

$$T_g = (N)^{1/4} T_{eff}, \tag{6.17}$$

which may be compared to the value, given by equation (6.14), that applied when solar energy was deposited at the surface. Here, as before, N defines the number of atmospheric layers.

There is an alternate, perhaps more useful, way to think of the problem, illustrated schematically in Figure 6.17. The net upward flux of infrared energy at a level z of the atmosphere must equal the net downward flux of solar energy at the same level; otherwise energy would pile up or dissipate below. The net downward flux of solar energy at z is equal to the flux of solar energy incident at z from above, minus the flux of solar energy reflected from Earth below; it is equal to the rate at which solar energy is absorbed below z. The net upward flux of infrared energy per unit area in layer i, $I(i)$, is equal to the energy delivered by layer i to layer $i - 1$, minus that returned:

$$I(i) = \sigma(T_i)^4 - \sigma(T_{i-1})^4 \tag{6.18}$$

If the quantity of solar energy absorbed below level i is small, so also is the difference between T_i and T_{i-1} (a small value for $I(i)$ in equaion (6.18) implies that T_i must be approximately equal to T_{i-1}). This accounts for the difference between the models discussed above. The deeper the level at which solar energy is absorbed, the greater the rate at which temperature must decrease with increasing altitude in order to transfer the energy to the topmost layer, where it can be radiated into space. The deeper the level at which solar radiation is absorbed, the larger the impact on surface temperature.

6.6 Summary

We pointed out that the Sun emits radiation (light) as though it were a blackbody at a temperature of about 6000 K. The flux of radiant energy crossing a unit area of the surface of a blackbody is given by σT^4, where T defines the temperature of the blackbody and σ is the Stefan-Boltzmann constant. The flux of solar radiation crossing a unit area decreases as the square of the distance from the Sun, as sunlight propagates in all directions through space. The decrease in flux with distance reflects the fact that the quantity of energy crossing the surface of an imaginary sphere at distance R from the Sun is exactly equal to the quantity of energy leaving the Sun; absorption of sunlight in the essentially empty intervening space is negligible. The area of the surface of a sphere increases as the square of its radius. The flux of energy crossing the unit area of the surface of the sphere accordingly decreases in proportion to the square of its radius. The flux of solar energy at the orbit of Earth (the energy crossing a unit area normal to the direction of propagation of sunlight) is equal to 1370 watts m^{-2}. This quantity is known as the "solar constant."

Approximately 70% of the solar radiation intercepted by Earth is absorbed by Earth; the balance is reflected into space. The fraction of incident energy reflected into space depends on the color of the planet, on the planet's reflective properties as defined by its albedo. A perfectly absorbing black object has an albedo of zero; the albedo for a perfectly reflecting white object is equal to unity. Earth's albedo is equal to about 0.3. It is determined to a significant extent by reflection of sunlight from brightly colored clouds ubiquitously present in the atmosphere.

The spectrum of a blackbody, the variation of light emitted as a function of wavelength, depends on temperature. Cold objects emit primarily at long wavelengths; emission from hot objects is shifted to shorter wavelengths. The peak in emission of radiation by the Sun is centered in the visible portion of the spectrum, at a wavelength of about 5000 Å. The energy absorbed by the Earth from the Sun is sufficient to maintain Earth at an average temperature of about 255 K. That is to say, emission from a blackbody of the size and shape of Earth at a temperature of 255 K would be just sufficient to offset the total amount of energy absorbed from the Sun. Most of the solar energy captured by Earth is absorbed at the surface. This energy is subsequently transferred to the atmosphere through a combination of radiation and latent and sensible heat. Energy absorbed by Earth from the Sun is balanced by emission of infrared radiation into space, primarily from the atmosphere. Polyatomic molecules, such as H_2O, CO_2, CH_4, N_2O, and O_3—gases capable of absorbing and emitting infrared radiation—play a critical role in the transfer of energy from the surface into space. They serve as insulating agents, like glass in a greenhouse, shielding the surface from the bitter cold of outer space. Following the greenhouse analogy, these infrared-absorptive molecules are known as "greenhouse gases." A simple model for the greenhouse effect was presented in which the atmosphere was represented by a series of insulating blackbody layers. We showed that a single insulating layer would be sufficient to raise the temperature of the surface by 48 K.

We shall next explore in more detail the processes involved in transfer of energy from the near-surface region of Earth to higher levels of the atmosphere, from which

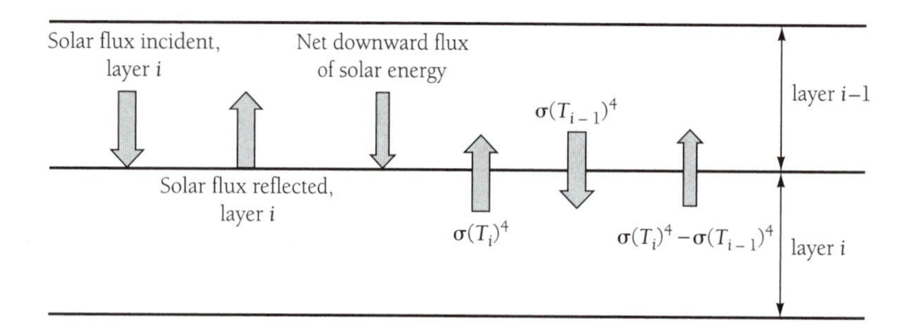

Figure 6.17 Schematic view of the flux balance for the layer model indicating exchange of energy across boundary between layers i and i−1.

radiation may be emitted directly to space. If transfer of energy from Earth's surface to the level of the atmosphere from which it can be radiated into space were accommodated solely by radiation, Earth's surface temperature would be much higher than it is today. As discussed in the next chapter, vertical motion (convective overturning of the lower layers of the atmosphere) has an important influence in cooling the surface and in transferring energy to higher altitudes. The greenhouse mechanism, as we shall see, is but one of a number of processes involved in regulating Earth's climate.

Vertical Structure of the Atmosphere

7

This chapter concerns the physical processes responsible for the variation of pressure, density, and temperature with atmospheric altitude. We introduce first, in Section 7.1, the barometric law, which accounts for the variation of pressure with altitude (assuming that we know the corresponding variation of temperature). An overview of the thermal structure of the atmosphere is given in Section 7.2. Specifically, Section 7.2 introduces the nomenclature used to describe the various atmospheric layers. The change of temperature with altitude under conditions where vertical motion takes place without exchange of heat with the surroundings is discussed in Section 7.3. In the absence of condensation, the decrease of temperature with altitude is given by a quantity known as the **dry adiabatic lapse rate**. Temperature decreases more slowly with altitude in the presence of condensation. The decrease of temperature with altitude is defined in this case by a quantity known as the **moist adiabatic lapse rate**. Vertical motion and the concepts of stability and instability are discussed in Section 7.4. Processes involved in the formation of clouds and precipitation are described in Section 7.5. Approaches to modeling the globally averaged thermal structure of the atmosphere are described in Section 7.6. Perspectives on the importance of latitudinal redistribution of energy as a result of the circulation of the atmosphere and ocean are discussed in Section 7.7. Section 7.8 provides a brief account of the challenge posed by current efforts to predict the climate of the future. Summary remarks are presented in Section 7.9.

7.1 The Barometric Law

Consider a volume of gas with a uniform horizontal cross-sectional area, A, confined to an altitude interval between z and $z + \Delta z$. Denote the pressure at z by $p(z)$, the pressure at $z + \Delta z$ by $p(z + \Delta z)$. Recall that pressure is the force associated with the transfer of momentum between molecules by collisions. Air above $z + \Delta z$ exerts a downward force on our sample volume of magnitude $p(z + \Delta z)A$ (remember that pressure measures the force per unit area). Similarly, air below z contributes an upward force of magnitude $p(z)A$. The net upward force on the volume due to pressure is given by $[p(z) - p(z + \Delta z)]A$.

In addition to the force due to pressure, air in the sample volume is subject to the force of gravity. The magnitude of the gravitational force is obtained by multiplying the mass contained in the sample volume by the gravitational acceleration, g. The mass in the volume is found by multiplying the **mass density**, ρ, by the volume, $A\,\Delta z$. The net downward force due to gravity is given by $\rho g A\,\Delta z$. The orientation of the different forces is illustrated in Figure 7.1.

Assume that the sample volume is at rest with respect to vertical motion, that the net vertical force on the sample volume is zero. In this case, balancing the upward force due to pressure with the downward force due to gravity, we find

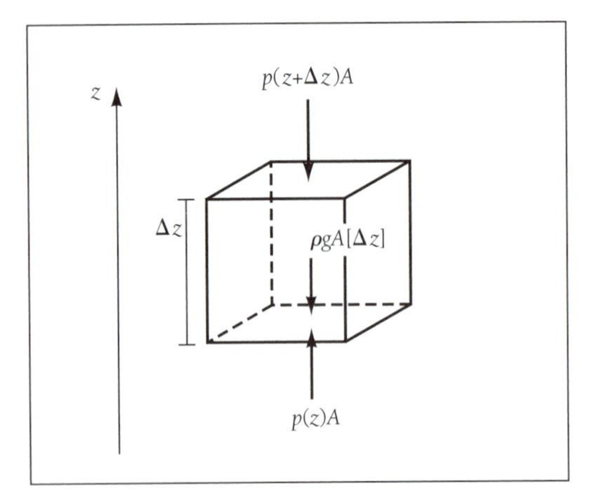

Figure 7.1 Forces acting on a parcel of air between heights z and z + Δz. The areas of the top and bottom of the parcel is A.

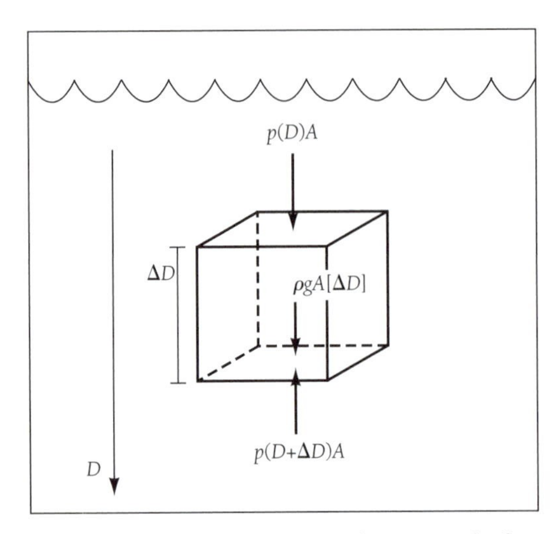

Figure 7.2 Same as Figure 7.1 but for a parcel of water between water depths of D and D + ΔD.

$$[p(z) - p(z+\Delta z)]A = \rho g A\,\Delta z \qquad (7.1)$$

Dividing both sides of equation (7.1) by A, we get

$$p(z) - p(z+\Delta z) = \rho g\,\Delta z \qquad (7.2)$$

Expressed in words, equation (7.2) states that the difference in pressure between altitudes z and $z + \Delta z$ is equal to the weight of air contained in a volume of unit horizontal cross-sectional area between z and $z + \Delta z$.

With minor rearrangement, equation (7.2) can be rewritten to give

$$p(z+\Delta z) = p(z) - \rho g\,\Delta z \qquad (7.3)$$

Assuming that the mass density, ρ, is known, equation (7.3) can be used to evaluate the pressure at $z + \Delta z$ in terms of the pressure at z. The equation confirms, as we would expect, that the pressure of the atmosphere decreases with height above the surface (to obtain the pressure at $z + \Delta z$, we *subtract* a quantity $\rho g\,\Delta z$ from the pressure at z).

We can apply a similar procedure to describe the variation of pressure as a function of depth in the ocean. The situation is illustrated in this case by Figure 7.2. Using D as the vertical coordinate to represent depth, we find

$$p(D + \Delta D) = p(D) + \rho g\,\Delta D \qquad (7.4)$$

The mass density of ocean water is essentially constant, independent of depth, and equal to about 1 g cm^{-3}. A column of water of depth 10 m with a cross-sectional area of 1 cm^2 contains a mass of 1×10^3 g (1 kilogram) roughly equal to the mass of the entire atmosphere above a square centimeter of Earth's surface. Pressure in the ocean increases by 1 atm (about 10^6 dyn cm^{-2}, or 1000 mb) with every 10 m increase in depth. This results in a particularly simple relation for pressure versus depth, as illustrated in Figure 7.3.

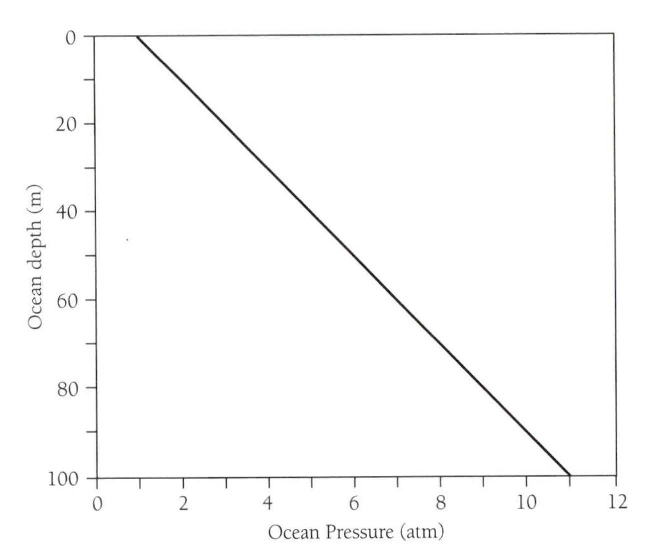

Figure 7.3 Pressure as a function of depth in the ocean.

Evaluation of the pressure-height relation for the atmosphere is complicated by the fact that the mass density varies with height; it decreases rapidly as a function of altitude. A gas such as the atmosphere is compressible. If the pressure is increased, the molecules are squeezed closer together, and the density of the gas increases. Conversely, if the pressure is reduced, the average separation between molecules increases, and the density drops. In contrast, molecules in a liquid are already quite densely packed; the liquid is comparatively incompressible. If you try to squeeze the molecules together, the effort to increase density is resisted by strong repulsive forces operating between neighboring molecules at small separations. It is similarly difficult, if you wish to drive them apart, to overcome the attractive forces that keep molecules together.

To evaluate pressure as a function of ocean depth, we were obliged to supplement equation (7.4) with an additional implicit constraint, the equation of state, which specified that ρ was constant, independent of depth. The analogous condition for a gas such as the atmosphere is provided by the perfect gas law. The perfect gas law was introduced in Chapter 2 as a relation between pressure, number density, and temperature (equation 2.17a). The mass density, ρ, is given in terms of the number density, n, by the equation

$$\rho = mn, \qquad (7.5)$$

where m is the average molecular mass of the atmosphere, defined as the number-weighted average of the masses of the individual constituents. Using the relation for n given by equation (7.5), the perfect gas law can be written as

$$p = \frac{k}{m} \rho T \tag{7.6}$$

The molecular mass of the average constituent of Earth's atmosphere, expressed in **atomic units** (or **au**: 1 au = 1.66 × 10^{-24} g) is 28.97 au, intermediate between the values for N_2 (28 au), composing 78% of the atmosphere, and O_2 (32 au), representing 21%. The value of m also reflects, in addition to these species, the presence of H_2O (18 au), Ar (40 au), CO_2 (44 au), and other less abundant compounds. Meteorologists prefer to replace k/m by a quantity R that they refer to as the **gas constant.** For m equal to 28.97 au (4.81 × 10^{-23} g), R has a value of 2.88 × 10^6 erg g^{-1} K^{-1}. We must carefully distinguish between the meteorologists' gas constant, R, and the universal gas constant used by chemists (as introduced in equation 2.17b), for which the same symbol was employed. As used in this chapter, R refers to the meteorologists' R. In an attempt to avoid confusion, we shall refer to the R used here as the *atmospheric gas constant,* distinguishing it from the universal gas constant. Substituting R for k/m in equation (7.6), the perfect gas law may be expressed in the form

$$p = R\rho T \tag{7.7}$$

We must emphasize again that, unlike the Boltzmann constant (k), the atmospheric gas constant is not a universal physical constant; its value depends on m, and as such, the value quoted here is specific for applications to Earth's atmosphere. If we wish to apply the analyses utilizing R as developed here to the atmosphere of another planet, Mars for example, we must remember to adjust the value of m and thus the value of R, as well.

The value of ρ in equation (7.2) is somewhat ill-defined. We introduced ρ such that the right-hand side of the equation specified the weight of a volume of air of unit horizontal cross-sectional area between z and $z + \Delta z$. Strictly speaking, ρ, as used in the equation, refers to the average mass density for air between z and $z + \Delta z$. If we used the value for ρ at z as the appropriate substitution for ρ in equation (7.2), we would tend to overestimate the weight of the air column between z and $z + \Delta z$ (since density decreases with altitude). Conversely, if we substituted for ρ in the equation using the density at $z + \Delta z$, we would tend to err in the opposite direction. If Δz is very small, the error in either case is negligible. Replacing ρ by $\rho(z)$, dividing both sides of equation (7.2) by Δz, and assuming that Δz is small, we find

$$\frac{dp}{dz} = -\rho(z)g, \tag{7.8}$$

where dp/dz defines the value of $[p(z + \Delta z) - p(z)]/\Delta z$ in the limit as Δz tends to zero. Readers familiar with calculus will recognize dp/dz as the derivative of p with respect to z.

Substituting for ρ in equation (7.8), and using the equation of state (7.7), we can rewrite (7.8) in the form

$$\frac{dp}{dz} = -\frac{pg}{RT} \tag{7.9}$$

Equivalently, (7.9) may be rearranged and expressed in the form

$$\frac{dp}{p} = -\frac{dz}{H}, \tag{7.10}$$

where

$$H = \frac{RT}{g} \tag{7.11}$$

The quantity H plays an important role in the physical description of the atmosphere. It has dimensions of length and is referred to in what follows as the atmospheric **scale height.** Scale height provides a measure of the effective thickness of the atmosphere; a gas with a vertical extent of H, possessing uniform density equal to the density at z, would have the same mass as the total mass contained in the actual atmosphere above z. The magnitude of H is about 8 km for temperatures representative of conditions in the lower atmosphere of Earth. Equation (7.10) implies that a given fractional change in p is associated with the same fractional change in z expressed in units of H; pressure may be expected to decrease by about 10% if altitude is increased by an amount equivalent to 10% of H, about 0.8 km.

Integrating (7.10) from a reference altitude z_0 to z, we find

$$\log_e(p(z)) - \log_e(p(z_0)) = -\int_{z_o}^{z} \frac{dz}{H}, \tag{7.12}$$

or

$$\log_e\left(\frac{p(z)}{p(z_0)}\right) = -\int_{z_o}^{z} \frac{dz}{H} \tag{7.13}$$

Raising both sides of (7.13) to the power e yields

$$p(z) = p(z_0)\exp\left(-\int_{z_o}^{z} \frac{dz}{H}\right) \tag{7.14}$$

If we assume that H is constant between z_0 and z, (7.14) may be rewritten in the simpler form

$$p(z) = p(z_0)\exp\left(-\frac{z - z_0}{H}\right) \tag{7.15}$$

Equation (7.14) defines what is known as the **barometric law** regulating the variation of pressure with altitude. The simpler expression (7.15) is exact for an **isothermal** (constant temperature) atmosphere. It may be used with minimal error to calculate a height profile for pressure if we apply the formula to relatively small-altitude intervals (i.e., for relatively small values of $z - z_0$), if T is represented by its average value between z_0 and z, and if we step vertically through the atmosphere, adjusting the value of $p(z_0)$ as we proceed.

Example 7.1: Given a sea-level pressure of 1013 mb and a scale height of 8 km, find the pressure at the altitudes 5 km, 10 km, and 20 km.

Answer:

$$p(z) = p(z_0)\exp\left(-\frac{z - z_0}{H}\right)$$

$$p(5) = p(0)\exp\left(-\frac{5-0}{H}\right)$$
$$= (1013 \text{ mb})\exp\left(-\frac{5}{8}\right)$$
$$= 542 \text{ mb}$$
$$p(10) = (1013 \text{ mb})\exp\left(-\frac{10}{8}\right)$$
$$= 290 \text{ mb}$$
$$p(20) = (1013 \text{ mb})\exp\left(-\frac{20}{8}\right)$$
$$= 83 \text{ mb}$$ ■

Equation (7.15), taking logarithms of both sides, can be expressed in the equivalent forms

$$\log_e(p(z)) = \log_e(p(z_0)) - \left(\frac{z-z_0}{H}\right) \tag{7.16}$$

and

$$z = z_0 - H\log_e\left(\frac{p(z)}{p(z_o)}\right) \tag{7.17}$$

It follows that a graph of $\log_e(p(z))$ versus z should resemble, more or less, a straight line. The variation of pressure as a function of z for a typical situation at midlatitudes is displayed on a log-linear plot in Figure 7.4. The slope of the curve at any point is equal to H and may be used to estimate local values of temperature, using the definition of H. The variation of T with height, corresponding to the pressure profile in Figure 7.4, is presented in Figure 7.5.

Equation (7.17) provides the basis for the function of an aircraft altimeter. It allows the altitude of the aircraft above sea level to be evaluated using barometric measurements of ambient pressure in combination with estimates for H based on local measurements of temperature. This information may be electronically incorporated in the altimeter to provide an immediate digital display for z. Values of z obtained in this fashion are known as **barometric altitudes.**

In its various forms, the barometric law can be readily converted to yield an equation for the variation of density with height. Substituting for p in equation (7.14) and using the perfect gas law, we find

$$\rho(z) = \frac{T(z_0)}{T(z)}\rho(z_0)\exp\left(-\int_{z_0}^{z}\frac{dz}{H}\right), \tag{7.18}$$

or, using (7.15) for the isothermal case where T and consequently H are independent of altitude,

$$\rho(z) = \rho(z_0)\exp\left(-\frac{z-z_0}{H}\right) \tag{7.19}$$

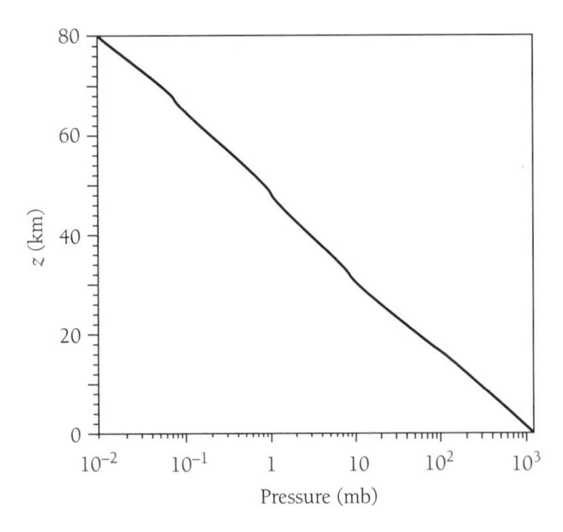

Figure 7.4 Pressure as a function of altitude in the atmosphere. Note that the relation between pressure and height is approximately linear if pressure is plotted on a logarithmic scale. Source: Farmer et al. 1987.

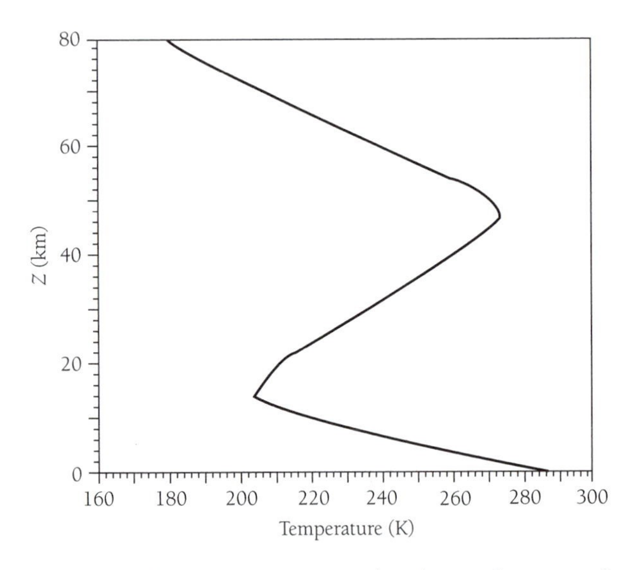

Figure 7.5 Temperature versus altitude in the atmosphere. Temperatures were taken from measurements made by the Atmospheric Trace Molecule Spectroscopy (ATMOS) Experiment flown on the space shuttle in May 1985 (Farmer et al., 1987). Data refer to 30°N. Temperatures shown here were used to construct the pressure profile shown in Figure 7.4. Source: Farmer et al. 1987.

Equations governing the variation of number density with altitude have the same form as (7.18) and (7.19), with ρ replaced by n. Mass and number densities decrease exponentially with height in the atmosphere, as does pressure, in contrast to the situation in the ocean where ρ and n are constant. It is the decrease of density with altitude that is responsible for the faster-than-linear drop of pressure with height in the atmosphere; the distinction between pressure

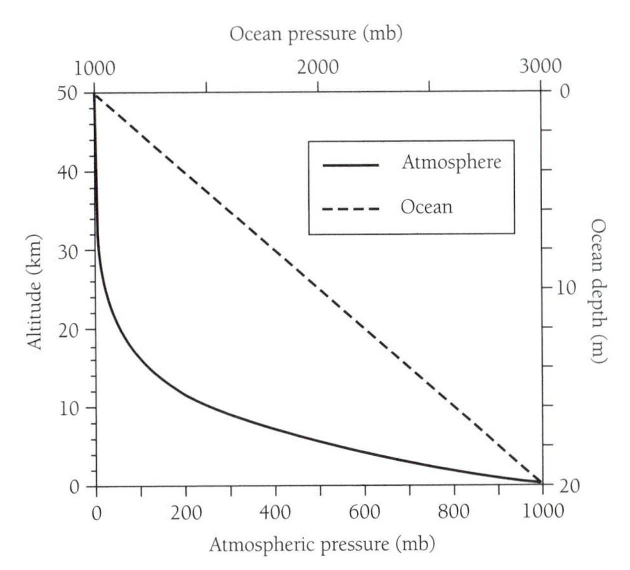

Figure 7.6 Contrast in pressure profiles for the atmosphere and ocean.

profiles for the atmosphere and ocean is ultimately associated with differences in the compressibilities of the two fluids. The contrast for the two systems is schematically illustrated in Figure 7.6.

7.2 Layers of the Atmosphere

Figure 7.5 indicates that temperature initially decreases with height above the surface at an average rate of about 7 K km^{-1}. It reaches a minimum value of about 216 K (–57°C) at an altitude near 11 km for midlatitudes. Temperature increases with height above about 20 km, reaching a maximum of about 270 K near 50 km, decreasing to a secondary minimum near 85 km, where the temperature is close to 180 K. The temperature increases again above 85 km, climbing steadily to a value of between 750 K and 2000 K, depending on the phase of activity of the Sun. (As previously discussed, the output of energy from the Sun is variable at the shortest wavelengths, which play a dominant role in the energy budget of the atmosphere at high altitude. Variations in solar luminosity at short wavelengths exhibit a period of approximately 11 years, referred to as the **solar cycle**.)

The lowest region of the atmosphere, up to the first temperature minimum, is known as the **troposphere**, a name derived from the Greek words *tropos,* meaning "turning," and *spaira*, meaning "ball." The troposphere is relatively unstable, a consequence, as we shall see, of the decrease of temperature with altitude. Air in the troposphere is poised to overturn, to convect, much like water in a kettle heated from below. Most of our planet's weather is confined to the troposphere. The upper boundary of the troposphere, the altitude corresponding to the temperature minimum, is known as the **tropopause**.

The atmosphere above the tropopause, where temperature increases with altitude, is known as the **stratosphere**, from the Latin word *stratus*, meaning "stretched out" or "layered." Vertical motions are strongly inhibited in the stratosphere as a consequence of the increase of temperature with altitude; an air parcel attempting to rise becomes rapidly denser than the air it seeks to displace and is driven back to its point of origin. The temperature maximum near 50 km marks the upper boundary of the stratosphere, known as the **stratopause.** We shall see that most of the world's O_3 is contained in the stratosphere.

The region of decreasing temperature above the stratopause is known as the **mesosphere** (from the Greek word *mesos*, meaning "middle"). The upper boundary of the mesosphere, the **mesopause**, is defined by the temperature minimum near 85 km. The region of increasing temperature above the mesopause is named the **thermosphere,** in recognition of the fact that this region is characterized by the highest temperatures observed anywhere in the atmosphere. Densities are low enough in the thermosphere to permit satellites to orbit Earth for extended periods of time. The thermosphere is also the site of the **ionosphere**, a region of the atmosphere including a significant number of free electrons liberated from parent atoms and molecules by absorption of energetic short-wavelength solar radiation. The presence of free electrons permits the propagation of radio signals over large distances. Thousands of kilometers above the surface, the atmosphere slowly merges with the **interplanetary medium.**

7.3 Adiabatic Motion in the Vertical

The rate at which temperature drops with altitude can be estimated using a simple application of thermodynamics, as follows. Consider a unit mass of atmosphere at height z. Suppose a quantity of heat, ΔQ, is added to this mass. The supplied heat can be used either to increase the internal energy (temperature) of the air or to do work, in particular to allow the volume occupied by the unit air mass to expand in the face of the constraining force of pressure. The **internal energy** of a gas depends on its temperature and on the number of molecules composing the gas (the average molecule contributes kinetic energy equal to $\frac{3}{2}\,kT$: see Section 2.6). Suppose that the internal energy changes by an amount ΔE and that the volume occupied by the sample air mass increases by an amount ΔV. The work done to change the volume from V to $V + \Delta V$ is given by $p\Delta V$, as indicated in Figure 7.7.[1] The **first law of thermodynamics**, a statement of the law of conservation of energy, requires that the heat supplied to the sample air mass must equal the change in internal energy plus the work done against external forces. Expressing this relation in mathematical terms,

$$\Delta Q = \Delta E + p\Delta V \tag{7.20}$$

or, equivalently,

$$\Delta Q = \left(\frac{\Delta E}{\Delta T}\right)\Delta T + p\Delta V \tag{7.21}$$

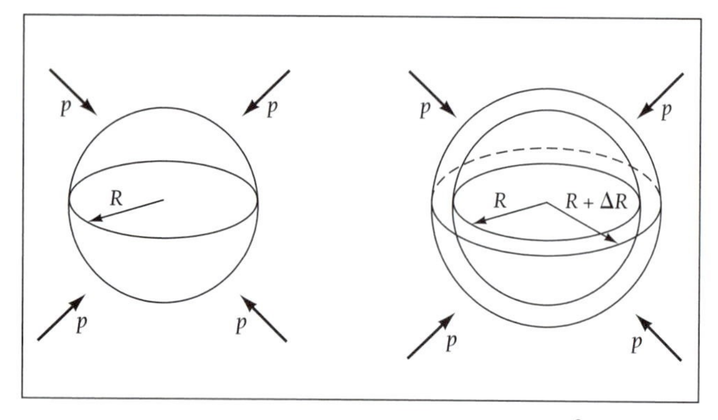

Figure 7.7 Illustrating the work done by a gas corresponding to a change in volume.

Suppose that heat is supplied to unit mass under conditions where the volume is held fixed. In this case we can interpret $\Delta E/\Delta T$ as a measure of the heat required to change the temperature of unit mass, with volume held fixed. We refer to this quantity as the **specific heat at constant volume**, denoted by the symbol c_v.

$$c_v = \left(\frac{\Delta E}{\Delta T}\right), \tag{7.22}$$

Those familiar with calculus will recognize that in the limit for vanishingly small changes of temperature,

$$c_v = \left(\frac{dE}{dT}\right), \tag{7.23}$$

where (dE/dT) defines the derivative of E with respect to T.

Substituting for $\Delta E/\Delta T$ in (7.21) using (7.22) allows us to write

$$\Delta Q = c_v \Delta T + p\Delta V \tag{7.24}$$

Equation (7.24) includes three variables: p, T, and V. These quantities are not independent. They are related through the perfect gas law (equation 2.17a), which can be written in the form

$$pV = RT, \tag{7.25}$$

where we have multiplied both sides of (7.7) by V and have set the product ρV (the mass contained in V) equal to 1, since our analysis refers to a system of unit mass. For small changes in the pressure, volume, and temperature of the gas, denoted by Δp, ΔV, and ΔT, respectively, the departures are related by

$$(\Delta p)V = -\,p\,\Delta V + R\Delta T, \tag{7.26}$$

where we have used[2]

$$\Delta(pV) = (\Delta p)V + p(\Delta V) \tag{7.27}$$

Using (7.25), equation (7.26) may be rearranged to give

$$\Delta V = \frac{R\Delta T}{p} - \frac{RT\Delta p}{p^2} \tag{7.28}$$

And using (7.28) to substitute for ΔV in (7.24), we obtain

$$\Delta Q = (c_v + R)\Delta T - \frac{RT\Delta p}{p} \tag{7.29}$$

It may be shown (by setting $\Delta p = 0$ in (7.29)) that the quantity $(c_v + R)$ is equal to the **specific heat at constant pressure**, c_p; meaning, it is the heat required to raise the temperature of unit mass by one degree at constant pressure ($\Delta p = 0$). Further, using the barometric law (equation 7.9), we can write

$$\frac{\Delta p}{p} = -\frac{g\,\Delta z}{RT} \tag{7.30}$$

Substituting using (7.30) in equation (7.29), the first law of thermodynamics for the atmosphere assumes the form

$$\Delta Q = c_p \Delta T + g\Delta z, \tag{7.31}$$

or, equivalently,

$$\Delta T = -\frac{g}{c_p}\Delta z + \frac{\Delta Q}{c_p} \tag{7.32}$$

The preceding analysis involves a great deal of mathematical manipulation. The individual steps are relatively straightforward. It is appreciated, though, that it may be difficult in wading through the details to distinguish the forest from the trees. A more compact presentation of the analysis is given in the accompanying box.

There are three possible sources of heat for a rising air mass. Heat can be communicated to the air mass from the neighboring air by radiation or as a result of molecular collisions (conduction). Alternatively, the air mass may gain heat internally as a result of phase changes of water. Radiation and conduction are inefficient processes; to transfer significant heat by either mechanism requires an elapsed time of days to weeks. We shall be interested here primarily in vertical motion taking place on much shorter time scales, minutes to hours. In this case, the influence of radiation and conduction may be neglected. If the temperature of the rising air mass reaches the dew point or the frost point, however, significant quantities of heat can be rapidly added as a consequence of the condensation of H_2O.

We distinguish between vertical motion occurring in the absence of condensation and that taking place in its presence. In the absence of condensation, the heat supplied is negligible ($\Delta Q = 0$) and the temperature lapse rate derived from (7.32) assumes the simple form

$$\frac{\Delta T}{\Delta z} = -\frac{g}{c_p} \tag{7.33}$$

We refer to the rate of change of temperature with altitude defined by (7.33) as the **dry adiabatic lapse rate**. The term

Table 7.1 Adiabatic Lapse Rates of Planetary Atmospheres

Planet	*Primary atmospheric composition*	*g* (*cm sec* $^{-2}$)	c_p (*erg* g^{-1} K^{-1})	*Adiabatic lapse rate* (*K* km^{-1})
Venus	CO_2	888	8.3×10^6	10.7
Earth	N_2, O_2	981	1.0×10^7	9.8
Mars	CO_2	373	8.3×10^6	4.5
Jupiter	H_2	2620	1.3×10^8	20.2

adiabatic, as used in thermodynamics, identifies a process taking place in the absence of exchange of heat with the surroundings ($\Delta Q = 0$). Values of g, c_p, and dry adiabatic lapse rates computed for atmospheres of a number of planets of the solar system, including Earth, are presented in Table 7.1. Lapse rates are quoted here as positive numbers. Remember that use of the word *lapse* presupposes that temperature must decrease with altitude. The dry adiabatic lapse rate for Earth is equal to 9.8 K km^{-1}.

If condensation occurs, heat is released and ΔQ is positive. Under such circumstances, the decrease of temperature with altitude is less than would be the case for a dry adiabatic process, as indicated by equation (7.32). We refer to the decrease of temperature with altitude for a rising air mass undergoing condensation as the *moist adiabatic lapse rate*. The significance of condensation as a source of heat for the rising air mass depends on the quantity of H_2O available for condensation; therefore, it depends ultimately on temperature. The moist adiabatic lapse rate is about 3°C km^{-1} for a temperature of 40°C near the surface. It increases to about 6°C km^{-1} as temperature drops to the freezing point, 0°C. Values for the moist adiabatic lapse rate as functions of temperature and altitude are presented in Table 7.2. (Although the moist adiabatic lapse rate is variable, we will use a value of 6°C km^{-1} in most of our examples.)

We can account qualitatively, without resort to the mathematical analysis above, for the drop of temperature with altitude for a rising air mass. As an air parcel rises, it moves into a regime of lower pressure. The perfect gas law requires that its volume increase to compensate for the drop in pressure. To increase its volume, the molecules in the sample air mass must work on the surrounding fluid; they must expend energy to move the constraining external molecules out of the way. In the absence of a heat source, this energy must be drawn from the internal energy of the

Box 7-1

Summary of Analysis Involved in Derivation of Adiabatic Lapse Rate

Equation of State

$$p = nkT$$

$$p = R\rho T \quad (R = \text{atmospheric gas constant})$$

$$pV = RMT \quad (M = \text{mass of air volume } V)$$

Barometric Law

$$\frac{dp}{dz} = -\rho g = \left(\frac{p}{RT}\right) g = -\frac{p}{H}$$

$$\frac{dp}{p} = -\frac{dz}{H}$$

$$\log \frac{p(z)}{p(z_0)} = -\frac{z - z_0}{H}$$

$$p(z) = p(z_0)\exp\left(-\frac{z - z_0}{H}\right)$$

Adiabatic Lapse Rate

$$\Delta Q = \Delta E + p\Delta V$$

$$\Delta Q = c_v \Delta T + p\Delta V$$

$$pV = RT \quad (M = 1)$$

$$(\Delta p)V + p(\Delta V) = R\Delta T$$

$$p\Delta V = R\Delta T - (\Delta p)V$$

$$p\Delta V = R\Delta T + g\Delta z,$$

(using the barometric law, setting $\rho V = 1$)

$$\Delta Q = (c_v + R)\,\Delta T + g\Delta z,$$

$$\Delta Q = c_p \Delta T + g\Delta z$$

$$\Delta Q = 0 \rightarrow \frac{dT}{dz} = -\frac{g}{c_p}$$

■

Table 7.2 Moist Adiabatic Lapse Rate (°C km^{-1}) for Different Temperatures and Pressures

Pressure (mb)	*Temperature (°C)*		
	−40	*0*	*40*
1000	9.5	6.4	3.0
600	9.3	5.4	2.6
200	8.6	3.4	2.0

sample air mass. The temperature of the rising air mass must fall as a consequence. If condensation occurs, part of the energy needed to expand the volume of the air mass can be supplied by latent heat; the demand for internal energy is reduced as a consequence, and the lapse rate is less under moist conditions than would be the case if the adjustment were to occur in the absence of a phase change for water.

7.4 Vertical Motion

Vertical motion can occur under a variety of conditions in the atmosphere. The simplest case arises when air encounters an obstruction, such as a mountain range aligned perpendicular to the flow. To flow past the obstacle, air is forced to rise. As it rises, it cools. If the moisture content is sufficiently high, the temperature drops to the dew point, and clouds form. We refer to such clouds as **topographic** or **mountain wave clouds.** As air crosses the mountain, it sinks. Sometimes, if atmospheric conditions are right, air flowing past a mountain rebounds and proceeds to oscillate in altitude, following a wave-like pattern, analogous to water in a pool set in motion by a falling stone. Clouds frequently appear downwind of a mountain, delineating high points in an often time-stationary pattern referred to as **mountain lee waves.**

Vertical motion can also occur in conjunction with convergence of air in a low pressure system. Air tends to spiral inward toward the center of a low pressure system, as discussed in Chapter 8. Needing somewhere to go, the converging air rises, its temperature often decreasing to the dew point. For this reason, low pressure systems are frequently accompanied by unsettled weather. Conversely, high pressure systems are associated with descending motion and usually accompanied by clear, cloud-free conditions and warm weather.

Vertical motion is also characteristic of conditions in the vicinity of a weather front, a region identified by an interaction of air masses with distinctly different thermal properties. An advance of warm air, forcing cold air to retreat, defines what is known as a **warm front** (Figure 7.8a). The density of warm air is less than that of cold. As a consequence, warm air rises, slowly riding over the cold air underneath. Warm fronts are frequently delineated by bands of clouds, accompanied by widespread precipitation, either rain or snow. Similarly, a situation with cold air overtaking warm air defines a **cold front** (Figure 7.8b). The cold air slips below, pushing the warm air aloft; the boundary between the cold and warm air is usually steeper for a cold front as compared to a warm front. Cold fronts are often associated with showers and thunderstorms.

Vertical motion can take place on relatively small spatial scales as a result of what is referred to as a **local instability.** Suppose a volume of air, temporarily warmer than air in its environment, evolves. The pressure force on the warm air is the same as that on the air in its vicinity. However, its density is less, and, as a consequence, the force of gravity is insufficient to balance the upward push exerted by pressure.

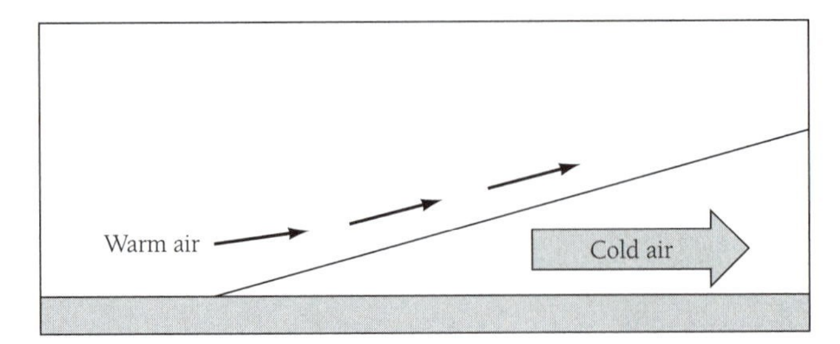

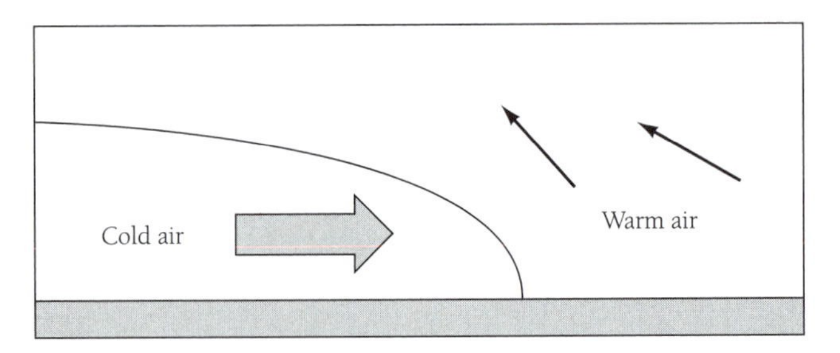

Figure 7.8 In the vicinity of a warm front, warm air advances, rising over the denser cold air that it pushes away. In a cold front, advancing cold air pushes in underneath the warm air; again the less dense warm air is forced to rise.

The air becomes buoyant; it begins to move vertically (see Figure 7.9).[3] The temperature of the moving air mass, in the absence of condensation, will decrease with altitude according to the dry adiabatic lapse rate, equation (7.33); alterna-

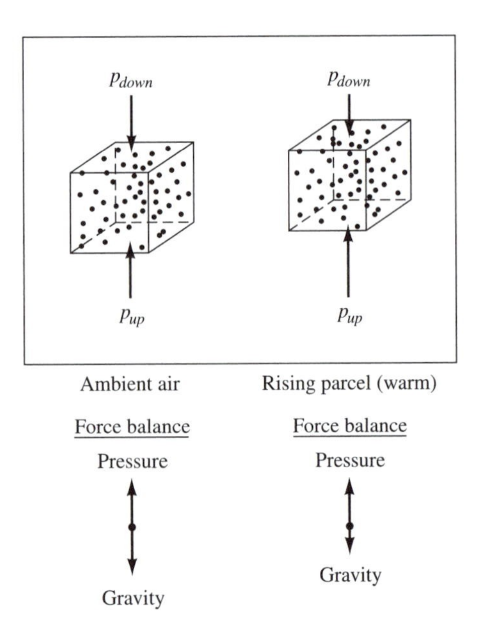

Figure 7.9 Buoyancy of a rising air parcel. The pressure forces acting on the warm parcel are the same as the forces acting on a similar volume of colder ambient air. The density of hot air is less than the density of air in the environment.

tively, with condensation, it will fall off with altitude consistent with the moist adiabatic rate. If the temperature of the rising air mass is higher than the temperature of the ambient air it displaces, it will remain buoyant and continue to rise. Otherwise, if its temperature is less than the temperature of the environment, its density will be higher than the density of the environment, and it will be forced to sink.

Atmospheric stability or instability depends on the gradient of temperature with altitude. If temperature decreases with altitude at a rate faster than the dry adiabatic limit, vertical motion will tend to feed on itself. The atmosphere in this case is said to be **unstable.** If the decrease of temperature with altitude is less than the limit defined by the dry adiabatic lapse rate but larger than the moist adiabatic value, the atmosphere is **conditionally unstable.** Spontaneous vertical motion can occur, but whether it does or not depends on the available supply of condensable H_2O. When the decrease of temperature with altitude is less than the moist adiabatic lapse rate, the atmosphere is **absolutely stable**, and vertical motion is strongly inhibited.

The dew point of a rising air mass drops by about 2°C per km as a result of the decrease in partial pressure of H_2O. In the absence of condensation, the decrease in the partial pressure of H_2O is regulated by the barometric law, equation (7.14). As noted earlier, the temperature of air rising under dry adiabatic conditions is expected to drop by almost 10°C per km. It follows that if instability extends to sufficient altitude, the temperature of the rising air mass will eventually decrease to the dew point. A cloud may be anticipated to form when the temperature reaches the dew point. Once condensation begins, the subsequent drop in temperature of the rising air mass follows the lapse rate defined by the moist adiabatic limit. Clouds can extend to great heights, especially in the tropics, where they occasionally penetrate the lower stratosphere, that is, to altitudes in excess of 20 km. Water may be present in clouds in the form of liquid or ice, depending on the temperature reached by the ascending, condensing air mass.

Figure 7.10 illustrates an unstable environment. Such a situation arises on a hot summer day, when intense warming near the surface leads to an unstable temperature profile, extending in this case to an altitude of 1.2 km. Suppose that the dew point near the surface is 27°C, corresponding to a relative humidity of 62%. Rapid vertical motion in the lower atmosphere results in a vigorous mixing of air, maintaining a constant relative abundance of H_2O up to the level where condensation commences. Figure 7.10 shows the temperature and dew point of the rising air mass as functions of height. The temperature of the ascending air reaches the dew point at an altitude of 1.0 km. We expect a cloud to form at this level. The temperature of the rising air mass is higher than the temperature of the ambient environment at the condensation level. The resulting buoyancy allows vertical motion to proceed to a higher altitude. The subsequent temperature decrease at higher altitude follows the moist adiabatic gradient, with relative humidity fixed at 100% (i.e., the air is saturated with respect to H_2O vapor).

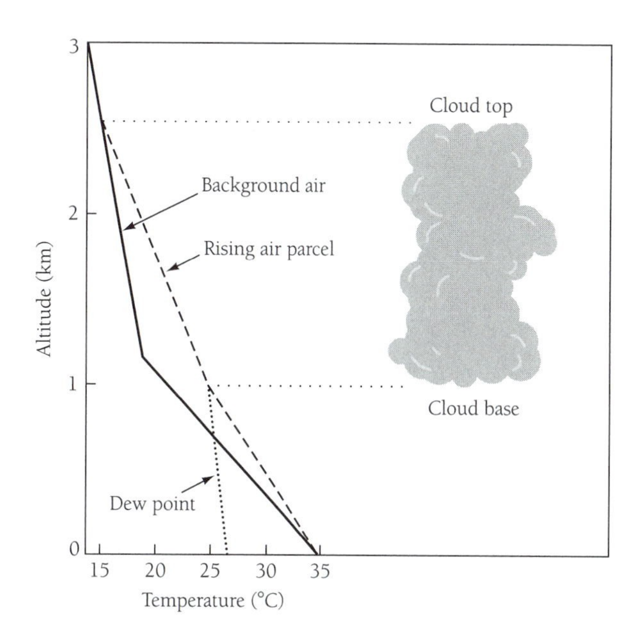

Figure 7.10 Ascent of a moist air parcel (dashed line). The temperature of the ambient air is indicated by the solid line. As the parcel begins to rise, its temperature drops, following the dry adiabatic lapse rate. Water vapor begins to condense when the temperature of the rising parcel falls to the dew point. This level corresponds to the base of the cloud. The parcel is still warmer than the environment. It continues to rise, following in this case the moist adiabat. The top of the cloud corresponds to the level where the temperature of the parcel is equal to the temperature of the ambient atmosphere.

In this case we anticipate clouds extending to an altitude of about 2.5 km, at which point the temperature of the rising air approximately equals the temperature of the background atmosphere. Upward vertical motion is suppressed since the rising air parcel is now cooler (less buoyant) than the surrounding atmosphere. We expect the cloud to exhibit a relatively sharp top where the rising air reaches neutral buoyancy. Its temperature in this case equals that of the background atmosphere.

7.5 Condensation, Clouds, and Precipitation

Condensation from vapor to liquid is normally initiated in the atmosphere by deposition of H_2O on surfaces of preexisting particles known as **condensation nuclei.** Direct production of liquid water from a pure vapor phase, known as **homogeneous nucleation**, is inefficient; growth of droplets under such conditions requires relative humidities significantly higher than 100%. There is usually no shortage of particles in the lower atmosphere to serve as condensation nuclei. Some of these particles, known as **Aitken nuclei**, are very small, with radii less than 0.2 μm (1 μm (**micron**) equals 10^{-4} cm). Others have radii ranging up to values as large as about 1 μm. The efficiency of individual particles as sites for condensation depends both on their size and chemical composition. Particles containing materials that readily dissolve in water, such as salt or acids, are especially

Table 7.3 Terminal Velocity of Differently Sized Particles

Radius (μm)	*Terminal velocity* (cm sec^{-1})	*Concentration* (number cm^{-3})	*Type of particle*
0.1	0.0001	1000	Condensation nuclei
10	1.0	10	Typical cloud droplet
50	27	1.0	Large cloud droplet
100	70	0.1	Large cloud droplet, or drizzle
1000	650	0.001	Typical raindrop
2500	900	0.0001	Large raindrop

effective. We refer to these particles as **hygroscopic** (water seeking). Salt particles formed by evaporation of water from sea spray are particularly effective as condensation nuclei in marine environments. Aerosols composed of sulfate and nitrate play an important hygroscopic role in continental regions and are especially effective in the vicinity of major sources of air pollution (see Chapters 17 and 18). The concentration of condensation nuclei in relatively clean regions of the atmosphere is typically about 10^3 cm^{-3}. In polluted environments, concentrations may exceed this background by as much as a factor of 10^2.

If the condensing water is equally distributed over all the condensation nuclei, the size of the average cloud particle may be estimated using the relation

$$M = N\rho\left(\frac{4}{3}\pi r^3\right), \qquad (7.34)$$

where M (g cm^{-3}) represents the mass of condensible water vapor, N (cm^{-3}) denotes the number density of condensation nuclei, ρ (1 g cm^{-3}) represents the mass density of liquid water, and r (cm) denotes the radius of the average cloud droplet. The factor $\frac{4}{3}\pi r^3$ defines the volume occupied by an individual spherical particle of radius r. With M equal to 5×10^{-6} g cm^{-3} and N equal to 10^3 cm^{-3}, we calculate that the radius of an average cloud particle should be about 10.6 μm. To efficiently scatter light, a water droplet must have a size comparable to, or larger than, the wavelength of the scattered light. The cloud forming in this example should be readily visible. It is more difficult to devise a mechanism that would allow it to rain, a mechanism permitting an average cloud particle to grow large enough to fall.

The speed of a falling raindrop is defined by what is known as the **terminal velocity.** Terminal velocity is identified as the velocity at which the drag on a falling particle, due to air resistance, exactly balances the force of gravity. Drag tends to hold the particle up; gravity pulls it down. The ratio of surface area to volume is important in determining the significance of drag compared to gravity. This ratio decreases as a function of increasing particle size (the ratio of the two quantities varies inversely with the radius). As a consequence, to generate sufficient drag to balance gravity, large particles are required to move faster than small ones. Table 7.3 presents terminal velocities as a function of particle radius. The radius of a typical raindrop is about 1000 μm. It has a fall velocity of about 6.5 m sec^{-1} and includes an amount of water comparable to that contained in approximately 10^6 cloud particles. It is a serious challenge to account for production of particles in the atmosphere large enough to fall; it is much simpler to explain the presence of clouds.

Even in a cloud there is some dispersion of particle sizes. Two factors contribute to this phenomenon. First, the initial growth of particles involves only a small fraction of the most hygroscopic of the available condensation nuclei. Second, large particles, at least in the early stages of growth, tend to grow faster than small, reflecting the dependence of the equilibrium vapor pressure on particle size. Water molecules are more tightly bound to a liquid with a flat surface than to a small particle with a curved surface; there are more molecules to hold the individual molecule of the liquid in place on a large flat surface, as schematically illustrated in Figure 7.11. Neither of these processes, however, can account for the production of particles in the size range observed in a rain shower.

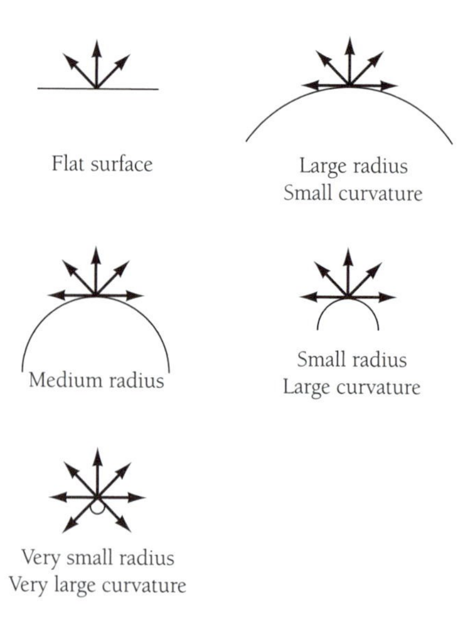

Figure 7.11 Illustrating the dependence of saturated vapor pressure on curvature. Molecules leave curved surfaces more easily than they can escape from flat surfaces.

There are two mechanisms that may account for the growth of particles large enough to fall. One involves coalescence of large numbers of small particles in the liquid phase. The second would trigger precipitation by formation of a small number of ice nuclei subsequently growing by accretion of water from the liquid phase. The ice mechanism is thought to play an important role at high and middle latitudes, where clouds frequently extend into regions of the atmosphere at temperatures below the freezing point. The liquid coalescence process must dominate elsewhere.

The mechanism by which small particles of liquid coalesce to form large drops is complex and not well understood. To initiate coalescence, a liquid drop must collide with a neighbor and stick to it. The collision rate depends on the relative velocity of the colliding drops. The velocity differential is small if the drops have a similar size and mass. Initial growth requires a dispersion of particle sizes and thus of velocities. A significant upward air speed serves to enhance the residence time of large particles in a cloud, allowing additional time for the accretion of smaller drops. In this manner, individual drops may grow large enough to precipitate. However, the theory is qualitative rather than quantitative. In this sense it is unsatisfactory. It is difficult to define the probability for coalescence (growth) rather than fragmentation (breakup) for a collision of a particular pair of particles. The extent to which electrical charging of particles may play a role in determining the collision rate or the chance of coalescence is unclear. Finally, the probability of accretion depends on details of the variation in the velocity field on very small spatial scales. Theories accounting for small-scale spatial and temporal variability in the velocity field are either lacking or regrettably deficient.

Ice crystals provide nearly ideal catalysts for the production of large particles. If the atmospheric temperature falls below 0°C, it is possible on purely thermodynamic grounds for liquid droplets to freeze. Formation of ice is not automatic, however; as we saw in Chapter 4, water can exist as supercooled liquid in the atmosphere to temperatures as low as −40°C. Ice crystal production requires a very specific molecular arrangement. This can occur most readily on a suitable preexisting embryonic nucleus, but it is relatively difficult for it to take place either directly from the gas phase or in a small liquid droplet composed of essentially pure H_2O. Once an ice crystal forms and achieves a critical size, it can grow rapidly under conditions where the temperature falls below the freezing point. The presence of ice crystals can trigger additional ice formation through collisions with supercooled drops by a process known as **contact nucleation.** Ice crystals grow preferentially from the gas phase since the vapor pressure of water in equilibrium over ice is less than that over liquid. Fragile ice crystals can fragment, resulting in an explosive growth of the number of ice crystals available to serve as sites for further condensation.

The ice nucleation mechanism may play an important role in production of precipitation, even in summer. Ice particles melt when they fall into regions of the atmosphere where the temperature is above 0°C, reaching the surface under such conditions as rain. If the underlying atmosphere is cold enough, precipitation falls to the surface as snow or, less commonly, as large ice particles we identify as hail.

There have been a number of attempts to induce precipitation by seeding clouds containing supercooled water with particles of silver iodide, taking advantage of the fact that the structure of silver iodide is superficially similar to that of water ice. These experiments are controversial, and it is difficult to gauge their success. There is no simple way to show that precipitation, when it occurs, is caused by the act of seeding, that it would not have naturally occurred in the absence of seeding.

The seeding of clouds for rain production raises a number of difficult legal issues. An enhancement of rain (or snow) as a result of seeding in one region may cause a reduction of precipitation in another. If operators of ski resorts practice seeding to improve conditions on the slopes of mountains in Colorado, can farmers in Kansas sue for damages if it appears that there might be an associated reduction in precipitation downwind? In the absence of a detailed understanding of the interplay of cause and effect, it is clearly difficult to establish conclusive guilt or innocence for any particular circumstance. Until we better understand the atmosphere, it would seem prudent for purposeful intervention to be kept to a minimum.

A cloud's visual appearance provides an excellent indication of the type of vertical motion taking place and of the condensation process underway in the region occupied by the cloud. There are four general cloud types, known by the Latin names introduced by the British naturalist Luke Howard in the early nineteenth century. Puffy clouds, formed by local convective instability, are termed **cumulus** (meaning *heap*). Cumulus clouds often evaporate as fast as they form. New clouds appear as rapidly as old clouds dissipate, tracking the ever fluctuating pattern of vertical motion. Low, gray, sheetlike clouds, often covering the entire sky, are known as **stratus** (meaning *layer*) and are usually associated with a large-scale (wide-area) uplift of moist air. **Cirrus** (meaning *curl of hair*) is the name used to identify wisps of cloud at high altitude, usually moving from west to east at the mid latitudes of the Northern Hemisphere, tracking the prevailing winds. Temperatures in a cirrus cloud are normally below the freezing point; clouds of this type are composed primarily of small ice crystals. **Nimbus** (meaning *rain cloud*) is the term used to describe a cloud undergoing active precipitation. Precipitating clouds usually exhibit a threatening black appearance near their bottoms, reflecting low levels of light transmitted by thick concentrations of clouds above. Individual clouds may display features of several different cloud types. They are identified in this case as composites with names such as **cumulonimbus**, **nimbostratus**, and **stratocumulus**.

7.6 Models for the Thermal Structure of the Atmosphere

Results from a pair of models incorporating a relatively complete representation of the complex details of the interaction of infrared and solar radiation with the atmosphere are presented in Figure 7.12. The models describe what is known as the **radiative equilibrium structure of the atmosphere.**

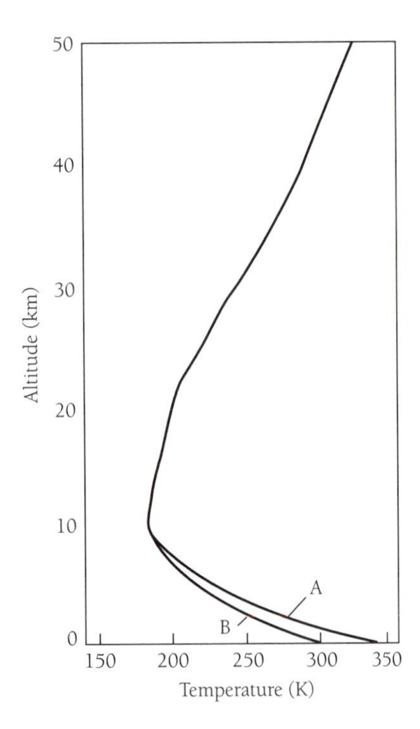

Figure 7.12 Results from a radiative equilibrium model for the variation of temperature with altitude. Curve A is for clear skies and Curve B is for cloudy skies. Source: Manabe and Moller 1961.

Model A is restricted to a consideration of absorption and emission of radiation by gaseous constituents of the atmosphere. Model B allows for the effects of clouds. Temperatures in both models decrease rapidly above the surface, falling to values near 190 K at an altitude of about 10 km. As noted earlier, we identify the region of large temperature lapse near the surface as the troposphere. Based on the layer model discussed above, we would expect the temperature above the tropopause, the top of the troposphere, to assume a constant value equivalent to what was identified earlier as the skin temperature of the planet. The increase of temperature with altitude above the troposphere in the models presented here reflects the influence of the absorption of ultraviolet radiation by O_3. The models in Figure 7.12 terminate at an upper boundary of 50 km; had they extended higher, they would have exhibited a temperature maximum at a level we would associate with the stratopause and, at higher altitudes, a region where temperature would decrease with altitude, a zone we would identify as the mesosphere.

The temperature lapse rate for the troposphere indicated by the models in Figure 7.12 is much larger than the lapse rate observed in the real atmosphere. Indeed, the lapse rate in the models exceeds the dry adiabatic limit. In this sense the models are unrealistic. If the lapse rate was as large as indicated in Figure 7.12, the atmosphere would be unstable: heat would be vertically transported by convection, and the lapse rate would decrease to a value close to the adiabatic limit. Convection acts to reduce the temperature near the surface, while increasing temperatures aloft. Figure 7.13 illustrates the manner in which this is accomplished.

The Figure depicts a parcel of air originating at the surface and initially undergoing a small vertical displacement. In the absence of condensation, its temperature is constrained to follow the dry adiabatic limit, as indicated by trajectory A. As it rises, the parcel is lighter than the air in its environment. Consequently, it is buoyant and continues to move. Suppose it rises to an altitude of 10 km, where it merges with the background atmosphere. The parcel arriving at 10 km is warmer than the air previously present at this level; it contributes a source of heat as it mixes with its new environment and is thus responsible for an increase in local temperature. Transfer of air by convection is a two-way process, however. Rising air in one environment is replaced by sinking air elsewhere. Suppose that the parcel rising from the surface to 10 km is replaced by a parcel sinking from 10 km. As before, assume that vertical motion proceeds adiabatically. The descending parcel is colder, and consequently denser, than the air through which it moves. As it sinks, it follows the temperature profile indicated by trajectory B. When it eventually merges with the background atmosphere, adding cold air to warm, it is responsible for a reduction in ambient temperature. Exchange of air by adiabatic motion between the surface and 10 km therefore results in a temperature increase at 10 km, with an associated temperature decrease at the surface.

Convective overturning of the atmosphere (vertical exchange of mass, as depicted in Figure 7.13) is accompanied by a net upward transport of heat (energy), when the lapse rate of temperature exceeds the adiabatic limit, as is the case for the radiative equilibrium temperature profiles presented in Figure 7.12. As we have seen, vertical redistribution of heat by convective motion results in a reduction in the ambient lapse rate. It acts to establish a lapse rate close to, though slightly larger than, that appropriate in the limit of

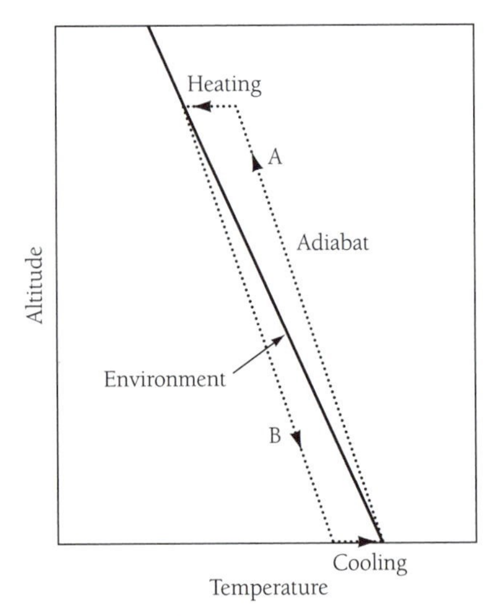

Figure 7.13 The effect of convection on an unstable temperature profile, illustrating the fact that the ambient atmosphere is heated by rising air, cooled by descending air.

strictly adiabatic motion. The lapse rate must be slightly superadiabatic to ensure that excess heat absorbed at the surface is transported vertically upward through the atmosphere. Solar energy deposited below any given level of the atmosphere in the purely radiative models discussed above (as summarized, for example, by the results in Figure 7.12) is removed solely by upward transfer of energy in the form of radiation at infrared wavelengths. A large temperature difference is required to carry the necessary flux of infrared radiation. The temperature difference will be smaller, when we allow for transport of energy by convection; the lapse rate needed to supply a given flux of energy by convection is significantly less than that required to drive a comparable flux of energy by radiation. Therefore it is clear that realistic models for the thermal structure of the atmosphere must allow for convective heat transfer.

In the simplest case, this is accomplished by assuming that, if the lapse rate calculated on the basis of radiative equilibrium is unstable, it should be replaced by the lapse rate corresponding to either the dry or the moist adiabatic limit. This is equivalent to assuming that, in the presence of instability, vertical heat transport is effected primarily by convection and that under such circumstances the gradient of temperature with altitude is indistinguishable from that appropriate in the adiabatic limit. We refer to models derived in this manner as **radiative-convective models.** Since lapse rates obtained under assumptions of radiative equilibrium are largest where the density of absorbing gas is highest, where transfer of energy by infrared radiation is least efficient, it is not surprising to find that the adiabatic adjustment in radiative-convective models is typically applied near the ground. A frequent procedure in one-dimensional models (where attention is focussed exclusively on transport of energy in the vertical) is to replace the lapse rate in the unstable region with the lapse rate observed to apply on average in the real atmosphere, about 6.5 K km^{-1} for the lower troposphere. Results obtained for such a model are presented in Figure 7.14.

One-dimensional models have been extensively used to assess the impact of the changing composition of the atmosphere on climate. They have the advantage that they can incorporate a relatively complete description of the radiative properties of the atmosphere (the transfer of energy associated with the absorption and emission of infrared radiation by H_2O, CO_2, O_3, CH_4, N_2O, SF_6, and the industrial chlorofluorocarbons, for example, and further allowing for effects of clouds) in a comparatively simple computer model. However, they have the disadvantage that the most important mechanism for heat redistribution in the lower atmosphere, overturning of air associated with convective instability, is treated in a qualitative rather than a quantitative fashion. A complete account of the vertical redistribution of energy near the surface requires not only a model for transport of heat on the small spatial scales characteristic of convection it must also allow for transport associated with the large-scale circulation of the atmosphere.

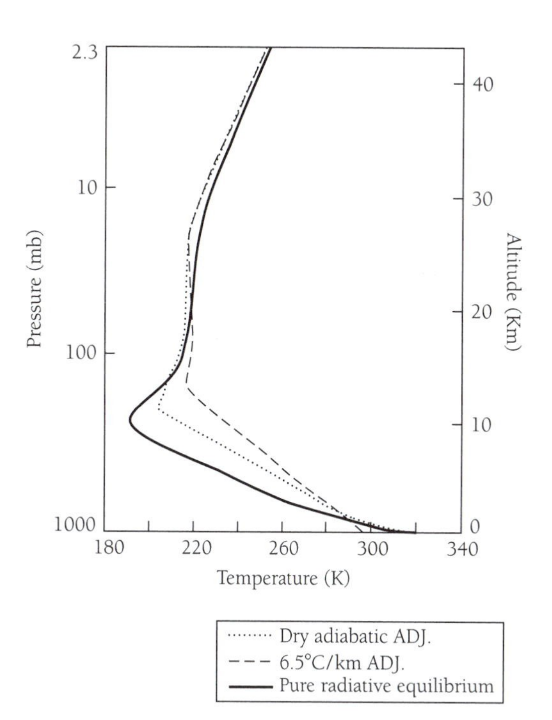

Figure 7.14 Comparison of results from a purely radiative equilibrium model of the atmosphere (solid line) with results from models in which the lapse rate of temperature was constrained not to exceed the dry adiabatic limit (dotted line) or a lapse rate of 6.5°C km^{-1} (dashed line). Source: Manabe and Strickler 1964.

7.7 Radiative Balance as a Function of Latitude

A perspective on the significance of the large-scale circulation is presented in Figures 7.15–7.17. The curve labeled R_s in Figure 7.15 summarizes estimates for the annually averaged net radiative balance at the surface. As a function of latitude,

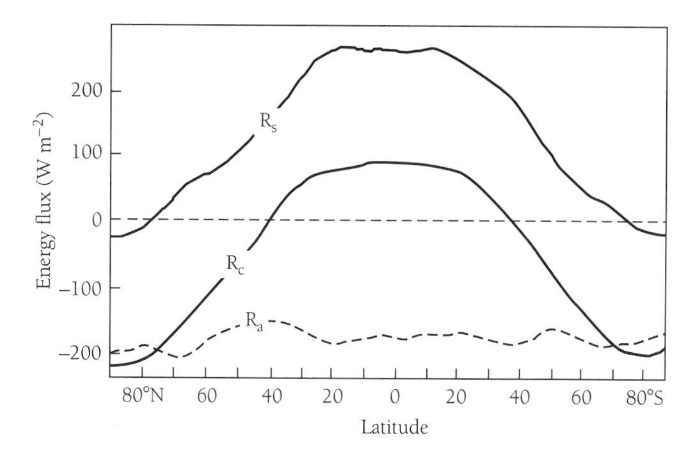

Figure 7.15 The annually averaged latitudinal distribution of radiation balances for Earth's surface (R_s), the atmosphere (R_a), and the combination of the atmosphere-surface system (R_c). Source: Sellars 1965.

it accounts for absorption of diffuse and direct solar radiation by the surface (a net gain), allowing at the same time for absorption and emission of energy in the infrared (a net loss). The radiant energy balance is positive for the surface, except at the highest latitudes, where, over the course of a year, the surface gains more energy by absorption of visible light than it loses by emission into the infrared. A portion of the excess radiant energy absorbed by the surface is used to evaporate H_2O; the balance is transferred to the atmosphere by conduction. The radiant energy balance of the atmosphere is negative at all latitudes, as indicated by the curve labeled R_a in Figure 7.15. The radiative deficit of the atmosphere is made up by energy transferred by conduction from the surface and by latent heat released in the precipitation of H_2O. The balance of radiant energy for the composite surface-atmosphere system is given by R_c in Figure 7.15.

The radiant energy balance of the composite surface-atmosphere system is positive for latitudes below about 40°, negative for those above. That is to say, the surface-atmosphere system gains more energy by absorption of sunlight than it loses by emission of infrared radiation for latitudes below 40°. The reverse situation applies for latitudes above 40°. Were it not for the moderating influence of the atmosphere and the ocean, and their ability to transfer heat efficiently from low to high latitudes, Earth would be much warmer at the equator, much colder at the poles. The flux of heat required to maintain the present observed latitudinal distribution of temperature (to resolve the latitudinal imbalance implied by R_c in Figure 7.15), averaged over the course of a year, is indicated in Figure 7.16. Part of the required redistribution of heat is accomplished by the atmosphere through bulk movements of air with the balance accomplished by the ocean, with the ocean being especially important at higher latitudes.

The results in Figure 7.15 assume that the capacity of the surface-atmosphere system to store heat is negligible on a global scale over the course of a year. The net quantity of solar energy annually absorbed by Earth should closely equal the flux of energy globally radiated at infrared wavelengths into space. Any imbalance must be taken up by the ocean (the heat capacities of the atmosphere and solid surface are trivial by comparison). Assume that the volume of ocean water interacting with the surface over an annual cycle is equivalent to the volume of water present to a depth of 20 m, about 7.2×10^{21} cm^3 (obtained by multiplying the surface area of the ocean, 3.6×10^{18} cm^2, by the depth of the annually interactive water, 2×10^3 cm). The mass of water in this volume is equal to 7.2×10^{21} g (the density of water is 1 g cm^{-3}). To change the water temperature by 0.2 K, a maximum estimate for the change in the temperature of the global ocean surface that might escape detection averaged over a year, would require the addition (or subtraction) of about 1.4×10^{21} calories of heat (the specific heat of water is 1 cal g^{-1} K^{-1}), equal to about 4×10^2 cal cm^{-2} averaged over the entire surface of Earth. This would correspond to about 0.6% of the average surplus of radiant energy at the surface implied by the results in Figure 7.15 (7.2×10^4 cal cm^{-2}). It would be difficult to accommodate an imbalance in the radiative budget of the global surface-atmosphere system much larger than this on an annual scale.

A slightly different perspective on the energetics of the climate system is presented in Figure 7.17. The data displayed here summarize results from satellite measurements of the total flux of outgoing planetary (mainly infrared) energy, together with estimates of the flux of solar (mainly vis-

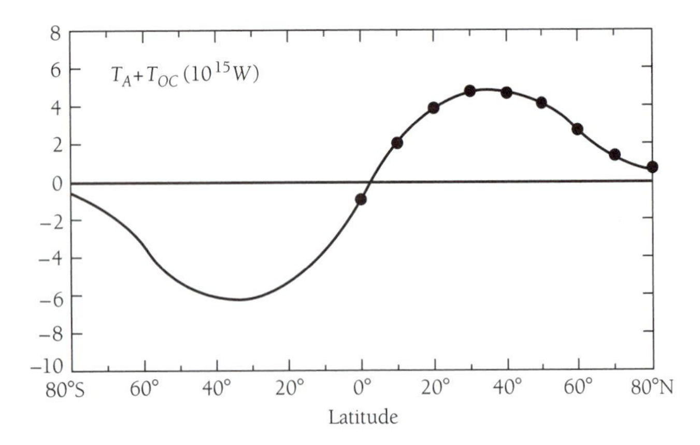

Figure 7.16 The annually averaged meridional transport of energy by the combination of the atmosphere and ocean, as a function of latitude, in units of 10^{15} W. Note that positive values represent northward transport, and negative values represent southward transport. Source: Peixoto and Oort 1992.

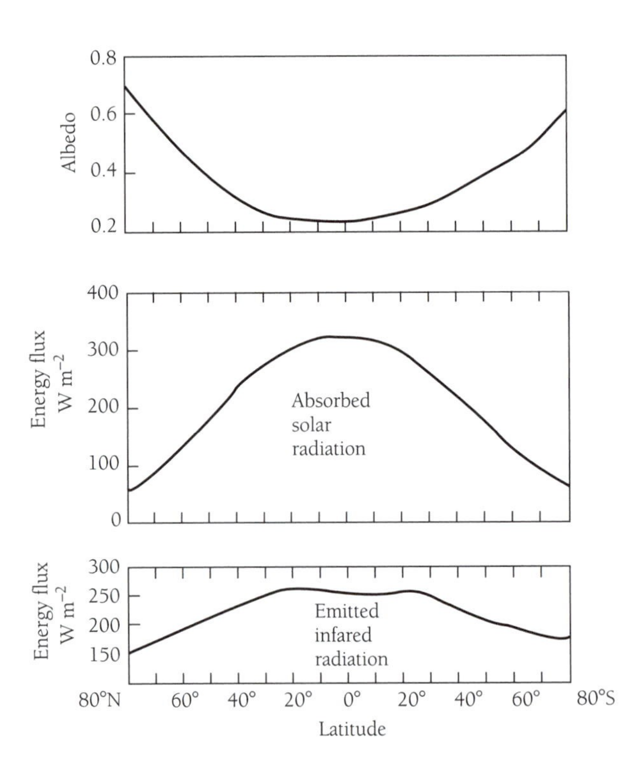

Figure 7.17 Variation of the average albedo of Earth as a function of latitude (upper panel), the flux of energy absorbed from the Sun (middle panel), and the energy emitted in the infrared (lower panel). Source: Peixoto and Oort 1984.

ible) radiation absorbed by the planet, presented in both cases as functions of latitude. Also indicated is the latitudinal variation of the planetary albedo. All quantities in Figure 7.17 refer to averages over longitude and over a full year. The radiative energy deficit poleward of about 40° is clearly evident, complementing the surplus at lower latitudes, as illustrated in Figure 7.15. The rise in albedo at high latitudes is a result of the high reflectivity of continental snow and seasonal sea ice.

7.8 Modeling the Changing Climate

Despite their obvious deficiencies, one-dimensional models have played an important role in focusing the scientific debate on the potential importance of the warming expected to occur as a consequence of the increase in the concentration of greenhouse gases such as CO_2, CH_4, N_2O, SF_6, and the industrial chlorofluorocarbons. However, it is essential that we recognize the limitations and uncertainties inherent in a one-dimensional formulation. Results obtained with one-dimensional models are overwhelmingly sensitive to the procedure used to correct the temperature profile obtained from purely radiative considerations to allow for the influence of convective overturning of air in the radiatively unstable layer near the ground. The results in Figure 7.14 indicate that the radiative profile for temperature in the contemporary atmosphere (i.e., the temperature profile calculated as a function of altitude, assuming that transfer of energy by radiation is the only mechanism available for vertical redistribution of heat) should be unstable on average for altitudes less than 10 km. Doubling the concentration of CO_2 is predicted to extend the domain of instability to a slightly higher altitude, about 10.3 km. Extrapolating to the surface in both cases using a fixed lapse rate of 6.5 K km^{-1} suggests that an increase in CO_2 by a factor of two should lead to an increase in surface temperature of about 2 K. Had we used a lapse rate of 6.7 K km^{-1} for the doubled CO_2 case, retaining a lapse rate of 6.5 K km^{-1} for the present atmosphere, the predicted increase in surface temperature would have risen to 4K. Conversely, had we adopted a reduced value for the lapse rate with doubled CO_2, say 6.3 K km^{-1}, our model would have predicted no change in surface temperature. It is obviously important to develop a better understanding of the physical processes that regulate the near-surface lapse rate. This understanding must be incorporated as fully as possible in models used to investigate effects of the changing composition of the atmosphere on Earth's heat budget.

It should be emphasized that variations in the concentration of H_2O play an important role in determining the atmospheric response to an increase in the concentration of greenhouse gases such as CO_2, one comparable indeed to the impact induced by the primary agent of change. We expect significant changes in both the concentration and height profile of H_2O as a consequence of a change in atmospheric temperature. The physical processes regulating the distribution of H_2O in the lower atmosphere are complex, similar to those involved in the dynamical redistribution of heat, complicated additionally by effects of precipitation. Complexity on this scale is beyond the scope of one-dimensional models. A customary response is to treat the adjustment of H_2O using a recipe rather than a physical model. Commonly, the profile of H_2O is fixed on the basis of observations for the present atmosphere and it is assumed that relative humidity is conserved, at least at lower altitudes, as the atmosphere is subjected to change. An increase in CO_2 tends to raise the level at which the radiatively computed temperature profile becomes unstable, thus extending the altitude domain over which convective adjustment must be applied, resulting in an increase in surface temperature (assuming that the lapse rate for the perturbed environment is the same as for the normal atmosphere). Higher temperatures, with constant relative humidity, result in an increase in H_2O concentration. Instability is transferred to higher levels of the atmosphere, extending the zone of convective instability and leading to an additional increase in surface temperature. The sequence provides an example of what is known as **positive feedback**; effects of a perturbation, once applied, are amplified by the subsequent response. The problem is that the specific feedback presupposed here critically depends on our assumption that the relative humidity remains fixed. One could imagine a situation in which the relative humidity might in fact decrease, a consequence of enhanced precipitation in a warmer atmosphere, for example. In this case, the feedback associated with H_2O would be negative; changes in H_2O would act to oppose changes induced by the postulated increase in CO_2. Lacking a quantitative model for H_2O and in the absence of a realistic treatment for heat transport by atmospheric motion, it seems only prudent that we treat predictions based solely on one-dimensional models of the atmosphere with a healthy degree of skepticism. They emphasize, however, the need for further, more realistic investigations into the prospects for future climate change.

7.9 Summary

The decrease of atmospheric pressure with altitude is described by the barometric law. We have shown that pressure exponentially decreases with altitude in the atmosphere, in part reflecting the decrease in mass of the overlying atmosphere, in part the drop of density (the mass of air contained in a unit volume) with pressure, as required by the perfect gas law. Pressure drops by a factor of *e* over an altitude interval given by a quantity known as the scale height. The scale height in the lowest layer of the atmosphere, the troposphere, has a value of about 8 km.

Temperature decreases with altitude in the troposphere, reaching a minimum at the tropopause. Temperature increases with height above the tropopause in the stratosphere, rising to a maximum at the stratopause. An increase of temperature with altitude in the stratosphere is caused by absorption of ultraviolet sunlight by ozone. Temperature decreases with altitude in the mesosphere above the stratopause dropping to a secondary minimum at the mesopause. Temperature increases again at higher altitudes in the thermosphere. An increase of temperature with altitude in the thermosphere is caused by absorption of short-wavelength solar radiation. Absorption of this short-wave radiation is responsible for the presence of the partially ionized region of the upper atmosphere known as the ionosphere.

We have introduced the concept of stability. If temperature decreases too rapidly with altitude, we showed that the atmosphere will be unstable. A small vertical displacement can feed upon itself. If the temperature of an upwardly moving air parcel is greater than the temperature of the background air through which the parcel is moving, the density of mass in the parcel will be less than the density of mass in the air it displaces. The weight of the parcel will be insufficient in this case to balance the upwardly directed force imparted by the vertical gradient of pressure. The parcel will experience a net upward force, leading to an acceleration of its initial motion. Similarly, if the increase in temperature of a downwardly moving parcel is less than the increase in temperature of the air through which it is moving, the density of mass in the parcel will be greater than the density of mass in the air it displaces. As before, the imbalance of forces will lead to an acceleration of the postulated initial motion (downward in this case).

The gradient of temperature with altitude marking the boundary between stable and unstable air is defined by what is known as the adiabatic lapse rate. The adiabatic lapse rate identifies the change of temperature with altitude that would result if an air parcel were to move vertically in the absence of exchange of heat with its surroundings. For dry air, the adiabatic lapse rate has a value of 9.8 K km^{-1}. If the decrease of temperature with altitude exceeds the adiabatic limit, the atmosphere is said to be unstable. In this case the profile of mass density with altitude ensures that upwardly moving air will move higher while downwardly moving air will move lower. An overturning of air will result in a vertical redistribution of heat, driving the temperature profile of the atmosphere toward the adiabatic limit.

Phase changes of water can cause instability to set in under conditions where the lapse rate of temperature with altitude is less than the value referenced above for dry air. An upwardly moving parcel can gain heat as a result of the condensation of water vapor. The decrease of temperature with altitude for adiabatic motion will therefore be less in the presence of condensable water vapor than for dry air. We refer to the decrease of temperature with altitude accounting for possible phase changes in water as the moist adiabatic lapse rate. As before, the term *adiabatic* is used to emphasize that motion takes place without an exchange of heat with the surroundings. The lapse rate for moist adiabatic motion is a sensitive function of temperature. It also depends on the available supply of water vapor; if relative humidity is less than 100%, adiabatic motion is constrained to follow the dry adiabatic profile. If the decrease of temperature with altitude is greater than the moist adiabatic limit but less than the dry adiabatic limit, the atmosphere is said to be conditionally unstable (it would be unstable if the water vapor content was high enough to ensure condensation). If the decrease of temperature with altitude is less than the moist adiabatic limit, the atmosphere is stable and vertical motion is suppressed.

We have shown that convective overturning of the atmosphere plays an important role in carrying heat from the surface to atmospheric levels where it can be radiated into space. We have described the procedure and limitations of the method used to account for the influence of convective instability in simple models for the one-dimensional thermal structure of the atmosphere. Finally, we have noted that a more realistic treatment of the energy budget and thermal properties of the atmosphere must account for transfer of heat from the tropics to higher latitudes. This transfer is accommodated by motions of the atmosphere and ocean to be discussed in the following chapters.

Horizontal Motion of the Atmosphere

8

Radiative-convective models provide a reasonable representation of the globally averaged state of the surface-atmosphere system. They are much less reliable as analogues for the condition of the environment at any given latitude. Temperatures obtained on the basis of radiative-convective models for the tropics are too high, while temperatures calculated for temperate and higher latitudes are too low. The missing ingredient, as we saw in Section 7.7, relates to the influence of atmospheric motion, with an additional contribution due to transport of heat by the ocean. Significant quantities of heat are transferred by bulk motions of the atmosphere and of the ocean from low to high latitude, as was indicated, for example, in Figure 7.16. An understanding of climate presumes an understanding of the factors regulating the spatial redistribution of heat by atmospheric and oceanic motions.

Adjustment of the radiative state of the atmosphere, associated for example with changes in the concentration of a greenhouse gas such as CO_2, may be expected to alter not only the average temperature but also the patterns of circulation of the atmosphere. How might the climate change in response to such a disturbance? Radiative-convective models can provide, we hope, an indication of the change in temperature expected for Earth as a whole, but such predictions are of limited use to the policymaker seeking answers to more specific questions. In the presence of an enhanced concentration of CO_2, should we anticipate a shift in the position of Earth's major climatic or ecological zones, a change, for example, in the areal extent or geographic location of deserts? How might changes in climate affect the supply of fresh water, or the capacity of Earth to produce food? Should we expect an increase or decrease in the incidence of violent storms, in the frequency of hurricanes for example? To address these issues we need a physical model for the dynamics of the atmosphere and a complementary model for the ocean. At a minimum we need to understand the processes that regulate the transport of heat and moisture from one region of Earth to another, an appreciation for the factors that regulate the connections between climate and latitude.

This chapter concerns the physical mechanisms responsible for what is referred to as the **general circulation of the atmosphere**, defined for present purposes as the long-term average state of motion of the air. (The circulation of the ocean is discussed in Chapter 9.) The pattern of winds associated with the general circulation at the surface has been known, at least in broad outline, for almost three hundred years, ever since the golden age of exploration, when navigators such as James Cook (1728–1779) traveled the seven seas in search of adventure and trade, produced the first comprehensive outlines of our planet, and recorded (incidentally) the ever fluctuating state of its weather. A conceptual model for longitudinally and time-averaged circulation is presented in Figure 8.1. According to this picture, winds in the tropics blow predominantly from the northeast in the Northern Hemisphere and from the southeast in the Southern Hemisphere. Winds at midlatitudes (30–60°) blow mainly from the west or southwest at midlatitudes in the Northern Hemisphere, from the west or northwest at comparable latitudes

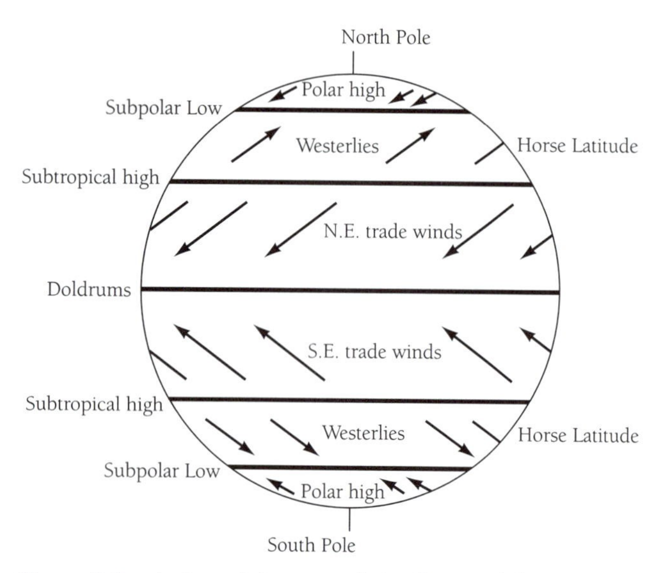

Figure 8.1 A view of the general circulation of the atmosphere showing the direction of prevailing winds at the surface.

in the Southern Hemisphere. Winds at high latitudes (60–90°) carry air from polar regions to lower latitudes, from a generally easterly direction in both hemispheres.[1]

The system of easterly winds in the tropics played an important role in early navigation, providing fast and reliable assistance to ships sailing from Europe to the New World. Merchants referred to the helpful winds on the outward trip as the "Trade Winds." The favored path for the return journey, despite dangers and the often devastating tolls taken by storms, was to the north, where the prevailing winds were from the west. Separating the zone of easterlies from westerlies was a region where winds were generally slight and temperatures unrelentingly hot under cloudless skies, where it was not uncommon for ships to be becalmed for days or even weeks. Samuel Taylor Coleridge (1772–1834) described the region graphically in his "Rime of the Ancient Mariner":

> All in a hot and copper sky,
> The bloody Sun, at noon,
> Right up above the mast did stand,
> No bigger than the Moon.
> Day after day, day after day,
> We stuck, nor breath nor motion;
> As idle as a painted ship
> Upon a painted ocean.
> Water, water, every where,
> And all the boards did shrink;
> Water, water, every where,
> Nor any drop to drink.

This area is known as the **Horse Latitudes** (around 30° both north and south), reputedly because horses were often killed on ships in this region to provide food and conserve water.

The picture of the general circulation displayed in Figure 8.1 is at best schematic. The actual situation is considerably more complicated, as indicated by maps of surface winds for the months of January and July, presented in Figures 8.2 and 8.3. The results given here are intended to illustrate the pattern of the surface flow expected when winds are averaged over many years for the months of January and July. The results demonstrate the importance of high and low pressure systems in guiding the motion of air, particularly at mid and high latitudes. Air in the Northern Hemisphere tends to circulate clockwise around regions of high pressure, in the opposite direction around zones of low pressure. (Reasons for this behavior are discussed below.) The pattern is reversed for the Southern Hemisphere. The highs and lows in Figures 8.2 and 8.3 are associated with particular geographic features. They tend to remain relatively fixed in position over the course of a season. The high pressure system over the Soviet Union in January, for example, is caused by the intense cooling and consequent sinking of air over the Asian continent during winter. The analogous lows over the Aleutian Islands and Iceland are associated with the relatively warm waters of the North Pacific and Atlantic.

A map of surface pressure for any given day in January or July, indeed for any time of year, would reveal high and low pressure systems in addition to the semipermanent features apparent in Figures 8.2 and 8.3. Unlike the features in the figures, these systems are mobile and dynamic. They appear and disappear over the course of a few days and can travel vast distances in the brief period between their births and decay. Averaged over the course of a month, disturbances associated with mobile highs and lows tend to cancel at any given location, leaving minimal traces of their presence on long-term average charts such as those displayed in Figures 8.2 and 8.3. In combination with stationary systems, however, as we shall see, they play an important role in the global transport of heat and are responsible for much of the weather—the day-to-day variable manifestations of climate—experienced at mid and high latitudes.[2]

The general circulation defines the complex response of the atmosphere to unequal rates at which solar radiation is absorbed over time and position and to related imbalances in rates at which energy is emitted to space in the infrared portion of the spectrum. As previously noted, there is a surplus of radiant energy (visible plus infrared) in the tropics, with a corresponding deficit at higher latitudes. Imbalances can also arise on much smaller spatial scales, between land areas and the surrounding ocean, for example.

We begin our account of the dynamics of the atmosphere in Section 8.1 with a qualitative description of the circulation arising near the sea in summer as a consequence of differences in the responses of land and water to **insolation** (exposure to the Sun's rays) over the course of a day. These differences give rise, as we shall see, to a shallow circulation (one confined to low altitude) with on-shore breezes at the surface during the day and off-shore breezes at night. The sea (or land) breeze problem provides a useful introduction

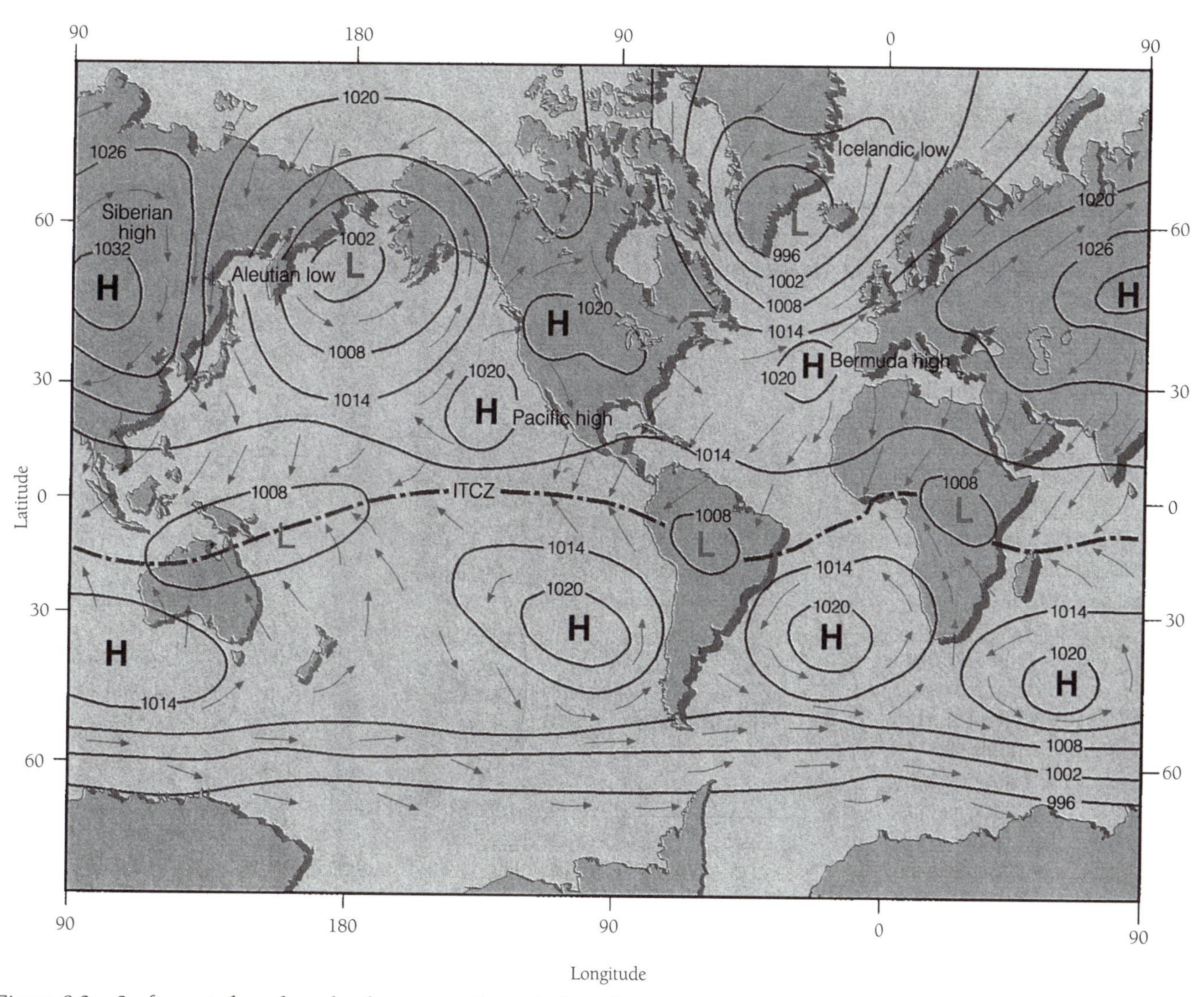

Figure 8.2 Surface winds and sea level pressures for typical conditions in January. The heavy dashed lines shows the position of the intertropical convergence zone (ITCZ), the approximate location of the meteorological equator. Source: Ahrens 1994.

to the circulation of the global atmosphere driven by differences in the absorption of heat at low as compared to high latitudes. An early model for the general circulation of the atmosphere is presented in Section 8.2. The role of rotation and its effect on atmospheric motion is discussed in Sections 8.3–8.6. We introduce in Sections 8.3 and 8.4 the Coriolis force, and in Section 8.6 the concept of geostrophy, in which the horizontal gradient of pressure acting on an air parcel is balanced primarily by the Coriolis force. The role of friction is discussed in Section 8.7. Further perspectives on the circulation of the atmosphere are presented in Section 8.8 as a prelude to a discussion in Sections 8.9–8.11 of the angular momentum budget of the atmosphere and the importance of transport by eddies. Summary remarks are presented in Section 8.12.

8.1 The Land-Sea Breeze Problem

Imagine a situation in which weather conditions near the land-sea boundary are initially uniform: there is no significant variation in the horizontal field of atmospheric pressure. Over land, solar energy is absorbed only by the surface. When absorbed by the sea, however, solar energy is distributed over a depth of several meters. Thus land temperature rises rapidly during the day and cools equally quickly at night, while ocean temperature responds more slowly, staying relatively constant over the course of a day.

As the temperature rises at the surface of the land during the day, the temperature of the air above the ground rises, as well. The pressure at the surface, determined by the mass of the overlying atmosphere, initially remains constant. Pressure, temperature, and density are constrained to satisfy the perfect gas law (equation 7.7). With pressure fixed, the increase in temperature at ground level over land is compensated by a decrease in density. To accommodate the decrease in density, air expands and the atmosphere rises at every level above the surface. This establishes a pressure gradient in the horizontal direction above the surface: pressure is higher at any given level over the warm ground than over the sea.

The variation of pressure with altitude may be calculated using the barometric law, (equation 7.15). The scale height, given by equation (7.11), is larger over land during

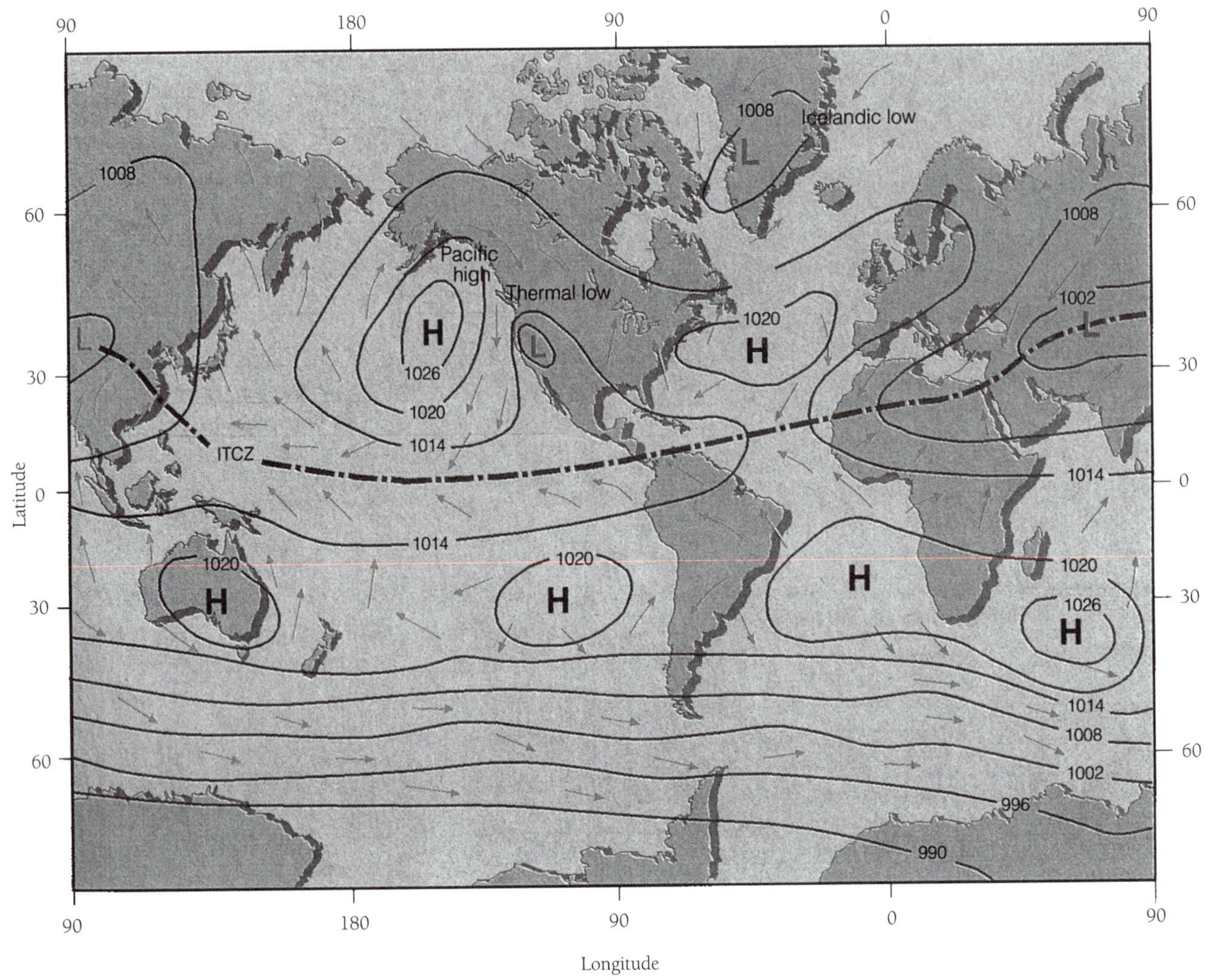

Figure 8.3 Surface winds and sea level pressures for typical conditions in July (as in Figure 8.2). Source: Ahrens 1994.

the day than over the sea, as a consequence of the higher temperature. The difference in the rate at which pressure drops with altitude over land and sea is illustrated schematically in Figure 8.4.

Example 8.1: What is the difference in pressure between two parcels of air at 500 m if one parcel, over land, is at 300 K and the other parcel, over the sea, is at 285 K? Assume that the temperature over the altitude range 0–500 m is the same as the value at the surface.

Answer: First find the appropriate scale heights over land (l) and sea (s).

$$H_l = \frac{RT_l}{g} = \frac{(2.87 \times 10^6 \text{ erg g}^{-1} \text{ K}^{-1})(300 \text{ K})}{980 \text{ cm sec}^{-2}}$$
$$= 8.79 \times 10^5 \text{ cm}$$

$$H_s = \frac{RT_s}{g} = \frac{(2.87 \times 10^6 \text{erg g}^{-1} \text{ K}^{-1})(285 \text{ K})}{980 \text{ cm sec}^{-2}}$$
$$= 8.35 \times 10^5 \text{ cm}$$

Now use the scale heights to find the two pressures.

$$\text{Pressure aloft} = (\text{surface pressure}) \left(\exp\left[\frac{-\text{height}}{\text{scale ht}}\right] \right)$$

$$P_l(500\text{m}) = 10^6 \text{ dyn cm}^{-2} \exp\left[\frac{-0.5 \text{ km}}{8.79 \text{ km}}\right]$$
$$= 9.45 \times 10^5 \text{ dyn cm}^{-2}$$

$$P_s(500\text{m}) = 10^6 \text{ dyn cm}^{-2} \exp\left[\frac{-0.5 \text{ km}}{8.35 \text{ km}}\right]$$
$$= 9.42 \times 10^5 \text{ dyn cm}^{-2}$$

$$\Delta P = P_l - P_s = 9.45 \times 10^5 \text{ dyn cm}^{-2} - 9.42 \times 10^5 \text{ dyn cm}^{-2}$$
$$= 3.0 \times 10^3 \text{ dyn cm}^{-2}$$

■

An air parcel at any given height above the surface in the transition zone between land and sea experiences a net force pushing it toward the sea during a summer day, as illustrated in Figure 8.5. The magnitude of this force de-

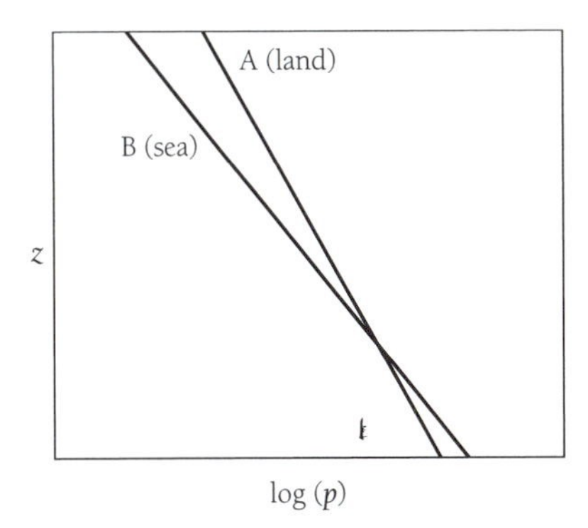

Figure 8.4 Schematic illustration of the variation of pressure with altitude. Curve A applies to the warmer conditions characteristic of air over land in the summer. Curve B applies to colder air over the ocean.

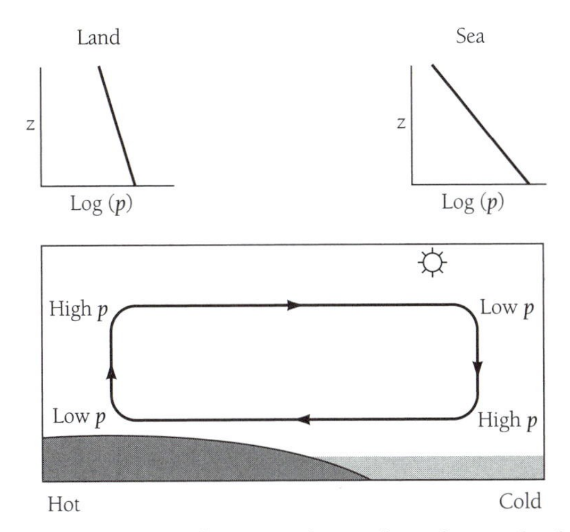

Figure 8.5 Airflow near the seashore during the day in summer.

pends on the pressure gradient, the change in pressure per unit distance. The pressure gradient aloft sets up an airflow from land to sea. The air mass over any given area of land decreases, and as a consequence there is a decrease in pressure near the ground. At the same time, as additional mass moves over the ocean, there is a corresponding increase in surface pressure in this region. A pressure gradient is established near the surface in the opposite sense to the gradient aloft. Air at the surface experiences a net force due to pressure, driving an airflow from sea to land. A circulation loop is set up, with rising motion over the land, an outflow of air from land to sea aloft, sinking motion over the sea, and a return flow toward land near the surface. An observer at the surface experiences this circulation as a cool breeze blowing off the sea toward the land during the day. The circulation reverses at night, as illustrated in Figure 8.6. The sea breeze at the surface by day is replaced by a land breeze at night.

The thermal circulation outlined above, driven by the temperature difference between land and ocean, plays an important role in the summertime meteorology of the State of Florida. Often, there is an inflow of air from the sea to the land at the surface during the day, from the Atlantic to the

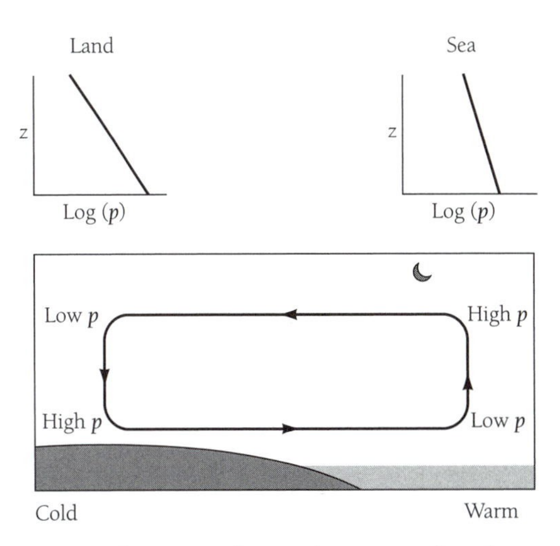

Figure 8.6 Airflow near the seashore at night when air over the sea is warmer than air over land. Compare the pressure profiles to those in Figure 8.5.

east and the Gulf of Mexico to the west, as indicated in Figure 8.7. Air drawn from off the sea has a high-moisture content. As it moves over land, converging from both west and east, it rises, setting off a bank of thunderstorms frequently extending along the entire axis of the state during early afternoon. A vertical perspective on the circulation, a cut through the atmosphere from west to east, is presented in Figure 8.8. Convergence of surface air and the subsequent rising motion over land can give rise to heavy daytime thundershowers.

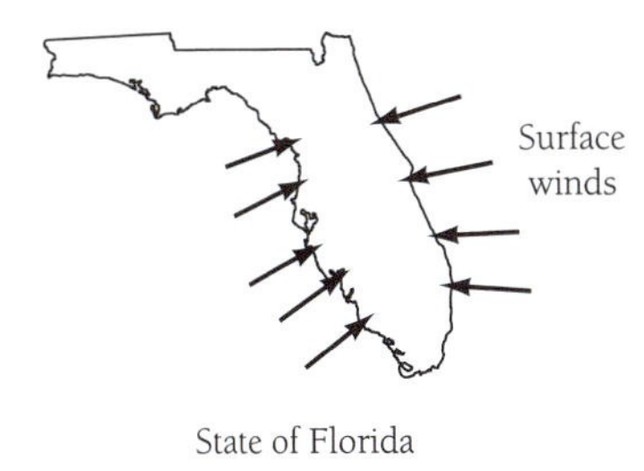

Figure 8.7 Pattern of onshore surface winds for Florida during summer.

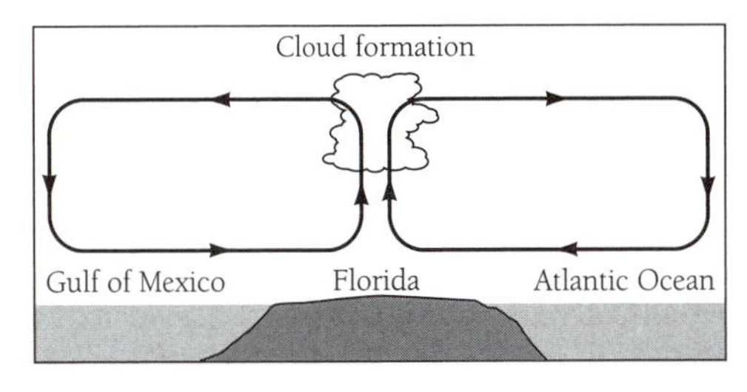

Figure 8.8 Vertical west-east cross section of the daytime circulation during summer in Florida. Rising air over land is supplied with abundant H_2O from the surrounding ocean, resulting in frequent daytime thunder showers.

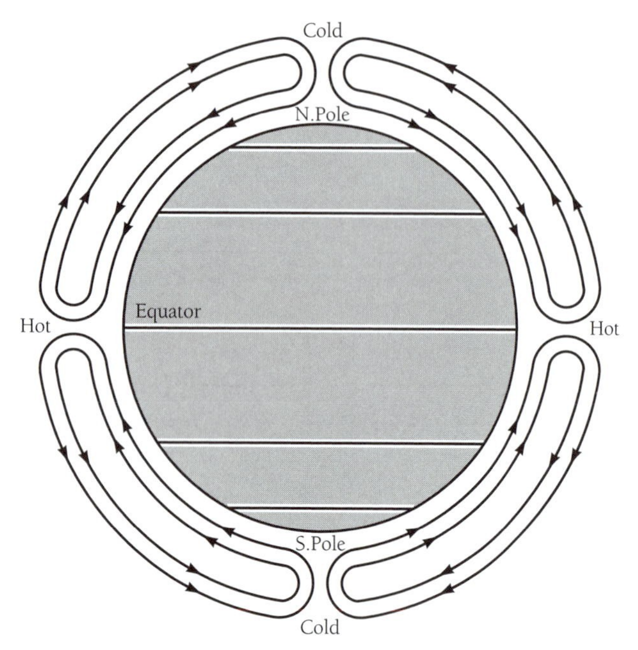

Figure 8.9 Hadley's view of the general circulation of the atmosphere.

8.2 Hadley's Model for the General Circulation

The first serious attempt to account for the general circulation of the atmosphere was presented by the English meteorologist George Hadley in 1735. Hadley began his paper with a bold statement: "I think the causes of the General Trade-Winds have not been fully explained by any of those who have wrote on that Subject." Building on an earlier paper by the astronomer Edmund Halley (1686), who presented a detailed description of the trade winds observed in three oceans, Hadley sought to explain the westerly winds at the surface at midlatitudes and the easterlies at low latitudes as natural consequences of the preferential absorption of solar radiation near the equator. He envisaged global circulation as a large-scale analogue to the sea breeze circulation outlined above. He assumed that air would rise over the tropics, split toward the north and south aloft, and return to the surface at higher latitudes in both hemispheres. The circulation would be completed by a return flow to the tropics at the surface in both hemispheres, as illustrated in Figure 8.9.

Hadley supposed that air, as it moved, would tend to conserve its absolute speed; in other words, air speed, as measured by an observer fixed in space, would appear to remain constant. The conservation of absolute speed is not a physical law, and, as we shall see, this assumption led to erroneous conclusions. However, the contrast between rotating and inertial frames of reference did lead to important ideas about circulation worth exploring here in some detail.

Recall that points on Earth's surface are not fixed in space but are always in a state of rapid motion, from west to east, as a consequence of planetary rotation. Speeds of rotation are largest at the equator: a body fixed to Earth at the equator moves a distance equal to the circumference of Earth, 2π times the radius of the planet (6.378×10^8 cm), over the course of a day (8.64×10^4 sec), corresponding to a speed, as measured by a space-fixed observer, of

$$v_r = \frac{2(3.14)(6.378 \times 10^8 \text{cm})}{8.64 \times 10^4 \text{ sec}} = 4.64 \times 10^4 \text{ cm sec}^{-1} \quad (8.1)$$

Here the numerator defines the distance moved in a day, or the circumference; the denominator is equal to the length of a day measured in seconds. Imagine touching a magic marker to a spinning globe. The circle traced at the equator is larger than the circle that would be traced at any other latitude. Figure 8.10 should convince you that the distance traveled in a day by a fixed point on Earth is $2\pi R(\cos \lambda)$, where λ represents the latitude. Magnitudes of rotational velocities for different latitudes are given in Table 8.1 as v_r.

Consider an air parcel, initially stationary with respect to Earth's equator, that begins to move northward. Following Hadley, assume that the absolute speed of the parcel remains constant, meaning equal to the rotation speed of the planet at the equator, or 4.6×10^4 cm sec^{-1} measured in the west-east direction. As the parcel moves northward, it begins to drift eastward: its eastward velocity is larger than the velocity of the solid Earth below. An observer on Earth would interpret the parcel's relative motion as a westerly wind. The speed of the wind at latitude λ would be given by the difference between the rotation speed of Earth at the equator and the rotation speed at the local latitude. Values of the wind speed computed in this fashion are included in Table 8.1 as v.

Example 8.2: Calculate the expected wind speed (as perceived by an observer rotating with Earth), based on Hadley's concept of constant absolute speed for a parcel moving from the equator to 60°N or S.

Answer: First, calculate the absolute speed of the parcel stationary at the equator. As shown in equation (8.1), it equals 4.6×10^4 cm sec^{-1}.

Next, calculate the absolute speed of a parcel stationary at 60°N or S.

$$v_s = \frac{2(3.14)(6.378 \times 10^8 \text{ cm})(\cos 60°)}{8.64 \times 10^4 \text{ sec}}$$

$$= 2.3 \times 10^4 \text{ cm sec}^{-1}$$

Finally, find the difference between the rotation speed at the equator and that at 60°N or S.

$$(4.6 \times 10^4) - (2.3 \times 10^4) = 2.3 \times 10^4 \text{ cm sec}^{-1}$$

This is over 800 km per hour! ■

As seen in this example, air reaching high latitudes in Hadley's scheme would have extremely strong flows eastward, and, in the absence of friction, the winds at the surface would be a mirror image of the flow aloft. As the air returns to the equator at the surface, the strength of the westerly wind would decrease steadily as it moves to

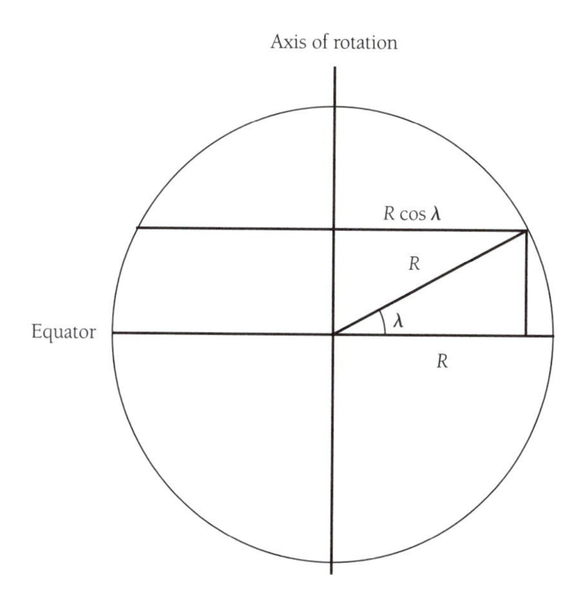

Figure 8.10 Geometry used to determine the circumference of Earth at latitude λ.

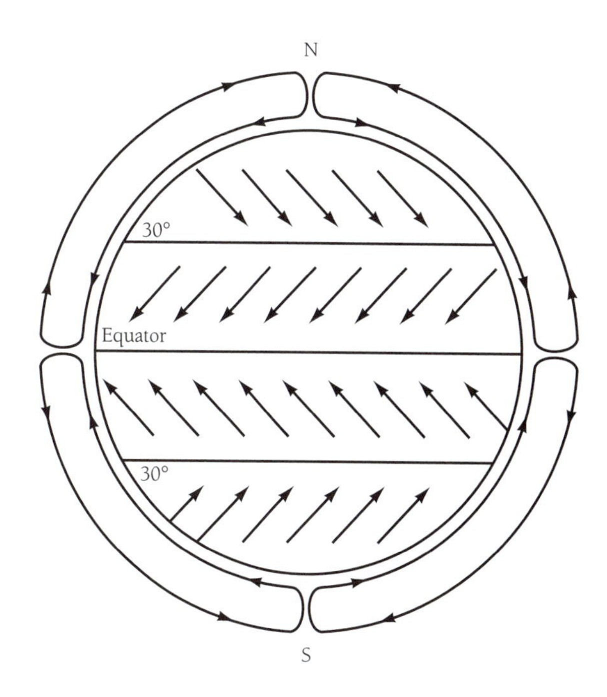

Figure 8.11 Hadley's circulation showing the direction of the prevailing winds at the surface with an exaggerated view of the circulation in the vertical. Source: Peixoto and Oort 1992.

Table 8.1 Computation of wind speeds

λ	v_r (cm sec^{-1})	v (cm sec^{-1})
0	4.64×10^4	0
15	4.49×10^4	1.5×10^3
30	4.03×10^4	6.1×10^3
45	3.29×10^4	1.4×10^4
60	2.32×10^4	2.3×10^4
75	1.20×10^4	3.4×10^4
90	0	4.6×10^4

NOTE: *v_r = magnitudes of rotational velocity at latitude λ; v = speed of the wind at latitude λ, assuming conservation of absolute speed*

lower latitudes. In this case, the model would account for the presence of westerly winds at midlatitudes. It would fail, however, to address the primary problem of interest to Hadley: the existence of easterly trade winds in the tropics.

Hadley resolved the dilemma by noting that friction introduced by transfer of momentum from air to ground would be expected to reduce the magnitude of the surface wind speed. He suggested that friction could cause the speed of the surface westerly wind to drop to zero at a latitude of about 30°. Prevailing winds at lower latitude would switch then to an easterly direction, reflecting the inability of air on its return journey to the equator to keep up with the increasing speed associated with west-east rotation of the underlying planet. A schematic illustration of Hadley's circulation is presented in Figure 8.11.

Hadley's contribution to our understanding of the general circulation of the atmosphere is remarkable, especially when we recall that his paper appeared more than 250 years ago. He offered the first detailed description of circulation arising as a result of differential heating. He correctly pointed to the importance of friction in limiting the magnitude of wind velocities near the surface. He provided the first plausible physical explanation for the general features of the circulation observed not only at the surface but also aloft. His paper, predating observations of the winds aloft, successfully predicted at least the eastward direction of the upper level winds. In this sense, Hadley may be credited with the first successful weather forecast!

His paper may be criticized on grounds that it failed to account for complications associated with the presence of continents and oceans and that it did not allow for seasonal variations in the absorption of solar radiation. A more serious error, perhaps, was his assumption that air should tend to conserve the magnitude of its absolute velocity. It would have been preferable to suppose that motion should be constrained to conserve angular momentum (see Section 2.10). Conservation of angular momentum, as previously noted, requires that the rotational speed of a skater increase as the arms are drawn in. For the same reason, the absolute value of the west-east speed of an air parcel should increase, rather than stay constant, as the parcel approaches the axis of rotation of the planet in moving to higher latitude. Hadley's model can be corrected in a relatively straightforward way to allow for conservation of angular momentum rather than absolute velocity. As we shall show, constraints on the west-east velocity imposed by the tendency of motion to conserve angular momentum may be treated by introducing a new effective force, the **Coriolis force**, named in honor of the nineteenth-century French physicist Gaspard Coriolis (1792–1843).

8.3 Angular Momentum and the West-East Coriolis Force

Denote the absolute value of the speed of an air parcel in the west-east direction at latitude λ by $v_a(\lambda)$. Suppose that the mass of the air parcel is equal to 1 gram. The magnitude of the angular momentum vector is directed to the north, aligned with the axis of rotation of the earth. The angular momentum of the parcel at latitude λ, $L(\lambda)$, is given by

$$L(\lambda) = \left(v_a(\lambda)\right) R \cos \lambda \qquad (8.2)$$

With $v_a(\lambda)$ expressed in cm sec^{-1} and the radius of Earth, R, in cm, L has units g cm^2 sec^{-1} (remember, we are considering a parcel with a mass of 1 g: as we saw in Chapter 2, the magnitude of the angular momentum vector is proportional to the mass of the parcel). Here $R \cos \lambda$ defines the distance separating the parcel from Earth's rotational axis at latitude λ (see Figure 8.10).

Imagine that the parcel is initially at rest with respect to Earth's equator. The absolute value of the west-east speed is given by in this case

$$v_a(0) = \frac{2\pi R}{(8.64 \times 10^4 \text{ sec})} \qquad (8.3)$$

Note that this equation is the same as (8.1). However, we indicate the velocity of the particle here by $v_a(0)$ rather than v_r to emphasize that we are dealing with the absolute wind speed rather than the speed relative to motion of the solid body; the air parcel is considered as though it were attached to the ground below it, meaning it is assumed to be stationary with respect to Earth. With R given in centimeters (6.378×10^8 cm), the speed calculated using (8.3) has units of cm sec^{-1}.

As discussed in Chapter 2, the quantity $2\pi/(8.64 \times 10^4 \text{ sec})$ in (8.3) is known as the **angular velocity**, written as Ω. It defines the rate at which the angular position of a particle changes with time, as illustrated in Figure 8.12. The angular velocity for Earth is equal to 7.3×10^{-5} sec^{-1}. As indicated by equation (8.3) (see also Section 2.11), the speed corresponding to solid body rotation at the equator is given by

$$v_r(0) = \Omega R \qquad (8.4)$$

At every other latitude, the solid body rotation is slower, with speeds given (see v_r in Table 8.1) by

$$v_r(\lambda) = \Omega R \cos \lambda \qquad (8.5)$$

$R \cos \lambda$ is always less than or equal to R, and as λ approaches 90° (the poles), $\cos \lambda$ approaches zero. Hence, the speed for solid body rotation decreases from the equator to the poles.

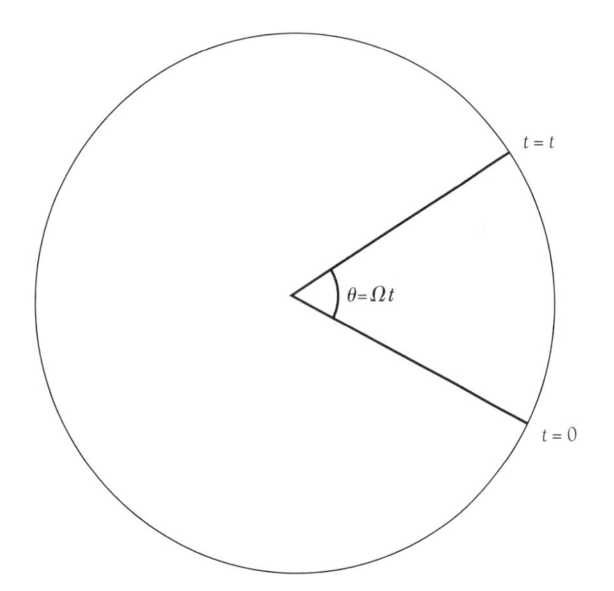

Figure 8.12 Change in the angular position of a rotating parcel with time. For a parcel rotating at angular velocity Ω, the angle θ traced out in time t is equal to Ωt.

Example 8.3: Find the angular momentum of an air parcel with a mass of 1 g at rest with respect to Earth at the equator (latitude 0°).

Answer: Recall that

$$L(\lambda) = v_r(\lambda) R \cos \lambda \times 1 \text{ g} \qquad (8.2)$$

Then

$$L(0) = v_r(0) R \cos(0)(1 \text{ g}) \qquad (8.6)$$
$$= v_r(0) R (1 \text{ g})$$

$$L(0) = \Omega R^2 (1 \text{ g}) \qquad (8.7)$$
$$= (7.3 \times 10^{-5} \text{ sec}^{-1})(6.38 \times 10^8 \text{ cm})^2 (1 \text{ g})$$
$$= 2.97 \times 10^{13} \text{ g cm}^2 \text{ sec}^{-1},$$

where the subscript r is included to emphasize that the quantity refers to an air parcel stationary with respect to Earth (or, solid body rotation). ■

Suppose that the air parcel with a mass of 1 g, which we assumed to be initially at rest with respect to Earth's equator, begins to move northward, as would be the case for air in the upper branch of the Hadley circulation. Its angular momentum at the equator would be given by equation (8.7). For conservation of angular momentum to hold, the value of L at λ must equal the initial value appropriate for solid-body rotation at the equator.

Thus, the air parcel at other latitudes must develop a velocity with respect to Earth. Using the value for $v_r(0)$ in (8.4) and equating (8.2) with (8.7), we find,

$$L(\text{latitude } \lambda) = L(\text{equator})$$

$$v_a(\lambda) R \cos \lambda = \Omega R^2 \qquad (8.8)$$

$$v_a(\lambda) = \frac{\Omega R}{\cos \lambda} \qquad (8.9)$$

Example 8.4: Find the absolute speed of this air parcel at latitude 45°.

Answer:

$$v_a(\lambda) = \frac{\Omega R}{\cos \lambda}$$

$$= \frac{(7.3 \times 10^{-5}\ \text{sec}^{-1})(6.38 \times 10^{8}\ \text{cm})}{(0.707)}$$
$$= 6.58 \times 10^{4}\ \text{cm}^2\ \text{sec}^{-1}$$ ■

Values of $v_a(\lambda)$ are given for different latitudes in Table 8.2. This value of $v_a(\lambda)$ reflects a combination of Earth's velocity and the air parcel's velocity with respect to Earth. Table 8.2 also includes values for the corresponding westerly wind speed (the magnitude of the west-east velocity as measured by an observer fixed with respect to Earth), denoted simply as $v(\lambda)$, obtained by subtracting from $v_a(\lambda)$ the speed, $v_r(\lambda)$, associated with planetary rotation given by (8.5):

$$v(\lambda) = v_a(\lambda) - v_r(\lambda) \tag{8.10}$$

$$v(\lambda) = \frac{\Omega R}{\cos \lambda} - \Omega R \cos \lambda \tag{8.11}$$

The results in Table 8.2 indicate that an air parcel moving northward from the equator and conserving angular momentum will tend to pick up speed eastward as it moves to higher latitudes. Viewed by an observer on Earth, the particle appears to turn to the right of its direction of motion. According to Newton's law of motion, equation (2.1), a change in velocity implies an acceleration and must be accompanied by a force. There is no obvious force we can point to as responsible for the apparent eastward drift of the northward moving air parcel. Fortunately, there is a straightforward trick that can be used to account for the complications arising for motion in a rotating frame. This is done by allowing for an additional force, the Coriolis force mentioned above. The simple expression of Newton's law of motion for an inertial frame is preserved for a rotating frame, when we allow for effects of the Coriolis force. The following discussion is intended to elucidate the nature and form of the Coriolis force.

As we have seen, an air parcel moving northward from the equator appears to turn eastward. Similarly, an air parcel moving southward toward the equator would appear to reverse the path followed by the particle moving northward; it would turn westward. The effect of rotation in both cases is to cause the air parcel to turn to the right with respect to its direction of motion: eastward for the particle moving northward, westward for the particle moving southward. This defines the first important property of the west-east Coriolis force: *it acts to the right of the direction of motion in the Northern Hemisphere.*

We can apply the same argument for **meridional** (north-south) motion in the Southern Hemisphere. An air mass moving southward from the equator acquires a component of eastward velocity with respect to Earth; it turns to the left with respect to its direction of motion. An air parcel moving northward in the Southern Hemisphere drifts to the west; again it appears to turn to the left. *The Coriolis force in the* **zonal** *(west-east) direction acts to the right of the direction of motion in the Northern Hemisphere, to the left in the Southern.*

Consider an air parcel traveling southward across the equator, moving from the Northern to the Southern Hemisphere. The Coriolis force on the parcel acts westward in the Northern Hemisphere, eastward in the Southern. There is no reason to expect an abrupt change in the force as the parcel crosses the equator. In fact, there is a smooth variation in the magnitude and direction of the force with latitude. The change in direction crossing the equator occurs because the west-east Coriolis force vanishes at the equator; its magnitude there exactly equals zero.

Now consider how we might expect the zonal (west to east) component of the Coriolis force to vary with meridional (south to north) velocity. If the magnitude of the absolute west-east speed at latitude λ_1 is given by $v_a(\lambda_1)$ and if angular momentum is conserved, the absolute value of the speed at latitude λ_2 may be obtained by equating the values for angular momentum at the two latitudes using the appropriate forms of equation (8.1):

$$L(\lambda_1) = L(\lambda_2) \tag{8.12}$$

$$v_a(\lambda_1) R \cos \lambda_1 = v_a(\lambda_2) R \cos \lambda_2 \tag{8.13}$$

It follows that

$$v_a(\lambda_2) = v_a(\lambda_1) \frac{\cos \lambda_1}{\cos \lambda_2} \tag{8.14}$$

Using the expression for the solid body rotation speed given by equation (8.5), equation (8.14) can be rewritten to yield an expression relating the zonal wind speed at λ_2 to the corresponding quantity at λ_1. First, rewrite equation (8.10)

$$v_a(\lambda) = v(\lambda) + \Omega R \cos \lambda \tag{8.15}$$

and substitute into equation (8.14)

$$v(\lambda_2) + \Omega R \cos \lambda_2 = [v(\lambda_1) + \Omega R \cos \lambda_1] \frac{\cos \lambda_1}{\cos \lambda_2} \tag{8.16}$$

$$v(\lambda_2) = [v(\lambda_1) + \Omega R \cos \lambda_1] \frac{\cos \lambda_1}{\cos \lambda_2} - \Omega R \cos \lambda_2 \tag{8.17}$$

The change in zonal velocity (as measured by an observer fixed on Earth) estimated using (8.17) is independent of the magnitude of the meridional velocity; it depends only on the magnitude of the initial zonal velocity and the values of the initial and final latitudes. If the meridional speed is large, the change in zonal speed will occur rapidly; the time required for an air parcel to move from latitude λ_1 to λ_2 is inversely proportional to the magnitude of the meridional wind speed. We wish to interpret the change in magnitude

Table 8.2 Values for absolute speed ($v_a(\lambda)$) and for the west-east speed ($v(\lambda)$) as measured by an observer fixed to Earth at latitude λ

λ	$v_a(\lambda)$	$v(\lambda)$
0	4.65×10^4	0
15	4.81×10^4	3.2×10^3
30	5.37×10^4	1.3×10^4
45	6.58×10^4	3.3×10^4
60	9.32×10^4	7.0×10^4
75	1.80×10^5	1.7×10^5

NOTE: *Speeds are quoted in units of cm sec^{-1} and are calculated assuming that the air parcels move northward from the equator conserving angular momentum.*

of the zonal wind as the result of a force operating in the zonal direction in the rotating coordinate system. Force, according to Newton's law of motion, is proportional to the *rate of change* of velocity (see equation 2.1). It follows that the *magnitude of the west-east Coriolis force should be proportional to the magnitude of the meridional velocity.*

The change in zonal wind directly depends on the speed of planetary rotation. This is most easily seen by considering an air parcel in solid body rotation at λ_1, moving to λ_2. Setting $v(\lambda_1)$ equal to zero in (8.17), we find

$$v(\lambda_2) = [0 + \Omega R \cos \lambda_1] \frac{\cos \lambda_1}{\cos \lambda_2} - \Omega R \cos \lambda_2 \quad (8.18)$$

$$v(\lambda_2) = \Omega R \left[\frac{(\cos\lambda_1)^2}{\cos\lambda_2} - \cos \lambda_2 \right]$$

It follows that the zonal component of the Coriolis force should be proportional to the angular velocity describing the rate of rotation of the planet, Ω. This completes our qualitative discussion of the zonal component of the Coriolis force. We conclude that the force should be proportional to the meridional velocity and to the value of Ω; in addition, the magnitude of the force must vanish at the equator. A rigorous analysis indicates that the force per unit mass satisfying these constraints is given by

$$F_c(w \rightarrow e) = 2\Omega v_m \sin \lambda, \quad (8.19)$$

where F_c $(w \rightarrow e)$ denotes the west-to-east Coriolis force and v_m denotes the magnitude of the meridional wind speed defined as positive for motion directed toward higher latitude. Recall that the acceleration is force per unit mass. Technically, the right-hand side of equation (8.19) is the Coriolis acceleration, or, equivalently, the force applied to unit mass.

8.4 Centrifugal Force and the North-South Component of the Coriolis Force

We turn our attention now to forces influencing motion in the meridional direction on a rotating planet. Consider a particle moving at constant speed, v, on the circumference of a circle of radius, R. For the particle to continue in circular motion, it must experience a force directed toward the center of the circle. This is most easy to visualize by supposing that the particle is attached to a string being whirled around, as discussed in Chapter 2. The force toward the center of the circle is provided by the tension in the string. If this were the only force operating on the particle, we might expect, by Newton's law of motion, that the particle would accelerate toward the center of the circle. The tension in the string is opposed, however, by a force of equal magnitude in the opposite direction to the force exerted by the string. This is the force you feel if you try to ride a bicycle at high speed on a circular track; you have a tendency to fall off the bike to the outside of the track if you fail to compensate by leaning to the inside. The force that tends to drive a particle in circular motion away from the center of its circular path is known as the **centrifugal force.** (This force was introduced in Chapter 2.) The centrifugal force per unit mass, F_{centr}, for circular motion is proportional to the square of the absolute speed and inversely proportional to the radius of the circle:

$$F_{centr} = \frac{v_a^2}{R} \quad (8.20)$$

Consider an air parcel of unit mass moving in the west-east direction at latitude λ. The speed of the parcel in the inertial system, v_a, includes a component due to planetary rotation given by (8.4), in addition to a contribution associated with motion relative to Earth, represented by a zonal wind of speed v. The parcel is moving in a circle of radius R cos λ, as indicated in Figure 8.10. The centrifugal force on the parcel is given by

$$F_{centr} = \frac{(\Omega R \cos \lambda + v)^2}{R \cos \lambda}$$

$$= \Omega^2 R \cos \lambda + 2\Omega v + \frac{v^2}{R \cos \lambda} \quad (8.21)$$

Evaluating F_{centr} with the wind speed set equal to 10^3 cm sec^{-1}, with λ taken equal to 60°, we find that the three terms in (8.21) have rather different magnitudes; they are equal to 1.7 (term 1, $\Omega^2 R \cos \lambda$), 0.15 (term 2, $2\,\Omega\, v$) and 3.1×10^{-3} (term 3, $v^2/R \cos \lambda$), expressed in units of cm sec^{-2}.

Little error is introduced by neglecting the third term. The first term is independent of the magnitude of the zonal wind; it depends simply on latitude and on the angular velocity and radius of the planet. The second term defines the component of the centrifugal force associated with motion relative to Earth.

The first term in (8.21) describes the centrifugal force associated with the rotation of the solid planet and is usually incorporated in the acceleration of gravity by modifying the value of g. The centrifugal force due to rotation of the solid body is responsible for a difference in the equatorial, as compared with the polar, radius of Earth: the equatorial radius is larger than the polar radius by about 21.3 km, as a result of the dependence of the centrifugal force on latitude.

For present purposes we are primarily interested in the force due to motion of an air parcel relative to Earth. The magnitude of this force is given, according to (8.21), by

$$F^r_{centr} = 2\Omega v, \quad (8.22)$$

where the superscript *r* is included to emphasize that the force in (8.22) refers to motion relative to Earth. The centrifugal force due to zonal motion operates in the direction indicated in Figure 8.13. The component of the force southward is obtained using the cosine rule discussed in Chapter 2: by multiplying the magnitude of the force given in (8.22) by the cosine of the angle separating the southerly direction from the actual orientation of the force, indicated as θ in Figure 8.13. The angle θ is the complement of the latitude angle λ; meaning, θ equals $(\pi/2 - \lambda)$. Using an elementary trigonometric identity, $\cos(\pi/2 - \lambda) = \sin \lambda$, the southerly component of the centrifugal force due to west-east relative motion may be expressed in the form

$$F^r_{centr}(n \rightarrow s) = 2\Omega v \sin \lambda \quad (8.23)$$

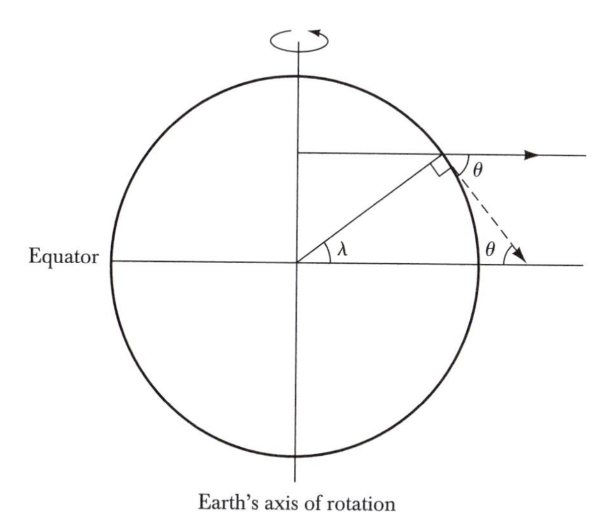

Figure 8.13 The direction of the centrifugal force at latitude λ for a particle rotating eastward. The component of the force southward, shown by the dashed line, is obtained by multiplying the magnitude of the force by $\cos\theta = \sin\lambda$.

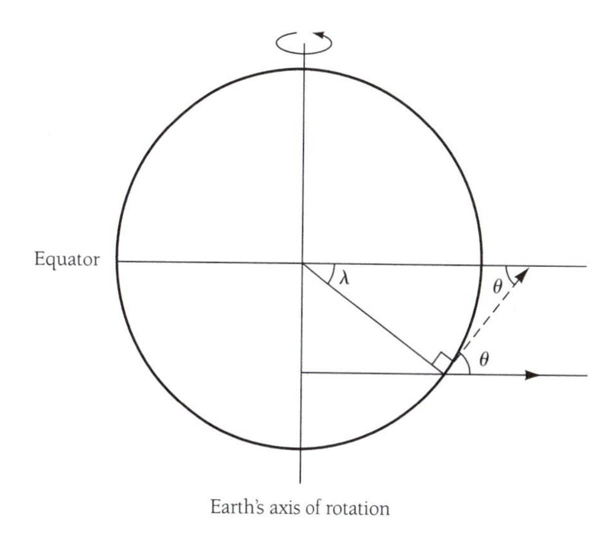

Figure 8.14 Same as in Figure 8.13 but for the Southern Hemisphere.

Note that the expression for the north-south component of the centrifugal force due to west-east relative motion is identical to the expression obtained earlier for the zonal component of the Coriolis force, with the meridional speed in (8.19) replaced by the zonal speed in (8.23). The meridional component of the centrifugal force associated with zonal motion is equivalent to the meridional component of the Coriolis force.

Suppose that the zonal motion represents an easterly wind, rather than the westerly wind treated above. The analysis is identical for this case, with v replaced by $-v$; the meridional component of the Coriolis force for an easterly wind operates northward. The Coriolis force, both for meridional and zonal motion, tends to deflect air parcels to the right in the Northern Hemisphere. It is easy to show that it acts to deflect parcels in the Southern Hemisphere to the left, as indicated for zonal motion in Figure 8.14 and as was explicitly demonstrated for meridional motion previously.

The functional form of the Coriolis force is identical for zonal and meridional motion. The force operates in both cases at right angles to the direction of motion: to the right in the Northern Hemisphere, to the left in the Southern. The vector describing horizontal motion in an arbitrary direction at latitude λ may be written as a vector sum of the appropriate components for meridional and zonal motion. It follows that the Coriolis force for horizontal motion in an arbitrary direction should operate, as it does for meridional and zonal motion separately, at right angles to the direction of motion—to the right in the Northern Hemisphere, to the left in the Southern—and that the magnitude of the force is given by $2\Omega v \sin\lambda$, where v defines the magnitude of the velocity vector (the wind speed).

$$F = 2\Omega v \sin\lambda \tag{8.24}$$

Expressed in vector notation, the Coriolis force is equal to $2\underline{\Omega} \times \underline{v}$.

8.5 Hadley's Circulation Revisited

Table 8.2 summarizes values for the zonal wind speed computed as a function of latitude for air moving northward in the upward branch of the Hadley circulation. It was assumed, in constructing the Table, that air was initially at rest with respect to Earth's equator and that angular momentum was conserved as the air moved northward. The zonal wind speed rapidly increases with latitude, reaching values comparable to the speed of sound at a latitude of about 40°. It would be impossible to maintain a wind speed higher than about a tenth of the sound speed. The Hadley circulation terminates in the present atmosphere at a latitude of about 30°. Air sinks to the surface and the return flow to the tropics takes place near the ground, deflected to the west by the Coriolis force. This accounts for the phenomenon of the trade winds. Hadley was successful in his primary objective to explain "the causes of the General Trade-Winds." His explanation for the midlatitude westerlies is less satisfactory but in no way detracts from the significance of his contributions to meteorology. His achievement is all the more astounding when we recall that Hadley was a contemporary of Newton, that he preceded Coriolis by more than a century, and that his analysis was carried out using qualitative reasoning without benefit of complex mathematics.

8.6 The Concept of Geostrophy

Our discussion to this point has emphasized the role of four basic forces in regulating atmospheric structure and dynamics. The vertical structure of the atmosphere reflects a balance between the vertical pressure gradient pushing air upward and the force of gravity pulling it downward. This balance is expressed through the barometric law, equations (7.14) and (7.15). Motions in the horizontal plane are regulated primarily by the horizontal pressure gradient and by the Coriolis force, with a contribution near the surface from the force of friction. The wind resulting from a balance between the pressure gradient and Coriolis force is known as the geostrophic wind (from *geo*, meaning "earth" and *strophic*, meaning "turning"). The geostrophic wind provides an excellent approximation to the real wind at mid to high latitudes (remember that the Coriolis force vanishes at the equator), for motions of moderately large spatial scale (in the absence of sharp gradients in pressure or temperature, or exceptionally high-wind speeds) at altitudes sufficiently far above the surface such that effects of friction may be neglected (a few kilometers or higher). We now discuss how an air parcel, initially at rest, may be expected to move in response to a pressure gradient and how its motion evolves to achieve the state of geostrophic balance.

Consider a parcel of unit volume in the Northern Hemisphere, with a gradient of pressure directed from left to right in a horizontal plane, as indicated in Figure 8.15a. Pressure to the left is higher than pressure to the right. As a result, the parcel is subject to a net force to the right with an associated acceleration, as illustrated in Figure 8.15b. As the parcel begins to move, it experiences the Coriolis force and is turned to the right with respect to its initial direction of motion (i.e., it turns toward the reader, as shown in Figure 8.15c). The parcel rotates until it moves at right angles to its initial direction of motion, as illustrated in Figure 8.15d. The Coriolis force now opposes the force associated with the pressure gradient. The parcel adjusts its speed such that the Coriolis force exactly cancels the force associated with the pressure gradient. The net force on the parcel vanishes, the acceleration drops to zero, and the parcel moves toward the reader at a constant speed determined by the magnitude of the left-right pressure gradient.

Consider a hypothetical spatial variation of pressure, as illustrated for a fixed altitude in Figure 8.16. A parcel of air that is initially stationary will first tend to move in the direction of lower pressure (toward the center of the low) at right angles to the pressure contour on which it is initially located. As it moves, it turns to the right in the Northern Hemisphere; it begins to circle in a counterclockwise direction around the low. In geostrophic balance, motion follows contours of constant pressure. Where contours of constant pressure are relatively far apart, motion is slow (the speed required for the Coriolis force to balance the pressure gradient is low); where contours are closely spaced, speeds are high. The pattern of the pressure field provides an instantaneous snapshot of the field of motion. Air in the Northern Hemisphere moves maintaining high pressure to the right of the direction of motion. The airflow is in a clockwise direction around a low in the Southern Hemisphere, with high pressure to the left. Flow around a low pressure system is said to be **cyclonic.**

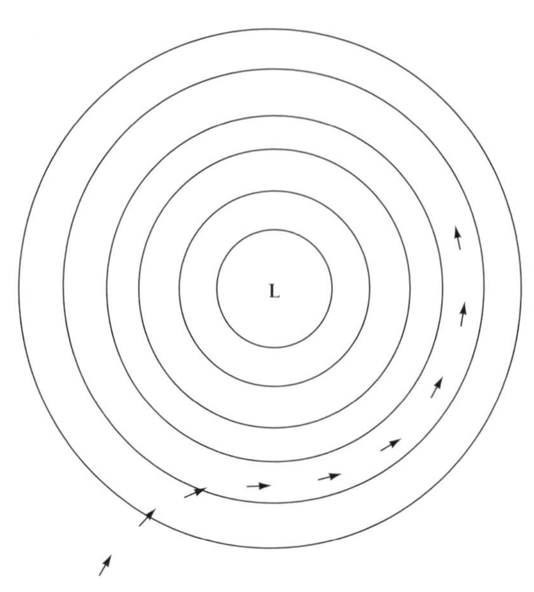

Figure 8.16 Track of an air parcel in the vicinity of a low pressure region in the Northern Hemisphere. The parcel is initially at rest but then adjusts to the pressure gradient force and the Coriolis force to achieve geostrophic balance.

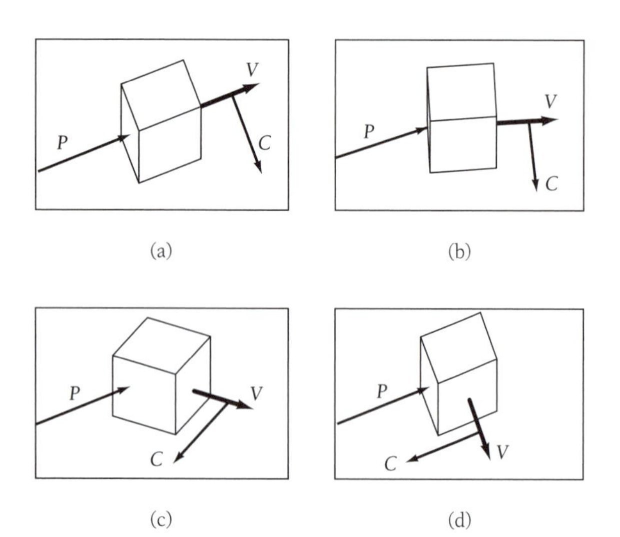

Figure 8.15 Effect of the Coriolis force on a particle moving initially in the Northern Hemisphere in the direction of the pressure gradient (top left). The Coriolis force turns the particle progressively to the right as it picks up speed. The final, force-balanced, geostrophic condition is illustrated at the bottom right.

Figure 8.17 illustrates the pattern of flow around a high pressure system. An air parcel that is initially stationary first tends to move away from the high, as indicated. Turning to the right in the Northern Hemisphere under the influence of the Coriolis force, it proceeds to move clockwise around the high, keeping high pressure to the right and following contours of constant pressure in the same fashion as for motion around the low. The pattern is reversed in the Southern Hemisphere: motion around a high is in a counterclockwise direction in this case, with high pressure to the left. The circulation of air around a high is said to be **anticyclonic.**

Meteorological data are presented most frequently not as maps of pressure at a fixed altitude (as in Figures 8.2 and 8.3) but as contours *giving altitudes corresponding to fixed values of pressure.* A typical situation is illustrated in Plate 3 (top)(see color insert). Heights of the 500 mb surface for 7 A.M. (EST) on 13 March 1993, varied from a low of 5160 m over northeastern Canada to a high of about 5800 m over the southern

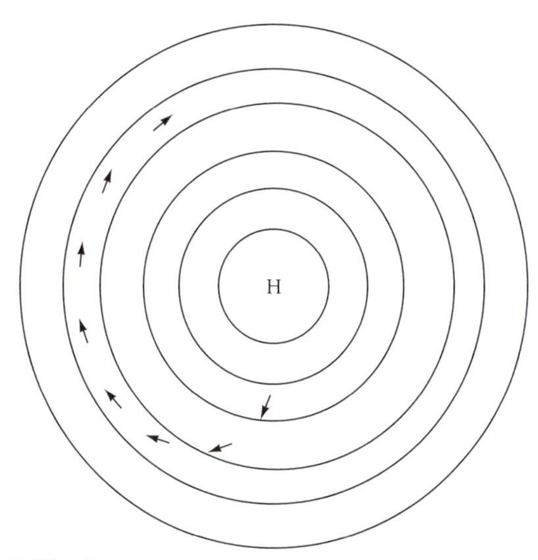

Figure 8.17 Same situation as in 8.16, except that the parcel is in the vicinity of a high pressure region in the Northern Hemisphere.

United States. A high value for altitude on this map would correspond to a region of high pressure on a map displaying variations in pressure at fixed altitude, as may be seen in the schematic profiles of pressure versus altitude in Figure 8.18. Suppose that the profile of pressure versus altitude is given for one location by curve 1, while the profile over a neighboring region is summarized by curve 2. A map giving heights corresponding to a pressure of p_1 would indicate that the pressure surface was high (altitude z_1) over region 1, low (z_2) over region 2. At the same time, as anticipated, the map of pressure at altitude z_1 would indicate a region of high pressure (p_1) over 1, with low pressure (p_2) over region 2.

We can think of the contours in Plate 3 (top) as defining a complex, three-dimensional surface. The pressure is constant everywhere on the surface, equal to 500 mb. Altitudes vary in a regular fashion as indicated; there are high regions to the west and east, with a depression in the central portion of the continent. The total relief amounts to about 500 m. Meteorologists, using an obvious geometric analogy, refer to elongated high and low regions on a map defining heights of the pressure field for a given value of pressure as *ridges* and *troughs*. As required by the barometric law, temperatures are relatively high in air underlying a ridge, low in the vicinity of a trough. Localized highs, such as the ridges in Plate 3 (top), would be associated with anticyclonic motion, while cyclonic motion would dominate in the vicinity of regional lows or depressions.

Geostrophic wind is directed along contours of constant height on a fixed pressure surface. The wind speed is proportional to the gradient of the height field, just as it is proportional to the pressure gradient on a constant-height map. The more closely bunched the contours in a map (such as presented in Plate 3), the larger the gradient of the pressure-height field and the higher the speed of the geostrophic wind. Wind speeds, expressed in **knots** (1 knot = 1 nautical mile per hour = 1.85 km per hour), are included in Plate 3, using a compact notation employed by meteorologists as explained in Figure 8.19. Winds on a constant-pressure surface blow in such a direction as to maintain ridges to the right in the Northern Hemisphere and to the left in the south, similar to the orientation with respect to pressure for geostrophic motion on a constant-height surface as discussed above.

It is relatively easy to calculate the speed of the geostrophic wind for a fixed height, given information on the magnitude of the pressure gradient and the mass density ρ. Assume that the situation depicted in Figure 8.20 applies to the Northern Hemisphere. The line ABC is drawn as normal (perpendicular) to the pressure contour at B and represents a distance of 100 km. The pressure falls by 4 mb in moving from A to C. Assume that the drop in pressure is linear with distance along AC: the gradient of pressure at B equals 4×10^{-7} mb cm^{-1}, obtained by dividing the total change in pressure from A to C by the distance from A to C. This defines the magnitude of the pressure gradient $\Delta p/\Delta s$, where Δp denotes the change in p taking place over distance Δs, expressed in units of mb cm^{-1}. Converting to *cgs* units (1 mb equals 10^3 dyne cm^{-2}), the force on a cubic centimeter due to the pressure gradient equals 4×10^{-4} dyne cm^{-3}.

The Coriolis force on the same cubic centimeter has magnitude $2\rho\Omega\ v_g \sin\lambda$ (equation 8.24), where v_g is the wind speed and λ is latitude as before. Equating the pressure force with the Coriolis force gives

$$\frac{\Delta p}{\Delta s} = 2\Omega\rho\ v_g \sin\lambda, \tag{8.25}$$

where the subscript g is included to emphasize that (8.25) reflects the assumption of geostrophy.

Solving for v_g, we find

$$v_g = \frac{1}{2\Omega\rho\sin\lambda}\left(\frac{\Delta p}{\Delta s}\right) \tag{8.26}$$

Example 8.5: Find the speed and direction of the geostrophic wind at point B in Figure 8.20, for a latitude of 30° N, with ρ taken as 7×10^{-4} g cm^{-3}, as appropriate for a pressure of about 500 mb.

Answer:

$$v_g = \frac{4 \times 10^{-4}\ \text{dyn cm}^{-3}}{2(7.3 \times 10^{-5}\ \text{sec}^{-1})(7 \times 10^{-4}\ \text{g cm}^{-3})(5 \times 10^{-1})}$$
$$= 7.8 \times 10^3\ \text{cm sec}^{-1} \tag{8.27}$$

The speed of the geostrophic wind is 78 m sec^{-1}, oriented along the pressure contour at B in the direction indicated by the arrow. If the configuration of the pressure contours in Figure 8.20 were applied to the Southern rather than to the Northern Hemisphere, the wind speed would be the same; the velocity would be reversed, however, and the wind would blow from right to left. ■

8.7 Effects of Friction

Consider now the effect of friction on the circulation around a high pressure system near the surface, as illustrated in Figure 8.21a. In the absence of friction, air in the

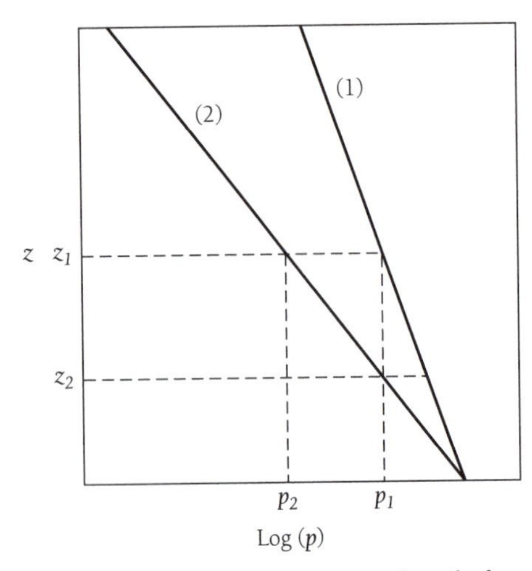

Figure 8.18 Profiles of pressure versus altitude for two different positions, labeled (1) and (2). The heights of the surface corresponding to a pressure of p_1 is z_1 at (1) and z_2 at (2). Pressure at altitude z_1 is higher at (1) where it is equal to p_1, than it is at (2), where it is equal to p_2.

Northern Hemisphere would describe a clockwise path around the high, following a contour of constant pressure. Orientations of the pressure gradient and Coriolis forces for a representative air parcel are indicated in the diagram. Friction leads to a decrease in the magnitude of the wind speed, with an associated reduction in the magnitude of the Coriolis force. Due to the weakening of the Coriolis force by friction, the geostrophic balance is disrupted. This favors the pressure force pushing the parcel in the direction of the gradient. The parcel spirals away from the center of the high, describing the path schematically indicated by the dashed line.

A similar correction occurs for frictional motion around a low, as illustrated in Figure 8.21b. In this case, friction causes the parcel to spiral inward. A typical trajectory for the Northern Hemisphere is indicated by the dashed line. Convergence of air toward the center of the low tends to reduce the magnitude of the pressure gradient, building up mass toward the center of the low and cutting off the flow. This tendency may be offset, however,

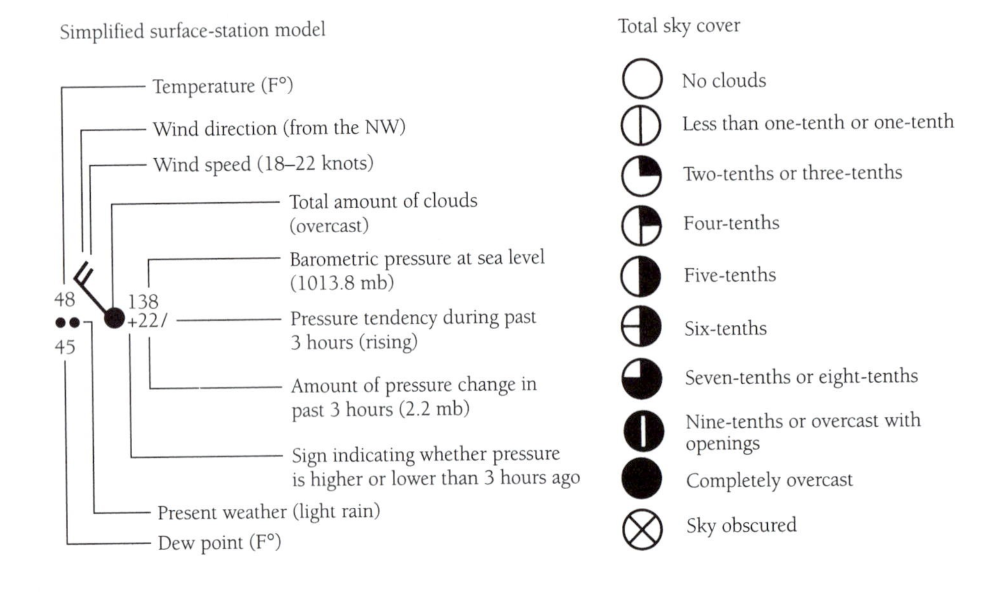

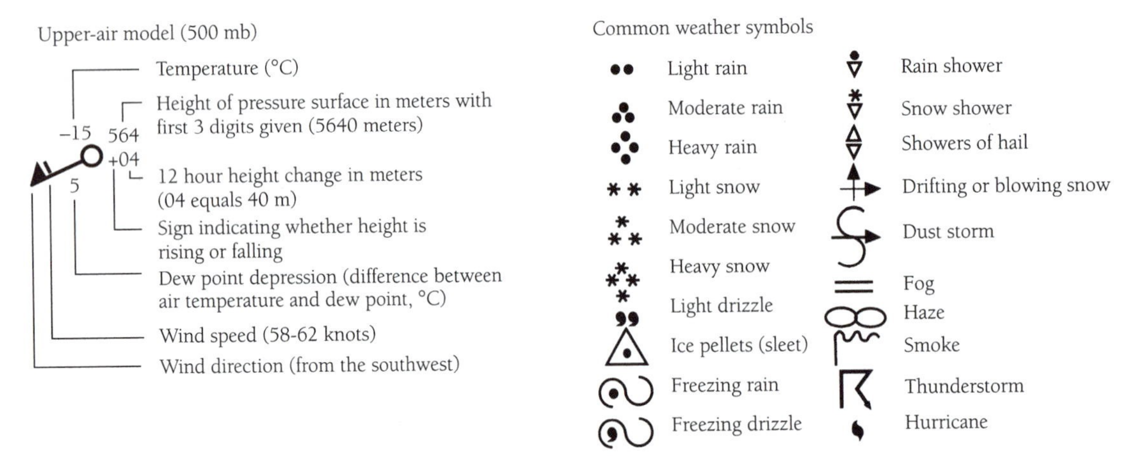

Figure 8.19 Notation employed by meteorologists on their maps. Source: Ahrens 1994.

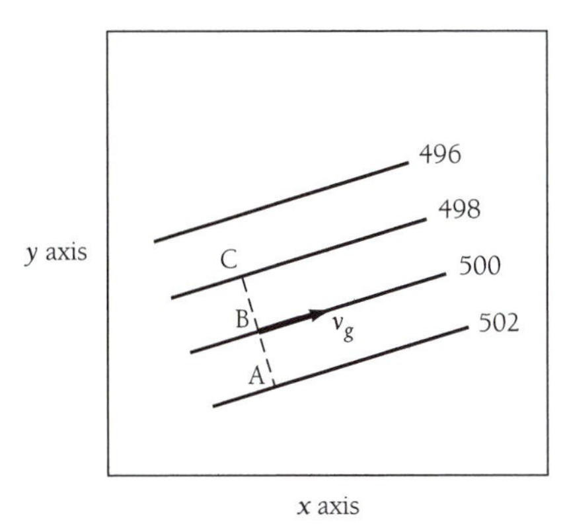

Figure 8.20 Schematic map of pressure surfaces for an altitude of about 6 km. It is assumed that the x and y axes measure distances to the east and north, respectively.

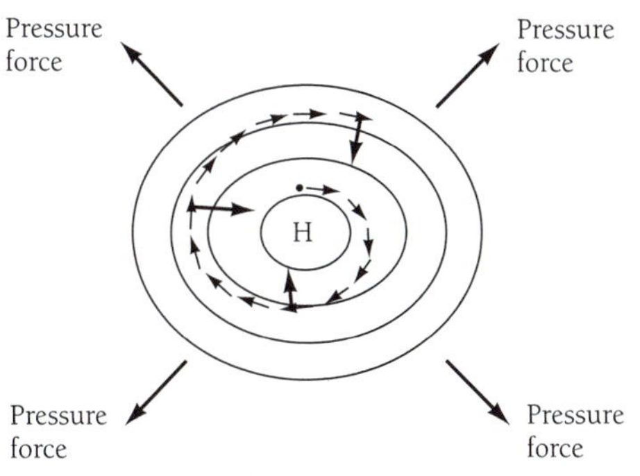

Figure 8.21a Schematic illustration of the trajectory of an air parcel at the surface in the Northern Hemisphere in the vicinity of a high pressure system, showing the effect of friction. Direction of Coriolis force indicated by arrows on trajectory.

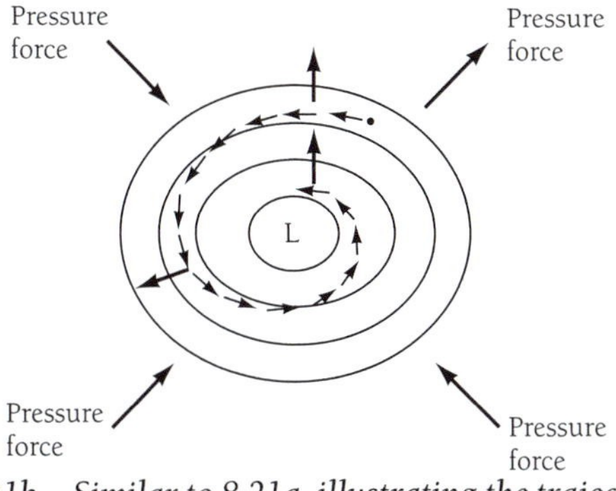

Figure 8.21b Similar to 8.21a, illustrating the trajectory of an air parcel at the surface in the Northern Hemisphere in the vicinity of a low pressure system. Direction of Coriolis force indicated by arrows on trajoectory.

by passage of a low pressure system aloft. In this case, air drawn to the low will be forced to rise, resulting in unsettled weather. If the low pressure system aloft is well developed, and if the moisture content of the surface air is relatively high, the rising motion can result in intense precipitation. Hurricanes provide an extreme example of this phenomenon. The change in pressure associated with Hurricane Bob, a violent storm that caused major damage to coastal areas of New England in August 1991, is illustrated in Figure 8.22. Conversely, passage of a high pressure system aloft will cause the atmosphere to descend, with an outflow of surface air. Descending air is normally dry. Temperatures will rise in this case as a result of adiabatic compression. High pressure systems are usually associated with warm weather and clear skies, especially during summer.

8.8 Further Perspectives on the General Circulation

As we have seen, Hadley's model, modified to account for conservation of angular momentum rather than absolute speed, provides a reasonable representation of the general circulation of the atmosphere in the tropics, for latitudes less than 20–30°. It accounts for the trade winds but fails to provide an explanation for the prevailing surface westerlies at midlatitudes. Figure 8.23 presents an alternate model proposed more than a century later, in 1856, by an English schoolteacher named William Ferrel.

Ferrel's model envisages a circulation with three distinct loops or cells in each hemisphere. The picture in Figure 8.23 indicates a circulation system in the tropics similar to the model proposed by Hadley for the globe. The tropical cells extend to the region of the Horse Latitudes, at about 30°. The cells at intermediate latitudes, 30–60°, however, have air revolving in a sense opposite to the flow in the tropics. Air rises on the poleward branch of the intermediate cell. It flows southerly in the upward branch in the Northern Hemisphere, returning to the surface at the southern margin. The loop is completed by northward transport near the ground. The high latitude (60–90°) cells are characterized by sinking motion near the

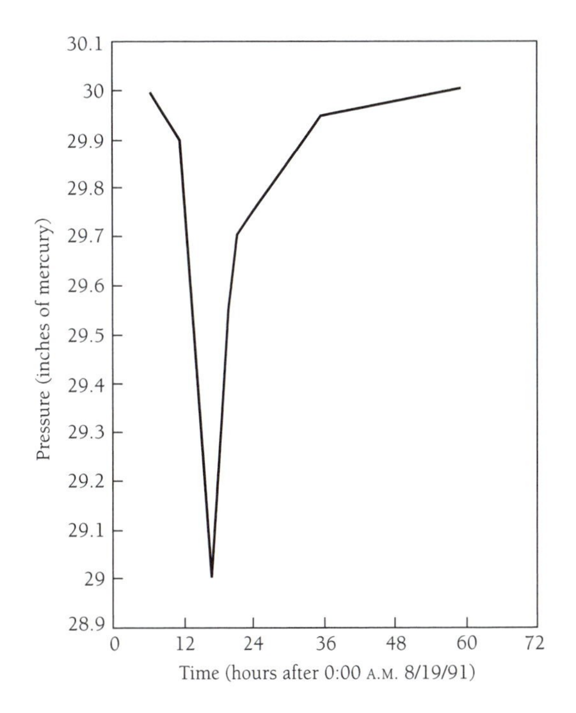

Figure 8.22 Surface pressure vs. time recorded at Ipswich, Massachusettes, during the passage of Hurricane Bob. Data courtesy of Quinn Sloan.

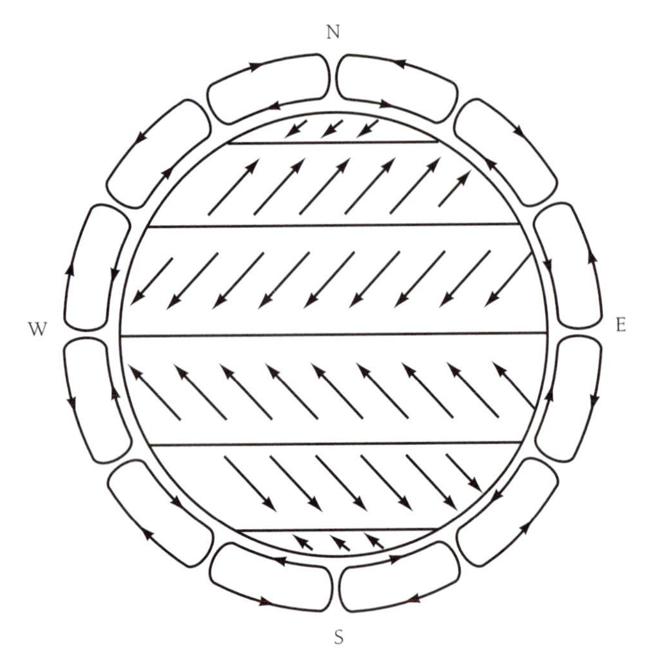

Figure 8.23 Model of the general circulation proposed by Ferrel (1856), illustrating directions of prevailing winds at the surface, with an exaggerated perspective on the vertical components of the motion field.

poles, with circulation in the same sense as for the tropics. Sharp thermal gradients mark the boundaries between the high temperatures of the tropics, the moderate temperatures at midlatitudes, and the cold temperatures near the poles. The boundary between the intermediate and polar cells is known as the **polar front**. The boundary between the tropical and midlatitude cells is associated with the midlatitude **jet stream.**

Ferrel's model was based on an attempt to provide a rationale for three-dimensional features of the general circulation observed at the surface. His model accounts, for example, for the prevailing surface westerlies at midlatitudes. These arise as a consequence of eastward deflection of winds in the lower branch of the intermediate cells. There are problems, however, with the model, particularly with the intermediate cell. The surface westerlies exert an eastward torque on the solid planet (see Section 2.10). This must be accompanied by a transfer of angular momentum from the atmosphere to the surface. A loss of angular momentum by the atmosphere at midlatitudes must be balanced by a supply from other regions, most likely from the tropics. But transport of angular momentum in the intermediate cell is from high to low latitudes, rather than in the reverse direction, as required to balance the sink at the surface. Ferrel's model predicts that the winds aloft at midlatitudes should blow easterly due to effects of the Coriolis force on the equatorward flow. Observations, summarized in Figures 8.24 and 8.25, indicate, however, that westerlies are dominant everywhere except for a limited region in the tropics. There is also a problem with

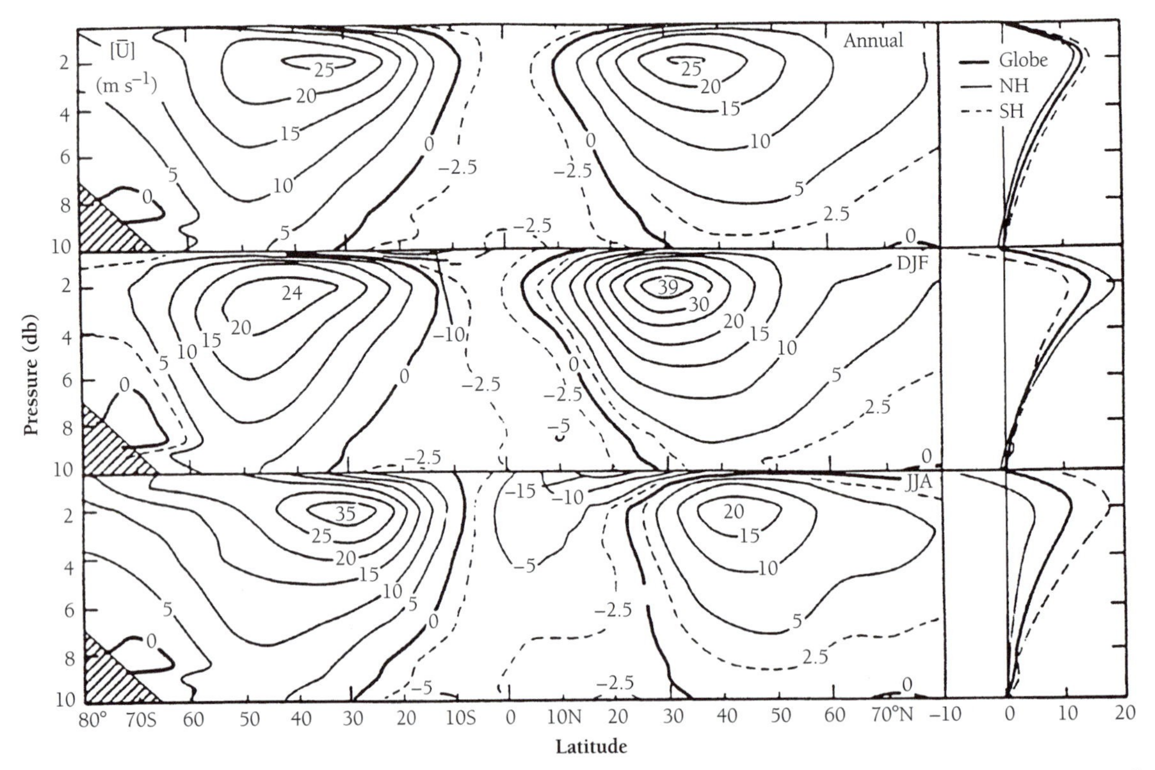

Figure 8.24 Three cross-sectional views of zonal (east-west) winds, showing zonal mean wind velocities (i.e., average wind velocity at each latitude). Wind velocities are given in m/sec^{-1}, showing only the zonal components of the wind speed. Positive values imply that the wind is from the west, negative values imply that wind is from the east. Note that altitude is expressed in terms of pressure. One decibar (db) equals 100 millibars (mb). The top panel shows mean annual values, the middle panel shows mean values for December-January-February (DJF), and the bottom panel shows mean values for June-July-August (JJA). Source: Peixoto and Oort 1992.

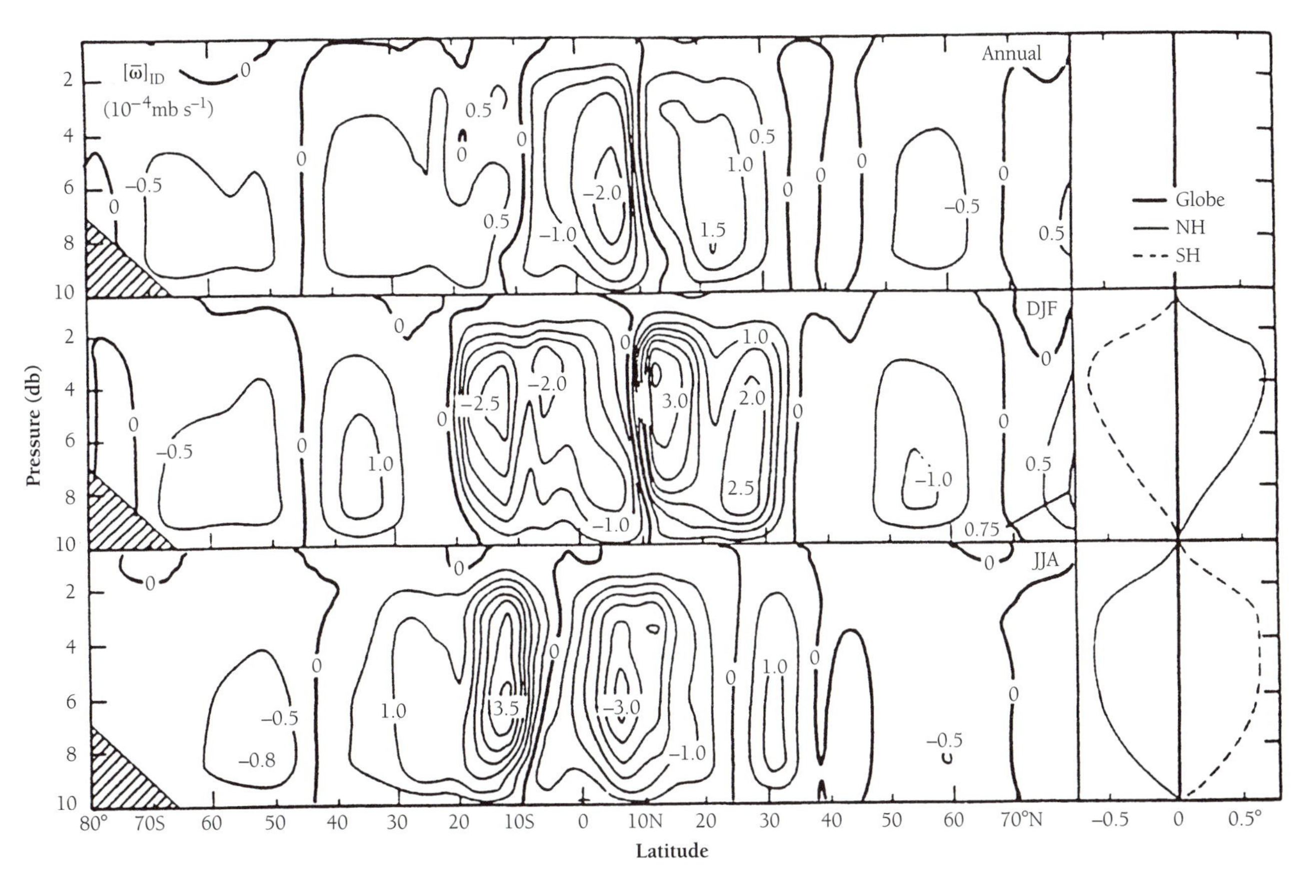

Figure 8.25 Three zonal-mean views of the mass stream function, showing streamlines of air motion in 10^{10} kg sec^{-1}. The depicted circulation cells are called the "mean meridional circulation." The top panel shows mean annual values, the middle panel shows mean values for December-January-February (DJF), and the bottom panel shows mean values for June-July-August (JJA). Note that the top graph is a result of averaging together rather different summer and winter patterns. This somewhat idealized picture resembles the Ferrel circulation seen in Figure 8.24. Source: Peixoto and Oort 1992.

the transport of heat. Rising motion at high latitudes, accompanied by sinking of air at low latitude, would imply a net transport of heat by the intermediate cell from high to low latitude. This is clearly inconsistent with the current understanding of the atmospheric radiative budget, as summarized in Chapter 7. The difficulty is ultimately linked to limitations inherent in the models presented to date, limitations associated primarily with their assumption that the circulation can be represented in terms of a zonal average. Important transports of angular momentum, and of heat and other dynamic variables, can arise due to departures from zonal symmetry as a consequence of longitudinally dependent features of the circulation. We refer to longitudinally variable characteristics of the circulation as **eddies**.

8.9 Transport by Eddies

Suppose that the time-averaged value of the meridional velocity at latitude λ and longitude θ is given by $v(\lambda, \theta)$. We write $v(\lambda, \theta)$ as a sum of two terms: one $[\bar{v}(\lambda)]$ defining the average of $v(\lambda, \theta)$ with respect to time and longitude, the other $[v'(\lambda, \theta)]$ providing a correction to account for the time-averaged value of $v(\lambda, \theta)$ at θ:

$$v(\lambda, \theta) = \bar{v}(\lambda) + v'(\lambda, \theta) \tag{8.28}$$

Assume, for the moment, that the averages over time in (8.28) refer to means with respect to a season. The quantity $\bar{v}(\lambda)$ is the meridional analogue to the time- and longitudinally averaged zonal velocity presented in Figures 8.24 and 8.25. By definition, $\bar{v}(\lambda)$ specifies the average speed with which air is transported across latitude over the course of a season; it is the seasonally averaged meridional speed at latitude λ. On the other hand, $v'(\lambda, \theta)$ represents the time averaged variance of v at (λ, θ) with respect to the zonally (longitudinally) averaged quantity $\bar{v}$.

Consider the simple variation of $v(\lambda, \theta)$ with θ indicated in Figure 8.26a. The time-averaged speed of the meridional wind is set equal to 2 m sec^{-1}, except in the regions denoted by A and B. The northward speed is assumed equal to 3 m sec^{-1} in region A. The wind is taken to blow southward in region B at a speed of 1 m sec^{-1}. We suppose that region A is three times more extensive than region B. It follows that the speed of the average wind associated with the combination of regions A and B is identical to the average for the meridional wind at other longitudes, 2 m sec^{-1} to the north.[3]

Corresponding values of $v'(\lambda, \theta)$ are given in Figure 8.26b. We identify v' as the velocity associated with eddies, or departures from the zonal mean of 2 m sec^{-1}. The contribution of eddies to the transport of air across a latitude circle is exactly zero, by definition. Excess transport northward in

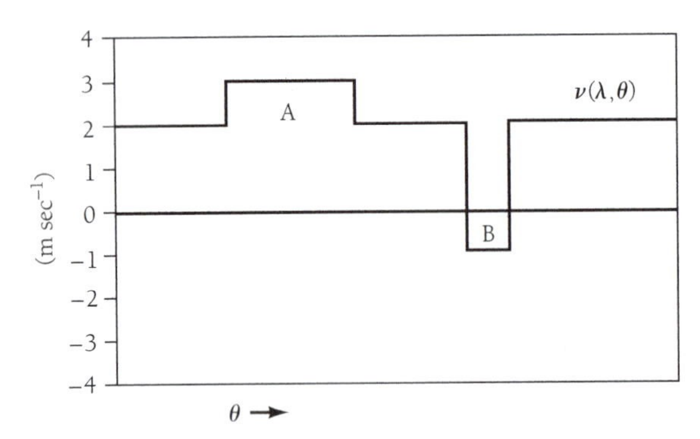

Figure 8.26a The average meridional velocities across a latitude band $[v(\lambda)]$ is $2m\ sec^{-1}$ to the north. Variation of meridional velocity as a function of longitude (θ) at latitude λ. The northward excess of velocity in region A is precisely compensated by a southward excess in B.

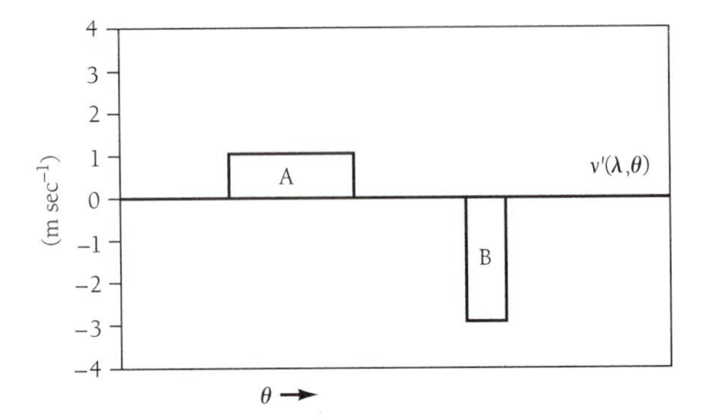

Figure 8.26b Velocity $[v'(\lambda,\theta)]$ corresponding to the difference between the actual velocity and the longitudinal mean as a function of longitude.

region A is balanced by a compensatory return flow in region B. The net transport of mass across a latitude circle is completely specified by the speed of the zonally averaged meridional wind, $\bar{v}(\lambda)$. However, this is not the case for transport of other quantities, such as heat or internal energy, for example.

Suppose that the temperature south of latitude λ is equal to T_1, while northward temperature is given by T_2. Northward transport of internal energy by the eddy at A is proportional to $v'_A T_1$, where v'_A denotes the magnitude of the eddy velocity at A. Internal energy is transferred southward by the eddy at B. The net flux of heat northward associated with the combination of the eddies at A and B flux is proportional to $(3v'_A T_1) + (v'_B T_2)$, where the 3:1 weighting reflects the relative spatial extents of the individual eddies. With our choice of values for v'_A and v'_B, the net northward flux of internal energy by eddies is proportional to $(3T_1 - 3T_2)$. If, as seems reasonable, the temperature southward, T_1, is larger than the temperature northward, T_2, we may anticipate a net transport of internal energy northward due to eddies.

The fluctuations in the meridional velocity shown for regions A and B in Figure 8.26 provide examples of what are known as **standing eddies**, or departures from the time-averaged zonal circulation persisting at particular longitudes, even when data are averaged over periods as long as a season. Standing eddies are often associated with specific geographic features. They can arise, for example, as a result of thermal contrasts between land and sea as a function of longitude at a particular latitude. The Indian Monsoon, induced by intense heating of the Indian subcontinent in summer and by inflow of moist air from the surrounding ocean, provides a useful example of a large-scale standing eddy. In addition to standing eddies, there are also fluctuations in velocity that are more ephemeral. We refer to these more rapid variations as **transient eddies.** They are associated, for example, with the passage of weather systems, high and low pressure systems moving rapidly through particular regions. Seasonally averaged maps show little trace of transient eddies; northward fluctuations in velocity occur just as often as southward fluctuations at particular locations, and net contributions from northward and southward motions tend to cancel when data are averaged over a period as long as a season. Transient eddies, nonetheless, can account for significant transport of dynamical variables such as heat and angular momentum and must be recognized as playing an important role in the general circulation of the atmosphere.

Contributions of transient eddies, stationary eddies, and mean circulation to meridional transport of sensible heat are summarized in Figure 8.27.[4] Note the dominant role of transport by eddies in both hemispheres for both winter and summer at latitudes higher than about 30° (Figure 8.27, panels b and c). Transport by the mean circulation plays an important role in the tropics (Figure 8.27d). Transport by the mean circulation is directed southward over the latitude band 30°N to about 18°N during Northern Hemispheric winter. The direction of transport by the mean circulation switches northward over the latitude band 30° N to about 18°N during Northern Hemispheric summer. The large seasonal swing in the direction of transport by the mean circulation in the tropics as indicated in Figure 8.27d reflects the influence of the Indian, or South Asian, monsoon. The monsoonal flow is northeasterly during Northern Hemispheric winter, southwesterly during Northern Hemispheric summer. Averaged over the course of a year, transport of sensible heat in the tropics is directed equatorward from the hot desert regions characteristic of the descending loops of the Hadley circulation. Accounting for all forms of energy (sensible heat, latent heat, and potential energy), net transport of energy over the course of a year is directed away from the equator in both hemispheres.

8.10 The Angular Momentum Budget of the Atmosphere

A discussion of the angular momentum budget of the atmosphere is necessary for a refined understanding of atmospheric circulation. Hadley's and Ferrel's models were deficient because of their failure to account for a balanced angular momentum budget. Taking the conservation of angular momentum into account is the key to comprehending why westerlies predominate in mid- to high latitudes. Toward this end, we must answer the following questions: what are the sources and sinks of atmospheric angular momentum, how is angular momentum transferred from sources to sinks, and how is this transfer of angular momentum associated with the observed wind patterns?

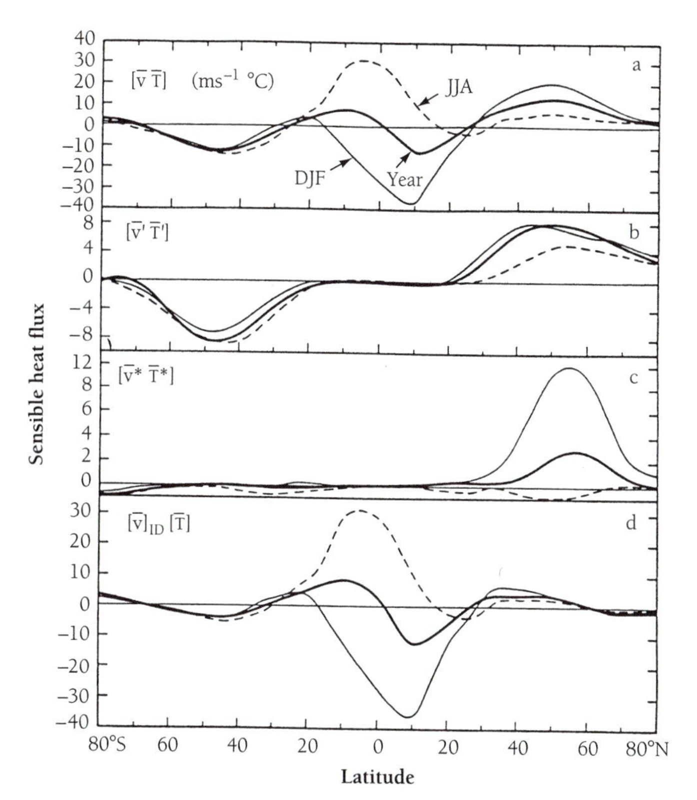

Figure 8.27 The northward flux of sensible heat, averaged over all latitudes and altitudes. The horizontal axis shows latitudes from 80°S to 80°N, and the vertical shows the heat flux in units of m sec^{-1} °C. The four panels show transport by (a) all motion, (b) transient eddies, (c) stationary eddies, and (d) mean meridional circulation. Dashed lines refer to months June-August; light solid lines are for December-February; solid lines define annual averages. Source: Peixoto and Oort 1992.

Exchange of angular momentum between the atmosphere and surface is accomplished primarily by torques imposed by the force of friction, with an additional component due to torques associated with west-east gradients in atmospheric pressure operating on north-south–aligned mountain chains such as the Rockies and the Andes. Friction acts to reduce the speed of the surface wind, transferring linear momentum (see Chapter 2) from the atmosphere to the liquid or the solid planet with which it is in contact. Frictional loss of linear momentum by the atmosphere is accompanied by a corresponding gain of the same by the rest of the planet. A change in linear momentum, by Newton's law of motion, implies a force. The force of friction on the air opposes the direction of the wind (friction causes a decrease in the speed and consequently a decrease in the linear momentum of the air measured with respect to its direction of motion). The corresponding force on Earth is in the direction of the wind.

Defining angular momentum in the direction of planetary rotation (eastward) as *positive,* it follows that an eastward force implies a positive torque and acts to increase angular momentum. A westward force implies a negative torque and acts to decrease angular momentum. If the wind is westward (i.e., easterly), the force of friction on the air is eastward; the atmosphere gains angular momentum at the expense of the solid-liquid planet. An easterly wind near the surface exerts a westward frictional torque on the planet: the angular momentum of the planet's solid and liquid region decreases as a consequence, resulting in a decrease in the rate of rotation, with a corresponding increase in the length of the day. Fortunately, this is balanced elsewhere by westerly winds that are responsible for a reduction in the angular momentum of the atmosphere, with a compensatory increase in the planet's angular momentum. The angular momentum of Earth as a whole—atmosphere plus solid and liquid components—is conserved. Angular momentum is communicated to the atmosphere from the underlying planet in one region, removed in another. Sources and sinks of angular momentum must balance separately over time on a global scale for the atmosphere, ocean, and solid planet. An imbalance, maintained for any appreciable time, would result in a detectable change in the length of the day.[5]

Estimates of the net eastward torque exerted on the atmosphere as a function of latitude by surface friction are presented in Figure 8.28. Note the influence of the lower branch of the Hadley circulation, the easterly trade winds at latitudes below 30°: the planet imparts an eastward net torque to the atmosphere in this region. In the region of the midlatitude

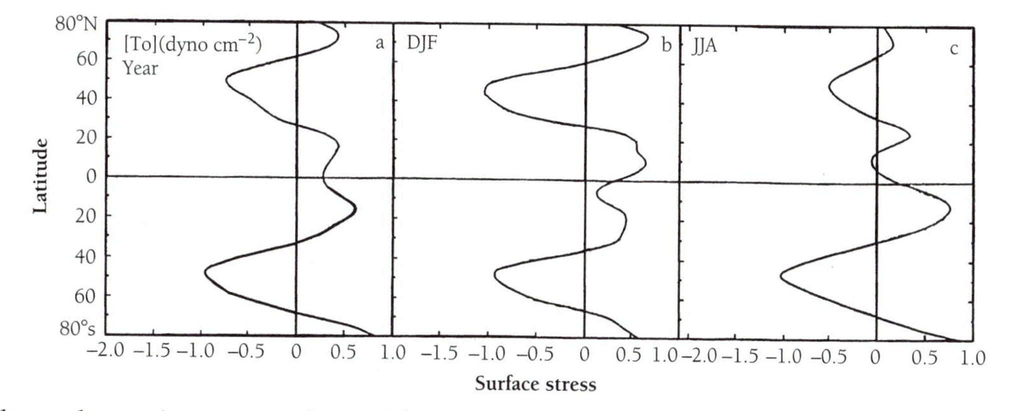

Figure 8.28 The zonal-mean (average at each latitude) stress imposed on the atmosphere by land plus ocean, expressed in dyn cm^{-2}, for the whole year (panel a), December through February (DJF) (panel b), and June through August (JJA) (panel c). Source: Peixoto and Oort 1992.

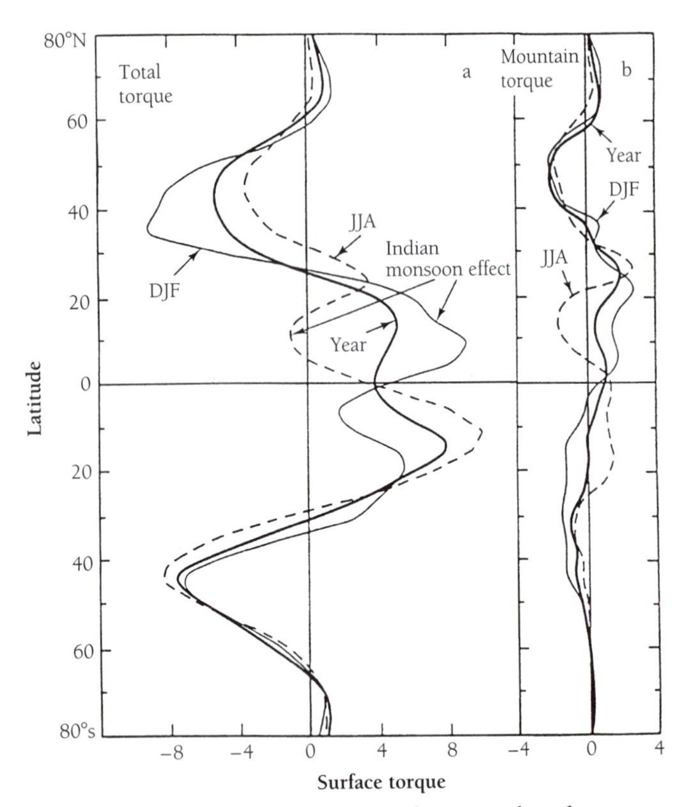

Figure 8.29 The torque exerted on the atmosphere by oceans and land at different latitudes (left-hand panel). The right-hand panel gives the torque exerted by mountains. Results are integrated over 5°C latitude bands and expressed in units of 10^{18} kg m^2 sec^{-2}. Separate curves refer to summer (June-August, JJA), winter (December-February, DJF), and annual mean. Source: Peixoto and Oort 1992.

surface westerlies, the atmosphere experiences a torque in the opposite direction, westward, and angular momentum is lost. Again, we may think of the Earth's surface as both the source and sink of atmospheric angular momentum (Figure 8.29). The atmosphere gains angular momentum from the planet at low latitudes, associated with the lower branch of the Hadley cell; it returns the surplus angular momentum to the planet at midlatitudes, associated with the lower branch of the theorized Ferrel cell. This requires a northward transfer of angular momentum by the atmosphere.

Integrated over zonal belts (longitude) for a range of latitudes and seasons, the torque associated with mountains appears to be oriented in the same general direction as the torque due to friction (Figure 8.29). Figure 8.29 also shows that the torque associated with mountains amounts to as much as 20% of the frictional torque for latitudes between 40° and 60° in the Northern Hemisphere.[6]

The picture of the circulation presented by Ferrel is conceptually useful but in some respects misleading. The circulation at mid- and high latitudes is governed by complex eddies rather than by an organized cell (or cells) as suggested in Figure 8.23. It is important to distinguish between the role of the mean (Hadley) circulation at low latitudes and the more complex effects of eddies at higher latitudes. The low latitudes are characterized by a cell-like movement of an air parcel, meaning that surface winds travel in a direction opposite to the winds aloft. Surface winds may blow southwestward while winds aloft may blow northeastward. At upper latitudes, the surface winds blow largely in the same zonal direction as the winds aloft, although at different speeds.

Estimates for the net northward transport of zonal momentum by the atmosphere (expressed as the product of the northward and eastward wind speeds) are presented in Figure 8.30. Net transport by the composite of all motions is summarized in panel (a). Transport by transient eddies, standing eddies, and mean meridional motion is indicated in panels (b)–(d).

Note the importance of transport by eddies for all latitudes. One can think of angular momentum as accumulating in the atmosphere in the tropics in the region of the easterly trade winds. It is expended at higher latitudes in the region of the surface westerlies. The bulk of the latitudinal redistribution of angular momentum by the atmosphere takes place in the upper troposphere, as indicated in Figure 8.31. The largest contribution originates from the vicinity of the midlatitude jets, from the region where the zonal wind is a maximum near the high-latitude edges of the Hadley regime (see Figures 8.24 and 8.25). Transport of angular momentum differs from transport of heat in that angular momentum is exchanged more readily between different air parcels.

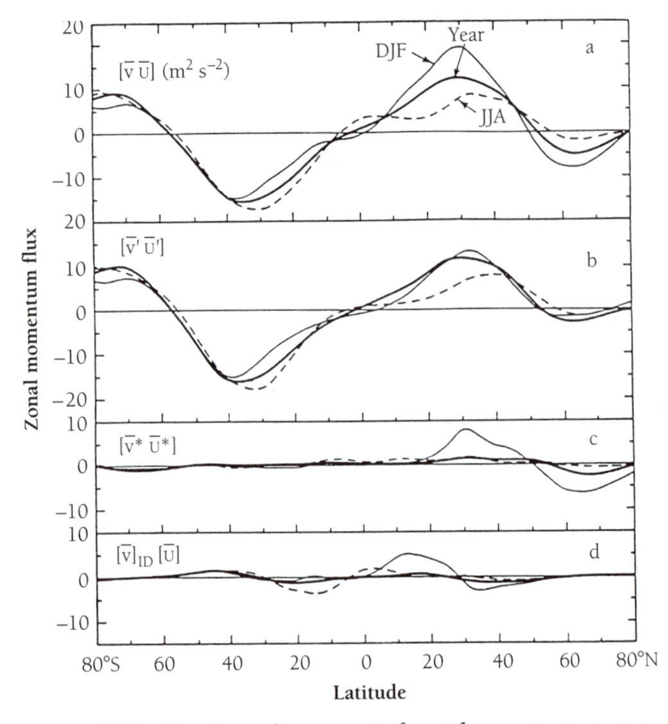

Figure 8.30 Northward transport of angular momentum averaged over latitude and altitude. Note that southward transport is represented by negative values. Panel (a) shows transport by all motion, panel (b) transport by transient eddies, panel (c) transport by stationary eddies, and panel (d) transport by mean meridional circulation. Data are presented in units of m^2 sec^{-2}. Seasonal and annual means are indicated as in Figures 8.28 and 8.30. Source: Peixoto and Oort 1992.

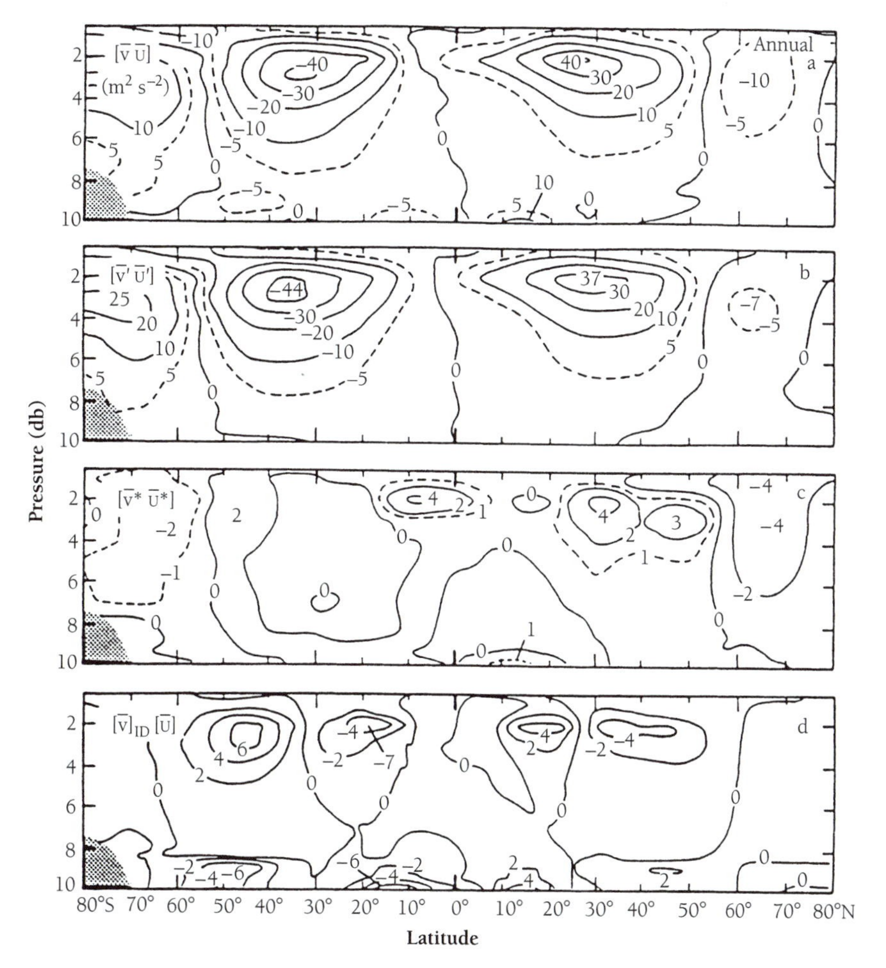

Figure 8.31 Annual average zonal-mean cross sections of the northward flux of angular momentum by (a) all motions, (b) transient eddies, (c) stationary eddies, and (d) mean meridional circulation. Units are m² sec⁻².

The question remains as to the physical mechanism by which surplus angular momentum is transferred to higher latitudes. Exchange of angular momentum between different atmospheric regions is accomplished primarily through torques that contort the zonal flow of air. In the Northern Hemisphere, the torques add angular momentum (an eastward force) to northward moving parcels and remove angular momentum (a westward force) from southward moving parcels. These torques are associated with zonal gradients of pressure. Thus, pressure gradients on a given latitude are ultimately linked to the existence of eddies.

The meridional transfer of angular momentum must be achieved through a wavelike pattern such that mass is conserved. Not every wavelike pattern will suffice, however, as some patterns are more efficient than others in transferring angular momentum. For example, the sinusoidal wave pattern in Figure 8.32 is not efficient in transporting angular momentum.

Example 8.6: Calculate the net angular momentum transferred from 30°N to 35°N by a 1 g air parcel traveling in one sinusoidal wavelength (refer to Figure 8.32). The west-to-east component of velocity for points A, B, and C are all 30 m sec⁻¹.

$$V_A(30°) = V_B(35°) = V_C(30°) = 30\text{m sec}^{-1}$$

Answer: The net angular momentum transferred is zero. The angular momentum transferred northward from point A to point B exactly equals the angular momentum transferred southward from point B to point C. ■

A more typical configuration of transient high and low pressure systems is illustrated in Figure 8.33. The stream lines of the flow depict the instantaneous directions of the

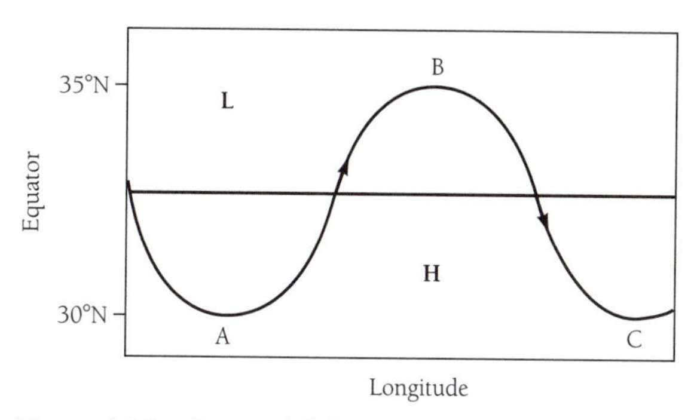

Figure 8.32 Sinusoidal fluctuations of zonal winds with associated highs and lows. This type of flow is inefficient in transporting angular momentum.

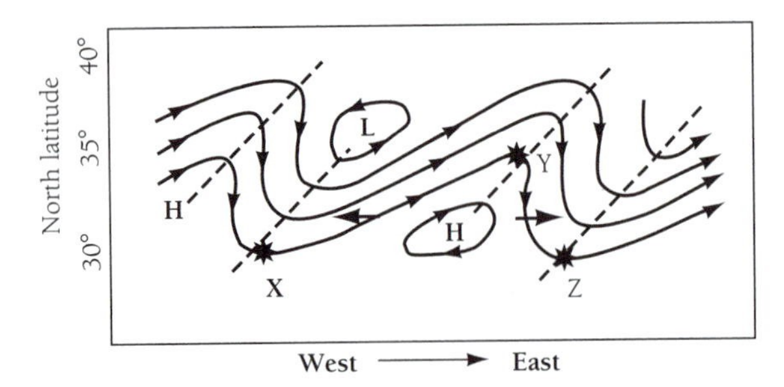

Figure 8.33 A typical fluctuation in the zonal motion illustrating the role of eddies in transporting angular momentum. In the Northern Hemisphere, note the strong easterly component of the flow northward as compared to the weak component of zonal motion associated with meanders of flow southward. Source: Peixoto and Oort 1992.

wind velocity. Note that the flow follows a clockwise pattern around the high and a counterclockwise pattern around the low, as expected for geostrophic balance. Furthermore, note that the air moving northward has a strong westerly component: its angular momentum, defined by the magnitude of the wind velocity eastward, is relatively high. The air moving southward has a relatively low angular momentum: the westerly component of the wind velocity in this case is small.

Example 8.7: Calculate the net angular momentum transferred from 30° to 35° by a 1 g air parcel (refer to Figure 8.33). Assume west-east velocities differ at points X,Y, and Z as follows:

$$V_X(30°) = V_Z(30°) = 30 \text{ m sec}^{-1}$$

$$V_Y(35°) = 0$$

Further, assume that the wind between points Y and Z has no zonal component, meaning it blows straight southward.

Answer: Recall that angular momentum is defined as

$$L(\lambda) = V(\lambda)R\cos(\lambda)$$

$$L_X(30°) = V(30°)R\cos(30°)$$

$$= (30\text{m sec}^{-1})(6.378 \times 10^6 \text{ m})(0.5)$$

$$= 9.6 \times 10^7 \text{ m}^2 \text{ sec}^{-1}$$

$$L_Y(35°) = V(35°)R\cos(35°)$$

$$= 0$$

■

Note that the angular momentum is transferred northward, while none is transferred southward. This type of eddy is extremely efficient in transferring angular momentum. Transport of angular momentum from low to high latitude by eddies yields strong westerly flows and weak easterly flows, accounting for the predominance of westerly winds in the middle troposphere at nonequatorial latitudes.

The meandering stream lines in Figure 8.33 are associated thus with significant northward transport of angular momentum. This is the case even though the flow pattern in Figure 8.33 is associated with little, or no, latitudinal transfer of mass. Southward transport of mass at Y is almost exactly balanced by northward transport at X.

Northward-moving eddies experience on average a westward pressure gradient while southward eddies are subject on average to an eastward pressure gradient. Thus, air parcels moving north at X experience a negative torque associated with the component of the pressure force operating westward. They lose angular momentum as they move. The angular momentum lost in this fashion is communicated to the air through which the parcels move. The surrounding air, then, has an increased flow from west to east. Similarly, parcels moving southward at Y are subjected to an eastward pressure force. Angular momentum is extracted from the ambient air in this case as a result of the positive west-east torque experienced by the southward-moving parcels. The angular-momentum surplus acquired on the southward excursion is returned to the ambient atmosphere on the flow's subsequent northward meander.

A similar mechanism operates in the Southern Hemisphere. Transport of angular momentum from low to high latitudes by eddies in both hemispheres plays a dominant role in the global redistribution of angular momentum. It provides the source required to balance the atmosphere's loss of angular momentum to the surface at midlatitudes associated with the negative torque exerted on the atmosphere by the planet in conjunction with the prevailing surface westerlies in this region. Westerly winds are ultimately required at midlatitudes to balance the negative torque exerted on Earth by the surface easterly trade winds at low latitudes. In this sense, eddies are an essential component of the global circulation of the atmosphere. They play a dominant role in the heat and angular momentum budget of the atmosphere at all latitudes poleward of about 20°.

8.11 The Origin of Eddies

Horizontal motions of the atmosphere are predominantly in the zonal (east-west) direction. Eddies typically arise as a result of instabilities in the zonal flow. A kink, a meridional excursion, develops and proceeds to propagate as a wave, similar to the pattern displayed in Figure 8.33. Instabilities are commonly linked to the north-south variation of pressure surfaces that typically slope downward from low to high latitudes. Adjustment of the slope of pressure surfaces can result in the release of significant amounts of energy. This provides the fuel for the growth of eddies. Their subsequent evolution is strongly influenced by rotation and constraints imposed by the need to conserve angular momentum. Densities can significantly vary along surfaces of constant pressure, due to variations in temperature. Eddies feeding on variations of density on constant-pressure surfaces are known as **baroclinic eddies.** Much of the transient atmospheric motion is associated with baroclinic eddies and is ultimately related to the tendency of the atmosphere to adjust to minimize the density variation on constant-pressure surfaces.

8.12 Summary

Atmospheric structure and temperature are controlled by absorption of solar radiation, mainly in the visible portion of the spectrum, by redistribution of energy absorbed from the Sun through the complex system of winds and eddies associated with the general circulation, and by emission of radiant energy into space, primarily in the infrared region of the spectrum. Solar radiation is most intensely absorbed at low latitudes. Emission of infrared radiation into space is more evenly distributed, such that the net radiative budget of the atmosphere is positive at low latitudes (more energy is absorbed from the Sun than is emitted in the infrared), negative at high latitudes. The radiative surplus at low latitudes is balanced by transport through the general circulation. Sufficient energy is transported by atmospheric and oceanic motions from low to mid and high latitudes of the atmosphere, with an important additional contribution, as will be discussed later, from the ocean to offset the radiative deficit in these regions.

Three distinct dynamical regimes can be identified. The thermal circulation described by Hadley plays an important role in the tropics. Air rises in the region of maximum heating, moving to higher latitudes aloft under the influence of a thermally maintained gradient in pressure. As it moves to higher latitude, air drifts eastward under the influence of the Coriolis force, seeking to conserve angular momentum. Instabilities in a primarily zonal flow give rise to eddies, assisting in the latitudinal transfer of angular momentum and heat, extending the tropical regime to a latitude of about 30° under present climate conditions. In the absence of eddies, the tropical regime would be confined to latitudes below about 20° and tropical temperatures would be most likely higher than today, with temperatures at mid- and high latitudes correspondingly colder.

The high-latitude boundary of the tropical, or Hadley, regime is marked by a strong westerly jet in the middle troposphere. The speed of the jet depends on the meridional pressure gradient, as required by geostrophy. The meridional pressure gradient is in turn linked to the meridional (south-north) gradient in temperature. The more efficient the transport of heat and angular momentum by eddies, the more extensive the tropical regime and the smaller the contrast in temperature between low and high latitudes. Ice ages may correspond to times when the tropical regime was relatively confined. Equable climates (occasions when Earth was uncommonly warm) may represent a response to unusually efficient transport by eddies, with an associated expansion of the tropical regime.

The upper-latitude boundary of the tropical, or Hadley, regime is characterized by the sinking motion of the atmosphere. Since water is removed by precipitation as air rises and cools in the ascending branch of the Hadley cell at the equator, the atmosphere in the descending branch is unusually dry. Temperatures at the surface are high as a result of adiabatic compression. This is the region occupied by the great desert areas on land and the location of the Horse Latitudes so eloquently described by Coleridge for the ocean. Air returns to the equator through the lower branch of the Hadley cell. As it moves southward (northward in the Southern Hemisphere), it is deflected westward by the Coriolis force, accounting for the existence of the trade winds.

Transport of heat and angular momentum is mainly effected by eddies at midlatitudes, particularly during winter. Eddies are manifest at the surface by passage of the high- and low pressure systems that play a dominant role in midlatitude weather. Geostrophic balance implies that the flow of air around a low pressure system, or cyclonic motion, should proceed in a counterclockwise direction in the Northern Hemisphere (high pressure is maintained to the right of the direction of motion). The sense of the flow is reversed for the Southern Hemisphere. Friction leads to a reduction in wind speeds near the surface. This causes air to converge toward the center of the low, spiralling inward and rising as it crosses **isobaric** (constant pressure) surfaces. Low pressure systems are associated frequently with unsettled weather, with precipitation often the consequence of the cooling of moisture-laden air as it rises. Conversely, high pressure systems are the harbinger of clear, cloud-free conditions, with descending motion dominant at the center of highs, accommodating frictionally driven divergence (or outward flow) of surface winds.

Cyclones, anticyclones, and frequent change are dominant characteristics of the weather at midlatitudes. Angular momentum transported by the atmosphere to high latitudes is dissipated by friction associated with prevailing westerly winds at the surface, in combination with torques relating to pressure gradients operating on north-south–aligned mountain chains. Surface westerlies at midlatitudes are an essential feature of the general circulation; they are required to balance oppositely directed torques imposed on the surface by the easterly trade winds at low latitudes. The indirect cell proposed by Ferrel for the mean circulation at midlatitudes is an idealization at best. It is difficult, indeed impossible, to divine a pattern as simple as the Ferrel cell in a realistic map of the weather. It exists, if at all, as a long-term average of the circulation, after dominant contributions from eddies have been eliminated. Mean circulation plays a trivial role in the transport of heat and angular momentum at all latitudes, except in the tropics. The meteorology of mid- and high latitudes is dominated by eddies.

The transition from midlatitudes to the polar region is marked, especially during winter, by a sharp temperature gradient, an associated pressure gradient, and a strong, geostrophically maintained, largely zonal wind reaching maximum speeds in excess of 100 mph (sometimes as high as 200 mph) at an altitude of about 10 km above the surface. The jet stream provides important assistance to aircraft flying eastward, reducing scheduled travel times from North America to Europe by as much as an hour with respect to the westward return journey. The jet stream undergoes frequent meanders, resulting in rapid surface temperature changes in regions near the margin of the polar environment. It is not uncommon for the jet stream to stick in a particular configuration for periods as long as a week or more. Regions on the poleward side of the jet stream are then subjected to an unrelenting cold snap, while areas on the equatorward side enjoy prolonged balmy temperatures. Such a condition occurred over the United States between the 15–20 April 1976, as illustrated for 17 April in Plate 3 (bottom) (see color insert). The western portion of the country was unseasonably cold as the southward meander of the jet brought frigid air from Northern Canada as far south as Northern Mexico. South of the jet stream, warm air invading from the subtropics resulted in record-high temperatures over much of the central and eastern United States. Temperatures rose as high as 88°F in Chicago, while records were broken in Boston, as temperatures climbed into the high 90s. Temperatures dropped by as much as 50°F over a 24-hour period a few days later, as tropical air derived from south of the jet stream was replaced by polar air from the north. Average temperatures for winter in cities such as Boston can vary significantly from year to year under the influence of apparently random migrations of the jet stream. Instabilities, or eddies, forming on the edge of the jet, have an important influence on weather and precipitation for most regions of the world at mid to high latitudes in winter.

Weather on the high-latitude side of the polar front is unrelentingly cold in winter. As is the case at midlatitudes, eddies play a dominant role. Even the concept of a mean circulation is problematic: large interannual variability is the rule rather than the exception, especially for the Northern Hemisphere. The picture of the general circulation displayed for high latitudes in Figure 8.23 is even more idealized than the view presented for midlatitudes. An appreciation of climate at high latitudes, as at midlatitudes, demands an understanding of eddies. Eddies represent the signal, not simply the noise, of the climate system for much of the world.

Dynamics of the Ocean

9

Chapter 8 presented an introduction to the dynamics of the atmosphere. We saw that atmospheric motions are driven by a combination of factors: unequal rates of heating between low and high latitudes; differences in the thermal response of land and sea to seasonally varying inputs of solar radiation; the importance of phase changes in water as sources and sinks for thermal energy; and the role of planetary rotation in modulating the response of the atmosphere to pressure gradients and other forces serving to set the air in motion. This chapter is intended to provide a complementary introduction to the dynamics of the ocean.

As we shall see, the motions of the ocean's surface layers arise primarily in response to stresses imposed by the large-scale circulation of the atmosphere. Air blowing over the ocean's surface acts to set its water in motion. As water begins to move, it experiences the effects of the Coriolis force deflecting it to the right of its initial direction of motion in the Northern Hemisphere, to the left in the Southern Hemisphere. In addition to the forces represented by the wind-imposed stress at the surface and by the Coriolis effect, oceanic water is subject to a frictional force introduced by the motion of surface water across the more slowly moving water underneath. The frictional force acts in a direction opposite to the direction of the water's motion. Figure 9.1 diagrammatically represents the forces acting on surface water. As a consequence of the combined influence of the three primary forces acting on the water—wind stress, friction, and the Coriolis effect—the surface water tends to move in a direction inclined at an angle of about 45° to the right of the wind's direction in the Northern Hemisphere, to the left of it in the south.

The tendency of water to move to the right of the wind velocity in the north was noted first, from observations of the drift of icebergs, by the Norwegian oceanographer Fritjof Nansen (1861–1930). The quantitative theory describing this result was developed by the Swedish physicist V. W. Ekman (1874–1954). He observed that the movement of surface water induced by the wind should result in a motion of the

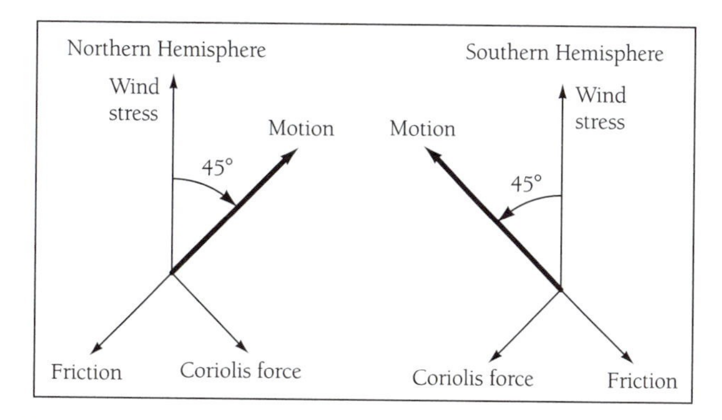

Figure 9.1 Diagrammatic representation of forces acting on surface water.

underlying water. The magnitude of the velocity imparted to the underlying water decreases, however, with depth, and the importance of the resulting Coriolis force should decline accordingly (the magnitude of the Coriolis force is proportional to the speed of the flow). There is thus a tendency for successively deeper layers to drift further to the right in the Northern Hemisphere. As a function of depth, the velocity vector of the water spirals to the right of the surface wind, eventually turning to the point where it is oriented opposite to the wind, as illustrated in Figure 9.2. Integrated over depth, we expect a net drift of water at right angles to the direction of the surface wind stress: to the right in the Northern Hemisphere, to the left in the Southern Hemisphere. Water is set in motion such that the depth-integrated impact of the Coriolis force on the water is precisely equal, but opposite, to the wind-imposed stress at the surface. As we have seen, the role of the Coriolis effect is small near the equator and increases in influence toward the poles. It follows that, to balance the effect of a given wind stress at low, as compared to high, latitudes, we need to set a larger volume of water in motion; other factors being equal, the influence of the wind extends to the greatest depths at low latitude. For a wind velocity of 10 m sec^{-1} at a latitude of 10°, the Ekman layer reaches a depth of about 100 m. The layer is only half as deep at a latitude of 45°.

The surface circulation of the ocean, that is, the major current systems, are described in Section 9.1. Vertical motion is discussed in Section 9.2. The role of the ocean in distributing heat—its influence on climate—is treated in Section 9.3.

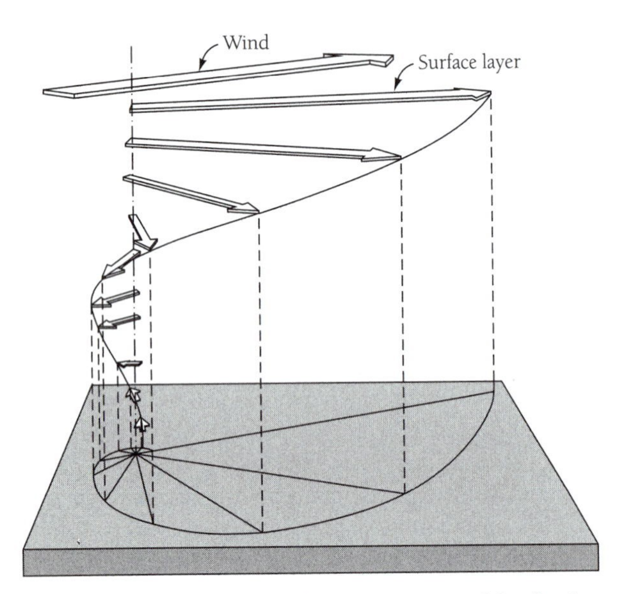

Figure 9.2 The Ekman spiral. As a function of depth, the velocity vector of water spirals to the right of the wind. Source: Gross 1990.

9.1 Surface Currents

The circulation of the surface ocean at midlatitudes is dominated by basin-scale quasi-circular patterns of motion, with water rotating clockwise in the Northern Hemisphere, counterclockwise in the Southern Hemisphere, as illustrated in Figure 9.3. These large-scale features, known as **gyres**, are driven by the prevailing easterlies at low latitudes and by the prevailing westerlies at higher latitudes, steered by the high pressure systems characteristic of the descending loops of the Hadley circulation (see Figures 8.2 and 8.3). The western, northward branch of the North Atlantic gyre is known as the **Gulf Stream**. The Gulf Stream is supplied from the south by the **North Equatorial Current**, which flows from east to west across the ocean basin in the tropics, driven by the easterly trade winds. The North Equatorial Current turns to the northwest as it reaches the South American coast, incorporating a component of water supplied by the **South Equatorial Current**. A portion of the Gulf Stream flow originates thus in the Southern Hemisphere. As we shall see, this cross-equatorial transport of surface water is balanced by a return flow of water in the depths. The deep water supplying this return originates mainly in the Norwegian Sea, where intense cooling of surface waters in winter causes densities to rise to the point where the water column becomes unstable (densities of surface waters are higher than densities of waters underneath), resulting in convective overturning and rapid mixing of the water column, analogous to what happens in the atmosphere when the lapse rate of temperature exceeds the adiabatic limit (see also Sections 7.3 and 7.4).

Water flow is highly concentrated on the western margins of the ocean gyres. The Gulf Stream, for example, attains speeds of up to 10 km h^{-1}. It represents an intense stream of warm water approximately 100 km wide, extending to a depth of about a kilometer. It transports a volume of water approaching 100 Sverdrups (Sv) as it turns to the northeast, leaving the coast of North America in the vicinity of Cape Hatteras.[1] The westward intensification of flow in the major ocean gyres results from a combination of the influences of friction and planetary rotation. As the flow progresses to higher latitudes, it tends to turn eastward, driven by the Coriolis force. Transfer of angular momentum to onshore water traveling in the opposite direction, and ultimately to the bordering land mass, acts to offset this tendency of the water to deviate eastward, permitting the flow to persist toward higher latitudes. The analogue of the Gulf Stream in the North Pacific is known as the **Kuroshio Current**. Western boundary currents in the Southern Hemisphere, such as the **Brazil Current**, the **Agulhas Current** (off West Africa), and the **East Australia Current**, are less clearly defined, reflecting a more unfavorable distribution of land masses required to shed angular momentum as the flows progress to higher latitudes.

The Gulf Stream becomes more diffuse as it moves northeastward across the Atlantic, merging into the **North Atlantic Drift**. As it approaches Europe, the North Atlantic

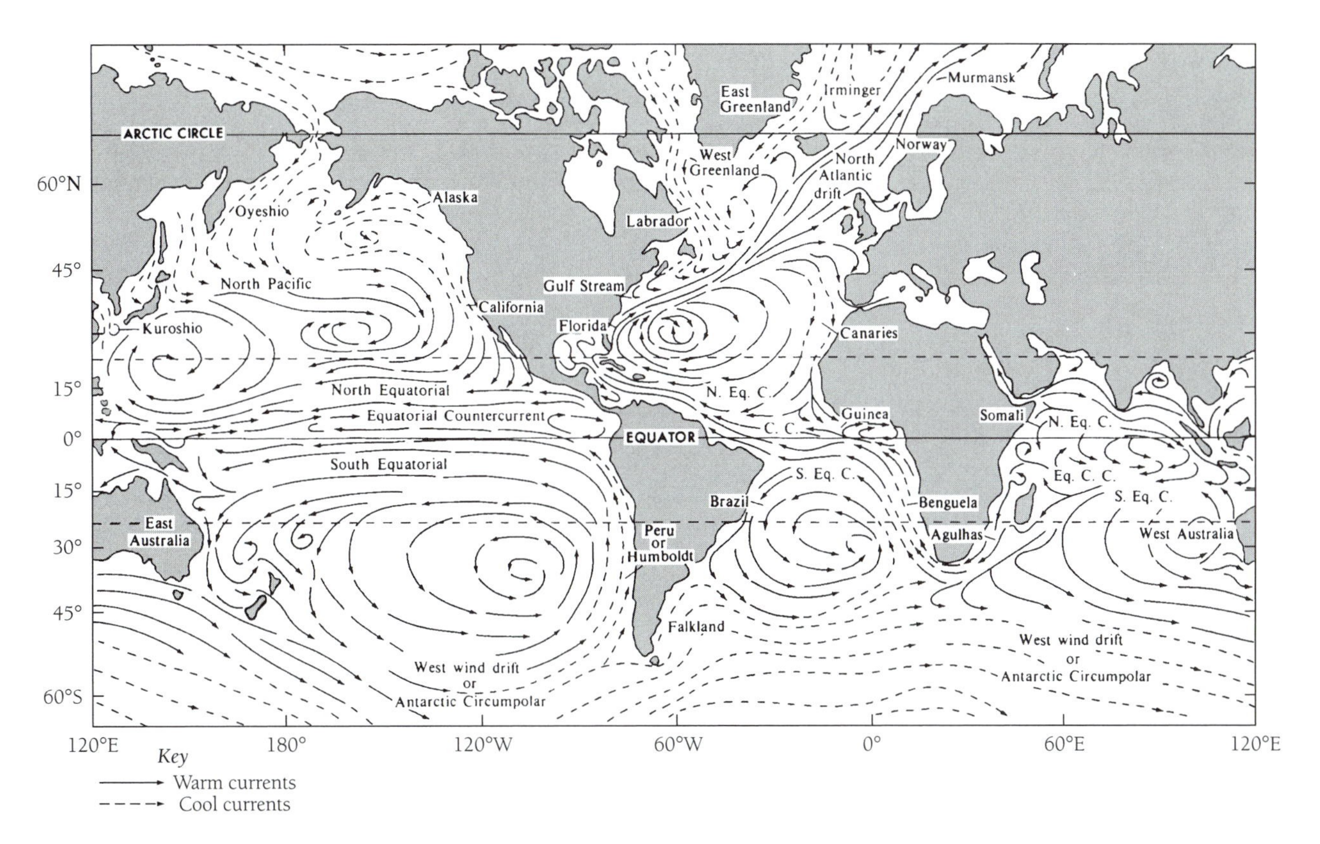

Figure 9.3 Major current systems of the world's oceans. Warm currents are indicated by solid lines, cold by dashed lines. Source: Tolmazin 1985.

Drift splits into two streams, one moving northward, bringing warm water and a temperate influence to the climates of the British Isles and Norway. The second branch flows southward, forming the **Canary Current** and carrying cool northern water back toward the equator. The cold **California Current** represents the analogue to the Canary Current for the Pacific, accounting for the return flow of waters to the equatorial region for the North Pacific gyre. The northward loop of the North Atlantic Drift, driven by the prevailing westerlies, is responsible for a supply of warm water to the Arctic Ocean. Water returns from the Arctic Ocean mainly in the western region of the Norwegian Sea, off the eastern coast of Greenland in the cold **East Greenland Drift**. A component of this water describes a counterclockwise loop around the Labrador Sea and Baffin Bay before flowing southward along the eastern coast of Canada and the northeastern coast of the United States, reaching as far south as Massachusetts during summer and all the way to North Carolina during winter. This component of the North Atlantic circulation is known as the **Labrador Current.**

There is an important distinction between the circulations of water represented by the North Atlantic and North Pacific gyres. The northward transport of warm water extends to a much higher latitude in the Atlantic as compared to that in the Pacific. Differences in the geometry of the basins provide a partial explanation for this difference: the connection between the Atlantic and Arctic Oceans is relatively unrestricted; in contrast, the connection between the Pacific and Arctic Oceans is limited by the comparatively narrow opening afforded by the Bering Straight separating Asia and North America. There is a further reason for the difference, as we shall see below. Water sinks from the surface to the depths in the North Atlantic, notably in the Norwegian and Labrador Seas, requiring an inflow of water to maintain balance. In contrast, surface waters of the North Pacific are relatively fresh (low in salt), and, even under extremes of winter cooling, densities are too low to permit surface waters to sink to significant depths.

Surface flows in the equatorial region arise largely as a consequence of the easterly trade winds. The dominant features for the Pacific are the North Equatorial Current (NEC), flowing westward between 20°N and 8°N and carrying approximately 20 Sv, the South Equatorial Current (SEC) between 2°N and 10°S, moving in the same direction with a flow that is almost twice as large as the NEC, and the **North Equatorial Countercurrent** (NECC), flowing eastward in the opposite direction, between the NEC and the SEC and carrying about 20 Sv. A fourth current, the **Equatorial Undercurrent** (EUC), straddling the equator between 2°N and 2°S, carries a significant volume of water (about 20 Sv) over a depth interval ranging from about 20 to 150 m. The EUC is thought to originate in an undercurrent of saline water moving northwestward along the northern coast of Papua New Guinea. The NECC represents a response to the buildup of water in the western region of the tropical Pacific as a consequence of the westward flows of the NEC and SEC. The ocean surface is higher in the west than in the east, driving an eastward return flow in the NECC. During the large El Niño episode of 1982–1983, transport in the NECC increased by approximately a factor of two; eastward moving water dominated surface transport in the equatorial Pacific over the entire region from 10°N to 5°S. We should note that the equatorial current systems in the Pacific are distributed somewhat asymmetrically with respect to the equator (the NEC extends to 20°N, while the SEC is confined to a region north of about 10°S). This reflects the fact that the intertropical convergence zone (ITCZ)—effectively the meteorological equator—is located on average to the north of the equator, reflecting the larger concentration of land in the Northern Hemisphere. Figure 9.4 gives a depiction of surface currents for the Pacific, offering a more detailed perspective than the global view presented in Figure 9.3.

The southern edge of the South Pacific gyre is defined by the **Antarctic Circumpolar Current** (ACC) flowing eastward and driven by the prevailing westerlies. The ACC feeds the eastern branch of the South Pacific gyre off the western coast of South America forming the **Peru Current** (also known as the **Humboldt Current**) and also contributes to the eastern branch of the South Atlantic gyre as the **Benguela Current** flowing northward off the western coast of Africa. Typical of flows on the eastern margins of ocean basins (in contrast to western currents), the Peru and Benguela Currents are relatively slow and diffuse.

9.2 Vertical Structure of the Ocean and the Formation of Deep Water

As we have seen, ocean surface circulation is largely determined by the transfer of momentum from the atmosphere. Sunlight, the dominant source of energy for the ocean, is absorbed mainly in the upper few meters of the water column. Its influence is distributed, however, across a much larger depth range (down to 100 m or so) as a consequence of mixing of water masses induced by surface winds. As a result, temperatures are relatively constant with depth across this upper mixed region of the ocean. Since solar energy is distributed over a comparatively large volume of water, surface water temperatures vary little from day to night. Even the seasonal swing of temperature at the ocean surface is small, much less than observed on land (the heat capacity of ocean water is much higher than that of continental rocks, as noted previously). Temperatures decline rapidly with depth below the surface's mixed layer. The region where the temperature gradient is greatest is known as the **thermocline** (from *thermo,* meaning "heat" and *cline,* meaning "slope"). The thermocline extends from a depth of about 200 to about 1000 m at mid and low latitudes, respectively, and separates the relatively warm surface water from the cold water that dominates properties in the depths of the ocean. Averaged over Earth, surface water accounts for about 2% of the total volume of the ocean. Approximately 18% of the ocean is included in the transitional thermocline zone. Cold water,

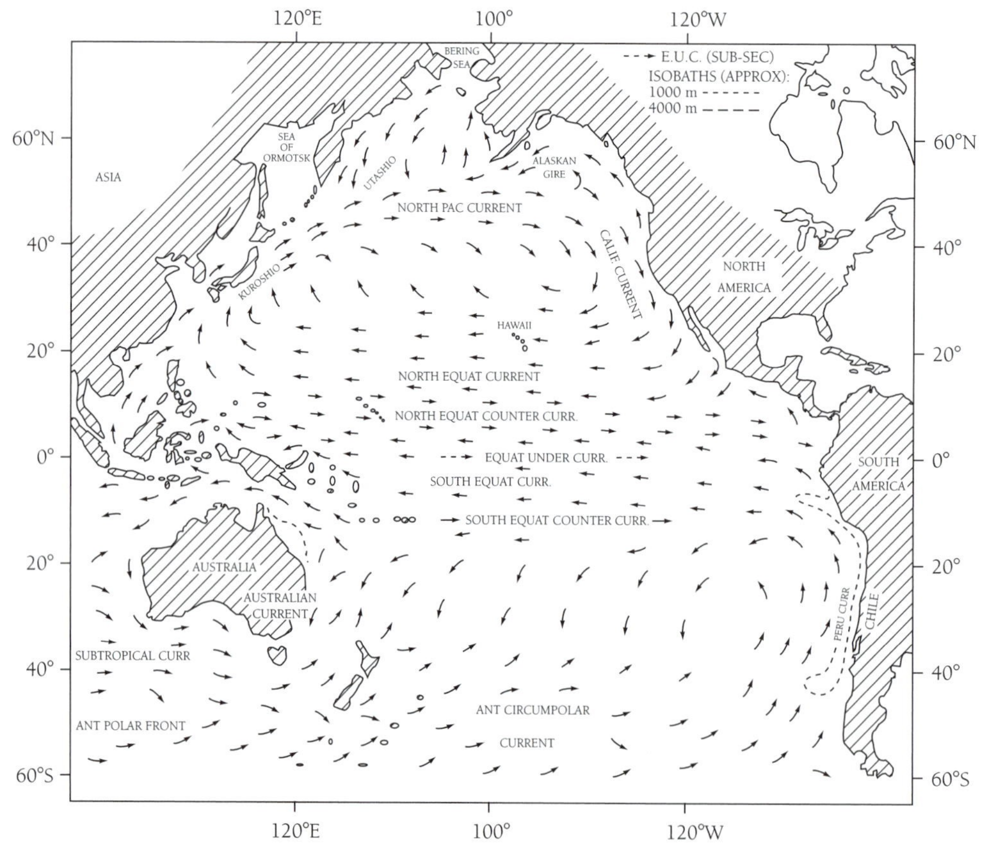

Figure 9.4 Surface currents of the Pacific. Source: Pickard and Emery 1990.

with an average temperature of about 3°C, accounts for by far the largest fraction (about 80%) of the total mass of water in the ocean.

The distribution of water in the different layers is schematically illustrated in Figure 9.5. Note that the distinction between surface and deep waters tends to disappear at high latitude; the thermocline rises to the surface at latitudes higher than about 40°. This behavior reflects the influence of deep convection (vertical overturning) taking place in the ocean, primarily at high latitudes during winter. The ocean's vertical stability depends on the contrast in density between the surface and depth. Surface water density can increase as a result of either cooling or an increase in salinity. Convection sets in when the density of surface waters is larger than (or comparable to) the density of waters below, a condition arising in selected regions of the North Atlantic and in the southern oceans during winter. Convection at high latitudes provides the dominant source of deep water for the world's oceans, accounting for the pervasive cold temperatures of waters in the depths.

To this point we have focused on temperature as the primary variable distinguishing the various vertical regions of the ocean. A more realistic discussion would emphasize contrasts in density. The density of surface waters is generally less than the density of waters in the region of the thermocline, while waters of highest density characterize the properties of the ocean's depths. Adopting a density-based perspective, the thermocline is identified closely with the **pycnocline** (from *pycno,* meaning "density"), the region where the gradient of density with depth is greatest. The rapid increase of density with depth in the vicinity of the pycnocline, from low above to high below, and the resulting stability, ensures that the pycnocline provides a serious barrier to vertical motion; waters at the surface are essentially isolated from waters in the depths across most of the ocean.

The density of seawater is a function primarily of temperature and salt content, with an additional, smaller dependence on pressure associated with the effects of compression. At a fixed temperature, an increase in salinity results in an increase in density. The response of density to a change in temperature at fixed salinity is more complicated. It depends on the salinity of the water, as illustrated in Figure 9.6.

For freshwater, density increases with declining temperature, reaching a maximum at a temperature of 4°C. A freshwater lake, as it cools during autumn and winter, arrives at a point, typically during early autumn, where the density of surface water is higher than the density of water underneath.[2] Convection sets in and continues, driven by a further increase in density as the water cools to the point of maximum density (4°C). With additional cooling (below 4°C), surface waters become lighter than waters underneath (their density decreases), and convection is suppressed. Eventually, surface water reaches the freezing point (0°C). The lake freezes from the top downward. The bulk of the water in the lake, however, if it is deep enough, remains above the freezing point throughout winter.

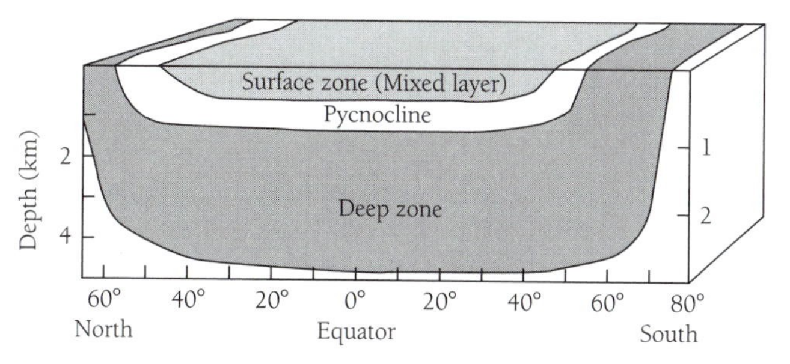

Figure 9.5 Vertical stratification of the ocean. Source: Gross 1996.

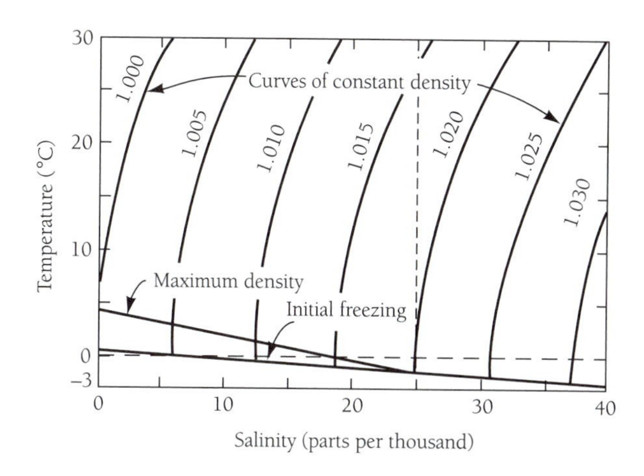

Figure 9.6 Relationship between temperature and density at fixed salinity. Source: Gross 1996.

The situation is different for the ocean. For typical values of salinity, ocean density increases steadily with declining temperature (in contrast to fresh water, which, as we have seen above, has a maximum in density at a temperature several degrees above the freezing point). Convection is triggered when the density of surface water exceeds the density of the water below. It continues until surface temperatures reach the freezing point, typically, for the ocean, at a temperature of about −1.8°C.[3] The entire water column (or at least the upper convective portion) must cool to close to −1.8°C before ice can begin to form in the ocean.

Plate 4 (top) (see color insert) illustrates the distribution of salinity over surface waters of the ocean. For a particular ocean, maximum values of salinity are observed in the subtropics, in the vicinity of the subtropical highs where the air is hot and dry. (Remember that the highs are associated with moisture-depleted descending loops of the Hadley circulation.) The rate at which fresh water is lost from the ocean to evaporation in this region exceeds the rate at which it is returned by precipitation. Evaporation of water from the subtropical ocean contributes to an excess of precipitation over evaporation at high latitudes, accounting for the decrease in salinity observed toward higher latitudes, as indicated in Plate 4. Salinities in the South Atlantic are similar to those in the South Pacific, reflecting the fact that both oceans are fed by a common source, the Antarctic Circumpolar Current. Salinities in the North Atlantic, however, are significantly higher than those in the North Pacific.

The difference in salinities between the North Atlantic and North Pacific plays a critical role in determining the locus of formation, and the direction of flow, of the cold deep waters pervading the world's oceans. Surface water salinities in the North Atlantic are such that, when waters cool during winter, they develop densities high enough to sink to the bottom at least in some regions. The North Atlantic provides the primary source of deep water for the ocean today. Deep water formed in the Norwegian, Greenland, and Labrador Seas flows south, circling Antarctica, picking up an additional contribution from waters sinking from the surface near Antarctica, before entering the Pacific and Indian Oceans. We expect a slow, diffuse return of water from the depths to the surface. The flow of deep, density-driven currents in the ocean defines what is known as the **thermohaline circulation.** A return flow of water to the surface from the depths is required for reasons of continuity: to maintain a more or less constant volume of deep water. It plays a critical role in determining the structure of the thermocline. In the absence of a steady upward flow of cold water, heat would diffuse down from the warm ocean above and it would be impossible to maintain the thermocline as a relatively stable feature of the vertical structure of the ocean.[4]

As will be discussed in Chapter 10, the gradient of salinity between the North Atlantic and the North Pacific requires a net transfer of fresh water from the former to the latter. This transfer most probably proceeds through the atmosphere. It implies that the rate at which water evaporates from the Atlantic must exceed the rate at which it returns by precipitation and by runoff in rivers. At first blush this result seems unexpected. After all, several of the largest rivers in the world, notably the Amazon, Orinoco, Zaire, Mississippi, and Saint Lawrence, flow into the Atlantic. In contrast, an excess of precipitation plus runoff over evaporation must apply for the Pacific. It appears that the difference in the salinity budgets of the North Atlantic and North Pacific may be due in part to transport of moisture by easterly trade winds blowing from the Atlantic to the Pacific over Central America, in part to transport by the westerlies across Eurasia (see Chapter 10). The low salinity wedge observed in waters off the west coast of central America may be attributable in large measure to the influence of vapor transported from the Atlantic across Central America.

Columbia University oceanographer Wallace Broecker coined the phrase the "Conveyor Belt" to describe the flow of water associated with the thermohaline circulation. The deep branch of the Conveyor Belt begins its lengthy traverse of the world's oceans when cold saline water sinks from the surface in the North Atlantic. It flows southward on the first leg of its journey toward Antarctica, where it turns eastward, circling the Antarctic continent before entering the Indian and Pacific Oceans. The branches of the deep Conveyor Belt in the Indian and Pacific Oceans are loaded not only with waters from the Atlantic but also with waters sinking in the cold surface environment of Antarctica.[5] Contributions to the overall global production of deep water from the North Atlantic and Antarctica are thought to be comparable. The upper loop of the Conveyor Belt is represented by a return flow of surface water from the Indian and Pacific Oceans to the Atlantic by way of the Antarctic Circumpolar Current. Figure 9.7 presents a pictorial representation of the Conveyor Belt.

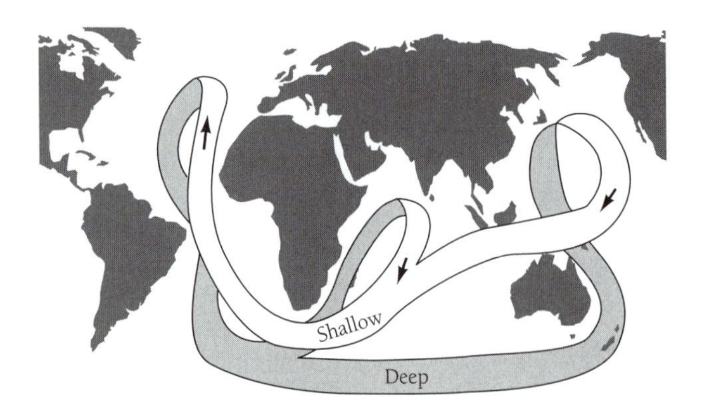

Figure 9.7 The Conveyor Belt. Source: Broecker 1987.

The North Atlantic has a powerful and direct influence on the climate of Western Europe. It works by drawing warm Gulf Stream water northward into the Norwegian and Greenland Seas as an extension of the North Atlantic Drift. Before it can achieve densities sufficient to sink to the bottom, the water must cool by about 8°C. Heat given up by the ocean is transferred to the atmosphere. Broecker points out that the contribution of the North Atlantic to the heat budget of the atmosphere amounts to nearly 30% of the total solar energy absorbed by the atmosphere over the Atlantic for the entire region north of the Straits of Gibraltar. He makes a persuasive case for the importance of the North Atlantic to the climate of Western Europe. It is unclear, though, whether the deep-water formation process is essential to the linkage. Warm water is steered northward by the prevailing westerly winds and guided by the low pressure system typically present in the atmosphere over Greenland. A more detailed analysis would be required to define the extent to which the northward flow of warm water depends on the rate at which surface water is transferred to the depths by the downward branch of the Conveyor Belt. It is conceivable that a stronger flow at the surface in the southward-moving Labrador Current could compensate for a reduction in the flux of water carried by the deep branch of the Conveyor Belt.

In any event the distinction may be moot. Production of deep water in the North Atlantic would appear to be an unavoidable consequence of the buildup of salinity in surface waters. Excess salinity resulting from the net transfer of water vapor from the Atlantic to the Pacific is balanced in the present environment by a net export of salt from the Atlantic to the Pacific and Indian Oceans. In the absence of a deep water removal mechanism, an export of salinity from the Atlantic would be suppressed and the salt content—and thus the density—of surface waters in the North Atlantic would increase. Eventually, the contrast in density between surface and deep water would rise to the point where convective overturning was inevitable.

As will be discussed in Chapter 10, an impressive body of evidence attests to the importance of the Conveyor Belt for climate. An increase in the supply of fresh water for the North Atlantic that occurred during the terminal stages of the last ice age is thought to have been responsible for the cold climatic interval known as the **Younger Dryas**, which interrupted a period of general warming beginning about 5,000 years earlier and marking the end of the last ice age.[6] The influence of the Younger Dryas was most pronounced in and around the North Atlantic, an association leading Broecker to suggest that a cessation of deep water formation in the North Atlantic may have provided the trigger for the transition to the Younger Dryas from the warm interval preceding it, known as the **Bolling Allerod.** The climatic fluctuation represented by the Bolling Allerod–Younger Dryas transition is but the last of a number of similar oscillations in climate occurring at irregular intervals throughout the last ice age. Broecker attributes these transitions to what he describes as modal shifts in the operation of the North Atlantic Conveyor Belt.

As noted earlier, the return flow of water from the depths to the surface (the diffuse, globally distributed, upward branch of the Conveyor) has an important influence on the structure and location of the thermocline. An increase in the water flow of the Conveyor would be expected to cause a shift in the location of the thermocline toward the surface, increasing the volume of cold, deep water at the expense of relatively warm water near the surface. Conversely, a decrease in the strength of the Conveyor Belt would cause the thermocline to drop, leading to an increase in the volume of warm water. The Conveyor Belt provides an important mechanism for transport of heat by the ocean from low to high latitudes. It is associated also with significant transfer of heat from one hemisphere to another. For example, there is a net transport of heat from the South to the North Atlantic today associated with the northward flow of warm water at the surface required to supply the cold return flow at depth. If the North Atlantic is the dominant source of deep water, the Conveyor is responsible for net transfer of heat from the southern to the Northern Hemisphere. If deep water was formed primarily in the southern ocean, the flow of heat would be directed opposite, from north to south.

9.3 Vertical Motions

We now turn our attention to a discussion of processes responsible for vertical motion in the near-surface region of the ocean. Net vertical displacement of water arises as a consequence of either convergence or divergence of the horizontal motion field. A net horizontal inflow of surface water requires that water must sink, or downwell, in order to maintain continuity. Conversely, a net outflow at the surface, implies that water must rise, or upwell, to maintain a balance. Regions of strong upwelling are associated with environments of high biological productivity, reflecting the importance of the supply of nutrients from below.[7]

Plate 4 (bottom) (see color insert) presents a composite satellite image of ocean color. Wavelengths of satellite sensors were selected to provide maximum sensitivity to chlorophyll, the primary pigment involved in marine photosynthesis. While maps of ocean color are not precisely equivalent to maps of primary productivity, they provide a useful indication of regions where biological production is high, where nutrients are supplied to the surface efficiently by upwelling, or of regions where biological production is low, where vertical motion is characterized by downwelling, and where the supply of nutrients is suppressed. Plate 4 (bottom) provides a clear indication of the importance of coastal upwelling, notably off California, Peru, and northwest Africa. Bands of high chlorophyll are also observed along the equator, in the southern ocean near Antarctica, and over extensive regions of the North Atlantic and North Pacific. The interiors of ocean gyres are associated with downwelling and stand out as regions of exceptionally low productivity, oceanic analogues to deserts on dry land.

Vertical motion is especially vigorous in coastal areas when prevailing winds blow parallel to the shore. As noted above, the stress imposed by the wind induces a flow of water at right angles to the direction of the wind: to the right in the Northern Hemisphere, to the left in the Southern Hemisphere. If the flow is directed onshore, downwelling must occur; otherwise water would pile up against the shore. If the flow is directed away from the land, subsurface waters rise to the surface to replace water driven out to sea; upwelling is the result in this case. The different situations are schematically illustrated in Figure 9.8, assuming a perspective appropriate for the Northern Hemisphere.

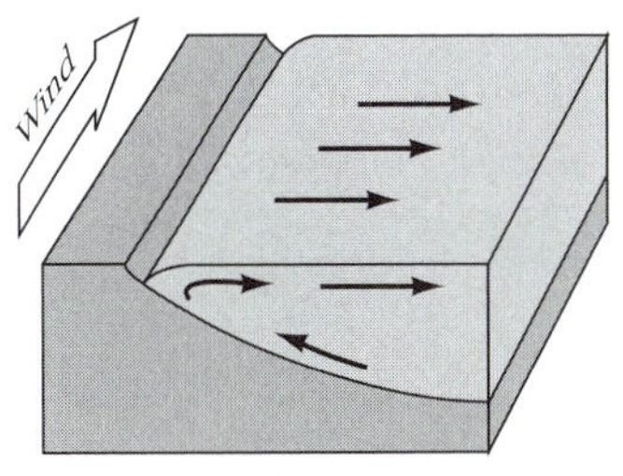

a. Upwelling

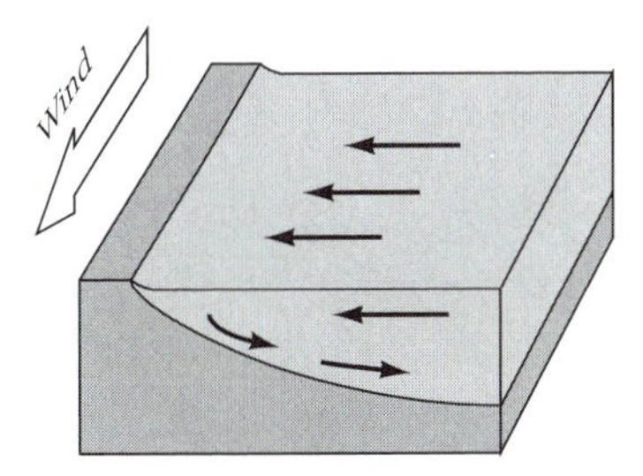

b. Downwelling

Figure 9.8 Upwelling (a) and downwelling (b) in the Northern Hemisphere. Source: Gross 1990.

Two quite distinct climatic regimes are observed in the tropics today. One is characterized by a strong upwelling off the Peruvian coast. Temperatures in the eastern tropical Pacific are lower than those in the west by up to 6°C. In the second regime, upwelling is suppressed and temperatures are relatively constant across the tropical Pacific Ocean. The average temperature of surface waters in the tropical Pacific is significantly lower in regime one than in regime two. The warm phase (regime two) is known as **El Niño**; the cold phase (regime two), in its most extreme configuration, is referred to as **La Niña.**[8]

Upwelling is driven in the cold phase by the southeasterly trade winds aligned fortuitously with the southeast to northwest trend of the Peruvian coast line (remember that, in the Southern Hemisphere, water is driven off shore, to the left of the direction of the wind). The thermocline is lifted to the surface as a consequence of the resulting upwelling, carrying an abundant supply of nutrients to the surface and fueling one of the world's richest fisheries. The trade winds, as they blow across the tropical ocean, cause warm water to pile up westward (near Indonesia). As a consequence, the thermocline deepens in the west and the ocean surface level rises to a point where it is eventually several meters higher in the west than in the east. This configuration is ultimately unstable. Eventually the system collapses. Warm water flows back across the ocean, reducing the gradient in temperature and the west-east contrast in thermocline depth. Redistribution of warm water, schematically shown in Figure 9.9, caps off the supply of cold, nutrient-rich water to the surface off the Peruvian coast and the fishery collapses, triggering the transition from the cold to the warm phase of tropical climate and signalling the onset of El Niño.

As was seen in Plate 4 (bottom), a band of high chlorophyll extends along the equator in all three major ocean basins. The high chlorophyll content of the equatorial ocean reflects the importance of upwelling driven by horizontal divergence of water at the surface. The divergence of the horizontal flow at the surface arises as a consequence of the easterly trade winds that cause surface waters to move to the south (left) below the equator, to the north (right) above. Divergence and upwelling are also associated with the region between the North Equatorial Current and the North Equatorial Countercurrent (at a latitude of about 10°N in the Pacific), whereas the area between the North Equatorial Countercurrent and the South Equatorial Current (at about 5°N) is associated with convergence and downwelling. The role of the Coriolis force in determining whether motion is convergent or divergent (up or down) is summarized for the tropical Pacific in Figure 9.10.

The equatorial undercurrent, which, as noted earlier, straddles the equator and carries water from west to east over a range of depths between about 20 and 150 m, is associated with convergence of water towards the equator (see Figure 9.10). Upwelling of water to the surface at the equa-

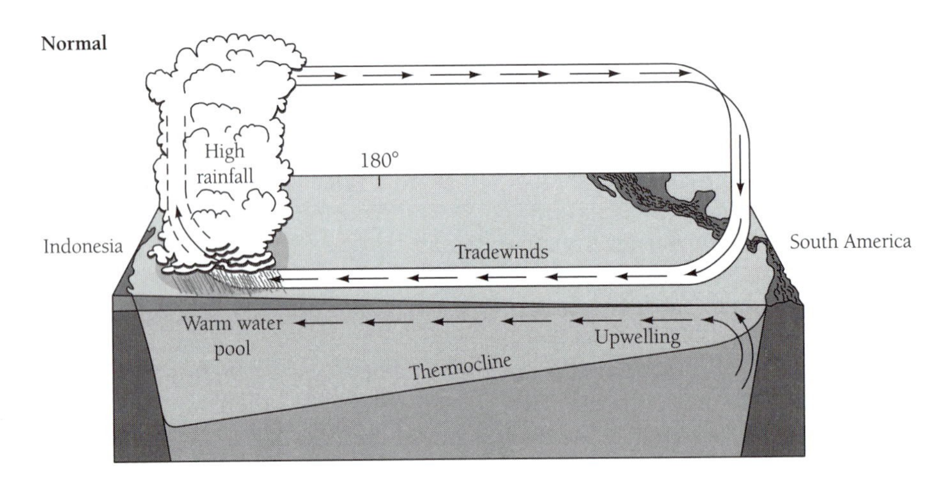

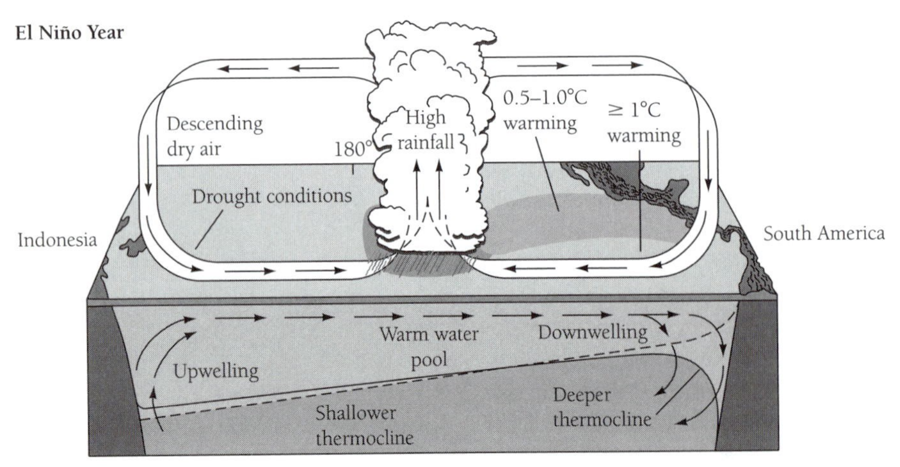

Figure 9.9 Comparison of El Niño versus normal conditions in the Pacific Ocean. Source: Skinner and Porter 1995.

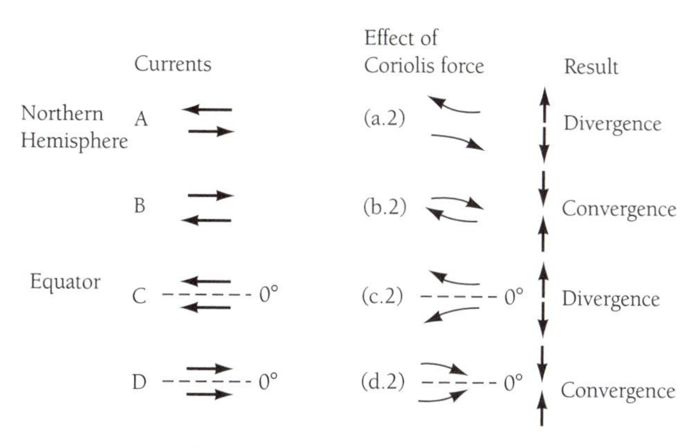

Figure 9.10 The role of the Coriolis force in determining consequences for either convergence (downward motion) or divergence (upward motion). Source: Pickard and Emery 1990.

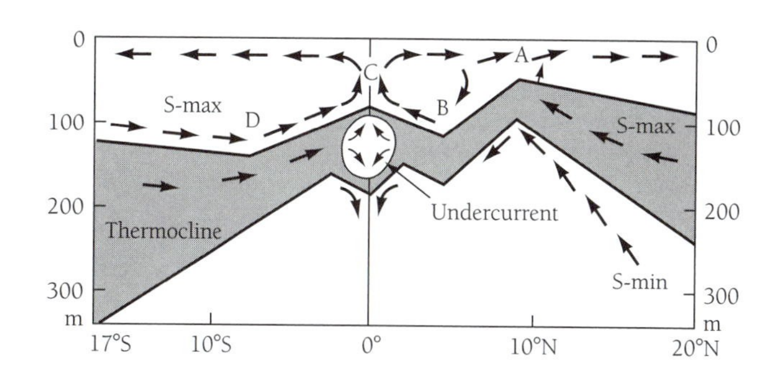

Figure 9.11 Patterns of meridional and vertical motion. Notice the effects of the divergent/convergent motion due to the Coriolis force: A, B, C, and D correspond to the same letters in Figure 9.10. Source: Pickard and Emery 1990.

tor is supplied in part by horizontal convergence associated with the undercurrent. Convergence and divergence of waters in different regions of the tropical ocean result in a complex pattern of meridional and vertical motion schematically illustrated in Figure 9.11. Note that when the vertical component of the motion is directed toward the surface, the isotherms are tilted upward. When the motion is directed down, the isotherms are oriented in the opposite sense.

The low-chlorophyll content of waters in the interior of the ocean gyres reflects the importance of horizontal convergence of surface water. Water is deflected to the interior of the gyres by the Coriolis force, resulting in subsidence. The circulation of the ocean is a mirror image of atmospheric flow. Air descends in the vicinity of the subtropical high pressure systems. As a result of friction, as discussed in Section 8.7, there is a net outflow of air from the center of the high. The frictional force responsible for the cross-isobaric flow is associated with transfer of linear momentum from the atmosphere to the ocean. It is precisely this frictional interaction that is responsible for the wind-driven component of the ocean's circulation; the force imparted by the atmosphere to the ocean is equal and opposite to the force applied to the atmosphere by the ocean. Outflow of air from the center of the high pressure region is associated with water inflow. Vertical motion is directed downward both in the atmosphere and in the ocean. The atmosphere at the center of the high pressure system is hot and generally cloud-free. The ocean at the surface in the interior of the gyres is nutrient-deprived and biologically unproductive, as indicated in Plate 4 (bottom).

As will be discussed in more detail in Chapter 11, carbon emitted to the atmosphere by the combustion of fossil fuels is derived from organic matter originally stored in sediments. To restore equilibrium, this carbon must be returned to the sediments. Before doing so, however, it must pass through the ocean. Carbon is taken up by the ocean mainly in the interior of the ocean gyres where, as seen above, the direction of vertical motion is favorable, meaning downward. (There is some uptake also in regions where deep water forms at high latitude, but this is generally less significant than uptake in the gyres, reflecting the larger surface area of the gyres.)

9.4 Summary

We have seen that the circulation of the ocean's surface is driven primarily by transfer of linear momentum from the atmosphere. Responding to a combination of surface stress, friction with underlying water, and the Coriolis effect, surface water tends to move in a direction inclined at an angle of 45° to the direction of the prevailing winds: to the right in the Northern Hemisphere, to the left in the Southern. The influence of wind stress extends to a maximum depth of about 100 m. The velocity of flow induced by the wind declines with depth, while the direction of water flow rotates with respect to the direction of the wind (in a clockwise sense in the Northern Hemisphere, counterclockwise in the Southern). Motion of water in the wind-driven surface layer traces out a characteristic pattern known as the Ekman spiral. Averaged over depth, the direction of flow of the surface layer is oriented at right angles to the direction of the wind stress. The force imposed by the wind is balanced by the Coriolis force arising as a result of the consequent bulk motion of surface water.

The wind-driven circulation of the ocean at midlatitudes is characterized by basin-scale quasi-circular features known as gyres. Warm water is transported from low to high latitudes in western portions of ocean basins, with eastern return flows. Flow in western loops of gyres is more focused than in the eastern, a consequence of the transfer of angular momentum from ocean water to the underlying solid planet. This transfer is effected primarily in the bordering coastal zone.

Circulation of waters in the gyres has an important influence on transport of heat from the tropics to higher latitudes. A portion of the water leaving the tropics in western boundary currents returns to the tropics as a component of water flow at intermediate depths of the ocean, in a region known as the thermocline. The thermocline separates the thin layer of relatively warm, low-density water at the surface from the much larger volume of cold, dense water accounting for the bulk of the water in the ocean. Intermediate water is formed as a result of the convergence of surface water in the interiors of the gyres. It returns to the surface in the tropics as a result of the divergence of surface water in the equatorial zone induced by the prevailing easterly trade winds.

Deep water is formed at high latitude, with a major contribution in the present environment from the North Atlantic. Formation of deep water depends on the contrast between the density of surface and deep water. The importance of the North Atlantic as a source of deep water today is a consequence of the relatively high salinity of surface water in the Atlantic as compared to that of the Pacific. The contrast in salinity between the Atlantic and Pacific results from a net transfer of water vapor from the Atlantic to the Pacific through the atmosphere. Deep water is also formed in the southern ocean near Antarctica. The salt content of the southern component of deep water is less than that of the deep water originating in the north. Circulation of deep water today follows a track described as the Conveyor Belt, originating in the North Atlantic and extending into the Pacific and Indian Oceans after passing around Antarctica. As we shall see, the mode of operation of the Conveyor Belt may have been different during glacial time. The flow of deep water may have originated then in the south rather than in the north. It is likely that deep water flowed north in the Atlantic during glacial time, returning to the south at shallow and intermediate depths. We shall argue in Chapter 10 that differences in modes of ocean circulation may have played an important role in the variety of climates that distinguished Earth in the past. A key question concerns the nature of changes that might arise in the future in response to changes in the concentration of greenhouse gases.

Climate through the Ages

10

Climate has varied dramatically over the course of Earth's 4.6 billion-year history. This chapter intends to provide an introduction to the range of climates that have visited Earth in the past. Our hope is that a careful survey of the past might educate our intuitions concerning the future. Our central premise is that the climate system is complex, involving a subtle array of processes both internal and external to the atmosphere—a suite of interactions coupling the atmosphere to the biosphere, cryosphere, and ocean, and implicating, on a sufficiently long time scale, the lithosphere.[1] An inclusive, mathematical description of the climate is beyond the present state of knowledge. Our objective here is to use enlightened studies of the past to elucidate processes that might be influencing climate today, to provide a context for the potential changes in future climate that may be driven by actions for which we are directly responsible.

Our current challenge is to predict the response of the climate system to an increasing burden of the greenhouse gases introduced earlier in Chapter 5. As indicated there, the concentration of CO_2 has risen from its pre-industrial value of about 280 ppmv to a level of more than 360 ppmv today, due largely to emissions associated with the burning of fossil fuels. It is expected to climb over the next few decades to values more than twice those of the pre-industrial era. An even larger increase, by almost a factor of three, has been observed for CH_4, the second-most abundant carbon-bearing gas in the atmosphere. And the concentration of N_2O, the longest-lived nitrogen compound in the atmosphere after N_2, is increasing at a rate fractionally similar to that for CO_2. There are greenhouse gases in the atmosphere, the industrial chlorofluorocarbons (CFCs) for example, for which there are no natural analogues. Chemical reactions triggered by industrial CFCs and analogous brominated gases have resulted in changes in both the height-integrated abundance and altitude distribution of O_3, particularly at high latitudes. All of these gases—CO_2, CH_4, N_2O, SF_6, the CFCs, and O_3—play a role in regulating the transmission of infrared radiation from Earth's surface into space. Together with water, in both its condensed and vapor forms, they provide the dominant contributions to the contemporary greenhouse.

There is general agreement that the direct effect of an increase in CO_2 and other greenhouse gases should be to warm the atmosphere, especially near the surface (its climate forcing is qualitatively positive).[2] A warmer atmosphere can accommodate a larger abundance of H_2O vapor, the most important greenhouse agent in the atmosphere today. Most models predict that an increase in H_2O should amplify the warming induced by CO_2 and other greenhouse gases—meaning the feedback on climate should be positive. But the issue is not without controversy. The greenhouse effect is sensitive to the abundance of water vapor in the upper troposphere. Richard Lindzen, a meteorologist at MIT, points out (Lindzen 1990) that H_2O is delivered to upper levels of the troposphere primarily by transport associated with convective systems (violent storms, mainly in the tropics) of relatively small spatial scale. Precipitation in these systems serves to dry the air as it rises, providing an important means for heat transfer from the surface to upper levels of the troposphere. It is due to this precipitation that the relative abundance of H_2O vapor in the upper troposphere is appreciably less than it is near the ground. Lindzen suggests that an increase in the concentration of CO_2 could serve to amplify the strength of convection. Cumulus systems could penetrate to higher, colder levels of the atmosphere. As a consequence, a relatively larger fraction of their initial burden of H_2O could be removed by precipitation. The abundance of H_2O in the upper troposphere would fall in this case, rather than rise, as a result of an increase in CO_2. Reduction in the concentration of H_2O in the middle and upper tropospheres would lead to a decrease, rather than an increase, in surface temperature. According to this view, the response of H_2O to an increase in CO_2 and other greenhouse gases should result in a negative, rather than positive, feedback: the change in H_2O should serve to offset warming induced by the increase in CO_2.

The difficulty with this suggestion is that the abundance of H_2O in the upper troposphere is not determined simply by the supply of that compound in vapor form. If it were, the abundance of H_2O in the upper troposphere of the contemporary system would be much less than presently observed. The resolution of this dilemma lies in the fact that water is transported to the upper troposphere not simply in vapor form but to a significant extent also as ice and liquid. Subsequent evaporation of its condensed phase provides the dominant source of H_2O to the upper troposphere. The supply of H_2O to upper altitudes is regulated by complex microphysical processes operating in individual cumulus towers and in the systems of towers characteristic of intense convective activity. Accurate simulation of the relevant physics requires exceptionally high spatial and temporal resolution, capabilities beyond the current, or even prospective, state of the art of three-dimensional general circulation models (GCMs) of the atmosphere.

Clouds have a dual impact on climate. An increase in cloud cover tends to cool the planet by raising the albedo, resulting in an enhanced reflection of incident solar radiation. At the same time, clouds contribute to the greenhouse effect, trapping infrared radiation that would otherwise readily escape into space.[3] Measurements from the Earth Radiation Budget Experiment (ERBE) reveal that clouds are responsible for a net cooling of Earth today. Cooling attributed to clouds registered 16.6 W m^{-2} on a globally averaged basis during July 1985, reflecting a difference between the short-wave cooling of 46.7 W m^{-2} and the long-wave heating of 30.1 W m^{-2}. Similar results were obtained for April 1985, October 1985, and January 1986. Radiative forcing (positive and negative) associated with clouds in the present environment is large compared to the magnitude of the forcing expected to arise as a consequence of enhanced levels of greenhouse gases: direct forcing associated with a doubling of CO_2, for example, is estimated at about 4 W m^{-2}. It is clear that any accurate assessment of the climate system's net radiative forcing induced by enhanced levels of greenhouse gases will require a precise treatment of the response by clouds and water vapor to different levels of gases such as CO_2.

Feedbacks on the climate system involving H_2O vapor and clouds (provoked for example by an increase in CO_2) can take effect almost instantly. Additional rapid interactions can be identified relating to changes in soil moisture affecting transfer of H_2O between the surface and atmosphere, and to variations in snow and ice cover, both on land and at sea, resulting in regional modification of Earth's albedo. Feedbacks operating on longer time scales (decades to centuries) are likely to involve shifts in oceanic circulation. Shifts in the ocean's circulation may be expected to result in changes in weather patterns, accompanied by shifts in vegetation, with implications for the albedo and heat capacity of the surface and for the exchange of H_2O between surface and atmosphere. A comprehensive model of climate accounting for these feedbacks requires a realistic simulation of the time-varying, coupled atmosphere-ocean-biosphere-cryosphere system, a task far beyond the range of current capability.

There are serious problems with even the atmospheric module of such a coupled system, while difficulties in modeling the ocean and biosphere-cryosphere are still more severe. As discussed above, a number of the key processes regulating the distribution of H_2O vapor, the production of precipitation, and the formation and dissipation of clouds involve physical processes operating on scales beyond the resolution of even the most elaborate current GCMs. A variety of strategies is used to deal with this problem. Small-scale processes are parameterized usually in terms of larger-scale variables predicted by the model. Invariably, GCMs are tuned to provide reasonable agreement with selected features of the present climate. The scientific problem involved in predicting future climate relates to the difficulty in testing the credibility of models for different atmospheric configurations. There are reasons to question the reliability of the simulations by GCMs for even the present climate. P. H. Stone and J.S. Risbey, colleagues of Richard Lindzen, point out that GCM predictions of meridional heat transport for the present atmosphere differ from observations by as much as a factor of two (observed fluxes are typically less than those obtained using models). They offer convincing evidence that modeled results are sensitive to the particular schemes adopted to simulate subgrid-scale processes.

It is apparent that we should view detailed results from models attempting to simulate the response of climate to contemporary increases in the concentration of greenhouse gases with caution. Our major reservation concerns the possibility that important feedbacks may be omitted from models. Uncertainty, of course, is a two-edged sword. One could argue that warming over the past century has been comparatively minor, but this should not be taken as a guarantee that changes in the future may not be more serious. There is little, we will argue, to support Lindzen's suggestion that an increase in greenhouse gases such as CO_2 should trigger a reduction in the abundance of H_2O in the upper troposphere and thus effectively cancel the direct-warming impact of the initial perturbation. The weight of the observational evidence presented below suggests that the climate is exceptionally variable. This would appear to fly in the face of the hypothesis that water should serve as an efficient climatic thermostat.

We begin our retrospective in Section 10.1 with a summary of changes in climate that have taken place over the past 150 years. Building on this base, we present a more expansive view, covering the past 20,000 years, in Section 10.2. Our perspective is extended back to about 3 million years B.P. in Section 10.3. We open a 130-million-year window in Section 10.4 and we extend our review back to the earliest epochs of Earth's history in Section 10.5. We present our summary remarks in Section 10.6.

10.1 The Past 150 Years

Gabriel Fahrenheit's invention of the thermometer in 1714 enabled scientists to carry out relatively straightforward measurements of temperature for the first time. Early applications of Fahrenheit's device for air temperature measurements were spotty. We can, however, construct a useful record of changes in global temperature for the past 150 years by using archival data from a wide range of stations around the world.

Obtaining such a record is no easy task. As we know, there are systematic differences in temperature between day and night. The amplitude of the diurnal change in temperature is typically greater in continental, as compared to marine, environments. Temperatures in cities are usually higher than in surrounding rural areas, reflecting intensive energy use and waste heat disposal by closely spaced households and industries. The diurnal temperature range in urban environments is normally different (less) than in surrounding rural areas as a consequence of the tendency of buildings to retain heat during the night and to warm up more slowly during the day. Most important, efforts to develop a record of geographically distributed (global or hemispheric) changes in surface temperature are complicated by the lack of data for the relatively large fraction of Earth's surface occupied by oceans and by the scarcity of measurements for the sparsely populated Southern Hemisphere.

Consider the challenge of constructing a long-term record for a particular location. Suppose that data for surface temperature are more or less continuously available, say, twice a day for the past 150 years. Most likely, this record will have been taken in an urban environment. It is probable that in the early years measurements were made at a convenient location within the city. More recently, the recording station may have been relocated to a site outside the city, an airport for example. Temperatures at the airport may be less on average than temperatures in the city. More subtly, there may have been changes in the time of day at which measurements were taken. The analyst must be aware of these changes in measurement protocol; otherwise inconsistencies will confound the record. Data must be corrected, or adjusted, to ensure a consistent basis for the study of long-term trends.

The most extensive record of temperature changes for marine environments comes from shipboard measurements of surface water. In the early years these measurements were taken by dipping a thermometer into a bucket of seawater lifted onto the deck of the ship. If the water temperature was measured immediately, we might expect this procedure to provide an accurate record of changes in sea surface temperature. Typically, though, there was a delay between lifting the water to the ship's deck and the time when the measurement was taken. In the interim, the temperature of the water would have relaxed to the temperature of the ambient air. To correct for this effect, the analyst must know something about the insulating properties of the bucket and make assumptions concerning the delay associated with a typical measurement. Was the bucket made of poorly insulating material, such as canvas, or a better-insulating material, such as wood? In recent years, shipboard measurements of surface water temperature have been taken using water continuously drawn on board as a cooling agent for the ship's engines. Again, there is a need to apply corrections to obtain a consistent long-term record since cooling water is taken from a region typically deeper than that sampled by buckets.

Given these complications, it is clear that the analysis of long-term trends in temperature must be carried out with painstaking attention to detail. This task has been implemented with skill over the past several decades by a group at the Climatic Research Unit of the University of East Anglia in England, led by P. D. Jones. Their work formed the basis for much of the analysis of recent climate trends summarized by the Intergovernmental Panel on Climate Change (IPCC)[4] and is extensively adopted for purposes of the following discussion.

A summary of trends for spatially averaged values of temperature at the land's and ocean's surface for the Northern Hemisphere from 1861 to 1994 is given in Figure 10.1a. Analogous data for the Southern Hemisphere and the globe are displayed in Figures 10.1b and 10.1c, respectively. A comparison of globally averaged trends distinguishing between land and ocean is given in Figure 10.1d. The data in Figure 10.1 are presented as differences with respect to averages for the 1961–1990 time period. They provide clear evidence of the unusual warmth for the past several decades. Emphasizing the significance of this result, the 1995 IPCC report concluded that Earth's average temperature is most likely higher now than it has been for at least the past 500 years.

Warming did not proceed uniformly over the past 150 years. The data in Figure 10.1 indicate that temperatures (between 1860 and 1910) were essentially constant for the Northern Hemisphere, the Southern Hemisphere, and the globe. They increased by about 0.4°C between 1910 and 1940. There are indications of a small decrease (by about 0.1°C) between 1940 and 1970. The increase resumes after 1970, climbing by about 0.3°C between 1970 and 1994. Patterns in temperature changes observed for the Northern and Southern Hemispheres and for the globe are remarkably similar. Lending confidence to the analysis, trends obtained

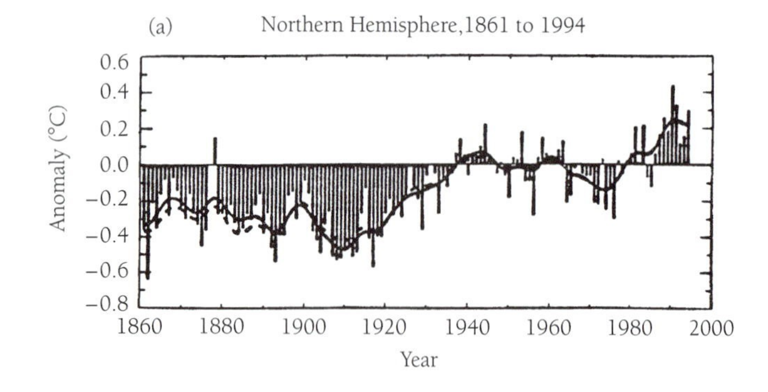

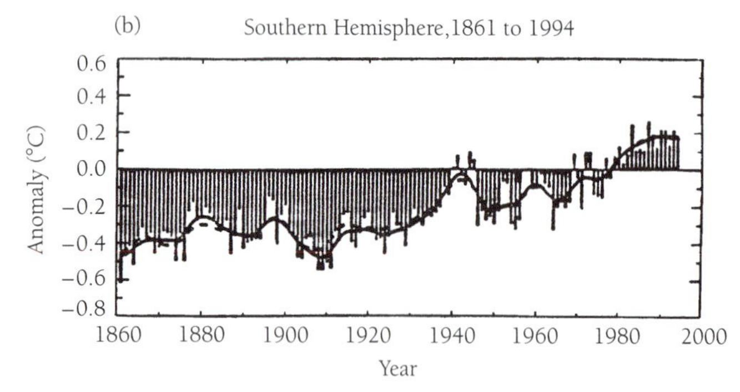

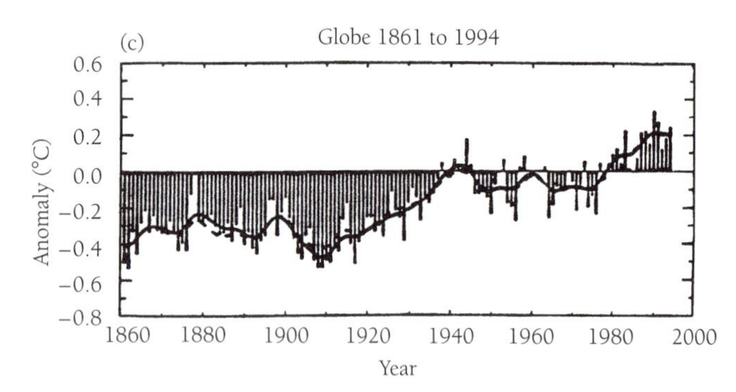

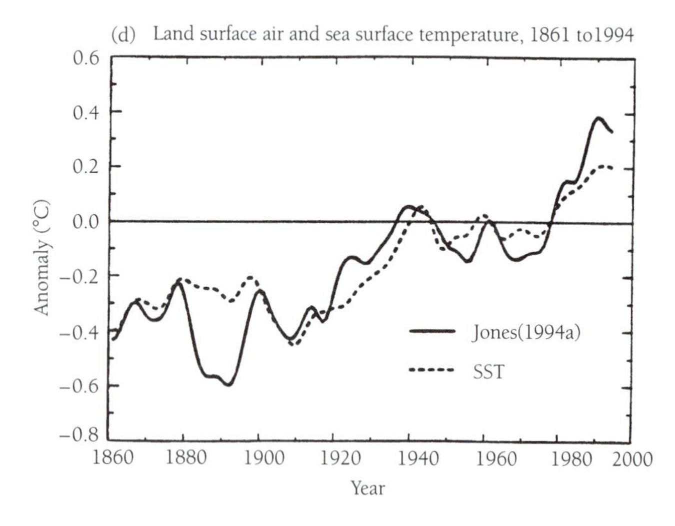

Figure 10.1 Combined annual land-surface air and sea-surface temperature anomalies (°C) from 1861 to 1994, relative to 1961 to 1990, for (a) Northern Hemisphere, (b) Southern Hemisphere, (c) Globe, and (d) Global land-surface air temperature (solid line) and sea surface temperature (SST, broken line). Source: IPCC 1996.

for sea surface temperature are qualitatively similar to those derived for land.

The data in Figure 10.1 provide an opportunity to test our understanding of the factors responsible for the changes in contemporary climate. As previously noted, the key question is whether the recently observed trends should be considered natural or, alternatively, an indication of changes due to the buildup of greenhouse gases. Results from a number of models considered in the first IPCC assessment are presented in Figure 10.2. The models are distinguished by sensitivities assumed for the globally averaged (equilibrium) change in surface temperature expected as a result of a doubling of CO_2.[5] As noted by the IPCC, the warming implied by the data in Figure 10.1 is actually *less* than we would have anticipated based on existing models for the climate's response to a doubling of CO_2. Results from some 17 coupled atmosphere-ocean general circulation models (AOGCMs) surveyed by the IPCC in 1995 indicated sensitivities for a doubling of CO_2 in a range of 2.1–4.6°C. Taken at face value, the comparisons in Figure 10.2 imply that either the sensitivity of climate to a doubling of CO_2 is lower than implied by models (perhaps as small as 1°C) or the approach to equilibrium is unexpectedly slow or factors other than the buildup of greenhouse gases have contributed significantly to the changes observed over the past 150 years.

The IPCC, in its second assessment, published in 1996, suggested that the discrepancy could be resolved by taking into account the cooling introduced by sulfate aerosols formed by the oxidation of sulfur dioxide released as a byproduct of the combustion of fossil fuels (see Chapter 18). Sulfate aerosols reflect light in the visible portion of the spectrum. Therefore they contribute directly to a local increase in the planetary albedo and a consequent reduction in the absorption of solar radiation. They have an additional, potentially important, indirect effect. Condensation of water in the atmosphere (as indicated in Chapter 7) does not normally proceed by a direct transfer from gas to the condensed phase. Rather, it is mediated by what are known as condensation nuclei, by particles with chemical properties favoring an absorption of water. Sulfate aerosols play an important role as condensation nuclei. The higher the abundance of condensation nuclei, the larger the number of cloud droplets that should result from a given condensation event; a given amount of liquid water or ice should be distributed over a greater number of cloud particles. The reflective properties of the cloud would be accordingly enhanced, resulting in a further tendency to cooling.

Results from a representative study (Mitchell et al. 1995) of the combined effect of the increase in greenhouse gases and the potential impact of sulfate aerosols are presented in Figure 10.3. The agreement between modeled and observed trends in globally averaged temperatures is clearly much improved over the comparisons displayed in Figure 10.2. It should be emphasized, though, that the treatment of the sulfate effect is subject to considerable uncertainty. In particular, the model adopted here did not allow for the indirect effects of sulfate noted above. Further, it is doubtful that the model

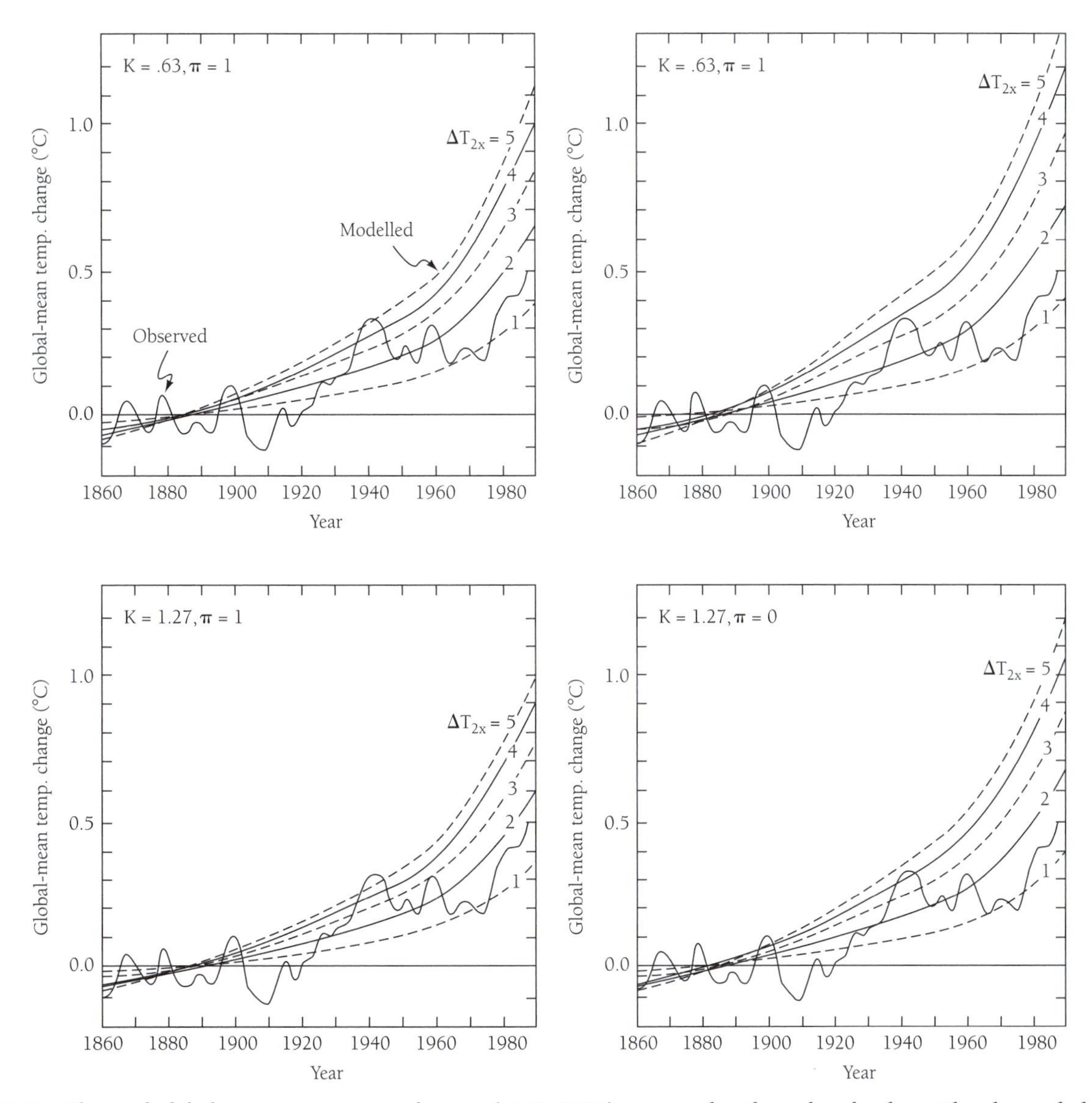

Figure 10.2 Observed global-mean temperature changes (1861–1989) compared with predicted values. The observed changes are smoothed to show decadal and longer time-scale trends more clearly. Observed and modeled data are both adjusted to have a zero mean over 1861–1900. The four panels show results using models with various values for the parameters K and π. Different curves reflect different sensitivities with respect to doubled CO_2 as indicated. Source: IPCC 1990.

could account for the temperature trends observed in the Southern Hemisphere, where emissions of anthropogenic sulfur should be relatively small but where trends in temperature are nonetheless similar to those observed in the north (cf. Figure 10.1). It is apparent, nonetheless, that sulfur emissions could have a significant impact on the climate, at least locally. They could contribute, for example, to the lack of a clearly established trend for temperatures over the continental United States. Indeed, as noted by Henderson-Sellers (1989), there is evidence of an increase in cloud cover over North America during the past century, with the largest increase observed between 1930 and 1950, roughly coincident with the period of modest cooling implied for the Northern Hemisphere in Figure 10.1a.

The results in Figure 10.3 were obtained by using a coupled atmosphere-ocean general circulation model distinguished by a relatively high sensitivity to forcing from greenhouse gases (5.2°C for a doubling of CO_2 in equilibrium). Results from a number of simpler models, termed *upwelling diffusion-energy balance models* by the IPCC,[6] are presented in Figure 10.4. The models in Figure 10.4 explored a range of sensitivities to CO_2-doubling, varying from 1.5 to 4.5°C. Allowing for effects of changes in greenhouse gases, sulfate aerosols, and solar radiation (Figure 10.4c), an optimal fit to the observational record is achieved assuming sensitivity for doubling in the range 3–4°C.

The analysis reported here should not be taken as a serious constraint on the *actual* sensitivity to CO_2. A rigorous constraint would require a more complete investigation of all of the factors contributing to climate change over the past 150 years. To accomplish this more demanding task, for example, we would need to understand, and incorporate in our model, a realistic simulation of climatic changes arising due to long-term, century-scale variations in oceanic circulation. We would need to understand the factors responsible

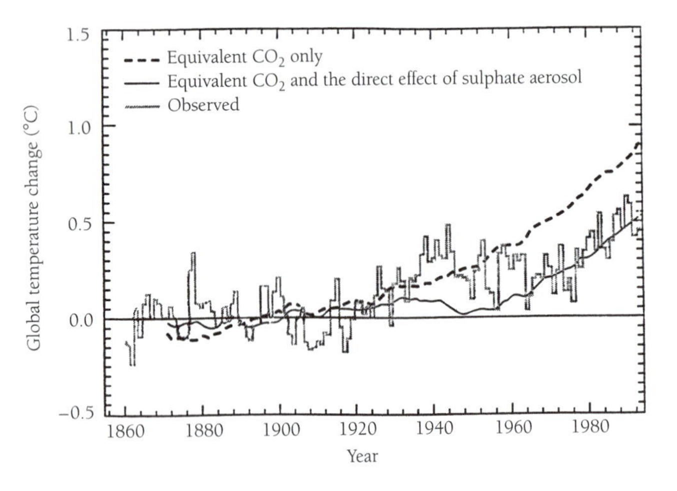

Figure 10.3 Simulated global annual mean warming, 1860–1990, allowing for increases in equivalent CO_2 only (dashed curve) and allowing for increases in equivalent CO_2 and the direct effects of sulfates (solid curve). The anomalies are calculated relative to 1880–1920. Source: IPCC 1996.

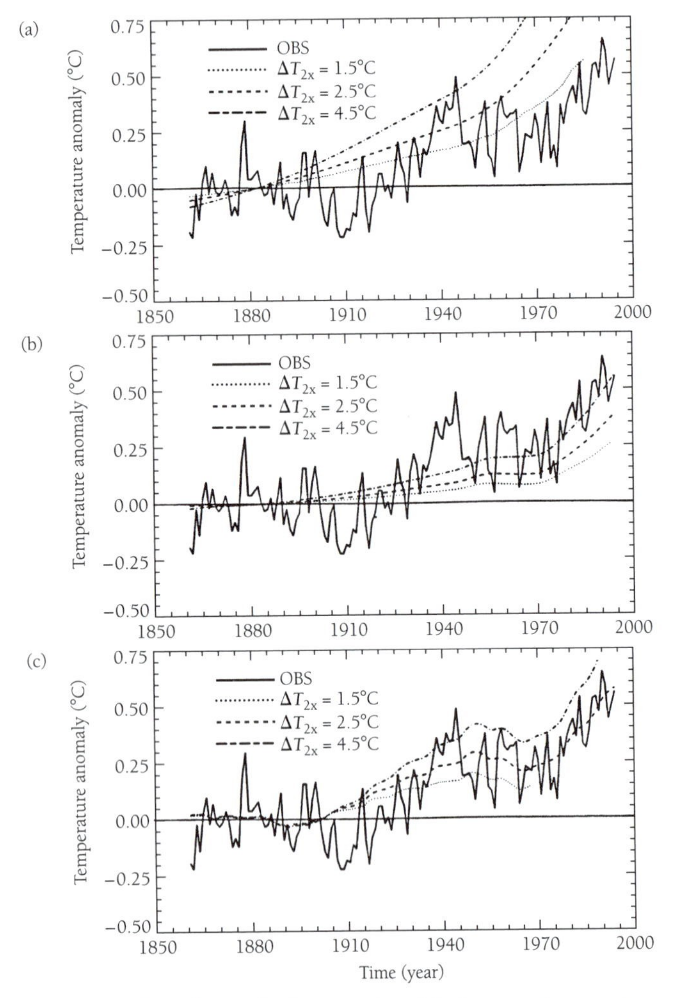

Figure 10.4 Observed changes in global-mean temperature, 1861–1994, compared to simulations using an upwelling diffusion-energy balance model: (a) greenhouse gases alone, (b) greenhouse gases and aerosols, and (c) greenhouse gases, aerosols, and an estimate of solar irradiance changes. Source: IPCC 1996.

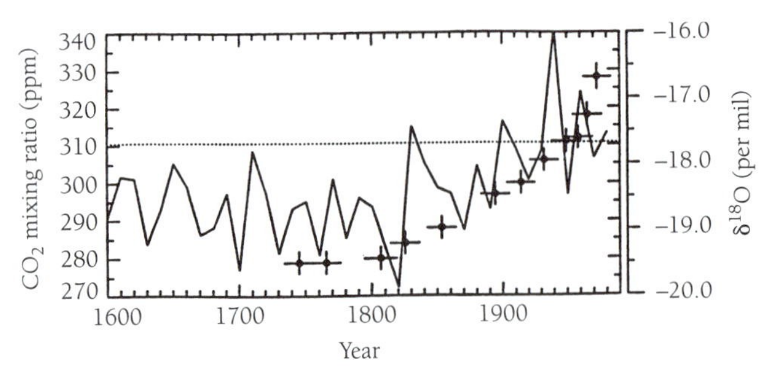

Figure 10.5 Variations in CO_2 concentrations inferred from measurements of gas recovered from ice at Siple Station, Antarctica (indicated by +) compared with values of $\delta^{18}O$ for ice from a core drilled at Huascarán, Peru (solid line, a surrogate for tropical temperature). Source: McElroy 1994.

for the Little Ice Age. And we would need to simulate how natural changes in oceanic circulation may have been altered by climatic changes introduced by rising levels of greenhouse gases and sulfate aerosols.

The relatively constant temperatures of the first 50 years of the record in Figure 10.1 could reflect a balance between warming driven by a rise in the concentration of greenhouse gases and a tendency toward cooling associated with the terminal stages of the Little Ice Age. As indicated in Figure 10.5, the end of the Little Ice Age is roughly coincident with the beginning of the modern rise in the concentration of CO_2. Absent the rise in CO_2, it is possible that the Little Ice Age might have persisted. We could speculate, in the absence of an anthropogenic source of CO_2, that temperatures today could be as cold as, or even colder than, they were in 1860. The implication is that climate may be still more sensitive than we believe to even small changes in CO_2.

Suppose, on the other hand, that the end of the Little Ice Age was a natural phenomenon; temperatures could have been as warm as they are today, even in the absence of the stimulus from additional CO_2. The inference in this case is that concerns over effects of higher levels of greenhouse gases in the future could be overstated.

On a purely empirical basis, evidence for warming induced by human activity must be considered ambiguous. In judging whether to take the issue seriously, we must be guided by our confidence in the credibility of models and by the intuitions we glean from informed studies of the past. We return to this issue later in Chapter 19.

10.2 The Past 20,000 Years

The transition from the last glacial maximum (LGM) to the present interglacial began about 19,000 years before the present (19 kyr B.P.): this is well documented in a remarkable historical record of the sea level obtained by Fairbanks (1989) in his pioneering study of coral in sediments off Barbados. Further detail is provided by measurements of the isotopic composition of carbonate shells of organisms preserved in deep sea sediments and in ice preserved at a variety of locations ranging from Greenland to Antarctica (Johnsen et al.

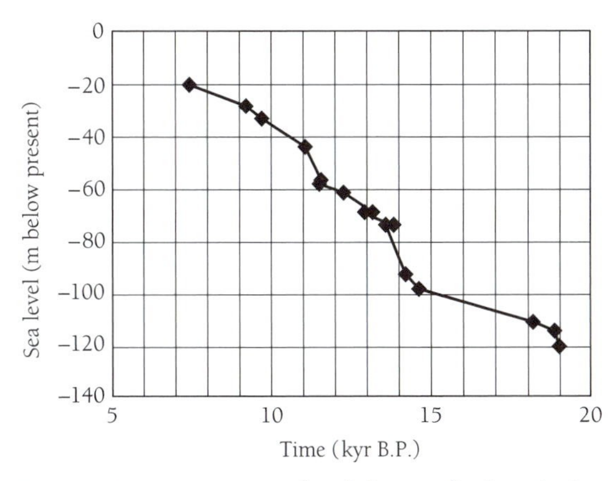

Figure 10.6 Rise in sea level during the last deglaciation.

1992; Jouzel et al. 1987). As indicated in Figure 10.6, the transition was marked by a slow but steady initial rise in sea level, averaging approximately 20 m in about 4000 years between 19 kyr B.P. and 15 kyr B.P. The rise accelerated rapidly at about 14 kyr B.P., climbing at an average of close to 4 meters per century between 14.2 and 13.7 kyr B.P. The subsequent increase was more gradual, with the rate of increase falling to a minimum at about 11.5 kyr B.P. during a cold period known as the Younger Dryas (discussed in more detail below). It picked up again about 11 kyr B.P., roughly coincident with the end of this neoglacial cold snap.

The recovery from the LGM, as recorded in ice cores, is summarized in Figure 10.7. The isotopic composition of snow or rain depends on the temperature at the point of the vapor's origin. Light isotopes of water evaporate more readily than heavy, reflecting differences in equilibrium vapor pressures. The contrast in vapor pressures decreases with increasing temperature. As a consequence, precipitation in cold regions is characteristically lighter than in warm.[7] The data in Figure 10.7, encompassing measurements of the isotopic composition of water in ice cores drawn all the way from Central Antarctica to Greenland and including the tropics, reflect this trend. Note that warming in Antarctica, as indicated by the data for Vostok and Byrd, begins at about 20 kyr B.P. Warming appears in the tropics roughly four thousand years later and is observed essentially simultaneously in Greenland.[8] The data from Vostok suggest that warming in the tropics and in Greenland is associated with modest cooling in Antarctica. A sharp pulse of warming is seen both in the tropics and in Greenland beginning at about 15 kyr B.P. This is followed by an abrupt climatic reversal, a resumption of near glacial conditions persisting for about two thousand years and setting in at about 13 kyr B.P. The cold snap ends abruptly, in both Greenland and the tropics. Counting annual layers in the Greenland core suggests that the final cold-to-warm transition may have taken place in a time interval as brief as 20 years!

Thompson et al. (1998) refer to the warm-cold-warm transition observed in the tropical and Greenland cores between 14 and 11.6 kyr B.P. as the **deglaciation climate reversal (DCR)**. It appears to be related to the Younger Dryas period, defined by a major shift in vegetation type over an

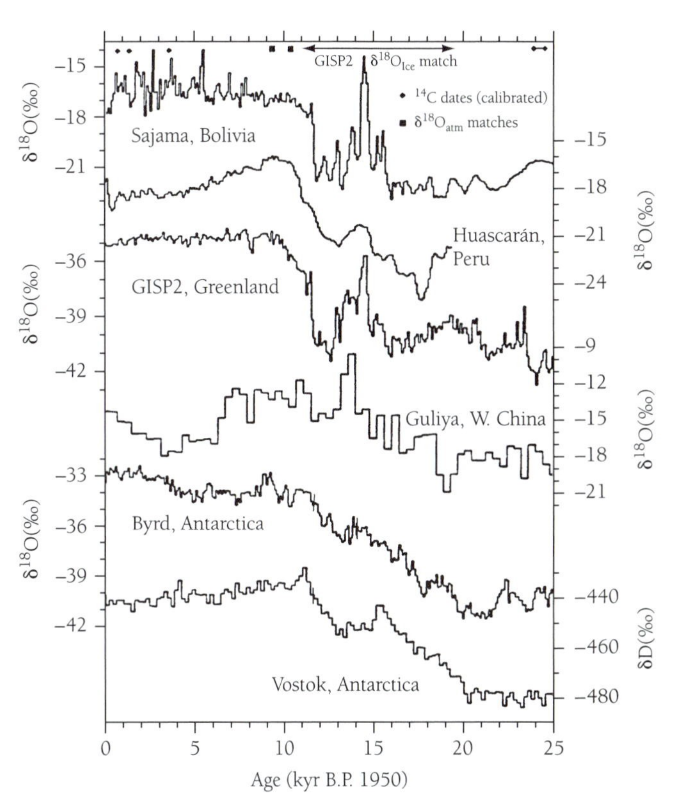

Figure 10.7 Interhemispheric comparison of stable isotope records from ice cores. Source: Thompson et al. 1998.

extensive region around the North Atlantic (forests were replaced by tundra in a time interval of 100 years or less). Like the Younger Dryas, the DCR ends abruptly. Indications of the DCR are notably absent in data for Antarctica presented in Figure 10.7. In contrast, it appears that cooling in the tropics and Greenland may have been temporally associated with a resumption of the warming trend in Antarctica.

Large-scale changes in climate, as manifest by recurrent ice ages and interglacials, are usually attributed to temporal variations in the seasonal input of solar radiation. Earth's rotational axis is tilted at an angle of about 23.5° today with respect to a perpendicular drawn through the ecliptic, the plane defined by Earth on its orbit around the Sun. This tilt is primarily responsible for the seasonal variation of solar radiation received at any given latitude over the course of a year. The axis maintains a more or less constant direction with respect to the fixed stars as Earth progresses around its orbit. During northern winter, for example, the Northern Hemisphere is tilted away from the Sun. During Northern Hemispheric summer, it is tilted towards the Sun. Solar radiation is obliquely incident in the Northern Hemisphere during winter. As a consequence, a given surface area of the Northern Hemisphere receives less energy from the Sun during winter than it does during summer; days are shorter and nights are longer. At the winter solstice, 21 December, latitudes higher than 66.5°N are dark 24 hours a day. In contrast, at the summer solstice, 21 June, the Sun never sets inside the Arctic Circle. Days and nights are equal in duration everywhere at the equinoxes, 21 March and 22 September. The geometry responsible for this situation is illustrated in Figure 10.8.

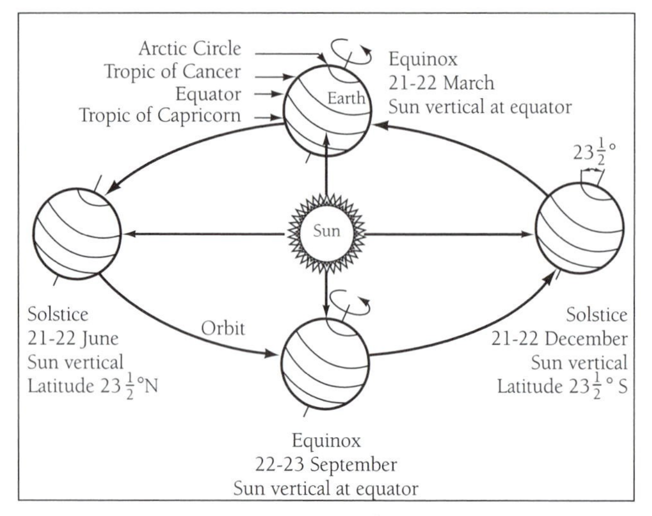

Figure 10.8 Seasons. Source: Anthes.

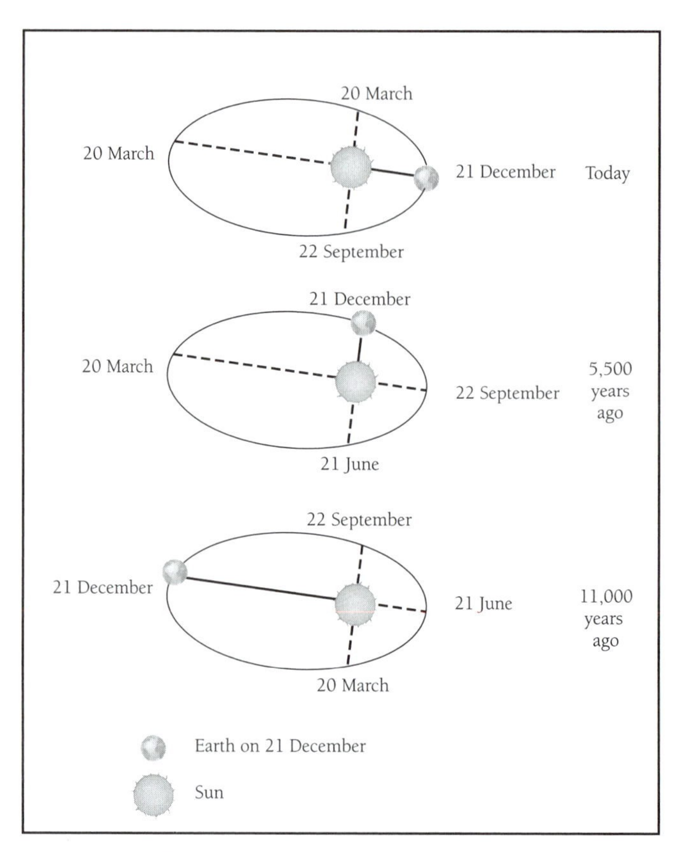

Figure 10.9 Precession of the equinoxes. Source: Imbrie and Imbrie 1978.

As hinted above, the direction of the rotational axis is not precisely constant with respect to the fixed stars. Due to gravitational interactions with other planets, the tilt of the rotational axis (defined with respect to a direction perpendicular to the ecliptic) varies from a minimum of 22° to a maximum of 25° and back, over a period of 41 kyr. We are currently at about the midpoint of the range heading toward lower values. In addition to its tendency to bob up and down, the rotational axis undergoes slow precession, describing a circle on the celestial sphere once every 26 kyr. If we think of Earth as a spinning top, the top is not only spinning but it is also wobbling. Viewed from the vantage of the fixed stars, the North Pole appears to rotate in space, tracing out a circle on the celestial sphere once every 26 kyr. At the same time, the center of the circle appears to oscillate back and forth repeating this pattern once every 41 kyr.

Earth traces out an ellipse as it moves on its annual cycle around the Sun. The degree of ellipticity—the departure from circularity of the orbit—is defined in terms of *eccentricity*. At present, the eccentricity of Earth's orbit is 0.06 (a circular orbit is characterized by an eccentricity of 0, a highly eccentric orbit by an eccentricity approaching 1.[9] The ellipse is not fixed in space. Like the axis of rotation, it precesses with respect to the fixed celestial reference frame, rotating around one focus of the ellipse, describing a complete cycle every 105 kyr. Axial and orbital precession combine to cause an apparent change in the position of Earth on its orbit around the Sun for any given season. A complete cycle around the ellipse takes about 22 kyr (more precisely, it fluctuates with associated periods of 23 kyr and, to a lesser extent, 19 kyr). That is to say, if we were to mark the position of Earth on its orbit at a particular season today, spring equinox for example, the position would appear to drift around the orbit with time, returning to the current position roughly 22 kyr from now. This effect is referred to as the **precession of the equinoxes**. During Northern Hemispheric winter today, Earth is closer to the Sun than it is for Northern Hemispheric summer (or Southern Hemispheric winter). As illustrated in Figure 10.9, approximately 11 kyr from now (or 11 kyr ago), the positions are reversed: winter in the Southern Hemisphere corresponds to the time when Earth is closest to the Sun, winter in the Northern Hemisphere, when it is furthest away.

Axial tilt, most commonly referred to as **obliquity**, and equinoctial precession are both important in influencing the flux of solar radiation received on any given day of the year at any selected latitude. The effect of obliquity is greatest at higher latitudes, while equinoctial precession is most important in the tropics and subtropics.[10] There is a third effect influencing the seasonal pattern of solar radiation received by Earth: long-term variations in ellipticity. The flux of radiation received at any given latitude for any given season is modulated by changes in ellipticity, primarily on time scales of 100 and 400 kyr. Ellipticity-related changes are small compared to changes induced by variations in obliquity and equinoctial precession. Changes in insolation for northern spring, summer, autumn, and winter are displayed for selected latitudes for the past 100 kyr in Figures 10.10–10.13.

The theory most commonly advanced to account for the presence of ice ages and interglacials is attributed to the Serbian astronomer Milutin Milankovitch (1941). According to this theory, an increase (or decrease) in energy received from the Sun during summer should favor recession (or expansion) of land-based ice.[11] Changes in ice cover would be amplified by positive feedback associated with changes in local albedo. Exposure of dark rock to sunlight should accelerate recession of the ice sheet. Conversely, expansion of ice cover should result in an increase in albedo favoring further

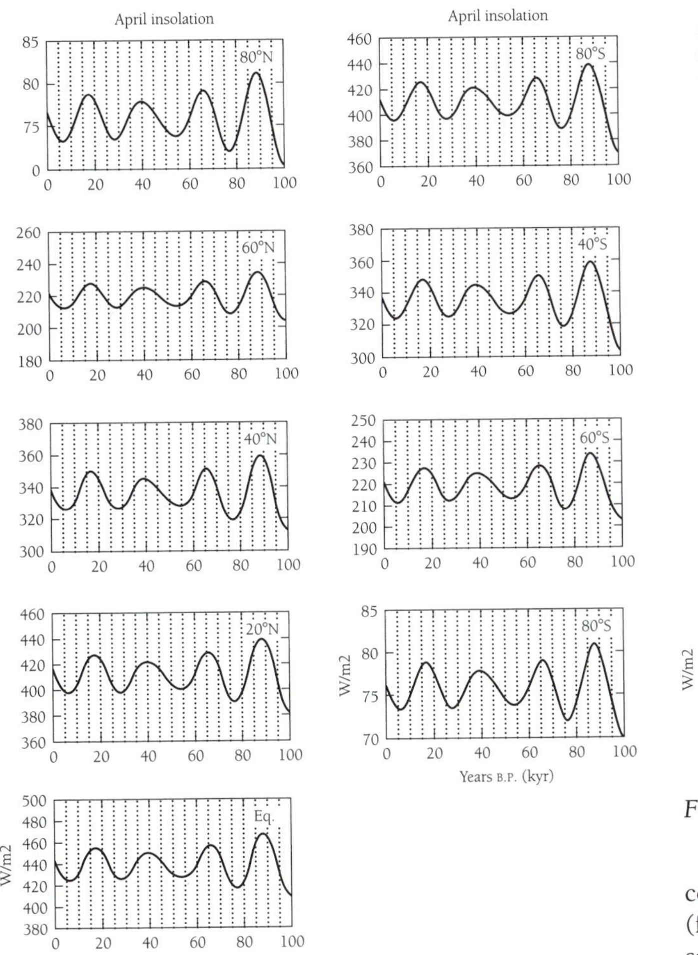

Figure 10.10 Insolation in April during the last 100 kyr.

extension. We would expect the Milankovitch effect to operate primarily on the large time-varying continental ice sheets of the Northern Hemisphere. In this case, it would appear that climatic changes should originate primarily in the north. As noted above, however, the first and most dramatic signs of the end of the last ice age show up in Antarctica. Indeed, there is evidence that changes in the climate of Antarctica preceded changes in Greenland over an extensive portion of the last ice age, from 47 to 23 kyr B.P., typically by between 1 and 2.5 kyr (Blunier et al. 1998). Warming in Antarctica marking the end of the ice age was accompanied, however, by a rise in sea level, as indicated in Figure 10.6. It is reasonable to assume that changes in sea level are induced primarily by changes in the volume of land-based ice. Since the ice sheets of the glacial epoch are located mainly in the Northern Hemisphere, it would appear likely, therefore, that warming at the end of the ice age may have set in essentially simultaneously in both hemispheres.[12] The absence of a detectable signal of this warming in Greenland may reflect inertia imposed by the presence of the large Laurentide Ice Sheet covering North America.

The glacial-interglacial transition is also associated with major changes in the concentrations of atmospheric CO_2 and CH_4, as summarized in Figures 10.14 and 10.15. The

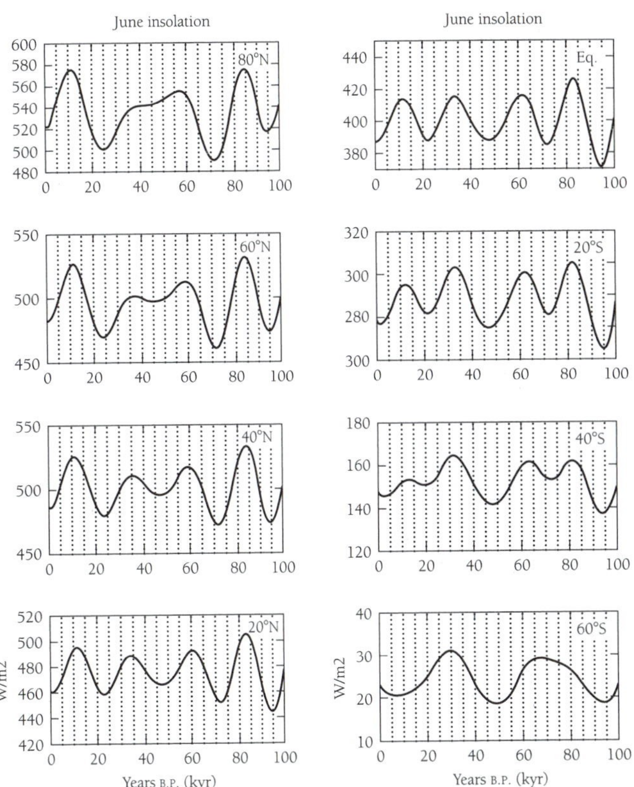

Figure 10.11 Insolation in June during the last 100 kyr.

concentration of CO_2 increased by approximately 80 ppmv (from 200 to 280 ppmv) over an interval of about 10 thousand years. The increases in CO_2 and CH_4 began at about the same time, essentially coincident with both the onset of warming in Antarctica and with the beginning of the rise in sea level indicated in Figure 10.6. The initial rise in CH_4 was relatively modest. The rate of increase picked up significantly at about 15 kyr B.P., roughly coincident with the increase in temperature recorded in both the Sajama and GISP2 cores and with the decrease in temperature implied for Antarctica by the data from Vostok (Figure 10.7). The onset of the Younger Dryas cold period at about 13 kyr B.P. is associated with a decrease in CH_4. It is accompanied by an essentially simultaneous decrease in tropical temperatures, as indicated by the isotopic data from Sajama, and by a resumption of warming in Antarctica, as implied by the results from Vostok. The abrupt end of the Younger Dryas at about 11.6 kyr B.P., as indicated in the data from GRIP2, is associated with a recovery of CH_4 and with an essentially contemporaneous increase in tropical temperatures, as indicated by the data from Sajama.[13]

Accounting for the complex nature of the recovery from the LGM, as defined by the data in Figures 10.6, 10.7, 10.14, and 10.15, poses a considerable challenge. There are a number of specific issues. We must explain why warming is observed first in Antarctica. We must account for the significant delay in the appearance of the first signs of warming in the tropics and Greenland. We must explain why the period of most rapid warming in the tropics and Greenland appears to be associated with an interruption of the warming trend

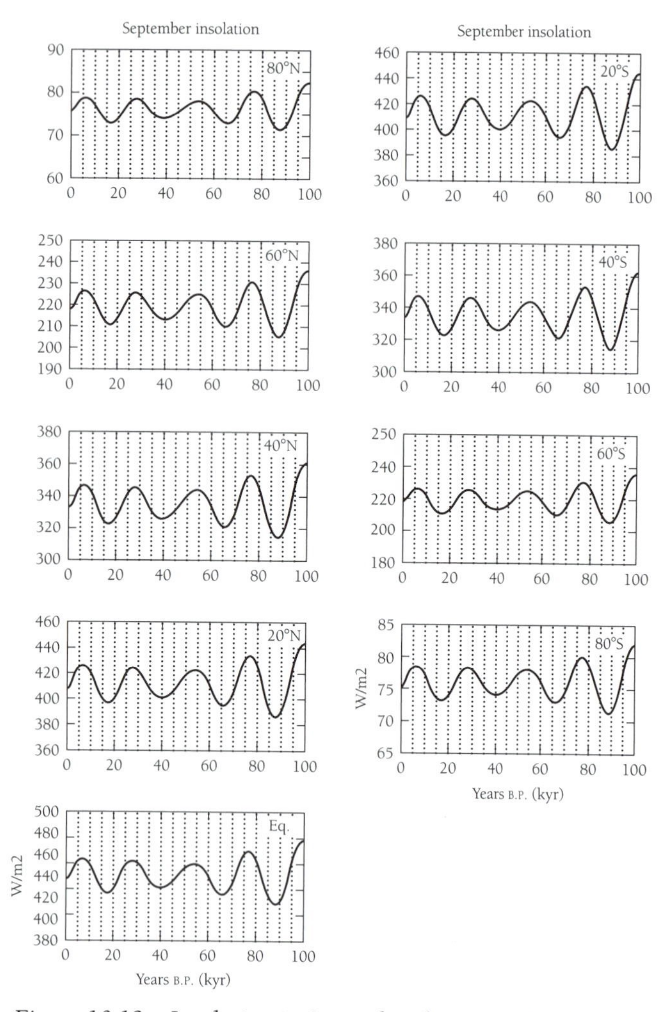

Figure 10.12 *Insolation in September during the last 100 kyr.*

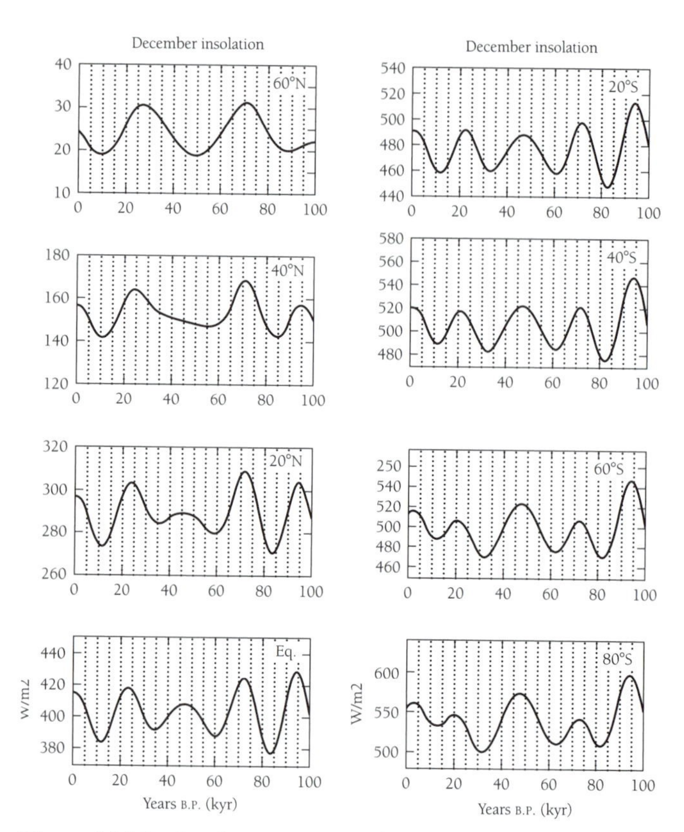

Figure 10.13 *Insolation in December during the last 100 kyr.*

in Antarctica. We must also account for the return to near glacial conditions defined by the Younger Dryas and for the apparently global nature of the phenomenon, as indicated by the DCR signal in the core from Sajama and by a variety of other spatially distributed climate proxies. We must account for the rapidity of the warming marking the end of the Younger Dryas. Finally, we must explain the trends observed in CO_2 and CH_4.

We suggest that the temperature changes evidenced by the Antarctic ice cores may reflect changes in the spatial extent of sea ice in the ocean surrounding the Antarctic continent. We should point out that this hypothesis is relatively untested; as noted in the preface, it has not as yet been presented in the peer-reviewed literature. Figure 10.16 illustrates variations of sea ice with season for the contemporary Arctic and Antarctic Oceans. Sea ice in the Southern Ocean reaches its maximum extent today in September when it expands to a latitude close to 60°S. It was almost twice as extensive during the LGM and there are indications that summer melt-back was considerably reduced at that time (Cooke and Hays, 1982). Cooling of the oceanic surface associated with the formation of sea ice ultimately depends on the temperature of the overlying atmosphere. This is regulated in turn by heat transport, primarily by wind systems originating at lower latitudes. We hypothesize that the flux of solar energy incident at about 40°S during early September may provide a reasonable proxy for the extent of Southern Ocean sea ice during winter. In this case, the data presented in Figure 10.12 would imply that conditions for expansion of sea ice would have been optimal near the peak of

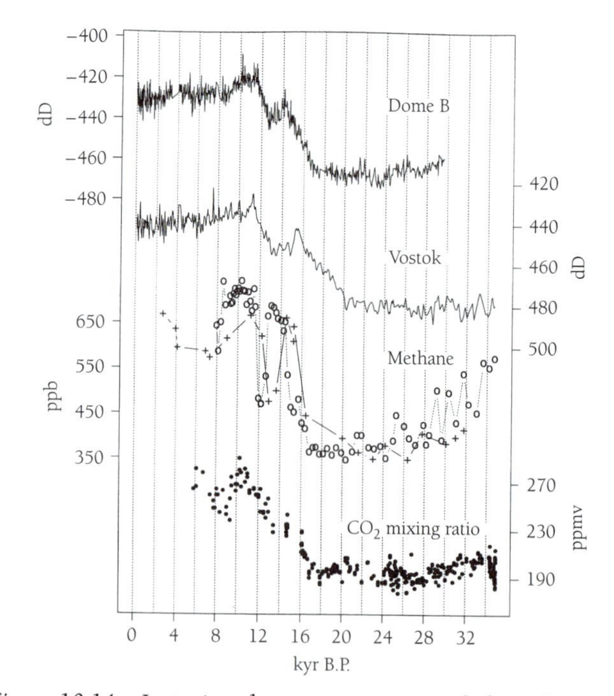

Figure 10.14 *Isotopic paleotemperature records from the Dome B and Vostok Antarctic cores and ice-core records of atmospheric greenhouse gas concentrations. Source: Hausman 1997.*

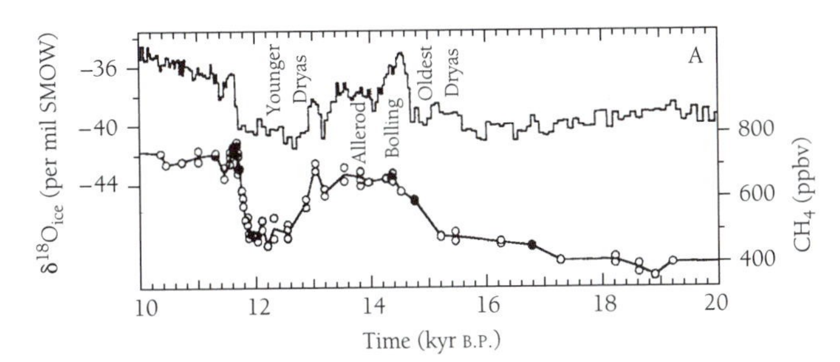

Figure 10.15 Expanded plot of GISP2 methane (lower) and $\delta^{18}O_{ice}$ (upper) records between 10 and 20 kyr B.P. *From* Science *1996. Source: Brook, et. al. 1996.*

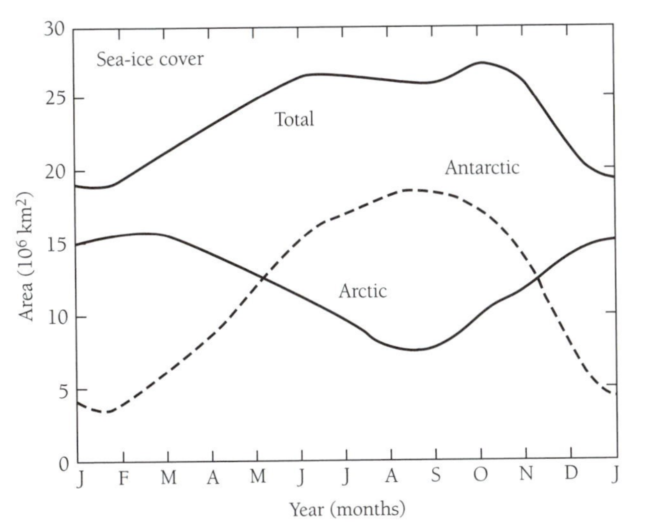

Figure 10.16 Annual cycle in sea ice cover in units of 10^6 km^2 for the Arctic, Antarctic, and the globe. Source: Peixoto and Oort 1992.

the LGM. They suggest that ice should have subsequently receded, evolving by about 10 kyr B.P. to a configuration approximating current, with a minimum at about 5 kyr B.P. Sea ice acts as an insulating blanket, cutting down the flux of heat from the ocean to the atmosphere. As the ice cover of the Southern Ocean is reduced during winter, we might expect an increase in evaporation accompanied by a significant heat transfer from the ocean to the atmosphere. We suggest that the warming implied by the isotopic data from the Antarctic ice cores may be attributed to this phenomenon. We propose that changes in winter/early spring–insolation at midlatitudes of the Southern Hemisphere impacting the spatial extent of sea ice in the Southern Ocean would provide a plausible explanation for the otherwise puzzling evidence that major, orbitally related shifts in global climate appear to originate in the south rather than the north. We further suggest that changes in the circulation of deep waters in the Atlantic Ocean initiated by changes in winter insolation in the Southern Hemisphere can provide the mechanism by which a climate signal triggered first in the south is eventually communicated to the north.

As was already discussed in more detail in Chapter 9, the circulation of waters in the deep ocean critically depends on the salinity budget of the North Atlantic. The salt content of the Atlantic is significantly higher today than that of either the Pacific or Indian Oceans. The buildup of salinity in the Atlantic reflects the fact that the rate at which water is evaporated from the Atlantic exceeds the rate at which it is returned by precipitation and by runoff in rivers. Ultimately this imbalance may be attributed to the presence of high mountain ranges extending from north to south over much of the Americas. The Rocky Mountains in the north and the Andes in the south provide important barriers blocking the transfer of moisture from the Pacific to the Atlantic at midlatitudes, where the prevailing winds blow from west to east. The barrier is less important in Central America, where the prevailing trade winds blow easterly. Wallace Broecker estimates that the Atlantic loses the equivalent of 0.35 Sv of water to the Pacific and Indian Ocean at the present time as a consequence of transport of water vapor through the atmosphere. Loss of water from the Atlantic results in a buildup of salt. The salt content of surface waters in the North Atlantic is high enough today such that waters, as they cool during winter, achieve densities high enough to sink to the bottom, feeding the deep Atlantic loop of the global Conveyor Belt. Salt removed by the lower branch of the Conveyor is replaced by input of lower salinity water at shallower depths, as schematically illustrated in Figure 10.17.

The overall salinity budget of the Atlantic may be described in terms of a simple pair of equations, one defining a balance in terms of inputs and outputs of water, the second reflecting an analogous constraint for salt:

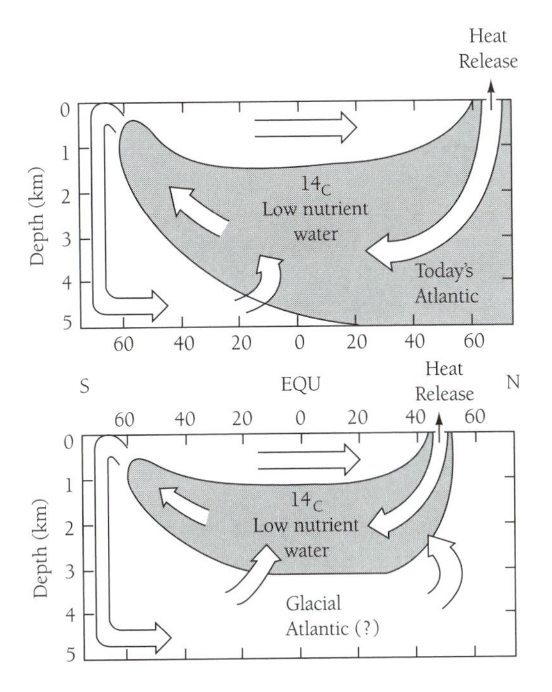

Figure 10.17 Diagram showing the suggestion, based upon the records of cadmium kept in shells of bottom-dwelling foraminifera, that there was conveyor-like circulation in the glacial ocean but that it did not penetrate to as great a depth. Source: Broecker 1995.

$$F_{p\to a} = F_e + F_{a\to p} \quad (10.1)$$

$$F_{p\to a}\,S_{p\to a} = F_{a\to p}\,S_{a\to p} \quad (10.2)$$

Here $F_{p\to a}$ and $F_{a\to p}$ define the fluxes of water entering the Atlantic from the Pacific and exiting the Atlantic to the Pacific, respectively; F_e is the flux of water removed from the Atlantic as a consequence of an excess of evaporation over precipitation plus runoff; and $S_{p\to a}$ and $S_{a\to p}$ refer to the salinities of waters associated with $F_{p\to a}$ and $F_{a\to p}$. Equations (10.1) and (10.2) may be combined as follows:

$$\frac{S_{p\to a}}{S_{a\to p}} = \frac{F_{a\to p}}{F_{p\to a}} = \frac{F_{p\to a} - F_e}{F_{p\to a}} = 1 - \frac{F_e}{F_{p\to a}} \quad (10.3)$$

It is convenient to specify water fluxes in units of Sverdrups (Sv).[14] Salinities are expressed usually in terms of the number of grams of salt contained in a kilogram of water, written as ‰ (the average salinity of ocean water is about 35 ‰). Broecker estimates that the net flux of water lost from the Atlantic by evaporation today is equivalent to about 0.35 Sv. The flux of water in the lower branch of the Atlantic Conveyor Belt responsible for removal of the excess salinity produced by evaporation is about 20 Sv. To maintain water balance, equation (10.1) requires that $F_{p\to a} = (F_e + F_{a\to p}) = 20.35$ Sv. Substituting for F_e and $F_{p\to a}$ in (10.3), it follows that $S_{p\to a}/S_{a\to p} = 0.99$. The salinity of water in the lower branch of the Atlantic Conveyor today is about 34.9 ‰. The salinity of the waters entering the Atlantic must therefore average about 34.3 ‰. Broecker suggests that the contemporary Atlantic salt budget is maintained in part by input of low-salinity waters from the Southern Ocean, in part by flow through the Bering Straits, with the former more important by about a factor of two.

The circulation of the Atlantic Ocean, as illustrated in Figure 10.17, was significantly different during glacial times. Evidence from studies of the chemical composition of shells of organisms preserved in deep-sea sediments indicates that in glacial times the deep Atlantic was dominated to a much larger extent by waters originating in the south.[15] The North Atlantic accounted for a source of intermediate rather than deep water. Equation (10.3) suggests a possible explanation for the change in circulation apparent for the Atlantic under glacial conditions. As indicated by the following examples, a decrease in the flux of water exported from the Atlantic through the atmosphere, possibly combined with an increase in the flux of deep water originating in the Southern Ocean, could have led to a reduction in the salinity of surface waters in the North Atlantic. A consequent decrease in the density of high-latitude waters at the surface relative to deep water, as compared to today, could have made it impossible for deep water to form at high latitudes in the glacial ocean. Put another way: in the face of a higher flux of cold, relatively fresh water from the south, the excess of salinity produced by a lower rate of net evaporation in glacial times may have been balanced by a production of intermediate water. There would have been no need to invoke the intense deep circulation that is required to offset production of excess salinity today.

Example 10.1: The average depth of the ocean today is about 3800 m. Sea level during glacial times was lower than today by about 120 m. The total salt content of the ocean is essentially constant: the associated time constant for salt is measured in millions, or even in tens of millions, of years. Assuming that the relative distribution of salt in the glacial ocean was the same as today, the salinity everywhere would be higher by a factor of 3800 / 3680 = 1.03. Assume that the salinity of waters entering the Atlantic during glacial times was higher than today by this factor, meaning that $S_{p\to a} = 35.3$ ‰. Calculate the salinity of water leaving the Atlantic during glacial times, assuming that the rate at which water was lost to the Atlantic by net evaporation was lower then than at present by about 30 %. Take $F_e = 0.25$ Sv. Assume that the flux of water leaving the glacial Atlantic was the same as today, $F_{a\to p} = 20$ Sv.

Answer: From Equation (10.3) it follows that

$$\frac{S_{p\to a}}{S_{a\to p}} = 1 - \frac{F_e}{F_{p\to a}} = 1 - \frac{0.25}{20.25} = 0.988$$

Thus,

$$S_{a\to p} = (0.988)^{-1}(35.3) = 35.73 \text{ ‰}.$$

In this case, the difference between the salinities of waters entering and leaving the Atlantic is equal to about 0.4 ‰ as compared to 0.6 ‰ today. ■

Example 10.2: Suppose now that the rate at which water was exchanged between the Atlantic and Pacific was somewhat higher in glacial times than today. This could reflect, for example, the influence of appreciably stronger wind-driven current systems in the Southern Ocean. We assume that production of deep water in the vicinity of Antarctica was the major source of deep Atlantic water in glacial times. As suggested by Figure 10.17, the circulation of the Atlantic was essentially reversed in glacial times, as compared to today. Deep water formed primarily in the south. It flowed northward, returning to the Southern Ocean as a component of a slightly more saline flow at shallower depths. Assume that the flux of water leaving the Atlantic was higher than today by about 25 %: meaning $F_{a\to p} = 25$ Sv. Assume that loss by net evaporation was the same as in Example 10.1: that is, $F_e = 0.25$ Sv. Assuming a value for the salinity for water entering the Atlantic of 35.3 ‰ (as in Example 10.1), calculate the salinity of waters leaving as required to maintain salt balance.

Answer:

$$\frac{S_{p\to a}}{S_{a\to p}} = 1 - \frac{F_e}{F_{p\to a}} = 1 - \frac{0.25}{25.25} = 0.990$$

Thus,

$$S_{a \to p} = (0.990)^{-1}(35.3)$$
$$= 35.66 \text{ ‰}.$$

In this case, the contrast in salinities would be reduced further by about 17 ‰. ■

Thus, we suggest that the transition from glacial to interglacial conditions was initiated by an increase in winter- and early spring–insolation in the Southern Hemisphere, resulting in a decrease in the supply to the North Atlantic of cold deep water originating in the Southern Ocean. As the flux of cold dense water declined, demands imposed on intermediate water to accommodate removal of excess salinity produced by net evaporation would have increased. Gradually, the dense, extremely cold (about –1°C), relatively fresh waters entering the Atlantic Basin from the south would have been displaced by warmer, more saline waters from the north. As cold southern waters receded to the south, and as warmer, more saline northern water began to dominate the composition of the deep North Atlantic, the contrast in density between surface and deep would have decreased. Matters would have eventually evolved to the point where cooling in winter would have permitted surface waters at high latitudes to sink to the bottom, as they do today. The circulation of the Atlantic would thus complete the transition from its glacial to its interglacial mode. Under glacial conditions, deep waters enter the Atlantic from the south, flowing northward, balanced by a southerly flow at shallower depths. In the interglacial mode, the flow of deep water is from north to south, balanced by flow of shallower water in the opposite direction.

The details of the transition from the glacial to the interglacial mode would depend in part on the strength of the flow of deep water originating in the south, in part on the tendency for salt to accumulate in surface waters in the north. The latter would reflect a complex balance between the rate at which freshwater was removed by net evaporation and the rate at which it was replaced by runoff associated with melting of Northern Hemispheric ice sheets. As indicated by the following examples, runoff would have been relatively unimportant in the early stages of the glacial-to-interglacial transition. It is likely, however, to have played a much larger role later. There are reasons to believe that an increase in runoff associated with rapid melting or release of waters held in large glacial lakes[16] may have triggered a reversal of circulation from the interglacial to glacial mode beginning at about 13 kyr B.P. As suggested by Broecker, such a reversal may have been responsible for the onset of the cold conditions characteristic of the Younger Dryas.

Example 10.3: Sea level increased by close to 20 m over a period of 4000 years in the early stages of deglaciation. Assuming that most of this extra water entered the oceans primarily in the Atlantic, estimate the corresponding value for the flux measured in units of Sv.

Answer: The oceans cover an area of 3.6×10^{18} cm^2. An increase in ocean depth of 1 cm yr^{-1} would require an input of water equivalent to 3.6×10^{18} cm^3 yr^{-1} . Noting that

$$1 \text{ Sv} = 10^{12} \text{ cm}^3 \text{ sec}^{-1}$$
$$= (3.14 \times 10^7)\, 10^{12} \text{ cm}^3 \text{ yr}^{-1}$$
$$= 3.14 \times 10^{19} \text{ cm}^3 \text{ yr}^{-1},$$

it follows that a 1 cm yr^{-1} rise of sea level would require an input flux of water given by

$$F = 3.6 \times 10^{18} \text{ cm}^3 \text{ yr}^{-1}$$
$$= (3.6 \times 10^{18} / 3.14 \times 10^{19}) \text{ Sv}$$
$$= 0.115 \text{ Sv}$$

A rise of sea level of 20 m in 4000 years is equivalent to

$$2000 / 4000 \text{ cm yr}^{-1} = 0.5 \text{ cm yr}^{-1}$$

Expressed in Sv, the additional flux of water to the ocean is given by

$$F = (0.5)(0.115) \text{ Sv} = 0.06 \text{ Sv}$$ ■

Example 10.4: Sea level rose by an average of about 2.8 m per century between 14.7 and 13.7 kyr B.P., climbing to a rate of close to 4 m per century between 14.2 and 13.7 kyr B.P. Estimate in units of Sv the magnitudes of the corresponding fluxes of water discharged to the ocean.

Answer: A rate of rise of 2.8 m per century (2.8 cm yr^{-1}) requires an additional flux of water to the ocean of

$$F = (2.8)(0.115) \text{ Sv}$$
$$= 0.32 \text{ Sv}$$

A rate of 4 m per century would require

$$F = (4)(0.115) \text{ Sv}$$
$$= 0.46 \text{ Sv}$$ ■

The data on sea level presented in Figure 10.6 and the ice-core evidence summarized in Figure 10.7 suggest that the period of rapid warming setting in about 14.7 kyr B.P. must have been essentially global in scale. The resolution of the ice-core data is insufficient to resolve the question of whether this warming was initiated first in the tropics spreading subsequently to higher latitudes, or the reverse. As discussed here, the reversal in climate at about 14 kyr B.P.— the change from warm to cold that ushered in the Younger Dryas—was almost certainly triggered by a shift in the North Atlantic Conveyor's mode of operation. This was most probably prompted by an increase in runoff from continents during the preceding warm period (see Example 10.4). Cold conditions of the Younger Dryas were not confined to the region around the North Atlantic, however. The ice-core data in Figure 10.7 indicate that the Younger Dryas–related cooling in Greenland was accompanied by essentially simultaneous cooling in the tropics. Sea surface temperatures in the Sulu Sea (121°E, 8.9°N) declined by as much as 3°C (Kudrass et al. 1991). Apparently synchronous changes were observed in the position of the polar front in the western North Pacific (Kallel et al. 1988), in the oceans off southern Portugal (Bard et al. 1987) and Argentina (Heusser and Rabassa 1989), in the Gulf of Mexico (Kennett, Elmstrom, and Penrose 1985), in the equatorial Atlantic (Pastouret et al. 1978), and in the Bengal Fan (Duplessy, Be, and Blanc 1981). It is doubtful that such large-scale changes could be attributed simply to a fluctuation in the circulation of the North Atlantic. We need a mechanism by which a change in North Atlantic circulation can be communicated efficiently, in less than a few decades, to alter the climate of Earth on an essentially global scale.

Our thesis is that a switch in Atlantic circulation from its glacial to interglacial mode prompted the warming at 14.7 kyr B.P. Following Broecker, we propose that the Younger Dryas cold period was triggered by a temporary reversal of circulation to the glacial mode. The decrease in runoff during the Younger Dryas period, as evidenced by the slowdown in the rise of sea level, allowed salinity to build up once again in the North Atlantic. This caused the circulation to return to its interglacial mode, permitting a resumption of the long-term warming trend. The critical challenge is to identify a mechanism to explain how a change in North Atlantic circulation was amplified to result in an important global change in climate, or at least a significant change in the climate of the tropics. Note that the abundance of ^{18}O relative to ^{16}O in ice sampled by the Sajama core (Figure 10.7) dropped during the Younger Dryas to a level close to the value that applied prior to 14.7 kyr B.P. The obvious interpretation is that the transitions into and out of the Younger Dryas epoch were accompanied by large, essentially synchronous changes in temperature not only in the North Atlantic but also in the tropics.

A wealth of data suggests that, during the LGM, surface temperatures in the tropics were about 5 K colder than today.[17] It is not easy to account for this result. The energy budget of the tropics depends on a balance of energy absorbed from the Sun, energy radiated into space, and energy transported to the extratropics by a combination of the atmosphere and ocean. The challenge posed by the cold temperatures of the tropical LGM is even more sharply defined if we focus on the factors that control the energy budget of the surface mixed layer of the Pacific, a region accounting for almost half of the total area of the tropics. These depend primarily on the net flux of radiant (visible plus infrared) energy absorbed at the surface, on the rate at which energy is expended in the evaporation of water, and on the rate at which heat is carried out of the tropics by oceanic circulation. As previously discussed in Chapter 9, the latter reflects a balance between the upwelling of relatively cold water flowing into the tropics from the subtropics and midlatitudes at intermediate depths (along the thermocline) and outflow of warmer water at the surface. Fielding Norton, in research for his Ph.D. degree at Harvard University (Norton 1995), concluded that there were two possible explanations for the cold temperatures of the glacial tropical Pacific: either cloud cover was more extensive, resulting in a significant decrease in sunlight reaching the surface, or ocean circulation was more vigorous, causing a larger flux of cold water to upwell to the surface. The second of these possibilities was viewed as more plausible.

Even today, the ocean plays an important role in the energy budget of the tropics. A portion of the incident solar energy is used to heat cold water upwelling from below. As previously discussed in Chapter 9, upwelling in the tropics is maintained by Ekman divergence driven by the easterly trade winds. Warm water is exported from the tropics to the subtropics, mainly in the western branches of the subtropical gyres with a compensating return flow of cold water entering the region primarily along isopycnal (constant density) surfaces defining the thermocline. Heat equivalent to an average of about 40 W m^{-2} is transferred by the ocean from the tropics to midlatitudes today (Hsiung 1986). There is little doubt that the strength of the upwelling of cold water in the tropics was enhanced during the LGM. The velocity of the trade winds averages about 6 m sec^{-1} at present. During the LGM it was significantly higher (about 8 m sec^{-1}). The strength of upwelling varies as the square of the wind speed. It follows, with our assumptions, that the rate of upwelling in glacial times should have been higher than today by a factor of $(8/6)^2 = 1.78$. Presumably the flow of subsurface water supplying this upwelling was accordingly enhanced.

The difference in temperature between waters entering and leaving the tropics today is about 6.1K. A slightly larger difference (about 6.9 K) combined with the more vigorous circulation noted here, would have resulted in a doubling of the heat flux removed from the tropics by the ocean during glacial times. Today, the ocean and atmosphere are comparably important in the removal of excess heat from the tropics. A stronger circulation of the ocean in glacial times, driven by higher wind speeds, could account for the appreciably colder temperatures of the tropical LGM.[18]

We suggest that the warming of the tropics at 14.7 kyr B.P., as indicated by the Sayama ice core data, arose as a consequence of a shift in the circulation of the Pacific from the more vigorous mode envisaged for the LGM to the more sluggish circulation characteristic of today. Specifically, we speculate that the export of vapor from the Atlantic responsible for the change in Atlantic circulation could have resulted in a decrease in salinity of Pacific surface waters: essentially a mirror image of the change invoked for the Atlantic. The decrease in Pacific salinity could have led to a reduction in the source of cold water sinking from the surface in the extratropics. The supply of cold water to the tropics could have accordingly declined. Temperatures in the tropics would have rapidly risen as the moderating influence of the cold-water input was reduced. Studies of bomb-derived ^{14}C in waters of the contemporary central tropical

Pacific indicate that the circulation carrying cold subsurface water to the tropics today has a time constant of about 5 years (Quay, Stuiver, and Broecker 1983). It would appear that the sequence envisaged here could account for the essentially simultaneous changes in climate observed for Greenland and Sayama.

In summary, we propose that the warming in Greenland, which marked the end of the ice age at 14.7 kyr B.P., was associated with a change in the operation of the Conveyor, prompted by a buildup of salinity in the North Atlantic. This buildup may have been facilitated by a reduction in the supply of cold, relatively fresh water from the Southern Ocean. Warming in the Atlantic led to an increase in evaporation, an increase in transfer of water vapor to the Pacific, a decrease in the salinity of Pacific surface waters, a consequent reduction in the supply of cold subsurface waters to the tropics, and thus to an increase in tropical temperatures. Subsequent changes in atmospheric circulation, fueled by an increase in heat transfer from a warmer tropical ocean to the atmosphere, extended the warming influence from the tropical Pacific to a global scale.

Rapid melting of continental ice, and eventually the breech of Lake Agassiz, ensured that the warming of the Bolling-Allerod period was short-lived. The salinity of the North Atlantic declined once again, prompting the Conveyor to revert to its glacial mode. The North Atlantic cooled, triggering the Younger Dryas cold snap. Transfer of vapor to the Pacific decreased, and the sequence of cause and effect invoked for the warming at 14.7 kyr B.P. ensured that the cooling of the North Atlantic was communicated to the tropical Pacific and eventually to much of the rest of the world (induced in this case by an increase in the salinity of waters in the extratropical Pacific). Input of meltwater to the Atlantic decreased as Earth cooled during the Younger Dryas. Salinity increased accordingly and the Conveyor reverted to its warm interglacial mode. The resulting increase in evaporation and freshening of waters in the Pacific ensured that warm conditions would extend once again to the tropics and hence to much of the globe, signaling the final demise of the ice age.

As suggested by the authors of the original paper (Brook, Sowers, and Orchardo 1996), the changes in CH_4 indicated in Figure 10.15 provide a useful proxy for the changes inferred for the climate of the tropics. The abundance of CH_4 increased from about 500 ppbv at approximately 15 kyr B.P., climbing to about 700 ppbv during the Bolling-Allerod. It fell back to 500 ppbv during the Younger Dryas, returning to higher levels (about 750 ppbv) at the end of the Younger Dryas. Severinghaus et al. (1998) showed that the rise in CH_4 at the end of the Younger Dryas set in over a period of less than 30 years following the first indication of warming in Greenland at 11.64 kyr B.P. Their data provide persuasive evidence that the changes in climate in the North Atlantic and the tropics were essentially contemporaneous. Warming of the tropics, and presumably an increase in rainfall as required to stimulate enhanced production of CH_4, trailed warming in Greenland by no more than a few decades. To our mind, the mechanism advanced above provides the only plausible hypothesis advanced so far to account for this behavior. We repeat, however, as previously noted, that the mechanism has yet to experience the scrutiny afforded by publication in a peer-reviewed journal.

The data in Figure 10.14 suggest that the rise in CO_2 at the end of the last ice age began at essentially the same time as the first indication of warming in Antarctica. The concentration of CO_2 was about 200 ppmv during the LGM. It climbed over a period of approximately 8 kyr to about 280 ppmv and remained at close to this level until recently, when human activity initiated the modern rise persisting today (see Figures 5.2 and 5.3). The low level of CO_2 in the atmosphere during glacial times and the indication of a smaller terrestrial biosphere imply that a significant quantity of carbon must have been transferred from the atmosphere and terrestrial biosphere to the ocean. Models for the glacial ocean suggest that the low level of CO_2 in the glacial environment was due to two primary factors: colder temperatures at the surface and a circulation resulting in less efficient ventilation of waters at the depths.[19] It is tempting to attribute the early rise in CO_2 at the end of glacial times to an increased ventilation of deep water in the Southern Ocean as sea ice receded: to precisely the set of circumstances responsible for the beginning of the warming in Antarctica. Warming, particularly in the tropics, would have contributed to an additional release of CO_2 after 15 kyr B.P. Details of the recovery of CO_2, among other factors, would be sensitive, however, to changes in biological productivity associated with changes in rates of upwelling, changes in the areal extent of high-nutrient waters exposed to the surface at high latitudes, and changes in the input of alkalinity to the ocean associated with changes in the rates of the weathering of continental rocks.[20]

The end of the Younger Dryas at 11.64 kyr B.P. was associated with a brief interval of rapid warming in the North Atlantic followed by a longer interval of more gradual warming lasting to about 9 kyr B.P., as indicated by the ice core data in Figure 10.7. Figure 10.18 presents a record, inferred using a variety of lithostratigraphic and paleobotanical techniques, of the temperature change over the past 10 kyr for Western Norway in the region of the Jostedalsbreen ice cap (Nesje and Kvamme 1991). Fluctuations of glaciers observed for the same region are summarized in Figure 10.19. It is tempting to attribute the advance of the Norwegian glaciation known as the **Erdalen** to an additional oscillation of the North Atlantic circulation, similar to that responsible for the Younger Dryas but apparently more modest. Sea level climbed rapidly between 10 and 7 kyr B.P., reaching to within about 15 m of its present value at the end of this period (Fairbanks 1989). The subsequent increase of sea level was more gradual, suggesting that the ice sheets had receded to close to their present interglacial configuration by 7 kyr B.P. Warmest temperatures over the past 10 kyr occurred during the period known as the **Hypsithermal**, from about 8 to 5 kyr B.P. The Jostedalsbreen ice cap disappeared completely at this time. Temperatures

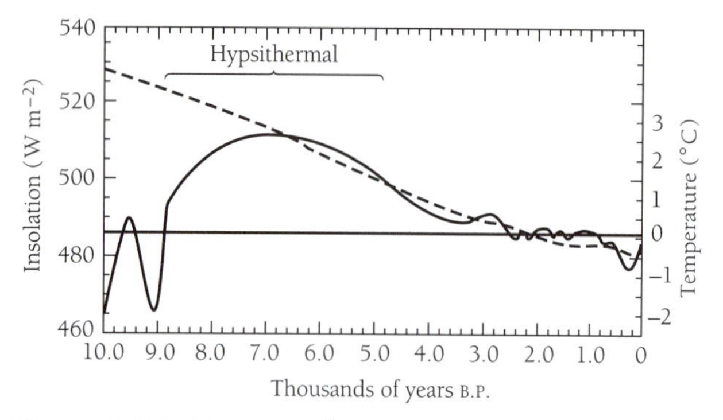

Figure 10.18 Variation of temperature (solid curve) and daily insolation at 60°N for summer solstice in the Jostedalsbreen region of Norway (62°N, 7°E), based on a variety of lithostatigraphic and paleobotanical techniques. Source: McElroy 1994.

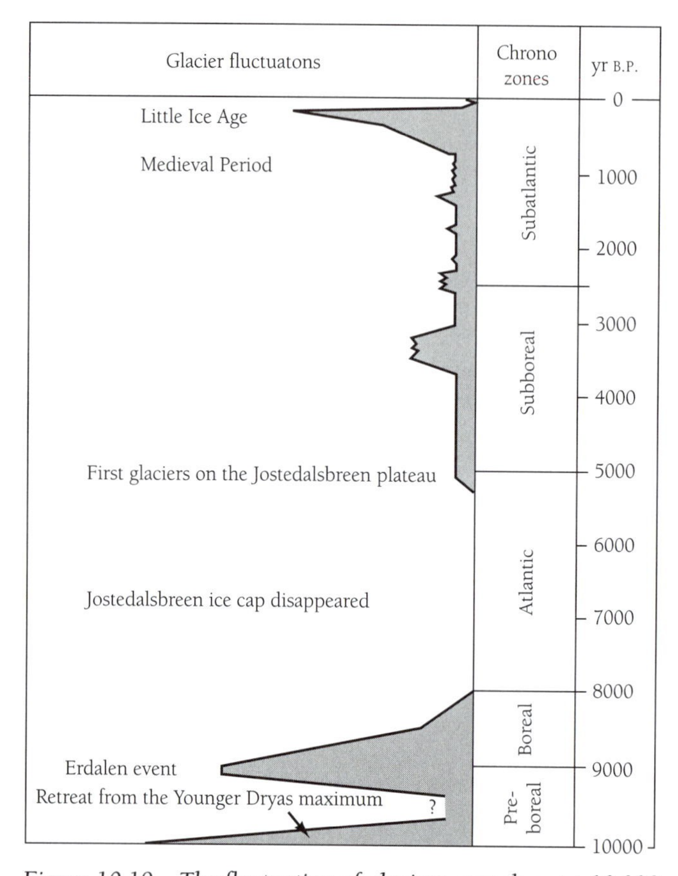

Figure 10.19 The fluctuation of glaciers over the past 10,000 years in the Jostedalsbreen region of Norway (62°N, 7°E), based on a variety of lithostatigraphic and paleobotanical techniques. Source: Nesje and Kwamme 1991.

have declined more or less steadily since, with the ice cap reappearing at about 5 kyr B.P., expanding to its maximum extent since the Erdalen during the Little Ice Age.

Figure 10.18 includes a curve showing the variation of insolation with time for the summer solstice at 60°N. It is intriguing to note the similarity in the behavior of the temperature and insolation curves for the period subsequent to about 6.5 kyr B.P. If the insolation curve is taken as a surrogate for temperature, the Little Ice Age appears as a relatively modest negative excursion with respect to a long-term cooling trend. If physical significance is attached to the similarity between the trends in temperature and insolation after 6.5 kyr B.P.—and in the absence of a quantitative model this may be dubious—we must account for the quite different behavior observed in the earlier record (before 7 kyr B.P.). The earlier trend could reflect feedbacks associated with the demise of the global ice sheets associated, for example, with changes in albedo and/or related changes in oceanic circulation induced by input of meltwater from the continents.

The Little Ice Age was a period of generally cold temperatures, specifically cold winters, lasting from about A.D. 1550 (or perhaps from as early as A.D. 1250) to about A.D. 1850, with extremes between about A.D. 1580 and 1700 (Fairbridge 1987). It was marked by a significant decline in sea surface temperatures both in the North Atlantic (Fairbridge 1987) and in the North Pacific (Yoshino and Xie 1983). The impact appears to have been global, with important changes reported not only for Europe (Pettersson 1912; Weikinn 1965; Lindgren and Neumann 1981; Lamb 1984) but also for North America (Ludlum 1966; Catchpole and Ball 1981; Baron 1982), China (Wang and Zhao 1981), Japan (Yamamoto 1972), and South America (LaMarche 1975). There is evidence for a worldwide advance of mountain glaciers, with snow lines declining by about 100–200 m at midlatitudes and by as much as 300 m in the equatorial Andes (Porter 1975, 1981; Hastenrath 1981; Broecker and Denton 1989). As previously noted, the end of the Little Ice Age was abrupt. It appears that it ended globally and essentially synchronously in the latter half of the nineteenth century (Broecker and Denton 1989).

It is tempting to attribute the Little Ice Age to a fluctuation in the circulation of the Pacific analogous to that invoked earlier to account for the variations of tropical climate observed between 15 and 11 kyr B.P. Data on the accumulation of snow in the Quelccaya ice cap of Peru (14°S) (Thompson et al. 1985), reproduced in Figure 10.20, indicate persistent anomalies in precipitation between about A.D. 1200 and A.D. 1900. A general increase in precipitation was observed from about A.D. 1200 to about A.D. 1650 , followed by a decline from A.D. 1650 to about 1850. Precipitation at Quelccaya at the present time is sensitive to conditions in the Pacific as they affect details of the circulation of the tropical atmosphere. The region receives most of its precipitation from the east during local summer. Atlantic inflow of air is reduced during an El Niño, as the strength of the trade winds is diminished; indeed, annual accumulation rates at Quelccaya provide an excellent proxy for El Niño events, as illustrated in Figure 10.21 (Thompson, Mosley-Thompson, and Arnad 1984).

We suggest that the long-term changes in precipitation at Quelccaya evident in Figure 10.20 may reflect oscillatory behavior on a century time scale in the operation of the Pacific circulation, analogous to the oscillatory behavior postulated for the Atlantic by Broecker, Bond, and Klas (1990a) and by Birchfield and Broecker (1990). Relatively saline waters at the surface of the midlatitude or subtropical Pacific would allow

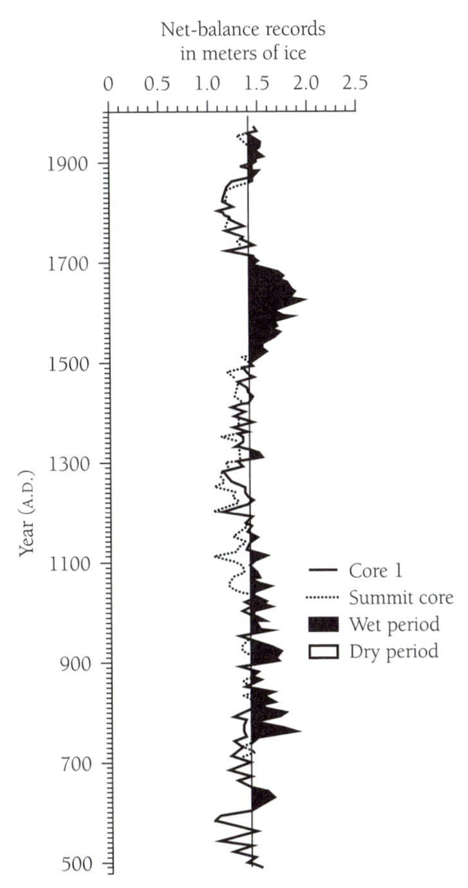

Figure 10.20 Reconstructed rate of precipitation, based on annual accumulation rates, for two ice cores in the Quelccaya ice cap (14°S, 71°W, 5160 m elevation). Extended periods of aridity and moistness are indicated. Source: Thompson et al. 1985.

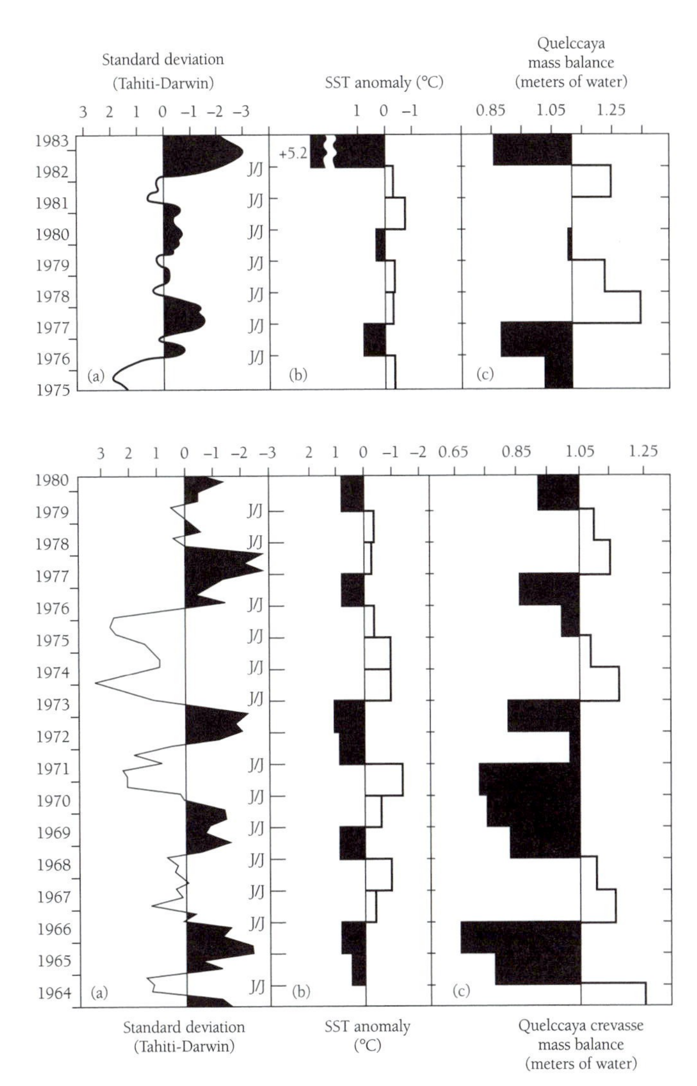

Figure 10.21 Time variation of the atmospheric pressure anomaly at sea level between Tahiti, Society Islands (18°S, 150°W), and Darwin, Australia (12°S, 131°E), the annual sea-surface temperature anomaly at Puerto Chicama, Peru (7°S, 79°W), and the annual accumulation of the Quelccaya ice core (in equivalent meters of water). The upper panel is the summit of Quelccaya; the lower panel is for a crevasse 1 km east of the summit. El Niño events correspond to negative excursions in the Tahiti-Darwin pressure anomaly and positive excursions in the sea-surface temperature anomaly. Source: Thompson, et. al. 1984.

the ocean circulation to operate in its strong mode, promoting a cooling of tropical surface waters, a reduction in atmospheric heating over the ocean, and a related intensification in the strength of the surface trade winds. Stronger trade winds could allow more efficient transfer of H_2O vapor from the Atlantic to the Pacific, over the Panama Isthmus, for example. This would tend to offset the postulated increase in salinity, resulting in a weakening of the oceanic circulation with a return to warm conditions in the tropics. The trade winds would accordingly weaken, choking off the supply of excess H_2O vapor from the Atlantic. It is unclear whether the recovery from the Little Ice Age was simply a natural response of this hypothetical oscillator or whether it might have been prompted in part by the post-industrial increase in the concentration of greenhouse gases. The changes are essentially synchronous, as illustrated in Figure 10.5.

The results in Figure 10.18 suggest that the Little Ice Age was but the last of a series of climatic fluctuations punctuating the long-term trend of cooling over the past 7 kyr. The record of the last 150 years must be examined in this context. As previously discussed, assessment of the extent to which recent warming should be attributed to the buildup of greenhouse gases presupposes knowledge of how the climate would have evolved in the absence of this buildup. On a purely empirical basis, we must conclude that the sensitivity of climate to an increase in the concentration of greenhouse gases could be either higher or lower than implied by current models. Lonnie Thompson returned to Quelccaya a few years ago seeking to update the record of climate summarized in Figures 10.20 and 10.21. The cores from which these data were derived were drilled in 1983. To his surprise, he found that surface snow had melted in the interim. Meltwater had percolated downward, eliminating the record of the past previously preserved in the ice. There was no indication of surface melting over the entire 1500-year record preserved in the earlier core. It is clear, therefore, that the climate of the recent past has been unusual, at least for Quelccaya. What it portends for the future is unclear but scarcely reassuring.

10.3 The Past 3 Million Years

Earth has been cold more often than it has been warm over the past 3 million years of its history. Analysis of the isotopic composition of the shells of organisms preserved in the sediments of the deep sea provides an invaluable record of the change of climate over this period (Emiliani 1955; Shackleton 1967; Broecker and Van Donk 1970; Shackleton and Opdyke 1973; Hays, Imbrie, and Shackleton 1976; Imbrie et al. 1984). The relative abundance of ^{18}O and ^{16}O in shells depends both on temperature and on the isotopic composition of the oxygen in the water from which the shells were formed. The isotopic composition of oxygen in seawater reflects primarily the volume of water withdrawn from the ocean to form the continental ice sheets. The shells thus record a complex combination of local water temperature and global ice volume.[21]

The variation of $\delta^{18}O$ in benthic foraminifera[22] for the past 2 million years is displayed in Figure 10.22. Note the distinctly different pattern observed for the past 800 kyr as compared to the first million years of the record. The climate of the past 800 kyr is marked by prolonged intervals of cooling (trends towards higher values of $\delta^{18}O$), interrupted by brief episodes of warming (indicated by sharp sawtooth transitions to lower values of $\delta^{18}O$). Cold intervals over the past 800 kyr persist for about 100 kyr on average. The transition from maximum cold to maximum warm conditions (known as the **glacial termination**) takes place over times that are typically as short as about 10 kyr. In contrast, the data in Figure 10.22 suggest that the climatic rhythm was distinctly different a million or so years ago. The dominant 100-kyr period of the recent record is missing, replaced by variations on much shorter time scales. For the first segment of the record, from 2.0 to 1.2 million years (myr) B.P. (as indicated by domain A in Figure 10.22), climate responds mainly to changes in seasonal insolation associated with variations in obliquity (a period of 41 kyr). While the obliquity influence remains dominant, the precession effect (a period of about 22 kyr) begins to show up in the intermediate portion of the record (1.2 to 0.65 myr B.P.; segment B). Effects of obliquity and precession are apparent throughout the record, but the dominance of the 100-kyr signal is comparatively recent. Note that the appearance of the 100-kyr signal is accompanied by a major expansion of continental ice sheets, as reflected by the significant increase in maximum values of $\delta^{18}O$ observed over the past 800 kyr.[23]

An expanded, smoothed view of the changes in $\delta^{18}O$ for benthic foraminifera over the past 650 kyr is presented in Figure 10.23. For comparison, the Figure includes a summary of changes in sunlight expected for 65°N in July. Note that terminations generally occur when insolation during summer at high northern latitudes is at a local (temporal) maximum. Cold periods persist on average for about 100 kyr. They typically include five precession cycles, but the number is not fixed. Termination II, for example, may have taken place following termination III on either the fourth or sixth peak in high-latitude northern-summer insolation associated with the precession cycle (the uncertainty reflects ambiguity in assignment of termination III to a specific individual peak in the data for $\delta^{18}O$ displayed in Figure 10.23).

A more detailed summary of variations of $\delta^{18}O$ in benthic foraminifera for the past 130 kyr, obtained from analysis of a core from the eastern tropical Pacific, is presented in Figure 10.24. An interpretation of these data in terms of changes in bottom-water temperature and ice volume is summarized in Figure 10.25 (Broecker 1995). Using the convention introduced by Cesare Emiliani, the penultimate interglacial period is designated as isotope stage 5 (subdivided subsequently into stages 5a–5d). The glacial period is distinguished by isotope stages 2 (late), 3 (intermediate), and 4 (early). The last termination is designated as isotope

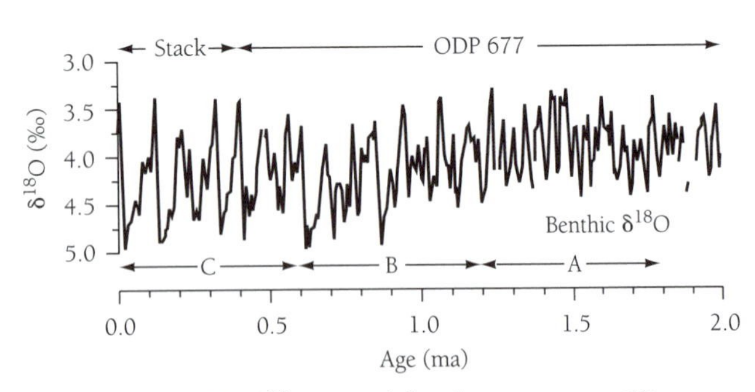

Figure 10.22 The $\delta^{18}O$ record for the past two million years, showing glaciation cycles. The data are plotted on an inverted scale: values of $\delta^{18}O$ decrease with height on the vertical axis. We choose this form of presentation to emphasize the associated changes in climate. Small values of $\delta^{18}O$ (peaks in the Figure) reflect interglacial conditions; high values of $\delta^{18}O$ (minima in the Figure) indicate periods of maximum glaciation. Source: Imbrie et al. 1984.

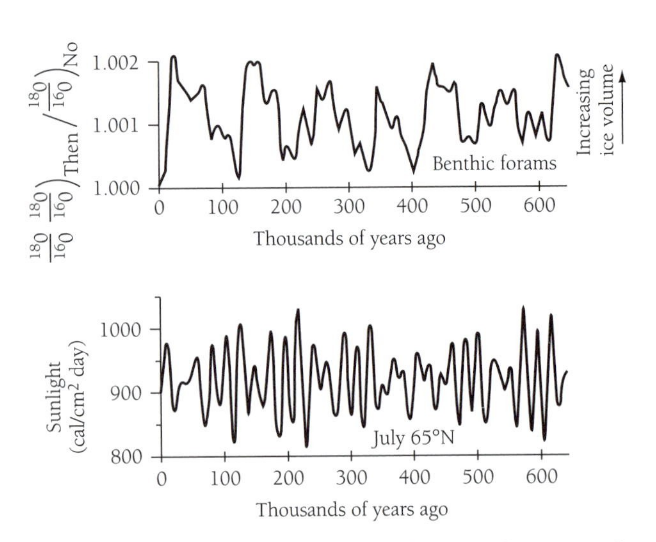

Figure 10.23 Comparison between the ice-volume record and the July solar-radiation record for 65°N. Source: Broecker 1995.

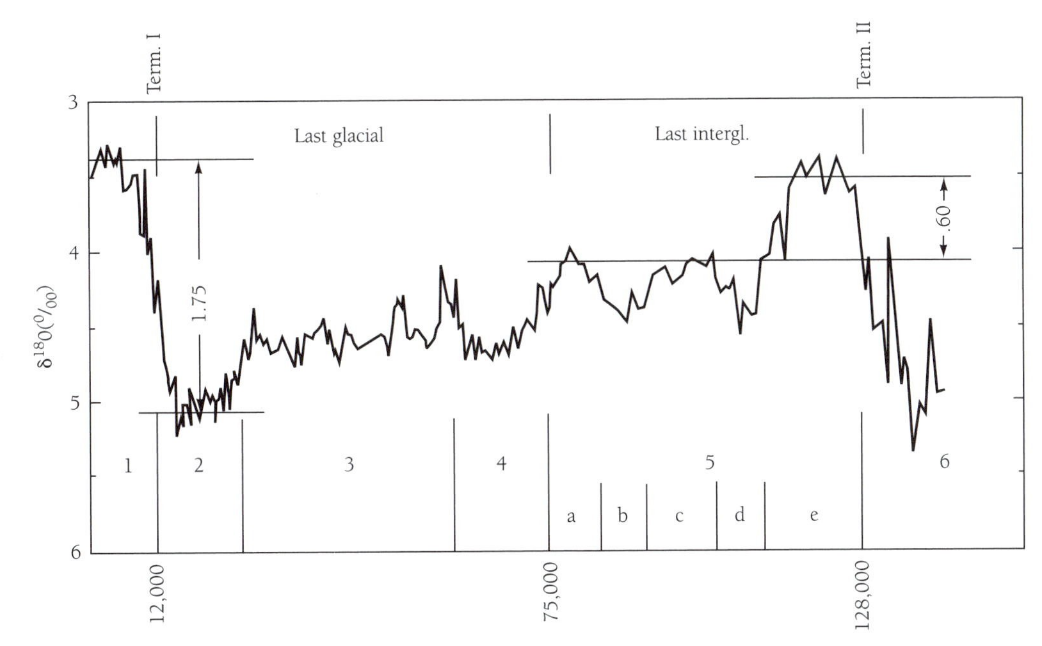

Figure 10.24 The $\delta^{18}O$ record for benthic foraminifera from a deep-sea core from the eastern part of the equatorial Pacific, as well as isotope stage numbers and generally accepted ages for the major boundaries. Source: Broecker 1995.

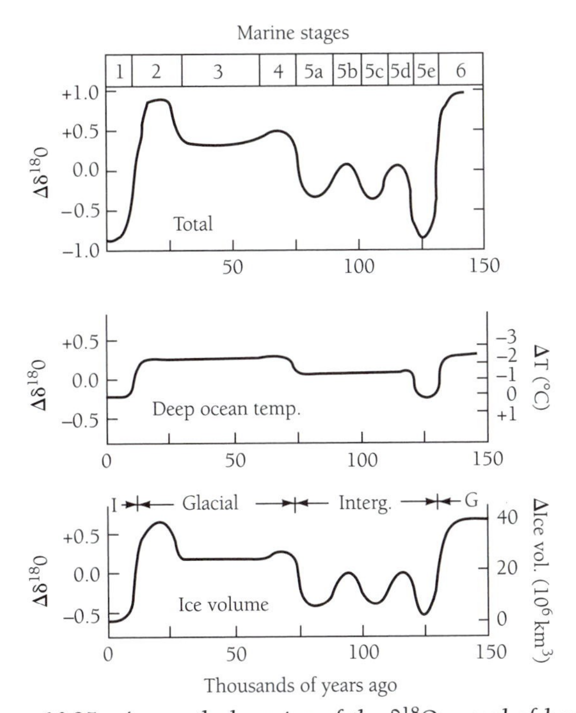

Figure 10.25 A smoothed version of the $\delta^{18}O$ record of benthic foraminifera from the deep Pacific (top), an estimate of the part of the $\delta^{18}O$ signal related to changes in deep-sea water temperature (middle) and an estimate of the part of the signal that can be attributed to changes in ice volume (bottom). Source: Broecker 1995.

stage 1. Broecker's interpretation of the $\delta^{18}O$ data suggests that the temperature of bottom water declined in two steps, through a total of about 2°C, over the course of the last ice age. Ice volume increased by about 4×10^7 km^3 over this interval. It fluctuated during stage 5, varying through a range equal to about 25% of the total glacial-to-interglacial change. The ice sheets grew rapidly during the transition between stages 5 and 4. They were relatively static during stages 4 and 3, advancing to their maximum extent at the transition between isotope stages 3 and 2.

Measurements of $\delta^{18}O$ from the GISP2 Greenland ice core are presented in Plate 2 (top) (see color insert) (Brook, Sowers, and Orchardo 1996). The Figure also includes a composite of measurements of CH_4 taken from the GISP2, GRIP, and Vostok ice cores. Changes in sunlight expected for June at latitudes of 60°N and 20°N were indicated in Figure 10.11. It is apparent that the large-scale variability of CH_4 correlates well with variations in summer insolation for the Northern Hemisphere.[24] It is clear, however, that an equally plausible correlation could be established with variations in sunlight for a variety of other seasons and latitudes. In particular, the changes observed for CH_4 are consistent with the hypothesis advanced earlier to account for the rise in CH_4 at 14.7 kyr B.P. We suggested then that the increase in CH_4 during the last deglaciation reflected an increase in tropical temperatures prompted by a transition of the Conveyor from its glacial to its interglacial mode. This transition, we proposed, was initiated by a decrease in the flow of cold deep water from the Southern Ocean. The decrease in the strength of the flow of southern deep water, we suggested, reflected a decrease in the areal expansion of sea ice during winter. We proposed that changes in insolation in September at 40°S could be used as a proxy to simulate orbitally driven changes in sea ice in the Southern Ocean. A comparison of the September 40°S insolation curve with the CH_4 data lends support to the suggestion that the large-scale changes in CH_4 could be paced by orbitally driven changes in sea ice in the Southern Ocean. Note the tendency for peaks in CH_4 to occur a few thousand years after sea ice (as inferred from the September

40°S insolation proxy) has reached its maximum extent: precisely the behavior observed during the most recent deglaciation.

Viewed at higher temporal resolution, it is clear that the variability of CH_4 is more complex than indicated by the data in Plate 2 (top) (see color insert). A comparison of $\delta^{18}O$ and CH_4 for the GISP2 core is presented in Figure 10.26, (also included in Plate 2 bottom). Taking the $\delta^{18}O$ record from Greenland as a reasonable record of the variation of climate in the North Atlantic, it is clear that the climate in this region was exceptionally variable during the last ice age. Peaks in $\delta^{18}O$, referred to here as **interstadials**, known also as **Dansgaard-Oeschger events**, denote times when the climate was unusually warm.[25] There is an intriguing pattern to these fluctuations: warming is exceptionally rapid; subsequent cooling, particularly early in the ice age, is comparatively more gradual. Warming in Greenland is normally associated with an increase in CH_4, suggesting that the changes in North Atlantic circulation imputed to account for the changes in temperature in Greenland may almost inevitably be responsible for sympathetic changes in the climate of the tropics. The pattern of variation observed for Dansgaard-Oeschger events is similar to that observed for the transition from the Bolling-Allerod to Younger-Dryas periods during the last deglacial epoch. It seems likely that the physical mechanism invoked previously to account for the latter may also be appropriate for the former. According to this scenario, warming in Greenland would be triggered by an excess of evaporation over runoff and precipitation in the North Atlantic, causing the circulation to temporarily revert from its glacial to its interglacial mode. A subsequent increase in melting and runoff would cause the circulation to revert to its glacial condition.

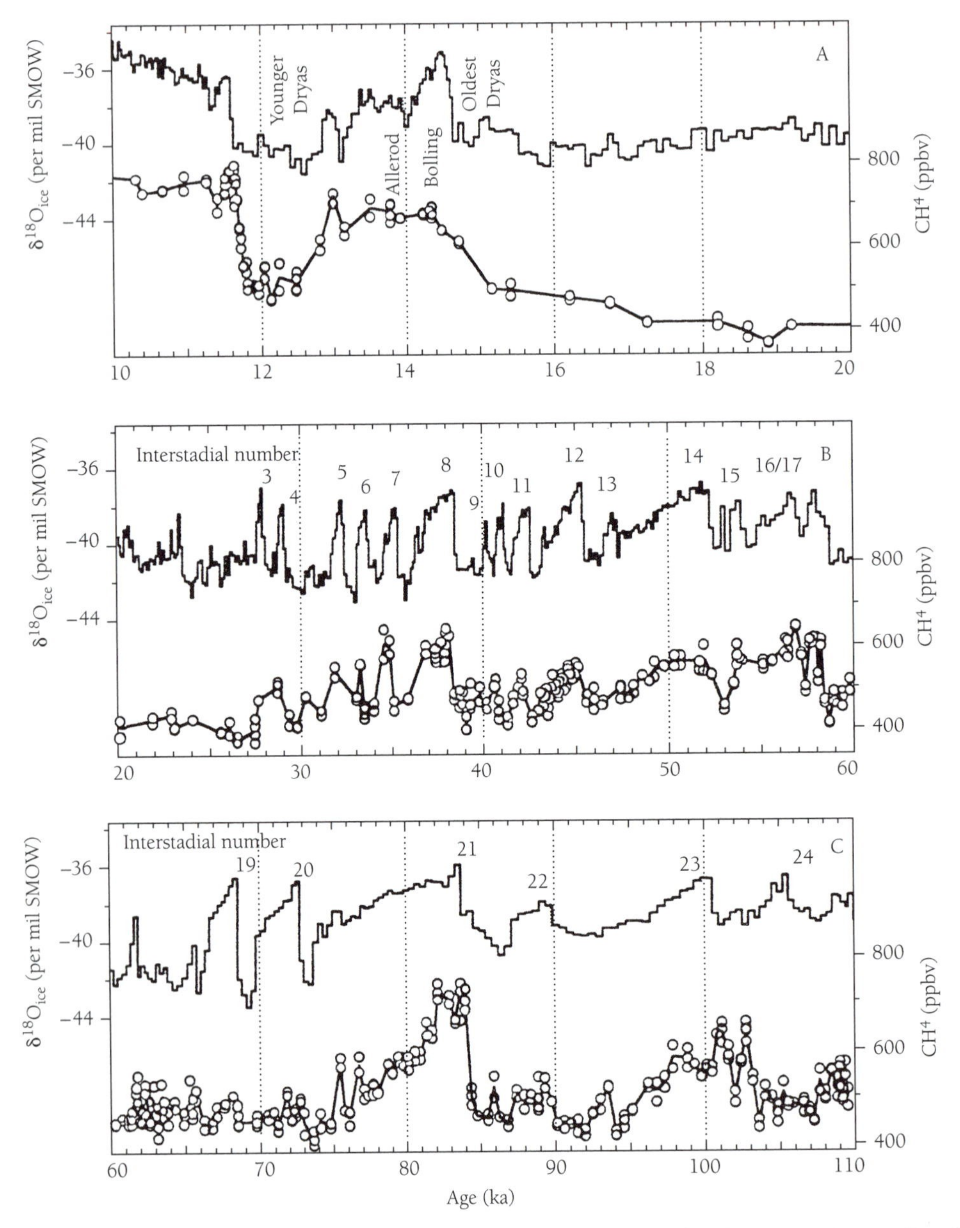

Figure 10.26 Variations of CH_4 and $\delta^{18}O$ obtained from analysis of the GISP2 Greenland ice core. Source: Brook, Sowers, and Orchardo 1996.

Changes in tropical climates would be induced by sympathetic changes in the circulation of Pacific waters.

It is interesting to note that Dansgaard-Oeschger events are most commonly observed during marine isotope stages 3 and 4 (between about 65 and 30 kyr B.P.), when overall changes in continental ice volume are relatively small (see Figure 10.25). A total of 24 Dansgaard-Oeschger events are observed between 20 and 110 kyr B.P., indicating that the glacial climate was variable not only on orbital time scales but also on time scales ranging from 10 to less than 1 kyr. As suggested earlier, the relatively modest fluctuations in climate observed during the Holocene (the Medieval Optimum and the Little Ice Age, for example) may represent muted analogues to the much larger variations observed during the Ice Age.

It is difficult to account for the dominance of the 100-kyr signal in the $\delta^{18}O$ record of climate for the past 800 kyr. We suggested that the sequence of events that led to the end of the last ice age began with an increase in late winter early spring insolation at midlatitudes of the Southern Hemisphere. We offered a mechanism to account for why changes in climate appear to arise initially at high latitudes of the Southern Hemisphere. We discussed how warming in the Southern Hemisphere would be communicated, with about a 4-kyr delay, first to regions round the North Atlantic and from there almost immediately to the tropical Pacific and to the rest of the globe. This sequence should be repeated, however, several times over a 100-kyr interval. The question is why it does not always and inevitably lead to the demise of the ice age. Why is it, at least over the past 800 kyr, that the ice sheets have been able to survive four to six of these events? It appears that the ice sheet is vulnerable to catastrophic collapse only when it has grown to some critical size. We need a mechanism to account for this behavior. To confound the challenge, we must be able to explain why the 100-kyr signal was recently dominant but effectively absent throughout the Pliocene (5.1 to 1.66 myr B.P.) and the early part of the Pleistocene (1.66 to 0.8 myr B.P.).

A number of writers have suggested that the 100-kyr signal is ultimately regulated by the dynamics of the ice sheet (Weertman 1976; Oerlemans 1980; Pollard, Ingersoll, and Lockwood 1980; Birchfield, Weertman, and Lunde 1981; Pollard 1982, 1983a, 1983b, 1984; Peltier and Hyde 1984; Hyde and Peltier 1985, 1987). According to this view, the collapse of the ice sheet is triggered by an insolation-driven retreat of ice into a surface depression caused by isostatic sinking of continental crust beneath the ice as it reaches its maximum thickness and greatest southern extent.[26] The most elaborate treatment of the evolution of the ice sheet (Peltier and Hyde 1984; Hyde and Peltier 1987) provides a satisfactory description of the changes in $\delta^{18}O$ observed over the past 0.5 myr or so. It accounts, in particular, for the rapid demise of the ice sheet observed during terminations, a phenomenon attributed in this model to a delay in isostatic rebound compared with the time required for breakup and melting of the ice sheet at its southern margin. The model adopts an empirical formula to define the rate at which snow accumulates as a function of height and latitude and accounts for a strong nonlinear dependence of the rate of ablation on the height of the ice sheet. It is driven by realistic variations in summer insolation. The 100-kyr period results from a particular choice of parameters to define rates for accumulation and ablation of ice.

There is a second consideration of potential relevance to the 100-kyr issue. It is possible that the ice sheet may be unusually stable during the early stages of its growth. Suppose that the insolation regime evolves to the phase favoring a weakening of the southern-source deep water. Imagine that the Conveyor then switches to its interglacial mode. Suppose further, however, that the meltwater in this case drains directly into the North Atlantic (in the early stages of its development, it is likely that the ice sheet is confined to relatively high northern latitudes). This could cause the circulation to revert essentially spontaneously to its glacial condition. The ice sheet would remain frozen in place. Growth of the ice sheet would resume as insolation conditions and ocean circulation revert to their glacial mode. It is possible that the change in insolation favoring warming may be effective in melting the ice sheet only when the ice sheet has advanced sufficiently far southward, where meltwater is able to drain to the Gulf of Mexico rather than directly (through the Saint Lawrence, for example) into the North Atlantic. Under these circumstances, an excess of salinity in the North Atlantic could be maintained for a longer period. The warming phase would persist, and the climate system would have time to evolve through the complex sequence of adjustments observed during the most recent termination.

Providing an explanation for the 100-kyr signal is but one of the problems to be confronted in accounting for the pattern of climatic change over the past several million years. Equally challenging, we must explain why the 100-kyr period emerged only during the past 0.8 myr, why it was absent during the Pliocene and Early Pleistocene. Clark and Pollard (1998) suggest that the transition from a dominant 41-kyr period to one of 100 kyr reflects a change in the nature of the underlying crustal material supporting the ice sheet. According to this proposal, the region of North America occupied by the Laurentide ice sheet was covered, up to the mid-Pleistocene, by a thick layer of unconsolidated regolith.[27] This loose, relatively mobile, basal material was incapable of supporting a thick ice sheet. With time, the regolith was removed, largely as a consequence of the scouring effect of the ice sheets. According to the Clark-Pollard proposal, it was effectively eliminated by the mid-Pleistocene. Only then was it possible to build on solid bedrock the thick ice sheets characteristic of glacial epochs over the past 0.8 myr. The Clark-Pollard model assumes that the ice sheets of the Pliocene and Early Pleistocene covered essentially the same area as the ice sheets of the later Pleistocene; they differed primarily in their thickness.

Barendregt and Irving (1998) present a somewhat different picture. They argue that the geographic coverage of ice sheets was quite different during the Pliocene and early Pleistocene. According to their reconstruction, summarized in Figure 10.27, the middle region of the continent was essentially free of ice during the early periods of glaciation. Continental ice cover consisted of two distinct ice sheets: one centered in the west (the Cordilleran ice

(a)

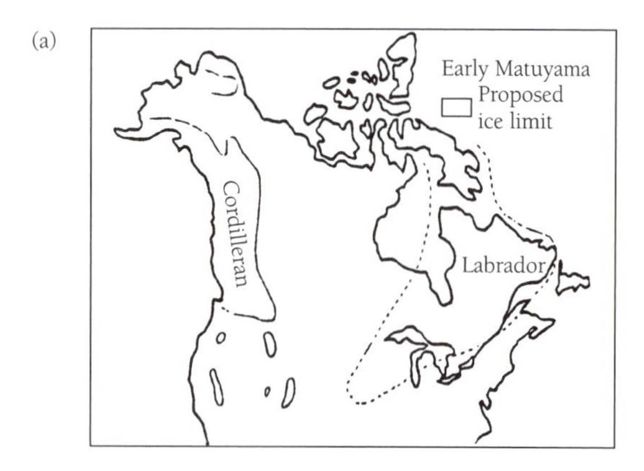

(b)

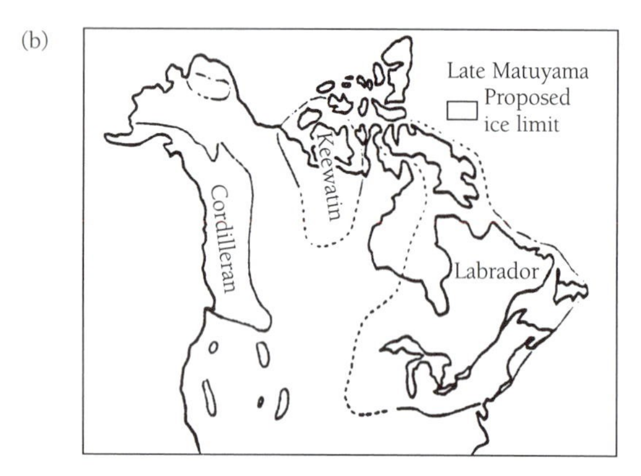

(c)

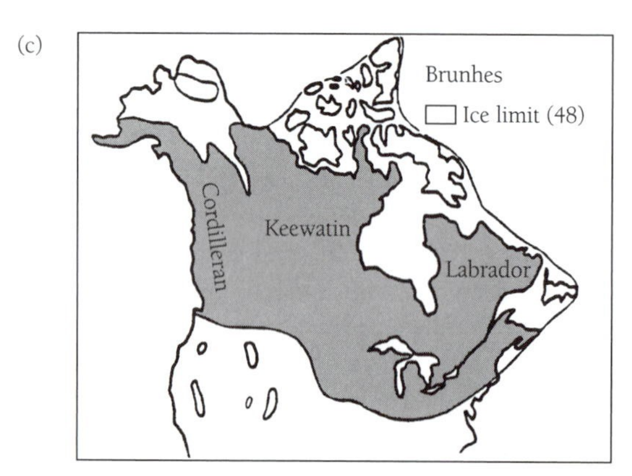

(d)

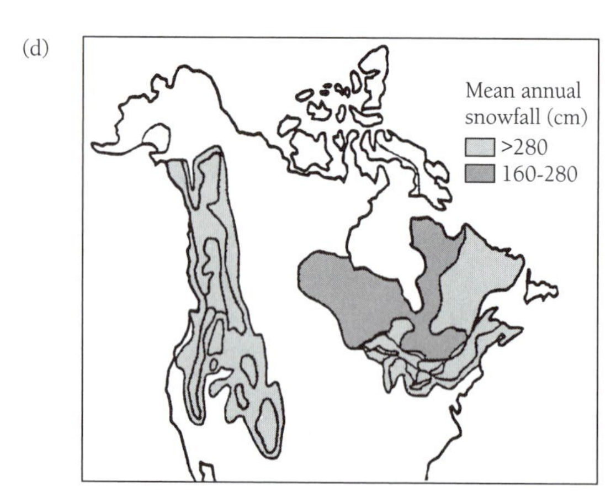

Figure 10.27 Proposed maximum ice distribution during (a) Early Matuyama, (b) Late Matuyama, and (c) Brunhes. (d) Present-day mean annual snowfall. Source: Barendregt and Irving 1998.

sheet), the other in the east (the Labrador ice sheet). With the passage of time, these ice sheets converged, joined by a third ice sheet originating in Keewatin in northern Canada. Barendregt and Irving offer no explanation for the change in ice cover that set in between 1.5 to 0.8 myr B.P. It would seem possible, however, to combine elements of the Clark-Pollard and Barendregt-Irving pictures to produce a composite view of climatic change over the past few million years. A thick blanket of regolith in the central portion of northern Canada during the early years could have made it difficult to develop a persistent ice sheet in this region prior to about 1.5 myr B.P. The regolith could have been removed with successive advances of the Labrador ice sheet permitting the Keewatin ice sheet to spread southward, eventually joining up with the Labrador ice sheet advancing from the northeast. A higher average level of CO_2 could also have contributed to the different rhythm of climate observed prior to 1.5 myr B.P. Closure of the previously open seaway connecting the Pacific and Atlantic (through what is now the Panamian isthmus) could also have had an impact.[28] Despite important recent advances, it is clear that a comprehensive theory of climatic change sufficient to account for the variety of climates observed over the past several million years is not yet at hand.

10.4 The Past 130 Million Years

We turn attention now to a discussion of the complex climatic changes marking the transition from the relatively warm conditions of the Cretaceous to the comparatively cold environments of the more recent past. In what follows, it will be convenient to refer to time intervals by using geologists' terminology for distinguishing different geological eras, periods, and epochs. A summary of such geological nomenclature and an associated time line is presented in Figure 10.28.

A record of changes in the temperature of the deep tropical Pacific for the past 130 myr, inferred from measurements of the oxygen-isotopic composition of benthic foraminifera, is presented in Figure 10.29. Interpretation of the isotopic data is subject to uncertainty, particularly for the most recent part of the record, where corrections must be made to account for bulk changes in the isotopic composition of ocean water due to the presence of continental ice. The temperature of deep-ocean water is taken usually as an indication of the temperature of surface water during winter at high latitudes. Despite the uncertainties implicit in a detailed interpretation of the record, it is difficult to escape the conclusion that important changes in climate have occurred over the past 130 myr. The implications of the data in Figure 10.29 for the climate at high latitudes are particularly striking. They imply that oceanic temperatures at high latitudes dropped from a high of about 18°C during the Cretaceous to a value near 0°C today.

The decrease in high-latitude temperature did not proceed uniformly. Temperatures in the early Eocene (65–40 myr B.P.) were almost as high as in the mid-Cretaceous

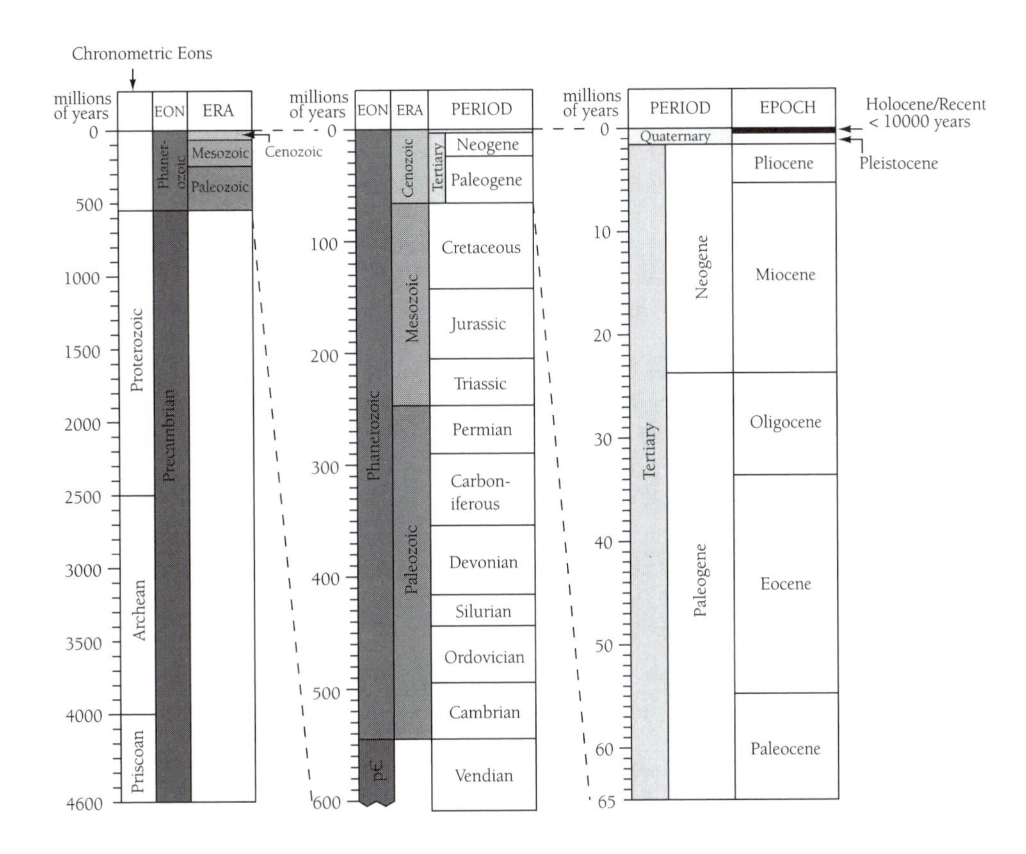

Figure 10.28 The geologic time scale. Source: Andrew MacRae 2000, using data from Gradstein and Ogg (1996) and Harland et al. (1990).

(80 to 120 myr B.P.). Significant cooling marked the late stages of the Eocene and much of the Oligocene. The Oligocene may be taken as marking a transition from the warm, ice-free world of the Cretaceous and Eocene to the cold, glacial world of the Pliocene and Pleistocene. The first signs of Antarctic glaciation appeared at about 40 myr B.P. Both the eastern and western Antarctic ice sheets were fully established by 10 myr B.P. The timing associated with the formation of the Greenland ice sheet is uncertain. It is usually assumed to have formed at about the same time that glaciers began to spread to midlatitudes in the Northern Hemisphere, at about 3 myr B.P. Evidence is lacking, however, and it is possible that initial development of the ice sheet on Greenland occurred much earlier.

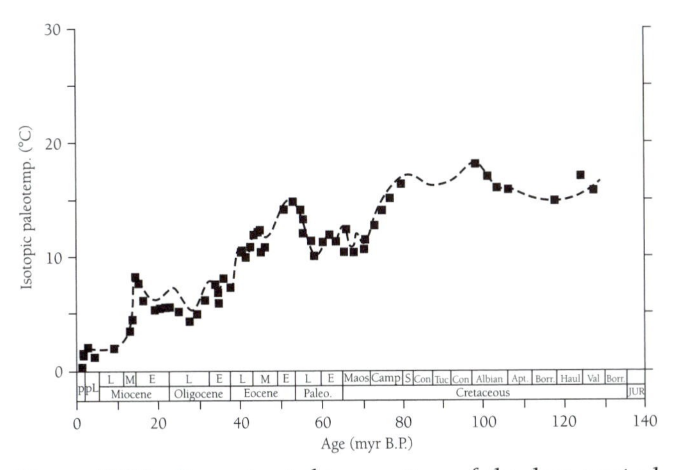

Figure 10.29 Reconstructed temperature of the deep tropical Pacific Ocean over the past 140 myr based on the oxygen-isotopic composition of benthic foraminifera. Source: Douglas and Woodruff 1981.

The temperate environments of the Cretaceous and the early Eocene pose a particular problem for climate models. It is especially difficult to account for the fact that temperatures were mild during winter at high latitudes even in regions far removed from the moderating influence of the warm ocean. Frost-intolerant vegetation was common at Spitsbergen (paleolatitude 79°N) during the Eocene (Schweitzer 1980). Alligators (Dawson et al. 1976) and flying lemurs (McKenna 1980) are observed in deposits of comparable age from Ellesmere Island (paleolatitude 78°N), while data from central Asia indicate the presence of palm trees in this region during the Cretaceous (Vakhrameev 1975). As recently as 5 myr B.P., the forest-tundra boundary extended to latitudes as high as 82°N, some 2500 km north of its present location, occupying regions of Greenland now perpetually covered in ice (Funder et al. 1985; Carter et al. 1986).

The evidence for mild winter weather in the interior of continents at high latitudes associated with what are known as the **equable** climates of the Cretaceous and early Eocene led Farrell (1990) to suggest that these conditions may have developed as a result of a significant poleward expansion of the tropical Hadley circulation. As previously discussed in Chapter 8, the transport of heat from the tropics to higher latitudes is restricted by the influence of the Coriolis force that deflects poleward-moving air eastward. More efficient exchange of angular momentum with either the surface or higher latitudes could allow the Hadley circulation to expand in latitude. Other factors influencing the efficiency of meridional heat transport by the Hadley circulation include the height of the tropopause, the time scale for radiative relaxation, and the magnitude of the latitudinal temperature gradient that would apply if the atmosphere were in radiative equilibrium (i.e., if meridional heat transport was negligible such that energy absorbed from the Sun was locally balanced by the emission of infrared radiation into space).

A high tropopause would provide a deeper troposphere as a conduit for heat transport. The more stable the atmosphere, the larger the heat content of air carried by the upper branch of the circulation. The longer the time scale for radiative relaxation, the more effective the circulation in transporting heat. The smaller the latitudinal temperature gradient corresponding to radiative equilibrium, the lower the demand imposed on the circulation to maintain a specified latitudinal gradient of temperature. According to Farrell (1990), the efficiency of heat transport by the Hadley circulation may be specified in terms of a dimensionless number, Γ, given by

$$\Gamma = \frac{S\tau_R}{\delta_H \tau_A}, \tag{10.4}$$

where S defines the mean static stability of the troposphere, τ_R denotes the time constant for radiative relaxation, δ_H specifies the fractional change of the radiative temperature with latitude, and τ_A refers to the time constant for dissipation of angular momentum. Farrell (1990) estimates a value of Γ for our present climate of 0.5. He concludes that an increase in Γ by about a factor of eight would be required to account for the equable climates of the Cretaceous and early Eocene.

The warm climate of the Cretaceous is usually attributed to an enhanced greenhouse effect associated with higher levels of CO_2. Concentrations of CO_2 inferred by Freeman and Hayes (1992) from measurements of the isotopic composition of carbon in porphyrins extracted from sedimentary organic material are displayed in Figure 10.30.[29] They suggest that concentrations of CO_2 in the Cretaceous were about three times higher than today. A higher concentration of CO_2 in the Cretaceous is consistent with expectations based on the geochemical model presented by Berner, Lasaga, and Garrels (1983).[30] According to this model, higher concentrations of CO_2 in the Cretaceous arise as a consequence of increased volcanism associated with a faster rate of sea floor–spreading. An increase in carbon emissions associated with enhanced tectonic activity must be balanced initially by an accelerated weathering of CaO from silicate rocks and ultimately by an increase in the rate at which carbon is sequestered in sediments (mainly as $CaCO_3$). We would expect a temporal increase in

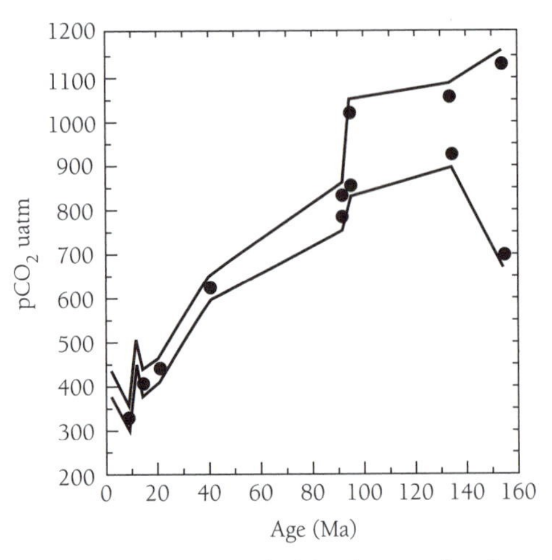

Figure 10.30 Reconstructed CO_2 during the Cretaceous and Eocene. Source: Freeman and Hayes 1992.

the rate of volcanism to result in a temporal increase in the concentration of atmospheric CO_2. Conversely we would expect an increase in the rate of weathering to lead to a reduction in CO_2. The decrease in CO_2 following the Eocene may have been prompted by both a decrease in the rate of sea floor–spreading and an increase in weathering related to an uplift of fresh continental material associated, for example, with the formation of the Tibetan plateau and the Himalayas, when India collided with the Asian continent approximately 50 myr ago (Raymo, Ruddiman, and Froelich 1988). It appears that, by the time of the Miocene, levels of CO_2 had declined to values lower than values applicable in the more recent past. Concentrations of CO_2 inferred from measurements of the $\delta^{13}C$ of alkenones (Pagani, Freeman, and Arthur 1999) indicate a slow secular increase in CO_2 between 14–9 myr B.P., with concentrations stabilizing at more recent levels by about 9 myr B.P.

It is unclear whether concentrations of CO_2 implied by the data in Figure 10.30 would be sufficient to account for the expansion of the Hadley regime advocated by Farrell (1990) as an explanation for the equable climates of the Cretaceous and early Eocene. A high level of atmospheric CH_4 could make an important additional contribution, as suggested by Sloan et al. (1992). A portion of the methane released at the surface is oxidized in the stratosphere, providing a source of high-altitude H_2O (in addition to CO_2). The abundance of H_2O is exceptionally low in air entering the stratosphere at the present epoch (a few parts per million) limited by vapor pressure restrictions imposed by the cold temperature of the tropical tropopause (the entry point of air to the stratosphere). The abundance of H_2O increases (it approximately doubles) as air moves upward and poleward in the stratosphere today, reflecting the additional source contributed by oxidation of CH_4. Concentrations of H_2O reach levels high enough in the polar regime to permit formation of thin, high-altitude clouds known as **polar stratospheric clouds** (see Chapter 15 for a discussion of the important role these clouds play in the chemistry of stratospheric O_3). Sloan et al. (1992) suggest that the source of

CH_4 may have been much higher in the early Eocene. Consequent production of high concentrations of H_2O in the stratosphere could have resulted in formation of dense, optically thick clouds in the polar stratosphere. They speculate that greenhouse screening imposed by these clouds could account for the mild temperatures of the winter polar regime during the early Eocene.

Data supporting this conjecture were presented recently by Katz et al. (1999), who reported evidence for an explosive release of CH_4 56 myr ago from a reservoir of gas initially bound as clathrate (a crystalline material in which gas is encased in a cage of frozen water molecules) in sediments 500 m below the sea floor off the coast of Florida. Clathrates are stable under conditions where pressures are high and temperatures are low (the melting condition depends on a combination of P and T). The idea is that large quantities of CH_4 may be produced in anoxic sediments of coastal regions underlying regions of unusually high biological productivity, where the supply of oxygen is insufficient to offset the supply of organic material. Under appropriate conditions of P and T, it may be converted to clathrate. Clathrate is unstable at shallow depths, where pressures are too low. It is also unstable at greater depths, where the flux of heat from Earth's interior causes temperatures to rise above the melting point. It may be stable, however, at intermediate depths. Katz et al. (1999) suggest that the destabilization of the clathrate reservoir off the coast of Florida 56 myr ago occurred as a result of an increase in the temperature of ocean water in contact with sediments in this region. Warming triggered the release of CH_4, causing a slumping of overlying sediments and leading eventually to an explosive release comparable to an eruption of an underwater volcano. They suggest that this was not an isolated phenomenon, that similar releases must have occurred elsewhere around the world to account for the global reduction of $\delta^{13}C$ observable in sedimentary carbonates. They estimate that clathrates were responsible for a source of atmospheric CH_4 as large as 2×10^{12} tons. Norris and Ruhl (1999) show that the bulk of this release must have taken place over a period of no more than a few thousand years.

It is not our intention to imply that variations in the concentrations of greenhouse gases such as CO_2 and CH_4 were solely responsible for the range of climates that evolved over the interval marking the transition from the Cretaceous to the Pleistocene. This was a period distinguished by important rearrangements in the position of continental landmasses and in the geometry of major ocean basins. We have already alluded to some of these circumstances: the closure of the Panama Straight and the encounter of India with Asia, for example. Other important factors include an increase in the size of both the North and South Atlantic Oceans (resulting from the westward drift of the Americas), separation of Australia from Antarctica, the opening of the Drake Passage, formation of major north-south–aligned mountain chains in western regions of the Americas, and the elimination of the passage (the Tethys Sea) where water previously could circulate freely around the globe in the tropics (obstructed only by a narrow peninsula in southeast Asia). It is likely that these tectonically derived changes were accompanied by important variations in modes of oceanic circulation and by significant, related variations in patterns of atmospheric circulation. A comprehensive theory must allow for all of these factors, accounting at the same time for changes in the radiative properties of the atmosphere. On time scales of a hundred million years or so, Earth's climate is regulated by a complex, interactive interplay of processes involving the atmosphere, biosphere, and ocean, implicating in addition the dynamics of the solid planet.

10.5 From Earliest Times to the Cretaceous

The weight of the evidence suggests that, with a few notable exceptions, Earth's climate was relatively mild over the earliest periods of the planet's history. This result is surprising in that astrophysical models suggest that the output of energy from the Sun should have been much less than today, by as much as 20% in the Proterozoic. If solar luminosity were to drop by as much in the present environment, we would expect Earth to plunge into a deep freeze. The dilemma of a warm Earth in the presence of an environment of low solar energy input is referred to in the climate literature as the **faint Sun paradox**. A resolution of this paradox is usually attributed to a high concentration of greenhouse gases, notably CO_2, in the early atmosphere.

A high concentration of CO_2 in the atmosphere of the early Earth is not surprising. We would expect tectonic activity to have been more vigorous at that time. Also, the supply of continental material to react with CO_2 would have been more limited. We would expect levels of CO_2 in the early atmosphere to have been variable, responding to fluctuations in volcanic emissions and in the supply of material for weathering. Temporary drops in the level of CO_2 could account for occasional cold snaps. The first evidence of continental glaciation appears in the geologic record at about 2.5 billion years before the present (2.5 byr B.P.). Recent attention has focused, however, on a spectacular series of climate fluctuations that developed during the Neoproterozoic, between 750 and 580 myr B.P.

On at least four occasions over this period, it appears that Earth moved into a deep freeze, a condition referred to as a **Snowball Earth.** The term defines an environment in which Earth was frozen over from pole to pole, where ice was ubiquitous, present year round, even in the tropics at sea level. The name was coined by Joseph Kirschvink, a geomagnetist at the California Institute of Technology (Kirschvink 1992). Recent work by Paul Hoffman, Dan Schrag, and colleagues at Harvard University, based on analysis of measurements of the $\delta^{13}C$ of carbonates, has demonstrated how this remarkable state may have evolved (Hoffman et al. 1998).

In their work, Hoffman and colleagues studied a sequence of Neoproterozoic glacial deposits in Namibia. These researchers showed that when the deposits formed, they were situated at sea level at a latitude of about 12°S (glaciation extended into the tropics). Mantle-derived carbon has an isotopic composition corresponding to a $\delta^{13}C$ value of about –5‰. Carbonates deposited in marine sediments today exhibit a $\delta^{13}C$ value close to zero. The relatively high $\delta^{13}C$ value for contemporary carbonates is offset by a much smaller $\delta^{13}C$ value for organic matter (about –25‰) such that the average $\delta^{13}C$ value for material delivered to sediments (carbonates plus organics) is approximately equal to the $\delta^{13}C$ value for carbon in the mantle from which the sedimentary material is derived (about –5‰). Hoffman and colleagues observed that $\delta^{13}C$ values for sedimentary carbonate were unusually high (+5–9‰) in the period preceding the onset of glaciation. They dropped precipitously (to values as low as –5‰), as the system evolved into the glacial world. They recovered to about –3‰ by the end of the glacial epoch, before resuming their decline, reaching values as low as –6‰. The terminal stage of the glacial-interglacial transition was distinguished by a steady increase in $\delta^{13}C$ to values close to those observed today (0‰). The sedimentary sequence marking the glacial epoch is overlain by a thick layer of carbonate rocks. Hoffman and colleagues estimate that these so-called cap carbonates were deposited at a rate as high as 40 cm yr^{-1}.

The elevated $\delta^{13}C$ values for carbonates formed during the interval preceding the transition to glacial conditions implies that the fraction of carbon deposited in sediments in organic form (as compared to carbonates) was significantly higher during this period than today. The relative portions of carbon in organic matter and carbonates shifted in favor of carbonates immediately prior to the onset of glaciation. The rate at which carbon is incorporated in sediments as a component of organic matter is ultimately limited by the rate at which phosphate is supplied to the ocean by the weathering of phosphate minerals on land. A decrease in the burial of carbon in organic form relative to carbonates is attributable to a reduction in supply of phosphate relative to calcium (obtained from the weathering of calcium-bearing silicate minerals). The decrease in phosphate may have developed as a consequence of the high rates of supply required to maintain the unusually high rates at which carbon was sequestered in organic form in sediments during the preceding time period. Alternatively, we might suppose that an increasing fraction of the P delivered to the ocean as phosphate during the preglacial period may have been sequestered in sediments in inorganic form, as phosphorite for example. According to the present understanding (Froelich et al. 1982), phosphorite is formed as a by-product of the oxidation of organic matter by sulfur-reducing bacteria. Phosphorus contained in organic matter is first converted to phosphate and subsequently transformed to phosphorite. In the process, sulfate is reduced to sulfide and precipitated in sediments as iron sulfide. An increase in production of phosphorite could reflect the evolution of more favorable environments for the formation of that mineral during the pre-glacial period, related perhaps to the development of extensive continental shelves associated with the breakup of the supercontinent Rodinia. A decline in the $\delta^{13}C$ of carbonates at the end of the glacial period could be attributed to an increase in the weathering of calcium-bearing silicate rocks exposed to exceptionally acidic precipitation, as discussed below. The final recovery of $\delta^{13}C$ in carbonates could reflect a gradual increase in the supply of phosphate with a recovery of the marine ecosystem to the point where proportions of carbon (deposited as carbonates and organic matter) were comparable to values applicable over much of Earth's history (a ratio of 4:1 for carbon in carbonates relative to organic matter). The sensitivity of the $\delta^{13}C$ of carbonates to changes in relative rates of burial of organic matter and carbonates is illustrated for a number of hypothetical scenarios in Table 10.1.

Hoffman et al. (1998) proposed that the Snowball Earth was triggered by a precipitous drop in CO_2, most probably resulting from a decrease in tectonic activity. Levels of CO_2, they suggested, decreased to the point where the glacial condition spread to planetary scale, globally capping the ocean with a thick layer of surface ice. The ocean was effectively isolated from the atmosphere. The climate of this glacial Earth would have been exceptionally arid. The concentration of CO_2 was not, however, fixed at the low level that triggered the Snowball. It began a slow, steady increase in response to continued input from continental volcanoes. Volcanic emissions would have also resulted in a buildup of sulfur oxides and other gases in the atmosphere. Eventually, levels of CO_2 recovered to the point where greenhouse-induced warming caused sea ice to melt. Exchange of gas between the atmosphere and ocean resumed. Evaporation of water from a relatively warm ocean amplified the greenhouse effect of CO_2 (we would expect the temperature of water below the surface ice layer to have remained higher than the temperature of the overlying atmosphere). Transfer of CO_2 from the atmosphere to the ocean could have initially resulted in accelerated dissolution of carbonate. The impact of this dissolution may have been muted, however, since additions of CO_2 to the ocean by subsurface volcanism over the prolonged period, when the ocean was effectively isolated from the atmosphere, could have depleted the potential supply of dissolvable sedimentary carbonate. Subsequent mixing of cold, deep, alkaline water with warm surface water would have resulted in precipitation of carbonate minerals in surface waters. Weathering of continental rocks (including carbonates) by a reinvigorated hydrological cycle could have provided an ad-

Table 10.1 Sensitivity of the $\delta^{13}C$ in sedimentary carbonates to assumptions made concerning the relative importance of carbonates and organic matter as sinks for carbon emitted by volcanoes (as CO_2).

Carbonate fraction (%)	*Organic fraction (%)*	*$\delta^{13}C$, carbonates*
60	40	+5‰
70	30	+2.5‰
80	20	0‰
90	10	–2.5‰
95	5	–3.75‰

ditional source of alkalinity and may have been responsible for the explosive growth of cap carbonates observed at the end of the glacial period. Under the conditions envisaged by Hoffman and colleagues, the intensity of weathering in the post-glacial world would have been enhanced by unusually corrosive levels of acidity in rain resulting from the high levels of CO_2 and acidic sulfur accumulating in the atmosphere over the period when the atmosphere was effectively isolated from the ocean.

The study by Hoffman and colleagues has provided a fascinating insight into extremes of climate visiting Earth in its distant past. We may expect new insights to emerge as the techniques they pioneered are applied to other periods of Earth's history.

10.6 Summary

We began this survey of past climates with a description of changes observed over the past 150 years. We noted that the global average temperature increased over this interval by about 0.7°C. The increase did not proceed uniformly. It developed over two intervals the first lasting from 1910 to 1940, the second beginning in about 1970 and persisting today. We pointed out that it is difficult to attribute the climatic changes observed over the recent past to any single factor. Increases in CO_2 and other greenhouse gases may have been responsible for warming. Sulfate aerosols produced as by-products of fossil fuel consumption may have been responsible for offsetting cooling, particularly over industrial regions of the Northern Hemisphere. Should the unusual warming observed over the past several decades be attributed primarily to the effects of the buildup of greenhouse gases such as CO_2? Can the trends recently observed be extrapolated as an indication of even warmer climates in the future? Answers to these questions require an understanding of how climate would have evolved in the absence of disturbances introduced by human activity. We need to understand the factors responsible for the climatic variations observed in the pre-industrial era. The current relatively warm period could reflect a natural recovery from the Little Ice Age, an analogue to the warm climate prevailing at the turn of the last millennium (the Medieval warm period). Alternatively, the recovery from the Little Ice Age may have developed as a direct response to the buildup of gases such as CO_2.

Ice ages have dominated Earth's climate over the past several million years. Variations of climate and CO_2 are strongly correlated, at least for the past 450 kyr for which we have direct measurements of CO_2 (from ice cores). Changes in CO_2 may have had an influence on climate. It would be difficult to argue, however, that the climatic changes were *driven* by changes in CO_2. We suggest that the opposite may have been the case: that changes in CO_2 arose largely as a response to climatic changes and related changes in the chemistry and biology of the ocean. This does not negate the hypothesis that changes in CO_2 may have *contributed* to the observed changes in climate.

Changes in orbital parameters (axial tilt, axial precession, orbital precession, and orbital eccentricity) have an important influence on the seasonal and spatial pattern of solar radiation incident on Earth (the so-called Milankovitch effect). Convincing evidence exists that astronomically driven changes in insolation are responsible for much of the variability of climate observed over the past several million years. The conventional Milankovitch theory suggests that continental ice cover should expand during a period when summer insolation is trending lower at high latitudes in the Northern Hemisphere. Conversely, ice cover should tend to decrease when summer insolation increases. The Milankovitch theory attributes the climatic response to a positive feedback drawing on changes in albedo driven by changes in the ice cover. Observational data suggest that major climatic changes in the Northern Hemisphere are preceded by significant climatic change in the Southern Hemisphere, as evidenced by ice cores from Antarctica. We suggested a mechanism to account for this feature of the climate system. We proposed that climatic changes in the Northern Hemisphere may be initiated by adjustments in the circulation of deep water in the Atlantic, prompted by insolation-driven changes in the seasonal expansion of sea ice in the Southern Ocean. The albedo feedback mechanism favored by Milankovitch could play an important role in the subsequent evolution of climate in the Northern Hemisphere. The timing of ice ages and interglacials may be paced, however, from the south rather than the north.

We summarized evidence indicating that climatic variations in the North Atlantic (in Greenland, for example) are rapidly communicated to the tropics. Warming in Greenland at the end of the Younger Dryas period was accompanied by essentially simultaneous warming in the tropics, as indicated by observations of a concurrent increase in the concentration of CH_4. The adjustment of the climate system observed at the end of the Younger Dryas took place over an interval of a few decades or less. We suggested that the increase in tropical temperature at the end of the Younger Dryas may have been triggered by a change in the circulation of shallow waters in the tropical Pacific (a tropical mini–Conveyor Belt). Water sinks from the surface in the interior of the subtropical gyres during winter. It enters the tropics as a flow of relatively cold water in the vicinity of the thermocline. Divergence driven by the easterly trade winds causes this cool water to upwell to the surface. The circulation loop is completed by poleward transport of relatively warm surface water in western branches of the subtropical gyres. We suggested that the flow of waters carried by the mini–Conveyor Belt could adjust in response to variations in the salinity of surface waters in the subtropical Pacific induced by changes in the rate at which water in vapor form is transferred from the Atlantic. Evaporation of H_2O from the Atlantic would be expected to increase as the Atlantic Conveyor Belt reverted to its interglacial (warm) mode at the end of the Younger Dryas. Warming in the Atlantic could have thus led to a decrease in the strength of the Pacific mini-Conveyor with an associated reduction in heat transport from the tropics to the subtropics by the ocean. Changes in the climate of the North Atlantic could be rapidly communicated in this manner to the tropical Pacific and subsequently extended to a global scale as the atmospheric circulation adjusted to the altered thermal state of the tropics.

Observed variations in the concentration of CH_4 over the course of the last ice age suggest that the observed rearrangements of the climate system, in conjunction with the Younger Dryas transition, may not be unusual. We suggested that they may have persisted in muted form into the Holocene. The changes in climate exemplified more recently by the Medieval warm period and by the Little Ice Age may be associated with similar, more modest fluctuations in the operation of the tropical mini–Conveyor with related variability in the flow of deep water carried by the Atlantic Conveyor. It is important to develop a perspective on changes in oceanic circulation possibly underway today. Our survey of the past offers compelling evidence that the climatic system can accommodate to a variety of different states and that it can rapidly evolve from one state to another. It would be instructive to search for changes in the circulation of tropical Pacific waters. A change in the average depth of the tropical thermocline could be a harbinger of an important climate change in the future. An open, but important, question is whether such a change could be triggered by a warming of surface ocean waters during winter at midlatitudes, induced by the contemporary increase in the concentration of greenhouse gases.

Major climatic variations over the past 800 kyr exhibit a characteristic periodicity of about 100 kyr. This is usually attributed to changes in insolation related to variations in the ellipticity of Earth's orbit around the Sun. The associated variations in insolation are small, however, and it has been difficult to devise a mechanism to account for how such small changes could be amplified to exert such a major influence on global climate. An alternate theory attributes the 100-kyr climate periodicity to factors relating to the stability of continental ice sheets. We view this explanation as more plausible. The 100-kyr signal only recently emerged as a dominant feature of the climate record. The earlier record is distinguished by climatic variations occurring predominantly on the time scales associated with major changes in insolation, specifically those corresponding to variations in obliquity (a period of 41 kyr) and precession (a period of 22 kyr). Growth of extensive ice sheets on the continents of the Northern Hemisphere is a relatively recent phenomenon. We suggested that the growth of the ice sheets may have been facilitated by the closure of the land bridge between North and South America. The relative isolation of Atlantic waters from waters in the Pacific resulting from the formation of the Panama Isthmus would have permitted a significant gradient to develop between the salinities of waters in the Atlantic and Pacific. Prior to closure of the land bridge, the Gulf Stream may have taken a more southerly track across the Atlantic. Transport of warm water to high latitudes would have been reduced accordingly. We speculated that the presence of warm water at high latitudes in the Atlantic during winter may have been critical to the growth of the extensive continental ice sheets that developed over the past 800 kyr.

This chapter was motivated from the outset by a question of whether the study of past climates could provide useful insights into climatic changes possibly developing in the future in response to higher concentrations of greenhouse gases. Our survey of the past offers irrefutable evidence of the variability of climate on a wide range of space and time scales. The changes observed during the Younger Dryas period indicate that globally significant adjustments can occur over periods as brief as a few decades. We attributed these changes to variations in the circulation of tropical Pacific waters. The evidence for equable climates in the Cretaceous and Eocene, the suggestions that higher concentrations of CO_2 and CH_4 may have contributed to these conditions, and the evidence of major climatic adjustments in the Precambrian offer the most persuasive support for the importance of greenhouse gases. It is unlikely, however, that studies of the past can provide on their own an unambiguous basis for a forecast of the future. For this we must rely on models. Studies of the past offer a means to test and refine models. They are also valuable in that they may draw attention to important processes and interactions that might otherwise be ignored.

The Carbon Cycle

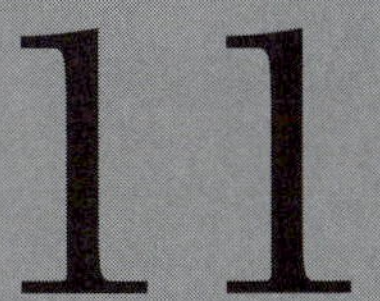

Carbon dioxide, CO_2, accounts for about 355 parts per million (ppm) by volume of our current atmosphere. The concentration is rising steadily by about 2 ppm per year, a consequence largely of industrial and domestic consumption of **fossil fuels**: coal, oil, and natural gas. Measurements of the concentration of CO_2 in the atmosphere have been available for a few locations (Mauna Loa, Hawaii, and the South Pole, for example) on an almost continuous basis since 1958, due largely to the dedicated efforts of C. D. Keeling, a geochemist at the Scripps Institute of Oceanography in California. Keeling's research was designed to provide a record of the changing concentration of CO_2 in regions of the atmosphere far removed from industrial sources. An Hawaiian mountaintop in the middle of the Pacific Ocean provides an ideal baseline for the Northern Hemisphere, whereas the United States research station at the South Pole serves a similar function for the Southern Hemisphere. Keeling's data, displayed as weekly averages of daily measurements of CO_2 concentration at Mauna Loa and the South Pole, are presented in Figures 11.1 and 11.2, respectively.

Two features of the results in Figures 11.1 and 11.2 merit comment. First, the measurements show indisputable evidence for a long-term increase in the concentration of atmospheric CO_2. The increase is roughly proportional to the rate at which CO_2 is added to the atmosphere by the burning of fossil fuels. Estimates for the fossil fuel–related source of CO_2 are presented in Figure 11.3. Emission of CO_2 from the combustion of fossil fuels grew from 2.47×10^9 tons C per year (1 ton = 10^6g) in 1959 to 5.08×10^9 tons C per year in 1982: cumulative emissions over this period

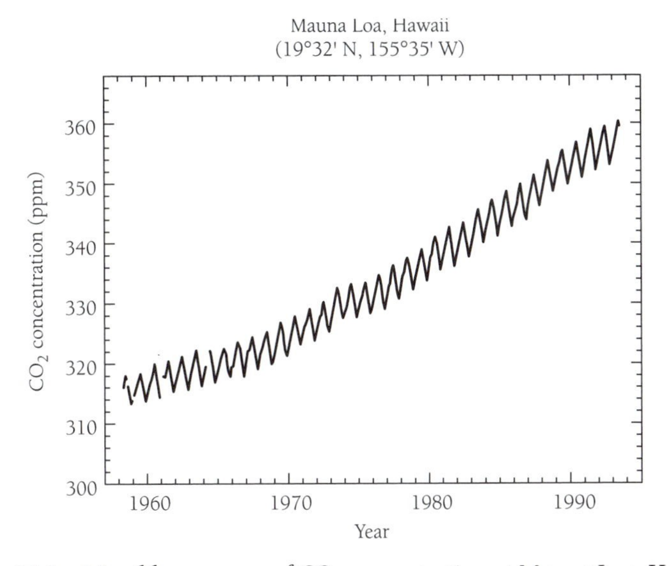

Figure 11.1 Monthly averages of CO_2 concentration at Mauna Loa, Hawaii, since 1958. Source: Keeling et al. 1989.

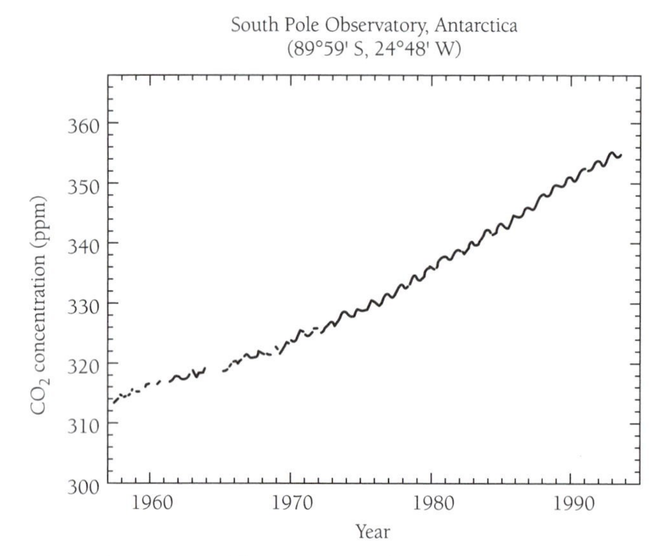

Figure 11.2 Monthly averages of CO_2 concentration at the South Pole since 1958. Source: Keeling et al. 1989.

amounted to 91.2×10^9 tons C.[1] The concentration of atmospheric CO_2 rose over the same interval by 24.3 ppm, implying an increase in the concentration of CO_2 in the atmosphere of approximately 1 ppm for each 3.3×10^9 tons C added by burning fossil fuels. The quantity of carbon added to the atmosphere from the fossil sources between 1959 and 1982 would have been sufficient to increase the concentration of atmospheric CO_2 by 43 ppm.[2] It follows that only 56% of the carbon emitted due to combustion of fossil fuels has remained in the atmosphere. The balance, as we shall see, must have been transferred either to the ocean or to the **terrestrial biosphere**, or it must have been sequestered in soils. We use the term *biosphere* here to refer generally to the living portion of the terrestrial system and primarily to the carbon content of trees and plants. Despite their overriding importance in the cycling of chemical elements, animals and microbes account for a negligibly small fraction of the total carbon content of the biosphere. The term *soil* as used here refers to the composite reservoir of carbon stored below ground.

The second notable feature of the results in Figures 11.1 and 11.2 is the sawtooth pattern evident for the variation of CO_2 over the course of a year. The abundance of CO_2 in the atmosphere at Mauna Loa decreases during spring and summer, rising during autumn and winter. This pattern reflects the role of photosynthesis as a sink for atmospheric CO_2 during spring and summer, and the balancing effects of respiration and decay returning CO_2 to the air during autumn and winter. The amplitude of the seasonal signal is even larger at Point Barrow, Alaska, as indicated in Figure 11.4, reflecting the closer proximity of this site to a greater concentration of deciduous plants. It is relatively small at the South Pole, as might be expected given the remote nature of this station and the comparative absence of vegetation, especially deciduous plants, in the Southern Hemisphere. The data in Figures 11.1, 11.2, and 11.4 provide unambiguous evidence for the importance of vegetation and soils in the budget of atmospheric CO_2.

This chapter concerns the processes regulating the concentration of CO_2 in the atmosphere. Ultimately we aim to understand the fate of CO_2 introduced to the atmosphere by the burning of fossil fuels. We begin, in Section 11.1, with an account of what is known as the **carbon cycle**, the merry-go-round by which carbon is shuffled back and forth between the atmosphere and biosphere, biosphere and soil, soil and atmosphere, soil and ocean, and the atmosphere and ocean. We discuss how carbon is eventually removed

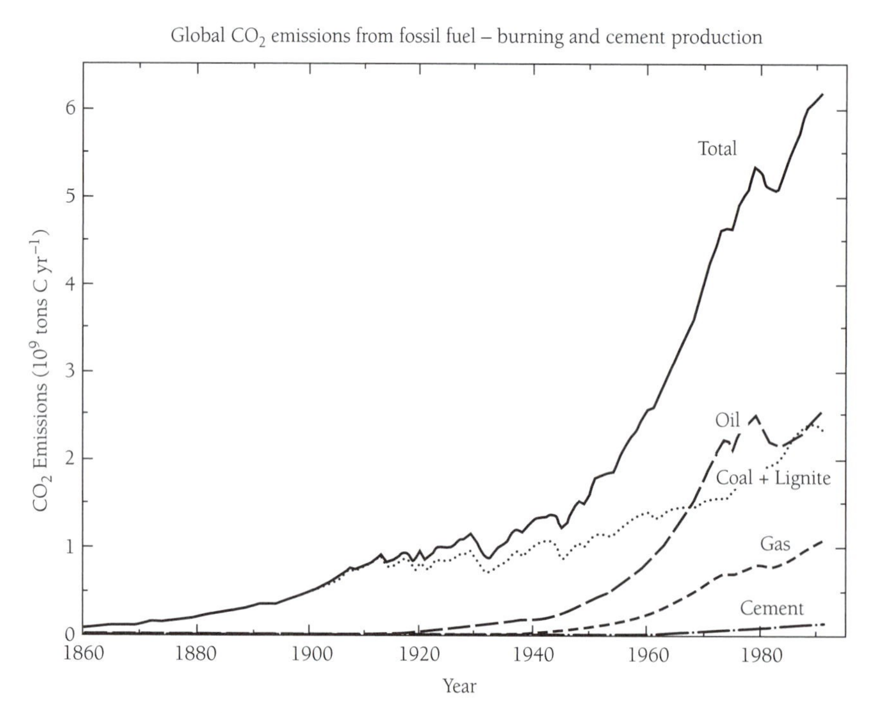

Figure 11.3 Production of CO_2 associated with the combustion of fossil fuels.

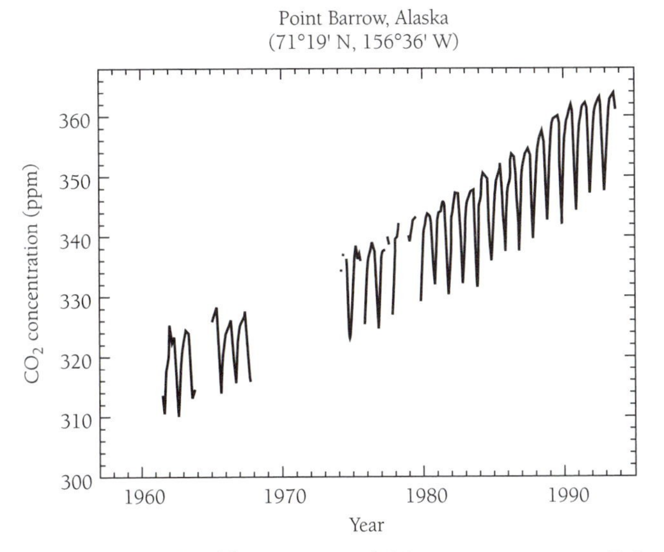

Figure 11.4 Monthly averages of CO_2 concentration at Point Barrow, Alaska, since 1974. Source: Keeling et al. 1989.

from the atmosphere-ocean-biosphere-soil system to sediments, and how it is returned to the atmosphere by uplift and erosion of sediments, and by volcanism. The discussion in Section 11.1 provides a perspective on the disturbance introduced by mining and combustion of fossil fuels. In its larger context, we see the influence of the consumption of fossil fuels as an acceleration of the rate at which carbon is transferred from sediments to the atmosphere. To restore the pre-industrial equilibrium, carbon must be returned to sediments. Section 11.1 sets the stage for a more focused discussion in Section 11.2 of the ocean's capacity to absorb excess quantities of CO_2 produced by the combustion of fossil fuels. We show that the rate at which **anthropogenic** carbon (i.e., carbon associated with human activity) is returned to the sediment is limited to a significant extent by the ocean's capacity to absorb additional CO_2. We discuss the contemporary budget of CO_2 in Section 11.3. The specific issue addressed here concerns the role of the biosphere as a source, or sink, for CO_2: deforestation in the tropics provides an important source of atmospheric CO_2, whereas regrowth of vegetation, mainly at midlatitudes of the Northern Hemisphere, provides a compensatory sink. We show how measurements of O_2, in combination with measurements of CO_2, can be used to distinguish between the relative contributions of the ocean and the biosphere to the contemporary budget of CO_2. We develop projections for growth of CO_2 in the future in Section 11.4. We present summary remarks in Section 11.5.

11.1 The Carbon Cycle

Carbon is transferred from the atmosphere to the biosphere by **photosynthesis**, the process by which plants use solar energy to transform H_2O and CO_2 to carbohydrates, releasing O_2 by the net reaction:

$$h\nu + CO_2 + H_2O \rightarrow CH_2O + O_2 \qquad (11.1)$$

The oxidation state of the carbon atom is changed in reaction (11.1) from the value +4 appropriate for CO_2 to the value 0 characteristic of organic matter (see Section 3.2). We say that carbon is *fixed* in (11.1) by photosynthesis. The fixing agents are the green plants that carpet Earth in diverse environments, ranging from the relative abundance of closed-canopy tropical rain forests to the comparative dearth of desert scrub and arctic tundra.

The solar energy captured in (11.1) provides the fuel for most of life on Earth. Animals, including ourselves, depend on carbon fixed by (11.1) for life-sustaining nourishment. We survive by virtue of energy extracted by the oxidation of carbon in the food we ingest. When we eat and breathe, we convert carbon in organic form to CO_2; we inhale O_2 and exhale the oxidation product CO_2, recovering in this fashion a portion of the solar energy absorbed in (11.1). The net reaction is represented by

$$CH_2O + O_2 \rightarrow CO_2 + H_2O + \text{energy}, \qquad (11.2)$$

essentially reversing (11.1). We refer to the process summarized by (11.2) as **respiration**.

Respiration is important not only for animals but also for plants. Plant leaves respire: so do their roots. The excess of the production rate of organic matter by plants as a result of (11.1) over and above the rate at which organic matter is consumed by respiration (11.2) is termed **net photosynthesis**, in contrast to **gross photosynthesis**, which defines the overall rate at which carbon is fixed by reaction (11.1).

Carbon fixed by photosynthesis is converted to a variety of chemical forms by plants and animals. As lignin and cellulose, it constitutes the structural material of the woody parts of trees and plants. In different arrangements, it appears in the roots, leaves, flowers, and fruits of plants, and in the body parts of organisms as large as the elephant or as small as the microbe responsible for decomposition of organic matter in soil. The chemical formula "CH_2O" in reactions (11.1) and (11.2) provides an umbrella description for the diversity of organic species in nature; its utility lies in the fact that it correctly accounts, on the average, for the oxidation state of the composite of these species. Carbon, absorbed by green plants from the atmosphere as CO_2, is reduced by photosynthesis, with the associated release of O_2. In conjunction with oxidation by respiration, it is returned into the atmosphere as CO_2. Production of O_2 by photosynthesis is offset on a global scale by the loss associated with plant and animal respiration.

A picture of the terrestrial portion of the carbon cycle is presented in Figure 11.5. In this model carbon is distributed among six reservoirs representing the atmosphere, ground vegetation, non-woody parts of trees (foliage, flowers, fruits, and transient roots), woody parts of trees (branches, trunks, and most root material), detritus/decomposers (litter and organisms involved in its decomposition near the soil surface), and a compartment identified as *active soil* (taken to include portions of the soil carbon reservoir exchanging with the atmosphere on a time scale of a century or so). Proceeding from the real world to the idealization in Figure 11.5 clearly requires a leap of faith. All of the carbon in all of the wood in all

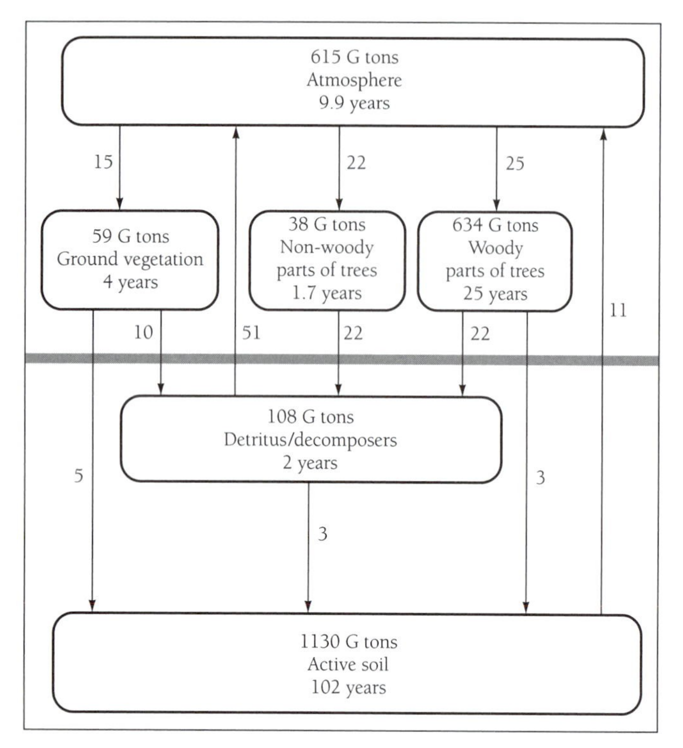

Figure 11.5 The terrestrial portion of the global carbon cycle. The biosphere is considered to be composed of three reservoirs: ground vegetation, non-woody parts of trees, and woody parts of trees. Soils are divided into two reservoirs: a short-lived compartment at the surface dominated by detritus, where decomposition by microorganism is particularly rapid, and a more persistent pool labeled active soil. *Carbon contents of individual reservoirs are indicated in units of 10^9 tons C (the content of the atmosphere is equal to 615×10^9 tons C). Also shown are lifetimes for carbon in individual reservoirs and rates for transfer between reservoirs (given in units of 10^9 tons C yr^{-1}). The data are designed to represent a hypothetical steady-state applicable to the pre-industrial environment. The mixing ratio of CO_2 in the atmosphere was taken equal to 290 ppm. Source: Emanuel et al. 1984.*

of the trees in the world is aggregated in a single unit; similar sweeping assumptions are made for carbon in the other reservoir units included in the Figure. One could imagine a finer distinction: a model accounting for characteristics of individual ecosystems, for example. The problem is that it would be difficult, if not impossible, to define the parameters of such a model, to specify contents and rates for exchange of carbon in the larger array of reservoirs. A model is no better than the data used in its definition. The model in Figure 11.5 was devised to study the response of the atmosphere–terrestrial carbon system to changes in land use (the deforestation associated with conversion of land to agriculture, for example) and the combustion of fossil fuels. The specific intent was to model the variation of the CO_2 concentration in the atmosphere over the past several centuries. In this sense, the model fulfills its purpose. The hope is that errors implicit in the definition of rates for carbon exchange between the atmosphere and individual ecosystems are minimized when reservoirs are combined into larger units. With better information, the model can be refined. In the meantime, it serves a useful function in focusing our thoughts on the manner and rate at which carbon is exchanged between the atmosphere and biosphere, biosphere and soil, and soil and atmosphere.

The model in Figure 11.5 includes estimates for the carbon content of individual reservoirs, together with data on rates at which carbon is exchanged between the different reservoirs; it is intended to represent the status of the terrestrial carbon cycle in a steady state prior to the large-scale disturbances introduced by modern changes in land use and combustion of fossil fuels. With the exception of the value quoted for the carbon content of the atmosphere, which assumes a mixing ratio for CO_2 of 290 ppm,[3] the data included in Figure 11.5 should be considered individually and collectively, to a greater or lesser extent, as uncertain.[4] Nonetheless, the contents and exchange rates given in Figure 11.5 can be used to obtain useful, or at least educational, estimates for the time a carbon atom spends on average in a particular reservoir or combination of reservoirs. The **residence time** for carbon in a particular reservoir is given by dividing the content of the reservoir by the rate at which carbon either enters or leaves the reservoir. Results are the same in either case, since the exchange rates in Figure 11.5 were defined such that budgets of the reservoirs were balanced; rates of production were constrained to equal rates of loss.

Example 11.1: Estimate the residence time for carbon contained in the woody parts of trees.

Answer: Woody parts of trees contain 634×10^9 tons carbon, according to the data in Figure 11.5. Carbon is delivered to the woody reservoir at a rate of 25×10^9 tons C yr^{-1}. The residence time, τ, is given by

$$\tau = \frac{634 \times 10^9 \text{ tons C}}{25 \times 10^9 \text{ tons C yr}^{-1}} = 25.36 \text{ yrs} \quad \blacksquare$$

Example 11.2: Estimate the average time that a carbon atom resides in the atmosphere before it is transferred by photosynthesis to the land-based biosphere.

Answer: The carbon content of the atmosphere, according to the data in Figure 11.5, is equal to 615×10^9 tons C. The total rate of photosynthesis, P, is obtained by combining the rates for photosynthesis by ground vegetation, non-woody parts of trees, and woody parts of trees.

$$P = \left[(15 \times 10^9) + (22 \times 10^9) + (25 \times 10^9)\right] \text{ tons C yr}^{-1}$$
$$= 62 \times 10^9 \text{ tons C yr}^{-1}$$

The residence time, τ, is given in this case by

$$\tau = \frac{615 \times 10^9 \text{ tons C}}{62 \times 10^9 \text{ tons C yr}^{-1}} = 9.92 \text{ yrs} \quad \blacksquare$$

Figure 11.5 indicates two paths for the return of carbon from the biosphere-soil system to the atmosphere. It can proceed by respiration either in the detritus/decomposer pool or in the active soil reservoir. Return associated with the respiration of plants and trees is incorporated in the definition of photosynthesis, that is, the data included in Figure 11.5 are intended to represent net, rather than gross, rates of photosynthesis. The relative importance of the separate soil compartments depends on the magnitudes of the associated transfer rates.

Example 11.3: Estimate the probability that a carbon atom should be returned to the atmosphere by respiration in the detritus/decomposer reservoir rather than by respiration in active soil.

Answer: The rate at which carbon is returned to the atmosphere from the detritus/decomposer reservoir is 51×10^9 tons C yr^{-1}, while the corresponding rate for active soil equals 11×10^9 tons C yr^{-1}. The total return rate is given by the sum of these rates, or 62×10^9 tons C yr^{-1}. The probability, P, that the return proceeds through the detritus/decomposer compartment is obtained by dividing the rate for transfer by this path by the rate for transfer through the combined soil reservoirs:

$$P = \frac{51 \times 10^9 \text{ tons C yr}^{-1}}{62 \times 10^9 \text{ tons C yr}^{-1}} = 0.82$$

It follows that the probability that carbon should be returned to the atmosphere through the active soil pool equals (1 – 0.82), or 0.18. ■

The lifetime for carbon in the active soil compartment is relatively long, 102 yrs, compared with 2 yrs in the detritus/decomposer pool. The cycling of carbon through the biosphere-soil system back to the atmosphere is comparatively rapid for 82% of the carbon atoms involved in photosynthesis (converting the probability obtained above to a percentage), relatively slow for the balance, 18%.

Example 11.4: Estimate the time a carbon atom resides as a constituent of the combined biosphere-soil system.

Answer: The content of carbon in the combined biosphere is obtained by summing the contents of ground vegetation, non-woody parts of trees, and woody parts of trees: $[(59 \times 10^9) + (38 \times 10^9) + (634 \times 10^9)]$ tons C = 731×10^9 tons C. The content of the soil compartment is given by the combination of the detritus/decomposer and active soil compartments: $[(108 \times 10^9) + (1130 \times 10^9)]$ tons C = 1238×10^9 tons C. The quantity contained by the combined biosphere-soil system is given by the sum of these numbers: $[(731 \times 10^9) + (1238 \times 10^9)]$ tons C = 1969×10^9 tons C. The rate at which carbon returns to the atmosphere is calculated by summing the rates for return from the separate soil compartments: $[(51 \times 10^9) + (11 \times 10^9)]$ tons C yr^{-1} = 62×10^9 tons C yr^{-1}. The residence time, τ, for carbon in the combined biosphere-soil system is given by

$$\tau = \frac{1969 \times 10^9 \text{ tons C}}{62 \times 10^9 \text{ tons C yr}^{-1}} = 31.76 \text{ yrs}$$ ■

A model describing the cycling of carbon within the ocean is presented in Figure 11.6. The ocean is split into four reservoirs: two (warm surface and cold surface) extending to depths of 100 m, are selected to describe surface regions of the ocean; a third, extending from 100 to 1100 m, represents **intermediate water**; the fourth, reaching to 3780 m, accounts for waters of the **deep ocean**. The surface reservoirs, labeled "warm" and "cold," distinguish regions of the ocean at low and high latitudes, respectively. The high-latitude cold-surface box is designed to simulate regions of the ocean where the surface supplies water to the depths as a result of the cooling of the surface layer during winter. Deep water is formed in limited regions of the world's oceans, mainly in the North Atlantic and Antarctic Oceans, as previously discussed in Chapter 9.[5] The cold surface reservoir in Figure 11.6 accounts for only 5% of the ocean's total surface area. The warm surface reservoir covers an area of 3.4×10^8 km^2, in contrast to the cold-surface reservoir, whose area extends only 1.8×10^7 km^2. Figure 11.6 includes data on the carbon content of the various oceanic reservoirs, together with

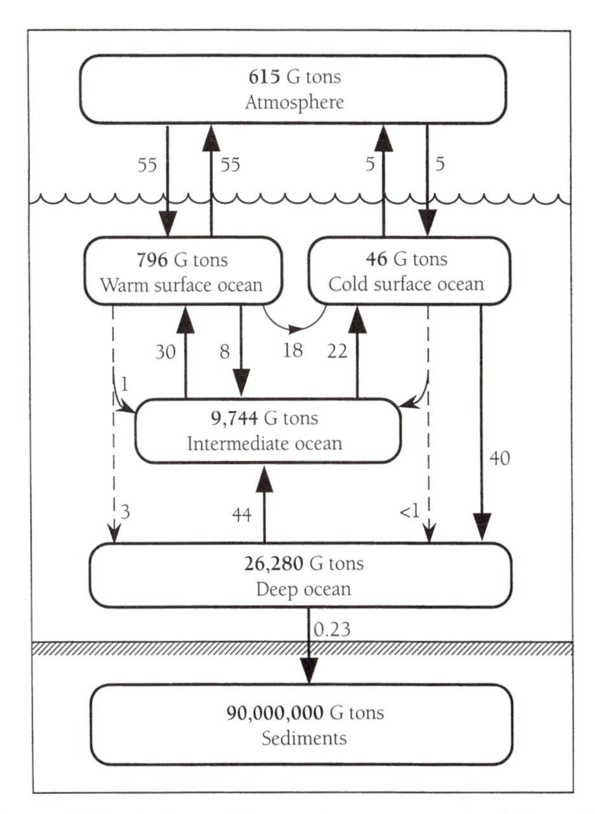

Figure 11.6 A view of the marine components of the carbon cycle. Reservoir contents are given in units of 10^9 C (1G ton = 10^9 tons). Transfer rates, indicated by arrows connecting reservoirs, are expressed in units of 10^9 tons C yr^{-1}. The dashed lines indicate transfer of carbon from the surface to intermediate and deep water associated with falling fecal material. Source: Ennever and McElroy 1985.

estimates for the rate at which carbon is exchanged between the different reservoirs.

As will be discussed in Section 11.2, most of the ocean's carbon is represented by inorganic species, such as the **bicarbonate ion** HCO_3^-. Living organisms account for a negligibly small fraction of the total carbon contained in the ocean. We should not interpret this to mean that the ocean is a desert; rates of photosynthesis in the ocean on a global scale are comparable to rates for photosynthesis on land.[6] The low value for the carbon content of living sea organisms as compared to land organisms reflects the relatively short life of the former (minutes to days) as compared with the persistence of the latter (years to centuries). Organic matter in the ocean is represented by a widely distributed diffuse component presumed to reflect the refractory residue (slowly decomposing body parts) of organisms that once lived in the sea. This component, referred to as **dissolved organic carbon**, accounts for a few percent of the carbon present in inorganic form. This is omitted in the budget summarized in Figure 11.6.

The abundance of carbon in the ocean, as indicated in Figure 11.6, is almost 60 times larger than that of the atmosphere. It exceeds the content of the combined atmosphere–terrestrial biosphere–soil compartment by more than a factor of fourteen. An appreciation for the cycle of carbon in the ocean is obviously prerequisite to an understanding of the distribution of carbon more generally on Earth. The picture presented in Figure 11.6 assumes flows of water between compartments, as summarized in Figure 11.7.

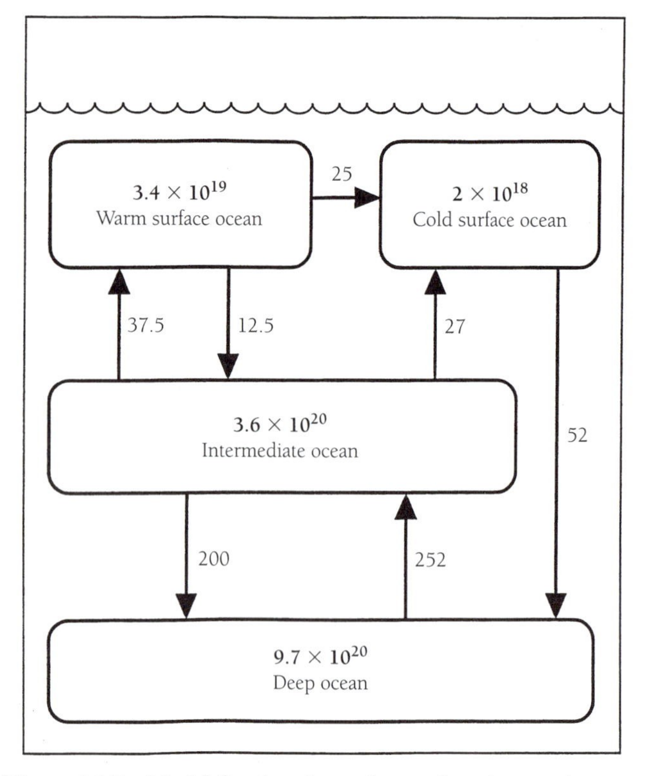

Figure 11.7 Model for the physical transfer of water between compartments used in constructing the data in Figure 11.6. Water content of individual reservoirs is given in units of liters. Transfer rates are in Sverdrups (1 Sv = 10^6 m^3 sec^{-1} = 3×10^{16} l yr^{-1}).

Rates for exchange of water between the different reservoirs in Figure 11.7 are quoted in units of Sverdrups (Sv), as introduced in Chapter 9 (1 Sv = 10^6 m^3 sec^{-1} = 10^{12} g sec^{-1}). The flow of water carried by the rivers of the world to the ocean equals about 1.2 Sv. The water flow from the surface to the depths, as indicated in Figure 11.7 (52 Sv), is equivalent to 43 times the flux carried by the sum of all rivers: more than 100 times the amount delivered by the largest river, the Amazon (0.5 Sv). Rates for exchange of water between the various reservoirs in Figure 11.7 are large, and so, too, are rates for exchange of carbon.

Figure 11.6 indicates a net flux of carbon from the intermediate pool to the warm surface of 22 × 10^9 (tons C yr^{-1}). Of this, 4 × 10^9 tons C yr^{-1} returns to intermediate levels and to depths associated with the decay of organic and carbonate components of fecal material. There is literally a shower of carbon from the surface to the underlying ocean, mediated by biological activity near the surface. The shower's strength depends on the rate at which nutrients are supplied to the surface.[7] The process has been described as a "biological pump," drawing carbon from the atmosphere, sequestering it in the deep sea. Changes in the efficiency of the biological pump are thought to play an important role in regulating fluctuations in atmospheric CO_2 concentrations associated with glacial-to-interglacial transitions in climate.

The ocean is not a closed system with respect to carbon, even for the hypothetical steady-state system envisaged here. Carbon is lost to sediments both in organic form and as a component of carbonate minerals. The corresponding fluxes are estimated at 0.08 and 0.15 × 10^9 tons C yr^{-1}, respectively. Production of calcium carbonate, $CaCO_3$, is associated with release of CO_2.[8] The organic pool incorporates a portion of the CO_2 formed by this process; the balance is released as CO_2 into the atmosphere, where it is consumed by the weathering of calcium-bearing rocks. The cycle is completed by the return of an equivalent amount of carbon to the ocean as HCO_3^- in rivers.

Figure 11.8. presents a picture of the overall carbon cycle, a composite of the data in Figures 11.5 and 11.6. For convenience, we have chosen to combine the carbon in ground vegetation, non-woody parts of trees, and woody parts of trees in a single biospheric reservoir, to aggregate carbon in detritus/decomposers and active soil in a single soil compartment, and to combine in a single unit carbon in the cold and warm water compartments of the surface ocean. Figure 11.8 explicitly represents the sedimentary reservoir.

Example 11.5: Estimate the lifetime for carbon in the combined atmosphere-terrestrial-ocean system.

Answer: The content of carbon in the aggregate atmosphere-biosphere-soil-ocean system is obtained by combining the abundances of the individual reservoirs:

$$\text{Total abundance} = 39{,}350 \times 10^9 \text{ tons C}$$

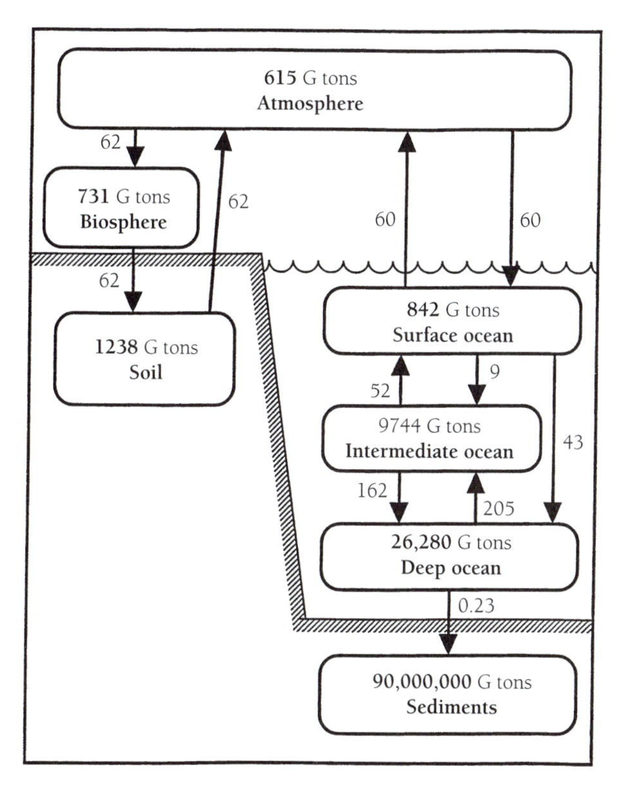

Figure 11.8 Composite model for the global carbon cycle, combining data in Figures 11.5 and 11.6. Reservoir contents are in units of 10^9 tons C; transfer rates are in 10^9 tons C yr^{-1}. Carbon is deposited in sediment both as $CaCO_3$ and as organic matter. There is a small release of CO_2 in steady state from the ocean; this source is employed in weathering of crustal rocks.

The rate at which carbon is withdrawn to sediments is equal to 0.23×10^9 tons yr^{-1}. The residence time, τ, is given therefore by

$$\tau = \frac{39,350 \times 10^9 \text{ tons C}}{0.23 \times 10^9 \text{ tons C yr}^{-1}} = 171{,}087 \text{ yrs} \qquad \blacksquare$$

Example 11.6: Estimate the residence time for carbon in the ocean.

Answer: The carbon content of the ocean equals to $36,866 \times 10^9$ tons C. Carbon atoms leave the ocean, transferring to either the atmosphere or sediments. The combined loss rate is given by $[(60 \times 10^9) + (0.23 \times 10^9)]$ tons C $yr^{-1} = 60.23 \times 10^9$ tons C yr^{-1}. The residence time, τ, is given by

$$\tau = \frac{36,866 \times 10^9 \text{ C}}{60.23 \times 10^9 \text{ tons C yr}^{-1}} = 612 \text{ yrs} \qquad \blacksquare$$

Example 11.7: Estimate the probability that a carbon atom in the ocean will be removed to sediments rather than released into the atmosphere.

Answer: The probability is determined by the relative magnitudes of the fluxes to the relevant reservoirs. The total removal rate is given by the sum of the separate contributions, 60.23×10^9 tons C yr^{-1}.

The rate for removal to sediments is equal to 0.23×10^9 tons C yr^{-1}. The probability, *P*, that a carbon atom is incorporated in sediment on a particular visit to the ocean is given by:

$$P = \frac{0.23 \times 10^9 \text{ tons C yr}^{-1}}{60.23 \times 10^9 \text{ tons C yr}^{-1}} = 0.0038 \qquad \blacksquare$$

Example 11.8: Estimate the average number of times a carbon atom visits the ocean before it is incorporated in sediments. Calculate the total time spent by a carbon atom in the ocean before it is incorporated in sediments.

Answer: On each visit, the probability that the atom should be incorporated in sediments is equal to the value derived above, 0.0038. The number of visits required to build this probability up to 1 equals $(0.0038)^{-1} = 263$. The total time, T, spent by a carbon atom in the ocean pending capture by sediments is obtained by multiplying the time per visit by the number of visits:

$$T = 611.9 \text{ yr} \times 263 = 160{,}930 \text{ yrs.} \qquad \blacksquare$$

The time the atom spends in the ocean, 160,930 yrs, may be compared with its integrated life in the combined atmosphere-ocean-biosphere-soil system: 171,087 yrs. It is clear that the average carbon atom spends most of its life in the ocean. It flits back and forth between the ocean and atmosphere many times, with brief but frequent visits to the terrestrial biosphere and soils. The carbon atoms that constitute our flesh, bones, and blood, had an earlier history in the atmosphere, biosphere, and ocean; we are in a real sense part of the stuff of Earth.

The residence time of a carbon atom in the combined atmosphere-ocean-biosphere-soil system is comparatively brief. Since life has existed on Earth for close to 4 billion years, it may be concluded that the sedimentary reservoir itself must be transitory. The residence time for a carbon atom in the sediment, estimated in the same fashion as for other reservoirs above, is about 390 million years. The typical carbon atom, then, has cycled more than 10 times through the sedimentary compartment over the course of geologic time. It is the motion of the crustal plates—the internal dynamics of Earth—that provides a solution to the dilemma. Sediments are transported by the crustal plates; over the course of time they are either uplifted (returned to the surface) or subducted (withdrawn to the mantle). In the latter case, they are cooked and carbon is vented to the atmosphere as a component of volcanoes and hot springs. In the former, carbon is exposed to the weather, where it can be eroded and returned to the more mobile environments of the atmosphere, biosphere, and ocean. Without this overturning of the crust, life might cease to exist on earth—all of the carbon in surface reservoirs would be tied up in the crust.

11.2 The Capacity of the Ocean as a Sink for Industrial CO_2

The preceding section sought to present a view of the carbon cycle in a steady state; sources and sinks were purposely balanced for the individual reservoirs included in Figures 11.5,

Table 11.1 Concentrations of the Major Cations and Anions in the Ocean

Positive Charge		
Cation	*mol l⁻¹*	*charge eq l⁻¹*
Na^+	0.470	0.470
K^+	0.010	0.010
Mg^{2+}	0.053	0.106
Ca^{2+}	0.010	0.020
Sum		0.606
Negative Charge		
Anion	*mol l⁻¹*	*charge eq l⁻¹*
Cl^-	0.547	0.547
SO_4^{2-}	0.028	0.056
Br^-	0.001	0.001
Partial Sum		0.604
$HCO_3^- + 2CO_3^{2-}$		0.002
Sum		0.606

NOTE: *A pH of 8 gives a concentration of H^+ equal to 10^{-8} mol l^{-1}, which is less by a factor of 10^5 than any of the values on the table.*

11.6, and 11.8. Our interest here is with the response of the system to a transient perturbation: release of CO_2 into the atmosphere associated with the combustion of fossil fuels. Specifically, our concern is with factors influencing the capacity of the ocean to absorb industrial CO_2.

The ocean contains a variety of electrically charged chemical species, as summarized in Table 11.1. We refer to the positively charged species as **cations**, the negatively charged compounds as **anions**. The most abundant cations are sodium (Na^+), potassium (K^+), magnesium (Mg^{2+}), and calcium (Ca^{2+}); the most common anions are chloride (Cl^-), sulfate (SO_4^{2-}), and bromide (Br^-).[9] The positive charge contributed by the cations listed in Table 11.1 is slightly larger than the negative charge supplied by the anions. The difference, 0.002 charge equivalents liter^{-1},[10] is compensated primarily by negative charge carried by the dominant forms of dissolved carbon, the bicarbonate and carbonate ions, HCO_3^- and CO_3^{2-} respectively. The excess of positive charge to be balanced by carbon is termed the **alkalinity**, given by

$$[Alk] = [HCO_3^-] + 2\,[CO_3^{2-}] \tag{11.3}$$

$$= [Na^+] + [K^+] + 2\,[Mg^{2+}] + 2\,[Ca^{2+}] - [Cl^-] - 2\,[SO_4^{2-}] - [Br^-]$$

(In this, and the subsequent, discussion, [X] denotes the concentration, expressed in mol l, of chemical species X.)

Carbon dioxide behaves as a weak acid in solution. The dissolution reaction may be written as

$$CO_2 + H_2O \rightarrow HCO_3^- + H^+ \tag{11.4}$$

The ocean's acidity is buffered to some extent by the presence of the carbonate ion. The proton liberated in (11.4) can react with CO_3^{2-} to form HCO_3^-.

$$H^+ + CO_3^{2-} \rightarrow HCO_3^- \tag{11.5}$$

The combination of (11.4) and (11.5) is equivalent to

$$CO_2 + CO_3^{2-} + H_2O \rightarrow 2\,HCO_3^- \tag{11.6}$$

The ability of the ocean to absorb CO_2 is ultimately limited by the availability of CO_3^{2-}.

The relative abundance of the various forms of inorganic carbon in solution is determined by conditions of chemical equilibrium. The equilibrium may be described by a series of chemically balanced reaction equations:

$$CO_2(g) \leftrightarrow CO_2(a), \tag{11.7}$$

$$CO_2(a) + H_2O \leftrightarrow HCO_3^- + H^+, \tag{11.8}$$

and

$$HCO_3^- \leftrightarrow CO_3^{2-} + H^+ \tag{11.9}$$

Here the labels *g* and *a* are used to distinguish CO_2 in the gas and aqueous phases, respectively. If the equilibrium constants for reactions (11.7) to (11.9) are denoted by α, K_1, and K_2, respectively, the partial pressure of $CO_2(g)$, denoted by pCO_2, and the concentrations of CO_2 (*a*), HCO_3^-, CO_3^{2-}, and H^+ are constrained to satisfy the relations

$$\alpha = \frac{[CO_2(a)]}{pCO_2}, \tag{11.10}$$

$$K_1 = \frac{[H^+]\,[HCO_3^-]}{[CO_2(a)]}, \tag{11.11}$$

and

$$K_2 = \frac{[H^+]\,[CO_3^{2-}]}{[HCO_3^-]} \tag{11.12}$$

Using (11.11) to substitute for $[H^+]$ in (11.12), we find

$$K_2 = \frac{K_1[CO_2(a)]\,[CO_3^{2-}]}{[HCO_3^-]^2} \tag{11.13}$$

Using (11.10) to substitute further for $[CO_2\,(a)]$ in (11.13), we obtain

$$\frac{[HCO_3^-]^2}{pCO_2[CO_3^{2-}]} = \frac{\alpha K_1}{K_2} \tag{11.14}$$

We write the combination $\alpha K_1/K_2$ as K'. Values of α, K_1, K_2, and K' are given as functions of temperature in Table 11.2.

Conversion of CO_3^{2-} to HCO_3^- associated with the introduction of additional CO_2 to the ocean is accompanied by a small increase in acidity. That this must be the case is immediately clear from a cursory examination of equation (11.12). Suppose that the concentration of CO_3^{2-} prior to the introduction of additional CO_2 is given by $[CO_3^{2-}]_0$. Following adjustment, assume that the concentration of CO_3^{2-} is equal to $x\,[CO_3^{2-}]_0$, where x is a number less than 1. The change in $[CO_3^{2-}]$ must be compensated by either an increase in $[H^+]$

Table 11.2 Values of Dissociation Constants as a Function of Temperature

T	α	K_1	K_2	K'
(°C)	10^{-2} mol (l atm)$^{-1}$	10^{-7} mol l^{-1}	10^{-10} mol l^{-1}	mol (l atm)$^{-1}$
0	6.33	6.28	3.50	113.8
5	5.25	7.12	3.99	93.6
10	4.42	7.93	4.67	75.0
15	3.77	8.68	5.52	59.3
20	3.26	9.36	6.52	46.8
25	2.85	9.95	7.60	37.4

or a decrease in $[HCO_3^-]$, or both (K_2 is a constant). But introduction of CO_2 leads to an increase rather than a decrease in $[HCO_3^-]$. Moreover, the concentration of HCO_3^- is much larger than that of H^+ (see Table 11.1); as a consequence, the fractional change in HCO_3^- is negligibly small. The decrease in $[CO_3^{2-}]$ is offset by a proportional increase in $[H^+]$; the revised value of the concentration of H^+ is given approximately by $[H^+]_0/x$, where $[H^+]_0$ is the concentration of H^+ prior to the introduction of the additional source of CO_2. Concentrations of H^+, CO_3^{2-}, and HCO_3^- adjust, with relative changes constrained by the equilibrium constant K_2.

Example 11.9: Verify that the change in $[CO_3^{2-}]$ associated with an input of CO_2 to the ocean is indeed compensated primarily by a change in $[H^+]$. Suppose that the concentrations of HCO_3^- and CO_3^{2-} are given initially by 1.8×10^{-3} mol l^{-1} and 2×10^{-4} mol l^{-1}, respectively. Consider a temperature of 293K with a pH of 8.23, $[H^+] = 5.9 \times 10^{-9}$ mol l^{-1}. Evaluate the changes in the chemistry of the water caused by the introduction of 2×10^{-5} mol l^{-1} of CO_2.

Answer: According to equation (11.6), we expect a reduction in $[CO_3^{2-}]$ of 2×10^{-5} mol l^{-1} with an increase of twice this magnitude in $[HCO_3^-]$ as a result of the addition of 2×10^{-5} mol l^{-1} of CO_2. The adjusted concentrations of CO_3^{2-} and HCO_3^- are given by

$$[CO_3^{2-}] = 1.8 \times 10^{-4} \text{ mol l}^{-1}$$

and

$$[HCO_3^-] = 1.84 \times 10^{-3} \text{ mol l}^{-1}$$

Equation (11.12) may be rewritten to define an expression for $[H^+]$ in terms of $[HCO_3^-]$, $[CO_3^{2-}]$, and K_2:

$$[H^+] = \frac{K_2[HCO_3^-]}{[CO_3^{2-}]}$$

Using the revised values for $[HCO_3^-]$ and $[CO_3^{2-}]$, with K_2 equal to 6.59×10^{-10} mol l^{-1} as appropriate for a temperature of 293 K, we find

$$[H^+] = \frac{(6.59 \times 10^{-10} \text{mol l}^{-1})(1.84 \times 10^{-3} \text{ mol l}^{-1})}{1.8 \times 10^{-4} \text{ mol l}^{-1}}$$

$$= 6.74 \times 10^{-9} \text{ mol l}^{-1}$$

corresponding to a revised value for pH given by

$$\text{pH} = -\log_{10}(6.74 \times 10^{-9})$$
$$= -[\log_{10}(6.74) + \log_{10} \times (10^{-9})]$$
$$= -(0.83 - 9.00)$$
$$= 8.17$$

In terms of the analysis above, x equals 0.9. We would have expected a revised concentration of H^+ of

$$[H^+] = \frac{[H^+]_0}{x} = \frac{5.9 \times 10^{-9}}{0.9} \text{ mol l}^{-1}$$
$$= 6.56 \times 10^{-9} \text{ mol l}^{-1}$$

The actual value is slightly larger, reflecting the small (2%) increase in $[HCO_3^-]$. ■

Exchange of CO_2 between the atmosphere and ocean depends on the difference between the partial pressure of CO_2 in the air and the partial pressure of CO_2 in air that has equilibrated with the surface water. If the value of pCO_2 for the ocean is larger than that for the atmosphere, we expect a net release of CO_2 from the ocean. Transfer of CO_2 from the ocean to the atmosphere continues until the pressure difference is eliminated, meaning until the partial pressure of CO_2 in the atmosphere equals that in the ocean. Similarly, if the partial pressure of CO_2 in the atmosphere exceeds that in the ocean, transfer of CO_2 takes place in the opposite direction, from air to sea. In order to understand the factors influencing the rate at which CO_2 is exchanged between the atmosphere and ocean, it is important to develop an appreciation not only for the processes responsible for changes in the concentration of CO_2 in the atmosphere but also for those resulting in variations of pCO_2 in surface water.

Figure 11.9 illustrates the change in the composition of ocean water expected as a consequence of a change in temperature. The concentration of total dissolved inorganic carbon ($\Sigma CO_2 = CO_2\,(a) + HCO_3^- + CO_3^{2-}$) is fixed in Figure 11.9 at a value of 2.05×10^{-3} mol l^{-1}, whereas alkalinity is set equal to 2.25×10^{-3} charge equivalents l^{-1}. An increase in temperature leads to a net conversion of HCO_3^- to $CO_2\,(a)$ and CO_3^{2-}: the adjustment is regulated by the temperature dependence of the equilibrium constants in (11.14). Conversion of HCO_3^- to CO_3^{2-} is associated with an increase in the concentration of H^+ (a drop in pH), with a concurrent rise in the value of pCO_2 and in the concentrations of $CO_2\,(a)$ and CO_3^{2-}.

Suppose that the water is in equilibrium initially with the atmosphere at a temperature T_0. The value of pCO_2 for the water then precisely equals the value of pCO_2 in the atmosphere, and the gas exchange rate equals zero. An increase in water temperature (assuming a fixed value for alkalinity) results in an increase in the value of pCO_2 for the

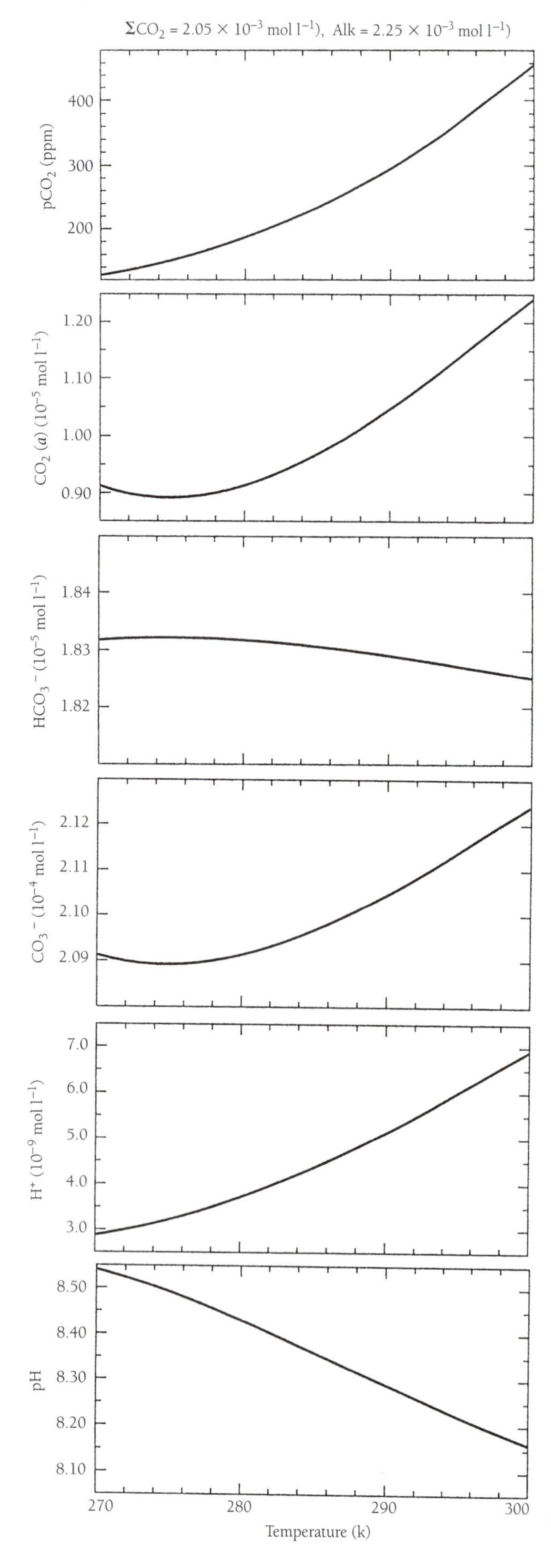

Figure 11.9 Changes in pCO_2 and in the concentrations of CO_2 (a), HCO_3^-, CO_3^{2-}, H^+, and pH as functions of temperature for surface waters of the ocean with $\sum CO_2 = 2.05 \times 10^{-3}$ mol l^{-1}, and alkalinity = 2.25×10^{-3} mol l^{-1}.

liquid, with associated transfer of CO_2 from the ocean to the atmosphere and a consequent decrease in the concentration of dissolved inorganic carbon. Conversely, a drop in water temperature would lead to a decline of pCO_2 in the liquid, with transfer of CO_2 in the opposite direction, from air to sea, resulting in an increase in the concentration of inorganic carbon in the ocean. The seasonal cycle of temperature in the surface ocean, ignoring effects of biology, should favor release of CO_2 from the ocean to the atmosphere during spring and summer, with a compensatory uptake during autumn and winter.

Figure 11.10 explores the response of the ocean to changes in the concentration of CO_2 in the atmosphere. As with Figure 11.9, we assume a fixed value for alkalinity, 2.25×10^{-3} charge equivalents l^{-1}. An increase in pCO_2, as indicated by reaction (11.6), results in a decrease in the concentration of CO_3^{2-}, with an associated increase in HCO_3^-. Acidity increases as expected: as previously noted, CO_2 is a weak acid. The fractional increase in total dissolved inorganic carbon is equal to approximately 10% of the fractional increase in pCO_2. The fractional change in pCO_2, expressed as a ratio with respect to the fractional change in total dissolved carbon, is known as the **Revelle factor**, a measure of the ocean's capacity to take up additional CO_2 from the atmosphere. The term honors Roger Revelle (1909–1991), an oceanographer, former director of the Scripps Institution of Oceanography, and founder of the Center for Population Studies at Harvard University, who played an influential role in modern studies of the fate and effects of industrial CO_2. With the assumptions in Figure 11.10, the Revelle factor increases from a value of about 10, corresponding to a level of pCO_2 equivalent to a mixing ratio of 200 ppm, to a value of about 13 for pCO_2 equal to 400 ppm. The capacity of the ocean to absorb an additional burden of CO_2 is limited by the supply of CO_3^{2-}. As Figure 11.10 indicates, the concentration of CO_3^{2-} declines at a rate proportional to the rise in pCO_2.

We are now in a position to estimate the fraction of CO_2 released by combustion of fossil fuels since the industrial revolution that persists in the atmosphere, as compared with the fraction incorporated into the ocean. Assume that the chemical composition of the ocean is uniform; that the concentration of CO_3^{2-} was equal to $[CO_3^{2-}]_0$ everywhere prior to the addition of the fossil source, and that it equals $[CO_3^{2-}]$ today. As discussed above, the change in the concentration of CO_3^{2-} provides a measure of the quantity of carbon added to the ocean. For each mol of CO_2 added to the ocean, we expect to observe a reduction of comparable magnitude in the concen-

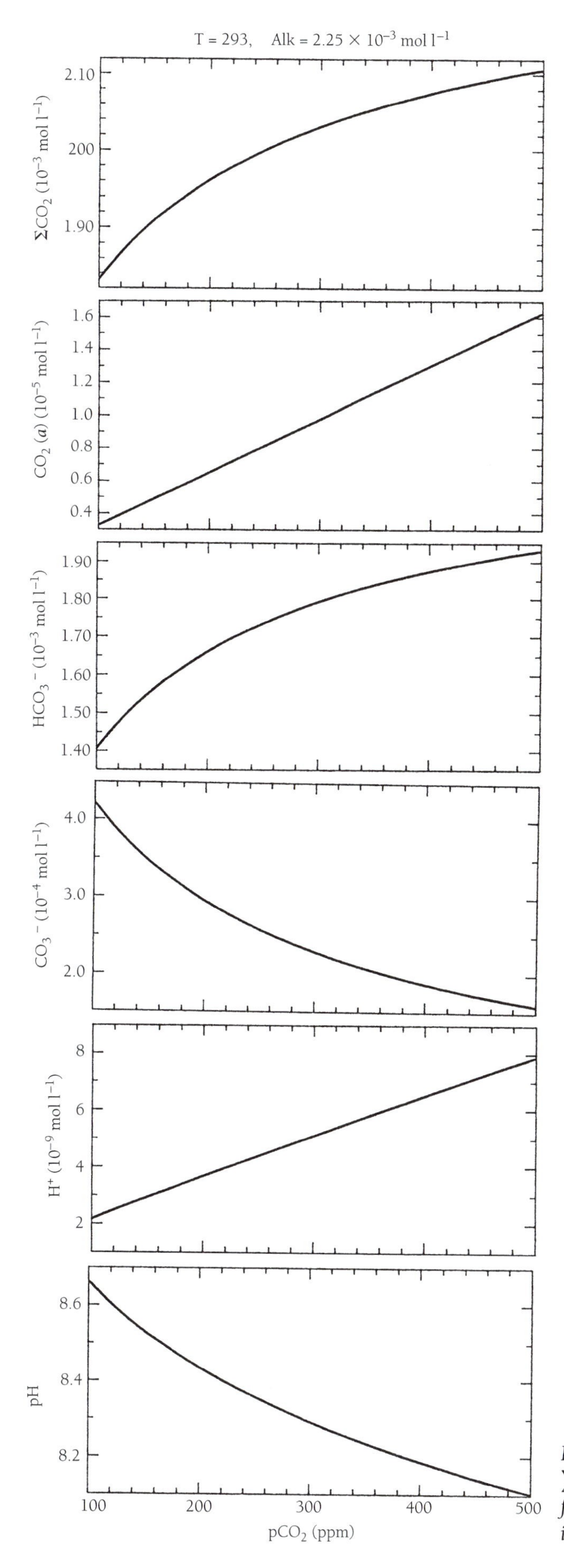

Figure 11.10 Changes in pCO_2 and in the concentrations of $\sum CO_2$, CO_2 (a), HCO_3^-, CO_3^{2-}, H^+, and pH as functions of pCO_2 for a temperature of the surface ocean equal to 293 K and alkalinity = 2.25×10^{-3} mol l^{-1}.

tration of CO_3^{2-}. The increase in the abundance of carbon contained in a unit volume of sea water is given by

$$(\Delta C)_{ocean} = \left([CO_3^{2-}]_0 - [CO_3^{2-}]\right) \qquad (11.15)$$

Assuming that the volume of ocean water exposed to the atmosphere since the industrial revolution (i.e., the quantity of water that has had the opportunity to interact directly with the atmosphere) is given by V, the net increase in the carbon abundance of the ocean is given by

$$(\Delta C)_{ocean} = V\left([CO_3^{2-}]_0 - [CO_3^{2-}]\right) \qquad (11.16)$$

The reduction in the concentration of CO_3^{2-} is offset by a proportional rise in the value of pCO_2.[11] If the initial partial pressure of CO_2 is given by $(pCO_2)_0$ with the present value equal to pCO_2, it follows that we may write the product $pCO_2\,[CO_3^{2-}]$:

$$(p_{CO_2})_0[CO_3^{2-}]_0 = p_{CO_2}[CO_3^{2-}] \qquad (11.17)$$

Using (11.17), to substitute then for $[CO_3^{2-}]$ in (11.16), we find

$$(\Delta C)_{ocean} = \frac{V[CO_3^{2-}]_0}{p_{CO_2}}\left(p_{CO_2} - (p_{CO_2})_0\right) \qquad (11.18)$$

Suppose that the mass of carbon initially present in the atmosphere is given by M_0. The mass of carbon in the atmosphere at any given time is, of course, proportional to the value of the partial pressure p_{CO_2}. It follows that the change in the mass of carbon in the atmosphere since the industrial revolution (beginning about 1800) is given by

$$(\Delta C)_{atmos} = \frac{\left(p_{CO_2} - (p_{CO_2})_0\right)}{(p_{CO_2})_0} M_0 \qquad (11.19)$$

The fraction of the total carbon added to the atmosphere-ocean system persisting in the atmosphere is given by

$$f_{atmos} = \frac{(\Delta C)_{atmos}}{(\Delta C)_{atmos} + (\Delta C)_{ocean}} \qquad (11.20)$$

Using (11.18) and (11.19) to substitute for $(\Delta C)_{ocean}$ and $(\Delta C)_{atmos}$, we find, after some algebraic manipulation,

$$f_{atmos} = \left[1 + \frac{(p_{CO_2})_0}{p_{CO_2}}[CO_3^{2-}]_0 \frac{V}{M_0}\right]^{-1} \qquad (11.21)$$

Example 11.10: Assume that the relative abundance, or mixing ratio, of CO_2 initially equaled 280 ppm, corresponding to a total atmospheric content, M_0, of 5×10^{16} moles. Assume an initial concentration of CO_3^{2-} equal to 200×10^{-6} mol l^{-1}. The volume of the ocean is equal to 1.33×10^{21} l. Calculate the value of f_{atmos} when the mixing ratio of CO_2 has risen to 355 ppm, the value appropriate today. Assume that the atmosphere is allowed to interact with the entire volume of ocean water.

Answer:

$$f_{atmos} = \left[1 + \frac{(p_{CO_2})_0}{p_{CO_2}}[CO_3^{2-}]_0 \frac{V}{M_0}\right]^{-1}$$

$$= \left[1 + \frac{280 \text{ ppm}}{355 \text{ ppm}} 200 \times 10^{-6} \text{ mol l}^{-1} \frac{1.33 \times 10^{21} \text{ l}}{5 \times 10^{16} \text{ mol}}\right]^{-1}$$

$$= \left[1 + (0.79)(2 \times 10^{-4})(2.7 \times 10^{4})\right]^{-1}$$

$$= (1 + 4.27)^{-1}$$

$$= 0.19$$ ■

If the entire ocean were to come to equilibrium with the atmosphere, a little less than 20% of the fossil fuel carbon would persist in the atmosphere. With our assumptions, the balance, 81%, would be incorporated in the ocean. The ocean, however, turns over slowly. Only about 10% of the ocean's total volume has had contact with the atmosphere since the industrial revolution. The actual fraction retained by the atmosphere is consequently much larger than the value implied by the example given here.

Example 11.11: Repeat the calculation above, assuming that the volume of the ocean equilibrating with the atmosphere equals to 1.33×10^{20} l.

Answer:

$$f_{atmos} = \left[1 + \frac{(p_{CO_2})_0}{p_{CO_2}}[CO_3^{2-}]_0 \frac{V}{M_0}\right]^{-1}$$

$$= \left[1 + \frac{280 \text{ ppm}}{355 \text{ ppm}} 200 \times 10^{-6} \text{ mol l}^{-1} \frac{1.33 \times 10^{20} \text{ l}}{5 \times 10^{16} \text{ mol}}\right]^{-1}$$

$$= \left[1 + (0.79)(2 \times 10^{-4})(2.7 \times 10^{3})\right]^{-1}$$

$$= (1 + 0.43)^{-1}$$

$$= 0.70$$ ■

The result obtained in this case is more realistic; as much as 70% of the carbon released by combustion of fossil fuels since the industrial revolution may persist in the atmosphere. Only 30% of the total has been transferred to the ocean. The ability of the ocean to take up extra CO_2 is limited both by the relatively small value of the concentration of CO_3^{2-} in solution and by the relatively sluggish nature of the vertical circulation of the ocean. Wallace Broecker refers to the quantity f_{atmos} as the **Keeling fraction**, honoring the contributions of C. D. Keeling, who is responsible (as noted above) for the remarkable record of the changing abundance of CO_2, as illustrated in Figures 11.1, 11.2, and 11.4.

The preceding examples fall seriously short of a quantitative model for the uptake by the ocean of CO_2 derived from fossil fuels. Note that nowhere in the examples did we consider the magnitude or history of the source of CO_2 associated with use of fossil fuels. Rather, we began with an assumption that the abundance of CO_2 in the atmosphere had increased from 280 ppm to 355 ppm and supposed in Example 11.10 that the atmosphere had come to equilibrium with the entire ocean, or with 10% of the ocean in Example 11.11. We shall now estimate the quantity of carbon that must be added to the combined atmosphere-ocean system to accommodate the conditions presupposed in these examples.

Example 11.12: Suppose that the abundance of CO_2 in the atmosphere has increased from 280 ppm to 355 ppm and that the atmosphere has come to equilibrium with the entire ocean, as assumed in Example 11.10. Calculate the quantity of carbon that must be added to the atmosphere-ocean system to account for this condition.

Answer: As indicated in Endnote 11-2, a mixing ratio of 355 ppm corresponds to an abundance of carbon in the atmosphere as CO_2 equivalent to 753.2×10^9 tons C. A mixing ratio of 280 ppm thus corresponds to a fraction equal to $\frac{280}{350}$, or 78.87%, of this amount: $(0.7887) \times (753.2 \times 10^9 \text{ tons C}) = 594.1 \times 10^9$ tons C. It follows that the carbon content of the atmosphere must increase by $(753.2 \times 10^9 - 594.1 \times 10^9)$ tons C $= 159.1 \times 10^9$ tons C as the mixing ratio of CO_2 climbs from 280 ppm to 355 ppm. Let M (tons C) denote the quantity of carbon added to the atmosphere-ocean system to bring about this result. According to Example 11.10, if the atmosphere were to come to equilibrium with the entire ocean, the fraction of the added carbon remaining in the air would be given by $f_{atmos} = 0.19$.

It follows that

$$f_{atmos} M = 159.1 \times 10^9 \text{ tons C}$$

Thus

$$M = (159.1 \times 10^9 \text{ tons C}) / f_{atmos}$$

$$= (159.1 \times 10^9 \text{ tons C}) / 0.19$$

$$= 837.4 \times 10^9 \text{ tons C}$$ ■

Example 11.13: Repeat the calculation above, assuming that the atmosphere achieves equilibrium with 10% of the ocean, as envisaged in Example 11.11.

Answer: In this case, $f_{atmos} = 0.70$.

Hence

$$M = (159.1 \times 10^9 \text{ tons C}) / 0.70$$

$$= 227.3 \times 10^9 \text{ tons C}$$

This is approximately equal to the quantity of carbon released by burning of fossil fuels since the beginning of the industrial revolution (from 1800 to about 1992). ■

The capacity of the ocean to absorb CO_2, as calculated here, is limited primarily by the abundance of dissolved CO_3^{2-}. For every molecule of excess CO_2 transferred to the ocean, a

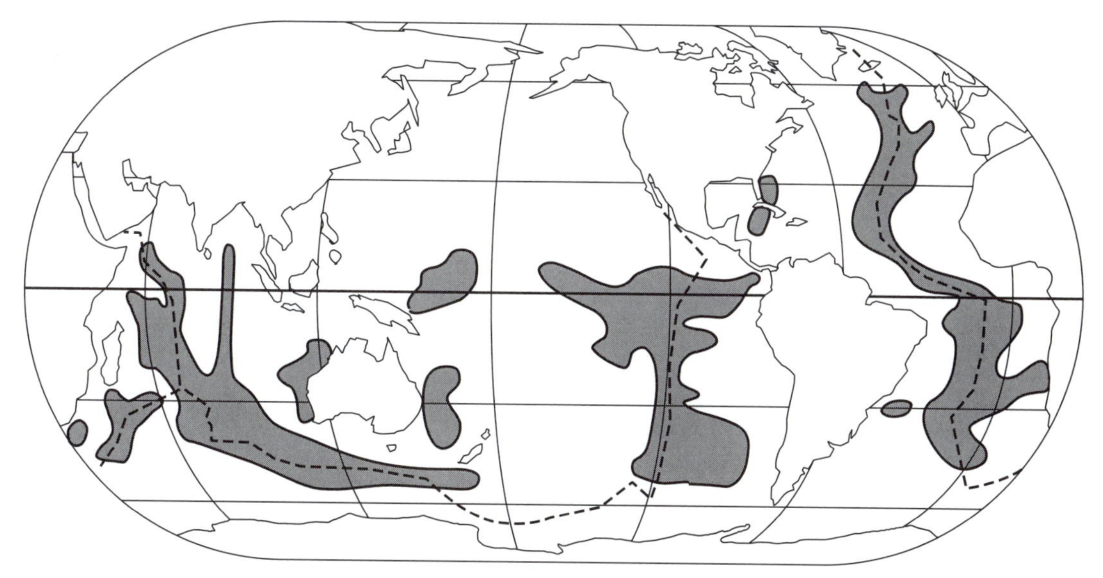

Figure 11.11 Map showing the distribution of calcite-rich sediments on the sea floor. Cross-hatched areas indicate regions with sediments containing more than 75% $CaCO_3$. This map is based on an Atlantic map published by Biscayne et al., a Pacific map published by Berger et al., and an Indian Ocean map published by Kolla et al. Source: Broecker and Peng 1982.

molecule of CO_3^{2-} is removed from solution: it is converted to HCO_3^- as summarized by reaction (11.6). The capacity of the ocean as a sink for fossil carbon would be enhanced if the abundance of CO_3^{2-} were higher. This could occur, for example, as a result of the dissolution of calcium carbonate (composed primarily of the mineral calcite) in sediments. The sediments in upper levels of the ocean, as illustrated in Figure 11.11, contain large quantities of $CaCO_3$; more than 80% of the total sedimentary mass is contributed by skeletal shells of organisms once living in the water column. In contrast, $CaCO_3$ is a minor component of sediments in the deep sea. The depth at which the transition from calcite-rich to calcite-poor sediment occurs is known as the **lysocline**.

The stability of calcite is controlled by the abundances of Ca^{2+} and CO_3^{2-}, as described by the reaction

$$CaCO_3(s) \leftrightarrow Ca^{2+} + CO_3^{2-} \quad (11.22)$$

In equilibrium with calcite, the concentrations of Ca^{2+} and CO_3^{2-} satisfy the relation

$$K = [Ca^{2+}]\,[CO_3^{2-}], \quad (11.23)$$

where K is the equilibrium constant (also known as the *dissolution constant*) for $CaCO_3$. The solubility of calcite increases with pressure and is greatest at low temperature; that is, other factors being equal, the value of K is largest in the cold deep ocean. The lifetime of calcium in the ocean is relatively long: about 10^6 yrs. As a consequence, Ca^{2+} is distributed more or less uniformly throughout the ocean. The abundance of CO_3^{2-}, on the other hand, is quite variable. It is relatively large near the surface and comparatively low in the depths as illustrated by the data in Figure 11.12. In practice, the stability of calcite in the ocean is determined mainly by variations in the abundance of CO_3^{2-}, which is controlled in turn by the production in surface waters and decay in deeper waters of organic matter and its associated carbonate.

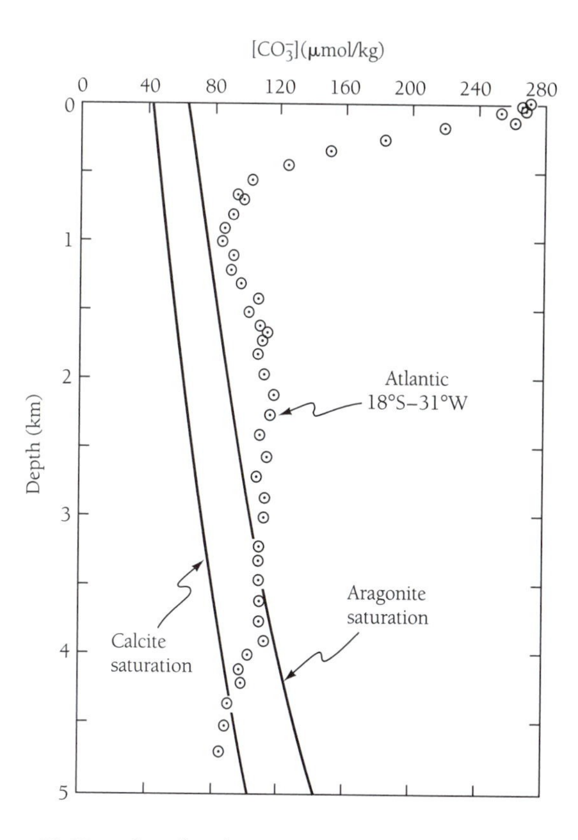

Figure 11.12 Plot of carbonate ion concentration versus water depth for a station in the western South Atlantic. Also shown are the saturation carbonate ion concentrations. They increase with depth mainly because of the pressure effect (calcium and carbonate ions occupy less volume when dissolved in sea water than when locked in $CaCO_3$). The concentration of carbonate ion remains fairly constant through the North Atlantic deep-water mass. It is higher in this mass than in the underlying wedge of the Antarctic Bottom Water and the overlying tongue of Antarctic Intermediate Water. Source: Broecker and Peng 1982.

The organic components of marine organisms include C, N, and P, in the approximate proportions 105:15:1 (expressed in terms of numbers of atoms).[12] The nutrients N and P are supplied mainly as nitrate and phosphate. For every four carbon atoms included in the organic fraction of the typical marine organism, an additional carbon atom is incorporated in the accompanying carbonate shell. Growth (or decay) of organisms results, therefore, not only in a change in the abundance of dissolved inorganic carbon but also in a change in alkalinity, as illustrated by the following examples.

Example 11.14: Estimate the change in the abundance of inorganic carbon and alkalinity if 4 units of carbon are incorporated in the organic fraction of typical Redfield marine organisms.

Answer: Let $[\Sigma CO_2]$ denote the abundance of inorganic carbon summed over all forms: CO_2, HCO_3^-, and CO_3^{2-}. Let [Alk] represent the alkalinity, as defined by equation (11.3). Indicate changes in $[\Sigma CO_2]$ and [Alk] by $\Delta[\Sigma CO_2]$ and Δ[Alk], respectively. Assuming that 4 units of carbon are incorporated in organic matter, it follows that 5 units of carbon must be removed from the inorganic pool (4 as organic matter, 1 as a component of shells). Thus

$$\Delta[\Sigma CO_2] = -5$$

The negative sign indicates a net loss of inorganic carbon.

Changes in alkalinity arise as a consequence of the production of $CaCO_3$ from Ca^{2+} and CO_3^{2-}. In addition, there is a change in alkalinity induced by conversion of NO_3^- (nitrate) to organic matter (electrically neutral). With our assumptions, production of $CaCO_3$, results in a reduction in alkalinity by 2 units. (Remember that Ca^{2+} contains a net charge of +2.) For each NO_3^- removed from the inorganic pool, the alkalinity increases by 1 unit. The number of NO_3^- ions eliminated by formation of 4 units of organic carbon is obtained by multiplying the fractional abundance of N relative to C by the number of carbon-atoms incorporated by organisms, that is, by (15/105) 4 = 0.57.

It follows that the change in alkalinity associated with production of 4 units of organic carbon is given by

$$\Delta[\text{Alk}] = -2 + 0.57$$
$$= -1.43$$ ■

As previously noted, the bicarbonate ion, HCO_3^-, is by far the most abundant component of inorganic carbon in the ocean, accounting for more than 90% of the total. The carbonate ion is the second most abundant contributor, such that the combination of HCO_3^- and CO_3^{2-} accounts for more than 99% of total inorganic carbon, ΣCO_2. To a good approximation, $[\Sigma CO_2]$ is given by summing the abundances of HCO_3^- and CO_3^{2-}:

$$[\Sigma CO_2] = [HCO_3^-] + [CO_3^{2-}] \qquad (11.24)$$

Note that alkalinity is defined as the net positive charge on all species exclusive of carbon; elimination of NO_3^- results in a net increase in the residual positive charge.

It is relatively straightforward to combine this relation with the definition of [Alk] (equation 11.3) to obtain an expression for $[CO_3^{2-}]$ in terms of $[\Sigma CO_2]$ and [Alk]. Using (11.24), we write

$$[HCO_3^-] = [\Sigma CO_2] - [CO_3^{2-}] \qquad (11.25)$$

$$[\text{Alk}] = [HCO_3^-] + 2\,[CO_3^{2-}]$$
$$= [\Sigma CO_2] - [CO_3^{2-}] + 2\,[CO_3^{2-}] \qquad (11.26)$$
$$= [\Sigma CO_2] + [CO_3^{2-}]$$

$$[CO_3^{2-}] = [\text{Alk}] - [\Sigma CO_2] \qquad (11.27)$$

For the conditions envisaged in Example 11.14, it follows that

$$\Delta[CO_3^{2-}] = \Delta[\text{Alk}] - \Delta[\Sigma CO_2]$$
$$= -1.43 - (-5) \qquad (11.28)$$
$$= 3.57$$

The change in $[CO_3^{2-}]$ may be qualitatively interpreted as a consequence primarily of the conversion of CO_2 to organic matter. A portion of the CO_2 taken up by photosynthesis is supplied as a by-product of the production of $CaCO_3$:

$$Ca^{2+} + 2HCO_3^- \rightarrow CaCO_3 + CO_2 + H_2O \qquad (11.29)$$

Removal of CO_2 leaves the water less acidic (remember, CO_2 is an acid). Incorporation of NO_3^- in organic matter results in a further drop in acidity (NO_3^- is neutralized by absorption of a proton). The drop in $[H^+]$ requires a net conversion of HCO_3^- to CO_3^{2-}, to maintain the equilibrium implied by (11.9).

We are now in a position to implement a more quantitative calculation of the change in $[CO_3^{2-}]$ associated with biological activity in surface waters.

Example 11.15: A typical value for the concentration of NO_3^- in deep water is about 3×10^{-5} mol l^{-1}. The biological productivity of surface water is fueled by upwelling of nutrient-rich water from below. During summer, biological activity is intense at mid and low latitudes and the concentration of NO_3^- is reduced to close to zero. Production of biological matter is then at a maximum, limited by the supply of nutrients (N and P are depleted more or less simultaneously). Estimate the change in $[CO_3^{2-}]$ resulting from complete uptake of nutrient as presupposed here.

Answer: We assume as above that C, N, and P are incorporated in the organic fraction in the proportions 105:15:1 and that one carbon atom is used to form $CaCO_3$ for every four included in the organic component. Formation of organic matter results in a net decrease of $[\Sigma CO_2]$ given by

$$\Delta[\Sigma CO_2] = \frac{-105}{15}\,[3 \times 10^{-5}] \text{ mol } l^{-1}$$

Production of carbonate results in further loss of inorganic carbon equal to 25% of this value. It follows that the net change in ΣCO_2 is given by

$$\Delta[\Sigma CO_2] = -(1.25)\frac{105}{15}\,[3 \times 10^{-5}]\ \text{mol l}^{-1}$$

$$= -2.63 \times 10^{-4}\ \text{mol l}^{-1}$$

The change in alkalinity represents the sum of the changes associated with uptake of nitrate and the growth of $CaCO_3$:

$$\Delta[\text{Alk}] = \left(3 \times 10^{-5} - \frac{2}{4}\left(\frac{105}{15}\right)3 \times 10^{-5}\right)\ \text{mol l}^{-1}$$

$$= -7.5 \times 10^{-5}\ \text{mol l}^{-1}$$

The factor 2 in the second term accounts for the double charge on Ca^{2+}. The net change in $[CO_3^{2-}]$ is given thus by

$$\Delta[CO_3^{2-}] = \Delta[\text{Alk}] - \Delta[\Sigma CO_2]$$

$$= (-7.5 \times 10^{-5} + 2.63 \times 10^{-4})\ \text{mol l}^{-1}$$

$$= +\,1.88 \times 10^{-4}\ \text{mol l}^{-1}$$ ■

The increase in $[CO_3^{2-}]$ near the surface as calculated here is similar to that shown in Figure 11.12.

11.3 The Contemporary Budget of Atmospheric CO_2

Section 11.1 presented an overview of the carbon cycle, an outline of how we might expect the element to be distributed in a steady state over its major potential reservoirs: the atmosphere, terrestrial biosphere, soils, ocean, and sediments. Implicit in the steady-state hypothesis was an assumption that carbon inputs and outputs from various compartments were in balance, meaning that abundances of carbon in individual reservoirs were fixed in time. It is clear, though, that this assumption is inappropriate for the contemporary environment. As we have seen, the concentration of atmospheric CO_2 has risen from about 315 ppm in 1958 to close to 360 ppm today (see Figures 11.1 and 11.2). Studies of gases trapped in polar ice (summarized in Figure 11.13) indicate that the modern increase began in the early part of the eighteenth century.[13] The abundance of CO_2 was relatively constant (at about 280 ppm) for more than 10,000 years, following the end of the last ice age up to the onset of the modern disturbance. The recent rise in CO_2 may be attributed, as we shall see, to complex influences of modern industrial and agricultural society. Emission of CO_2 associated with the burning of fossil fuels (see Figure 11.3)—transfer of carbon from the sedimentary reservoir to the atmosphere—played a major role but was not the only influence. As discussed below and illustrated by the following examples, at least a portion of the increase, especially in the early years, must be attributed to a transfer of carbon from soils and from the terrestrial biosphere to the atmosphere.

Example 11.16: C. D. Keeling estimates that consumption of fossil fuels, mainly coal, added 5.8×10^9 tons of carbon as CO_2 to the atmosphere over the period 1860–1890. Over the same interval, according to the data summarized in Figure 11.13, the concentration of atmospheric CO_2 increased by about 7 ppm.

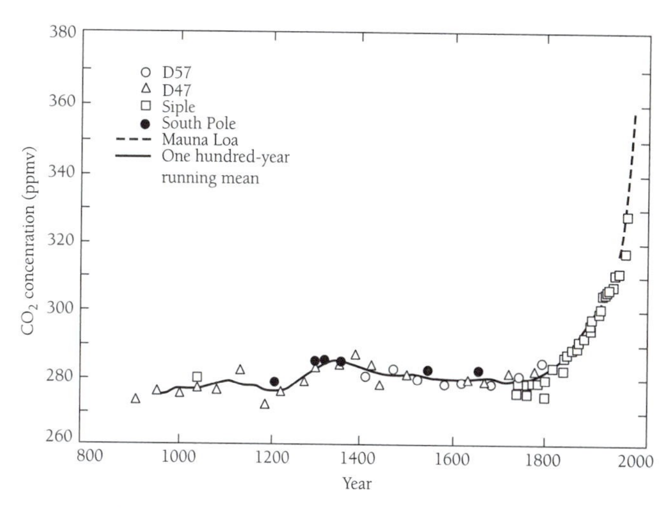

Figure 11.13 CO_2 concentrations over the past 1,000 years from ice-core records (D47, D57, Siple South Pole) and (since 1958) from the Mauna Loa, Hawaii, measurement site. The smooth curve is based on a hundred-year running mean. The rapid increase in CO_2 concentration since the onset of industrialization is evident and has closely followed the increase in CO_2 emissions from fossil fuels. Source: IPCC 1996.

Estimate the maximum contribution to the rise in CO_2 attributable to burning of fossil fuels between 1860 and 1890.

Answer: The analysis presented in Endnote 11–2 indicates that 1 ppm of CO_2 in the atmosphere corresponds to 2.12×10^9 tons C. The observed rise in CO_2 between 1860 and 1890 requires an input of carbon of at least $7 \times 2.12 \times 10^9$ tons C = 14.8×10^9 tons C, larger than the fossil fuel source by a factor of 2.56. ■

Example 11.17: Assume an airborne fraction (f_{atmos}) of 0.7, as estimated in Example 11.11. What would we expect the increase in concentration of CO_2 to be as a consequence of the input from fossil fuel consumption between 1860 and 1890?

Answer: We saw that the total input from fossil fuels was 5.8×10^9 tons C. Of that we would expect $0.7 \times 5.8 \times 10^9 = 4.06 \times 10^9$ tons C to remain in the atmosphere. That would imply an increase in atmospheric concentration of 4.06×10^9 tons C/(2.12×10^9 tons/ppm) = 1.92 ppm. ■

Example 11.18: Assume that the rise in CO_2 between 1860 and 1890 is largely due to the release of carbon from a combination of the soil and terrestrial-biosphere reservoirs. Assume further that the fraction of carbon emitted from this source remaining in the atmosphere (i.e., the airborne fraction) is 0.7. Estimate the quantity of CO_2 that must be provided by the biosphere-soil source to supplement the shortfall from fossil fuels in accounting for the rise in CO_2 between 1860 and 1890. Give the result in terms of the annual average source required to account for the ice-core data.

Answer: We saw in Example 11.16 that 14.8×10^9 tons C were added to the atmosphere between 1860 and 1890. And we saw in Example 11.17 that 4.06×10^9 tons C of the same came from fossil fuels. The difference, 10.74×10^9 tons C, must be attributable to emissions from the soil-biosphere system. Assuming an airborne fraction of 0.7, net emission from the biosphere-soil system must exceed the quantity remaining in the atmosphere by a factor of 1/(0.7). Net emission from the soil-biosphere system over the 30-year interval between 1860 and 1890 is given then by 10.74×10^9 tons C/ 0.7 = 15.34×10^9 tons C = 1.534×10^{10} tons C. This implies an average annual emission of $(1.534 \times 10^{10})/30 = 5.11 \times 10^8$ tons C per year. ■

In retrospect, it should come as no great surprise that a hundred years ago the transfer of carbon from the biosphere and soil to the atmosphere should have provided a larger source of atmospheric CO_2 than combustion of fossil fuels. Where it was available, wood provided a convenient and relatively inexpensive fuel for use both in domestic situations (for heating and cooking) and in industry (where it was employed extensively not only as a fuel but also as a source of essential industrial feedstock, supplying, for example, the large quantities of charcoal consumed in smelting ore). As forests were depleted in the early part of the eighteenth century, coal replaced wood as the primary fuel used in England and in much of Western Europe. The transition from wood to coal occurred almost 200 years later in the United States, in the early part of the twentieth century. It is an even more recent phenomenon in countries such as China and India. As fossil fuels are substituted for wood in developed societies, forests were often allowed to regrow. In the eastern region of the United States, for example, forests were destroyed in the eighteenth century to provide wood for industry and land for agriculture. Construction of the railroads in the nineteenth century provided access to much richer agricultural land in the Midwest. Plowing the relatively pristine prairies of the Midwest stimulated release of significant quantities of CO_2, as organic matter deposited over millennia was exposed to atmospheric oxygen.[14] Farming in the east declined as a consequence of the more efficient supply of agricultural products from the Midwest. Land was abandoned and allowed to slowly revert to forest.

This Section is concerned with the contemporary budget of atmospheric CO_2. We know that at least part of the rise in CO_2 can be attributed to emissions associated with combustion of fossil fuels. We also know that a portion of the carbon emitted by burning fossil fuels will remain in the atmosphere, although some fraction may be taken up by the ocean. A major uncertainty concerns the significance of the exchange between the atmosphere and a combination of the terrestrial biosphere and soil. Net exchange of carbon between the biosphere–soil system and the atmosphere reflects the aggregate influence of human activity on the global biosphere. Experience in the United States and Europe indicates that this influence can be extremely complex and difficult to quantify. In principle, the biosphere–soil system could represent either a source or a sink for atmospheric CO_2 on a global scale. Compounding the problem, it could provide a source in one geographic region, a sink in another. Deforestation in the tropics represents a source of CO_2; on the other hand, a regrowth of forests in previously deforested regions should contribute a sink. A warmer climate could promote enhanced growth of vegetation, thus a sink for CO_2; offsetting higher temperatures could enhance rates of decomposition of organic matter in soils, resulting in a source. Higher levels of CO_2 and industrial sources of fixed nitrogen and sulfur (see Chapter 12) could contribute to an increased growth of vegetation representing a sink for CO_2; the carbon to nitrogen ratio in plants, however, could change in response to a change in the availability of these essential elements and feedbacks in terms of exchange of carbon with the atmosphere could be altered significantly. These influences are complex: feedbacks with respect to atmospheric CO_2 are uncertain, even with respect to sign. We need to develop a sense of the factors regulating the exchange of carbon between the biosphere-soil system and the atmosphere today and an understanding of how these conditions may evolve in the future. Clearly, this exchange contributed a significant source of atmospheric CO_2 in the past: it is important that we define its contribution today and that we develop a sense of how it might vary in the future.

Given the heterogeneity of biosphere–soil systems and the diversity of natural and human-induced changes that can influence exchange of carbon between the biosphere–soil system and the atmosphere, this is a daunting task. An a priori, first–principles approach is beyond our current capability.[15] It would be helpful, however, to know what is happening today, at least on a global scale. If we could provide spatial information, even on a coarse scale, this would be a bonus and an incentive for further, more directed research. We describe an approach here that promises to at least address these objectives. Interestingly, the architect and first practitioner of the approach is another member of the Keeling clan, Ralph Keeling, son of the C. D. Keeling cited earlier for his contributions to modern studies of CO_2. His approach is based on a recognition that careful measurements of the abundance of O_2 in the atmosphere, coupled with measurements of CO_2, can provide a means to distinguish between the roles of the ocean-biosphere-soil system as a sink for fossil fuel–derived CO_2. Ralph Keeling, while a graduate student at Harvard, developed an instrument with the capability to measure the abundance of O_2 to the precision required to implement this strategy and subsequently demonstrated its potential in an important series of papers analyzing implications of the changes in O_2 occurring over the past several years. The following discussion shows how measurements of O_2 and CO_2 may refine our understanding of the contemporary budget of atmospheric CO_2.

Combustion of fossil fuels is responsible for both an increase in production of CO_2 and simultaneously a (predictable) decrease in O_2. The relative decrease in O_2 is largest if the fuel consumed is natural gas (CH_4), less for oil and least for coal.[16] On average, with the current mix of fossil fuels, accounting for production of cement, it is estimated that 1.39 mol of O_2 are removed from the atmosphere for each mol of CO_2 released by burning fossil fuels. Independent analysis suggests that uptake of CO_2 by the biosphere (photosynthesis) is associated with release of O_2 in relative proportions of 1 mol of CO_2 for each 1.1 mol of O_2 with similar proportions assumed to apply for release of CO_2 and consumption of O_2 by the biosphere and soils (the combined effects of respiration and decay). The net global source of CO_2 (the composite release due to transfer from the biosphere–soil system and burning of fossil fuels) will be distributed between the atmosphere and ocean in proportions that depend on the capacity of the ocean to absorb excess CO_2, measured in terms of the airborne-fraction parameter (f_{atmos}) introduced above. The change in O_2, however, is essentially confined to the atmosphere, reflecting the relatively low solubility of O_2 in water. Given a measurement of the change in O_2, knowing the change expected due to combustion of fossil fuels, we can derive an estimate for the contribution to the observed change due to oxidation (or reduction) of biosphere–soil organic carbon. Given a measurement of the change in the abundance of atmospheric CO_2, with information on the contribution to this change from the burning of fossil fuels and oxidation of organic carbon in soils and the biosphere, we can obtain empirical estimates for the quantity of carbon transferred from the atmosphere to the ocean and for the net exchange of atmospheric carbon with the biosphere and soil, as illustrated in Figures 11.14 and 11.15.

Variations in CO_2 are reported usually in terms of changes in the concentration of the gas expressed in parts per million by volume (mixing ratios in ppmv). For practical reasons,[17] changes in O_2 are given in terms of the fractional change in the ratio of the concentration of O_2 with respect to N_2, referenced with respect to the ratio of O_2 to N_2 in a standard (typically a sample of air taken in 1988, when Ralph Keeling began his program of sustained modern measurements of O_2). The standard is carefully maintained to ensure that relative measurements of O_2/N_2 are reliable. The change in O_2 relative to N_2 is usually expressed in terms of a quantity $\delta(O_2/N_2)$ (with units of per million or per meg) given by

$$\delta\left(\frac{O_2}{N_2}\right) = \frac{\left(\frac{O_2}{N_2}\right) - \left(\frac{O_2}{N_2}\right)_{ref}}{\left(\frac{O_2}{N_2}\right)_{ref}} \times 10^6,$$

where O_2/N_2 is the ratio recorded for the measurement while $(O_2/N_2)_{ref}$ is the ratio for air in the standard. If the abundance of N_2 is assumed to remain constant (as seems reasonable), it may be seen that $\delta(O_2/N_2)$ defines the fractional change (in parts per million) in the abundance of O_2 relative to the abundance of O_2 in the standard. Assuming a mixing ratio by volume for O_2 in dry air of 0.2095, it follows that a change in the O_2 mixing ratio of 1 ppmv would be equivalent to a change in delta of 1/0.2095 = 4.77 per meg (parts per million).

Emissions of CO_2 from combustion of fossil fuels between July 1991 and July 1994 would have been sufficient to change the mixing ratio of CO_2 by 8.65 ppmv.[18] The equivalent decrease in $\delta(O_2/N_2)$ would be given by (8.65) × (1.39) × (4.77) = 57.4 per meg. The observed decrease is 42.2 per meg. It follows that exchange of O_2 between the biosphere–soil system and the atmosphere must be responsible for a net increase in $\delta(O_2/N_2)$ of (57.4 – 42.2) = 15.2 per meg. This implies that the biosphere–soil system must have accounted for a net source of O_2 over this time period and thus for a net sink of CO_2. The equivalent sink for CO_2 (converting the change in $\delta(O_2/N_2)$ to an equivalent change in the mixing ratio of O_2 and accounting for the fact that 1.1 mol of O_2 are released for every mol of CO_2 taken up by the biosphere-soil) would be given by [(15.2) × (0.2095)]/(1.1) = 2.89 ppmv CO_2. The observed increase in CO_2 is 3.36 ppmv. Thus, the ocean must have accounted for removal of CO_2 equivalent to (8.65 – 2.89 – 3.36) = 2.39 ppmv.

A graphical summary of this analysis is presented in Figure 11.14. Point A denotes the (CO_2, O_2) combination measured at the beginning of the record, in July 1991. Point D summarizes observations for the end of the record in July 1994. The influence of fossil fuel combustion (production of CO_2, consumption of O_2) is represented by the segment A–B (slope = –0.15 ppmv/meg). The role of the ocean in providing a sink for CO_2 with minimal impact on O_2 is indicated by the horizontal line B–C. Finally, the path from points A to D is completed by an exchange between the atmosphere and the biosphere-soil system as described by C–D (slope = –0.19 ppmv/meg). Point C lies at the intersection of a horizontal line passing through B and a line of slope –0.19 ppmv/meg passing through the final state D. Figure 11.15 illustrates a hypothetical case for which the biosphere–soil system would provide a net source of CO_2.[19]

In summary, over the period mid 1991 to mid 1994 we added CO_2 to the atmosphere from combustion of fossil fuels sufficient to increase the atmospheric abundance by 8.65 ppmv, approximately 18.3 × 10^9 tons C. Of this, a quantity of CO_2 equivalent to 2.89 ppmv (6.13 × 10^9 tons C) was incorporated in either soils or the terrestrial biosphere,

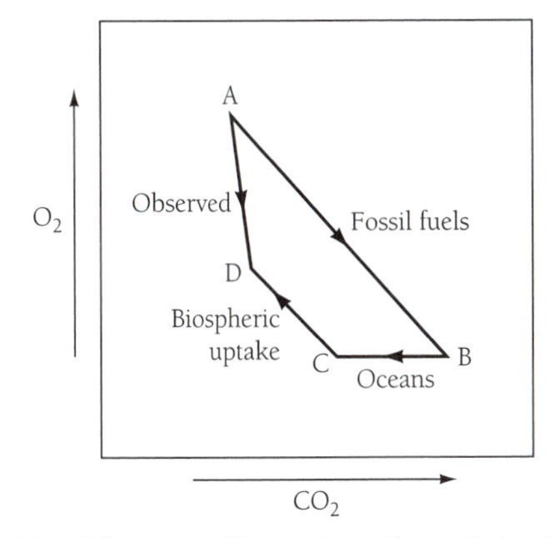

Figure 11.14 Schematic illustration of trends in CO_2 and O_2, 1991–1994.

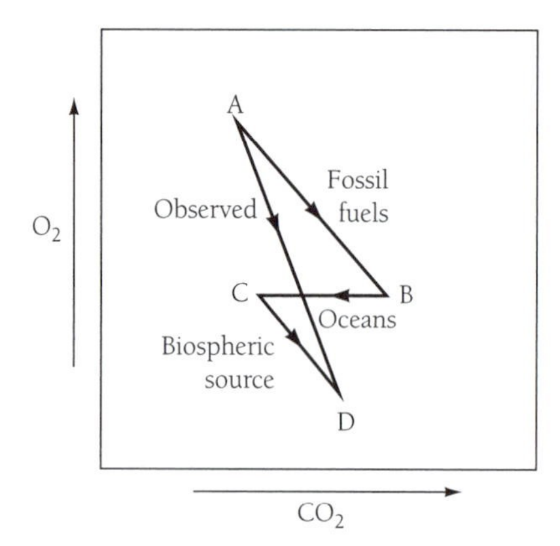

Figure 11.15 Hypothetical situation with a biospheric source of CO_2.

2.39 ppmv (5.07×10^9 tons C) was transferred to the ocean and 3.36 ppmv (7.12×10^9 tons C) remained in the atmosphere. Of the carbon added to the atmosphere by burning fossil fuels, 39% remained in the atmosphere, 33% was incorporated in a combination of soils and the biosphere, and the balance, 28%, was absorbed by the ocean. The analysis implies that 58% of net carbon added to the atmosphere (the fossil fuel source minus the biosphere-soil sink) over the period 1991 to 1994 remained in the atmosphere, whereas the balance was transferred to the ocean. On a global basis, it appears that in the early years of the 1990s the carbon content of soils and the terrestrial biosphere increased at an annual rate of close to 2×10^9 tons C per year.

From an analysis of gradients in CO_2 and O_2 observed between the Northern and Southern Hemispheres, R. F. Keeling and associates (1996) concluded that the growth of the biosphere–soil reservoir in the 1990s occurred primarily in the north. Transfer of air between the Northern and Southern Hemispheres is a relatively slow process. The associated time constant—as indicated, for example, by analysis of spatially distributed data for the industrial halocarbons—is about one year. Fossil fuels are consumed mainly in the industrial Northern Hemisphere. If net input of CO_2 to the Northern Hemisphere was dominated by fossil fuels, we would expect the abundance of CO_2 in the atmosphere of the Northern Hemisphere to exceed that in the Southern Hemisphere by approximately a year's worth of production. The abundance of O_2 would be correspondingly lower in the north. Observations indicate that the concentration gradient between the hemispheres is less than one would expect according to this scenario. The dilemma is resolved if we suppose that the net input of CO_2 (and net consumption of O_2) in the Northern Hemisphere was less than would have been expected as a consequence of fossil fuel consumption. The apparent reduction in the net northern CO_2 source (and the smaller decrease in O_2) can be explained if the global biosphere–soil sink for CO_2 (source of O_2) inferred from the analysis is associated primarily with growth of the biosphere–soil system in the north.

Direct (in situ) measurements of CO_2 and O_2 sufficient to carry out the analysis of the relative importance of the ocean and the biosphere–soil system in the global budget of atmospheric CO_2 are available only for the 1990s. In a remarkable *tour de force*, a group of scientists from the University of Rhode Island, Penn State University, the National Oceanographic and Atmospheric Administration, and the University of Colorado (Battle et al. 1996) succeeded in extending the study by Keeling et al. (1996) back to 1977 by measuring the composition of air in the upper unconsolidated (firn) layer of snow at the South Pole. Analysis of the firn data indicates that the biosphere–soil system played a much smaller role in the global budget of atmospheric CO_2 over the period 1977 to 1985 than it did in the 1990s; trends in CO_2 and O_2 observed in the earlier data can be straightforwardly attributed to the influence of fossil fuels without invoking a role for either the biosphere or soil. Therefore, it appears that the biosphere-soil system played a minor role in the global CO_2 budget in the 1977–1985 time frame, even though it was clearly important in the 1990s.

An interesting perspective on the factors influencing changes in the abundance of CO_2 since 1958 was presented in a paper published by C. D. Keeling and associates in 1995. Moreover, as indicated in Figure 11.16, they found that the trend in CO_2 from 1958 until 1980 could be explained, if fossil fuels provided the dominant source of CO_2 over this interval and if the air-borne fraction was taken as 56%. The airborne fraction they used is almost identical to the value inferred from the CO_2–O_2 observations for the 1991–1994 period (58%) when we allow in the latter case for uptake of carbon by the biosphere–soil system. Assuming that the ocean's capacity to take up excess CO_2 should be relatively constant, the obvious conclusion to be drawn from this analysis is that the role of the global biosphere–soil system, although important in the 1990s, must have been small throughout the period 1958–1985; the conclusions of C. D. Keeling and co–workers (1995) are consistent with those of Battle and co–workers (1996).

There is a simple resolution to the dilemma. The gradient in CO_2 between the Northern and Southern Hemispheres throughout the record is consistently less than would be expected if the input from combustion of fossil fuels was the only factor influencing this gradient. Moreover, as indicated in Figure 11.17, the gradient appears to increase in proportion to the source of CO_2 contributed by fossil fuels. We can account for all of the existing constraints if we assume a persistent sink for CO_2 associated with uptake of carbon by the biosphere–soil system at middle latitudes of the Northern Hemisphere. The analysis of R. F. Keeling and co–workers (1996) suggests that this sink contributed to a net annual removal of about 2×10^9 tons of carbon from the atmosphere in the early 1990s—roughly one-third of the input from fossil fuels. For most of the time, the midlatitude Northern Hemisphere sink is offset by a source of CO_2 of comparable magnitude in the tropics, most probably attributable to deforestation in some countries, including Brazil and Indonesia. The tropical source appears to have been much reduced in the early 1990s in the period covered by the observations of R. F. Keeling and co–workers (1996). Indeed, there are data to support this conjecture.

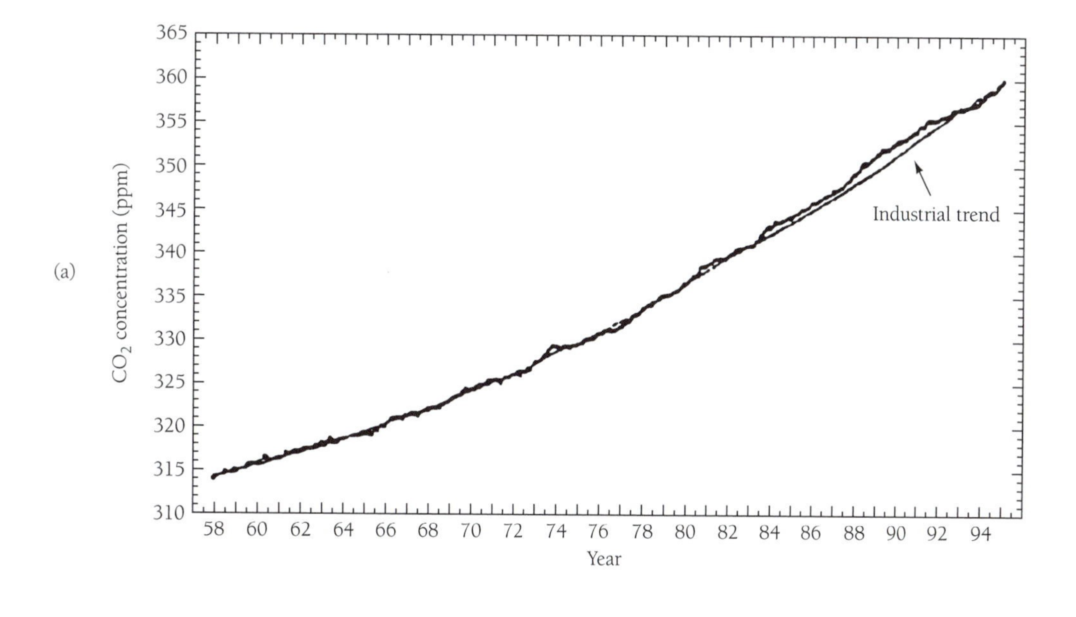

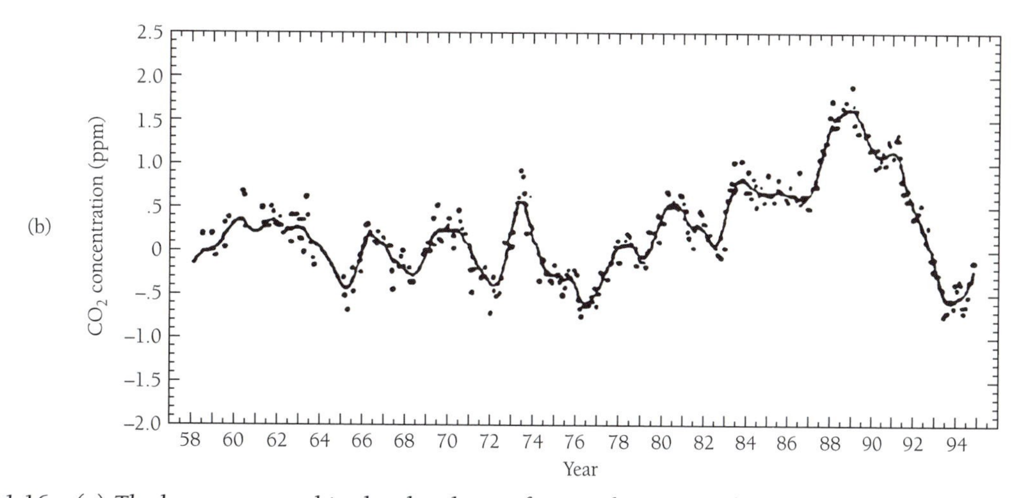

Figure 11.16 (a) The long-term trend in the abundance of atmospheric CO_2 obtained from an average of data from Mauna Loa, Hawaii, and the South Pole, after seasonal adjustment of monthly averages at both stations. The curve labeled "industrial trend" was derived with an assumption that combustion of fossil fuels and cement manufacturing were primarily responsible for changes in CO_2 and that 55.9% of emissions persist in the atmosphere. (b) The anomaly in CO_2 obtained by subtracting the industrialized trend curve in (a) from the observed, monthly mean CO_2 observations. Source: Keeling, et al. 1995.

The burning of forests in Brazil appears to have diminished in the early 1990s, largely because of changes in economic incentives, perhaps helped by sensitivities heightened by Brazil's role in hosting the Earth Summit in Rio de Janeiro in 1992. Anecdotal evidence suggests that the pace of tropical deforestation has increased more recently, beginning in the mid 1990s.

C. D. Keeling and co-workers (1995) used a combination of CO_2 data and measurements of the isotopic composition of carbon in CO_2 to isolate influences of the ocean-biosphere-soil system on the trends in CO_2 observed since 1978.[20] Between 1980 and 1989, they found that the concentration of CO_2 rose more rapidly than expected, given the earlier trend (that is, if fossil fuels had continued as the major source, we would have needed to assume a higher airborne fraction to account for the data in the 1980s). The anomalous rise ended in 1989. By 1994, CO_2 concentrations had reverted to the long-term behavior observed prior to 1980. As illustrated in Figure 11.18, adapted from C. D. Keeling and co-workers (1995), globally averaged surface temperatures rose steadily over the decade of the 1980s, peaking in 1990 and declining by several tenths of a degree after the 1991 eruption of the volcano on Mount Pinatubo. They concluded that, with reference to the long-term trend, there was a net cumulative increase of about 2.3×10^9 tons of carbon released from soils and the biosphere between 1980 and 1989, with a compensating increase of 3.5×10^9 tons of carbon taken up by soils and the biosphere between 1989 and 1991. Those numbers, when expressed in terms of mean–annual exchange rates, are small relative to the mean annual uptake by the biosphere of 2×10^9 tons of carbon per year inferred for the 1990s by R. F. Keeling and co-workers (1996): averaged over the period 1980–1991, the

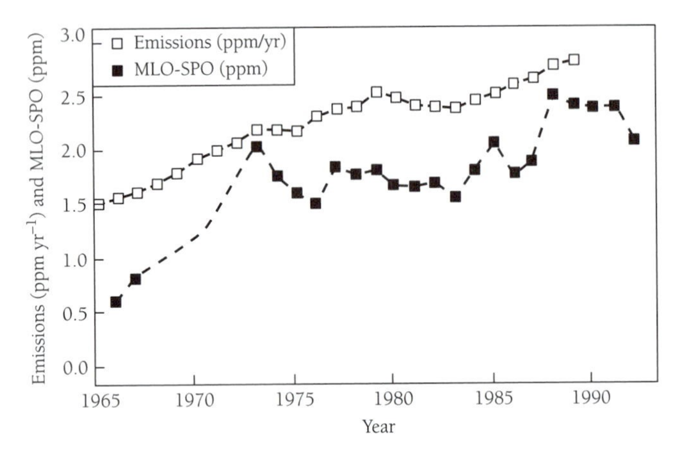

Figure 11.17 Emissions of CO_2 associated with combustion of fossil fuels (open circle), quoted in terms of the equivalent change in the abundance of atmospheric CO_2 in units of ppmv per year and a measure of the gradient in CO_2 between the Northern and Southern Hemispheres (closed square), expressed as the difference between concentrations of CO_2 measured at Mauna Loa (MLO) and South Pole (SPO) (ppmv).

anomalous biosphere-soil uptake inferred from the isotopic analysis amounts to less than 10^8 tons of carbon per year. The pattern reported by C. D. Keeling and co-workers (1995) is generally consistent with results based on the CO_2-O_2 analyses discussed above. They argued that higher ocean temperatures contributed to a cumulative reduction of 2.9×10^9 tons of carbon taken up by the ocean over the period 1980–1991; again, this is small relative to the annual mean uptake of 1.7×10^9 tons C per year inferred from the CO_2-O_2 analysis. There are additional indications in the isotopic data of an influence of El Niño.[21] Release of carbon from the tropical ocean is reduced as a consequence of suppressed upwelling during an El Niño. At the same time, during an El Niño it appears that tropical terrestrial environments are responsible for increased release of CO_2 associated most probably with related regional changes in tropical climate.

Observations of turbulent CO_2 exchange between the atmosphere and a deciduous forest in New England (Harvard Forest in central Massachusetts) provide independent support for the existence of a midlatitude Northern Hemispheric sink for CO_2 (Goulden et al. 1996). Net ecosystem exchange of carbon varied from a low of –1.4 tons of carbon per hectare per year in 1992–1993 (negative numbers indicating uptake by the biosphere) to a high of –2.8 tons of carbon per hectare per year in 1990–1991. Interannual variability was associated with changes in climatic conditions and was sensitive particularly to the length of the growing season (regulated primarily by temperatures in spring and early autumn), cloud cover during summer, drought during summer, snow depth, and other factors affecting soil temperatures during the dormant season. If we were to assume that Harvard Forest was representative of forests and woodlands in North America (obviously a leap of faith) and were to adopt an average of the observed range of values for uptake of carbon at Harvard Forest over the period 1990–1995, we would conclude that forests and woodlands in North America could be responsible for net regional uptake as large as 1.3×10^9 tons C per year, with comparable regional emissions of 1.5×10^9 tons C per year in 1991 contributed by the combustion of fossil fuels.

Additional evidence supporting a Northern Hemispheric sink for CO_2 comes from an analysis reported by Fan and co-workers (1998) of CO_2 measurements reported by a distributed network of 62 sampling stations (Masarie and Tans 1995). Using two distinct general circulation models for the atmosphere and a model to describe ecosystem exchange of carbon, they concluded that the combination of biosphere–soil systems in Eurasia and North America accounted for a net CO_2 sink of about 2×10^9 tons of carbon per year for the period 1981–1987, with somewhat larger removal in 1988–1992. The sink inferred in their analysis is remarkably similar in magnitude to the result derived above. Their study suggested that the sink is operative primarily in temperate regions of North America and in the

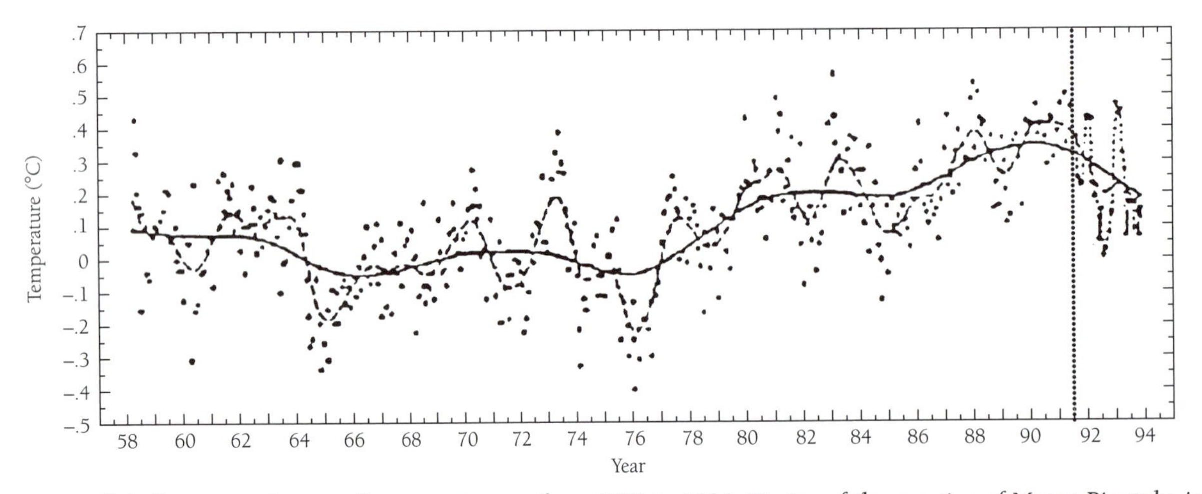

Figure 11.18 Globally averaged sea-surface temperatures from 1958 to 1994. Timing of the eruption of Mount Pinatubo is indicated by the dotted vertical line. From Keeling et al. 1995.

boreal forests of northern Eurasia. Consistent with the speculation above, they raised the possibility that the release of CO_2 associated with combustion of fossil fuels and cement manufacturing in North America may be significantly offset at present by an uptake by regional temperate forests.

In summary, we have shown that geographically distributed measurements of O_2 and CO_2, in combination with data on the isotopic composition of CO_2, may be used to place important constraints on the budget of atmospheric CO_2. A convincing body of evidence suggests the presence of an important and persistent sink for CO_2 associated with the biosphere-soil system in the Northern Hemisphere. The analysis indicates that the sink may accommodate as much as one-third of the carbon added to the atmosphere today by the burning of fossil fuels. The sink is most likely in regions of North America and Europe where forests were once abundant but where overuse resulted in their depletion (early in Europe, more recently in North America). As coal replaced wood as a primary fuel and as industrial practices evolved, forests were allowed to regrow. The sink for CO_2 at northern midlatitudes could be temporary and likely to diminish in importance as the biosphere–soil system approaches a new steady state, reflecting current mixed patterns of land use.

Soils at high latitudes represent an important reservoir for organic carbon, between 200 and 500 × 10^9 tons C (Masarie and Tans 1995; Gorham 1991; Dixon et al. 1994). The eddy correlation method used to study carbon exchange between the atmosphere and Harvard Forest has also been applied to the carbon balance of a mature black spruce forest in central Canada (Goulden et al. 1998). Decomposition of organic carbon in the soils of this system resulted in a small, though significant, net source of CO_2 to the atmosphere—0.4 tons of carbon per hectare per year—in 1994–1996. Emission rates increased by a factor of ten, however, as temperatures rose from −2°C to 5°C, raising the possibility that high-latitude soils could provide a much more important source of CO_2 in the future if climate were to significantly warm at high latitudes.

In forecasting upcoming trends in CO_2, it will be important to allow for feedbacks between the climate system and the complex suite of processes regulating the distribution of carbon over its dominant reservoirs—the atmosphere, soils, biosphere, and ocean. As indicated by the analysis of C. D. Keeling and co-workers (1995), a warmer ocean may provide a less-efficient sink for excess concentrations of atmospheric CO_2. It appears that the global biosphere–soil system played a relatively minor role in the budget of CO_2 over the past 40 years, with release from tropical ecosystems offset by regrowth of forests at northern midlatitudes. But the signature of the biosphere–soil system could change over the next few decades, as midlatitude forests approach a new steady state, completing their recovery from earlier disturbances. The global biosphere–soil system could switch from its current neutral role to become an important net source of CO_2 if warmer temperatures develop, stimulating an increase in emissions from soils at high latitudes.

The analysis indicates that for most of the past 40 years, tropical ecosystems contributed a net source of CO_2 to the atmosphere equal to as much as one-third of emissions associated with the global combustion of fossil fuels. This was balanced for the most part by uptake of a comparable quantity of CO_2 due to net growth of vegetation at midlatitudes of the Northern Hemisphere, most probably in regions deforested earlier but subsequently abandoned and allowed to regrow. The deforestation source appears to have declined in the early 1990s but has reappeared more recently. The importance of the deforestation source of CO_2 suggests that countries such as Brazil and Indonesia could bear as much responsibility as the United States and China for the contemporary buildup of atmospheric CO_2.

The suggestion that regrowth of forests in North America can provide a sink for CO_2 comparable to that associated with local consumption of fossil fuels, although plausible, is clearly speculative, based on current evidence. It merits further attention, however. If validated, it would complicate the task of those charged with developing a strategy to minimize the future growth of greenhouse gases in the atmosphere.

11.4 Projections for the Future

A model for gas transfer from the atmosphere to the ocean is illustrated in Figure 11.19. The model assumes that the partial pressure of a gas in the liquid phase at the air–water interface should be in equilibrium with the atmospheric concentration of the gas. Before air can enter the ocean, it must diffuse through a thin, essentially stagnant film of water at the surface. By definition, the net vertical velocity

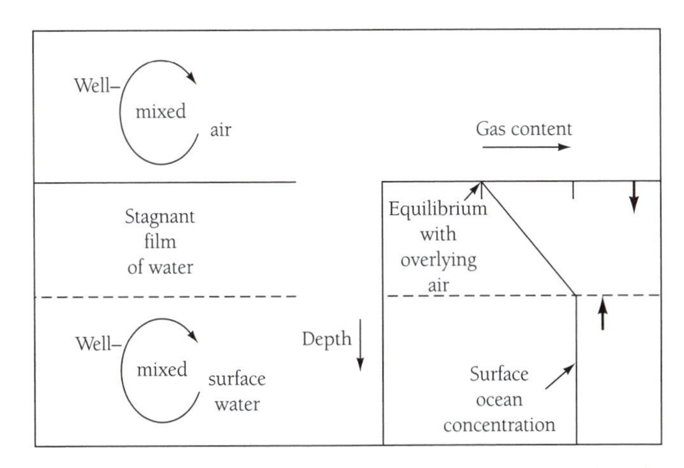

Figure 11.19 Schematic illustration of the model used to describe the oceanic uptake of gas. Concentrations of gases are assumed to be uniformly distributed in the atmosphere as a result of mixing. Mixing processes are assumed to be efficient in the ocean below the surface boundary layer, indicated in the Figure as a stagnant film. Transfer of gas from the water surface to the bottom of the boundary layer proceeds by molecular diffusion.

of water at the surface must equal zero (otherwise the water surface would move either upward or downward). Vertical motion is possible in the water column at some distance below the surface. Water moving upward in one region may be replaced by water moving downward elsewhere. If the concentration of a chemical species in the water moving downward is greater than the concentration in water moving upward, this overturning results in a net downward transfer of the chemical species into the liquid. Suppose that this vertical exchange involves transfer of water masses between levels of the ocean corresponding to depths Z and $Z + \Delta Z$. Suppose also that the concentrations of the chemical species at depths Z and $Z + \Delta Z$ are given by $c(Z)$ and $c(Z + \Delta Z)$, respectively. Assume further that the (downward directed) vertical motion responsible for mass exchange is associated with a velocity V. The net downward flux, F, of the chemical species resulting from this vertical overturning of water masses is given then by

$$F = Vc(Z) - Vc(Z + \Delta Z) \tag{11.30}$$

If we express the difference between the concentrations of the chemical species at Z and $Z + \Delta Z$ in terms of the derivative of c with respect to Z, then

$$c(Z) - c(Z + \Delta Z) = -\Delta Z \frac{dc}{dz} \tag{11.31}$$

Thus

$$F = -V\Delta Z \frac{dc}{dz} \tag{11.32}$$

Introducing a quantity K defined by

$$K = V\Delta Z, \tag{11.33}$$

it follows that the downward flux is given by

$$F = -K\frac{dc}{dz}, \tag{11.34}$$

where the velocity V refers to bulk (downward) vertical motion of water masses, the parameter K (introduced in 11.33) is known as the **eddy diffusion coefficient**. The depth ΔZ is known as the **mixing length**.

The formulation presented here assumes that fluctuating vertical motions (eddies) with velocity V are responsible for the transfer of the water column's properties over vertical-length scales given on the average by the mixing length ΔZ. A critical assumption is that rates of transfer may be expressed in terms of gradients with respect to the mean values of these properties.

As previously noted, bulk vertical motion of the water column is impossible in the immediate vicinity of the water's surface. Movement of a chemical species from the surface into the fluid's bulk is interrupted by collisions with water molecules. The length scale for vertical motion in this case (the analogue of the mixing length) is equal to the mean free path (the distance a molecule can move on average before its motion is interrupted by a collision with the water molecules through which it is trying to move). The flux is defined in this regime in terms of a diffusion coefficient appropriate for motion on a molecular scale (a product of velocity and length with velocity defined by the mean molecular speed and the length scale taken equal to the mean free path). The downward flux of a chemical species through the stagnant boundary layer is given thus by

$$F = -D\frac{dc}{dz}, \tag{11.35}$$

where D denotes the magnitude of the coefficient for molecular diffusion.

If the vertical scale of the boundary layer is given by λ, (11.35) may be written approximately in the form

$$F = \frac{-D}{\lambda}\left[c(\lambda) - c(0)\right], \tag{11.36}$$

where $c(\lambda)$ and $c(0)$ define the concentrations of the chemical species at the bottom and top of the boundary layer, respectively. The quantity D/λ has the dimensions of velocity. Writing

$$v = \frac{D}{\lambda}, \tag{11.37}$$

it follows that

$$F = vc(0) - vc(\lambda) \tag{11.38}$$

The quantity v is known as the **piston velocity**. We can think of a piston pushing chemicals into the ocean from the surface where the concentration is $c(0)$, drawing them out from the bottom of the surface film where the concentration is equal to $c(\lambda)$. Adopting a typical value for the molecular diffusion coefficient of 10^{-5} cm^2 sec^{-1} with a stagnant boundary-layer depth of 4×10^{-3} cm as indicated by a number of studies (e.g., Broecker and Peng 1982), it follows that the piston velocity is equal to about 2.5×10^{-3} cm sec^{-1}, or 2.2 m per day. Every day, the gas content of a column of water about 2 m thick is exchanged with the atmosphere.

The formalism described here may be used to model the change in atmospheric CO_2 expected to develop in the future in response to the continuing (accelerating) emissions of CO_2 associated with combustion of fossil fuels. An increase in emissions results in an increase in the CO_2 concentration in the atmosphere. The concentration of CO_2 in surface waters rises accordingly. The fractional change in the concentration of total dissolved inorganic carbon is less however than the change in CO_2; as previously noted in this chapter, electrically neutral CO_2 accounts for less than 1% of total dissolved inorganic carbon. The impact of a higher concentration of CO_2 in the atmosphere is transferred rapidly to the base of the stagnant boundary layer (the concentration of all dissolved carbon species adjusts to the higher level of CO_2). Transfer to deeper levels of the ocean proceeds at a rate determined by the strength of eddy mixing.

To apply the model to account for uptake of CO_2 by the ocean, we need to specify values for the diffusion coefficients, K. Values for K are selected based on analysis of data for selected chemical species serving as tracers of ocean motion. Measurements of the distribution of ^{14}C in the ocean are especially useful in this context. The testing of nuclear weapons in the atmosphere was responsible for a large (known) source of atmospheric ^{14}C in the 1950s and 1960s. An extensive database is available documenting the spread of this tracer into the ocean. Measurements of industrial CFCs (whose source again is known) provide an additional opportunity to calibrate models for transfer of chemical species from the atmosphere to the ocean. The results presented below used values of K consistent with analyses of data for both ^{14}C and CFCs.

Concentrations of CO_2 calculated for the period 1900–2100 are presented in Figure 11.20. Results indicated by the dashed line depict a sink for carbon associated with uptake by the biosphere. The magnitude of the sink is assumed to vary in proportion to the concentration of CO_2. The constant of proportionality was selected to ensure that the results of the model are in satisfactory agreement with observational data for CO_2 in the 1980s. The sink is intended to account for uptake of carbon by the biosphere at midlatitudes of the Northern Hemisphere, as discussed in Section 11.3. The potential importance of the sink was omitted in the model represented by the solid line in Figure 11.19. The dotted line summarizes results of a simulation that allowed for uptake of carbon by the biosphere until 2060, assuming that the biosphere is subsequently responsible for a small net source. The strength of this late twenty-first–century source was assumed to be proportional to the magnitude of the sink implicated in earlier uptake.

Emissions of CO_2 employed in this study were taken from IPCC Scenario A (IPCC 1992). Scenario A allows for a source due to deforestation in addition to the contribution from combustion of fossil fuels. According to the scenario, emissions of

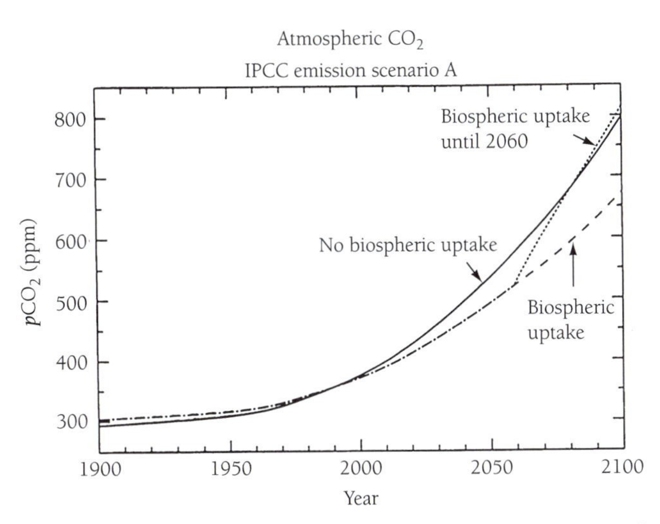

Figure 11.20 Concentrations of CO_2 (ppm) obtained using the model described in the text. Emissions from 1995 to 2100 are taken from IPCC Scenario A. Results indicated by the dashed line allow for uptake of CO_2 by the midlatitude Northern Hemispheric biosphere. The discontinuity at 2050 (the dashed line continued by the dotted line) reflects an assumption that the midlatitude sink switches to a net source subsequent to 2050. The influence of the midlatitude biosphere was omitted in the model summarized by the solid line.

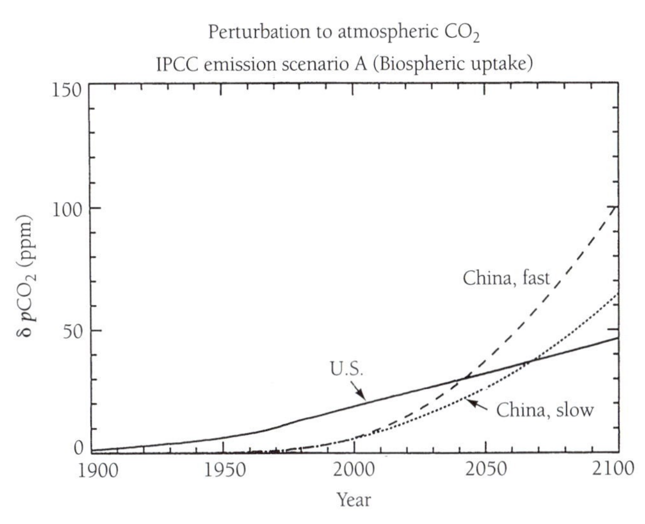

Figure 11.21 Contributions of the United States and China to past and future growth of CO_2. Emissions for the United States are consistent with assumptions in IPCC Scenario A. The figure accounts for two possible scenarios for future growth emissions by China, as described in the text.

CO_2 are projected to grow globally from a contemporary (1990) level of 7.4×10^9 tons C yr^{-1} to 20.3×10^9 tons C yr^{-1} by 2100. The IPCC projection includes an estimate for the source of CO_2 due to deforestation, taken as 1.3×10^9 tons C yr^{-1} in 1990, decreasing to 1.1×10^9 tons C yr^{-1} by 2025, dropping to close to zero by 2100. The results in Figure 11.20 suggest that the concentration of CO_2 may increase by as much as 150 ppm over the next 50 years (to a value close to 500 ppm) and that it may rise to levels in excess of 700 ppm by 2100.

Contributions to past and anticipated future growth of CO_2 arising as a result of consumption of fossil fuels in the United States and the Peoples Republic of China are summarized in Figure 11.21. The model implies that industrial sources of CO_2 in the United States have been responsible for an increase in global CO_2 over the past 100 years by about 20 ppm. Consumption of fossil fuels in the United States is expected to result in a further increase in CO_2 by about 30 ppm over the next 100 years, adding up to a net total impact of close to 50 ppm. The fast-growth model for China (per capita emissions of C in 2100 comparable to current values for the United States), represented by the dashed line in Figure 11.21, suggests that China may be responsible for an increase in global CO_2 by as much as 100 ppm by 2100. Even if economic growth is slow (per capita emissions by 2100 equal to roughly half of those for the contemporary United States), China's impact on global CO_2 is likely to exceed 60 ppm by 2100. It is difficult to escape the conclusion that China's contribution to the growth of atmospheric CO_2, already important, will soon exceed that of the United States, the world's largest current emitter.

11.5 Summary

The contemporary rise in the concentration of atmospheric CO_2 is dominated by the contribution from combustion of fossil fuels. As indicated in Section 11.3, deforestation in the tropics accounts for a significant source of CO_2, but this source is offset by a sink of comparable magnitude associated with uptake of carbon by vegetation at midlatitudes of the Northern Hemisphere. The tropical source was anomalously small in the early 1990s. It appears to have since resumed its earlier upward trend.

Approximately 50% of carbon emitted by burning fossil fuels since the dawn of the industrial revolution remains in the atmosphere. The balance has been transferred to the ocean. The capacity of the ocean to take up excess carbon, as discussed in Section 11.2, is limited by the relatively small abundance of carbonate ion available to neutralize the resulting acidity. Absent aggressive action to limit emissions, concentrations of CO_2 are expected to double over the course of the next century (Section 11.4). Models presented by the IPCC (1996) suggest that, if the increase in the concentration of CO_2 is to be limited to a doubling by 2100, it will be necessary to significantly slow the growth in emissions over the next 50 years and to reverse the current trend, to establish a net decrease over the latter half of the century. Concentrations of CO_2 are higher now than at any time over the past 450 kyr (see Chapter 5). They are likely to rise over the next few decades to levels not seen since dinosaurs roamed the surface of Earth (see Chapter 10).

Cycles of Nitrogen, Phosphorus, and Sulfur

12

Living organisms require a variety of chemical elements for healthy growth. Carbon, hydrogen, oxygen, nitrogen, phosphorus, and sulfur constitute what may be called the **macronutrients**, the major elements of living organisms. The minor elements—the **micronutrients**—include sodium, magnesium, potassium, calcium, iron, manganese, fluorine, chlorine, and bromine. Our discussion of the carbon cycle in the preceding chapter outlined the importance of carbon, hydrogen, and oxygen. Supplied in the form of CO_2 and H_2O, these elements furnish the essential ingredients for photosynthesis. This chapter introduces the biogeochemical cycles regulating the supply of the other macronutrients: N, P, and S.

We begin in Section 12.1 with a discussion of the nitrogen cycle. A major fraction of Earth's nitrogen resides in the atmosphere as N_2. In this form, however, the element is relatively inaccessible to biological organisms. Before it can be incorporated in biological tissue, it must be **fixed**; meaning it must be transformed from the relatively inert form of N_2 to more accessible compounds such as NH_4^+ and NO_3^-. A specific motivation in this Section is to understand the factors responsible for the rise in the abundance of N_2O over the past several centuries. As previously noted, N_2O is an important greenhouse gas. It also plays a critical role in the chemistry of the stratosphere. An increase in the abundance of N_2O may be expected to result in an increase in the global average temperature at Earth's surface, accompanied by a drop in the abundance of O_3 in the stratosphere. The concentration of N_2O is presently increasing at a rate roughly comparable to that of CO_2: while the increase in CO_2 may be attributed mainly to combustion of fossil fuels, more subtle influences, it appears, are responsible for the rise in N_2O. We shall argue that decomposition of human and animal waste, and microbial processes triggered by large-scale applications of chemical fertilizer, are the major factors affecting the contemporary rise in the abundance of N_2O. The global budget of atmospheric N_2O is currently out of equilibrium: the rate at which N_2O is emitted to the atmosphere is about 40% larger than the rate at which it is removed. It is clearly important that we understand the processes responsible for this imbalance.

The phosphorus cycle is the subject of Section 12.2. In contrast to the situation for nitrogen, the atmosphere plays a relatively unimportant role in the geochemistry of phosphorus. Phosphorus is transferred primarily by rivers flowing from the continents to the oceans, either in solution as phosphate or as a component of particulate matter. Not all of the phosphorus entering the ocean is available for uptake by the biota. A portion is relatively immediately immobilized in sediments. The abundance of phosphate plays an important role, however, in determining the overall biological productivity of the ocean. It thus affects the capacity of the ocean to take up carbon from the atmosphere and may also have an influence on biological productivity and on the uptake of carbon by specific terrestrial ecosystems. Phosphate can have a further indirect influence on nitrogen. Excessive supply of phosphate to fresh-water aquatic systems, associated, for example, with intensive applications of phosphate-based fertilizers, with high concentrations of domestic animals, or with use of phosphate-based detergents, can enhance production of fixed nitrogen by stimulating the growth of nitrogen-fixing organisms such as blue-green algae. A comprehensive appreciation for the influence of human activity on the atmospheric environment requires at least a peripheral understanding of the role of phosphorus.

The sulfur cycle is discussed in Section 12.3. In contrast to the case of phosphorus, the atmosphere is significantly involved in the redistribution of global sulfur. Important gaseous contributors to the sulfur cycle include the reduced species hydrogen sulfide (H_2S), carbonyl sulfide (COS), and dimethylsulfide ($(CH_3)_2S$). These species are oxidized in the atmosphere to SO_2, adding to the burden of atmospheric SO_2 contributed by the combustion of fossil fuels. Oxidation of SO_2 to H_2SO_4, mainly in the aqueous phase, plays an important role in the phenomenon of acid rain, as we will discuss in more detail in Chapter 18. In addition to its role as a source of acid rain, sulfur in the atmosphere contributes a source of sulfate aerosols, which serve as condensation nuclei and can contribute to an increase in cloudiness, particularly at low altitude. As previously noted in Chapter 10, sulfate aerosols, by enhancing the reflectivity of the atmosphere, can serve to reduce the flux of solar radiation reaching the surface, offsetting to some extent effects of warming contributed by an increasing burden of greenhouse gases such as CO_2, CH_4, and N_2O. We conclude with summary remarks in Section 12.4.

12.1 The Nitrogen Cycle and the Budget of N_2O

The abundance of nitrogen in different reservoirs is summarized in Table 12.1. Estimates of rates for transfer between these reservoirs are presented in Figure 12.1. As indicated in the Table, a major fraction of Earth's store of nitrogen resides in the atmosphere as N_2. The bulk of the available supply of fixed nitrogen is present as organic material in terrestrial biomass, soils and sediments and in inorganic form, mainly as NO_3^-, in soils and in the ocean. In contrast, a relatively small fraction of the total store of fixed nitrogen, less than 1 part in 10^9, resides in the atmosphere.

The data in Figure 12.1 were purposely selected to ensure that inputs to, and outputs from, major reservoirs are in balance. They are therefore intended to represent conditions for the pre-industrial environment. For example, the terrestrial system (biosphere plus soil) is postulated to annually receive a quantity of nitrogen equivalent to 240 Mt N, of which 200 Mt N is supplied by the fixation of atmospheric N_2, 30 Mt N by precipitation, and 10 Mt N by the recycling of N from sediments. Nitrogen is assumed to be removed from the terrestrial compartment by runoff in rivers (20 Mt N yr^{-1}), by volatilization of fixed N (30 Mt N yr^{-1}), and by **denitrification** (190 Mt N yr^{-1}), with an integrated loss constrained to equal the postulated source of 240 Mt N yr^{-1}.[2] The corresponding balance for the ocean assumes an input of 20 Mt N yr^{-1} from rivers, 20 Mt N yr^{-1} from fixation, and 10 Mt N yr^{-1} from pre-

Table 12.1 Nitrogen Content of Individual Terrestrial Reservoirs

Reservoir	*Content (g)*	*Notes*
Atmosphere	4×10^{21}	(a)
Land biomass	3×10^{16}	(b)
Humus	1×10^{17}	(c)
Soil inorganic	1×10^{16}	(d)
Ocean biomass	8×10^{14}	(e)
Ocean organic	2×10^{17}	(e)
Ocean nitrate	6×10^{17}	(f)
Ocean N_2	2×10^{19}	(f)
Sediments	2×10^{21}	(g)

Notes:

(a) The total mass of the atmosphere is equal to 5.3×10^{21} g. The mass of N is obtained by using the fact that the mass-mixing ratio of N is equal to 7.6×10^{-1}, corresponding to a volume mixing ratio of N_2 equal to 7.8×10^{-1}.

(b) The N content of terrestrial biomass was estimated assuming a carbon content of 7.3×10^{17} g C (Figure 11.8) with a C/N ratio of 30.

(c) The N content of humus (soil organic material) was calculated assuming a carbon abundance of 1.2×10^{18} g C (Figure 11.8) with a C/N ratio of 15.

(d) Adapted from Delwicke (1970), as amended by McElroy (1976).

(e) McElroy (1976).

(f) Vaccaro (1965).

(g) Holland (1978).

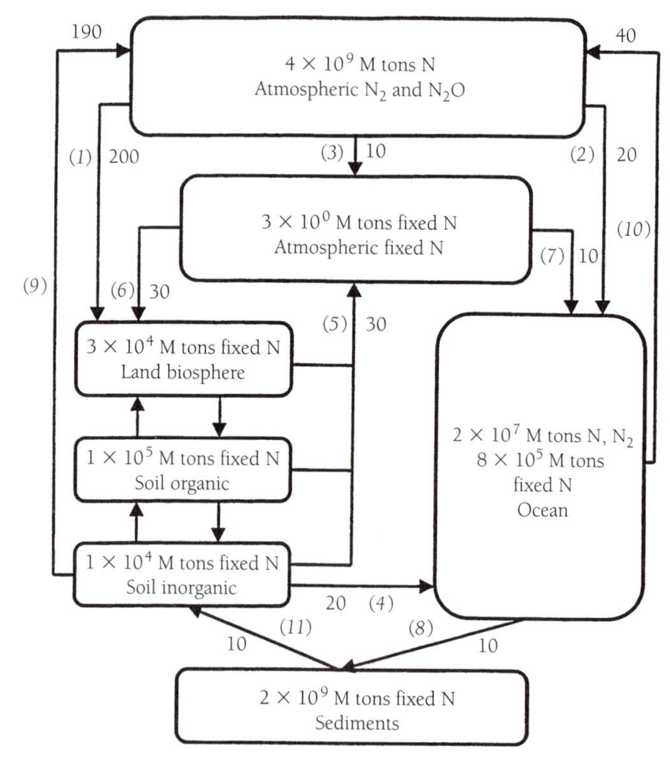

Figure 12.1 Overview of the global nitrogen cycle. Reservoir contents are expressed in units of MT N (10^{12} g N). Transfer rates are in units of MT N yr^{-1} (10^{12} g N yr^{-1}). Processes involved in specific transfer processes are numbered in parentheses (1)–(11). Specific processes are as follows: (1) terrestrial fixation, (2) marine fixation, (3) lightning, (4) transfer from land to ocean in rivers, (5) volatilization of fixed N from terrestrial sources, (6) precipitation on land, (7) precipitation on ocean, (8) incorporation in sediments, (9) denitrification on land, (10) denitrification in ocean, (11) return of fixed N mainly to terrestrial systems by uplift of sediments.

cipitation (total: 50 Mt N yr^{-1}), offset by a loss of 40 Mt N yr^{-1} to denitrification, with 10 Mt N yr^{-1} removed by incorporation in sediments (total: 50 Mt N yr^{-1}).

It should be emphasized that the transfer rates adopted here are all to a greater or lesser extent uncertain. A similar caveat must be applied to the values assumed for the abundances of N in specific reservoirs, with the notable exception of the result for the atmosphere and to a lesser extent the number adopted for the ocean. The summaries in Table 12.1 and Figure 12.1 are nonetheless useful. They provide a convenient framework, allowing us to focus on the large-scale, overall features of the global nitrogen cycle. In particular, they afford a useful starting point to assess the significance of specific disturbances associated with human activity, as may be illustrated by considering the case of fixation.

The fixation of nitrogen under natural conditions in terrestrial systems is accomplished to a large extent by bacteria acting in symbiotic association with selected plants. Particularly important is the relationship between bacteria of the genus *Rhizobium* and plants of the family *Leguminosia*. Neither the bacterium nor the plant acting alone has the capacity to fix nitrogen. Acting together, however, they are able to accomplish this formidable feat with remarkable ease. The relationship is said to be *symbiotic*. Energy equivalent to 226 kcal mol^{-1} is required to rupture the N:N triple bond in N_2. A portion of this energy may be recovered as nitrogen is converted to compounds such as NH_4^+ or NO_3^-. Fixation of nitrogen in an industrial chemical plant requires a major allocation of economic resources. It proceeds at temperatures in excess of 500°C, consuming large quantities of fossil fuels. The rhizobium-legume combination accomplishes a similar feat using sunlight as an energy source and operating under natural conditions without the need for extraneous inputs of either material or energy. A healthy leguminous crop can be remarkably effective in fixing nitrogen. Yields as large as 120 kg N per hectare per year are common in temperate regions, whereas yields as much as 3 times higher have been achieved in well-managed pastures. Alfalfas, clovers, and lupines are particularly effective as agents for symbiotic fixation of nitrogen. Yields from peas, beans, and peanuts are relatively more modest.

Nitrogen may be fixed also by certain free-living bacteria and by blue-green algae. The important bacteria include members of the heterotrophic classes *Azotobacter* and *Clostridium*,[3] in addition to the photosynthetic purple sulfur, green sulfur, and non-sulfur purple bacteria—*Thiorhodaceae*, *Chlorobacteriaceae*, and *Athiorhodaceae*, respectively. These organisms require rather specialized conditions for growth and proliferation and are not thought to play a major role in the contemporary budget of fixed nitrogen. Blue-green algae have the capacity to synthesize organic material from CO_2 and H_2O. They flourish prolifically in aquatic media, given an adequate supply of phosphate and other essential nutrients. They are especially effective in the tropics, where associated rates for fixation of nitrogen may be as high as 75 kg N per hectare per year. They play an important role also in the nitrogen budget of rice paddy fields and can dominate the nitrogen economy of lakes and rivers receiving excessive inputs of phosphate.[4] Like the free-living bacteria, blue-green algae are primitive and are thought to have played a significant role in the early evolution of life.

Human activity makes an important additional contribution to the budget of fixed nitrogen today. The success of modern agriculture depends in large measure on widespread use of increasing quantities of chemical fertilizer. The manufacture and application of nitrogen-based fertilizer contributed an extra source of fixed nitrogen to terrestrial systems in 1997 equivalent to 81 Mt N, of which approximately 25% was employed in China. Production of chemical fertilizer has risen from about 4 Mn N yr^{-1} in 1950 and has more than doubled over the past 25 years. The source of fixed nitrogen from the combustion of fossil fuels has grown at a comparable rate, reaching a level of about 20 Mt N yr^{-1} today. There are two components to the emission of fixed nitrogen associated with the combustion of fossil fuels: one reflects emission of fixed nitrogen contained in the fuel; the other involves fixation of atmospheric N_2 at high temperatures associated with combustion. Biomass burning, associated primarily with clearance of land for agriculture in the tropics, is responsible for an additional contemporary source of about 12 Mt N yr^{-1}. The present anthropogenic contribution to the global budget of fixed nitrogen equals at least 50% of the source attributed to natural agents, and there is little sign that the trend will be reversed in the foreseeable future. It should be emphasized that the anthropogenic component was omitted in the budget presented in Figure 12.1, which was intended to represent a hypothetical steady state that might have applied several centuries ago, prior to onset of the disturbances associated with more recent human activity.

Nitrogen is present in nature in a variety of oxidation states, as illustrated in Table 12.2. Its importance for the biosphere may be attributed in no small measure to this aspect of its chemistry. Ammonium (NH_4^+), nitrite (NO_2^-), and nitrate (NO_3^-) represent the major forms of inorganic nitrogen in soils and water systems. The process responsible for conversion of organic nitrogen to NH_4^+ is known as

Table 12.2 Oxidation status of N for common forms of nitrogen.

Species	*Organic N* *NH_4^+* *NH_3*	*N_2*	*N_2O*	*NO*	*NO_2^-*	*NO_2*	*NO_3^-*
Oxidation state	−3	0	+1	+2	+3	+4	+5

ammonification. Oxidation of NH_4^+ to NO_2^-, summarized by the bulk reaction

$$NH_4^+ + 3/2\,O_2 \rightarrow NO_2^- + H_2O + 2H^+, \qquad (12.1)$$

is an exothermic process: at a temperature of 298 K and a pressure of 1 atm, it is accompanied by the release of free energy equivalent to 51.8 kcal mol^{-1}. Conversion of NH_4^+ to NO_2^-, the first step of what is referred to as **nitrification**, is effected by bacteria known as *Nitrosomonas*. A different class of bacteria, *Nitrobacter*, is involved in the conversion of NO_2^- to NO_3^-:

$$NO_2^- + 1/2\,O_2 \rightarrow NO_3^- \qquad (12.2)$$

Reaction (12.2) is also exothermic. The energy released in this case is equal to 20.1 kcal mol^{-1}. Reaction (12.1) involves a change in the oxidation state of N from −3 to +3; the oxidation state is raised by two units in (12.2) from +3 to +5.

Plants can satisfy their requirements for nitrogen by using any combination of NH_4^+, NO_2^-, or NO_3^-. Not surprisingly, when it is available, they tend to prefer NH_4^+. Nitrogen in NO_2^- and NO_3^- must be reduced before it can be incorporated in plant tissue: from +3 to −3 in the case of NO_2^-, from +5 to −3 in the case of NO_3^-. The relevant transformations, identified as **assimilatory reduction**, are summarized by

$$\begin{aligned} &NO_2^- + 3/2\,CH_2O + 2H^+ \\ &\rightarrow NH_4^+ + 3/2\,CO_2 + 1/2\,H_2O \end{aligned} \qquad (12.3)$$

and

$$\begin{aligned} &NO_3^- + 2CH_2O + 2H^+ \\ &\rightarrow NH_4^+ + 2CO_2 + H_2O \end{aligned} \qquad (12.4)$$

The energy required to effect the reduction of nitrogen in (12.3) and (12.4) is supplied by oxidation to CO_2 of organic matter, represented here by CH_2O.

The processes summarized by reactions (12.1)–(12.4) provide an interesting illustration of the efficiency with which chemical elements, such as nitrogen, are used and reused by the biota. One organism's waste is another's essential nutrient. Nitrifying bacteria are able to grow and flourish extracting energy by oxidizing N; on the other hand, if plants were unable to utilize nitrogen in its oxidized form, the entire system would be in trouble. In the presence of an adequate source of O_2, nitrogen would accumulate as NO_3^- and organisms would be starved for nutritional N. Assimilatory reduction of NO_2^- and NO_3^- allows oxidized nitrogen to be reduced and reused by plants. Energy extracted by oxidation of plant tissue in turn allows nitrogen transformations to proceed in both directions of the oxidation–reduction line illustrated in Table 12.2. Ultimately, sunlight captured by photosynthesis is the fuel allowing this perpetual-motion machine to function.

Nitrogen fixation is balanced on a global scale by denitrification. Otherwise, as illustrated by the following examples, given sufficient time, the reservoir of nitrogen as N_2 in the atmosphere would be depleted and nitrogen would steadily accumulate in the ocean as NO_3^-.

Example 12.1: Estimate the lifetime for N in the atmosphere as N_2, assuming a global rate for N fixation equal to 220 Mt N yr^{-1}, as indicated by the data for terrestrial and marine fixation included in Figure 12.1.

Answer: The lifetime of atmospheric N is obtained by dividing the abundance of N in the atmosphere by the rate at which it is removed by fixation. Expressed in Mt N, the abundance is equal to 4×10^9 Mt N. The lifetime is given by

$$\tau = \frac{4 \times 10^9}{2.2 \times 10^2}\,yr = 1.8 \times 10^7 yr$$ ■

Example 12.2: Assuming that nitrogen fixed from N_2 is eventually converted to NO_3^- and delivered to the ocean, estimate the lifetime for oceanic NO_3^-. Assume that the fixation rate is the same as quoted above in Example 12.1: 220 Mt N yr^{-1}. As indicated in Table 12.1, the abundance of marine NO_3^- may be taken equal to 6×10^{17} g N.

Answer: The abundance of N as NO_3^- in the ocean expressed in Mt N is equal to 6×10^5. The lifetime is given therefore by

$$\tau = \frac{6 \times 10^5}{2.2 \times 10^2}\,yr = 2.7 \times 10^3\,yr$$ ■

The denitrification process may be summarized by the bulk reaction

$$\begin{aligned} &NO_3^- + 5/4\,CH_2O \\ &\rightarrow 1/2\,N_2 + 3/4\,H_2O + 5/4\,CO_2 + OH^- \end{aligned} \qquad (12.5)$$

Denitrification proceeds under anaerobic conditions. It provides an important respiratory path for bacteria when O_2 is in short supply. Reaction (12.5) is exothermic, releasing about 124 kcal mol^{-1}. Denitrification can also proceed by a path resulting in the production of N_2O. The bulk reaction in this case is given by

$$\begin{aligned} &NO_3^- + CH_2O \\ &\rightarrow 1/2\,N_2O + 1/2\,H_2O + CO_2 + OH^- \end{aligned} \qquad (12.6)$$

The details of the path for denitrification are not well understood. It appears that nitrate is reduced first to nitrite and that N_2O may be produced as a precursor to production of N_2. It is clear, though, that in nature, N_2O may be consumed if the concentration of O_2 is exceptionally low. In the core of the anoxic water observed off the coast of Peru, associated with the strong upwelling system normally present in this region (see Chapter 9), the concentration of N_2O drops essentially to zero. On the edge of the anoxic zone, where the concentration of O_2 is low but not too low, concentrations of dissolved N_2O are observed to be exceptionally high, many times those that should apply in equilibrium with the atmosphere.

Motivated by concerns that human intervention in the nitrogen cycle could result in a significant increase in emission of N_2O into the atmosphere, we initiated a major program of research at Harvard University in the late 1970s, seeking to develop a better understanding of the factors reg-

Plate 1

Period	Group** 1 IA 1A	2 IIA 2A	3 IIIB 3B	4 IVB 4B	5 VB 5B	6 VIB 6B	7 VIIB 7B	8 VIII 8	9 VIII 8	10 VIII 8	11 IB 1B	12 IIB 2B	13 IIIA 3A	14 IVA 4A	15 VA 5A	16 VIA 6A	17 VIIA 7A	18 vIIIA 8A
1	1 H 1.008																	2 He 4.003
2	3 Li 6.941	4 Be 9.012											5 B 10.81	6 C 12.01	7 N 14.01	8 O 16.00	9 F 19.00	10 Ne 20.18
3	11 Na 22.99	12 Mg 24.31											13 Al 26.98	14 Si 28.09	15 P 30.97	16 S 32.07	17 Cl 35.45	18 Ar 39.95
4	19 K 39.10	20 Ca 40.08	21 Sc 44.96	22 Ti 47.88	23 V 50.94	24 Cr 52.00	25 Mn 54.94	26 Fe 55.85	27 Co 58.47	28 Ni 58.69	29 Cu 63.55	30 Zn 65.39	31 Ga 69.72	32 Ge 72.59	33 As 74.92	34 Se 78.96	35 Br 79.90	36 Kr 83.80
5	37 Rb 85.47	38 Sr 87.62	39 Y 88.91	40 Zr 91.22	41 Nb 92.91	42 Mo 95.94	43 Tc (98)	44 Ru 101.1	45 Rh 102.9	46 Pd 106.4	47 Ag 107.9	48 Cd 112.4	49 In 114.8	50 Sn 118.7	51 Sb 121.8	52 Te 127.6	53 I 126.9	54 Xe 131.3
6	55 Cs 132.9	56 Ba 137.3	57 La* 138.9	72 Hf 178.5	73 Ta 180.9	74 W 183.9	75 Re 186.2	76 Os 190.2	77 Ir 190.2	78 Pt 195.1	79 Au 197.0	80 Hg 200.5	81 Tl 204.4	82 Pb 207.2	83 Bi 209.0	84 Po (210)	85 At (210)	86 Rn (222)
7	87 Fr (223)	88 Ra (226)	89 Ac~ (227)	104 Rf (257)	105 Db (260)	106 Sg (263)	107 Bh (262)	108 Hs (265)	109 Mt (266)	110 --- ()	111 --- ()	112 --- ()		114 --- ()		116 --- ()		118 --- ()

*Lanthanide Series	58 Ce 140.1	59 Pr 140.9	60 Nd 144.2	61 Pm (147)	62 Sm 150.4	63 Eu 152.0	64 Gd 157.3	65 Tb 158.9	66 Dy 162.5	67 Ho 164.9	68 Er 167.3	69 Tm 168.9	70 Yb 173.0	71 Lu 175.0
~Actinide Series	90 Th 232.0	91 Pa (231)	92 U (238)	93 Np (237)	94 Pu (242)	95 Am (243)	96 Cm (247)	97 Bk (247)	98 Cf (249)	99 Es (254)	100 Fm (253)	101 Md (256)	102 No (254)	103 Lr (257)

**Groups are noted by 3 notation conventions.

Periodic table of elements.

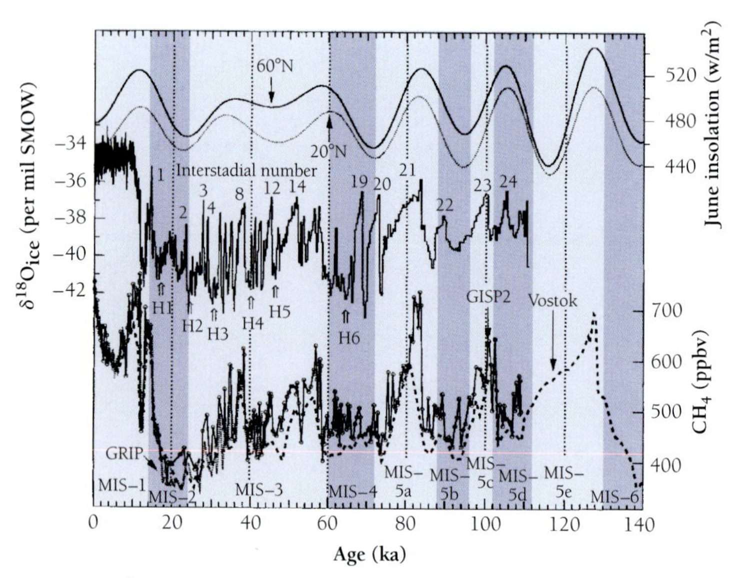

Variation of CH_4 (ppb) over the past 140 kyr. Source: Brook et al. 1996.

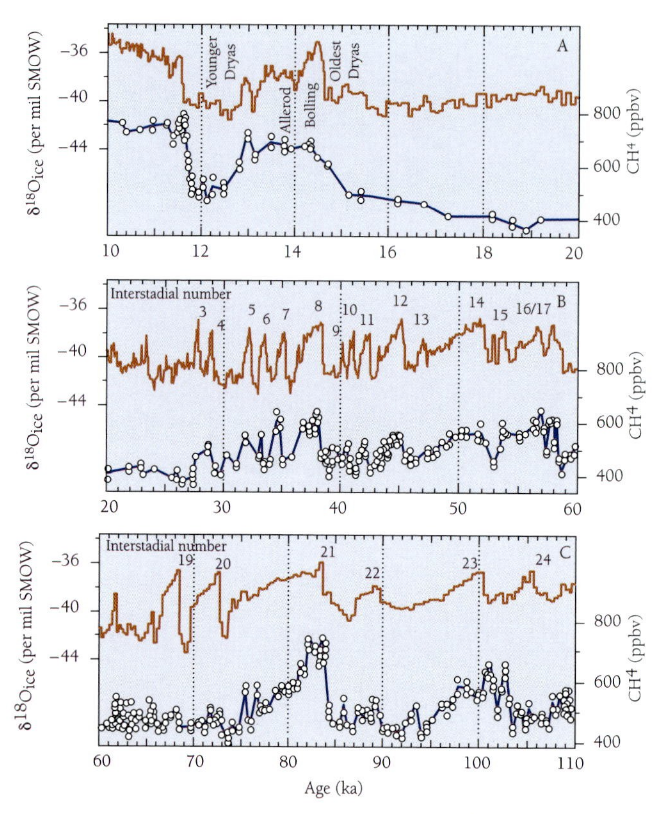

Comparison on a common time scale of GISP2, GRIP, and Vostok methane records, along with June insolation at 60°N and 20°N during the last 110,000 years. Source: Brook et al. 1996.

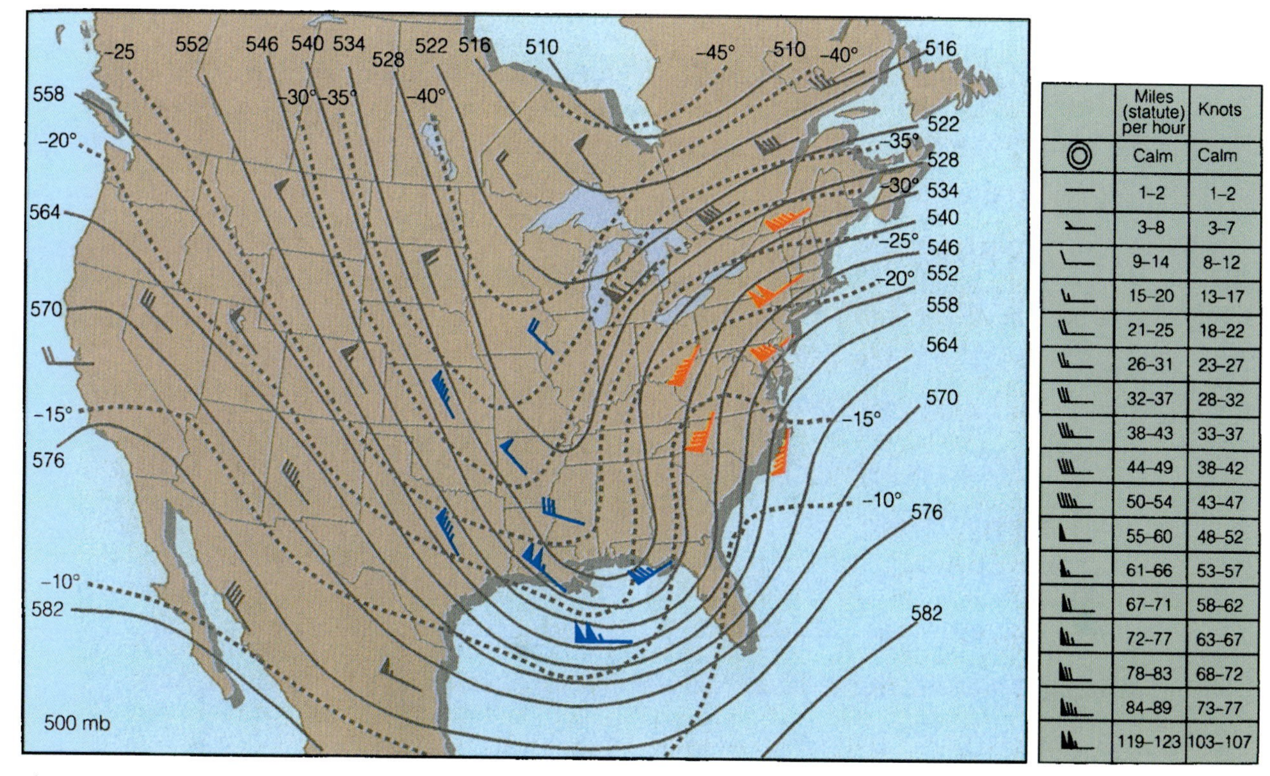

The 500 mb chart for 7 A.M. (EST), 13 March 1993. Heights are expressed in units of tens of meters (i.e., 560 corresponds to a height of 5600 m or 5.6 km). Dotted lines indicate regions of constant temperature (°C). Symbols for wind speed are defined in the right-hand panel. Source: Ahrens 1994.

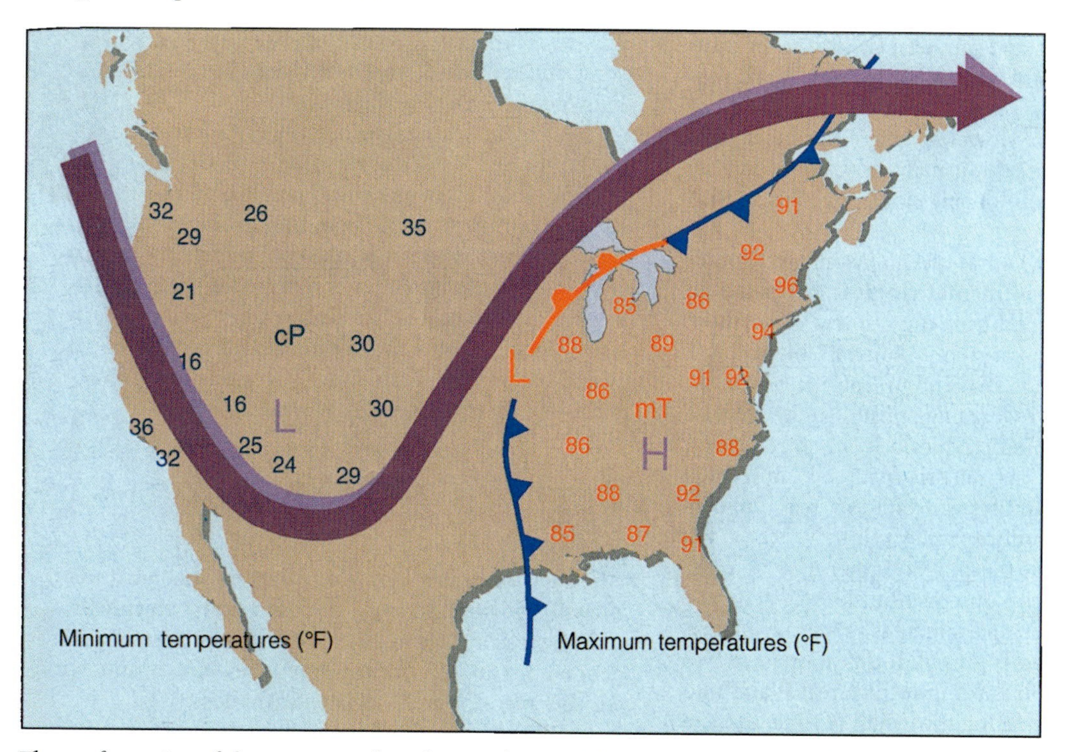

The configuration of the jet stream (broad, curved arrow) during an unusual weather condition observed 15–20 April, 1976. Numbers refer to temperatures (°F) observed at the surface. Note the extreme contrast between the (cold) western and (warm) eastern portions of the United States over this period. Source: Ahrens 1994.

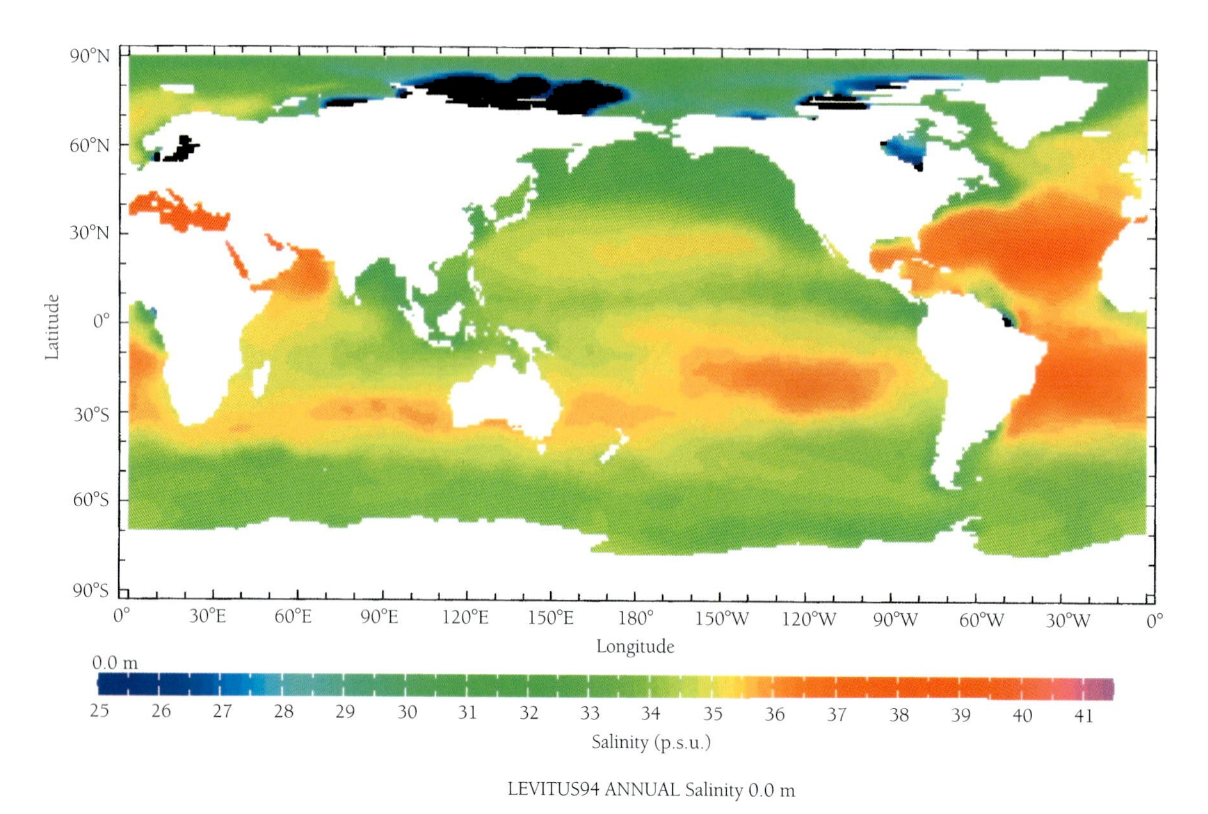

Color rendition of the distribution of surface salinity in the world's oceans. Note that the scale goes from 25 p.s.u. (practical salinity units) to about 41 p.s.u.

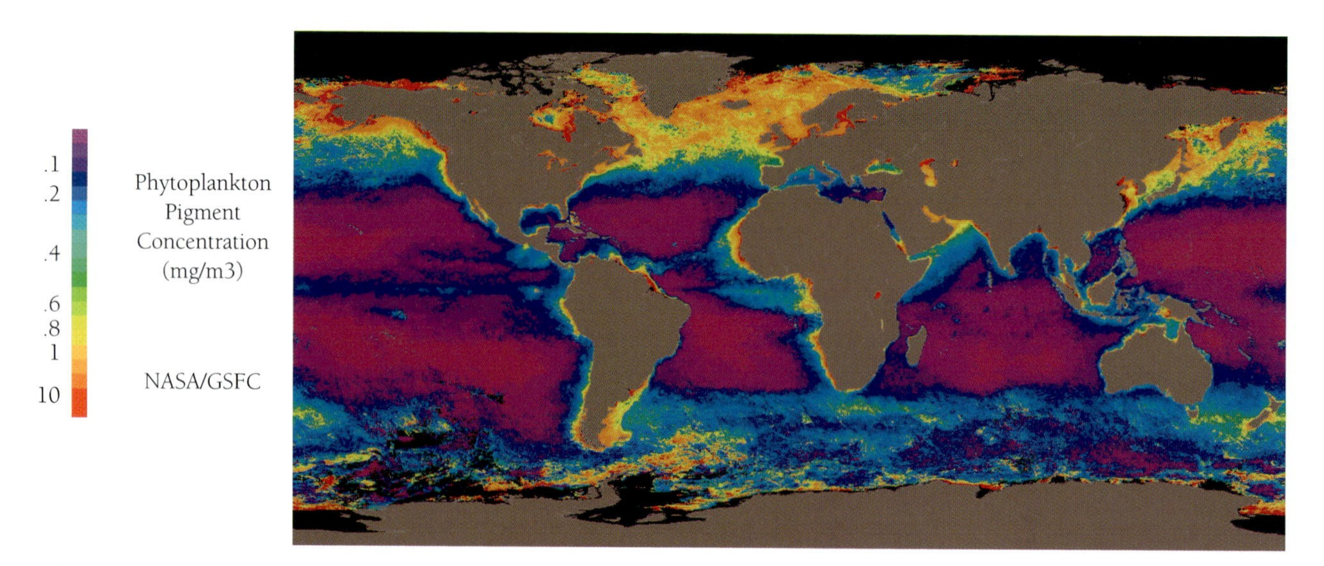

False-color satellite picture of ocean color. Source: ORBIMAGE (© Orbital Imaging Corporation and processing by NASA Goddard Space Flight Center).

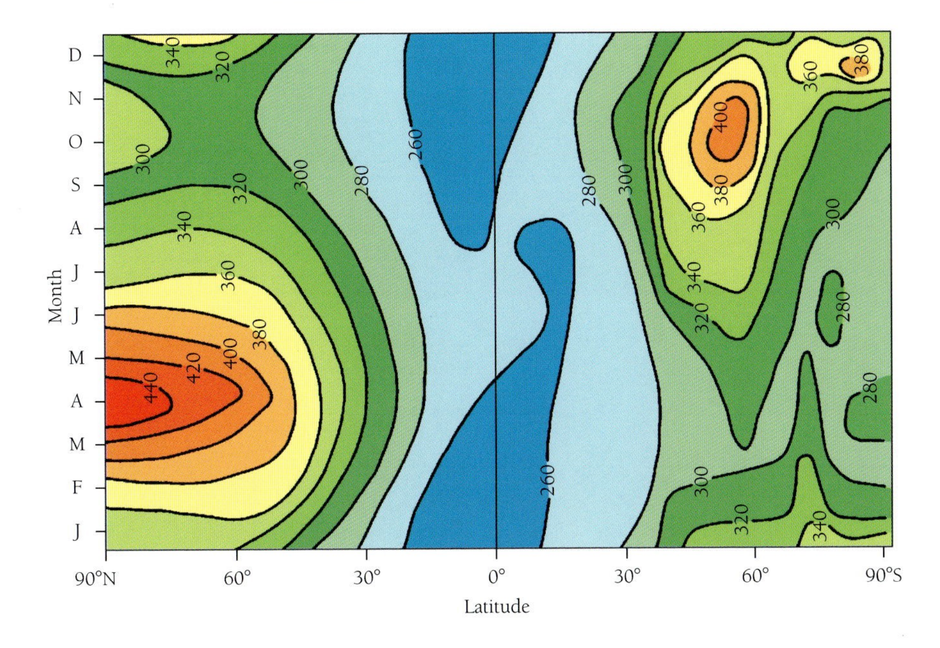

The global distribution of the column density of ozone (DU) as a function of time and latitude, for measurements obtained prior to the onset of the Antarctic ozone hole. Source: Deustch 1974.

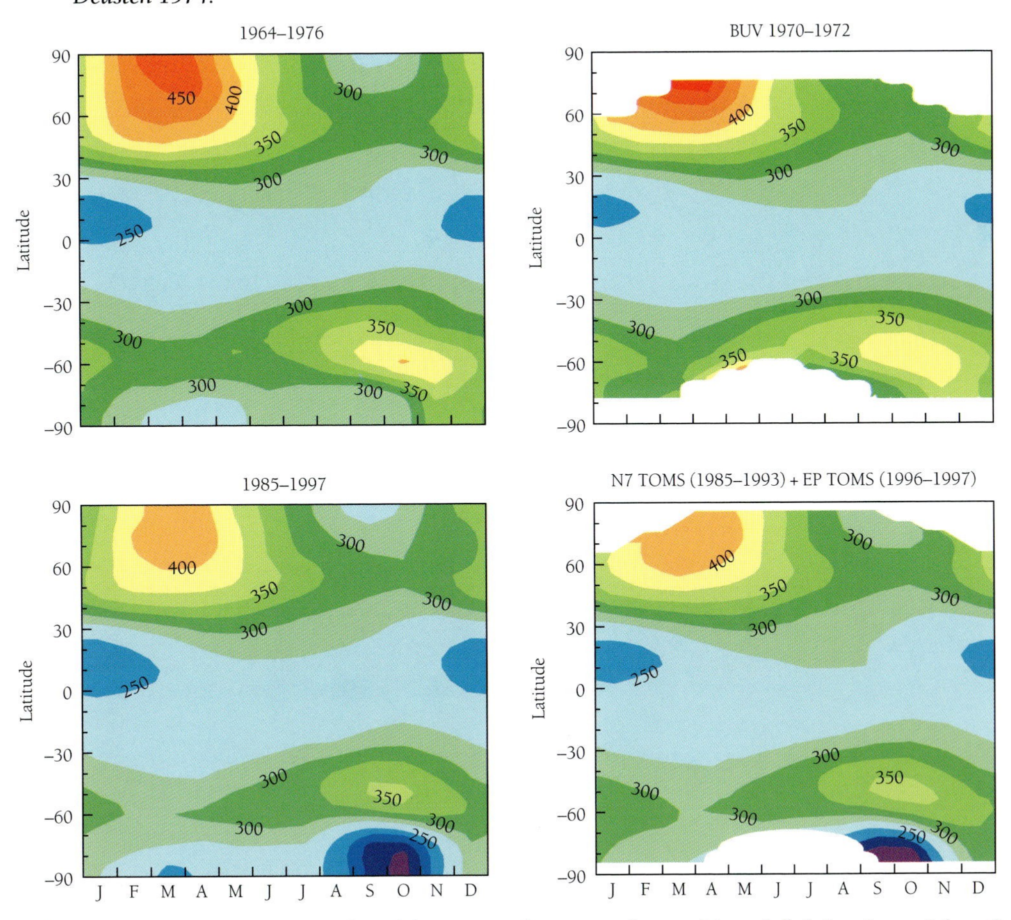

Column densities of O_3 (DU) inferred from a combination of ground-based (left-hand panels) and satellite-based (right-hand panels) observations. From upper left clockwise to lower left, data refer to time intervals 1964–1976, 1970–1972, 1985–1997, and 1985–1997. Source: WMO 1999.

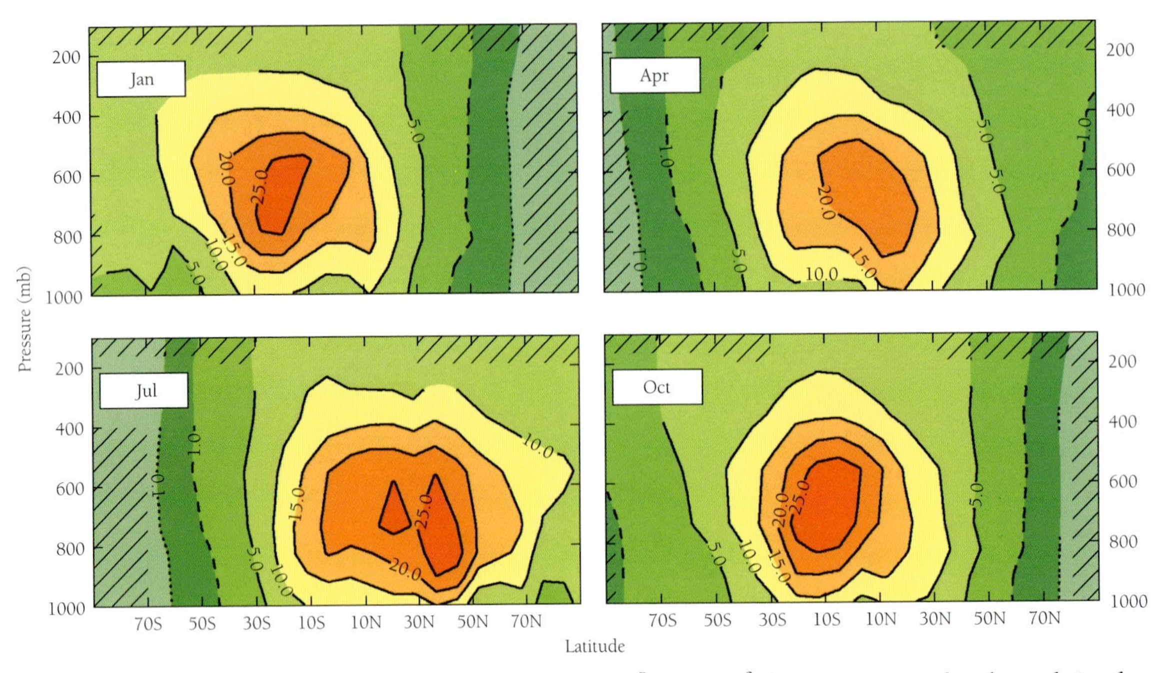

Zonally and monthly averaged concentrations of OH in 10^5 mol cm^{-3} for January, April, July, and October. Source: Spivakovsky et al. 2000.

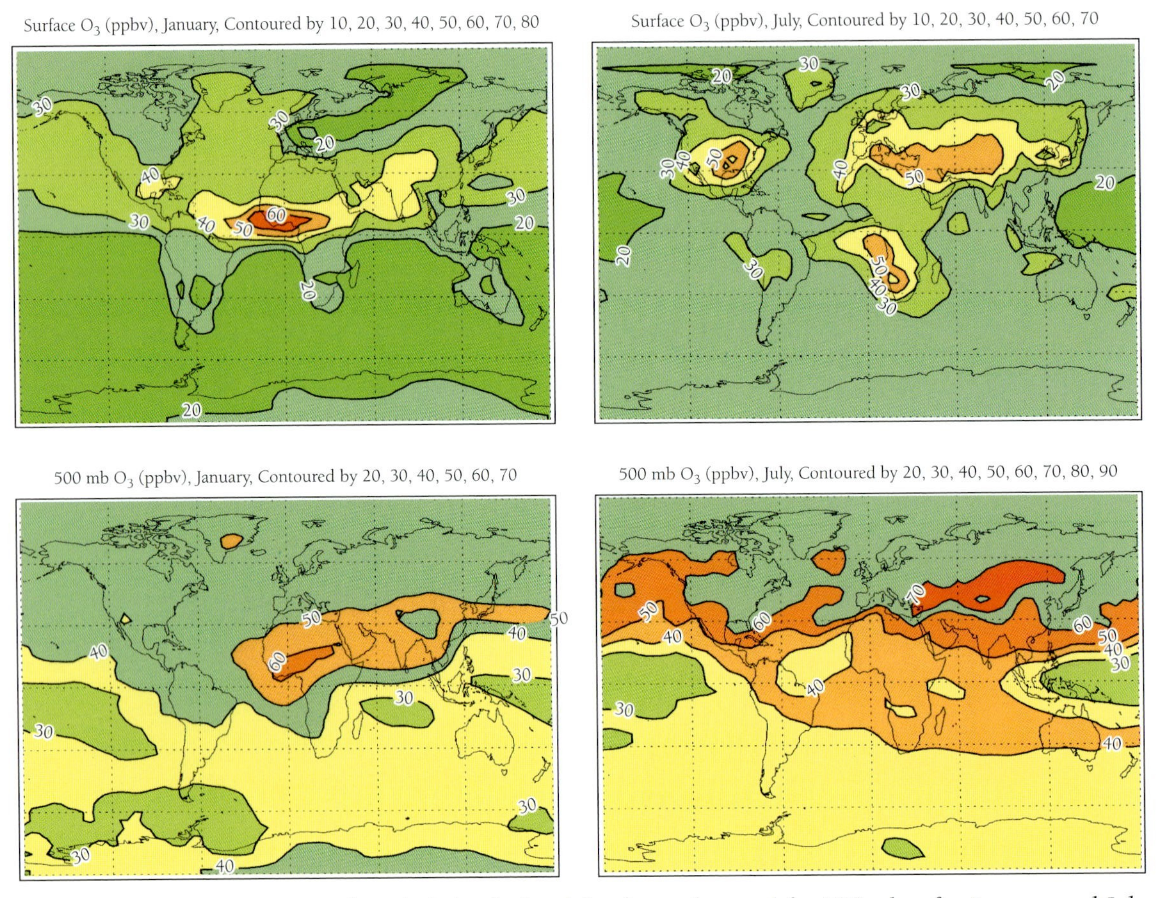

Monthly mean concentrations of O_3 (ppbv) calculated for the surface and for 500 mbar for January and July. Source: Wang et al. 1998.

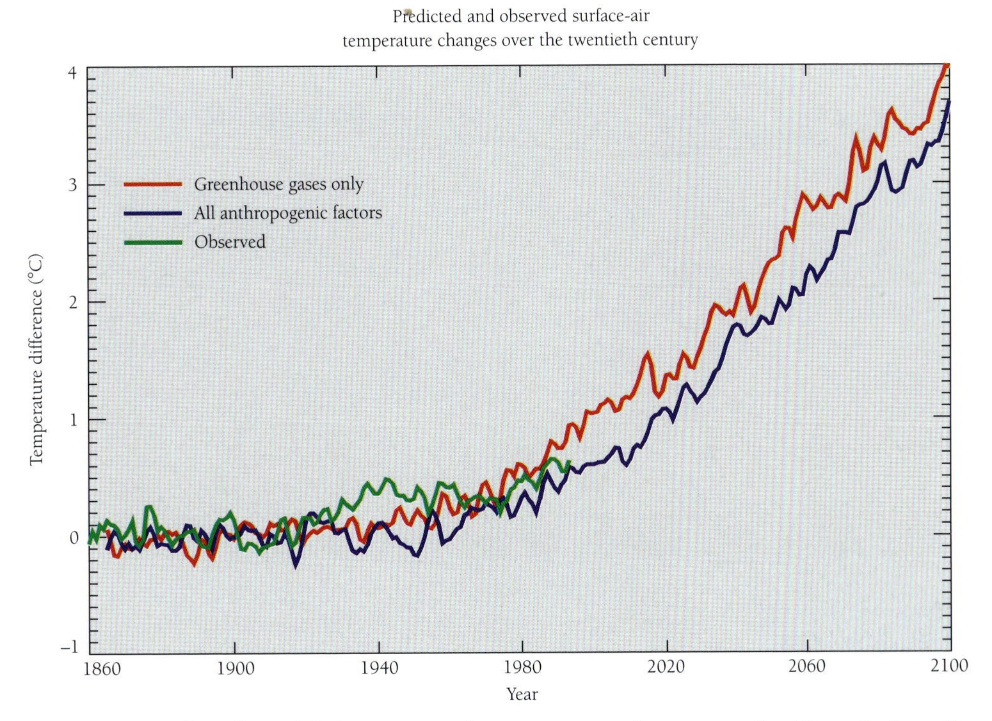

A comparison of trends in global average surface temperature (green curves) with results from the Hadley Centre model. Model results indicated in red indicate the impact of radiative forcing by greenhouse gases assuming growth according to the business-as-usual scenario to 2100. Results indicated in blue account both for greenhouse gases and sulfate aerosols. Source: Hadley Centre for Climate Prediction and Research.

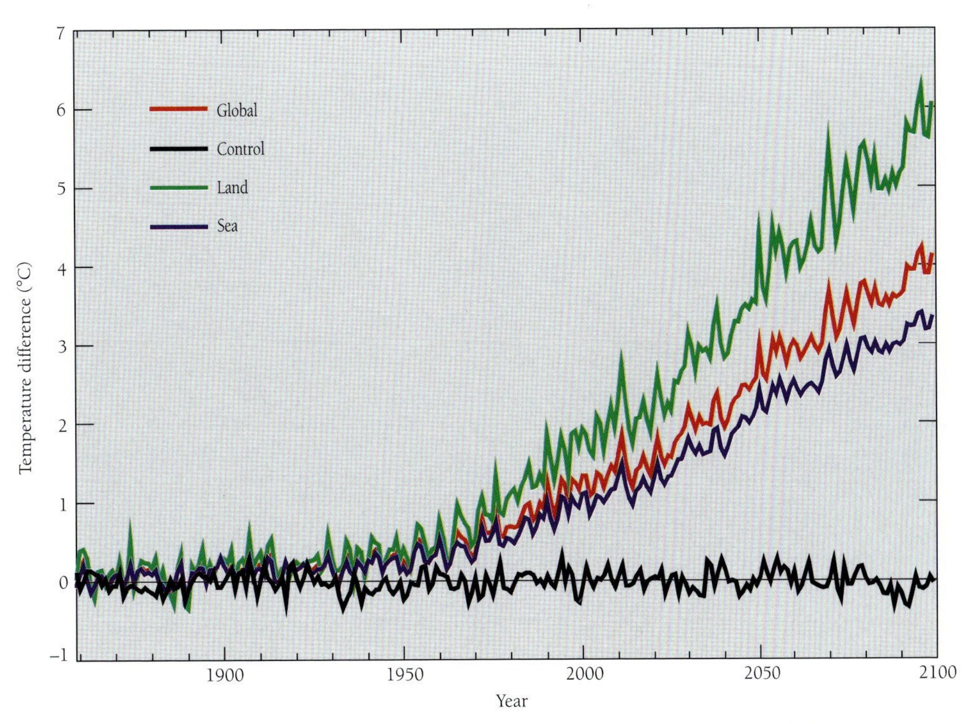

Predictions by the Hadley Centre model for changes in land (green), sea (blue) and global average surface temperatures (red), assuming growth in greenhouse gases according to the business-as-usual scenario. Results for the control run of the model are indicated in black. Source: Hadley Centre for Climate Prediction and Research.

Changes in surface temperature predicted for the 2050s using the Hadley Centre model assuming growth in greenhouse gases according to the business-as-usual scenario. Effects of sulfate aerosols were not included in the simulation. Source: Hadley Centre for Climate Prediction and Research.

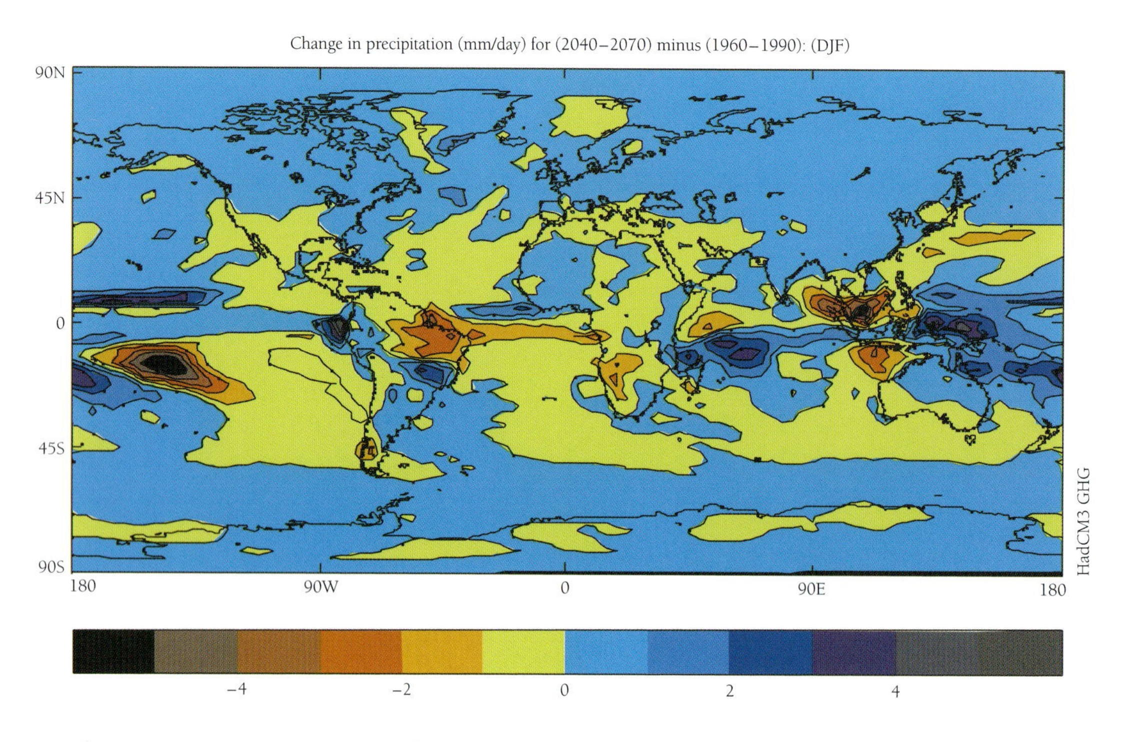

Changes in precipitation predicted for the period 2040–2070 relative to 1960–1990. Results based on the Hadley Centre model accounting for growth in greenhouse gases according to the business-as-usual scenario. Source: Hadley Centre for Climate Prediction and Research.

ulating emission of N_2O. Sampled systems ranged from the Peruvian upwelling system as noted above, to the open ocean and to selected rivers, including an intensive study of the Potomac River in the vicinity of Washington, D.C. Field observations were complemented by measurements of bacterially mediated transformations of nitrogen under pure-culture conditions in the laboratory. What developed from this work was a clear indication that significant quantities of N_2O are produced as by-products, not only of denitrification but also of nitrification. In retrospect, this may be viewed as not-too-surprising. After all, in terms of its oxidation state, N_2O stands as an intermediate between NH_4^+ and NO_2^-. We found that when oxygen levels are high, a predictable fraction of NH_4^+ consumed in the first stage of nitrification is converted to N_2O. When oxygen levels are low, yields of N_2O are higher. We concluded that nitrification provided the primary source of N_2O in the ocean and that sources associated with denitrification under anoxic conditions were probably small by comparison, in part reflecting the tendency for N_2O to be consumed in these environments. It follows that production of N_2O by the ocean should be linked rather directly to rates of primary productivity. It is likely that nitrification is also an important source of N_2O in terrestrial systems, though the evidence is more ambiguous in this case. Note that the first step in nitrification is associated with production of H^+. Both steps in the nitrification sequence involve consumption of O_2.

Photodissociation in the stratosphere provides the primary sink for atmospheric N_2O. The lifetime, set by a combination of photolysis and reaction with $O(^1D)$ (see Chapter 14), is estimated at 116 yr (McElroy and Jones 1996). The concentration of N_2O, globally averaged, was equal to about 307 ppb in 1988 (Figure 5.9), implying a sink of magnitude 12.5 Mt N yr^{-1}.[5] The concentration of N_2O is increasing in the current epoch at a rate of about 0.7 ppb yr^{-1} (IPCC 1996), equivalent to a growth in the atmospheric inventory of N-N_2O in 1988 of about 3.4 Mt N yr^{-1}. It follows that the global rate for production of N_2O in 1988 equaled about 15.9 Mt N yr^{-1}. Production exceeded loss of N_2O on a global basis in 1988 by about 27%.

The abundance of N_2O in the pre-industrial environment was about 280 ppb (see Figure 5.7). The global sink at that time averaged about 11.4 Mt N yr^{-1}.[5] If we assume that production and loss were in balance in the pre-industrial environment, it follows that the source of N_2O must have increased over the past several centuries by close to 40 %, from 11.4 Mt N yr^{-1} to 15.9 Mt N yr^{-1}. A summary of the budgets of atmospheric N_2O for both the pre-industrial and contemporary environments is presented in Table 12.3.

We assume as a working hypothesis that the pre-industrial budget provides a measure of the natural source of N_2O and that the increase in production in recent years may be attributed primarily to the influence of anthropogenic activity. Measurements of N_2O dissolved in surface ocean water, combined with estimates of rates for transfer of N_2O between the ocean and atmosphere, imply an ocean source for N_2O of about 4 Mt N yr^{-1} (Nevison, Weiss, and Erickson 1995). It follows that the natural component of the terrestrial source should be equal to about 7.4 Mt N yr^{-1} and that the anthropogenic source, presumably terrestrial, should correspond to about 4.5 Mt N yr^{-1}, implying an increase in the terrestrial source by about 60%, roughly comparable to the increase in fixation inferred above.

Table 12.3 Budgets of Atmospheric N_2O for the Pre-industrialized and Contemporary Environments

	Pre-industrialized	*Contemporary (1988)*
Concentration (ppb)	280	307
Inventory (Mt N)	1327	1455
Growth Rate (ppb yr^{-1})	0	0.7
Loss Rate (Mt N yr^{-1})	11.4	12.5
Production Rate (Mt N yr^{-1})	11.4	15.9
Production by ocean (Mt N yr^{-1})	4.0	4.0
Production by land (natural) (Mt N yr^{-1})	7.4	7.4
Production by land (anthropogenic) (Mt N yr^{-1})	0	4.5
Human and animal waste component (Mt N yr^{-1})	0	1.8
Agricultural component (Mt N yr^{-1})	0	2.7

Measurements of N_2O over wide regions of the ocean indicate that the concentration of N_2O in solution is strongly correlated with the deficiency of O_2 as measured by a quantity chemical oceanographers refer to as *apparent oxygen utilization* (AOU).[6] If we assume that consumption of O_2 reflects the combined effects of oxidation of organic carbon and nitrogen, and assuming further a ratio of C to N in typical marine organic material of 6.6 (see Chapter 11), the ocean measurements may be interpreted to indicate that one molecule of N_2O is produced for every 4700 molecules of carbon undergoing oxidation. This assumption provides an upper limit to the ocean source of N_2O associated with a particular rate of net primary productivity. If an appreciable fraction of reduced nitrogen is taken up by the biota before it is oxidized, we would be obliged to associate *a larger* rate of carbon oxidation with a given rate of production of N_2O.[7] A value for net primary production or organic carbon in the ocean equal to 20×10^9 tons C yr^{-1} (Elkins et al. 1978) would thus imply an upper limit to production of N_2O by the ocean of about 10 Mt N yr^{-1}. A source of 4 Mt N yr^{-1}, as assumed here, would require that either the rate for primary productivity is less than 20×10^9 tons C yr^{-1} or a significant fraction of the nutritional requirements of phytoplankton for nitrogen should be supplied by reduced forms of N, or both.

According to the budget presented in Figure 12.1, fixed nitrogen is supplied to terrestrial systems under natural conditions at a rate of about 200 Mt N yr^{-1}. The summary in Table 12.3 suggests that N_2O is produced under natural conditions in terrestrial environments at a rate of 7.4 Mt N yr^{-1}. These data imply that a little less than 4.0% of N added by fixation is converted to N_2O (actually 3.7% or 7.4/200, if the numbers are taken literally). The additional source of N_2O attributed to human activity was assessed at 4.5 Mt N yr^{-1} (Table 12.3). The anthropogenic source of fixed N was previously estimated at 113 Mt N yr^{-1} (representing contributions of 81, 20, and 12 Mt N yr^{-1} from chemical fertilizer, combustion of fossil fuels, and burning of biomass, respectively). Conversion of anthropogenically fixed nitrogen to N_2O with an efficiency equal to that inferred for natural environments (3.7%) would imply an anthropogenic source for N_2O of 4.2 Mt N yr^{-1}. Within the accuracy of the data assumed here, it would appear that the extra source of N_2O is simply attributable to acceleration of the nitrogen cycle in terrestrial systems driven by a higher, human-induced rate of application of fixed nitrogen.

The additional source of N_2O is most likely associated with processes involved in the human food chain, with transformations of N in agricultural systems and with the disposal of N included in human and animal waste. In contrast to the case of CO_2, combustion of fossil fuels makes a negligible contribution to the budget of N_2O, about 0.2 Mt N yr^{-1} according to the IPCC (1990). Human populations in the developed world consume and excrete N at an annual average per capita rate of about 5.4×10^3 g N yr^{-1} (National Academy of Sciences 1972). Adopting a somewhat lower value for the world as a whole, 4×10^3 g N yr^{-1}, but multiplying this value by a factor of four to allow for N involved in the animal part of the food chain, we estimate that human and animal waste is responsible for a contemporary global source of N of magnitude 90 Mt N yr^{-1}, roughly comparable to the quantity of N applied to agricultural systems as chemical fertilizer.[8]

Laboratory studies and field observations indicate a yield for N as N_2O produced by nitrification equal to about 0.3%, when the concentration of O_2 is close to equilibrium with the atmosphere (Elkins et al. 1978; Goreau et al. 1980). Yields of this magnitude were observed in 1978 under high-flow, high-oxygen conditions in a region of the Potomac River receiving large inputs of N as NH_4^+ from the sewage treatment plant at Blue Plains serving Washington, D.C. Yields in 1977, when flow rates were lower and when levels of O_2 were suppressed, were significantly higher, approaching 5% (Elkins et al. 1980). Higher yields in 1977 were attributed to an enhanced production of N_2O at low levels of O_2. We tentatively assign an average value of 2% to the yield of N_2O associated with global disposal of human and animal waste. The choice is somewhat arbitrary. We choose a value toward the high end of the range of yields observed for the Potomac, on grounds that O_2 concentrations are likely to be low in environments receiving large inputs of animal waste and that they are likely to also be depressed under conditions applying to disposal of human waste in the developing world. A yield of 2% would imply a source of N_2O from disposal of human and animal waste globally of 1.8 Mt N yr^{-1}. We caution that the actual value could be either higher or lower. There is a need for additional observations with a particular emphasis on the fate of nitrogen in animal waste and on the yield of N_2O associated with disposal of human waste in the developing world.

A source of 1.8 Mt N yr^{-1} of N_2O from waste disposal would imply a contribution of 2.7 Mt N yr^{-1} from other forms of human activity to account for the anthropogenic component of the N_2O budget given in Table 12.3. It is likely that a major fraction of this extra source of N_2O is associated with agricultural systems receiving large inputs of nitrogen-based fertilizer. The IPCC (1990) quotes a range of values for emission of N_2O from cultivated soils of 1.8 to 5.3 Mt N yr^{-1}, recommending a value of 3.5 Mt N yr^{-1}, slightly higher than the estimate adopted here.[9]

12.2 The Phosphorus Cycle

Phosphorus is transferred from land to ocean primarily as a component of river water. Ultimately, the phosphorus carried by rivers is supplied by the weathering of continental rocks. Estimates of preagricultural weathering, combined with measurements of the average P content of continental rocks, suggest a preagricultural flux of P from land to ocean of about 10 Mt P yr^{-1}. Human activity has been responsible for a significant acceleration of continental weathering in recent years. In addition, mining of high-content phosphate rocks has added to the phosphorus content of rivers, especially to those draining intensively cultivated agricultural lands.[10] The quantity of P carried by rivers today may exceed the long-term average by as much as a factor of three (Froelich et al. 1982). The key question concerns the fraction of this P that is available as a nutrient for uptake by the biota.

It appears that much of the phosphorus carried by rivers is biologically inaccessible: it is present as a component of refractory minerals incorporated without modification in estuarine, coastal, and marine sediments. For the flux of biologically available (reactive) P carried to the ocean by rivers in the preagricultural environment, Froelich et al. (1982) estimate a magnitude of 1.2 Mt P yr^{-1}, with an uncertainty of about ±50%. The fraction of reactive P directly delivered to the ocean, as compared to the portion incorporated in estuarine and coastal sediments, is unclear. Ultimately, though, it would seem reasonable to equate the flux of reactive P carried by rivers with the long-term average rate at which biologically useful P is supplied to the ocean.[11]

The dominant sinks for marine P involve (a) burial in sediments as a component of organic matter and (b) incorporation in sediments as a component of biogenically produced calcium carbonate. As discussed by Froelich et al.

(1982), the concentration of organic P is remarkably constant for a variety of sediments in the ocean, averaging about 100 ppm, apparently independent of the abundance of organic C. If carbon and phosphorus were included in organic material in sediments in proportion to Redfield ratios (see Chapter 11), we would expect a ratio of P_{org} to C_{org} of 1/106 (9.4×10^{-3}). Data from a variety of ocean locations indicate that the abundance of P in organic-rich sediments (environments where the abundance of organic C is higher than about 0.5%) is, relative to C, less than would be expected based on Redfield ratios, in some cases by as much as a factor of five. In contrast, the abundance of P is enriched, by as much as a factor of two, in organic-poor sediments. The factors responsible for this behavior are not well understood. The high abundance of organic P in organic-poor sediments could reflect differential retention of P-rich organic moieties when diagenesis proceeds under oxic conditions. The low abundance of P in organic-rich sediments, on the other hand, could reflect preferential loss of P (retention of C) when diagenesis proceeds under anoxic conditions. Froelich et al. (1982) estimate a removal rate of P in organic form from the global ocean equal to 0.48 Mt P yr^{-1}.

The abundance of P in $CaCO_3$ averages about 300 ppm. Froelich et al. (1982) summarize evidence indicating that the abundance of P in coccoliths (plants) is higher than in foraminifera (animals): 400 ppm as compared to about 50 ppm. The P content of calcite in sediments may be expected therefore, to vary, in proportion to the relative abundances of calcite supplied to sediments by coccoliths and foraminifera. The globally averaged content of P in carbonates (300 ppm) implies that coccoliths globally account for about 70% of sedimentary carbonate with the balance provided by foraminifera. Accounting for dissolution of carbonates below the lysocline, Froelich et al. (1982) calculated a global rate for removal of P in carbonates equal to 0.48 Mt P yr^{-1}, essentially the same as the rate for removal in organic matter.

There are three possible fates for organic P in sediments: it may be retained in organic form, it can be converted to phosphate and diffused through the sediment-pore water back to the overlying water column, or it can be converted to phosphate and subsequently precipitated as a phosphate salt. When the salt formed in this fashion is carbonate fluoropatite, it is referred to as **phosphorite**. The geologic record includes a number of giant deposits of phosphorite. It is clear that, at times in the past, phosphorite must have provided a major sink for marine phosphorus. Its role in the current environment appears to be more modest. Froelich et al. (1982) conclude that phosphorite accounts for less than 10% of the total contemporary loss of P from the ocean.

Chemical reactions associated with hydrothermal activity at oceanic spreading centers provide another potential sink for ocean P, as suggested first by Berner (1973). Hot water emanating from vents near ocean ridge crests includes high concentrations of reduced iron, which is oxidized on contact with sea water, resulting in precipitation of ferric hydroxide. Phosphate is scavenged from sea water in the process, leading to formation of metal-rich material that is added to the sediment. The P content of this material can be as high as 0.6% P. Froelich et al. (1982) estimate the sink for ocean P associated with this process at 0.13 Mt P yr^{-1}. Key elements of the phosphorus cycle are summarized in Table 12.4.

12.3 The Sulfur Cycle

The biogeochemical cycle of carbon was previously discussed in Chapter 12. The processes regulating supply of N and P were outlined in Sections 1 and 2 of the present Chapter. We turn our attention now to mechanisms regulating the availability of another of the chemical elements essential for life, sulfur.

We have seen that biological processes in the ocean play an important role in the cycles of C, N, and P. All three elements are removed from the ocean as they are incorporated as components of both organic and inorganic molecules in sediments. In the absence of a restorative process, supplies of C, N, and P essential for life would be depleted on relatively short time scales (compared with the tenure of life on the planet). The ocean's store of C would be eliminated on a time scale of about 10^5 years. The time for removal of P is comparable to that for C (8×10^4 yr according to the data in Table 12.4) while the lifetime of N is somewhat longer, about 8×10^5 yr.[12] As we shall see, sediments provide an important sink not only for C, N, and P but also for S.

Fortunately, sediments represent but a transitory sink for the life-essential elements C, N, P, and S. Material accumulating in sediments is laterally transported as a consequence of the motion of the giant plates that form Earth's crust. At plate boundaries, sedimentary material may be either uplifted or withdrawn into Earth. Uplifted material may be eroded and transferred from sediments to the atmosphere-ocean-biosphere-soil system initiating a further loop through the appropriate geochemical cycle; withdrawn material may be raised to high temperature. Volcanoes and hot

Table 12.4 Key Components of the Global Phosphorus Cycle (after Froelich et al. 1982).

Flux of reactive P carried by rivers from land to the ocean	1.2 Mt P yr^{-1}
Ocean P content	9.6×10^4 Mt P
Flux of P removed from ocean to sediments in organic form	0.48 Mt P yr^{-1}
Flux of P removed as component of carbonates	0.48 Mt P yr^{-1}
Flux of P out of the ocean including removal of P in organics, carbonates, losses at hydrothermal vents, and other minor-loss mechanisms	1.2 Mt P yr^{-1}

springs provide a significant route for return of material processed in this way. Volcanoes and hot springs are especially important in recycling C, N, and S.

A schematic representation of the sulfur cycle is presented in Figure 12.2. According to the data displayed here, sulfur is transferred from the ocean to sediments at a rate of 90 Mt S yr^{-1}, implying a lifetime for oceanic S of about 1.1×10^7 yr. Sulfur is moved from sediments to the atmosphere at a rate of 20 Mt S yr^{-1}, mainly as SO_2, as a consequence of emissions associated with volcanoes and hot springs. It is transferred to continental soils following the uplift of sediments and subsequently incorporated in the terrestrial biosphere at a rate estimated as about 70 Mt S yr^{-1}. Microbially mediated decay of organic material in soils and water bodies is responsible for emission of gaseous, chemically reduced forms of sulfur, notably H_2S and COS, providing an additional path for return of S into the atmosphere. The corresponding flux in this case is indicated at 30 Mt S yr^{-1}. Runoff from land delivers S to the ocean at a rate of 80 Mt yr^{-1}. Sulfur is transferred from the ocean into the atmosphere, mainly by emission of dimethyl sulfide (DMS). The rate in this case is given as 40 Mt S yr^{-1}. Reduced forms of S are oxidized in the atmosphere where they are ultimately converted to sulfuric acid (H_2SO_4), contributing to the phenomenon of acid rain, as we will discuss in more detail in Chapter 18. According to the data presented in Figure 12.2, the atmosphere contributes about 40 Mt S yr^{-1} to terrestrial systems, with a somewhat larger quantity of sulfur (50 Mt S yr^{-1}) supplied to the ocean.

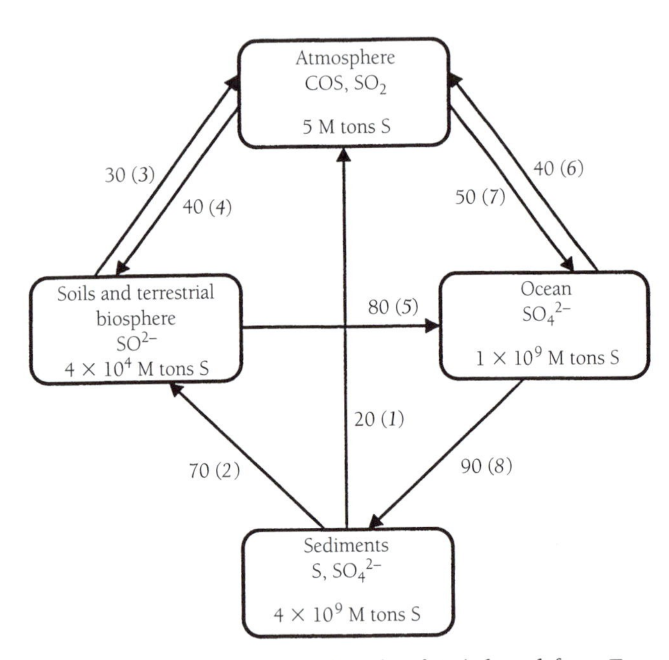

Figure 12.2 The geochemical cycle of S (adapted from Turco 1997). Reservoir contents are given in units of MT S (10^6 tons S). Exchange rates are in units of MT S yr^{-1}. Specific exchanges are as follows: (1) transfer of S from sediments to the atmosphere (mainly emissions associated with volcanoes and hot springs), (2) uplift of sediments, (3) emission from soils and the biosphere, (4) transfer from the atmosphere to soils and the biosphere (mainly as sulfate), (5) run-off in rivers to the ocean, (6) emissions from the ocean to the atmosphere (mainly as DMS), (7) return from the atmosphere to the ocean (mainly as SO_4^{2-}), (8) transfer from the ocean to the sediments.

Oxidation of chemically reduced sulfur in the atmosphere is initiated primarily by reaction with OH (for a more detailed discussion of the chemistry of OH, see Chapter 17). Lifetimes of reduced-sulfur species are generally brief (a few days, for example, in the case of H_2S), reflecting the fact that rates for relevant reactions with OH are typically rapid; removal of COS is an exception. The rate for reaction of COS with OH is comparatively slow: the lifetime for COS is relatively long as a consequence, about a year. Reflecting its relatively long lifetime, COS is distributed ubiquitously throughout the atmosphere: with a mixing ratio of about 0.5 ppb, it is the most abundant sulfur-bearing component of the atmosphere. Oxidation of COS provides a spatially distributed source of atmospheric SO_2 and consequently of H_2SO_4. Paths for oxidation of SO_2 to H_2SO_4 are discussed in Chapter 18.

For the most part, the sulfur emitted by volcanoes and hot springs is released into the lower atmosphere. Represented mainly by SO_2, it is rapidly converted in the atmosphere to H_2SO_4 (on a time scale of a day or less). It is removed by a combination of dry and wet deposition, providing a relatively localized source of sulfur for soils and for the biosphere in the region close to the source. Occasionally, however, the emission of S is associated with explosive, extremely energetic, eruptions in which volcanic debris is lofted to altitudes as high as 20 km, or more. The sulfur contributed by these volcanic eruptions is oxidized to H_2SO_4, which is converted to particulate form in the stratosphere, leading to production of large numbers of small sulfate particles. These particles are eventually widely dispersed throughout the stratosphere as a result of atmospheric motions. They are effective in scattering sunlight, especially at short wavelength. Their presence results in the development of spectacular purple-colored sunrises and sunsets observed at a variety of locations around the world for as long as several years following major eruptions. Absorption of sunlight by sulfate aerosols can lead to a significant increase in stratospheric temperature. They are also responsible for an increase in the planetary albedo, resulting in a small but detectable decrease in surface temperature. Sulfate aerosols can play a further role in stratospheric chemistry. Reactions on the surfaces of these particles can affect the abundance of key chemical components of the stratosphere, with the potential to alter the abundance and spatial distribution of O_3 (see Chapter 15).

The eruption of Mount Pinatubo in the Philippines in June 1991 is thought to have been responsible for a decrease in globally averaged surface temperature in 1992 of 0.3–.5°C, effectively canceling the rise attributed to the increase in the concentration of greenhouse gases over the past century (IPCC 1996). The importance of a volcanic eruption for climate depends most critically on the quantity of sulfur delivered to the stratosphere. From the point of view of its impact on climate, Mount Pinatubo was perhaps the most significant volcanic event of the century: the associated source of stratospheric sulfur is estimated at about 10 Mt S. It pales in significance, however, compared with the catastrophic eruption of Tambora in Indonesia in 1815: it is estimated that Tambora may have accounted for a source of stratospheric sulfur as much as 10 times larger than that from Mount Pinatubo. The influence of volcanic aerosols on climate and stratospheric chemistry is temporary, however. We expect the source of sulfur contributed to the stratosphere by a particular volcanic eruption to be depleted on a time scale of a few years as a consequence of the exchange of air between the stratosphere and troposphere. Effects of Mount Pinatubo are no longer significant: surface temperatures appear to have resumed the long-term trend toward warming that was apparent in the record prior to 1991.

Human activity has an important influence on the contemporary geochemical cycle of sulfur; mining and the combustion of fossil fuels are accelerating the rate at which sulfur is transferred from the sedimentary reservoir to the atmosphere, by as much as a factor of two at the present time. The source of S (mainly as SO_2) associated with fossil fuel combustion in the United States amounts to about 15 Mt S yr^{-1} today. On a global basis, fossil fuels, mainly coal, contribute a source of close to 100 Mt yr^{-1} of S to the atmosphere as SO_2. Application of sulfur-based fertilizers to agricultural systems is responsible for the mobilization of an additional 30 Mt S yr^{-1}. Combined with wastes from various industrial activities, fertilizers are responsible for a significant increase in the concentration of S in rivers draining agricultural and industrial regions, by close to a factor of two. It is difficult to quantify the environmental impact of this additional burden of S carried by rivers and streams into estuaries and coastal zones. There is little doubt, however, that concentrated emissions of SO_2 associated with fossil fuel combustion have an important negative impact on the health of people living in proximity to these emissions and that they are a major contributor to the phenomenon of acid rain, as we will further elaborate upon in Chapter 18.

12.4 Summary

Human activity has an important, in some cases a dominant, influence on the biogeochemical cycles responsible for the mobilization of N, P, and S and for the supply of these elements to important components of Earth's near-surface environment. We reached a similar conclusion previously in our discussion of C. For C, P, and S, the disturbance introduced by man reflects an accelerated transfer rate of these elements from sediments, either directly to the atmosphere or, in the case of P, to soils and the biosphere. In the case of N, the disturbance also involves N fixation by industrial processing of N_2 in the atmosphere, in conjunction with either the manufacture of N-based fertilizer or the combustion of fossil fuels.

The rise in N_2O concentration provides dramatic evidence for the influence of human activity on the nitrogen cycle. We suggested that the increase in N_2O is most likely due to the efforts expended to produce a food supply adequate to satisfy the needs of an ever expanding world population and with processes implicated in disposal of relevant organic waste. We discussed the microbially mediated reactions by which nitrogen is transformed through a suite of oxidation states, ranging from the reduced forms characteristic of organic nitrogen and NH_3, for example, to fully oxidized forms, such as NO_3^-. We noted that N_2O may be produced as an intermediate in the oxidation of nitrogen, by nitrification, or as a by-product of nitrogen reduction by denitrification. We suggested that agriculture, stimulated by ever increasing applications of nitrogen-based fertilizers, is likely to provide an important source of N_2O. Additional production was attributed to microbially mediated reactions associated with the disposal of human and animal waste.

The productivity of modern agriculture, and ultimately our ability to keep up with the ever growing demand for food, is due in no small measure to increases in productivity stimulated by applications of fertilizer, not only nitrogen but also phosphorus, sulfur, and other essential elements, such as potassium. Runoff of nutrients from agricultural systems can stimulate biological productivity in environments other than the agricultural systems to which the nutrients are originally applied—rivers and lakes, for example—with implications not only for carbon but also for nitrogen. We pointed out that excess quantities of P entering aquatic systems have the potential to change the ecology of these systems. By giving rise to blooms of blue-green algae, they can enhance the source of fixed nitrogen with implications not only for storage of carbon but also for production of important gaseous species, such as CH_4 and N_2O.

Combustion of fossil fuels is a significant source not only of nitrogen oxides but also of sulfur oxides, notably SO_2. Nitrogen oxides play an important role in the production of tropospheric O_3, as we will discuss in Chapter 17. Oxidation of nitrogen and sulfur oxides emitted to the atmosphere as by-products of fossil fuel combustion results in the production of nitric acid and sulfuric acid, the major ingredients of acid rain (see Chapter 18).

Stratospheric Ozone (I): Chapman's Formulation for the Chemistry of an Oxygen-Nitrogen Atmosphere

13

In terms of its abundance, **ozone**, O_3, is a minor constituent of the atmosphere. It accounts for less than 1 ppm (part per million by volume) of the total gaseous composition of the air. Most of the atmosphere's budget of O_3 is contained in the stratosphere, more than 15 km above Earth's surface. Despite its relatively low concentration, O_3 plays a critical role in regulating the environment for life on Earth. Its importance relates to its function in shielding the planet's surface from what would otherwise constitute a lethal dose of ultraviolet solar radiation (see 13.1). Radiation with wavelengths below 300 nm has the potential to fragment molecules of DNA, the essential constituent of all living tissue. In the absence of stratospheric O_3, Earth's surface would be exposed to solar radiation with wavelengths as short as 240 nm. It is generally believed that the synthesis of O_3 in the atmosphere was a prerequisite for the evolution of life at Earth's surface.

The past thirty years have witnessed a remarkable growth of interest in the chemistry of stratospheric O_3. Prior to 1971, studies of stratospheric O_3 were considered rather esoteric, confined to a relatively small community of atmospheric chemists and meteorologists. The transformation occurred as a consequence of concern that diverse forms of human activity could cause a reduction in Earth's protective shield of O_3, with an associated increase in the intensity of ultraviolet solar radiation reaching the planetary surface. The alarm was initially sounded in 1971 by a Californian chemist, Harold Johnston, and by a Dutch meteorologist, Paul Crutzen, in the context of a study of possible environmental consequences of a fleet of commercial supersonic transports (SSTs) under consideration by the United States government during the 1960s and early 1970s. The planes, as proposed, would fly mainly in the stratosphere, where the air is exceptionally stable (see Section 6.4) and where exhaust gases could persist for several years before reaching the troposphere, where they would be removed by precipitation and by deposition at the surface. Johnston and Crutzen suggested that **nitric oxide**, NO, emitted as a component of aircraft exhaust gas could cause an increase in the efficiency with which O_3 is removed from the stratosphere, resulting in a decrease in its abundance.

We can think of the controls on O_3 as loosely analogous to the situation illustrated in Figure 13.1, where water flows into a leaky barrel at a constant rate. The water rises to a level such that the rate at which it enters the barrel equals the rate at which it leaves. Imagine that an additional hole is punched in the side of the barrel. The water level sinks in this case

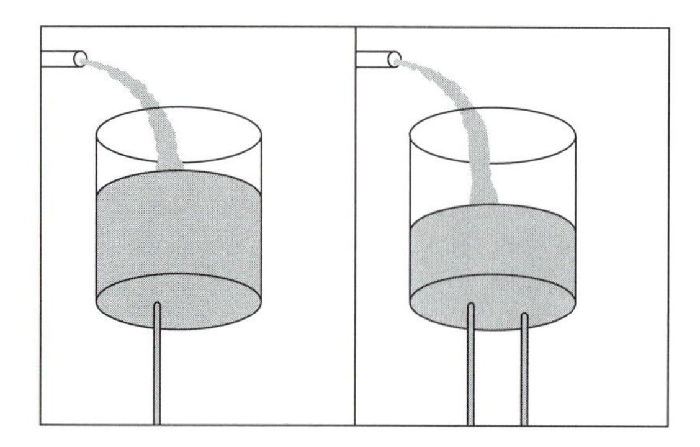

Figure 13.1 Illustrating the change in water level when another hole is punched in the barrel.

to find a new equilibrium, where again inflow and outflow are in balance; the water content of the barrel is less under this condition. The barrel in the analogy is the stratosphere: the water is ozone. Ozone is produced, as we shall see, at a constant rate by sunlight breaking apart molecular oxygen. Release of nitric oxide from the SST, according to Johnston and Crutzen, would be responsible for an additional leak in the barrel resulting in a decrease in the abundance of O_3.

Models of stratospheric O_3 in 1971 suggested that a fleet of 500 SSTs flying eight hours a day in the stratosphere, as proposed by the Boeing Company, could cause reductions in O_3 of a few percentage points. The expected change in O_3 might appear modest at first blush; public consciousness was raised, however, by the possibility that even small changes in O_3 could trigger significant increases in the incidence rate of human skin cancer. The relation between changes in O_3 and increases in skin cancer was first recognized by James McDonald, a meteorologist at the University of Arizona. McDonald was a controversial figure in 1971: he had achieved a level of notoriety in the 1960s as a result of work he had conducted on unidentified flying objects (UFOs). McDonald maintained that many of the official explanations advanced by the United States Air Force to account for specific sightings of UFOs were flawed. Despite this peculiar twist to his career, McDonald's credentials as a scientist were impeccable: earlier, he had made important contributions to our understanding of the physics of clouds. His achievement in defining a quantitative connection between changes in O_3 and the incidence rate of skin cancer (deduced from consideration of the dependence of these quantities on latitude) was insightful and normally would have earned McDonald kudos for an important contribution to environmental health. Unfortunately, although his ideas were extensively presented in the early 1970s at meetings concerned with environmental effects of supersonic aircraft, the record of his contribution is rarely acknowledged. (It is an inviolable, and indeed defensible, rule that credit for scientific discoveries must be earned not by talks orally presented at meetings but by papers published in peer-reviewed scientific journals.) McDonald died in 1972, without committing his ideas to writing. His last years were troubled: he found himself on the receiving end of bitter, often personal, attacks from advocates of the SST program. His foes did not hesitate to allude to his earlier involvement with UFOs; the extent to which the emotional stress imposed by the causes he espoused contributed to McDonald's untimely end (by suicide) is unclear. There is little doubt, though, that, together with Harold Johnston and Paul Crutzen, he played an important role in the decision of the United States government to suspend funding for the SST in 1972.

The SST issue had an important influence on the development of stratospheric science. It resulted in a major research program in the United States, funded by the Department of Transportation, with a complementary effort in Europe. The Europeans had an independent interest in the environmental impact of supersonic aviation, since the Concorde had just been introduced and since the Anglo-French consortium responsible for its development had ambitious plans (never realized) to market as many as several thousand of the aircraft. The SST issue spawned a new community of stratospheric scientists and was responsible for the development of a variety of innovative new techniques for the measurement of stratospheric gases using aircraft, balloons, rockets, and satellites, in addition to ground-based remote sensing. It led to an improved understanding of the physical, chemical, and biological processes responsible for control of O_3 under natural conditions; indirectly, it set the stage for the discovery of a new and even more serious threat to O_3, identified in 1974 by two chemists from the University of California at Irvine: Mario Molina and Sherwood Rowland.

The problem in this case involved a class of industrial chemicals known as **chlorofluorocarbons**, or CFCs. Developed in the 1930s, these compounds are comparatively inert and had long been considered environmentally benign. Molina and Rowland pointed out that CFCs would rise into the stratosphere, where they would decompose, providing a source of chlorine atoms. Chlorine atoms are even more effective than NO as agents for destruction of O_3; they punch an even larger hole in the metaphorical barrel. The CFC issue has dominated stratospheric research since 1974. The contributions of Paul Crutzen, Mario Molina, and Sherwood Rowland to our understanding of the impact of human activity on stratospheric ozone were recognized in 1995 with the award of the Nobel Prize in Chemistry.

We begin our discussion of the stratosphere here with a treatment of the photochemistry of an atmosphere composed solely of O_2 and N_2; the formulation is relatively straightforward and "classic" in the sense that it was developed more than 50 years ago by the British mathematician-physicist Sydney Chapman. As we shall see, Chapman's formulation involved only four reactions. Despite its simplicity, it provides a useful first-order description of the chemistry of stratospheric ozone, and it allows us to introduce chemical principles to be elaborated more fully in subsequent chapters: we will develop the importance of gas-phase reactions involving nitrogen, hydrogen, chlorine, and bromine radicals later in Chapter 14; we will elaborate the role of heterogeneous processes, reactions on particle surfaces, and their particular significance for the polar environment in Chapter 15; and we will present an integrated view of the combined influence of chemistry and dynamics in regulating both the abundance and latitudinal distribution of ozone in Chapter 16.

We begin by introducing the Chapman reactions in Section 13.1. The treatment of reactions triggered by absorption of sunlight is the subject of Section 13.2. We show in Section 13.3 how the Chapman reactions can be distinguished instructively using the concept of *odd oxygen*, defined as the combination of O and O_3. We introduce the notion of a chemical lifetime and illustrate by numerical examples how one would proceed to calculate abundances of O and O_3 as a function of altitude in the stratosphere. The casual reader may find the emphasis on algebraic manipulations in Sections 13.2 and 13.3 somewhat tedious: such manipulations are required, however, to obtain analytic expressions for the lifetimes and concentrations of the key species of the stratosphere (O, O_3, and odd oxygen). Insights inferred on the basis of the analysis developed here will prove useful in interpreting the more complex reaction schemes developed in subsequent chapters.

13.1 Chapman's Treatment for the Photochemistry of an Oxygen–Nitrogen Atmosphere

Our approach here will emphasize results appropriate for the limiting case of an atmosphere in **photochemical equilibrium**. That is to say, we will assume that, for any particular location and time, the concentration of photochemically active species, such as O and O_3, is set primarily by a locally applied balance between rates at which the species are produced and removed by photochemical reactions. Using the earlier analogy, we want to specify both the input of water to the barrel and the size of the hole and in this fashion determine the equilibrium depth of the water. We suppose that chemical species are born, live, and die in more or less the same place. We hope that we can ignore complications introduced by a lifetime of travel, by the influence of winds transporting species formed in one chemical regime to another, where the conditions of removal may be different.

The assumption of photochemical equilibrium leads to useful simplifications: the concentration of a chemical constituent is uniquely determined by specifying location and time, and we avoid the need to consider the complex history of an air parcel as it wanders around the world exposed to greater or lesser intensities of sunlight and chemical reactivity. Photochemical equilibrium should be particularly valid for species living for a relatively short time; the shorter the life of a species, the less likely it is that it will have time to venture into a significantly different chemical regime.

Rates for production and loss of photochemically active species depend on the intensity of sunlight and may be expected to vary with altitude, latitude, and time. They may significantly change over the course of a day, depending on the zenith angle of the Sun (as illustrated in Figure 13.2). If the Sun is high in the sky (small zenith angle), the path for solar radiation through the atmosphere is relatively short. Absorption of light, as solar rays pass through the atmosphere to a particular altitude, is at a minimum, and the intensity of sunlight reaching a given altitude is at a maximum, when the Sun is at its zenith.

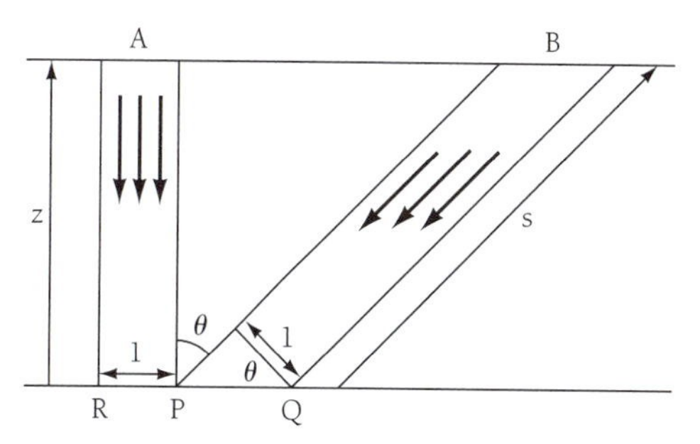

Figure 13.2 A schematic illustration of the fact that a larger number of molecules are intercepted by sunlight incident from a slanting direction. The solar zenith angle is given by θ. Case A illustrates radiation incident from the zenith; case B depicts radiation incident with zenith angle θ. Height is indicated by z; distance along the slant path by s.

As we shall see, O_3 is both formed and removed by reactions involving O. Variations in the abundance of O as a function of time and position are relatively large, reflecting the comparatively rapid response of O to changes in the flux of incident sunlight at altitudes of interest (the number of photons in relevant portions of the spectrum-crossing unit area that is perpendicular to the direction of light propagation in unit time, photons cm^{-2} sec^{-1}). Variations in the abundance of O_3 are more modest, given that its lifetime is typically measured in days or weeks, in contrast to minutes or hours for O. We choose here to ignore the relatively small changes in O_3 occurring over the course of a 24-hour period. Production and loss of O_3 can be described with adequate accuracy in terms of the diurnally averaged concentration of O. Since the chemistry of O_3, rather than O, provides the primary focus for this chapter, we choose to emphasize here a 24-hour, diurnal-average picture of stratospheric chemistry.

Production of O_3 is initiated when sunlight at ultraviolet wavelengths is absorbed by molecular oxygen, O_2. Radiation with wavelengths below 240 nm has sufficient energy to dissociate O_2, to fragment the molecule into its component oxygen atoms. Ozone is formed when the oxygen atom produced by this process is attached to O_2. Ozone is removed by either the absorption of sunlight (breaking the molecule apart into its components O and O_2) or reaction with an oxygen atom (forming two molecules of O_2). The chemistry of O_3 is comparatively simple for a hypothetical atmosphere formed solely of oxygen and molecular nitrogen. There are four important reactions in this case, as first noted by Chapman.

The Chapman reactions may be summarized as follows:

$$h\nu + O_2 \rightarrow O + O \tag{13.1}$$

$$O + O_2 + M \rightarrow O_3 + M \tag{13.2}$$

$$h\nu + O_3 \rightarrow O + O_2 \tag{13.3}$$

$$O + O_3 \rightarrow O_2 + O_2 \tag{13.4}$$

The short-hand notation employed in reactions (13.1)–(13.4) may be translated into everyday language as follows: (13.1) indicates a process in which a photon (hv) is absorbed by molecular oxygen (O_2), causing the molecule to fragment into its component atoms; (13.2) symbolizes a reaction in which an oxygen atom combines with a molecule of O_2 in the presence of an inert collision partner denoted by M (see below), resulting in the production of O_3; (13.3) denotes absorption of a photon by O_3, leading to dissociation (fragmentation) of O_3 with associated production of O and O_2; (13.4) indicates a reaction between O and O_3 in which the initial O atom captures an O atom from O_3, resulting in formation of two molecules of O_2.

Reactions (13.1) and (13.3) provide examples of what are known as **photolytic processes** (see Section 13.2). The rate of a photolytic reaction is defined as the number of reactions taking place in unit volume in unit time (with *cgs* units, $cm^{-3}\ sec^{-1}$). It is proportional to the intensity of radiation responsible for the reaction; it also depends on the **cross section** for absorption of the relevant radiation and on the concentration of species (targets) involved in the reaction.[1] This follows as a matter of common sense: the larger the number of photons available to carry out the reaction, the greater the cross section, and the higher the concentration of molecules with which photons can react, the larger the number of reactions we expect to take place.

The rate of reaction (13.1), call it $R_{13.1}$, may be written in the form

$$R_{13.1} = J_{13.1}\,[O_2], \qquad (13.5)$$

where $[O_2]$ denotes the concentration of O_2. In a similar manner, the rate for reaction (13.3) is given by

$$R_{13.3} = J_{13.3}\,[O_3] \qquad (13.6)$$

The quantities $J_{13.1}$ and $J_{13.3}$ account for the dependence of $R_{13.1}$ and $R_{13.3}$ on the intensity of the radiation field and on the magnitude of the relevant cross sections. With concentrations expressed in units of cm^{-3} and reaction rates given in units of $cm^{-3}\ sec^{-1}$ as appropriate for the *cgs* system, the values of $J_{13.1}$ and $J_{13.3}$ in (13.5) and (13.6) have units of sec^{-1}.

Reaction (13.2) provides an example of what is known as a **3-body reaction**. Formation of O_3 from O and O_2 involves a transition to a lower energy state. Just as energy was required to dissociate O_3 in (13.3), energy is liberated as O_3 is formed from O and O_2. It is similar to what happens if you climb a cliff; energy is expended to reach the top, but if you fall off, most of it can be recovered when you hit the ground below. The role of the third body, represented by M in (13.2), is to carry off part of the excess energy. We can think of the reaction as proceeding through a series of steps. First, the O atom and the O_2 molecule come together to form a loosely associated O-O_2 complex. The complex is unstable: it has too much energy. It can accommodate the excess energy for a while with internal motion (rotation and vibration), but eventually the surplus energy is converted to unstable vibration, and the system flies apart. The complex is stabilized (converted to a chemical form which can survive for an extended period of time) by collision with another molecule, M (any species that is available to collide with the complex and carry off its excess energy). Under the circumstances we expect that the rate for (13.2) should be proportional to the product of the concentrations of O, O_2, and M. At least to a first approximation, we can identify the density of M with the total density of air molecules at the altitude of interest. Denoting the proportionality constant by $k_{13.2}$, the rate, $R_{13.2}$, is given by

$$R_{13.2} = k_{13.2}\,[O]\,[O_2]\,[M] \qquad (13.7)$$

Reaction (13.4) is somewhat simpler. An oxygen atom encounters an ozone molecule. In the course of collision, it picks up one of the O atoms composing O_3. The reaction is exothermic (i.e., energy is liberated), since O_2 is much more stable than either O or O_3. In this case, much of the excess energy is carried off in the form of the products' kinetic energy. Both of the collision partners are converted to O_2. We expect that the rate for reaction (13.4) should be proportional to the product of the concentrations of O and O_3. Denoting the constant of proportionality by $k_{13.4}$, the rate for (13.4) is given by

$$R_{13.4} = k_{13.4}\,[O]\,[O_3] \qquad (13.8)$$

The k factors are known as **rate constants.** Considering the dimensions of the various quantities in (13.7) and (13.8), with the *cgs* system of units, we see that $k_{13.2}$ and $k_{13.4}$ have dimensions of $cm^6\ sec^{-1}$ (3-body reactions) and $cm^3\ sec^{-1}$ (2-body reactions), respectively. Rate constants for chemical reactions of interest in the atmosphere are usually determined on the basis of careful studies under controlled conditions in the laboratory; they typically vary as functions of temperature. For complex reactions, they may also change in response to variations in pressure and, in the case of 3-body reactions, may depend on the nature of the third body.

13.2 Evaluation of Rates for Photolytic Processes

As previously noted, rates for photolytic reactions depend on the intensity of the radiation field at the photolytically active wavelengths. In general, the intensity of the radiation field is a function of altitude. It may be expected to vary, as previously discussed, with the time of day, responding to changes in the solar zenith angle, as illustrated in Figure 13.2. The intensity of radiation reaching level z (*note:* z is measured from the ground) depends on the number of absorbing molecules, N (expressed in units of cm^{-2}), encountered by light as it propagates through the atmosphere from space to z and on the magnitude of the molecular cross section, Q (expressed in units of cm^2), the quantity defining the probability that the light is absorbed by an individual molecule. Propagation of light through an inhomogeneous medium, such as the atmosphere, is complex, with complications associated mainly with changes in the density of absorbing gases along the path followed by the light as it passes through the atmosphere from space to level z.

We propose to pause here in our treatment of stratospheric chemistry to consider in more detail the transmission of solar radiation through the atmosphere. We need to do this in order to be able to calculate values for the photolytic rate constants $J_{13.1}$ and $J_{13.3}$ as functions of altitude and solar zenith angle. It is convenient to begin with discussion of a simpler problem: the propagation of light through a medium of uniform (constant) density in the laboratory.

Consider the situation illustrated in Figure 13.3. Light (represented by parallel rays) enters a tube filled by a gas of uniform density n. Distance along the tube in the direction of the propagation of the light is measured by s. Light enters the tube at the point where $s = 0$. The flux of radiation in a particular wavelength interval at s (the number of photons crossing a unit area normal to the direction of light propagation at s in unit time, photons cm^{-2} sec^{-1}) is defined in terms of a quantity F(s). In a similar manner, the flux at $s + \Delta s$, a short distance along the tube in the direction of light propagation, is given by $F(s + \Delta s)$. The flux at $s + \Delta s$ differs from the flux at s by a quantity representing the number of photons absorbed between s and $s + \Delta s$. We may express this relation as an equation in words as follows:

The number of photons crossing unit area in unit time at s + Δs normal to the direction of propagation of light	=	The number of photons crossing unit area in unit time at s normal to the direction of propagation of light	−	The number of photons absorbed per unit time between s and s + Δs in a volume defined by unit cross-sectional area normal to the direction of light propagation	(13.9)

The number of photons absorbed per unit time in the volume between s and $s + \Delta s$ depends on three quantities: (1) the number of photons that enter the volume in unit time, (2) the number of molecules available to absorb the light, and (3) the size of the target or cross section offered by individual molecules. Expressed mathematically, the rate at which photons are absorbed between s and $s + \Delta s$ is given by

$$\text{Rate} = [F(s)] \times [n\,\Delta s] \times [Q], \qquad (13.10)$$

where the separate factors are identified with the terms in square brackets. The units for the quantities in this equation are as follows:

$$cm^{-2}\ sec^{-1} = (cm^{-2}\ sec^{-1})\ (cm^{-3}\ cm)\ (cm^{2})$$

$$cm^{-2}\ sec^{-1} = cm^{-2}\ sec^{-1}$$

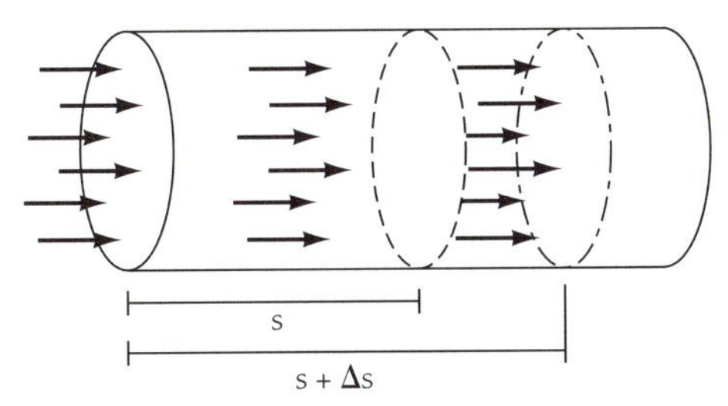

Figure 13.3 Radiation is assumed to enter a tube of unit cross-sectional area from the left. It passes down the tube with its travel distance measured by s.

As expected, the rate at which photons are absorbed between s and Δs has dimensions of cm^{-2} sec^{-1}. It defines the number of photons absorbed per unit time (sec) between s and $s + \Delta s$ in a volume of unit cross-sectional area (photons cm^{-2} sec^{-1}).

We are now in a position to translate the word equation (13.9) into mathematical terms. It is equivalent to

$$F(s + \Delta s) = F(s) - [F(s)(n\,\Delta s)Q], \qquad (13.11)$$

which may be rewritten as

$$\frac{F(s + \Delta s)}{F(s)} = 1 - (n\,\Delta s)Q \qquad (13.12)$$

Equation (13.12) indicates that the flux of radiation is reduced by a *fractional* amount equal to $(n\,\Delta s)Q = N(s, s + \Delta s)Q$ in passing from s to $s + \Delta s$, where $N(s, s + \Delta s) = n\,\Delta s$ is the number of absorbing molecules in a volume of unit cross-sectional area between s and Δs. This means that the fractional reduction in the flux of radiation over a particular path is simply determined by the cross-section for absorption of light and the number of absorbing molecules.

Suppose that the flux is reduced by a factor of two while traversing 10^{18} molecules contained in a column defined by a cross-sectional area of 1 cm^2. When the beam encounters an additional 10^{18} molecules cm^{-2} (i.e., 10^{18} molecules contained in a column of cross-sectional area 1 cm^2), the flux will be further reduced by a factor of two, and so on. More generally, suppose that the flux of radiation entering the tube at $s = 0$ is given by F_0. Imagine that the flux is reduced by a factor of two in propagating a distance L. It should be decreased thus by a factor of four in moving a distance $2L$. This behavior is described by a function known as an **exponential.** The variation of F with s is schematically illustrated in Figure 13.4.

The decrease of F with s is expressed by a relation of the form

$$F(s) = F_0 \exp[-as] \qquad (13.13)$$

The exponent of the exponential, as, determines the rate at which F falls off with s. The exponent of an exponential must be dimensionless. It follows that, since s has dimensions of cm, a must have dimensions of cm^{-1}.

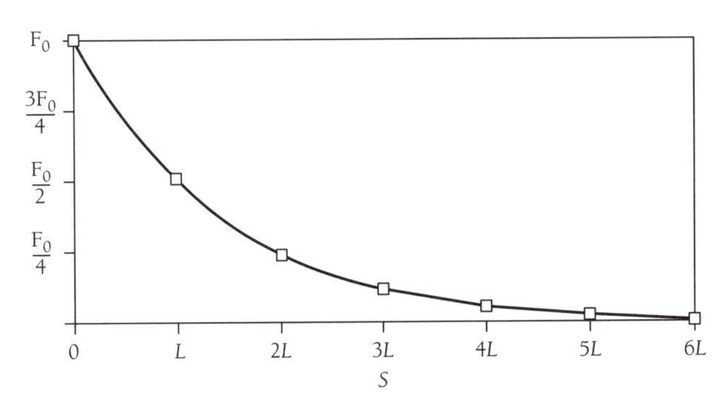

Figure 13.4 Situation assuming that the flux entering the tube in Figure 13.2 has an initial value equal to Fo, which is then reduced by a factor of two while moving a distance L down the tube. The Figure indicates the change in F with s, illustrating the exponential behavior of F(s).

Intuitively, we would expect the rate at which F falls off with s to depend on the magnitude of the cross section, Q. This suggests that we should write

$$as = bQs \tag{13.14}$$

Given that Q has dimensions of cm^2, it follows that b should have dimensions of cm^{-3}. It requires little imagination to conclude that b should be identified with the density of absorbing gas, n. Thus:

$$F(s) = F_0 \exp\,[-nQs] \tag{13.15}$$

The expression (13.15) for F(s) could have been derived more directly, without resort to the dimensional argument used here, by writing (13.11) in the form:

$$\frac{F(s+\Delta s) - F(s)}{\Delta s} = -F(s)nQ \tag{13.16}$$

Taking the limit as $\Delta s \rightarrow 0$, this becomes

$$\frac{dF}{ds} = -F(s)nQ \tag{13.17}$$

Writing (13.17) in the form

$$\frac{dF}{F} = -nQ\,ds \tag{13.18}$$

and integrating from s = 0 to s = s, we obtain

$$\log_e \frac{F(s)}{F_o} = -nQs \tag{13.19}$$

Taking exponentials of both sides of (13.19), we recover the solution for F(s) given by (13.15).

The quantity ns defines the number of molecules (cm^{-2}) in a volume of unit cross-sectional area (1 cm^2) between 0 and s. The dimensionless quantity nQs is known as the **optical depth** between 0 and s. The reduction in flux between 0 and s is determined by taking the exponential of the negative of the optical depth.

Note the origin of the exponential relation summarized by (13.15): it arises as a consequence of the assumption that the rate at which photons are absorbed between s and s + Δs should be proportional to the flux at s. If the rate at which photons are absorbed (cm^{-2} sec^{-1}) between s and Δs was simply proportional to the cross section and to the number of intervening molecules, the rate would be given by $c(n\,\Delta s)Q$, where c would indicate, in this case, a constant independent of F. It would follow then from dimensional considerations that the units of c should be the same as those for F(cm^{-2} sec^{-1}). The variation of flux with s would be given in this instance by

$$F(s + \Delta s) = F(s) - c(n\,\Delta s)Q \tag{13.20}$$

The difference in the behavior of the solutions described by (13.11) and (13.20) is displayed in Figure 13.5. For purposes of this example, we assumed values for the various parameters as follows: $n = 10^{18}$ cm^{-3}, $Q = 10^{-18}$ cm^2, $F_0 = 100$ cm^{-2} sec^{-1} and $c = 100$ cm^{-2} sec^{-1}. Figure 13.5 illustrates the difference between exponential and linear decay of F as a function of s.

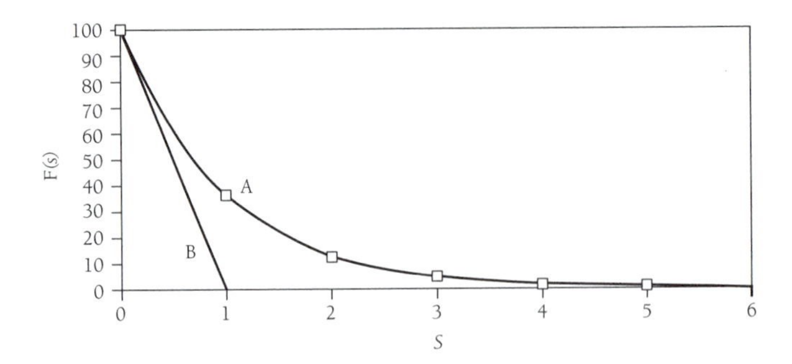

Figure 13.5 Comparison of F as a function of s, as computed using either (13.13), curve A, or (13.20), curve B. The flux entering at s = 0 is taken to equal 100 cm^{-2} sec^{-1}. The density of absorbing molecules is taken as 10^{18}cm^{-3}. Other parameters are as follows: Q = 10^{18} cm^2, c = 100 cm^{-2} sec^{-1}. The Figure illustrates the difference between exponential (curve A) and linear (curve B) variations of F with s.

We return now to a discussion of factors influencing the variation of rates for photolysis in the atmosphere as a function of altitude and time of day. The intensity of solar radiation exponentially decreases with transmission through the atmosphere. By analogy with the situation discussed above for a homogeneous medium, the exponent of the exponential regulating transmission of radiation to altitude z is given by ($-N(z)Q$), where Q is the cross section (cm^2) and N(z) is the number of absorbing molecules (cm^{-2}) encountered along the path from space to z. The quantity NQ, for *vertical incidence*, is equivalent to the *optical depth* introduced above.

The number of molecules along the path for light incident from a slant (off-vertical) direction increases as the inverse of the cosine of the solar zenith angle θ (considering, as before, absorption by molecules in a column with a cross-sectional area of 1 cm^2 aligned along the direction of incidence). This may be readily understood for the case where the density is constant along the direction of incidence. Referring to Figure 13.2 and noting that

$$PQ = \frac{1}{\cos\theta} PR = \frac{1}{\mu} PR, \tag{13.21}$$

we see that the volume occupied by the slant column is larger than that for the vertical column by a factor of $1/\mu$ where $\mu = \cos(\theta)$. The integrated density along the column is accordingly enhanced. It is easy to show that this result must hold quite generally. Denoting the integrated density for the slant column extending from altitude z to the top of the atmosphere by N(θ, 0$\rightarrow$s$\rightarrow\infty$), it follows that

$$\begin{aligned} N(\theta, 0\rightarrow s\rightarrow\infty) &= \int_0^\infty n(s)\,ds \\ &= \frac{1}{\mu}\int_0^\infty n(z)\,dz \\ &= \frac{1}{\mu} N(z), \end{aligned} \tag{13.22}$$

where we made use here of the relation $s = (1/\mu)z$ (see Figure 13.2) to affect the transformation of variable from s to z.

It follows that the flux of radiation incident at z at zenith angle θ in a particular wavelength band F(z) is given by

$$F(z) = F_0 \exp\left(-\frac{N(z)Q}{\cos\theta}\right), \tag{13.23}$$

where F_0 defines the flux at the top of the atmosphere, N(z) is the vertical column density of absorbing molecules, and Q is an appropriate, wavelength-averaged value of the absorption cross section. Defining the vertical optical depth for the atmosphere above z by

$$\tau(z) = N(z)\,Q, \tag{13.24}$$

it follows that

$$F(z) = F_0 \exp\left(-\frac{\tau(z)}{\mu}\right) \tag{13.25}$$

Suppose that more than one gas is involved in limiting the transmission of radiation through the atmosphere. Suppose there are two important absorbers. Denote the corresponding vertical-column densities by N_1 and N_2. Let Q_1 and Q_2 signify the average values of the cross sections for absorption of radiation by these species in a particular wavelength band. The vertical optical depth is defined in this case by $N_1Q_1 + N_2Q_2$. The flux of radiation reaching z at zenith angle θ is given by

$$F(z) = F_0 \exp\left(-\frac{N_1(z)Q_1 + N_2(z)Q_2}{\cos\theta}\right) \tag{13.26}$$

Again, writing

$$\tau(z) = N_1(z)\,Q_1 + N_2(z)\,Q_2, \tag{13.27}$$

we see that the flux at z is given by (13.25).

Rates for photolysis at altitude z are defined by the product of the flux of radiation F(z) ($cm^{-2}\ sec^{-1}$), at relevant wavelengths reaching z, and the cross section, Q (cm^2), for the particular photolytic reaction $J(z) = F(z)Q$. In practice, photolysis occurs over a range of wavelengths. For computational purposes, it is customary to divide the solar spectrum into a number of discrete intervals; absorption cross sections, determined on the basis of measurements in the laboratory, are associated with individual intervals, and J values are obtained by summing contributions to photolysis from all of the relevant spectral intervals. Values of J are obviously largest at noon: they decrease toward sunrise and sunset and are zero at night. Values of $J_{13.1}$ and $J_{13.3}$, averaged over a day and computed using observed concentrations of O_3 and O_2 for a latitude of 30°N at equinox, are presented in Figure 13.6. The gases most effective in limiting the transmission of radiation at the wavelengths involved in photolysis of O_2 and O_3—reactions (13.1) and (13.3)—are O_2 and O_3, with the latter more important for (13.3).

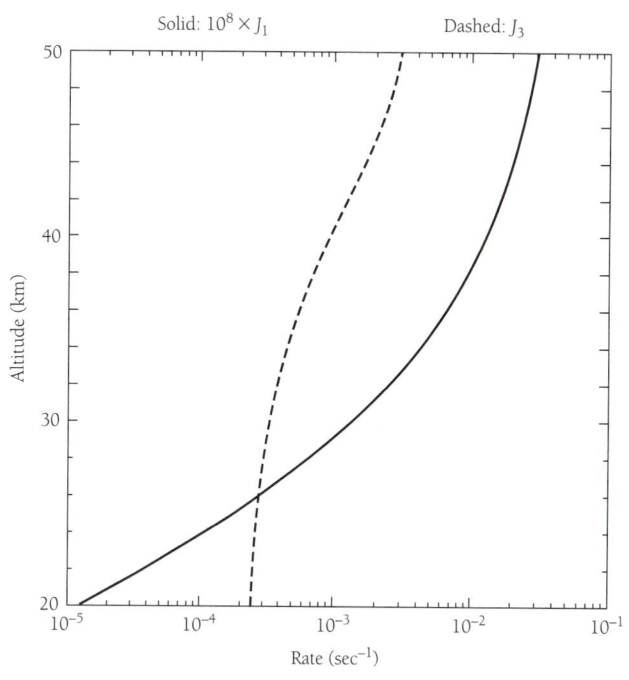

Figure 13.6 Diurnally averaged values for J_1 (solid line) and J_3 (dashed line) as functions of altitude for 30°N at equinox. The scale at the bottom of the Figure is for J_3; for J_1 multiply by 10^8. Values of J_1 and J_3 (expressed in sec^{-1}) given here were computed using observed values for $[O_3]$ as a function of altitude.

Example 13.1: Radiation between 250 and 350 nm is absorbed mainly by O_3. The vertical column densities for O_3 on a particular day at altitudes above 50 and 20 km are measured and found to have values of 5×10^{16} and $1.0 \times 10^{19}\ cm^{-2}$. The cross section for absorption of radiation by O_3 between 250 and 260 nm, denoted by Q, has an average value of $1 \times 10^{-17}\ cm^2$. Estimate the factor by which the solar radiation incident between 250 and 260 nm is reduced in passing from space to both 50 and 20 km. Assume that it is high noon in the tropics and that the Sun is at the zenith, i.e., $\theta = 0$.

Answer: The optical depth, $\tau(z)$, corresponding to altitude z is given by

$$\tau(z) = N(z)\,Q,$$

where N(z) is the vertical column density of the absorbing gas (O_3). Note that $\tau(z)$ is dimensionless. N(z) has dimensions of cm^{-2}, whereas Q has dimensions of cm^2. Using N(z) at 50 and 20 km and substituting for Q, we find:

$$\tau(50\ \text{km}) = (5.0 \times 10^{16}\ \text{cm}^{-2})(1 \times 10^{-17}\ \text{cm}^2)$$
$$= 5 \times 10^{-1} = 0.5$$

and

$$\tau(20\ \text{km}) = (1 \times 10^{19}\ \text{cm}^{-2})(1 \times 10^{-17}\ \text{cm}^2)$$
$$= 1 \times 10^2 = 100$$

The intensity of radiation reaching z from the zenith is reduced by a factor f(z) given by

$$f(z) = \exp(-\tau)$$

It follows that

$$f(50\text{ km}) = \exp(-0.5) = 0.61$$

and

$$f(20\text{ km}) = \exp(-100) = 3.7 \times 10^{-44}$$

The value of $f(20\text{ km})$ is, for all intents and purposes, equal to zero: an imperceptibly small fraction of the solar radiation incident between 250 and 260 nm penetrates to 20 km. ■

Example 13.2: For the same situation, imagine that it is 4:00 P.M. The solar zenith angle is now 60°. Estimate the factor by which the flux of sunlight is reduced in penetrating to 50km.

Answer: The reduction factor, $f(z)$, corresponding to altitude z and zenith angle θ is given by

$$f(z) = exp\left(-\frac{\tau}{\mu}\right),$$

where

$$\mu = \cos\theta$$

With $\theta = 60°$, $\mu = 1/2$, and $f(z) = \exp(-2\tau)$. It follows that

$$f(50\text{ km}, \theta = 60°) = \exp(-2 \times 0.5) = \exp(-1) = 0.368,$$

where the value of τ is adopted from the preceding example. ■

Example 13.3: The cross section for photodissociation of O_3 at wavelengths between 250 and 260 nm equals the cross section for absorption of radiation by O_3 in this spectral region ($Q = 10^{-17}\text{ cm}^2$). The solar flux at the top of the atmosphere in the wavelength interval 250 to 260 nm is equal to 10^{14} photons $\text{cm}^{-2}\text{ sec}^{-1}$. Estimate the contribution of this wavelength interval to the rates for photolysis of O_3 at 50 km for both $\theta = 0$ and $\theta = 60°$.

Answer: The flux of radiation, $F(z)$, reaching altitude z with zenith angle θ, is given by

$$F(z, \theta) = F_0 f(z, \theta),$$

where F_0 is the flux at the top of the atmosphere and $f(z, \theta)$ defines the reduction factors introduced in the preceding examples.

$$F(50\text{ km}, (\theta = 0) = (10^{14}\text{ photons cm}^{-2}\text{ sec}^{-1})(0.61) = 6.1 \times 10^{13}\text{ photons cm}^{-2}\text{ sec}^{-1}$$

$$F(50\text{ km}, (\theta = 60°) = (10^{14}\text{ photons cm}^{-2}\text{ sec}^{-1})(0.368) = 3.68 \times 10^{13}\text{ photons cm}^{-2}\text{ sec}^{-1}$$

It follows that the contributions of the selected wavelength interval to photolysis at 50 km, for $\theta = 0$ and 60°, are given by

$$J(50\text{ km}, \theta = 0) = (6.1 \times 10^{13}\text{ photons cm}^{-2}\text{ sec}^{-1})(10^{-17}\text{ cm}^2) = 6.1 \times 10^{-4}\text{ sec}^{-1}$$

$$J(50\text{ km}, \theta = 60°) = (3.68 \times 10^{13}\text{ photons cm}^{-2}\text{ sec}^{-1})(10^{-17}\text{ cm}^2) = 3.68 \times 10^{-4}\text{ sec}^{-1}$$

■

Example 13.4: At wavelengths between 200 and 210 nm, penetration of sunlight is limited both by O_3 and O_2. The cross sections for absorption by these gases in this wavelength interval have values of $Q_{O_3} = 2 \times 10^{-19}\text{ cm}^2$ and $Q_{O_2} = 10^{-23}\text{ cm}^2$. Assume, as before, that the vertical column density of O_3 above 20 km has magnitude 10^{19} cm^{-2}. Assume a vertical column density for O_2 above the same altitude of $2.5 \times 10^{23}\text{ cm}^{-2}$. Estimate, for a zenith angle of 60°, the factor by which the flux of sunlight between 200 and 210 nm is reduced in penetrating to 20 km.

Answer: The optical depth above 20 km is given in this case by

$$\tau(20\text{ km}) = Q_{O_3}N_{O_3}(20\text{ km}) + Q_{O_3}N_{O_2}(20\text{ km}),$$

where $N_{O_3}(20\text{ km})$ and $N_{O_2}(20\text{ km})$ are the column densities of O_3 and O_2, respectively. It follows that

$$\tau(20\text{ km}) = (2 \times 10^{-19}\text{ cm}^2)(10^{19}\text{ cm}^{-2}) + (10^{-23}\text{ cm}^2)(2.5 \times 10^{23}\text{ cm}^{-2}) = 2 + 2.5 = 4.5$$

The flux of radiation is reduced by a factor

$$f(20\text{ km}, \theta = 60°) = \exp\left(-\frac{\tau}{\cos 60°}\right) = \exp(-2\tau) = \exp(-9.0) = 1.23 \times 10^{-4}$$

■

Example 13.5: Using the information in Example 13.4, estimate the contribution of the interval 200 to 210 nm to photolysis of O_2, reaction (13.1). Assume that the cross section for dissociation of O_2 is equal to the cross section for absorption by O_2. The flux of solar radiation at the top of the atmosphere in this wavelength interval (F_0) is equal to 10^{13} photons $\text{cm}^{-2}\text{ sec}^{-1}$.

Answer: The flux of radiation reaching 20 km corresponding to zenith angle $\theta = 60°$ is given by

$$F(20\text{ km}, \theta = 60°) = F_0 \times f(z, \theta) = (10^{13}\text{ cm}^{-2}\text{ sec}^{-1})(1.23 \times 10^{-4})$$

The contribution to the photolysis rate of O_2 is given by

$$J_{O_2} = Q_{O_2}\, F(20\ km, \theta = 60^o)$$
$$= (10^{-23}\ cm^2)(1.23 \times 10^9\ cm^{-2}\ sec^{-1})$$
$$= 1.23 \times 10^{-14}\ sec^{-1}$$ ■

13.3 The Concept of Odd Oxygen and the Evaluation of Lifetimes for Chemical Species

It is convenient for many purposes to consider O and O_3 as a family of closely related chemical species: atmospheric chemists refer to the combination as **odd oxygen.** Focusing on odd oxygen rather than O and O_3 separately provides important insights into the chemistry of stratospheric O_3. It allows us to distinguish between reactions implicated in production and loss of odd oxygen—real sources and sinks of O_3—in contrast to reactions representing temporary rearrangements of O atoms. An O atom formed by (13.1) is converted almost immediately to O_3 by (13.2). It stays as O_3 for a short while before being reconstituted as O as a consequence of photolysis, reaction (13.3). Reactions (13.2) and (13.3) provide no net source or sink of odd oxygen: the species is destroyed by (13.3) just as rapidly as it is reformed by (13.2). An O atom flips back and forth many times between (13.2) and (13.3) over the course of its life in the atmosphere, before eventually finding an O_3 molecule with which it is removed as a result of reaction (13.4). To the extent that we are interested in odd oxygen as a whole, it is clear that reactions (13.1) and (13.4) are more significant than either (13.2) or (13.3). Reactions (13.1) and (13.4) are responsible for the overall production and loss of odd oxygen; they represent the *real* sources and sinks of O_3. Reaction (13.1) involves production of two elements of odd oxygen, introduced initially as O, while two elements of odd oxygen, O and O_3, are consumed by reaction (13.4). The bond joining the O atoms in O_2 is fractured in (13.1), and reconstituted in (13.4). Reactions (13.2) and (13.3) involve no net change in the concentration of odd oxygen: an atom of O is transformed to O_3 in the former case whereas the reaction is reversed in the latter. Reactions (13.2) and (13.3) play an important role in regulating the proportions of the separate forms of odd oxygen; they influence the abundance of odd oxygen as a whole by setting the relative abundances of O and O_3, controlling by this means the rate of reaction (13.4). A pictorial view of odd oxygen chemistry is presented in Figure 13.7.

Example 13.6: The importance of reactions (13.2) and (13.3) in affecting the overall abundance of O_3, may be illustrated as follows. Suppose for some altitude that the rate for dissociation of O_2 by (13.1), $R_{13.1}$, is equal to $10^6\ cm^{-3}\ sec^{-1}$. Let the abundance of O relative to O_3 (set by reactions 13.2 and 13.3) be given by a ratio of 10^{-5} (one atom of O for every 10^5 molecules of O_3). Suppose that the rate constant for (13.4), $k_{13.4}$, is equal to $10^{-13}\ cm^3\ sec^{-1}$. Calculate the abundance of O_3.

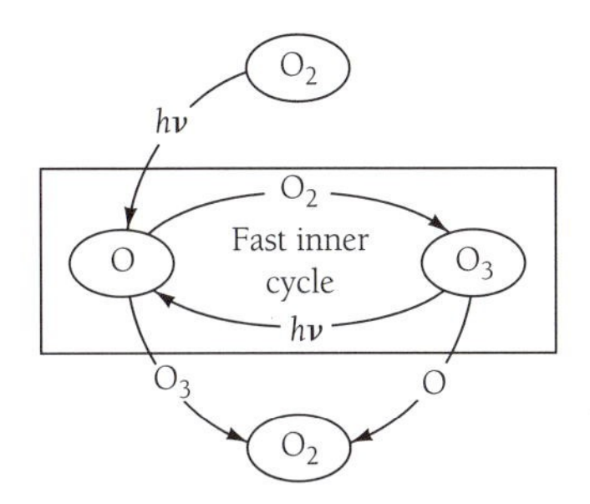

Figure 13.7 A pictorial representation of Chapman chemistry. Species of odd oxygen are indicated in the interior of the rectangle. Odd oxygen is formed by reaction (13.1) and removed by (13.4). Interconversion of O and O_3 occurs by (13.2) and (13.3).

Answer: The source of odd oxygen is given by twice the dissociation rate of O_2 (i.e., two atoms of odd O are formed for each dissociation event). Thus the source of odd O = $2 \times 10^6\ cm^{-3}\ sec^{-1}$. The rate for removal of odd O is equal to twice the rate for reaction (13.4), or $2R_{13.4}$:

$$Sink = 2\, k_{13.4}\, [O][O_3]$$
$$= 2 \times 10^{-13}\, (10^{-5}[O_3])\, [O_3]$$
$$= 2 \times 10^{-18}\, [O_3]^2$$

In this and the subsequent discussion, [X] denotes the concentration (cm^{-3}) of chemical species X. Equating source and sink:

$$2 \times 10^6 = 2 \times 10^{-18}\, [O_3]^2$$

It follows that.

$$[O_3]^2 = 10^{24}$$

Hence,

$$[O_3] = 10^{12}\ cm^{-3}$$ ■

Example 13.7: Repeat the calculation above, assuming a relative abundance of O with respect to O_3 of 10^{-3}.

Answer:
$$Sink = 2\, k_{13.4}\, [O][O_3]$$
$$= 2 \times 10^{-13}\, (10^{-3}[O_3])\, [O_3]$$
$$= 2 \times 10^{-16}\, [O_3]^2$$

Equating source and sink:

$$2 \times 10^6 = 2 \times 10^{-16}\, [O_3]^2$$

It follows that

$$[O_3]^2 = 10^{22}$$

Hence,

$$[O_3] = 10^{11}\ cm^{-3}$$ ■

The abundance of O_3 is reduced as a consequence of the higher relative abundance of O. Note that the abundance of odd oxygen essentially equals the abundance of

O_3 in both Examples 13.6 and 13.7: allowing for O introduces only a small correction. The abundance of O is equal to 10^{-5} $[O_3] = 10^{-5} \times 10^{12} = 10^7$ cm^{-3} in the case of Example 13.6. In the case of Example 13.7, it is given by 10^{-3} $[O_3] = 10^{-3} \times 10^{11} = 10^8$ cm^{-3}.

As noted above, the exchange of forms of odd oxygen between O and O_3 is efficient, compared with the production and removal by (13.1) and (13.4). This point is illustrated in Figure 13.8, which presents a comparison of rates for production of O by (13.1) and (13.3). The source of O due to (13.3) is larger, by several orders of magnitude over most of the stratosphere, than the rate for production by (13.1): the rate for production of O by (13.1) at 40 km, for example, is about 4.6×10^6 cm^{-3} sec^{-1}: by way of comparison, the rate for formation of O by photolysis of O_3, reaction (13.3), is almost 100 times larger, 4.4×10^8 cm^{-3} sec^{-1}. It follows that an O atom, once produced by (13.1), will traverse the inner loop in Figure 13.7 many times over its life, exchanging its label, or form of odd oxygen, from O to O_3 and back. It will cycle in this fashion close to 100 times at 40 km (4.4×10^8/ 4.6×10^6) before it is eventually removed by reaction with O_3 in reaction (13.4).

We can estimate the time an odd oxygen atom spends as O_3, τ_{O_3}, before cycling to O as a consequence of (13.3), by dividing the concentration of O_3 by the rate of reaction (13.3), as given by reaction (13.6).[3]

$$\tau_{O_3} = \frac{[O_3]}{R_{13.3}} = \frac{[O_3]}{J_{13.3}[O_3]} = (J_{13.3})^{-1} \tag{13.28}$$

Similarly, the time an odd oxygen atom spends as O before it is converted to O_3 by (13.2) is given by

$$\tau_O = \frac{[O]}{R_{13.2}} = \frac{[O]}{k_{13.2}[O][O_2][M]} = (k_{13.2}[O_2][M])^{-1} \tag{13.29}$$

The lifetime of O is short at low altitudes, where the densities of O_2 and M are large. Equation (13.29) indicates that the lifetime of O should rapidly increase with altitude, varying inversely as the square of the total density (the abundance of O_2 relative to total density M is essentially constant with altitude). The lifetime of O may be expected to increase by a factor of *e* when altitude is raised by an amount equal to approximately half a scale height, about 3 km. The lifetime of O_3 with respect to (13.3) varies more slowly as a function of altitude, responding simply to changes in $J_{13.3}$. Profiles for τ_{O_3} and τ_O are presented as functions of altitude in Figure 13.9.

Example 13.8: The probability of finding a member of the odd oxygen family as O rather than O_3 depends on the relative values of the lifetimes for O and O_3. The shorter the lifetime of O relative to O_3, the smaller the abundance of O relative to O_3. The lifetime of O at 40 km, as indicated in Figure 13.9, is equal to 0.76 sec, while the lifetime for O_3 (with respect to photolysis) is 1.1×10^3 sec. Use these data to calculate the abundance of O relative to O_3.

Answer: Equating the rates for reactions (13.2) and (13.3) we have:

$$K_{13.2}\ [O][O_2][M] = J_{13.3}\ [O_3]$$

It follows that

$$\frac{[O]}{[O_3]} = \frac{J_{13.3}}{k_{13.2}[O_2][M]},$$

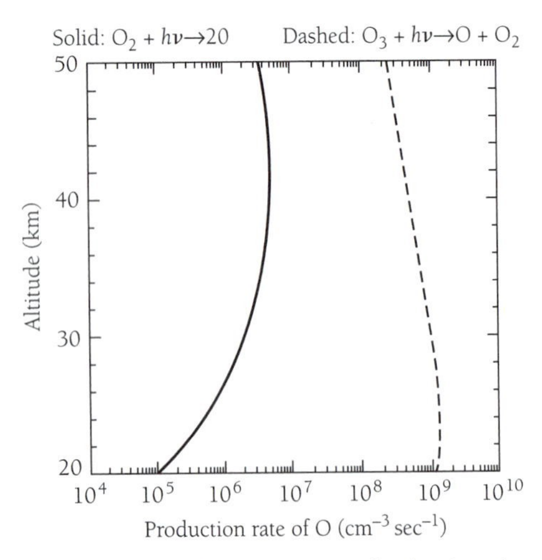

Figure 13.8 Rates for the production of O by (13.1) compared with production by (13.2), illustrating the dominance of recycling of odd O associated with photolysis of O_3.

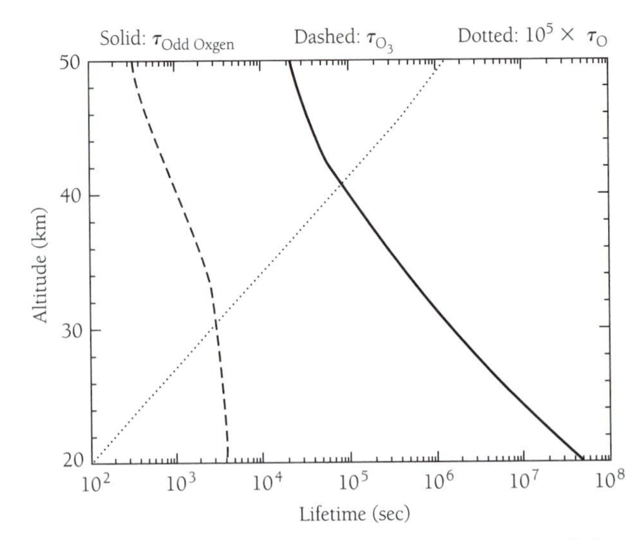

Figure 13.9 Lifetime for odd oxygen, $\tau_{OddOxygen}$, and for removal of O and O_3 by (13.2) and (13.3) τ_O and τ_{O_3}, respectively.

and using the definitions for τ_{O_3} and τ_O given by (13.28) and (13.29),

$$\frac{[O]}{[O_3]} = \frac{\tau_O}{\tau_{O_3}}$$

Using the values quoted for τ_O and τ_{O_3} at 40 km:

$$\frac{[O]}{[O_3]} = \frac{7.6 \times 10^{-1}}{1.1 \times 10^3} = 6.9 \times 10^{-4}$$ ■

Example 13.9: Repeat this calculation in (13.7) for 20 km where the lifetime for O and O_3 are 9.7×10^{-4} sec and 4.2×10^3 sec, respectively.

Answer: In this case,

$$\frac{[O]}{[O_3]} = \frac{9.7 \times 10^{-4}}{4.2 \times 10^3} = 2.3 \times 10^{-7}$$ ■

The residence time for odd oxygen as a whole (O + O_3) may be obtained by dividing the concentration of odd oxygen by the rate at which the composite species is produced by (13.1), or alternatively, by the rate at which it is consumed by (13.4). Results are identical for the case of photochemical equilibrium in which, by assumption, the rates for production and loss are equal. In general, the concentration of odd oxygen can be affected by transport, as discussed above. Depending on location, transport can account for either a net source or sink, compensating for an imbalance in rates for chemical production and loss. When the chemical time constant is short, the concentrations of odd oxygen can be estimated with adequate accuracy on the basis of photochemical equilibrium. We assume, somewhat arbitrarily, that photochemical equilibrium is justified if the chemical time constant is less than about a month, about 10^6 sec.[4] We choose to estimate the time constant for odd oxygen by using the rate for production given by (13.5), rather than the rate of loss expressed by (13.8). We make this choice simply for convenience: it is easier to proceed using (13.5) rather than (13.8), since the former depends on the quantities $J_{13.1}$ and $[O_2]$, which are readily available, whereas the latter requires information on the concentrations of both O and O_3. The distinction between chemical time constants based on rates for production rather than loss is inconsequential for present applications: our interests here are qualitative rather than quantitative; we intend to use the concept merely to delineate the approximate region of applicability of the assumption of photochemical equilibrium.

With our assumptions, the lifetime for odd oxygen is given by

$$\tau_{\text{Odd Oxygen}} = \frac{[\text{Odd Oxygen}]}{2J_{13.1}[O_2]} \quad (13.30)$$

The factor of two accounts for the fact that two elements of odd oxygen are produced by reaction (13.1). The abundance of O_3 is much larger than that of O throughout the stratosphere (see Examples 13.8 and 13.9). The distinction between O_3 and odd oxygen in (13.30) is consequently moot; without introducing significant error, we can replace the concentration of odd oxygen in (13.30) with that of O_3. The time constant for odd oxygen is then given by

$$\tau_{\text{Odd Oxygen}} = \frac{[O_3]}{2J_{13.1}[O_2]} \quad (13.31)$$

Results for $\tau_{\text{Odd Oxygen}}$ are included in Figure 13.9. The data for τ_{O_3}, τ_O, and $\tau_{\text{Odd Oxygen}}$ displayed here were computed using values for the concentration of O_3 measured as a function of altitude for 30°N under equinoctial conditions. The time constant for O is comparatively brief, declining from less than a minute at 50 km to fractions of a second below 40 km. We expect significant variations in the concentration of O over the course of a day in response to changes in the intensity of sunlight affecting the magnitude of $J_{13.3}$. The time for odd oxygen to achieve a steady state is much longer than that for O, ranging from about a day at 40 km to more than a year at 20 km. Changes with time in the concentration of odd oxygen, and consequently O_3, should be gradual, with relatively little variation expected over the course of a day. The concentration of O_3 at a particular altitude and latitude responds primarily to seasonal variations in the intensity of sunlight affecting values of $J_{13.1}$. Assuming a cutoff for the validity of the assumption of photochemical equilibrium at a time constant of 10^6 sec, we conclude that transport may be expected to influence the distribution of O_3 significantly for altitudes below about 30 km.

Changes in the concentration of O_2 associated with conversion of oxygen to O and O_3 are minor (the combined abundance of O plus O_3 is less than 10^{-5} that of O_2 throughout most of the stratosphere and mesosphere). Changes in O_2 at a particular altitude arise primarily in response to fluctuations in pressure associated with variations in atmospheric temperature (the mixing ratio of O_2 is effectively constant everywhere below about 100 km, with its value, 0.21, almost exactly equal to that at the ground; see Table 5.1).

Concentrations of odd oxygen reflects a balance between the rate at which it is produced by (13.1) and the rate at which it is consumed by (13.4). Expressing this balance in mathematical terms, setting the expression for $R_{13.1}$ given by (13.5) equal to that for $R_{13.4}$ given by (13.8), we find

$$k_{13.4}\,[O][O_3] = J_{13.1}\,[O_2] \quad (13.32)$$

Similarly, assuming that the relative abundances of O and O_3 are determined by reactions (13.2) and (13.3), as justified by the fact that τ_{O_3} and τ_O are appreciably less than $\tau_{\text{Odd Oxygen}}$ (interconversion of O and O_3 is much more rapid than loss of the compounds by reaction 13.4), we may write (equating the expressions for $R_{13.2}$ and $R_{13.3}$ given by reactions 13.7 and 13.6, respectively),

$$k_{13.2}\,[O][O_2][M] = J_{13.3}\,[O_3] \quad (13.33)$$

Equation (13.33) may be rearranged to provide an expression for [O] as a function of $[O_3]$:

$$[O] = \frac{J_{13.3}[O_3]}{k_{13.2}[O_2][M]} \quad (13.34)$$

Using (13.34) to substitute this expression for [O] in (13.32) gives

$$J_{13.1}[O_2] = \frac{k_{13.4}J_{13.3}[O_3]^2}{k_{13.2}[O_2][M]} \quad (13.35)$$

Equation (13.35) may be rearranged to yield

$$[O_3]^2 = \frac{k_{13.2}J_{13.1}[O_2]^2[M]}{J_{13.3}k_{13.4}} \quad (13.36)$$

Since the mixing ratio of $[O_2]$ (denoted by f_{O_2}) is approximately constant with altitude, we may write

$$[O_2] = f_{O_2}[M] \quad (13.37)$$

Using (13.37) to solve equation (13.36) for $[O_3]$, we find

$$[O_3] = \left(\frac{k_{13.2}J_{13.1}}{J_{13.3}k_{13.4}}\right)^{1/2} f_{O_2}[M]^{3/2} \quad (13.38)$$

Using (13.38) to substitute in equation (13.34) for $[O_3]$ and using (13.37) for $[O_2]$, we obtain the analogous expression for [O]:

$$[O] = \left(\frac{J_{13.1}J_{13.3}}{k_{13.2}k_{13.4}}\right)^{1/2} [M]^{-1/2} \quad (13.39)$$

Figure 13.10 presents concentrations of $[O_3]$ and [O] computed using (13.38) and (13.39) for 30°N at equinox. The Figure includes a comparison of values derived for $[O_3]$ using (13.38) to results obtained when observational data for O_3 are employed to evaluate the time constants given in Figure 13.9.

The variation of $[O_3]$ and [O] with altitude is determined primarily by the altitude dependence of $J_{13.1}$, $J_{13.3}$, and [M]: values of rate constants k are more or less constant with altitude. (Remember that J's values depend on the intensity of sunlight (from Example 13.1) and that this can vary rapidly with altitude when the optical depth is large).

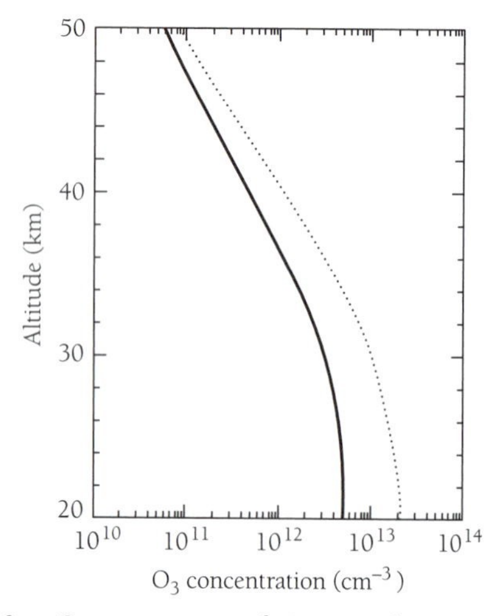

Figure 13.10a Concentration of O_3 as a function of altitude, using the Chapman scheme (dashed line), compared with observation (solid line).

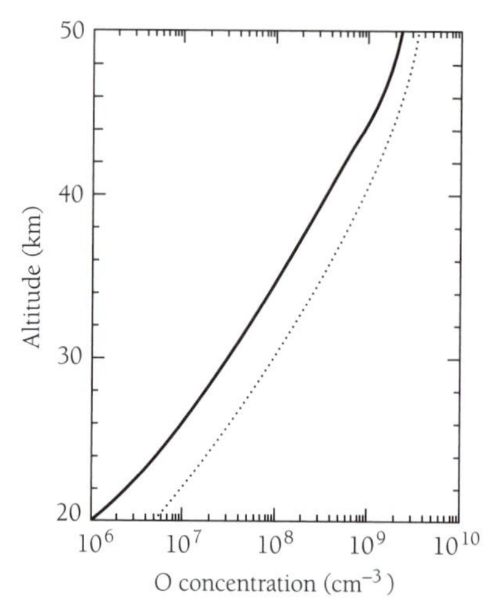

Figure 13.10b Concentration of O as a function of altitude using the Chapman scheme (dashed line), compared with the concentration of O computed using observed values of $[O_3]$ (solid line).

The concentration of O_3 depends mainly on $[M^{3/2}]$, which varies exponentially with z, and on the combination $(J_{13.1}/J_{13.3})^{1/2}$. Figure 13.11 illustrates the trend of these terms with altitude. Values of $J_{13.1}$ and $J_{13.1}$, and consequently $(J_{13.1}/J_{13.3})^{1/2}$, are essentially constant at high altitude, where absorption of incident solar radiation is minimal. The concentration of O_3 falls off with altitude at high elevation, with a rate determined mainly by the factor $[M^{3/2}]$. The decline in $[O_3]$ at low altitude is regulated primarily by the rapid drop in $J_{13.1}$, which essentially reflects the complete absorption at higher altitudes of radiation required to effect dissociation of O_2 (see Example 13.1). The decrease in $J_{13.1}$, and consequently $(J_{13.1}/J_{13.3})^{1/2}$, is more than sufficient to compensate for the increase in $[M]^{3/2}$ at low altitudes. The concentration of O_3 is small at high altitudes, reflecting the relatively low concentration of O_2 available as a source of odd oxygen in this region and the relatively high concentration of O available for its removal. It is small at low altitudes, reflecting the comparative absence of radiation with wavelengths sufficiently short to effect dissociation of O_2. It comes as no surprise, then, that the concentration of O_3 exhibits a maximum at an intermediate altitude, as illustrated in Figure 13.10. This explains the occurrence of an **ozone layer.** The concentration of O climbs monotonically with rising altitude, responding to the increase in both of the important terms in (13.39), $(J_{13.3} \times J_{13.1})^{1/2}$ and $[M]^{-1/2}$, as indicated in Figure 13.12.

13.4 Summary

The concept of *odd oxygen* offers a useful means to classify the relative roles of the four reactions involved in the Chapman scheme for the photochemistry of an idealized O_2-N_2 atmosphere. Photodissociation of O_2 (reaction 13.1) provides the primary source of odd oxygen for the stratosphere. In chemical equilibrium, the source of odd O contributed by (13.1) is balanced by the reaction of O with O_3, reaction

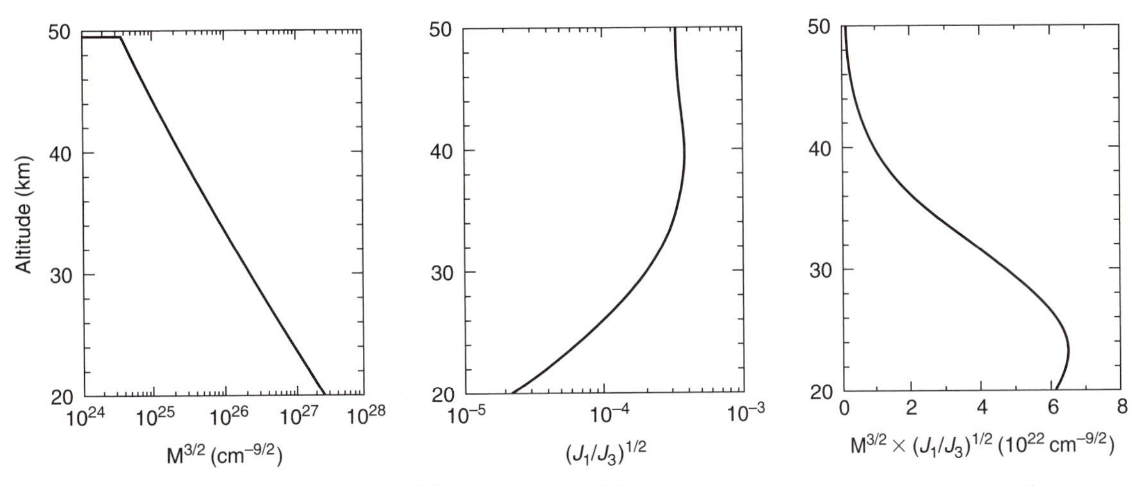

Figure 13.11 The variation with altitude of $[M]^{3/2}$ and $[J_{13.1}/J_{13.2}]^{1/2}$ and the product, the term regulating the variation of $[O_3]$ with altitude, as defined by reaction (13.38). Values of $J_{13.1}$ and $J_{13.3}$ are indicated for simplicity as J_1 and J_3, respectively.

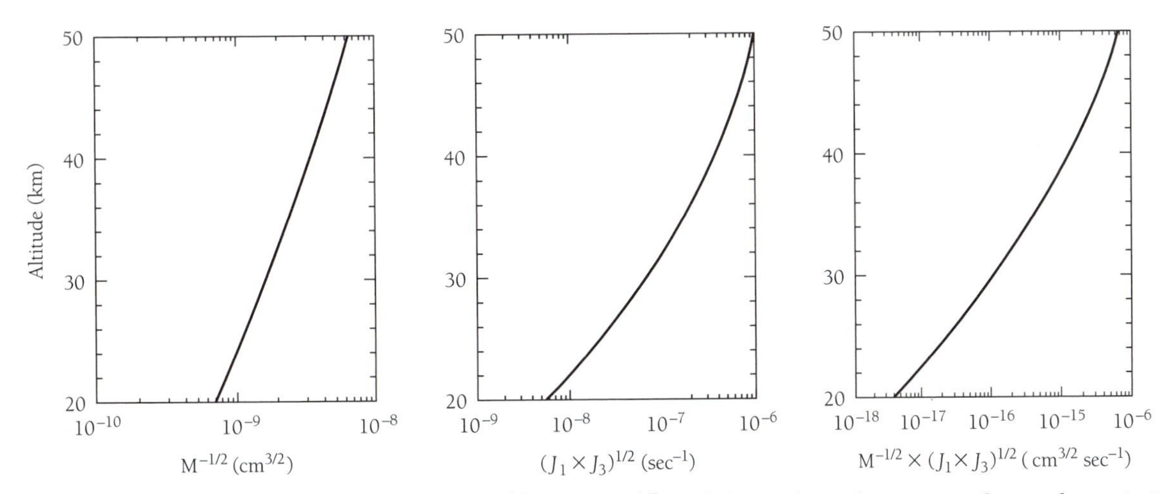

Figure 13.12 The variation with altitude of $[J_{13.1}/J_{13.2}]^{1/2}$ and $[M]^{-1/2}$, and the product, the term regulating the variation of $[O]$ with altitude, as given by reaction (13.39). Values of $J_{13.1}$ and $J_{13.3}$ are indicated for simplicity as J_1 and J_3, respectively.

(13.4). The relative abundances of O and O_3 are set primarily by the interchange reactions (13.2) and (13.3). Over most of the atmosphere, the abundance of O is small compared with the abundance of O_3. To an excellent approximation, the abundance of O_3 equals that of odd oxygen.

The radiation responsible for dissociation of O_2 is absorbed mainly at altitudes above about 20 km. This reflects the high abundance of O_2 and occurs despite the fact that the cross section for photodissociation of O_2 is relatively small. Photodissociation of O_3 persists to comparatively low altitudes as a consequence of the absorption of radiation at longer ultraviolet wavelengths. This contributes a source of O atoms at low altitude, ensuring that reaction (13.4) provides an efficient sink for odd oxygen in the lower stratosphere. The presence of this sink, combined with the fact that the source of odd oxygen from (13.1) is small at low altitudes, ensures that the abundance of odd oxygen, and thus O_3, has a maximum at intermediate altitudes.

The lifetime of odd oxygen rapidly increases with decreasing altitude, as a result of the decrease in the rate at which it is produced by photolysis of O_2 (or equivalently, the rate at which odd oxygen, mainly O_3, is removed by reaction with increasingly smaller concentrations of O). Below 30 km, it rises to values measured in months, or even years. The abundance and distribution of O_3 at low altitudes thus depends on the rate at which it is supplied by transport. The assumption of chemical equilibrium is valid only for altitudes above about 30 km.

Stratospheric Ozone (II): The Role of Radicals

14

Concentrations of O_3, computed with the idealized Chapman scheme, are greater than values observed over much of the middle region of the stratosphere, as illustrated in Figure 13.10. The discrepancy is due in large measure, as we shall see, to the neglect of paths for removal of odd oxygen over and above the single loss reaction included in the Chapman scheme, reaction (13.4). The focus in this chapter is on the role of **radicals.**[1] We shall indicate how the removal of odd oxygen can be accelerated in the presence of even trace quantities of species such as NO_2, OH, HO_2, ClO, and BrO. Understanding the role of radicals is key to understanding the effects of human activity on the stratosphere: the issue raised by Crutzen and Johnston—that operations of supersonic aircraft in the stratosphere could result in a significant reduction in the abundance of ozone—specifically targeted emissions of nitrogen oxides associated with the exhaust gases of aircraft; Molina and Rowland drew attention to the importance of chlorine radicals introduced into the stratosphere as by-products of the decomposition of industrial CFCs: a number of subsequent studies (Wofsy, McElroy, and Yung 1975; Yung et al. 1980) suggested that industrial activity could be responsible also for a significant source of bromine radicals. This chapter is intended to provide the chemical background and to outline the approaches used to assess the significance of these perturbations.

We begin in Section 14.1 with a discussion of the chemistry of nitrogen and hydrogen radicals, following with a treatment of the chemistry of chlorine and bromine radicals in Sections 14.2 and 14.3, concluding with a summary of the key elements of the presentation in Section 14.4.

14.1 Chemistry of an Oxygen-Nitrogen-Hydrogen Atmosphere

The stratosphere contains small but significant concentrations of nitrogen and hydrogen radicals, notably NO, NO_2, OH, and HO_2, and the chemistry of O_3 in the natural stratosphere is intimately linked to the chemistry of these compounds. Expanding the scope of our discussion to include nitrogen and hydrogen leads to a significant increase in complexity. The Chapman scheme, as we saw, involved four reactions; by the end of this Section, the number of reactions implicated in our description of stratospheric O_3 will have increased to twenty-six, and the number of species will have expanded from Chapman's three (O, O_3, and O_2) to more than fifteen. The approach followed here involves a logical, if complex, extension of the treatment developed in the previous chapter for a stratosphere assumed to be composed simply of O_2 and N_2. We continue to suppose that rates for chemical production and loss of individual species are in balance. We shall show that complications introduced in the expanded chemical scheme are more comprehensible if interpreted using a generalization of concepts developed in the Chapman formalism; only a small number of the additional reactions are directly implicated in the loss of odd oxygen. It will be convenient, as we shall see, to expand the definition of odd oxygen to include HO_2 and NO_2.

Radicals act as **catalysts** for the removal of odd oxygen. *Webster's New Collegiate Dictionary* defines a catalyst as "a substance that initiates a chemical reaction and enables it to proceed under milder conditions (as at a lower temperature) than otherwise possible." As we shall discuss below, nitric oxide (NO) allows reaction (13.4) to proceed faster ("under milder conditions") than would be the case in its absence. In this sense, NO serves as a catalyst for reaction (13.4). The key reactions are

$$NO + O_3 \rightarrow NO_2 + O_2 \tag{14.1}$$

followed by

$$NO_2 + O \rightarrow NO + O_2 \tag{14.2}$$

The sequence (14.1) followed by (14.2) is equivalent to O + $O_3 \rightarrow 2O_2$, or reaction (13.4). Nitric oxide is consumed in (14.1) but reformed in (14.2). An individual NO molecule, participating in reactions (14.1) and (14.2), can account for the removal of many molecules of odd oxygen over the course of its life in the stratosphere. As we shall see, the sequence of reaction (14.1) followed by (14.2) provides a dominant path for removal of odd oxygen in the contemporary stratosphere.

There is another important path for conversion of NO_2 to NO:

$$NO_2 + h\nu \rightarrow NO + O \tag{14.3}$$

Reaction (14.1), followed by (14.3) and (13.2), results in no net change in the abundance of odd oxygen. We refer to the combination of reactions in this case as a *do nothing*, or *null sequence* for odd oxygen.

The key to removal of odd oxygen involves the fate of NO_2. If conversion of NO_2 to NO occurs by way of (14.2), the NO $\rightarrow$ NO_2 $\rightarrow$ NO cycle is associated with the removal of two compounds of odd oxygen; if it proceeds by (14.3) the cycle involves no net loss of odd oxygen. Reaction (14.2) is identified as the **rate-limiting step** for the removal of odd oxygen by the NO-catalyzed sequence of reaction (14.1) followed by (14.2). The reactions responsible for interconversion of NO and NO_2 are schematically illustrated in Figure 14.1. The rate-limiting step for removal of odd oxygen is indicated by the heavy line connecting NO_2 to NO.

Rates (cm^{-3} sec^{-1}) for reactions (14.1)–(14.3) are given by

$$R_{14.1} = k_{14.1}\,[NO][O_3], \tag{14.4}$$

$$R_{14.2} = k_{14.2}\,[NO_2\,][O], \tag{14.5}$$

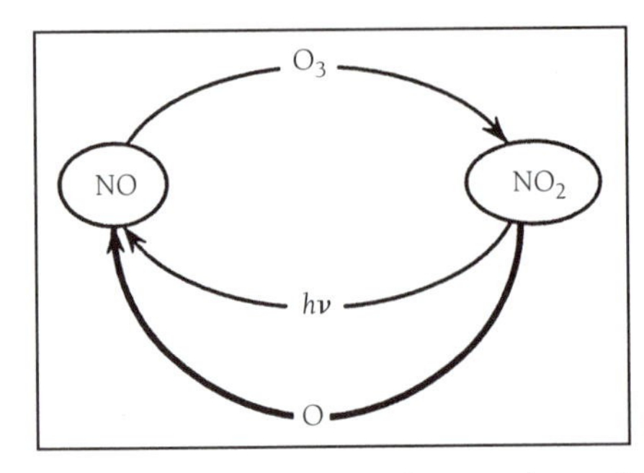

Figure 14.1 The interconversion of NO and NO_2 by reactions (14.1), (14.2), and (14.3). The heavy line indicates the conversion of NO_2 to NO by the rate-limiting step in odd oxygen removal (reaction 14.2).

and

$$R_{14.3} = J_{14.3}\,[NO_2\,], \tag{14.6}$$

respectively. Time constants for the conversion of NO to NO_2 by reaction (14.1), and for conversion of NO_2 to NO by reactions (14.2) and (14.3), are given by

$$\tau_{NO}\,(14.1) = (k_{14.1}\,[O_3])^{-1}, \tag{14.7}$$

$$\tau_{NO_2}\,(14.2) = (k_{14.2}\,[O])^{-1}, \tag{14.8}$$

and

$$\tau_{NO_2}\,(14.3) = (J_{14.3})^{-1} \tag{14.9}$$

Results for $\tau_{NO}(14.1)$, $\tau_{NO_2}(14.2)$, and $\tau_{NO_2}(14.3)$ are presented in Table 14.1 for altitudes of 40 km, 30 km, and 20 km for a latitude of 30°N at equinox.

Example 14.1: Use the results for $\tau_{NO_2}(14.2)$ and $\tau_{NO_2}(14.3)$ presented for altitudes of 40 km, 30 km, and 20 km in Table 14.1 to estimate the probability that the NO $\rightarrow$ NO_2 $\rightarrow$ NO cycle is completed by the path contributing to odd oxygen loss, rather than by the "do nothing" loop.

Answer: The probability that the loop is carried out using (14.2) rather than (14.3) depends on the rates for these reactions. The probability of carrying out the cycle by (14.2) is given by

$$P = \frac{R_{14.2}}{R_{14.2} + R_{14.3}}$$

$$= \frac{k_{14.2}[NO_2][O]}{k_{14.2}[NO_2][O] + J_{14.3}[NO_2]}$$

Table 14.1 Lifetimes for NO and NO_2 as Determined by Specific Reactions as Indicated by Reference Numbers in Parentheses

Time Constant by Specific Reaction	*40 km*	*30 km*	*20 km*
$\tau_{NO}(14.1)$	2.0×10^2 sec	6.0×10^1 sec	7.4×10^1 sec
$\tau_{NO_2}(14.2)$	7.2×10^2 sec	3.5×10^3 sec	1.1×10^5 sec
$\tau_{NO_2}(14.3)$	4.0×10^2 sec	2.0×10^2 sec	2.4×10^2 sec

$$= \frac{k_{14.2}[O]}{k_{14.2}[O] + J_{14.3}}$$

$$= \frac{\left(\tau_{NO_2}(14.2)\right)^{-1}}{\left(\tau_{NO_2}(14.2)\right)^{-1} + \left(\tau_{NO_2}(14.3)\right)^{-1}}$$

Values of P for 40 km, 30 km, and 20 km are given by

$$P(40\text{ km}) = \frac{(7.2 \times 10^2)^{-1}}{(7.2 \times 10^2)^{-1} + (4.0 \times 10^2)^{-1}}$$

$$= \frac{1.4 \times 10^{-3}}{1.4 \times 10^{-3} + 2.5 \times 10^{-3}}$$

$$= 0.35$$

$$P(30\text{ km}) = \frac{(3.5 \times 10^3)^{-1}}{(3.5 \times 10^3)^{-1} + (2.0 \times 10^2)^{-1}}$$

$$= \frac{2.86 \times 10^{-4}}{2.86 \times 10^{-4} + 5.0 \times 10^{-3}}$$

$$= 5.4 \times 10^{-2}$$

$$P(20\text{ km}) = \frac{(1.1 \times 10^5)^{-1}}{(1.1 \times 10^5)^{-1} + (2.4 \times 10^2)^{-1}}$$

$$= \frac{9.1 \times 10^{-6}}{9.1 \times 10^{-6} + 4.2 \times 10^{-3}}$$

$$= 2.2 \times 10^{-3}$$

The probability, P', that the cycle is traversed by the "do nothing" loop is given by subtracting these values from unity:

$$P'(40\text{ km}) = 1 - P(40\text{ km}) = 0.65$$

$$P'(30\text{ km}) = 1 - P(30\text{ km}) = 0.95$$

$$P'(20\text{ km}) = 1 - P(20\text{ km}) = 0.998$$ ■

As we have seen, odd oxygen is consumed when the cycle depicted in Figure 14.1 is completed by way of reaction (14.2) (recall that each loop of the cycle by this path is equivalent to reaction 13.4). We expect, therefore, that the removal rate of odd oxygen by NO catalysis should be represented by

$$L_{\text{Odd Oxygen}} = 2\,k_{14.2}\,[NO_2][O] \qquad (14.10)$$

This may be established as follows. Using the definition of odd oxygen introduced in Section 13.1, reactions (14.1) and (14.2) are associated with the removal of odd oxygen (O_3 in the former case, O in the latter), whereas (14.3) is involved with producing odd oxygen (specifically O).

$$L_{\text{Odd Oxygen}} = k_{14.1}[NO][O_3] + k_{14.2}[NO_2][O] - J_{14.3}[NO_2] \qquad (14.11)$$

In photochemical equilibrium, the rate of NO_2 formation by reaction (14.1) must equal the rate at which it is consumed by (14.2) and (14.3). Thus,

$$k_{14.1}[NO][O_3] = k_{14.2}[NO_2][O] + J_{14.3}[NO_2] \qquad (14.12)$$

Substituting this expression for $k_{14.1}[NO][O_3]$ in (14.11), we establish the identity of the relations (14.10) and (14.11) for $L_{\text{Odd Oxygen}}$:

$$L_{\text{Odd Oxygen}} = k_{14.1}[NO][O_3] + k_{14.2}[NO_2][O] - J_{14.3}[NO_2] \qquad (14.13)$$

$$= k_{14.2}[NO_2][O] + J_{14.3}[NO_2] + k_{14.2}[NO_2][O] - J_{14.3}[NO_2]$$

$$= 2\,k_{14.2}[NO_2][O]$$

The analysis above suggests an expansion of the definition of odd oxygen to include NO_2 in addition to O and O_3. It is clear that reaction (14.2) involves the loss of two forms of odd oxygen: O and NO_2. Odd oxygen is neither produced nor consumed by reactions (14.1) and (14.3). It is simply converted from one form to another: from O_3 to NO_2 in (14.1), from NO_2 to O in (14.3). As with (13.2) and (13.3), reactions (14.1) and (14.3) play an important role in regulating the relative abundance of the separate forms of odd oxygen; the loss of odd oxygen—the hole in the barrel previously discussed—is controlled by (13.4), extended now to account additionally for (14.2).

Nitrogen radicals are introduced into the stratosphere mainly by the reaction of nitrous oxide, N_2O, with an electronically excited, metastable form of O, denoted by $O(^1D)$.[2]

$$O(^1D) + N_2O \rightarrow NO + NO \qquad (14.14)$$

The label ^1D defines a specific excited state of the oxygen atom. The $O(^1D)$ oxygen atom is produced by photolysis of O_3 by solar radiation at wavelengths below about 300 nm:

$$h\nu + O_3 \rightarrow O(^1D) + O_2 \qquad (14.15)$$

Nitrous oxide is a relatively stable component of the atmosphere produced (as discussed in Chapter 12) by microbial processes in soils and the ocean. It is removed from the atmosphere primarily by photolysis in the ultraviolet portion of the light spectrum by

$$h\nu + N_2O \rightarrow N_2 + O, \qquad (14.16)$$

with additional, minor removal by (14.14) and

$$O(^1D) + N_2O \rightarrow O_2 + N_2 \qquad (14.17)$$

Nitrogen radicals are converted to HNO_3 and N_2O_5 by the reactions

$$OH + NO_2 + M \rightarrow HNO_3 + M \qquad (14.18)$$

and

$$NO_2 + NO_3 + M \rightarrow N_2O_5 + M, \qquad (14.19)$$

where, as before, M denotes the passive third body required to carry off the excess energy generated by these associative reactions (the role of M in reactions 14.18 and 14.19 is similar to that discussed earlier for reaction 13.2). The relatively unstable compound NO_3 is formed by reaction of NO_2 with O_3,

$$NO_2 + O_3 \rightarrow NO_3 + O_2, \quad (14.20)$$

and it is removed by photolysis, either by

$$h\nu + NO_3 \rightarrow NO + O_2 \quad (14.21)$$

or by

$$h\nu + NO_3 \rightarrow NO_2 + O \quad (14.22)$$

The hydroxyl radical, OH, which appears in (14.18), is produced by reaction of $O(^1D)$ with H_2O:

$$O(^1D) + H_2O \rightarrow OH + OH \quad (14.23)$$

Its abundance is controlled by (14.18), (14.23), and a series of reactions including

$$OH + O_3 \rightarrow O_2 + HO_2, \quad (14.24)$$

$$HO_2 + O_3 \rightarrow OH + O_2 + O_2, \quad (14.25)$$

$$NO + HO_2 \rightarrow NO_2 + OH, \quad (14.26)$$

$$HO_2 + HO_2 \rightarrow H_2O_2 + O_2, \quad (14.27)$$

$$OH + HNO_3 \rightarrow H_2O + NO_3, \quad (14.28)$$

$$h\nu + HNO_3 \rightarrow OH + NO_2, \quad (14.29)$$

and

$$h\nu + H_2O_2 \rightarrow OH + OH \quad (14.30)$$

It is convenient to once again expand the definition of odd oxygen to include HO_2 (in addition to O, $O(^1D)$, O_3, and NO_2). The expanded set of reactions provides additional catalytic paths for the removal of odd oxygen. For example, reaction (14.24) followed by (14.25), and (14.20) followed by (14.21) and (14.1), are equivalent to

$$O_3 + O_3 \rightarrow O_2 + O_2 + O_2 \quad (14.31)$$

These cycles are illustrated in Figures 14.2 and 14.3.

Some of the new loss mechanisms for odd oxygen are transitory: reaction (14.18), for example, represents a sink, but it is almost precisely balanced by the compensatory reaction (14.29). A summary of source and sink reactions for odd oxygen is given in Table 14.2. A list of reactions included in contemporary state-of-the-art models of stratospheric chemistry is given in Table 14.3.

In light of the importance of NO and NO_2 for removal of odd oxygen, it is convenient to recognize these species as a related pair. Atmospheric chemists refer to the combination NO + NO_2 as $\mathbf{NO_x}$. The composite of all of the reactive nitrogenous species emphasized here, NO + NO_2 + NO_3 + HNO_3 + 2 × N_2O_5, is referred to as $\mathbf{NO_y}$. The definition of NO_y is inclusive with respect to reactive nitrogenous species: it omits only the relatively stable compounds N_2 and N_2O. The hydrogenous species, OH + HO_2 + 2 × H_2O_2 + HNO_3, are

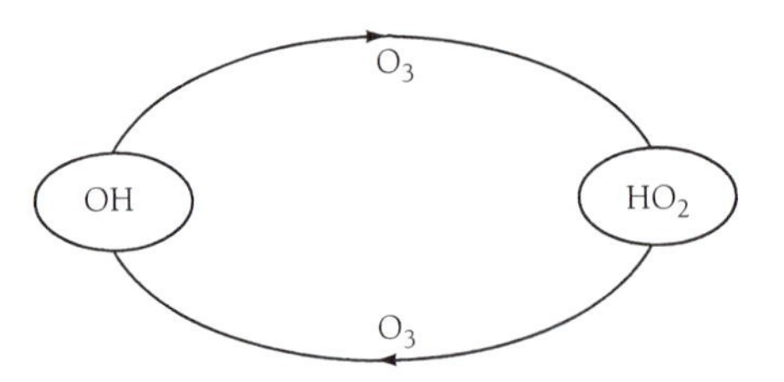

Figure 14.2 A path for reaction of O_3 with itself, catalyzed by OH (as discussed in the text), resulting in the formation of three molecules of O_2.

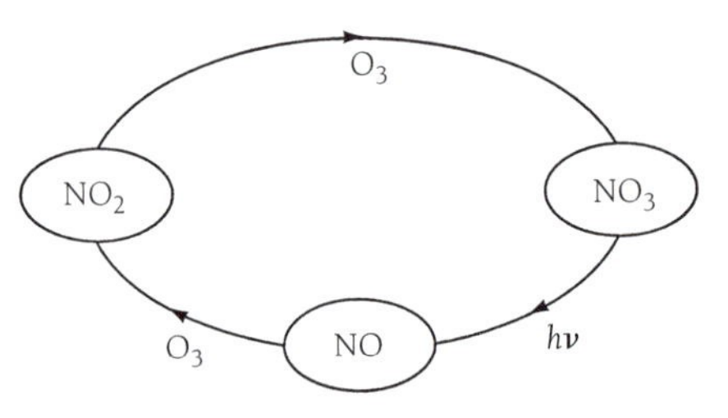

Figure 14.3 A path for reaction of O_3 with itself, catalyzed by NO_2 (as discussed in the text), resulting in the formation of three molecules of O_2.

Table 14.2 Source and Sink Reactions for Odd Oxygen

Odd Oxygen Production:
$O_2 + h\nu \rightarrow O + O$
$H + O_2 + M \rightarrow HO_2 + M$
Odd Oxygen Loss:
$O + O_3 \rightarrow O_2 + O_2$
$O + O + M \rightarrow O_2 + M$
$O + NO_2 \rightarrow NO + O_2$
$NO_2 + HO_2 \rightarrow HNO_2 + O_2$
$HO_2 + NO_2 + M \rightarrow HNO_4 + M$
$NO_2 + O_3 \rightarrow NO_3 + O_2$
$NO_3 + h\nu \rightarrow NO + O_2$
$O + HO_2 \rightarrow OH + O_2$
$O_3 + HO_2 \rightarrow OH + O_2 + O_2$
$H + HO_2 \rightarrow H_2 + O_2$
$OH + HO_2 \rightarrow H_2O + O_2$
$HO_2 + HO_2 \rightarrow H_2O_2 + O_2$
$Cl + HO_2 \rightarrow HCl + O_2$
$ClO + O \rightarrow Cl + O_2$
$ClO + ClO \rightarrow Cl_2 + O_2$
$ClO + HO_2 \rightarrow HOCl + O_2$
$ClO + ClO + M \rightarrow Cl_2O_2 + M$
$BrO + O \rightarrow Br + O_2$
$BrO + BrO \rightarrow 2Br + O_2$
$BrO + BrO \rightarrow Br_2 + O_2$
$BrO + HO_2 \rightarrow HOBr + O_2$
$ClO + BrO \rightarrow Cl + Br + O_2$
$ClO + BrO \rightarrow BrCl + O_2$

Table 14.3 Reactions Included in Contemporary Models of Stratospheric Chemistry

$O_2 + h\nu \rightarrow O + O$	$O_3 + h\nu \rightarrow O(^1D) + O_2$
$O_3 + h\nu \rightarrow O(all) + O_2$	$O(^1D) + M \rightarrow O + M$
$O(^1D) + O_2 \rightarrow O + O_2$	$O + O_2 + M \rightarrow O_3 + M$
$O + O_3 \rightarrow O_2 + O_2$	$O + O + M \rightarrow O_2 + M$
$O(^1D) + N_2O \rightarrow NO + NO$	$O(^1D) + N_2O \rightarrow N_2 + O_2$
$O + NO_2 \rightarrow NO + O_2$	$NO_2 + h\nu \rightarrow NO + O$
$O_3 + NO \rightarrow NO_2 + O_2$	$O + HNO_3 \rightarrow OH + NO_3$
$O + NO + M \rightarrow NO_2 + M$	$O + NO_2 + M \rightarrow NO_3 + M$
$O_3 + NO_2 \rightarrow O_2 + NO_3$	$H + NO_2 \rightarrow OH + NO$
$HO_2 + NO_3 \rightarrow OH + NO_2$	$NO_2 + OH \rightarrow HNO_3$
$HNO_3 + h\nu \rightarrow OH + NO_2$	$HNO_3 + OH \rightarrow H_2O + NO_3$
$NO + OH \rightarrow HNO_2$	$NO_2 + HO_2 \rightarrow HNO_2 + O_2$
$HNO_2 + h\nu \rightarrow OH + NO$	$HNO_2 + OH \rightarrow H_2O + NO_2$
$HO_2 + NO_2 \rightarrow HNO_4$	$HNO_4 \rightarrow HO_2 + NO_2$
$HNO_4 + h\nu \rightarrow OH + NO_3$	$HNO_4 + OH \rightarrow H_2O + NO_2 + O_2$
$NO_3 + h\nu \rightarrow NO_2 + O$	$NO_3 + h\nu \rightarrow NO + O_2$
$NO_3 + NO \rightarrow 2NO_2$	$NO_3 + NO_2 \rightarrow NO + O_2 + NO_2$
$NO_3 + NO_3 \rightarrow 2NO_2 + O_2$	$NO_2 + NO_3 \rightarrow N_2O_5$
$N_2O_5 \rightarrow NO_2 + NO_3$	$N_2O_5 + h\nu \rightarrow NO_2 + NO_3$
$NO + h\nu \rightarrow N + O$	$N + O_2 \rightarrow NO + O$
$N + O_3 \rightarrow NO + O_2$	$N + NO \rightarrow N_2 + O$
$N + NO_2 \rightarrow N_2O + O$	$NH_3 + OH \rightarrow NH_2 + H_2O$
$NH_2 + O_3 \rightarrow NO_X + \ldots$	$NH_2 + NO \rightarrow N_2 + \ldots$
$ClO + NO \rightarrow Cl + NO_2$	$ClO + NO_2 \rightarrow ClNO_3$
$N_2O + h\nu \rightarrow N_2 + O$	$Cl + HNO \rightarrow HCl + NO_2 + O_2$
$ClNO_3 \rightarrow ClO + NO_2$	$ClNO_3 + h\nu \rightarrow O + ClONO$
$ClNO_3 + h\nu \rightarrow Cl + NO_3$	$ClNO_3 + O \rightarrow ClO + NO_3$
$ClNO_3 + OH \rightarrow HOCl + NO_3$	$ClNO_3 + H_2O$ (aerosol) $\rightarrow HOCl + HNO_3$
$ClNO_3 + HCl$ (aerosol) $\rightarrow Cl_2 + HNO_3$	$N_2O_5 + H_2O$ (aerosol) $\rightarrow 2HNO_3$
$N_2O_5 + HCl$ (aerosol) $\rightarrow HNO_3 + ClNO_2$	$ClNO_2 + h\nu \rightarrow Cl + NO_2$
$NO + ClNO_3 \rightarrow ClONO + NO_2$	$ClONO + h\nu \rightarrow Cl + NO_2$
$Cl + NO_{23} \rightarrow ClONO$	$Cl + NO_2 \rightarrow ClNO_2$
$ClNO_3 + O \rightarrow ClONO + O_2$	$CH_3OO + NO \rightarrow RO + NO_2$
$NO + OClO \rightarrow NO_2 + ClO$	$O(^1D) + N_2 + M \rightarrow N_2O + M$
$O + NO_3 \rightarrow O_2 + NO_2$	$NO_3 + O_2 \rightarrow NO + O_2 + O_2$
$NH_2 + NO_2 \rightarrow N_2 + \ldots$	
$O(^1D) + H_2O \rightarrow OH + OH$	$O(^1D) + H_2 \rightarrow OH + H$
$O(^1D) + CH_4 \rightarrow OH + CH_3$	$O(^1D) + CH_4 \rightarrow H_2 + H_2CO$
$O + H_2 \rightarrow OH + H$	$CO_2 + h\nu \rightarrow CO + O$
$O + OH \rightarrow O_2 + H$	$O + HO_2 \rightarrow OH + O_2$
$O + H_2O_2 \rightarrow OH + HO_2$	$O_3 + H \rightarrow OH + O_2$
$O + H_2O_2 \rightarrow OH + HO_2$	$O_3 + H \rightarrow OH + O_2$
$O_3 + OH \rightarrow HO_2 + O_2$	$O_3 + HO_2 \rightarrow OH + O_2 + O_2$
$H_2O + h\nu \rightarrow H + OH$	$HO_2 + h\nu \rightarrow O + OH$
$H_2O_2 + h\nu \rightarrow OH + OH$	$H + O_2 + M \rightarrow HO_2 + M$
$H + HO_2 \rightarrow OH + OH$	$H + HO_2 \rightarrow H_2 + O_2$
$H + HO_2 \rightarrow H_2O + O$	$H + H_2O_2 \rightarrow OH + H_2O$
$H + H_2O_2 \rightarrow H_2 + HO_2$	$OH + OH \rightarrow H_2O + O$
$OH + HO_2 \rightarrow H_2O + O_2$	$OH + H_2O_2 \rightarrow H_2O + HO_2$
$HO_2 + HO_2 \rightarrow H_2O_2 + O_2$	$OH + H_2 \rightarrow H_2O + H$
$OH + CO \rightarrow CO_2 + H$	$OH + CH_4 \rightarrow CH_3 + H_2O$
$CH_4 + h\nu \rightarrow \ldots H_2CO$	$CH_3OO + HO_2 \rightarrow ROOH + O_2$
$CH_3OO + CH_3OO \rightarrow R_2O_2 + O_2$	$CH_3OOH + h\nu \rightarrow CH_3O + OH$
$CH_3OOH + OH \rightarrow RO + H_2O$	$H_2CO + OH \rightarrow HCO + H_2O$
$H_2CO + h\nu \rightarrow H + HCO$	$H_2CO + h\nu \rightarrow H_2 + CO$
$CH_4 + h\nu \rightarrow H_2 + \ldots$	

(continued)

Table 14.3 Reactions Included in Contemporary Models of Stratospheric Chemistry (continued)

$HCl + OH \rightarrow Cl + H_2O$	$HCl + O \rightarrow Cl + OH$
$HCl + H \rightarrow Cl + H_2$	$HCl + hv \rightarrow Cl + H$
$Cl + CH_4 \rightarrow HCl + CH_3$	$Cl + H_2CO \rightarrow HCl + HCO$
$Cl + H_2 \rightarrow HCl + H$	$Cl + HO_2 \rightarrow HCl + O_2$
$Cl + H_2O_2 \rightarrow HCl + HO_2$	$Cl + Cl + M \rightarrow Cl_2 + M$
$Cl_2 + hv \rightarrow Cl + Cl$	$Cl_2 + H \rightarrow HCl + Cl$
$Cl_2 + O \rightarrow ClO + Cl$	$Cl + O_3 \rightarrow ClO + O_2$
$ClO + O \rightarrow Cl + O_2$	$ClO + hv \rightarrow Cl + O$
$ClO + ClO \rightarrow Cl_2 + O_2$	$ClO + H \rightarrow OH + Cl$
$ClO + OH \rightarrow Cl + HO_2$	$ClO + HO_2 \rightarrow HCl + O_3$
$ClO + HO_2 \rightarrow HOCl + O_2$	$HOCl + hv \rightarrow OH + Cl$
$HOCl + OH \rightarrow H_2O + ClO$	$ClO + ClO \rightarrow OClO + Cl$
$OClO + hv \rightarrow ClO + O$	$OClO + O \rightarrow ClO + O_2$
$ClO + O_3 \rightarrow OClO + O_2$	$Cl + OClO \rightarrow ClO + ClO$
$ClO + OClO \rightarrow ClOO + ClO$	$ClO + ClO \rightarrow Cl + ClOO$
$ClO + ClO \rightarrow ClO_2O_2$	$Cl_2O_2 + hv \rightarrow Cl + ClOO$
$Cl_2O_2 + hv \rightarrow Cl + OClO$	$ClOO + M \rightarrow Cl + O_2 + M$
$ClOO + hv \rightarrow ClO + O$	$Cl + ClOO \rightarrow Cl_2 + O_2$
$Cl + ClOO \rightarrow ClO + ClO$	$Cl + O_2 + M \rightarrow ClOO + M$
$O(^1D) + O_3 \rightarrow O_2 + O_2$	$Cl_2O_2 + M \rightarrow ClO + ClO$
$O(^1D) + CCl_4 \rightarrow \ldots$	$O(^1D) + CFCl_3 \rightarrow \ldots$
$O(^1D) + CF_2Cl_2 \rightarrow \ldots$	$O(^1D) + HCl \rightarrow OH + Cl$
$ClO + OH \rightarrow HCl + O_2$	$CF_2Cl_2 + hv \rightarrow \ldots$
$CFCl_3 + hv \rightarrow \ldots$	$CCl_4 + hv \rightarrow \ldots$
$CH_3Cl + hv \rightarrow \ldots$	$CH_3Cl + OH \rightarrow \ldots$
$CH_3CCl_3 + hv \rightarrow \ldots$	$CH_3CCl_3 + OH \rightarrow \ldots$
$HBr + OH \rightarrow Br + H_2O$	$HBr + O \rightarrow Br + OH$
$HBr + hv \rightarrow H + Br$	$Br + HO_2 \rightarrow HBr + O_2$
$Br + O_3 \rightarrow BrO + O_2$	$BrO + O \rightarrow Br + O_2$
$BrO + NO \rightarrow Br + NO_2$	$BrO + O_3 \rightarrow Br + 2O_2$
$BrO + BrO \rightarrow {}_2Br + O_2$	$BrO + BrO \rightarrow Br_2 + O_2$
$BrO + hv \rightarrow Br + O$	$BrO + HO_2 \rightarrow HOBr + O_2$
$HOBr + hv \rightarrow Br + OH$	$HOBr + OH \rightarrow BrO + H_2O$
$BrO + NO_2 \rightarrow BrNO_3$	$BrNO_3 + hv \rightarrow BrO + NO_2$
$BrNO_3 + hv \rightarrow Br + NO_3$	$ClO + BrO \rightarrow Cl + Br + O_2$
$Br + H_2CO \rightarrow HBr + HCO$	$ClO + BrO \rightarrow OClO + Br$
$ClO + BrO \rightarrow BrCl + O_2$	$BrCl + hv \rightarrow Br + Cl$
$CH_3Br + hv \rightarrow \ldots$	$CH_3Br + OH \rightarrow \ldots$

combined in a family referred to as **HO_x**. Obviously the definitions of NO_y and HO_x are not exclusive: HNO_3 is simultaneously included as a member of both the NO_y and HO_x families.

Figures 14.4 and 14.5 present a pictorial view of the reactions important in regulating the concentrations of individual members of the NO_y and HO_x families. Time constants associated with individual gases are indicated for selected NO_y and HO_x species for altitudes of 40 km, 30 km, and 20 km in Tables 14.4 and 14.5. Mixing ratios for individual NO_y species, averaged over a 24-hour period, are depicted as functions of altitude in Figure 14.6, while comparable data for HO_x are given in Figure 14.7.

Rates for production and loss of odd oxygen, illustrating the significance of the nitrogen and hydrogen catalytic processes, are presented as functions of altitude in Figure 14.8. Reaction (13.1) provides the only source of odd oxygen. Note the dominant role of (14.2) in removal of odd oxygen for altitudes up to about 42 km. The Chapman reaction (13.4) is relatively unimportant, except at high altitude, and even there its contribution is dominated by reactions involving HO_x. The results shown here refer to conditions for 30°N at equinox and attest to the significance of nitrogen- and hydrogen-catalyzed paths for removal of odd oxygen. The results in Figure 14.8 were computed using observed values of O_3. The imbalance between computed net production and loss incidated for altitudes above about 42 km may reflect inaccuracies in assumed values for $[O_3]$ or errors in computed values for net source minus sink (or both).

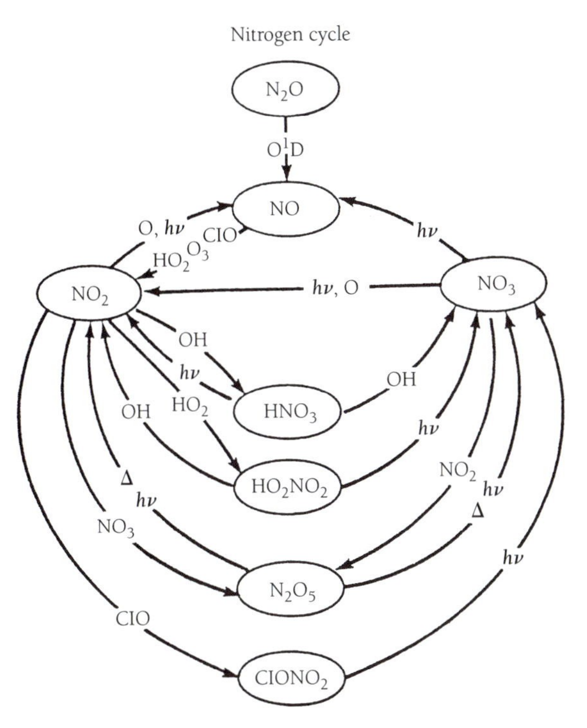

Figure 14.4 The complex paths by which exchange proceeds among members of the NO_y family. Production of NO_y is associated with reaction of $O(^1D)$ with N_2O.

Note that the budget of NO_y as presented above is unbalanced; reaction (14.14) provides a source, but there is no comparable, off-setting in situ chemical loss. Compounds of NO_y formed by (14.14) are removed, for the most part, by transport. There is a sink for NO_y at high altitudes associated with the reaction

$$N + NO \rightarrow N_2 + O \quad (14.32)$$

Reaction (14.32) is limited mainly by the lack of a significant source for N. The only known source of N for the stratosphere is

$$h\nu + NO \rightarrow N + O \quad (14.33)$$

Most of the NO_y produced by (14.14) is carried out of the stratosphere to the troposphere, where it is cleansed from the atmosphere by precipitation or by contact with the surface. Removal of N_2O is a one-way street; the gas, once destroyed, is not reformed in the atmosphere. In general, the mixing ratio of N_2O declines with altitude, whereas NO_y increases, reaching a maximum at about 35 km, as indicated in Figure 14.9 a–b.

The relative abundance of NO_y is related simply to that of N_2O, as illustrated in Figure 14.10. The linear relationship between NO_y and N_2O reflects the fact that an essentially constant fraction, about 18%, of the N_2O removed from the atmosphere is associated with the production of NO_y. A small mixing ratio of N_2O implies that a relatively large fraction of the available N_2O has been removed with associated, proportionally significant production of NO_y. Departures from the linear trend at low levels of N_2O reflect the importance of the loss mechanism (reaction 14.32) for NO_y at high altitudes. The photons required to effect reaction (14.33) have wavelengths near 200 nm and are available only at

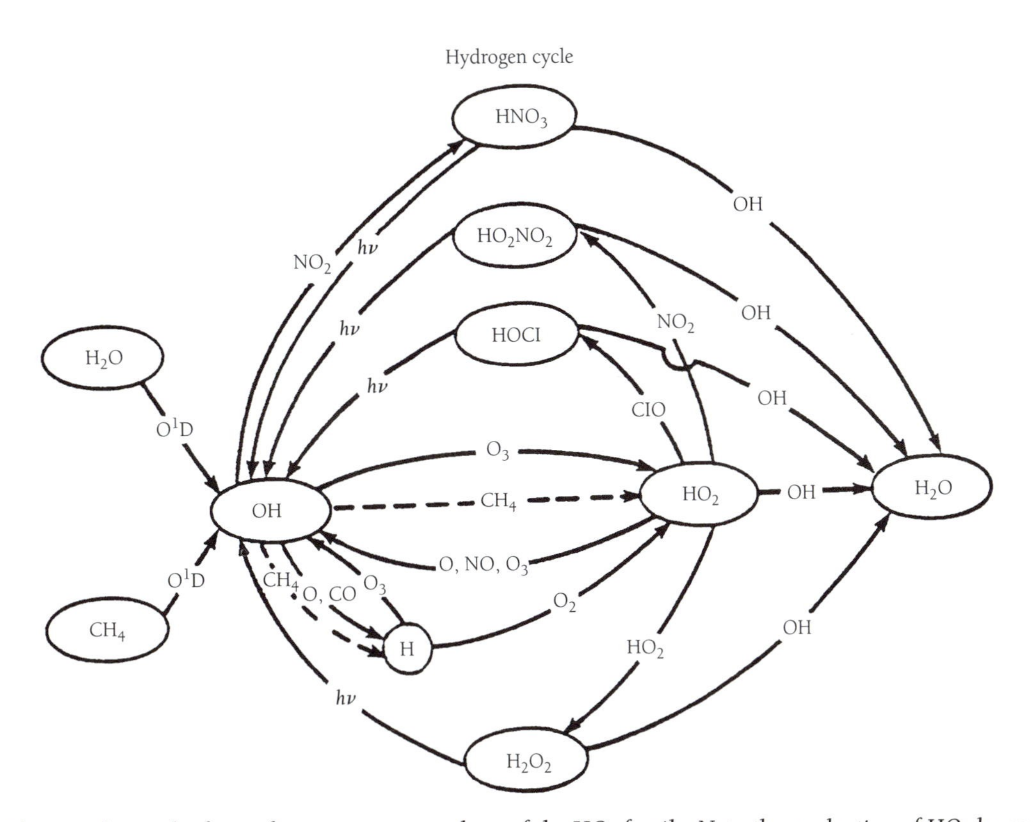

Figure 14.5 The complex paths for exchange among members of the HO_x family. Note the production of HO_x by reaction of $O(^1D)$ with H_2O and CH_4; HO_x is removed mainly by reactions resulting in the formation of H_2O.

Table 14.4 Photochemical Lifetimes (sec) for Selected NO_y Species

	40 km	30 km	20 km
NO	2.0×10^2	60.0	74.0
NO_2	2.6×10^2	1.9×10^2	2.4×10^2
NO_3	1.4×10^3	1.4×10^2	93.0
N_2O_5	2.2×10^4	4.8×10^4	1.0×10^5
HNO_4	1.6×10^4	3.7×10^4	1.6×10^5
HNO_3	3.5×10^4	1.5×10^5	1.8×10^6
$ClNO_3$	2.6×10^4	2.8×10^4	4.2×10^4

Table 14.5 Photochemical Lifetimes (sec) for Selected HO_x Species

	40 km	30 km	20 km
H	8.4×10^{-3}	4.0×10^{-4}	1.4×10^{-5}
OH	20.0	10.0	8.7
HO_2	17.0	39.0	74.0
H_2O_2	2.8×10^4	7.4×10^4	1.4×10^5
HOCl	5.2×10^3	4.0×10^3	4.5×10^3
HNO_4	1.6×10^4	3.7×10^4	1.6×10^5
HNO_3	3.5×10^4	1.5×10^5	1.8×10^6

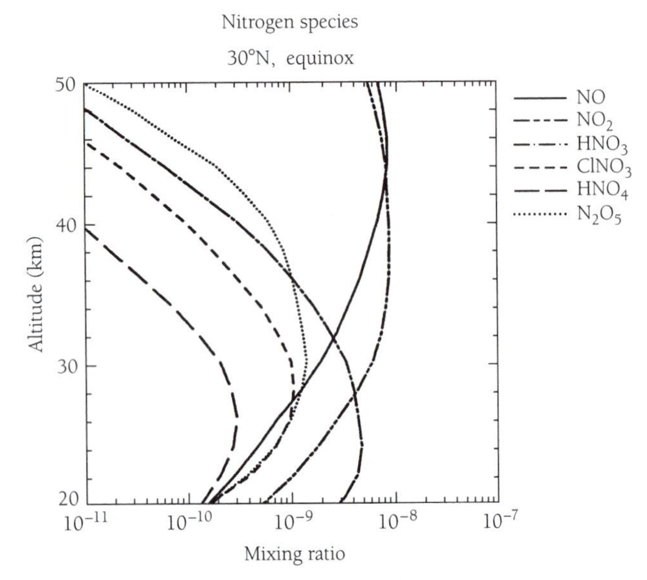

Figure 14.6 Diurnally averaged values for the mixing ratios of NO, NO_2, HNO_3, $ClNO_3$, HNO_4, and N_2O_5 at 30°N equinox.

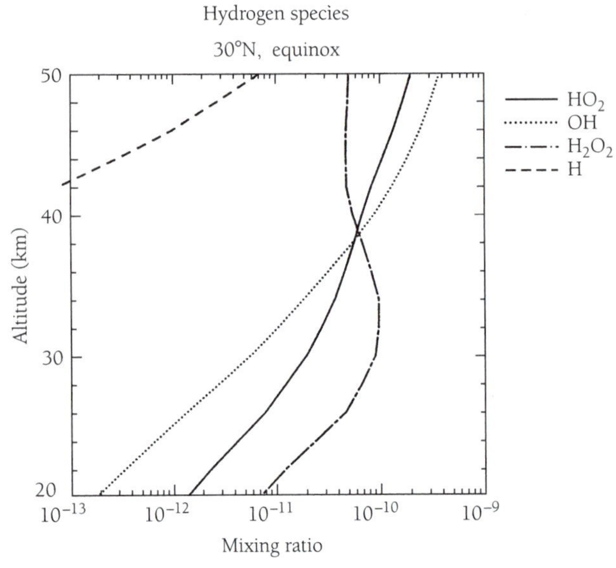

Figure 14.7 Diurnally averaged values for the mixing ratios of OH, HO_2, H_2O_2, and H at 30°N equinox.

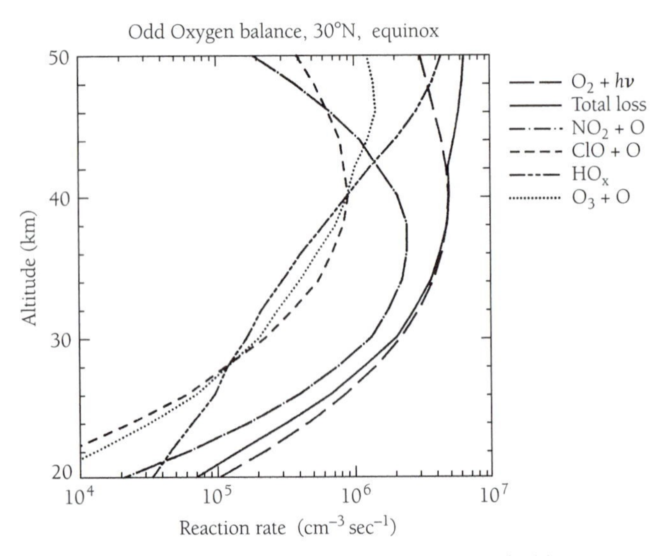

Figure 14.8 Rates for the production and loss of odd oxygen as functions of altitude for 30°N equinox.

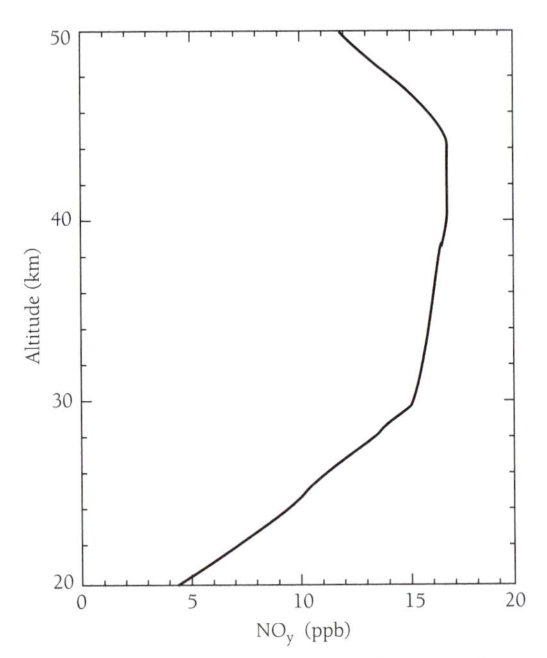

Figure 14.9a Mixing ratio of NO_y as a function of altitude.

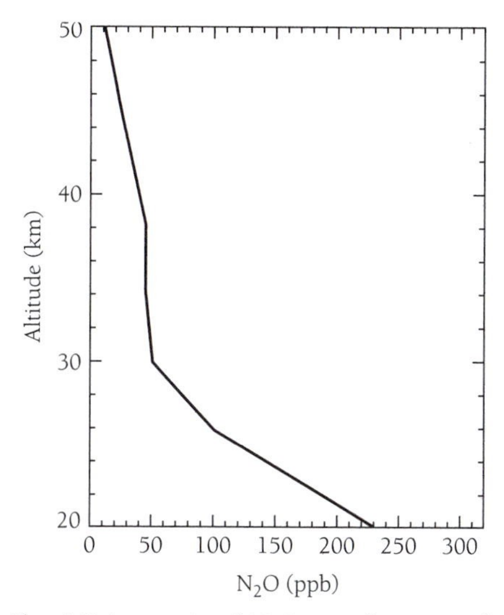

Figure 14.9b Mixing ratio of N_2O as a function of altitude.

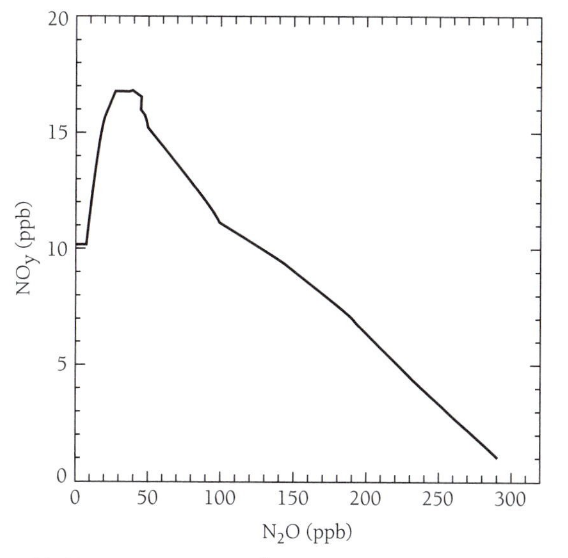

Figure 14.10 Mixing ratio of NO_y, expressed as a function of the N_2O mixing ratio.

altitudes above about 40 km. Reaction (14.33) is responsible for the decline in the mixing ratio of NO_y evident at high altitudes in Figure 14.9a.

14.2 The Chemistry of Chlorine

The atmosphere includes significant and, until recently, growing concentrations of halogenated molecules of industrial origin, the most important of which are the chlorofluorocarbons $CFCl_3$ and CF_2Cl_2 (trichlorofluoromethane and dichlorodifluoro-methane, more commonly known as CFC-11 and CFC-12, respectively).[3] A partial list of the more important compounds, with information on their primary industrial uses, is given in Table 14.6. The mixing ratio of chlorine contributed by these compounds to the atmosphere in 1992 amounted to about 3.7 ppb; it has rapidly risen over the past few decades, from a preindustrial background of about 0.6 ppb, supplied mainly by CH_3Cl (methyl chloride) produced by organisms in the ocean. Estimates of the concentration of atmospheric chlorine as a function of time since 1950, are presented in Figure 14.11.

Fully halogenated hydrocarbons, such as CFC-11 and CFC-12, are extremely long-lived in the atmosphere; they are removed mainly in the stratosphere by photolysis at relatively short wavelengths in the ultraviolet portion of the spectrum. Hydrogen-bearing species such as CH_3CCl_3 (methyl chloroform) and CH_3Cl, generically known as HCFCs, are also removed by photolysis. In addition, HCFCs can react with OH; significant loss can thus occur in the troposphere, and the lifetime of hydrogenated halocarbons is abbreviated accordingly. A summary of estimated lifetimes for a number of the more important halogenated gases is given in Table 14.6.

It is convenient to distinguish from the outset between chlorine contained in halocarbons—the form in which it is delivered to the stratosphere—and chlorine present in the variety of inorganic species formed as a consequence of halocarbon decomposition. Decomposition of chlorinated halocarbons

Table 14.6 Lifetimes and Industrial Sources and Origins for Major Chlorinated and Brominated Gases

Compound	*Chemical formula*	*Atmospheric lifetime (years)*	*Amount used worldwide, 1998 (million kg)*	*Major uses*
CFC-11	($CFCl_3$)	64.0	376.0	Refrigeration, foams
CFC-12	(CF_2Cl_2)	108.0	421.0	Foams, air conditioning, refrigeration
CFC-113	($C_2F_3Cl_3$)	88.0	247.0	Solvents
CFC-114	($CClF_2CClF_2$)	185.0	16.0	Foams, refrigeration
CFC-115	(CF_3CF_2Cl)	380.0	14.0	Refrigeration
Halon 1211	(CF_2ClBr)	25.0	7.0	Portable fire extinguishers
Halon 1301	(CF_3Br)	110	7.0	Total flooding–fire extinguishing systems
Carbon tetrachloride	(CCl_4)	67.0	87.0	Pesticides, solvents, feedstock
Methyl chloroform	(CH_3CCl_3)	8.5	666.0	Solvents, pesticides, aerosols

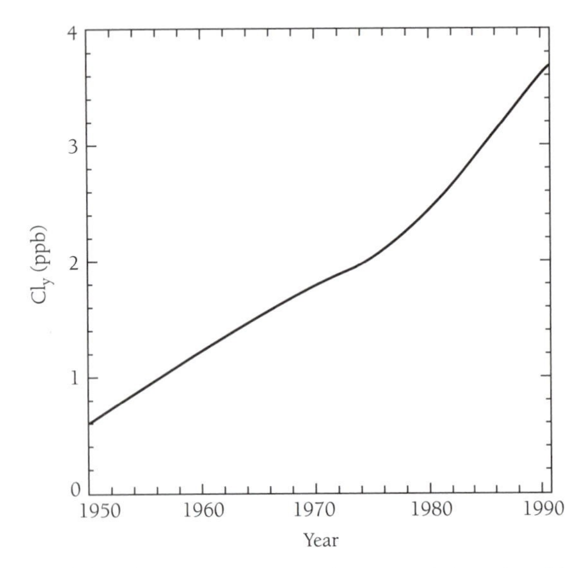

Figure 14.11 Mixing ratio as a function of time for chlorine contained in the form of organic molecules that penetrate into the stratosphere and contribute to the stratospheric budget of reactive chlorine.

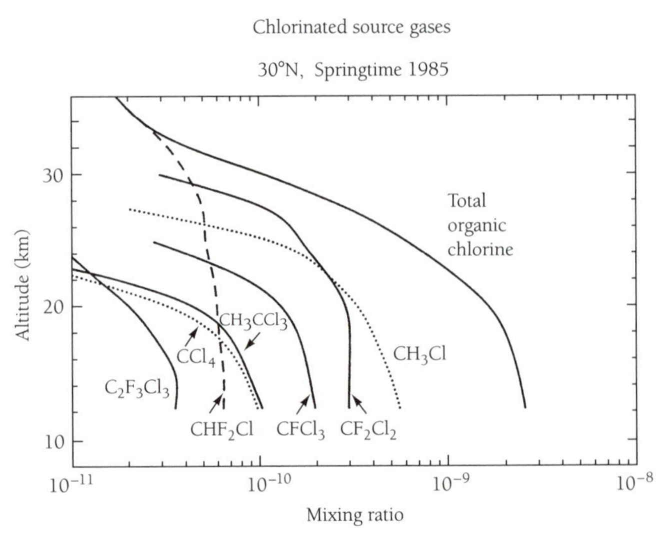

Figure 14.12 Observed concentration of chlorinated source gases as a function of altitude. Measurements of CH_3Cl, CF_2Cl_2, CHF_2Cl, and CCl_4 were obtained near 30°N between 30 April and 1 May 1985 from the ATMOS experience on board Skylab 3 (Zander et al., 1992). Measurements of CH_3CCl_3 and $C_2F_3Cl_3$ were obtained near 30°N during 1983 using balloon-borne instrumentation, and have been scaled by factors of 1.19 and 1.21, respectively, to reflect 1985 values. Source: Meteorological Association 1986.

provides the dominant source of inorganic chlorine in the stratosphere; important inorganic products in the stratosphere include Cl, ClO, HCl, HOCl, and $ClNO_3$. The chlorine composition of an air mass rising from the troposphere to the stratosphere is initially dominated by organic compounds such as CF_2Cl_2, $CFCl_3$, CH_3CCl_3, and CH_3Cl (we refer to these species as **organic**, since the constituent chlorine atoms are bonded to carbon). As the air moves to higher altitudes, it is exposed to sunlight at shorter wavelengths (a consequence of the reduction in the O_3 content of the overlying column). Photolysis rates increase and the composition of the air mass changes as chlorine is converted from organic to inorganic form. The more intense the exposure to short-wavelength ultraviolet solar radiation, the more complete the conversion from organic to inorganic species.

We expect the relative abundance of total chlorine in an air mass, the abundance measured without respect for chemical form, to remain essentially constant as an air mass moves around the stratosphere. In the absence of in situ sources or sinks, and in the absence of mixing with air masses of distinctly different composition, the abundance of chlorine should be fixed, equal to the value that applied when the air entered the stratosphere from the troposphere.[4] The chemical form of chlorine should be dominated by organic compounds at low altitude and by inorganic species at high altitudes. The decrease of the organic chlorine mixing ratio with altitude is illustrated for selected species in Figure 14.12, while the increase of inorganic chlorine with altitude is depicted in Figure 14.13.

Chlorine-catalyzed removal of odd oxygen is dominated by reactions involving the radicals Cl and ClO:

$$Cl + O_3 \rightarrow ClO + O_2 \quad (14.34)$$

and

$$ClO + O \rightarrow Cl + O_2 \quad (14.35)$$

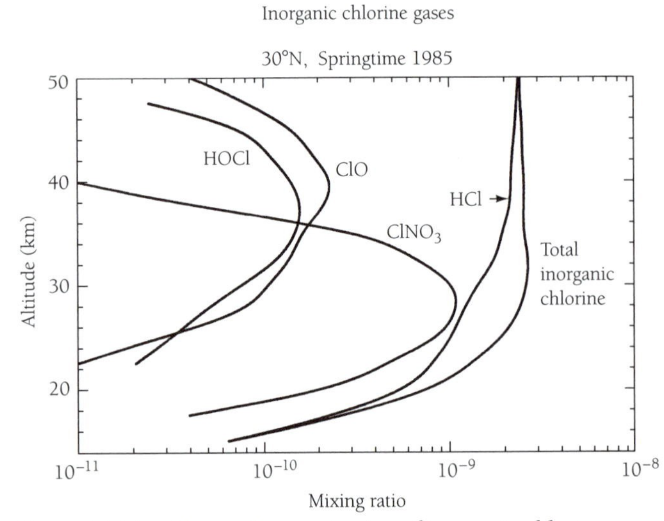

Figure 14.13 Observed concentration of inorganic chlorine gases as a function of altitiude. Measurements of HCl and $ClNO_3$ were obtained near 30°N between 30 April and 1 May 1985 from the ATMOS experiment on board Skylab 3 (Zander et al., 1992). Measurements of ClO (Waters et al., 1988) and HOCl (Chance, Johnson, and Traub, 1989) were obtained near 30°N by balloon-borne instrumentation during May of 1985 and 1988, respectively. Source: Waters 1988.

Reactions (14.34) and (14.35) for chlorine are analogous to (14.1) and (14.2) for NO_x: the result is equivalent in both cases to (13.4). The chlorine atom that enters (14.34) is recycled many times through (14.35). As with NO, a single chlorine atom can be responsible for the removal of many molecules of odd oxygen over the course of its lifetime in the stratosphere.

Just as it was convenient in our treatment of NO_x chemistry to expand the definition of odd oxygen to include NO_2, it is useful at this point to generalize the concept further to encompass ClO. This seems reasonable. The odd oxygen atom contained in O_3 is abstracted by Cl in (14.34); it disappears on the left-hand side of (14.34) only to reappear on the right as a component of ClO. Reaction (14.35) provides a sink for two species of odd oxygen (ClO and O). It represents the rate-limiting step for the removal of odd oxygen by chlorine catalysis, analogous to (14.2) for NO_x.

The abundance of Cl relative to ClO is controlled by (14.34), (14.35), and by

$$ClO + NO \rightarrow Cl + NO_2 \qquad (14.36)$$

The cycle regulating exchange of chlorine between Cl and ClO is illustrated in Figure 14.14. Denoting rate constants for (14.34), (14.35), and (14.36) by $k_{14.34}$, $k_{14.35}$, and $k_{14.36}$, respectively, and equating rates for production and loss of Cl, we find

$$k_{14.34}[Cl][O_3] = k_{14.35}[ClO][O] + k_{14.36}[ClO][NO] \qquad (14.37)$$

This expression may be rearranged to give

$$\frac{[Cl]}{[ClO]} = \frac{k_{14.35}[O] + k_{14.36}[NO]}{k_{14.34}[O_3]} \qquad (14.38)$$

As previously noted (see Figure 13.10b), the abundance of O is expected to increase as a function of increasing altitude. At sufficiently high altitude, we anticipate that the numerator on the right-hand side of (14.38) should be dominated by the first term; reaction (14.35) is then more important than (14.36) in cycling ClO back to Cl. The ratio [Cl] to [ClO] is given in this case by

$$\frac{[Cl]}{[ClO]} \approx \frac{k_{14.35}[O]}{k_{14.34}[O_3]} \qquad (14.39)$$

Chlorine atoms represent the dominant form of radical chlorine at altitudes above about 50 km. Conversion of ClO to Cl at intermediate altitudes is dominated by (14.36). In this case,

$$\frac{[Cl]}{[ClO]} \approx \frac{k_{14.36}[NO]}{k_{14.34}[O_3]} \qquad (14.40)$$

Reaction (14.34), followed by (14.36), involves no net loss of odd oxygen; it corresponds to a rearrangement of species within the odd oxygen family, with one of the components, O_3, converted to another, NO_2.

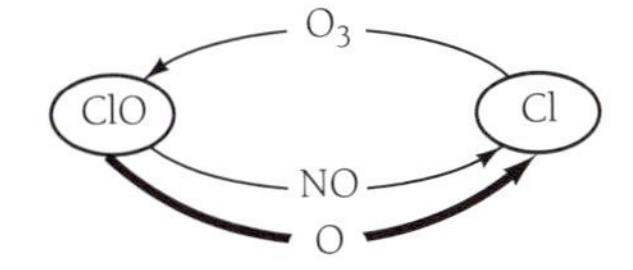

Figure 14.14 The interconversion of Cl and ClO by reactions (14.34), (14.35), and (14.36). Conversion of ClO to Cl by reaction (14.35), the rate-limiting step in odd oxygen removal, is indicated by the heavy line.

Example 14.2: The diurnally averaged concentrations of O, O_3, and NO at an altitude of 50 km are given approximately by:

$$[O] = 2.3 \times 10^9 \text{ cm}^{-3}$$

$$[O_3] = 6.3 \times 10^{10} \text{ cm}^{-3}$$

$$[NO] = 1.5 \times 10^8 \text{ cm}^{-3}$$

The rate constants $k_{14.34}$,$k_{14.35}$, and $k_{14.36}$ have values:

$$k_{14.34} = 1.2 \times 10^{-11} \text{ cm}^3 \text{ sec}^{-1}$$

$$k_{14.35} = 3.8 \times 10^{-11} \text{ cm}^3 \text{ sec}^{-1}$$

$$k_{14.36} = 1.7 \times 10^{-11} \text{ cm}^3 \text{ sec}^{-1}$$

Use the expression given by (14.38) to estimate the ratio [Cl]/[ClO].

Answer:

$$\frac{[Cl]}{[ClO]} = \frac{k_{14.35}[O] + k_{14.36}[NO]}{k_{14.34}[O_3]}$$

$$= \left(\frac{(3.8 \times 10^{-11})(2.3 \times 10^9) + (1.7 \times 10^{-11})(1.5 \times 10^8)}{(1.2 \times 10^{-11})(6.3 \times 10^{10})}\right)$$

$$\left(\frac{\text{cm}^3 \text{ sec}^{-1} \text{ cm}^{-3}}{\text{cm}^3 \text{ sec}^{-1} \text{ cm}^{-3}}\right)$$

$$= \frac{(8.74 \times 10^{-2}) + (2.55 \times 10^{-3})}{7.54 \times 10^{-1}}$$

$$= \frac{9.00 \times 10^{-2}}{7.54 \times 10^{-1}}$$

$$= 0.12$$ ■

Example 14.3: With the numbers quoted above for 50 km, estimate the importance of (14.35) relative to (14.36) and the fraction of the time reaction (14.34) is followed by (14.35), rather than by (14.36).

Answer: The rates ($\text{cm}^{-3} \text{ sec}^{-1}$) for reactions (14.35) and (14.36), $R_{14.35}$ and $R_{14.36}$, are given by

$$R_{14.35} = k_{14.35}[O][ClO]$$

and

$$R_{14.36} = k_{14.36}[NO][ClO]$$

The fractional efficiency of (14.35) relative to (14.36) is given by

$$f = \frac{R_{14.35}}{R_{14.35} + R_{14.36}}$$

$$= \frac{k_{14.35}[O][ClO]}{k_{14.35}[O][ClO] + k_{14.36}[NO][ClO]}$$

$$= \frac{k_{14.35}[\mathrm{O}]}{k_{14.35}[\mathrm{O}] + k_{14.36}[\mathrm{NO}]}$$

$$= \left(\frac{(3.8 \times 10^{-11})(2.3 \times 10^{9})}{(3.8 \times 10^{-11})(2.3 \times 10^{9}) + (1.7 \times 10^{-11})(1.5 \times 10^{8})} \right) \left(\frac{\mathrm{cm^{-3}\ sec^{-1}}}{\mathrm{cm^{-3}\ sec^{-1}}} \right)$$

$$= \frac{8.74 \times 10^{-2}}{(8.74 \times 10^{-2})(2.55 \times 10^{-3})}$$

$$= 0.97$$ ■

It follows that the rate for reaction of Cl with O_3 at 50 km is to an excellent approximation equal to the rate for the chlorine-catalyzed path for the reaction of O with O_3: 97% of the time reaction (14.34) is followed by (14.35). Approximately 3% of reactions of Cl with O_3 are followed by (14.36), resulting in the production of NO_2.

Reaction (14.2) is about three times more important than (14.3) at 50 km. Thus, production of NO_2 is usually followed by a reaction resulting in additional removal of odd oxygen (reaction 14.2 is responsible for the removal of the odd oxygen species NO_2 and O). More than 99% of Cl reactions with O_3 at 50 km result in the removal of odd oxygen. To an excellent approximation, the rate at which odd oxygen is removed at 50 km as a result of chlorine catalysis is equal to twice the rate for reaction (14.34).

Example 14.4: Estimate the lifetime of Cl with respect to (14.34) at 50 km and, for the same altitude, the lifetime of ClO with respect to the combination of reactions (14.35) and (14.36).

Answer:

$$\tau_{\mathrm{Cl}} = \frac{[\mathrm{Cl}]}{R_{14.34}} \left[\frac{\mathrm{cm^{-3}}}{\mathrm{cm^{-3}\ sec^{-1}}} \right]$$

$$= \frac{[\mathrm{Cl}]}{k_{14.34}[\mathrm{Cl}][\mathrm{O_3}]}\ \mathrm{sec}$$

$$= \frac{1}{k_{14.34}[\mathrm{O_3}]}\ \mathrm{sec}$$

$$= \frac{1}{(1.2 \times 10^{-11})(6.3 \times 10^{10})}\ \mathrm{sec}$$

$$= \frac{1}{7.56 \times 10^{-1}}\ \mathrm{sec}$$

$$= 1.32\ \mathrm{sec}$$

$$\tau_{\mathrm{ClO}} = \frac{[\mathrm{ClO}]}{R_{14.34} + R_{14.36}} \left[\frac{\mathrm{cm^{-3}}}{\mathrm{cm^{-3}\ sec^{-1}}} \right]$$

$$= \frac{[\mathrm{ClO}]}{k_{14.35}[\mathrm{ClO}][\mathrm{O}] + k_{14.36}[\mathrm{ClO}][\mathrm{NO}]}\ \mathrm{sec}$$

$$= \frac{1}{k_{14.35}[\mathrm{O}] + k_{14.36}[\mathrm{NO}]}\ \mathrm{sec}$$

$$= \frac{1}{(3.8 \times 10^{-11})(2.3 \times 10^{9}) + (1.7 \times 10^{-11})(1.5 \times 10^{8})}\ \mathrm{sec}$$

$$= 11\ \mathrm{sec}$$ ■

Chlorine radicals (Cl and ClO) are removed by reactions resulting in production of HCl, HOCl, and $ClNO_3$:

$$\mathrm{Cl + CH_4 \rightarrow HCl + CH_3,} \qquad (14.41)$$

$$\mathrm{ClO + HO_2 \rightarrow HOCl + O_2,} \qquad (14.42)$$

and

$$\mathrm{ClO + NO_2 + M \rightarrow ClNO_3 + M} \qquad (14.43)$$

Loss of chlorine radicals is not, however, permanent. The species HCl, HOCl, and $ClNO_3$ are recycled, with chlorine radicals reformed by reactions such as

$$\mathrm{OH + HCl \rightarrow Cl + H_2O,} \qquad (14.44)$$

$$h\nu + \mathrm{HOCl \rightarrow Cl + OH,} \qquad (14.45)$$

and

$$h\nu + \mathrm{ClNO_3 \rightarrow Cl + NO_3} \qquad (14.46)$$

An expanded pictorial representation of the chemistry of stratospheric chlorine is presented in Figure 14.15.

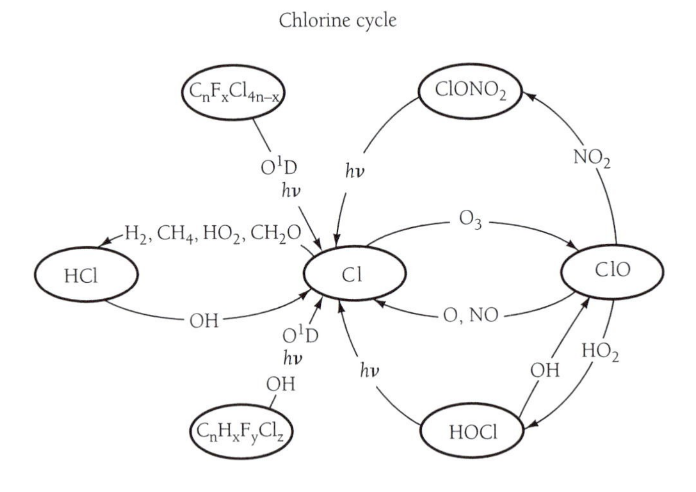

Figure 14.15 The complex paths by which exchange proceeds among members of the Cl_y family. Production of Cl_y is associated with reaction of source gases with $O(^1D)$ and their photolysis.

Example 14.5: The concentrations of Cl and ClO at 50 km are given approximately by:

$$[\mathrm{Cl}] = 2.9 \times 10^5\ \mathrm{cm^{-3}}$$

$$[\mathrm{ClO}] = 2.4 \times 10^6\ \mathrm{cm^{-3}}$$

The most important paths for removal of chlorine radicals at 50 km are provided by reactions (14.41) and (14.42). The respective rates are given approximately by

$$R_{14.41} = 1.7 \times 10^2\ \mathrm{cm^{-3}\ sec^{-1}}$$

and

$$R_{14.42} = 6.7 \times 10^1\ \mathrm{cm^{-3}\ sec^{-1}}$$

Estimate the lifetime of chlorine radicals at 50 km.

Answer: The concentration of chlorine radicals is given by the sum of the concentrations of Cl and ClO. The lifetime of chlorine radicals, τ, is obtained by dividing the concentration of radicals by the rate at which they are removed:

$$\tau = \frac{[\mathrm{Cl}] + [\mathrm{ClO}]}{R_{14.41} + R_{14.42}} \left(\frac{\mathrm{cm}^{-3}}{\mathrm{cm}^{-3}\ \mathrm{sec}^{-1}} \right)$$

$$= \frac{(2.9 \times 10^5) + (2.4 \times 10^6)}{(1.7 \times 10^2) + (6.7 \times 10^1)}\ \mathrm{sec}$$

$$= \frac{2.69 \times 10^6}{2.37 \times 10^2}\ \mathrm{sec}$$

$$= 1.14 \times 10^4\ \mathrm{sec}$$

■

Example 14.6: Estimate the number of odd oxygen pairs removed by a chlorine radical at 50 km before the radical is itself removed (by conversion to either HCl or HOCl).

Answer: The easiest approach to this problem involves a comparison of the rates for reaction of Cl with O_3 with the rate for removal of chlorine radicals; recall that the rate for removal of odd oxygen by chlorine catalyst is given to an excellent approximation by twice the value of $R_{14.34}$ (Example 14.3).

$$R_{14.34} = k_{14.34}\ [\mathrm{Cl}][\mathrm{O_3}]\mathrm{cm^{-3}\ sec^{-1}}$$

$$= (1.2 \times 10^{-11})\ (2.9 \times 10^5)\ (6.3 \times 10^{10})\ \mathrm{cm^{-3}\ sec^{-1}}$$

$$= 2.2 \times 10^5\ \mathrm{cm^{-3}\ sec^{-1}}$$

$$R_{14.41} + R_{14.42} = 2.37 \times 10^2\ \mathrm{cm^{-3}\ sec^{-1}}$$

The rate for (14.34) exceeds the combined rates for (14.41) and (14.42) by a factor:

$$f = \frac{R_{14.34}}{R_{14.41} + R_{14.42}}$$

$$= \frac{2.2 \times 10^5}{2.37 \times 10^2}$$

$$= 9.3 \times 10^2$$

It follows that a chlorine radical can catalyze the removal of about 1000 odd oxygen pairs over its life in the atmosphere at 50 km. (Recall that reaction 14.34, followed by either 14.35 or 14.36 and 14.2, removes two molecules of odd oxygen.) Moreover, this is not the end of the story; after a brief inactive residence as HOCl or HCl, the chlorine radical is recycled and is free once again to gobble up odd oxygen. ■

Example 14.7: HCl is the most abundant form of inorganic chlorine at 50 km. It is removed mainly by reaction with OH, or reaction (14.44). The rate constant for this reaction at 50 km is about $7.0 \times 10^{-13}\ \mathrm{cm^3\ sec^{-1}}$, whereas the concentration of OH is estimated at $8.8 \times 10^6\ \mathrm{cm^{-3}}$. Calculate the residence time for HCl.

Answer: The residence time is given by:

$$\tau_{\mathrm{Cl}} = \frac{[\mathrm{HCl}]}{R_{14.44}} \left(\frac{\mathrm{cm}^{-3}}{\mathrm{cm}^{-3}\ \mathrm{sec}^{-1}} \right)$$

$$= \frac{[\mathrm{HCl}]}{k_{14.44}[\mathrm{OH}][\mathrm{HCl}]}\ \mathrm{sec}$$

$$= \frac{1}{k_{14.44}[\mathrm{OH}]}\ \mathrm{sec}$$

$$= \frac{1}{(7 \times 10^{-13})(8.8 \times 10^6)}\ \mathrm{sec}$$

$$= \frac{1}{(6.16 \times 10^{-6})}\ \mathrm{sec}$$

$$= 1.6 \times 10^5\ \mathrm{sec}$$

Chlorine spends on average about 2 days as HCl at 50 km before it is converted back to radical form (Cl and ClO). The time a chlorine atom spends on average as HCl at 50 km is about 14 times as large, $(1.6 \times 10^5 / 1.144 \times 10^4)$, as the time spent as a radical (see Example 14.5). ■

Lifetimes of selected chlorine species are presented for a number of representative altitudes in Table 14.7. Results included here were obtained using the model previously employed for NO_y and HO_x (see Tables 14.4 and 14.5; Figures 14.6–14.7). They are intended to represent conditions for a latitude of 30° at equinox. Note that for all altitudes included in Table 14.7 the lifetimes of the chemically inactive forms of chlorine, the relatively stable species HOCl, $ClNO_3$, and HCl,[5] are long compared with those for the radical

Table 14.7 Photochemical Lifetimes (sec) for Selected Cl_y Species

	40 km	30 km	20 km
Cl	0.097	6.8×10^{-3}	2.8×10^{-4}
ClO	28.0	28.0	56.0
HOCl	5.2×10^{3}	4.0×10^{3}	4.5×10^{3}
$ClNO_3$	2.6×10^{4}	2.8×10^{4}	4.2×10^{4}
HCl	1.8×10^{5}	7.0×10^{5}	4.8×10^{6}

species, Cl and ClO. Lifetimes of radical and reservoir species are comparable for an altitude of about 60 km, with the former more persistent at higher elevation. As illustrated in Figure 14.16, hydrogen chloride (HCl) is the dominant form of inorganic chlorine at altitudes below 50 km.

Radicals account for approximately 10% of inorganic chlorine above 30 km, dropping to less than 1% below 20 km. The relatively small abundance of chlorine radicals at low altitudes may be attributed in part to the comparatively long lifetime of HCl (a consequence of the low abundance of OH) and in part to the increased importance of $ClNO_3$ (a result of the dependence on density M of the rate at which the compound is formed by the 3-body reaction 14.43). Rates for production and loss of HCl and $ClNO_3$ are displayed as functions of altitude in Figure 14.17. Complementing these data, rates for production and loss of OH are presented in Figure 14.18.

As illustrated in Figure 14.8, the loss of odd oxygen induced by chlorine-catalyzed chemistry proceeds mainly in the upper stratosphere, reflecting the distribution of ClO with altitude (see Figure 14.14). These results suggest that chlorine should account for about 12% of odd oxygen loss at

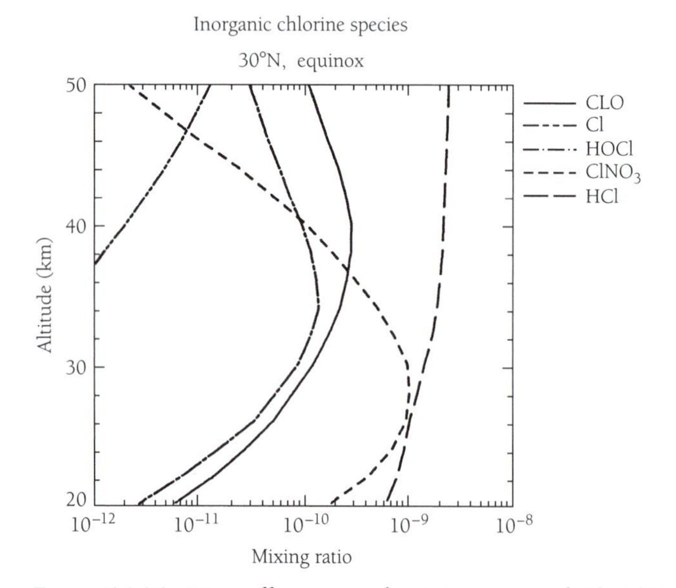

Figure 14.16 Diurnally averaged mixing ratios of Cl, ClO, HOCl, $ClNO_3$, and HCl as a function of altitude, calculated for 30°N at equinox.

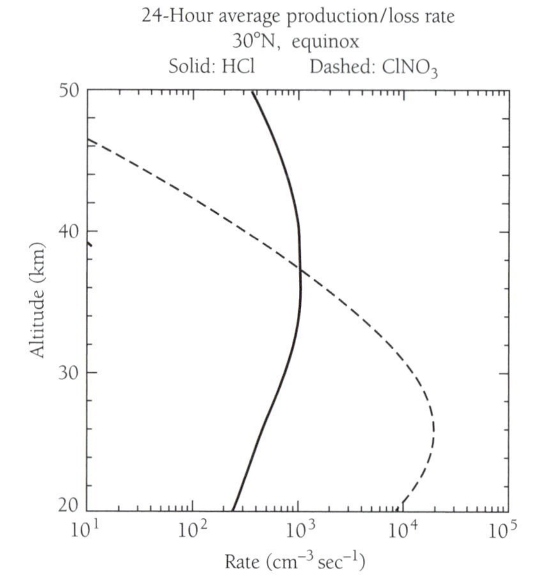

Figure 14.17 Diurnally averaged production rates for HCl and $ClNO_3$, calculated as a function of altitude for 30°N at equinox (production and loss rates are assumed to balance in the model).

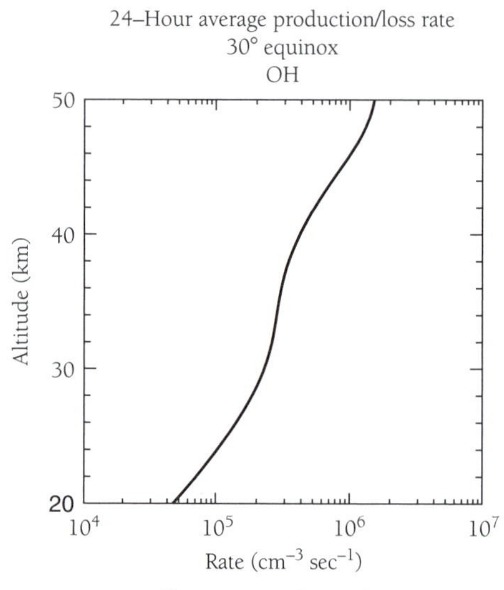

Figure 14.18 Diurnally averaged production rates for OH, calculated as a function of altitude for 30°N at equinox (production and loss rates are assumed to balance in the model).

50 km and for roughly 13% at 40 km, dropping to about 2% at 30 km and to about 0.1% at 20 km. Less than 1% of the O_3 column is located at altitudes above 50 km; a few percent is sited above 40 km, with roughly 20% above 30 km. It follows, according to the present analysis, that effects of chlorine-catalytic chemistry on the column density of O_3 should be relatively contained. In terms of the metaphor developed earlier, the leak introduced by chlorine should be confined to the upper portion of the barrel; most of the water in the container should be unaffected by the high-level leak.

The mixing ratio of chlorine in the stratosphere increased from about 2.5 to 3.6 ppb between 1980 and 1990.

Models accounting for the effect of reaction (14.35) on the balance of odd oxygen suggest that the column abundance of O_3 should have declined over the past few decades by between 1–2%. Observational data indicate that the actual change may be much larger, particularly at mid and high latitudes. Data from the TOMS satellite[6] suggest that reductions in the O_3 column over the decade of the 1980s may have been as much as 8% at a latitude of 40°N, and even greater at higher latitudes, particularly in the Southern Hemisphere. We address this problem below in the context of a discussion of the role of **heterogeneous chemistry** (the influence of reactions taking place on the surface of particles); as we shall see, heterogeneous processes play a crucial role in accounting for the dramatic loss of O_3 observed in recent years over Antarctica during spring. They may be implicated also in removal of O_3 in the Arctic and at midlatitudes in both hemispheres.

14.3 The Chemistry of Bromine

A variety of compounds contributes to the inventory of brominated gases in the stratosphere; bromine, like chlorine, enters the stratosphere mainly in organic form, as a component of molecular species ranging from hydrogen-rich gases, such as CH_3Br (methyl bromide) and CH_2Br_2 (ethylene dibromide), to fully halogenated compounds, such as $CBrClF_2$ (Halon 1211) and $CBrF_3$ (Halon 1301). The mixing ratio of bromine in the stratosphere is presently thought to lie in the range of 20–25 ppt. Methyl bromide, with a mixing ratio of 10–15 ppt, is produced by both natural and industrial processes and provides the single largest contribution to the global inventory of atmospheric bromine. Halons, used as fire extinguishers, represented the most rapidly growing component until recently; their use has now been proscribed and concentrations of the gases in the atmosphere have begun recently to level off. They account for about 6 ppt Br in the stratosphere at present; their contribution was increasing at a rate of between 15 and 20% per year until recently.[7]

The chemistry of inorganic bromine in the stratosphere is generally similar to that of chlorine, with some important differences: bromine is more reactive than chlorine. Compared with chlorine, a relatively larger portion of the inorganic bromine budget of the stratosphere is present in the form of the radicals Br and BrO, as compared with that contained in more stable reservoir species, such as HBr and $BrNO_3$. The difference is due mainly to the fact that the reaction of Br with CH_4 that forms HBr, the analogue of (14.41) for chlorine, is endothermic (i.e., energy must be supplied for the reaction to proceed); consequently, production of HBr is suppressed relative to that of HCl.

The important loss mechanisms for odd oxygen involving Br are

$$Br + O_3 \rightarrow BrO + O_2, \quad (14.47)$$

$$BrO + O \rightarrow Br + O_2,$$

and the sequence

$$\begin{aligned} Br + O_3 &\rightarrow BrO + O_2 \\ Cl + O_3 &\rightarrow ClO + O_2 \\ BrO + ClO &\rightarrow Br + Cl + O_2 \end{aligned} \quad (14.48)$$

In terms of its impact on odd oxygen, (14.47) is identical in effect to the NO_x–catalytic couplet, reaction(14.1) followed by (14.2), and to the analogous sequence involving chlorine, reaction (14.34) followed by (14.35). The sequence (14.48) is equivalent to the reaction of O_3 with itself, as summarized by (14.31). Figure 14.19 illustrates the chemistry of inorganic bromine. Figure 14.20 displays abundances of the more important brominated inorganic species, calculated using the chemical model summarized in Table 14.3. And Figure 14.21 presents a comparison, as a function of alti-

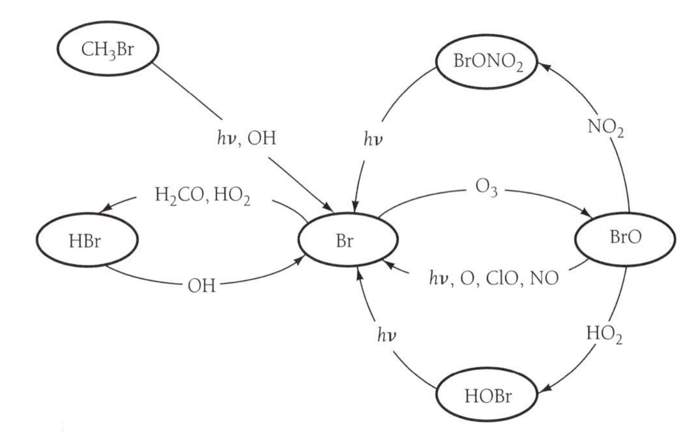

Figure 14.19 The complex paths by which exchange proceeds among the members of the Br_y family. Production of Br_y is associated with reaction of source gases with $O(^1D)$ and with photolysis.

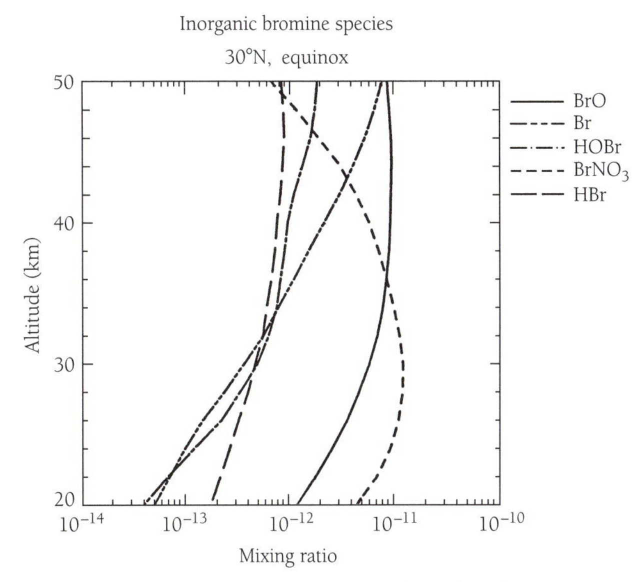

Figure 14.20 Diurnally averaged values for the mixing ratios of Br, BrO, HOBr, $BrNO_3$, and HBr as a function of altitude at 30°N at equinox.

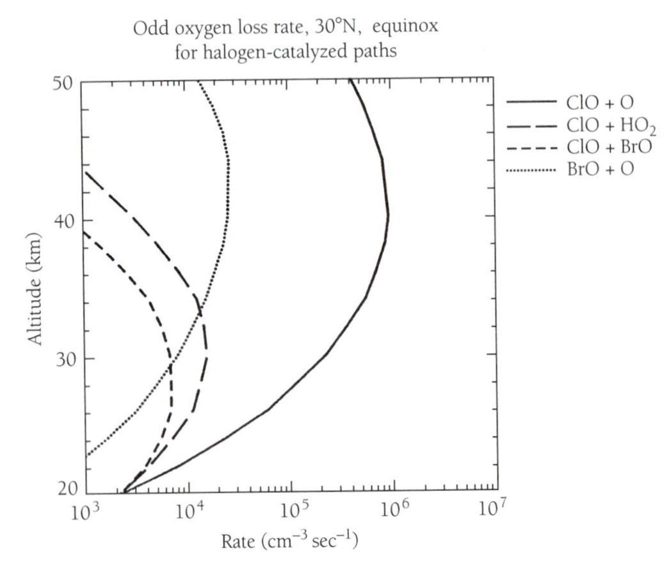

Figure 14.21 Rates for production and loss of odd oxygen due to halogen-catalyzed paths as a function of altitude at 30°N at equinox.

tude, of rates for several chlorine and bromine catalyzed paths for removal of odd oxygen (see Table 14.2).

14.4 Summary

Radicals play an important role in the chemistry of stratospheric ozone, serving as catalysts for the removal of odd oxygen. Under natural conditions, reactions involving oxides of nitrogen account for more than half of the total loss of odd oxygen. They are formed primarily as by-products of the decomposition of N_2O by reaction of $O(^1D)$ with N_2O, where $O(^1D)$ is produced by photodissociation of O_3. In this sense, microbial processes in soils and water, by regulating production of N_2O, exert an important influence on the abundance and distribution of ozone and thus on the flux of ultraviolet radiation reaching Earth's surface.

In addition to NO and NO_2, the stratosphere also contains important concentrations of hydrogen radicals, notably OH and HO_2, formed by reaction of $O(^1D)$ with H_2O. The abundance of nitrogen radicals depends on the abundance of hydrogen radicals, and vice versa; their chemistries are inextricably linked by a complex series of reactions, by which radicals are transformed to, and ultimately released from, more stable reservoir species, such as HNO_3. Catalytic reactions involving hydrogen radicals provide important paths for removal of odd oxygen and are especially significant at high and low altitudes (above 40 and below about 25 km).

Emission of relatively long-lived industrial gases containing chlorine and bromine atoms has had an important influence on the chemistry of the stratosphere in recent years. These gases penetrate to the stratosphere, where they decompose, releasing their constituent chlorine and bromine. They provide an important source of radicals that are notably effective as catalysts for the removal of odd oxygen. We found it useful to generalize the definition of odd oxygen introduced in Chapter 13 to include the radical species NO_2, HO_2, ClO, and BrO. In this manner, it was possible to focus on a relatively small set of (rate-limiting) reactions that are ultimately responsible for the removal of odd oxygen and thus for regulating the abundance of stratospheric ozone.

Stratospheric Ozone (III): Influence of Heterogeneous Chemistry

15

By the early 1980s there was general agreement that continued use and release of the longer-lived chlorinated and brominated gases into the atmosphere would result in small, though significant, losses of O_3. Simulations of stratospheric chemistry and dynamics suggested that, if the use of CFC-11 and CFC-12 were to continue at rates applicable in 1980, we might expect reductions in the column density of O_3, globally averaged, to eventually reach levels of 5–9%. The anticipated impact was less when models allowed, not only for changes in CFC-11 and CFC-12 but also for increases in the abundances of CO_2, CH_4, and N_2O. Reductions in the column density of O_3 predicted for the decade of the 1980s were small, about 1%.[1] It was thus with considerable surprise that the atmospheric community greeted the announcement in 1985 that major losses of O_3, approaching 40%, had been observed during spring over Antarctica dating back to the mid 1970s.

The paper, authored by J. C. Farman, G. G. Gardiner, and J. D. Shanklin of the British Antarctic Survey, reporting this startling result appeared in the 16 May 1985 issue of *Nature*. Figure 15.1 reproduces the key Figure

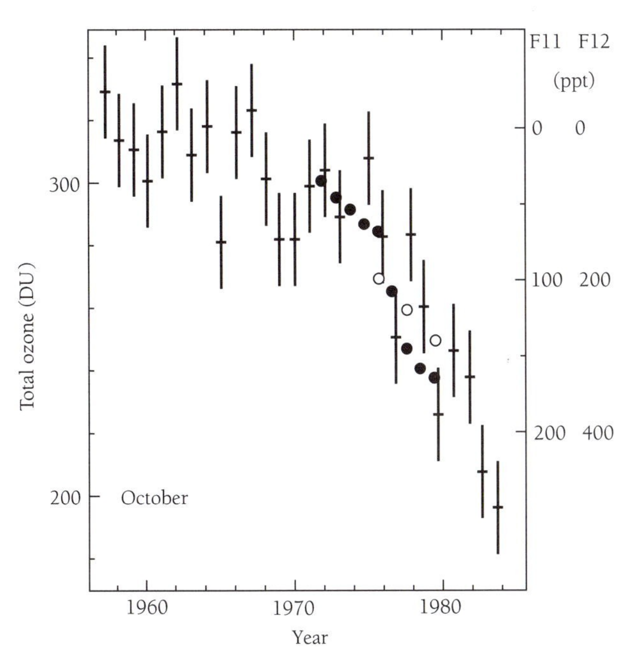

Figure 15.1 Column density of ozone (averaged for the month of October) and atmospheric burden of CFC-11 and CFC-12 as a function of years over Halley Bay, Antarctica (76°S, 27°W). Column density of ozone is shown by the vertical bars; units are on the left. Concentration of CFC-11 is shown by the closed circles; units are on the right (0–200). Concentration of CFC-12 is shown by open circles; units are on the right (0–400). Source: Farman, et al. 1985.

from the paper, illustrating the change with time of the average value of the column density of O_3 observed over Halley Bay on the Antarctic coast (76°S, 27°W) for the month of October dating back to 1957. Providing context for these data, measurements of the global and temporal distribution of the column density of O_3 are summarized in Plate 5 (top) (see color insert). Note that column densities are typically largest at high latitudes during spring; we would have expected, according to the results in Plate 5, to find column densities of O_3 over Antarctica during local spring of about 320 DU,[2] rising to about 380 DU with the approach of summer. The data in Figure 15.1 indicate that by 1984 the column density of O_3 over Halley Bay in October had dropped to an average of about 200 DU. As indicated above, the decline appears to have set in quite abruptly in the mid 1970s. In recent years, column densities have fallen over Antarctica to levels near 100 DU, corresponding to reductions in excess of 60%. The abundance of O_3 over Antarctica during spring is now less than it is at any time of year over the equator (see Plate 5).

As discussed above, the reductions in O_3 over Halley Bay were unexpected; they were soon confirmed, however, by independent observations, in particular by analyses of data from TOMS (see Endnote 14–6). Satellite observations provide a record dating back to 1978 and were influential in placing the results from Halley Bay in a larger spatial and temporal context: they indicated that the reduction in O_3 detected by Farman was associated with a relatively large region of the atmosphere, a system known as the **polar vortex**, a mass of cold, comparatively isolated, spinning air that forms over the polar region during winter. At its full extent the vortex occupies approximately 10% of the Southern Hemisphere.[3]

A variety of theories were advanced in quick order to account for what eventually became known as the **Antarctic ozone hole**. We have taken the unusual step here of reproducing, in Figures 15.2–15.7, title pages and abstracts of several of the more influential papers. They attest to the fast pace and the excitement of the research stimulated by the Farman paper—by the aggressive search for a theory to account for an entirely unexpected observation. An acceptable theory had to address a number of key issues: first, it had to explain why large reductions in O_3 appeared to be confined to high latitudes of the Southern Hemisphere during spring; second, it had to rationalize why the reductions appeared to have had a relatively abrupt onset during the 1970s; finally, it had to account for the height-dependence of the loss, that is, for the fact that the deficit in O_3 appeared to be most intense over the altitude range 10–20 km, as indicated in Figure 15.8.

What has emerged over the past ten years or so is a relatively complete understanding of the physical and chemical processes responsible for the ozone hole over Antarctica. The pace of progress has been astounding, a tribute to the breadth and depth of the scientific and technological talent engaged over this period in the quest to understand the scope and nature of the influence of human activity on stratospheric O_3.

We begin in Section 15.1 with a discussion of the importance of reactions taking place on the surfaces of particles formed under cold conditions in the high-latitude polar stratosphere. Specifically, we show that these reactions can result in an important source of chlorine and bromine radicals by providing a path for conversion of stable compounds, such as $ClNO_3$ and HCl, to more reactive species, such as Cl_2 and HOCl. Section 15.1 introduces the catalytic paths thought to be responsible for the low-altitude loss of O_3 observed over Antarctica. Section 15.3 discusses implications for the Arctic with low and mid-latitudes treated in section 15.4. Section 15.4 gives a summary of the key elements of this Chapter.

15.1 The Role of Polar Stratospheric Clouds

A number of the papers included in Figures 15.2–15.7 (specifically 15.2–15.5) proposed that high concentrations of either ClO or BrO could account for the rapid loss of O_3 observed during spring over Antarctica and that heterogeneous processes should play an important role in conditioning the stratosphere for this loss. The key first step in these studies invokes the transfer of chlorine from stable reservoirs, such as HCl and $ClNO_3$, to more reactive forms, such as Cl_2 and HOCl. The most important reaction involves rearrangement of HCl and $ClNO_3$ on the surfaces of particles that appear in the stratosphere when temperatures drop below about 196 K. Termed **Type 1 polar stratospheric clouds** (Type 1 PSCs), these particles are believed to form from a combination of HNO_3 and H_2O (Figure 15.4) included in an icy crystalline matrix in a molecular ratio of 1:3. Reflecting their composition, the particles are known as **nitric acid trihydrate**, or NAT.[4]

The overall reaction converting HCl and $ClNO_3$ to Cl_2 and HNO_3 may be summarized as follows:

$$ClNO_3 + HCl \rightarrow Cl_2 + HNO_3 \qquad (15.1)$$

We expect that nitric acid produced on the surface of NAT by (15.1) should be incorporated into the condensed phase (i.e., should be included as a component of NAT), whereas Cl_2 should be released into the atmosphere. Additional production of reactive chlorine, in this case HOCl, may arise as a consequence of the heterogeneous reaction of $ClNO_3$ with H_2O:

$$ClNO_3 + H_2O \rightarrow HOCl + HNO_3 \qquad (15.2)$$

We expect HOCl evolved in (15.2) to be released to the atmosphere, while the fate of the product HNO_3 may depend on temperature and on the chemical nature of the particles involved in the reaction.

There is compelling evidence from laboratory studies and indirect, but convincing, support from atmospheric measurements that reaction (15.1) proceeds rapidly on the surface of NAT. Given that (15.1) is efficient, we can understand, in retrospect at least, why this should be the case. Water on the surface of frozen particles such as NAT exhibits properties similar to those of a liquid; this behavior reflects the fact that molecules are more mobile on the surface than in the interior of particles, where they are constrained to a greater extent by

Reductions of Antarctic ozone due to synergistic interactions of chlorine and bromine

Michael B. McElroy, Ross J. Salawitch, Steven C. Wofsy & Jennifer A. Logan

Harvard University, Center for Earth and Planetary Physics, 29 Oxford Street, Cambridge, Massachusetts 02138, USA

The vertical column density of ozone observed in October over Antarctica has fallen precipitously over the past 10 yr. The concentration at Halley Bay (76° S 27° W), expressed conventionally in Dobson units (DU) (1 DU = 10^{-3} atmos. cm = 2.7×10^{16} molecules cm^{-2}), has dropped from about 300 DU in 1975 to <200 DU in 1984 (ref. 1). Values in 1985 were even lower, comparable with the lowest values recorded anywhere on Earth[2]. We suggest here that the loss of O_3 in Antarctica may be attributed to catalysis of O_3 recombination by a scheme in which the rate-limiting step is defined by the reaction $ClO + BrO \rightarrow Cl + Br + O_2$. Concentrations of NO_2 must be low and heterogenous reactions involving particles in the polar stratospheric clouds must be an important element of the relevant chemistry. Industrial sources make important contributions to the contemporary budgets of both BrO and ClO and are likely to grow significantly in the future.

Figure 15.2 Heterogenous reactions involving PSCs and the catalytic cycle involving BrO and ClO are emphasized as an important part of the chemistry leading to springtime removal of O_3 in the Antarctic. Source: McElroy et al. 1986.

molecular neighbors. Imagine that HCl dissolves in the surficial liquid of NAT (a coating probably no thicker than a few molecular layers), where it dissociates to form H^+ and Cl^-. Imagine that $ClNO_3$ is captured also by the liquid and that the products in this case are Cl^+ and NO_3^-. Suppose that the positively and negatively charged chlorine ions combine to form Cl_2, while H^+ and NO_3^- associate to form HNO_3. This sequence could account for the efficiency of (15.1). Further work is required to establish that this is indeed the mechanism for reaction of HCl with $ClNO_3$ in the presence of NAT.

Reactions (15.1) and (15.2) represent the first important steps in the transformation of chlorine from unreactive forms, such as HCl and $ClNO_3$, to radicals which eventually catalyze the removal of O_3. Reactions (15.1) and (15.2) can proceed in the dark of the polar night; synthesis of radicals requires sunlight. As sunlight returns to the polar environment in spring, the chlorine contained in Cl_2 and HOCl is converted to radicals by reactions such as

$$h\nu + Cl_2 \rightarrow Cl + Cl \tag{15.3}$$

and

$$h\nu + HOCl \rightarrow HO + Cl \tag{15.4}$$

Similar processes are effective in converting bromine from more stable reservoirs, such as $BrNO_3$ to Br and BrO.

The abstract in Figure 15.2 attributes loss of O_3 during spring over Antarctica to the sequence

$$\begin{gathered} Cl + O_3 \rightarrow ClO + O_2 \\ Br + O_3 \rightarrow BrO + O_2 \\ ClO + BrO \rightarrow Cl + Br + O_2, \end{gathered} \tag{15.5}$$

On the depletion of Antarctic ozone

Susan Solomon*, Rolando R. Garcia†, F. Sherwood Rowland‡ & Donald J. Wuebbles§

* National Oceanic and Atmospheric Administration Aeronomy Laboratory, Boulder, Colorado 80303, USA
† NCAR, Boulder, Colorado 80307, USA
‡ Department of Chemistry, University of California, Irvine, California 92717, USA
§ Lawrence Livermore Laboratory, Livermore, California 94550, USA

Recent observations by Farman *et al.*[1] reveal remarkable depletions in the total atmospheric ozone content in Antarctica. The observed total ozone decreased smoothly during the period from about 1975 to the present, but only in the spring season. The observed ozone content at Halley Bay was ~30% lower in the Antarctic spring seasons (October) of 1980–84 than in the springs of 1957–73. No such obvious perturbation is observable in other seasons, or at other than the very highest latitudes in the Southern Hemisphere, and the magnitude of the observed change there far exceeds climatological variability[2]. We present here balloonsonde ozone data[3,4] which show that these ozone changes are largely confined to the region from about 10 to 20 km, during the period August to October. We show that homogeneous (gas phase) chemistry as presently understood cannot explain these observed depletions. On the other hand, a unique feature of the Antarctic lower stratosphere is its high frequency of polar stratospheric clouds[5], providing a reaction site for heterogeneous reactions. A heterogeneous reaction between HCl and $ClONO_2$ is explored as a possible mechanism to explain the ozone observations. This process produces changes in ozone that are consistent with the observations, and its implications for the behaviour of HNO_3 and NO_2 in the Antarctic stratosphere are consistent with observations of those species there, providing an important check on the proposed mechanism. Similar ozone changes are obtained with another possible heterogeneous reaction, $H_2O+ClONO_2$.

Solomon et al., *Nature*, **321, 755, 1986**

Figure 15.3 Heterogenous reactions involving PSCs are emphasized as an important part of the chemistry leading to springtime removal of O_3 in the Antarctic. Source: Solomon et al. 1986.

which is equivalent to

$$O_3 + O_3 \rightarrow 3O_2 \qquad (15.6)$$

The paper highlighted in Figure 15.5 suggests a mechanism involving ClO that accommodates the same result (recombination of O_3 with itself as summarized by reaction 15.6):

$$ClO + ClO + M \rightarrow Cl_2O_2 + M$$

$$h\nu + Cl_2O_2 \rightarrow Cl + ClOO \qquad (15.7)$$

$$ClOO + M \rightarrow Cl + O_2 + M$$

$$2[Cl + O_3 \rightarrow Cl + O_2]$$

Additional loss can occur through the sequence

$$ClO + HO_2 \rightarrow HOCl + O_2$$

$$h\nu + HOCl \rightarrow OH + Cl \qquad (15.8)$$

$$OH + O_3 \rightarrow HO_2 + O_2$$

GEOPHYSICAL RESEARCH LETTERS, VOL. 13, NO. 12, PAGES 1284-1287, NOVEMBER SUPPLEMENT 1986

CONDENSATION OF HNO_3 AND HCl IN THE WINTER POLAR STRATOSPHERES

Owen B. Toon[1], Patrick Hamill[2], Richard P. Turco[3], Joseph Pinto[1]

[1]NASA Ames Research Center, Moffett Field, California
[2]San Jose State University, San Jose, California
[3]R and D Associates, Marina Del Rey, California

Abstract. Nitric acid and hydrochloric acid vapors may condense in the winter polar stratospheres. Nitric acid clouds, unlike water ice clouds, would form at the temperatures at which polar stratospheric clouds (PSCs) are observed and would have optical depths of the magnitude observed suggesting that HNO_3 is a dominant component of PSCs. ClO, N_2O_5 and $ClNO_3$ may react on cloud particle surfaces yielding additional HNO_3, HCl, and HOCL. In the vicinity of PSCs these reactions could deplete the stratosphere of photochemically active NO_x species. The sedimentation of PSCs may remove these materials from the stratosphere. The loss of vapor phase NO_x might allow halogen-based chemistry to create the ozone hole.

Figure 15.4 Condensation of HNO_3 on the surface of PSCs and subsequent sedimentation of PSC particles is emphasized as an important part of the chemistry leading to springtime removal of ozone in the Antarctic. Source: Toon et al. 1986.

$$Cl + O_3 \rightarrow ClO + O_2,$$

which is also equivalent to (15.6). Efficient removal of O_3 by reactions (15.5), (15.7), and (15.8) requires, as noted in Figure 15.2, that the "concentrations of NO_2 must be low." It appears indeed that the abundance of NO_2 in the Antarctic stratosphere is significantly depleted during the long cold winter. Otherwise, the radicals ClO and BrO would be bound up as $ClNO_3$ and $BrNO_3$. The sensitivity of ClO to the concentration assumed for NO_2 is illustrated by the following example:

Example 15.1: The concentration of O_3 exhibits a maximum over Antarctica at an altitude of about 18 km near the end of August (Figure 15.8). Removal of O_3 responsible for development of the ozone hole is most effective between 10–20 km. It appears to set in near the end of August or the beginning of September and is more or less complete by the early part of October.

Assume that the abundance of ClO relative to $ClNO_3$ is controlled over Antarctica by the reactions

$$ClO + NO_2 + M \rightarrow ClNO_3 + M \qquad \text{(A)}$$

and

$$h\nu + ClNO_3 \rightarrow ClO + NO_2 \qquad \text{(B)}$$

Denote the corresponding rate constants by k_A and J_B, respectively. Estimate the abundance of ClO relative to $ClNO_3$ at 18 km over Antarctica on 15 September, assuming NO_2 concentrations given by 2×10^9 cm^{-3}, 1×10^9 cm^{-3}, and 2×10^8 cm^{-3}, assuming rate constants k_A and J_B given by $k_A = 6.7 \times 10^{-31}$ cm^6 sec^{-1} and $J_B = 2 \times 10^{-5}$ sec^{-1}, respectively, and assuming $[M] = 2 \times 10^{18}$ cm^{-3}.

Answer: We may assume that the rate for production of $ClNO_3$ by (A) equals the rate at which it is removed by (B):

$$k_A\,[ClO][NO_2][M] = J_B\,[ClNO_3]$$

Thus

$$\frac{[ClO]}{[ClNO_3]} = \frac{J_b}{k_a[NO_2][M]}$$

$$= \frac{(2 \times 10^5\ \text{sec}^{-1})}{(6.7 \times 10^{-31}\ \text{cm}^6\ \text{sec}^{-1})[NO_2](2 \times 10^{18}\ \text{cm}^{-3})}$$

$$= \frac{1.5 \times 10^7\ \text{cm}^{-3}}{[NO_2]}$$

Corresponding to the three different values assumed for $[NO_2]$, equivalent to mixing ratios of 1 ppb, 0.5 ppb, and 0.1 ppb, we obtain

J. Phys. Chem. 1987, 91, 433–436

Production of Cl_2O_2 from the Self-Reaction of the ClO Radical

L. T. Molina and M. J. Molina*

Jet Propulsion Laboratory, California Institute of Technology, Pasadena, California 91109
(Received: April 22, 1986; In Final Form: September 15, 1986)

The species Cl_2O_2 has been generated in a gaseous flow system at 220–240 K by reacting Cl atoms with one of three different ClO precursors: O_3, Cl_2O, or OClO. The infrared spectra of the reactive mixture indicate that at least two different dimers are produced: a predominant form with bands centered at 1225 and 1057 cm^{-1} attributed to ClOOCl, and a second form with a band at 650 cm^{-1} attributed to ClOClO. The UV spectrum of the predominant form shows a maximum absorption cross section of $\sim 6.5 \times 10^{-18}$ cm^2/molecule around 270 nm, with a wing extending beyond 300 nm. The implications of these results for the chemistry of the stratosphere are discussed.

Assuming these concentration values, and assuming the rate constant for reaction 4 (the Cl_2O_2 formation reaction) to be in the $(1-5) \times 10^{-32}$ cm^6 $molecule^{-2}$ s^{-1} range (as discussed above), ClO would react with itself about as fast as with BrO. Furthermore, in the lower stratosphere over Antarctica the Cl_2O_2 photodissociation rate would be comparable, if not considerably faster than its thermal decomposition rate, due to the very low temperatures found at those latitudes (190–220 K). Hence, if our UV measurements are correct, and if the primary Cl_2O_2 photolysis products are Cl + ClOO, the cycle consisting of reactions 4–7 might indeed play a role in explaining the observed Antarctic ozone decline. Clearly, additional studies of the chemistry and photochemistry of Cl_2O_2 are needed to address this question.

Figure 15.5 The catalytic cycle involving ClO and ClO is emphasized as an important part of the chemistry leading to springtime removal of O_3 in the Antarctic. Source: Molina and Molina, 1987.

$$\frac{[ClO]}{[ClNO_3]} = 7.5 \times 10^{-3}, [NO_2] = 2 \times 10^9 \text{ cm}^{-3}$$

$$\frac{[ClO]}{[ClNO_3]} = 1.5 \times 10^{-2}, [NO_2] = 1 \times 10^9 \text{ cm}^{-3}$$

$$\frac{[ClO]}{[ClNO_3]} = 7.5 \times 10^{-2}, [NO_2] = 2 \times 10^8 \text{ cm}^{-3} \quad \blacksquare$$

The abstracts included in Figures 15.2–15.7 offer three general types of hypotheses to account for loss of O_3 over Antarctica during spring: one hypothesis (Figures 15.2–15.5) invokes high concentrations of ClO and/or BrO, requiring (implicitly) low concentrations of NO_2; a second (Figure 15.6) argues for rising motion transporting ozone-poor air from low altitudes into middle regions of the polar stratosphere; the third (Figure 15.7) proposes an important role for NO_x chemistry (reaction 14.2), suggesting that accelerated loss of O_3 in the polar region in recent years is a natural phenomenon caused by "enhanced formation of odd nitrogen during solar cycle 21." Measurements of NO_y, N_2O, O_3, ClO, and BrO with instrumentation carried into the Antarctic vortex during late winter and spring of 1987 by the ER-2 research aircraft[5] effectively eliminated the second and third of these explanations, while providing support for the first. The profile of N_2O as a function of altitude (or pressure) changed little as O_3 significantly decreased between 23 August and 22 September, as illustrated in Figure 15.9. If the second hypothesis were valid, we would have expected to observe an increase in the mixing ratio of N_2O for fixed values of pressure as elevated levels of N_2O were transported to higher altitudes (recall that we expect the relative abundance of N_2O to decrease on average with altitude). The abundance of NO_y was low (Figure 15.10), consistent with the first hypothesis but in conflict with the third. Finally, concentrations of ClO were high inside the vortex, with a clear negative correlation between concentrations of ClO and O_3 in mid September, when the rate of loss of O_3 was largest (Figure 15.11). Even the minor wiggles in the curves in Figure 15.11 display this behavior; increases (decreases) in the concentration of ClO are associated with decreases (increases) in the concentration of O_3.

The ER-2 data (Fahey et al. 1989, 1990) provide unambiguous evidence that not only NO_x but also NO_y is low in the Antarctic vortex. As discussed in Chapter 14, we expect concentrations of NO_y to vary in a predictable fashion as a function of N_2O, reflecting the fact that NO_y is formed as a by-product of N_2O decomposition. The standard relationship is summarized in Figure 14.10. Inside the Antarctic vortex (see Figure 15.10), concentrations of NO_y observed on 23 August 1987 fell significantly below the level expected, based on this relationship. The shaded area in Figure 15.10 indicates the quantity of NO_y inferred as "missing" from the vortex. Loss of NO_y from the vortex is attributed to incorporation of HNO_3 in

Are Antarctic ozone variations a manifestation of dynamics or chemistry?

Ka-Kit Tung*, Malcolm K. W. Ko†, José M. Rodriguez† & Nien Dak Sze†

* Massachusetts Institute of Technology, Cambridge, Massachusetts 02139, and Clarkson University, Potsdam, New York 13767, USA
† Atmospheric and Environmental Research, Inc., Cambridge, Massachusetts 02139, USA

Observations[1–3] reveal a large seasonal decrease in the column density of ozone during Antarctic early spring, followed by a rapid increase after October beyond its pre-spring value. Given the unique circumstances that exist in the Antarctic environment—relatively stable circumpolar vortex, cold air temperature achieved during polar night, and the increase in absorption of solar radiation by ozone as the Sun returns—we surmise the existence of a reverse circulation cell with rising motion in the polar lower stratosphere. The upwelling brings ozone-poor air from below 100 mbar to the stratosphere, possibly contributing to the observed ozone decline in early spring. At the same time, the Antarctic stratosphere might contain a very low concentration (<0.1 p.p.b.v. (parts per 10^9 by volume) of $NO_x(NO+NO_2)$, a condition that could favour a greatly enhanced catalytic removal of O_3 by halogen species[2,4,5]. We argue that heterogeneous processes and formation of OClO by the reaction $BrO + ClO \rightarrow OClO + Br$ before and after the polar night might help to suppress the NO_x levels during the early spring period. However, the dilution of the concentrations of the chlorine species by the upwelling may reduce the effectiveness of the photochemical removal of O_3.

Tung et al., *Nature*, 322, 811, 1986

Figure 15.6 A reverse-circulation cell leading to dynamical import of O_3-poor air to the stratosphere is emphasized as a possible cause of the springtime removal of O_3 in the Antarctic. Source: Tung et al. 1986.

particles growing to sizes large enough to fall from the stratosphere (for a discussion of the size of particles required for this phenomenon to occur, see Section 7.5). Most likely, these particles consist of a combination of NAT and water ice; specific requirements must be met to ensure that the particles are large enough to fall. If condensable water were equally distributed among all of the available nuclei, the average particle size would be too small to permit gravitational settling. Waibel et al. (1999) suggest that the removal of NO_y from the stratosphere begins with the production of a small number of ice crystals, formed most likely from supercooled droplets by a mechanism similar to that discussed in Section 7.5. They propose that HNO_3 is incorporated in these ice crystals, most probably as NAT, and that the particles rapidly grow to large size, probably at the expense of preexisting supercooled droplets. Ice crystals could capture both gaseous HNO_3 and NAT as they fall if the atmosphere is sufficiently cold; eventually the particles would melt or sublime and HNO_3 and H_2O would be returned to the gas phase. The phenomenon responsible for removal of NO_y from the stratosphere is known as **denitrification**. Return of NO_y from the particulate to the gas phase is referred to as **nitrification**.

The following examples illustrate the efficiency of reaction sequences (15.5) and (15.7) as sinks for ozone over Antarctica, assuming concentrations of ClO and BrO consistent with the measurements carried out by James Anderson

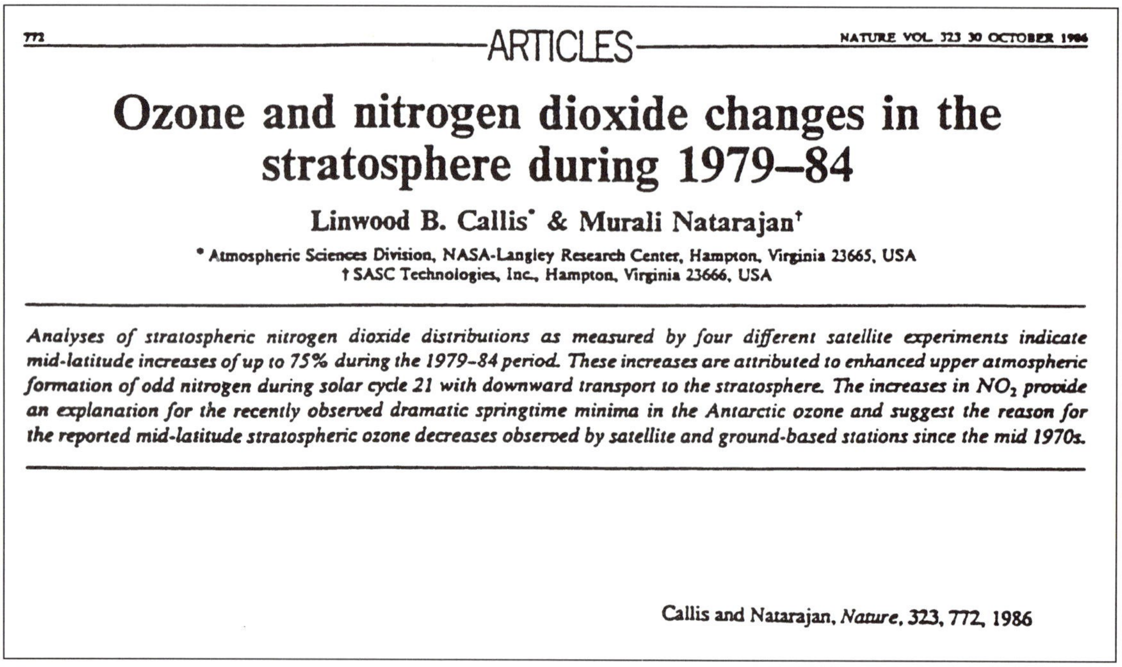

772 — ARTICLES — NATURE VOL. 323 30 OCTOBER 1986

Ozone and nitrogen dioxide changes in the stratosphere during 1979–84

Linwood B. Callis* & Murali Natarajan†

* Atmospheric Sciences Division, NASA-Langley Research Center, Hampton, Virginia 23665, USA
† SASC Technologies, Inc., Hampton, Virginia 23666, USA

Analyses of stratospheric nitrogen dioxide distributions as measured by four different satellite experiments indicate mid-latitude increases of up to 75% during the 1979–84 period. These increases are attributed to enhanced upper atmospheric formation of odd nitrogen during solar cycle 21 with downward transport to the stratosphere. The increases in NO_2 provide an explanation for the recently observed dramatic springtime minima in the Antarctic ozone and suggest the reason for the reported mid-latitude stratospheric ozone decreases observed by satellite and ground-based stations since the mid 1970s.

Callis and Natarajan, *Nature*, 323, 772, 1986

Figure 15.7 The catalytic cycle involving NO and NO_2, accelerated due to enhancements in NO_x associated with natural fluctuations in solar activity, is emphasized as a possible cause of the springtime removal of O_3 in the Antarctic. Source: Callis and Natarajan 1986.

and colleagues, using instruments flown into the Antarctic vortex on the ER-2 aircraft:

Example 15.2: Assume that the rate at which O_3 is removed by (15.5) equals twice the rate for reaction of ClO with BrO:

$$ClO + BrO \rightarrow Cl + Br + O_2 \qquad (A)$$

This is equivalent to an assumption that (A) is the rate-limiting step for the catalytic sequence (15.5). (Recall that two molecules of O_3, or odd oxygen, are removed each time the cycle is traversed.)

Calculate the rate at which O_3 is removed by (15.5) at 18 km over Antarctica. Densities of ClO and BrO and the rate constant for (A) may be taken as follows

$$[ClO] = 1.1 \text{ ppb} = 2.3 \times 10^9 \text{ cm}^{-3}$$

$$[BrO] = 8 \text{ ppt} = 1.6 \times 10^7 \text{ cm}^{-3}$$

$$k_A = 1 \times 10^{-11} \text{ cm}^3 \text{ sec}^{-1}$$

Answer:

$$R_A = k_A [ClO][BrO]$$

$$= (10^{-11})(2.3 \times 10^9)(1.6 \times 10^7) \text{ cm}^{-3} \text{ sec}^{-1}$$

The removal rate of O_3 is equal to

$$2 R_A = 7.4 \times 10^5 \text{ cm}^{-3} \text{ sec}^{-1}$$ ■

Example 15.3: Under the circumstances outlined in Example 15.2, estimate the rate at which O_3 is removed by (15.7).

Assume that the rate-limiting step in this case is

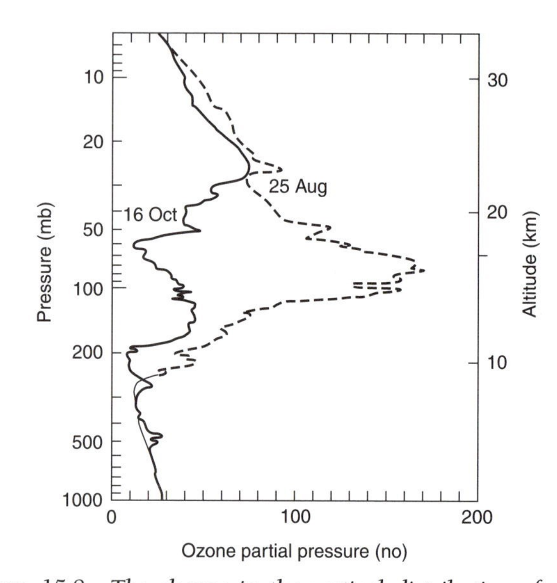

Figure 15.8 The change in the vertical distribution of ozone from 28 August to 16 October, demonstrating the temporal component of ozone loss over Antarctica. Source: Hoffman et al. 1987.

$$ClO + ClO + M \rightarrow Cl_2O_2 + M \qquad (B)$$

Suppose further that the rate constant is $k_B = 8 \times 10^{-32}$ cm^{-3} sec^{-1} and that $[M] = 2 \times 10^{18}$ cm^{-3}.

Answer:

$$2 R_B = 2 k_B [ClO]^2 [M]$$

$$= 2 (8 \times 10^{-32})(2.3 \times 10^9)^2 (2 \times 10^{18}) \text{ cm}^{-3} \text{ sec}^{-1}$$

$$= 1.7 \times 10^6 \text{ cm}^{-3} \text{ sec}^{-1}$$ ■

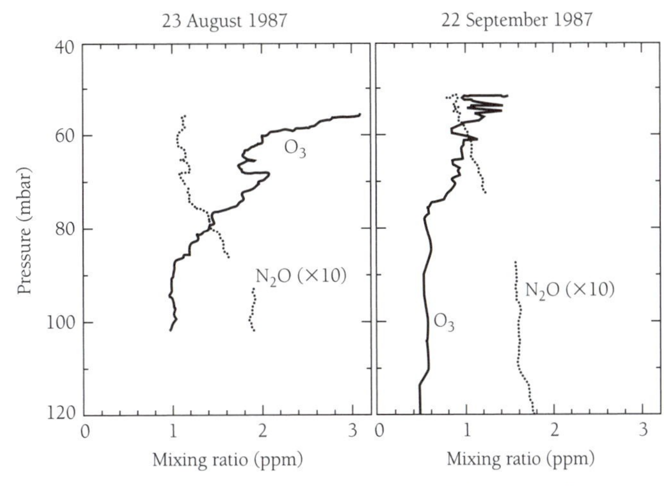

Figure 15.9 Vertical profile for O_3 and N_2O for the flight of 23 August and 22 September 1987. Data for N_2O have been multiplied by a factor of ten to facilitate display in the same scale as O_3. Measurements were obtained near 70°S, during the dive portion of the flights.

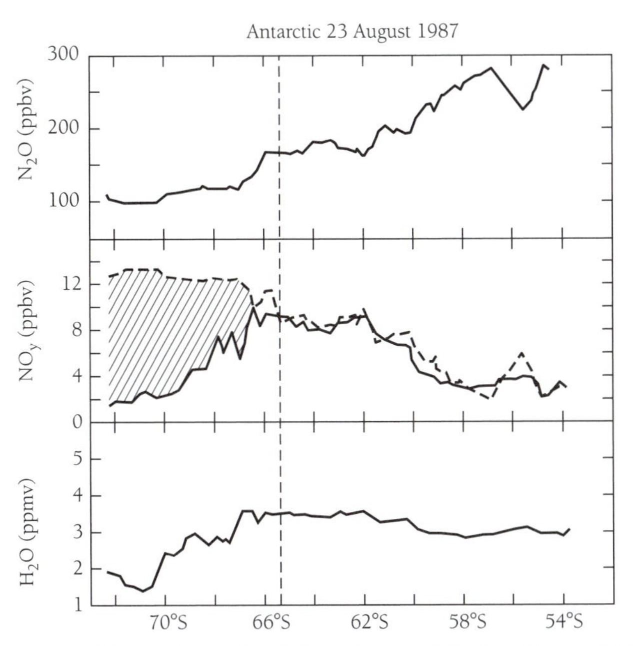

Figure 15.10 Latitudinal distribution of H_2O, NO_y, and N_2O for 23 August 1987. The shaded area in the middle graph depicts the loss of NO_y in the ozone hole. Source: Fahey et al. 1990.

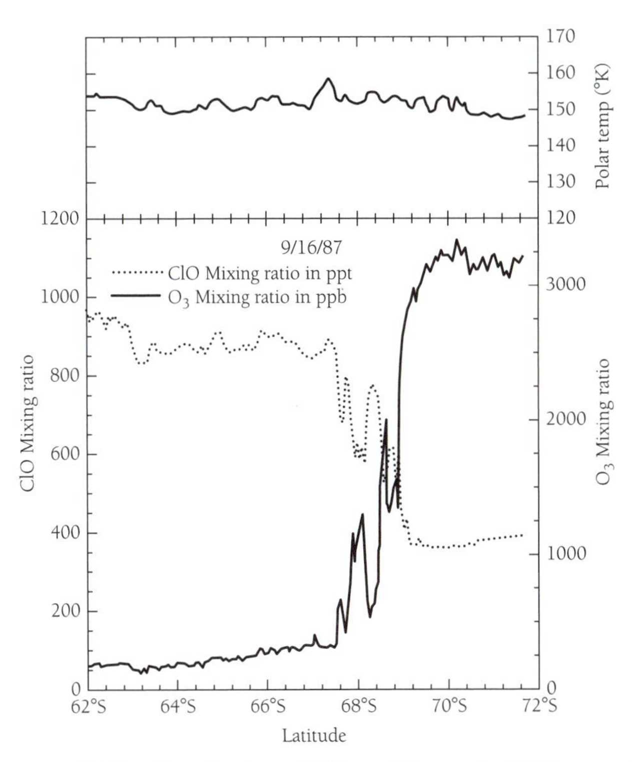

Figure 15.11 Distribution of ClO and O_3 as the ER-2 entered the ozone hole on September 16, 1987. Notice the strong anti-correlation between the two species. Source: Anderson et al. 1989.

Example 15.4: Using the results above, estimate the the loss of O_3 exhibited over the course of a day in mid September at 18 km. Assume that the sequences (15.5) and (15.7) are effective only during the sunlit portion of the day.

Answer: Net removal of O_3 by (15.5) is given by

$$L_{15.5} = (4.32 \times 10^4 \text{ sec})\ (7.4 \times 10^5 \text{ cm}^{-3} \text{ sec}^{-1})$$
$$= 3.2 \times 10^{10} \text{ cm}^{-3}$$

We assume here that the sunlit portion of the day is 12 hours (4.32×10^4 sec), as appropriate for the equinox. Loss by (15.7) is given by

$$L_{15.7} = (4.32 \times 10^4 \text{ sec})\ (1.7 \times 10^6 \text{ cm}^{-3} \text{ sec}^{-1})$$
$$= 7.3 \times 10^{10} \text{ cm}^{-3}$$

Total loss of O_3 over the day is equal to 1.05×10^{11} cm^{-3}. The mixing ratio of O_3 at 18 km in late August is about 2 ppm, equivalent to 4×10^{12} cm^{-3}. It follows that (15.5) and (15.7) can account for removal of half of the O_3 present at 18 km over Antarctica at the end of winter within a time interval of $(2 \times 10^{12}/1.05 \times 10^{11}) = 19$ days. ■

The aircraft measurements from 1987 provide unambiguous proof that loss of O_3 over Antarctica during spring is a consequence of reactions (15.5) and (15.7).[6] Trends observed for O_3 as a function of time from late August to mid September are compared in Figure 15.12 with trends calculated on the basis of (15.5) and (15.7), using measured concentrations of ClO and BrO. Approximately 70% of ozone removed over Antarctica in 1987 is attributed to (15.7), with most of the balance due to (15.5). Since the bulk of the chlorine and a significant fraction of the bromine budgets of the stratosphere are derived from industrial sources, the data in Figure 15.12 leave little doubt that human activity is responsible for the unexpected losses in O_3 observed in recent years over Antarctica.

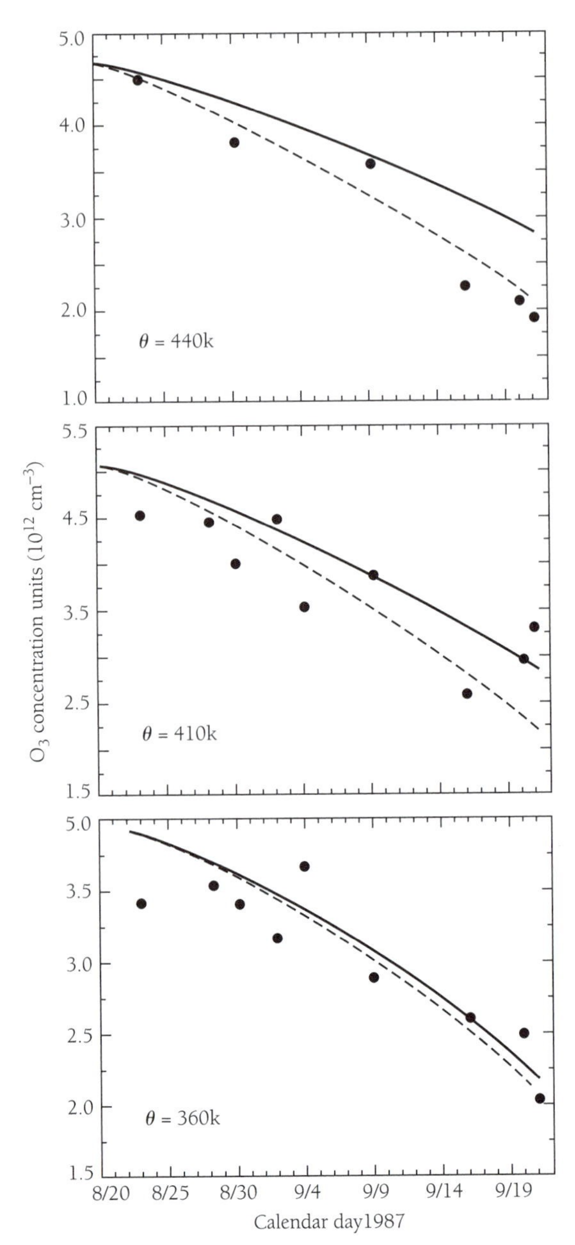

Figure 15.12 Model results demonstrating the importance of the ClO and BrO mechanisms for O_3 loss. The solid line only calculates the predicted loss due to ClO chemistry; the dashed line adds the effects of BrO chemistry. The points are from actual observations. Results in different panels reflect changes for different potential temperature surface with altitude decreasing from top to bottom. Source: Anderson et al. 1991.

15.2 The Chemistry of the Arctic Stratosphere

Armed with their newly acquired understanding of the Antarctic stratosphere, scientists turned their attention to the Arctic in 1989. The Arctic and Antarctic environments, although similar in general terms, significantly differ in detail. A vortex forms in the stratosphere over the Arctic during winter, just as it does over the Antarctic. The northern vortex is less stable, however, and less persistent than that in the south; it typically breaks up in mid February, as compared with mid October, or even later in the south (equivalent in terms of insolation or season to mid April, or later, in the north). Temperatures in the northern polar-vortex winter are generally higher than in the south but still low enough to permit formation of both Type 1 and Type 2 PSCs. The aircraft campaign mounted to study the northern polar stratosphere, the Airborne Arctic Stratospheric Expedition (AASE), covered the period from early January to mid February of 1989. Measurements were carried out using both the DC-8 and ER-2 aircrafts, the same planes employed in the AAOE mission. The primary objective of the AASE mission was to explore the chemical processes operating in the Arctic, to see how similar or different they were from those in the Antarctic. The important question to be answered, at least in the long term, is whether we might expect a hole to develop in O_3 over the Arctic, just as it did in the 1970s over Antarctica. The related issue is whether processes in the Arctic winter stratosphere could have an influence on O_3 more generally in the Northern Hemisphere; could they be involved in conditioning the atmosphere for accelerated loss of O_3 in the extra-polar environment, following breakup of the vortex? The mission was spectacularly successful in addressing the primary objective, defining the chemical properties of the winter polar stratosphere. With respect to the more challenging question of what might happen in the future, the jury is still out, but the issues are now defined and can be addressed with further work.

The AASE mission conclusively showed that the chemical processes responsible for loss of O_3 over Antarctica could also occur in the Arctic. Mixing ratios of ClO in excess of 1 ppb were observed in the Arctic during February, comparable to values observed earlier by AAOE for the Antarctic during September. Concentrations of ClO and BrO were high enough to drive rapid loss of O_3 in the Arctic stratosphere during February. It was estimated that about 10% of O_3 contained in the Arctic stratosphere between 16–20 km may have been lost during the winter of 1989, with approximately 40% of this loss attributed to the bromine-catalyzed cycle (reaction 15.5), the balance to the ClO dimer mechanism (reaction 15.7). As in the Antarctic, high concentrations of ClO in the Arctic are a consequence of low levels of NO_2, inhibiting the conversion of chlorine radicals to $ClNO_3$ and efficient conversion of chlorine in stable compounds, such as $ClNO_3$ and HCl, to ClO by reaction (15.1), followed by reactions (15.3) and (14.34). There are two important differences between the Arctic and Antarctic environments: first, levels of NO_y are relatively high in the Arctic winter (there is no evidence for extensive denitrification); second, as previously noted, the Arctic vortex breaks up relatively early. If the Arctic vortex should persist, and if denitrification should occur in the Arctic as in the Antarctic, we might expect major loss of O_3 to occur in the north as in the south. Assessment of this possibility requires a credible description of the mechanism for denitrification and a better understanding of the physical processes that control the vortex. An intriguing possibility is that loss of O_3 itself could contribute to a more persistent vortex by inhibiting warming resulting from absorption of sunlight by O_3. The fact that the southern vortex appears to have been more persistent in recent years when loss of O_3 was largest provides circumstantial support for this conjecture.

15.3. The Role of Heterogeneous Chemistry at Mid- and Low Latitudes

In Chapters 13 and 14 we developed a picture of stratospheric chemistry emphasizing processes that take place in the gaseous phase. The puzzles posed by the unexpected loss of O_3 observed in recent years over Antarctica and related studies of the chemistry of the Arctic (Section 15.2) forced an extension of this gaseous-phase perspective. We were obliged to account for reactions taking place on the surface of particles formed in the polar stratosphere, to allow in particular for the role of NAT in mediating the transformation of relatively stable forms of inorganic chlorine, such as $ClNO_3$ and HCl, to more reactive forms, such as Cl_2 (reaction 15.1), subsequently transformed into radicals, such as Cl and ClO, with analogous processes responsible for the production of high concentrations of Br and BrO. We return in this Section to stratospheric chemistry at mid and low latitudes. Specifically, our objective is to inquire whether heterogeneous processes could be important for these environments also. The evidence, as we shall see, is unequivocal; the influence of heterogeneous chemistry, at least in lower regions of the stratosphere, is ubiquitous.

Sulfate, present mainly as liquid particles, is an important component of stratospheric aerosols. Oxidation of SO_2 provides the primary source of stratospheric sulfate. As previously noted, oxidation of gaseous species, such as COS, is thought to maintain a relatively constant background level of stratospheric SO_2, an irreducible minimum episodically augmented by injection of large quantities of SO_2 associated with volcanic eruptions. Trends in the aerosol burden of the stratosphere since 1974 are illustrated in Figure 15.13. Contributions from major volcanoes such as El Chichon in 1982 and Mount Pinatubo in 1991 are clearly evident, as are inputs from a variety of smaller sources identified throughout the record. Volcanism has a dominant influence on the abundance of aerosols in the stratosphere most of the time: it has been especially important in recent years since the eruption of Mount Pinatubo.

Volcanically derived SO_2 is converted to sulfate aerosols in the stratosphere on a time scale of weeks. The resulting aerosols spread globally over the course of about a year, at least in lower regions of the stratosphere (altitudes below about 30 km), transported by seasonally varying stratospheric winds. The stratosphere recovers from the volcanic input slowly, with an *e-folding time* (defined as the time for the concentration to be reduced by a factor of *e*) for removal of sulfur from the stratosphere of several years, reflecting the relatively sluggish rate at which the stratosphere exchanges air with the troposphere (recall that vertical motion is inhibited in the stratosphere as a consequence of the stability imparted by the relatively strong positive temperature gradient with altitude). Sulfur is eliminated promptly from the

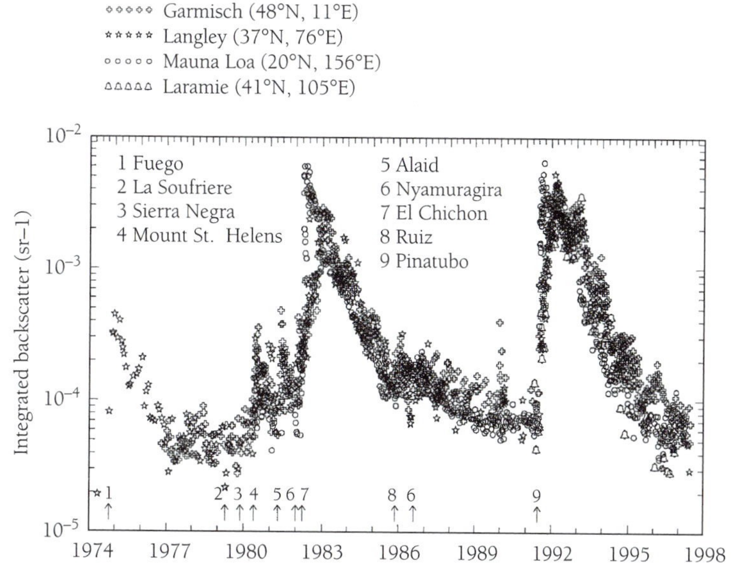

Figure 15.13 Integrated aerosol backscatter at a wavelength of 0.6943 μm from lidar measurements at Garmisch-Partenkirchen (48°N), NASA Langley (37°N), and Mauna Loa (20°N) for the period 1974–1998. Also indicated in the Figure are times corresponding to known volcanic eruptions, ranging from Fuego in 1974 to Mount Pinatubo in 1991. The Figure includes data from a number of locations as indicated. Source: World Meteorological Organization 1999.

troposphere, on a time scale of days to weeks, by either dry or wet deposition to the surface.[7]

Measurements with instrumentation deployed by the ER-2 aircraft provide a sensitive test of models for the stratosphere at mid and low latitudes, just as they did for the polar regime (Section 15.1). The primary limitation of the ER-2 data is imposed by the fact that the plane is restricted to fly at altitudes below about 20 km. This constraint is relatively unimportant for the polar environment, where the bulk of stratospheric O_3 is accessible to the ER-2. It is somewhat more limiting for lower latitudes, where most of stratospheric O_3 is located above the ceiling of the ER-2 (the tropopause is highest in the tropics, sloping downward toward the poles).

Figures 15.14 and 15.15 summarize results from an ER-2 mission flown from Moffett Field, California (37.4°N), on 18 May 1993. The plane took off at 7:40 A.M. local time and flew north at an altitude of about 18 km (60 mb) to a latitude of 58°N. It reached this location at 1:00 P.M. before reversing direction and returning to Moffett Field, where it landed at 3:30 P.M. The plane briefly descended to about 15 km at positions on the return leg corresponding to latitudes of 56.9°N and 47.7°N, permitting measurements to be taken locally over a small range of altitudes. Figure 15.14 summarizes the variety of latitudes, pressures, solar zenith angles (reflecting local time of day), and overhead column abundances of O_3 (obtained by combining data from an onboard spectrometer with measurements by TOMS) encountered by the aircraft over the course of the mission.

Results from model simulations are compared with measurements for OH, HO_2, ClO, and NO in Figure 15.15. Accounting for gaseous-phase processes only (Model A), results in a significant and systematic underestimate of abundances for OH, HO_2, and ClO, with a comparable overestimate for NO. The agreement between model and observation is notably better when we allow for the influence of heterogeneous chemistry, specifically for the reaction

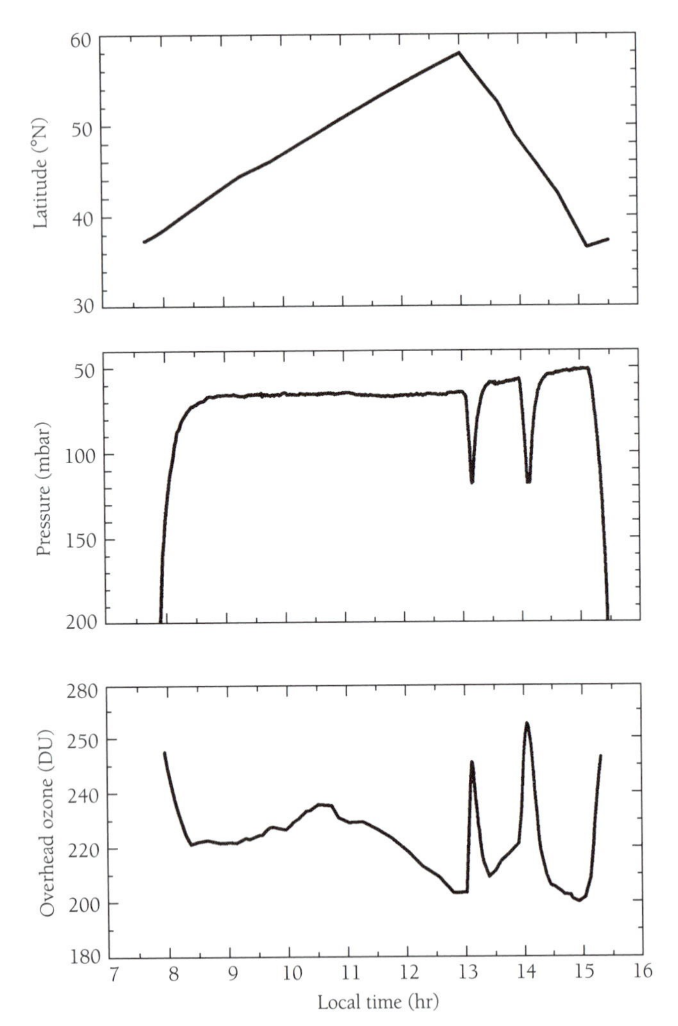

Figure 15.14 *Changes in latitude, pressure (height), and overhead column densities along the track of the ER2 on the flight from Moffet Field, California, on 18 May 1993. Abrupt changes in pressure near 13 hr and 14 hr reflect times when the aircraft adjusted altitude to obtain vertical profiles of relevant atmospheric parameters.*

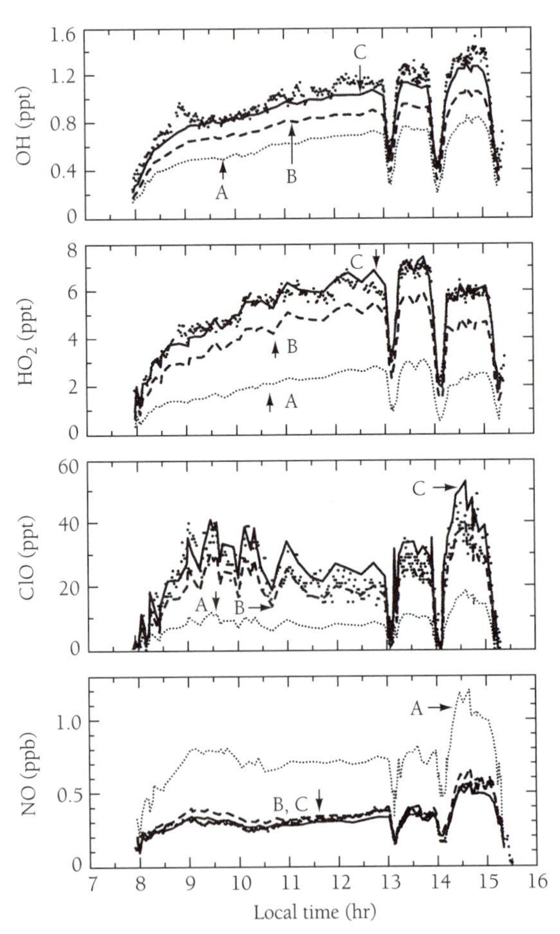

Figure 15.15 *Measurements of OH, HO_2, ClO, and NO along the track of the ER2 flight of 18 May 1993. Comparison of measurements with results of model simulations: A, gaseous-phase chemistry only; B, accounting for reaction (15.9); C, results from a study involving "tuning." Source: Salawitch et al., 1994.*

$$N_2O_5 + H_2O(l) \rightarrow 2HNO_3\ , \qquad (15.9)$$

where (l) indicates that the water molecule in (15.9) is present in the liquid phase. The impact of (15.9) depends on the abundance of aerosols, more precisely on the total surface area of aerosols available for reaction. The rate at which N_2O_5 is converted to HNO_3 by (15.9), $R_{15.9}$ (cm^{-3} sec^{-1}), may be expressed in terms of an effective rate constant, $k_{15.9}$ (sec^{-1}), such that

$$R_{15.9} = k_{15.9}\ [N_2O_5], \qquad (15.10)$$

where

$$k_{15.9} = \gamma\left[\frac{(SA)V_{N_2O_5}}{4}\right] \qquad (15.11)$$

Here γ measures the efficiency of the heterogeneous reaction (the fraction of collisions with particles resulting in reaction), *SA* denotes the total surface area contributed by particles included in unit volume, and $V_{N_2O_5}$ is the mean (average) speed of N_2O_5. Laboratory studies suggest a value for γ of about 0.06 for reaction (15.9), that is, an efficiency of about 6%.[8]

Reaction (15.9) results in net conversion of NO_x to the more stable form of NO_y, HNO_3. As illustrated by Model B, this leads to a reduction in abundances of NO and NO_2, as compared with values obtained with the gaseous-phase-only Model A. Lower abundances of NO_2 favor higher concentrations of ClO by lowering the rate at which ClO is converted to $ClNO_3$ by reaction (14.43); they have an additional impact on OH by reducing the efficiency of the major sink for hydrogen radicals, represented by conversion of OH and NO_2 to HNO_3, reaction (14.18). As a consequence, concentrations of OH and HO_2 are higher in Model B than in Model A, and agreement with observation is enhanced accordingly.

Further improvements in the agreement between model and observation can be achieved with minor adjustments in rates for particular gaseous-phase reactions, fine-tuning rates within the limits of uncertainties associated with specific reactions. The pattern of the discrepancies implied by the detailed comparison of the observations with Model B in Figure 15.15 suggests that the agreement between model and observation could be improved if the source of hydrogen radicals was slightly larger than that included in Model B (or equivalently, if the sink was less efficient). A larger source of hydrogen radicals would result in higher abundances of OH and HO_2, a decrease in NO_x induced by more efficient conversion of NO_2 to HNO_3 (reflecting involvement of higher concentrations of OH in reaction 14.18), and a modest increase in ClO, caused by the combination of more efficient conversion of HCl to Cl (reflecting the role of higher concentrations of OH in reaction 14.44) and less-efficient removal of ClO by formation of $ClNO_3$ (reflecting the influence of lower abundances of NO_2 in reaction 14.43). A modest "tuning" exercise results in excellent agreement between model and observation, as indicated by Model C in Figure 15.15.

The work summarized in Figure 15.15 provides unambiguous evidence for the significance of (15.9) at midlatitudes. Less definitively, it suggests that present models may underestimate the abundance of OH with implications for all of the other key radicals: HO_2, NO, NO_2, Cl, ClO, Br, and BrO. Recognition of the role of reaction (15.9) dictates an important shift in our perception of the relative importance of the various cycles responsible for the removal of O_3. The gaseous-phase-only model implies that catalytic reactions involving NO_x provide the dominant path for removal of odd oxygen, as illustrated in Figure 14.8. With inclusion of reaction (15.9), cycles involving HO_x, Cl_x, and Br_x assume larger roles.

The influence of (15.9) on the relative importance of the various paths for removal of odd oxygen for equinoctial conditions at 30°N is illustrated in Figure 15.16. Results in Figure 15.16a are analogous to those in Figure 14.8; meaning, effects of heterogeneous chemistry are omitted. Figure 15.16b allows for the impact of a relatively small burden of aerosols, similar to that present in the stratosphere at the time the ATMOS measurements were taken in May 1985. The effect of aerosol surface area on the fraction of odd oxygen loss attributable to the chlorine- and bromine-catalytic cycles is illustrated by the additional curves included in Figure 15.17b. The Figure explores a range of surface areas, from low values characteristic of conditions that might be considered background (that is, temporally removed from large volcanic eruptions), to intermediate values, to high values (reflecting disturbances attributable to a major eruption). Profiles of surface area as a function of altitude for these three conditions are displayed in Figure 15.17a. The low-value case (background) is taken to represent conditions in May 1985. The high-value case refers to September 1992 (approximately one year after the eruption of Mount Pinatubo). The intermediate case refers to September 1993. The fractions of odd oxygen loss attributable to combined influences of chlorine- and bromine-catalytic chemistry for these three conditions are displayed in Figure 15.17b. Results in Figures 15.16 and 15.17 indicate the enhanced importance of reactions involving HO_x, Cl_x, and Br_x, compared with NO_x, in the budget of odd oxygen when we account for the influence of (15.9). Note in particular the increased significance of the cycles involving Cl_x and Br_x. The influence of heterogeneous chemistry is even greater at high latitudes, where photolysis rates are lower and where partitioning of NO_x between NO and NO_2 favors NO_2 to a larger extent than for the conditions modeled in Figure 15.16. This is illustrated further in Figure 15.18 with a simulation of ATMOS results for 47°S.

As indicated above, the impact of heterogeneous chemistry at mid and low latitudes, in terms of its potential affects on O_3, is directly related to the role of reaction (15.9) in modifying the partitioning of NO_y between NO_x and HNO_3. Lower abundances of NO_2 result in a change in the relative importance of the catalytic cycles responsible for the removal of odd oxygen favoring HO_x, Cl_x, and Br_x over NO_x. The shift is directly attributable to changes in the abundance of NO_2. The response of NO_2, specifically the ratio of its

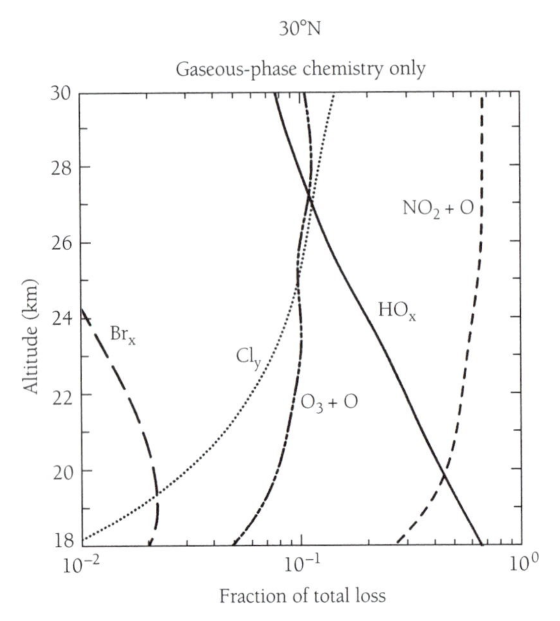

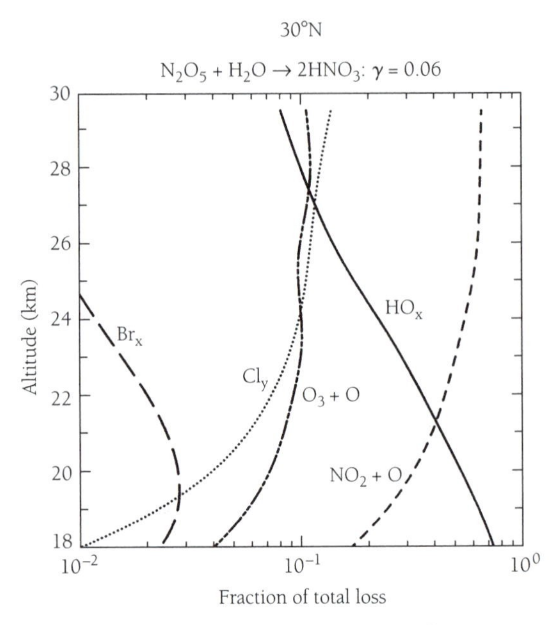

Figure 15.16 Contributions of various cycles to removal of odd oxygen for equinoctial conditions at 30°N. Panel a accounts only for gaseous-phase chemistry; panel b allows for the influence of reaction (15.9). From McElroy, et al., 1992.

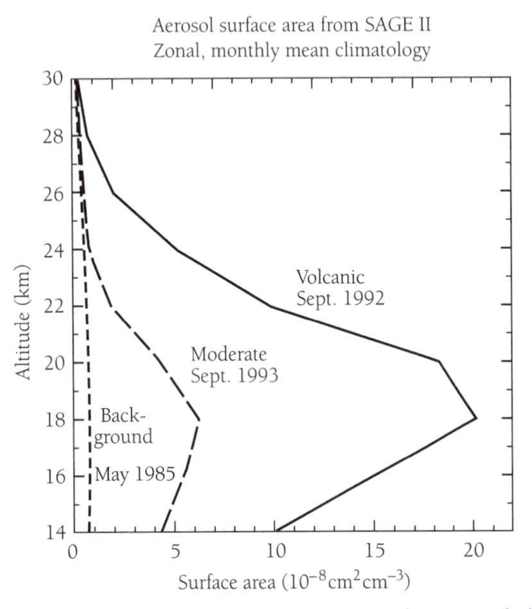

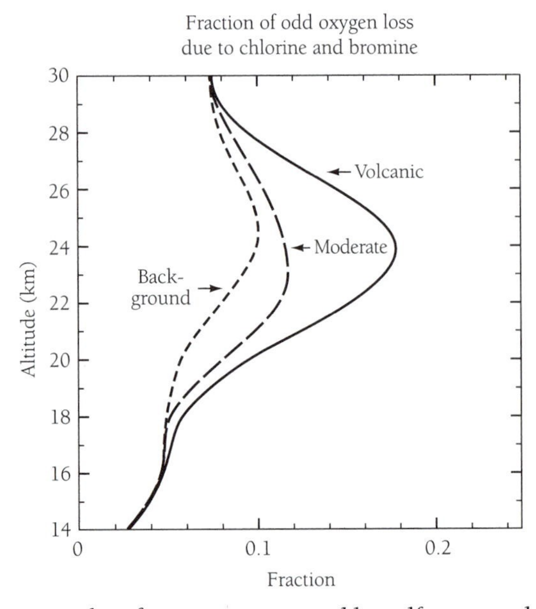

Figure 15.17 Panel a presents profiles as a function of altitude for the integrated surface area represented by sulfate aerosols. The separate profiles are intended to reflect conditions where surface area is low (May 1985), intermediate (September 1993), and high (September 1992). Fractional losses of odd oxygen attributable to chemical reactions catalyzed by chlorine and bromine for these three cases are presented in panel b.

abundance with respect to HNO_3, to an expansion of surface area (an increase in the rate for reaction 15.9), for an altitude of 20 km is illustrated in Figure 15.19. Note that the curve defining the variation of NO_2 with $k_{15.9}$ tends to be relatively flat for both low and high values of $k_{15.9}$. The behavior at low values of $k_{15.9}$ is easy to understand; if the rate for the heterogeneous reaction is slow, the abundance of NO_2 relative to HNO_3 is determined by processes in the gaseous phase. The pattern at high values of $k_{15.9}$ is a little more unexpected: it reflects the fact that conversion of NO_2 to HNO_3 is limited ultimately by the rate at which NO_2 is converted to N_2O_5—by the rate of the gaseous-phase process (14.19) rather than by (15.9).[9] Volcanoes, by adding sulfur to the stratosphere, can enhance the importance of heterogeneous chemistry, amplifying the impact of industrial chlorine and bromine on O_3; the impact is limited, however, eventually by restrictions imposed in the gaseous phase. According to the picture painted here, the direct

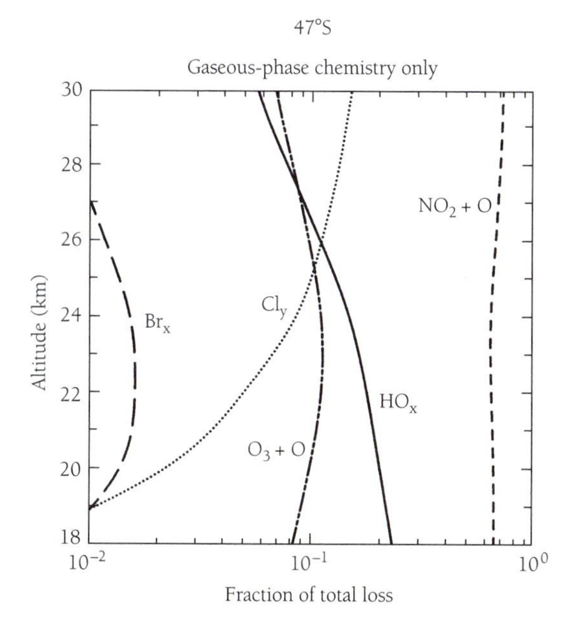

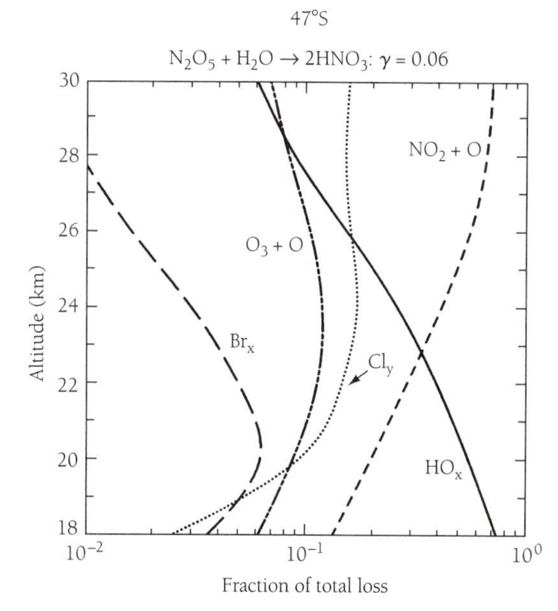

Figure 15.18 Similar to Figure 15.17 but for 47°S. Panel a, gaseous-phase only; panel b, accounting for reaction (15.9). Results included here are intended to simulate conditions appropriate for the measurements taken by ATMOS in 1985.

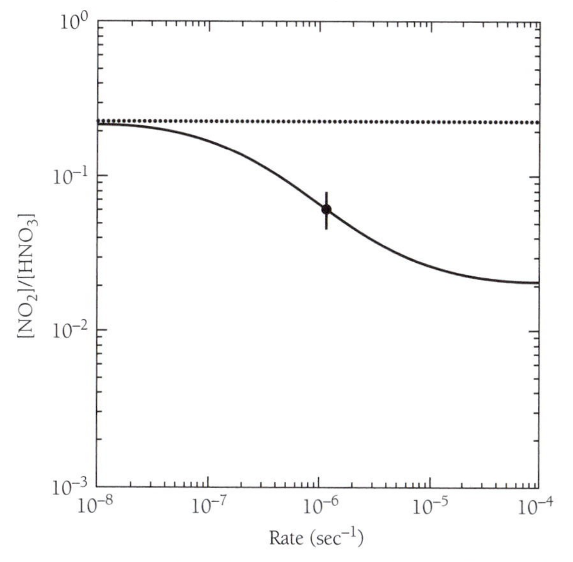

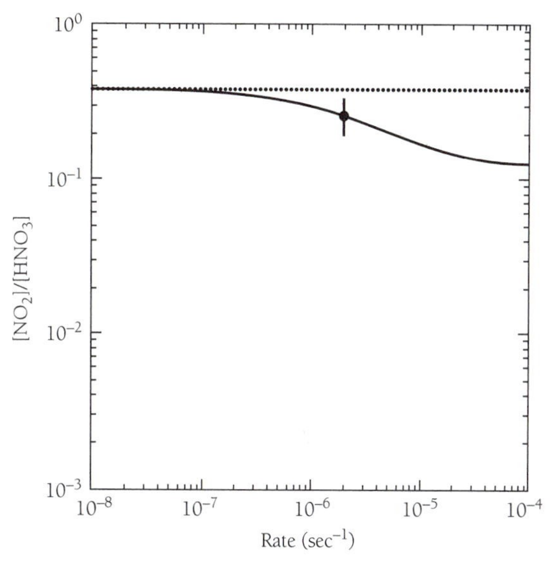

Figure 15.19 Model results for the ratio [NO$_2$]/[HNO$_3$] for an altitude of 20 km for latitudes of 30°N (panel a) and 47°S (panel b) as appropriate for analysis of the ATMOS observations taken in 1985. The dotted lines indicate results expected if we allowed only for gaseous-phase processes. The solid lines indicate the sensitivity of model results to the value assumed for $k_{15.9}$. The ratios measured by ATMOS are indicated by the solid circles, with one-sigma estimates of uncertainty defined by the vertical-error bars.

chemical effects on O_3 of a very large volcano should not be much larger than those observed for Mount Pinatubo.

15.4 Summary

The first essential step to production of a high concentration of chlorine and bromine radicals in the low-altitude polar stratosphere involves formation of NAT crystals. The threshold for this process is reached at a temperature of about 196 K and is commonly surpassed in both the Arctic and Antarctic environments. Reactions on the surfaces of NAT particles offer an effective means for conversion of stable compounds, such as $ClNO_3$ and $BrNO_3$, to less stable species, such as Cl_2, HOCl, and HOBr, paving the way for production of ClO and BrO as sunlight returns to the polar regime during local spring. Odd oxygen (O_3) is rapidly removed in the presence of these radicals, mainly by reactions (15.5) and (15.7), resulting in conversion of two molecules of O_3 to three molecules of O_2.

The abundance of NO_y was exceptionally low in the Antarctic environment sampled during the AAOE mission in 1987. Denitrification appears to be a common phenomenon in the winter stratosphere over Antarctica but may occur more sporadically in the Arctic. Cold temperatures persist later in spring in the Antarctic than in the Arctic, reflecting the higher stability of the vortex in the south (related presumably to the lower degree of wave activity in the Southern Hemisphere, as we shall discuss in Chapter 16). As a consequence, a high concentration of chlorine and bromine radicals can be maintained in the southern polar region later in spring than in the north, due to repetitive processing of $ClNO_3$ and HCl on NAT particles. In the absence of denitrification, chlorine and bromine radicals in the northern polar region are converted to more stable species, such as $ClNO_3$ and $BrNO_3$, as temperatures warm up during spring and as NO_2 is released by photodissociation of HNO_3. Loss of ozone by reactions (15.5) and (15.7) would be limited accordingly.

There is only one viable mechanism accounting for the phenomenon of denitrification: incorporation of HNO_3 in particles large enough to settle gravitationally carrying NO_y from high to low altitudes. The details of how this takes place remain unclear, however. The mass of H_2O and HNO_3 included in pure NAT particles is too small to permit gravitational settling to occur directly through this medium. A more likely possibility, suggested by Waibel et al. (1999), is that HNO_3 is included (most likely as NAT) in a small number of relatively large water-ice particles formed by selective nucleation of supercooled droplets of water. Observational evidence in support of this mechanism is reported by Dye et al. (1992). Lending further support to the hypothesis, Schreiner et al. (1999), using direct measurements with a balloon-borne instrument, showed that PSC particles present in the Arctic stratosphere at temperatures between 189–192 K contain both H_2O and HNO_3, with molar ratios H_2O/HNO_3 in excess of 10. Supercooled droplets might be expected to form as temperatures drop below the freezing point (about 188 K for the polar stratosphere). (A mechanism to account for selective nucleation of ice crystals was introduced in Section 7.5.) NAT-bearing ice crystals would melt or sublime as they fall to warmer regions at lower altitudes. This would be associated with release of H_2O and HNO_3, resulting in a local increase in the abundance of these species, a phenomenon referred to as *hydration* or *nitrification,* depending on whether the focus is on H_2O or HNO_3.

Ozone loss in polar regimes is likely to persist for some time in the future despite recently taken steps to reduce emission of industrial chlorinated and brominated gases. This relates to the fact that continuing release of greenhouse gases may be expected to result not only in an increase in tropospheric temperatures but also in a decrease in stratospheric temperatures. A decrease in temperatures in the polar regime, favoring an increase in occurrence of NAT particles and a more favorable environment for denitrification, could result in an enhanced efficiency of heterogeneous processes as a sink for ozone. Shindell, Rind, and Lonergan (1998) argued that temperatures in the Arctic stratosphere could decrease by as much as 8–10 K over the next few decades and that ozone loss in the Arctic would peak in about 2015. The problem could be exacerbated by increases in the abundances of CH_4 and N_2O, resulting in higher concentrations of H_2O and NOy in the stratosphere leading to an increase in the temperature of the threshold for formation of both NAT and water ice.

Heterogeneous reactions are expected to play a role also for the midlatitudinal stratosphere; the primary effect in this case involves conversion of NO_x to NO_y by reaction of N_2O_5 with H_2O on the surface of sulfate aerosols. This is likely to result in an increase in the importance of HO_x-catalyzed chemistry, as compared to NO_x as a sink for odd oxygen. The role of heterogeneous processes at midlatitudes will be sensitive to the magnitude of the input of sulfur from volcanic eruptions.

Stratospheric Ozone (IV): Influence of Dynamics

16

The preceding chapters have emphasized the chemical processes responsible for production, and loss, of stratospheric O_3. We noted that the lifetime of odd oxygen is long at low altitudes and high latitudes during winter (from months to years); in contrast, it is relatively short at high altitudes (approximately a month at 30 km, decreasing to less than a day above 40 km). If the lifetime of O_3 is short, we expect that the concentration of the gas should reflect primarily a local balance between the rate at which it is produced and consumed by chemical reactions; if the lifetime is long, we must allow for the influence of transport. When the lifetime is long, odd oxygen may be produced in one region and consumed under very different photochemical conditions elsewhere. This chapter intends to provide a perspective on the influence of transport (dynamics) on the distribution of chemical species in the stratosphere.

We begin in Section 16.1 with an overview of the physical processes regulating the circulation of air in the stratosphere. We illustrate principles with results from a 2-dimensional model in Section 16.2. We present summary remarks in Section 16.3.

16.1 Stratospheric Dynamics

As previously noted, atmospheric temperature increases with altitude in the stratosphere. The increase of temperature with altitude is due mainly to absorption of ultraviolet solar radiation by O_3. It is instructive for present purposes to focus on potential temperature rather than physical temperature: the potential temperature of an air mass is defined as the temperature that would apply if the air were adiabatically displaced (in the absence of heat exchange with its surroundings: see Chapter 7) to the surface at a standard pressure of 1000 mb. Potential temperature, θ, is given in terms of physical temperature, T, by the relation

$$\theta = T\left(\frac{1000}{p}\right)^{\kappa}, \tag{16.1}$$

where p is the local value of pressure expressed in millibars (κ for dry air has a value of 0.288).[7] Physical temperatures can change by as much as 10 K as the altitude of an air mass varies by as little as 1 km. The value of potential temperature, however, can be altered only by addition or removal of heat.

Contours illustrating the variation of potential and physical temperatures as a function of latitude and altitude are displayed in Figure 16.1. The data included here reflect an average over longitude for January 1993 (Holton et al. 1995). The heavy solid line denotes the approximate location of the tropopause outside the tropics; the tropopause in the tropical region is identified with the level of the atmosphere corresponding to a potential temperature of about 380 K. The tropopause slopes downward in the poleward direction dropping in altitude from about 15 km in the tropics to almost 7 km near the poles; it is intersected by successively lower values of potential temperature at higher latitudes.

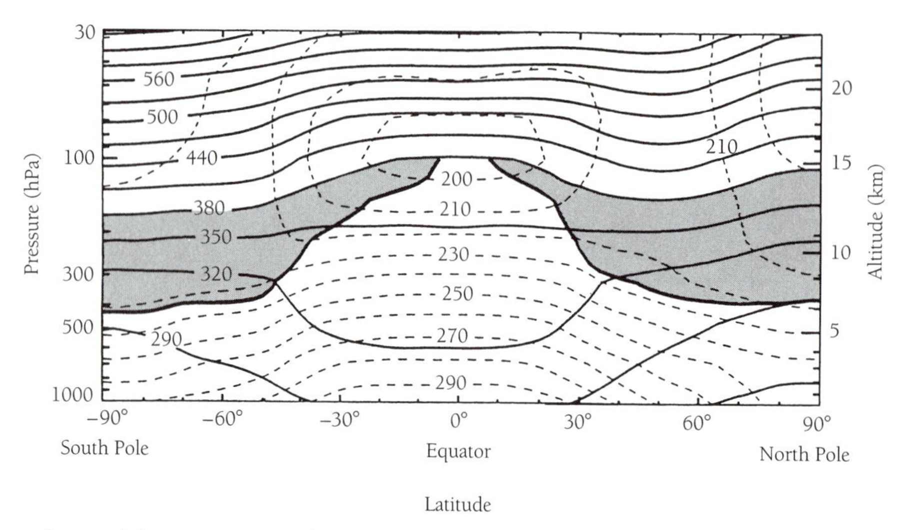

Figure 16.1 Contours of potential temperature as functions of pressure (height) and latitude for January 1993 (solid lines). Actual temperatures are indicated by dashed lines. The shaded area indicates a region where surfaces of potential temperature include both the stratosphere and the troposphere (the middle world). Source: Holton et al. 1995.

Contours defining potential temperatures in the lower stratosphere in the extratropical region (denoted by the shaded area in the Figure) are smoothly connected to contours identifying similar values of potential temperature in the troposphere at lower latitudes; in contrast, contours identified with potential temperatures in excess of 380 K are located exclusively in the stratosphere. Air in the region above the 380 K potential-temperature contour must cool before descending into the troposphere. Similarly, air with potential temperatures less than about 290 K must acquire heat before it can rise into the stratosphere; exchange of air between the stratosphere and troposphere can proceed adiabatically in the middle (shaded) region. The extent to which this exchange results in an irreversible transfer of air from the stratosphere to the troposphere is limited, however, by constraints imposed by more complex considerations of atmospheric dynamics.[2] Holton et al. (1995) refer to the region of the stratosphere where potential temperatures are greater than 380 K as the *overworld*. The zone where potential temperature contours lie exclusively in the troposphere is described as the *underworld*. By obvious extension, the region where potential temperature contours include both the stratosphere and troposphere is identified as the *middle world*.

Circulation of air in the overworld is driven by wavelike disturbances originating in the troposphere. Of particular importance are planetary scale disturbances, known as **Rossby waves**. These disturbances propagate vertically in a westward direction; they break, or dissipate, at middle levels of the stratosphere, applying a westward force to the background mean flow. The forcing associated with breaking Rossby waves is especially important in the subtropics in the winter hemisphere. The westward torque imposed on the predominantly eastward flow of background stratospheric air drives a secondary circulation: air begins to move poleward.[3] This poleward motion is supplied by upward motion of air in the tropics. The circulation loop is completed by downward motion at higher latitudes with a return flow at lower altitudes in the troposphere.

Following Holton et al. (1995), we may think of the breaking waves as applying suction, drawing air upward and outward from the tropics and depositing it at higher latitudes. The upward (adiabatic) motion in the tropics causes temperatures to drop below values that would apply in radiative equilibrium (the condition where temperatures are determined by a local balance between absorption and emission of radiation). Conversely, the downward motion at higher latitudes causes temperatures to rise above the radiative equilibrium standard. As a consequence, air in the tropics experiences net heating due to interaction with the radiation field; net cooling is associated with the region where air is descending at higher latitudes. On a globally averaged basis, the variation of temperature with stratospheric altitude may be represented to reasonable accuracy by a simple consideration of radiative balance.

The overturning of the stratosphere described here is referred to variously as the **wave-driven** or **diabatic circulation**. The latter terminology implies that vertical motion arises as a result of a departure from radiative equilibrium (as a consequence of net heating or cooling). According to the view presented here, the departure from radiative equilibrium is the *result* not the *cause* of the vertical motion. Stratospheric circulation is driven primarily by forcing associated with breaking Rossby waves, with an additional contribution, especially at higher altitudes, due to the dissipation of smaller-scale disturbances known as **gravity waves**.

The influence of the wave-driven circulation on temperature extends all the way to the tropical tropopause. The coldest temperatures are observed during Northern Hemispheric winter, when the wave-forcing is most vigorous (presumably related to differences in winter weather patterns ultimately associated with the higher concentration of land masses in the Northern Hemisphere). Temperatures at

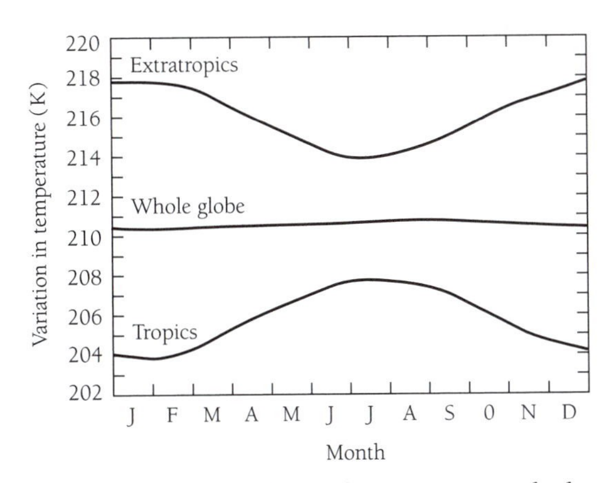

Figure 16.2 Annual variations of temperature in the lower stratosphere taken from space-based measurements over the period 1979–1991. Data are presented for the tropics (lower panel), the extratropical region poleward of 30°N and 30° S (upper panel), and the globe. Source: Holton et al. 1995.

the tropical tropopause drop to values as low as 188 K during January. The mixing ratio of H_2O vapor in air entering the stratosphere at that time is limited by saturation vapor considerations to a value of about 2.4 ppm. Temperatures of the tropical tropopause are higher at other seasons and the mixing ratio of H_2O vapor entering the stratosphere is increased accordingly. The seasonal variation of temperature in the lower stratosphere is illustrated for the tropics, the extratropics, and the globe in Figure 16.2. Warm temperatures in the tropics are associated with cold temperatures in the extratropics, and the reverse is also true, as expected based on the wave-forcing mechanism (directions of vertical motion are opposite for the tropics and extratropics). Temperatures for the globe as a whole show little variation with season.

Estimates for exchange of mass between the troposphere and stratosphere are presented in Table 16.1. As expected, the flux of mass entering the stratosphere is largest during the period of maximum wave-forcing, during Northern Hemispheric winter. It is smaller by about a factor of two during Northern Hemispheric summer (Southern Hemispheric winter). As indicated by the following example, the data in Table 16.1 imply a mean residence time of about 2 yr for air in the stratosphere above the 100 mb level.

Table 16.1 Estimates for Exchange of Mass between Troposphere and Stratosphere

	DJF	*JJA*	*Annual Mean*
Northern Hemisphere, midlatitudes	–81	–26	–53
Southern Hemisphere, midlatitudes	–33	–30	–32
Tropics	114	56	85

Estimates by Rosenlof and Holton (1993) for the flux of mass across the 100 mb surface (in units of 10^8 kg sec^{-1}) for the tropics and for midlatitudes of the Northern and Southern Hemispheres. Positive numbers imply that the flux is directed upward (from troposphere to stratosphere); negative numbers indicate that the flux is downward (from stratosphere to troposphere). Data are presented for the months of December–February (DJF) and June–August (JJA). Also included are estimates for annual means.

Example 16.1: Using the data in Table 16.1 for the annual average rate at which mass enters the stratosphere for the troposphere across the 100-mb surface in the tropics, estimate the lifetime for air in the stratosphere above 100 mb.

Answer: The mass, M, per unit area above a level in the atmosphere corresponding to a pressure p is given by (see equation 7.8)

$$M = \frac{p}{g}$$

With p expressed in dynes cm^{-2} and g in cm sec^{-2}, M, in g cm^{-2}, is given by:

$$M(g\ \text{cm}^{-2}) = \frac{10^5(\text{dynes cm}^{-2})}{9.81 \times 10^2(\text{cm sec}^{-2})}$$

Integrating over the total surface area of Earth (5.1×10^{18} cm^2), the total mass above the 100-mb surface is given by

$$M(g) = \frac{(5.1\times10^{18})(10^5)}{9.81\times10^2}$$
$$= 5.2 \times 10^{20} g$$
$$= 5.2 \times 10^{17}\ \text{kg}$$

The lifetime, τ, is given by

$$\tau = \frac{5.2\times10^{17}}{8.5\times10^1}\ \text{sec}$$
$$= \frac{5.2\times10^{17}}{(8.5\times10^1)(3\times10^7)}\ \text{yr}$$
$$= 2\ \text{yr}$$

■

A schematic illustration of stream lines describing the stratospheric flow of air under solstice conditions is presented in Figure 16.3. Air is drawn into the stratosphere from the troposphere by an upward flow in the tropics; it moves from there to the summer hemisphere. A portion is carried to higher altitudes under the influence of wave-forcing exerted mainly in the winter hemisphere. It subsequently moves to the winter hemisphere where it progressively sinks to lower altitudes. The flow pattern for the lower stratosphere depicted in the Figure reflects an upward projection of the Hadley circulation at lower altitudes.

A more symmetric flow pattern is expected to develop under equinoctial conditions. In this case also, air is transported upward in the tropics; it moves poleward in both

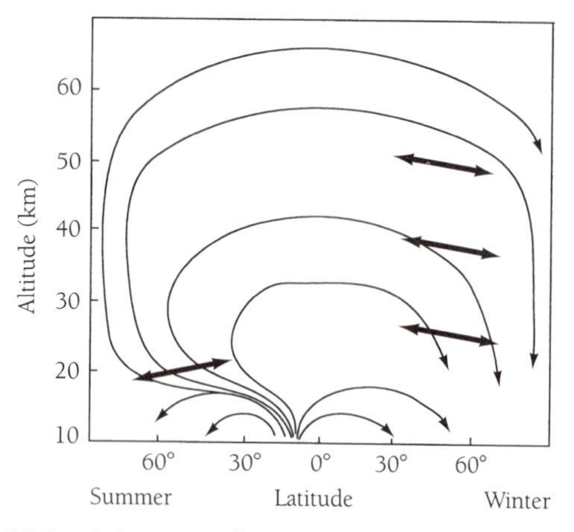

Figure 16.3 Schematic illustration of the circulation of the stratosphere under solstice conditions. The influence of mixing by eddy motions is indicated by the bidirectional arrows.

hemispheres, sinking to lower altitudes at higher latitudes, as illustrated in Figure 16.4.

The flow illustrated for the mean circulation in Figures 16.3 and 16.4 presumes that transport of chemical species in the stratosphere may be treated as 2-dimensional. The bidirectional arrows in the Figures indicate the potential importance of transport by eddies. Eddies are introduced in the 2-dimensional formulation to account for transport induced by variations in meridional flow as a function of longitude. These variations are most probably associated with fluctuations in the velocity fields of waves propagating upward from the troposphere. It is presumed that the motion associated with these waves is adiabatic. The transport of chemical species by eddies is expected to proceed, therefore, primarily along surfaces of constant potential temperature. The influence of eddies is confined to the extratropical region; transport, at least at low altitudes in the tropics, is associated primarily with seasonal variations of the mean flow.

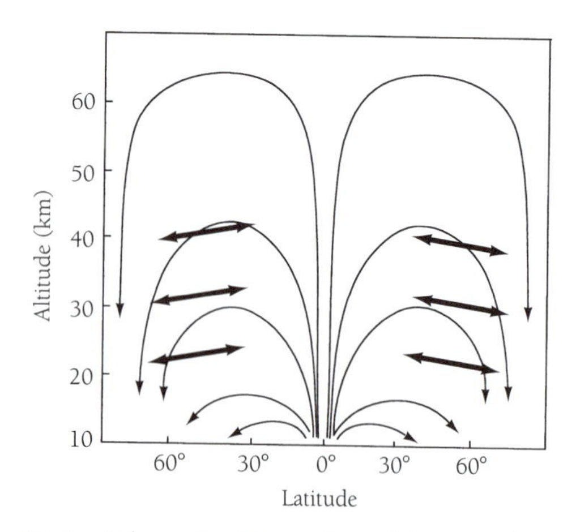

Figure 16.4 Schematic illustration of the circulation of the stratosphere under equinoctial conditions. As in Figure 16.3, the influence of eddy motions is indicated by the bidirectional arrows.

16.2 The Distribution of Ozone with Latitude

We turn our attention now to the processes that regulate the distribution of O_3 with latitude; results in this Section were obtained using a version of the 2-dimensional model of the stratosphere developed by Schneider et al. (1993). The effects of breaking planetary waves were treated using a diffusive formulation with diffusion coefficients selected based on analysis of observational data by Newman and Schoeberl (1986) and Newman et al. (1988). Gravity waves were assumed to exert a frictional drag on the zonal flow. The magnitude of the associated frictional force was taken to be proportional to the speed of the zonal wind (using the Rayleigh friction approximation); drag coefficents were assumed independent of latitude and season but were allowed to vary with altitude according to the prescription suggested by Holton (1982). The basic equations describing conservation of energy, momentum, mass, and chemical species were numerically solved in a dynamically consistent fashion. The model is *interactive* in the sense that heating rates and details of the circulation were determined using model-derived results for the distributions of O_3 and other chemical species. It includes a relatively complete representation of the relevant chemistry of the stratosphere (both gaseous phase and heterogeneous), as discussed in the preceding chapters.

The impact of radical-related chemistry on the height and latitude distribution of O_3 is illustrated in Figures 16.5–16.7. Data presented here refer to conditions at equinox. Figure 16.5 summarizes results from a simulation accounting only for the reactions included in the simple Chapman scheme (Chapter 13); Figure 16.6 allows for reactions involving hydrogen and nitrogen radicals (Chapters 14 and 15); the impact of chlorine radicals is illustrated in Figure 16.7, assuming concentrations of stratospheric chlorine roughly comparable to levels present in 1985. Peak values of the mixing ratio of O_3 for all three simulations are predicted to occur in the tropics at an altitude of about 35 km. Mixing ratios in the vicinity of the peak calculated using the Chapman formulation (Figure 16.5) are approximately twice as large as those obtained with models accounting for the influence of radicals (Figures 16.6 and 16.7): mixing ratios computed using the Chapman formulation below the peak are lower than those obtained with the more complete models. A similar pattern is observed at higher latitudes. The decrease in O_3 at low altitudes (and high latitudes) in the Chapman treatment results from a decrease in transmission of ultraviolet solar radiation. The reduction in O_3 at low altitudes is thus a direct result of the presence of higher abundances of O_3 at higher altitudes. Radiation that would be otherwise available to dissociate O_2 is absorbed by O_3 at higher altitudes. Its contribution to local production of odd oxygen is reduced accordingly.

The relative importance of the removal of odd oxygen by the catalytic cycles involving NO_x, HO_x, and ClO_x is illustrated in Figure 16.8. The Figure includes also a panel

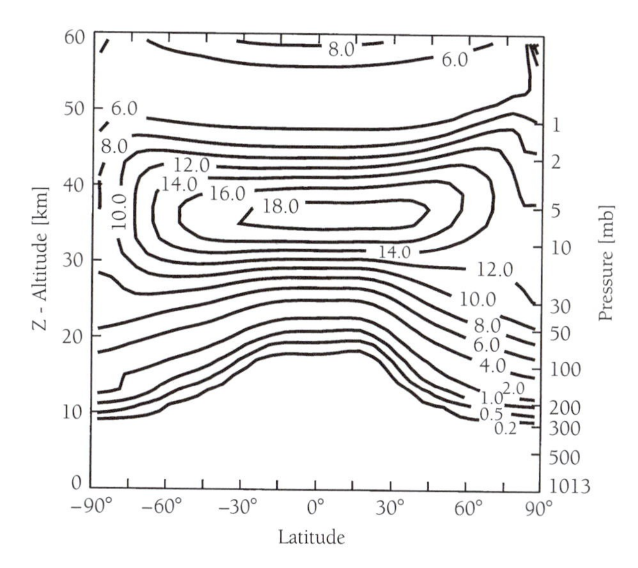

Figure 16.5 Distribution of O_3 as a function of latitude and altitude as obtained using the simple Chapman chemistry (neglecting effects of radicals). In the Figure concentrations are presented as mixing ratios expressed in units of ppm. Results refer to equinox.

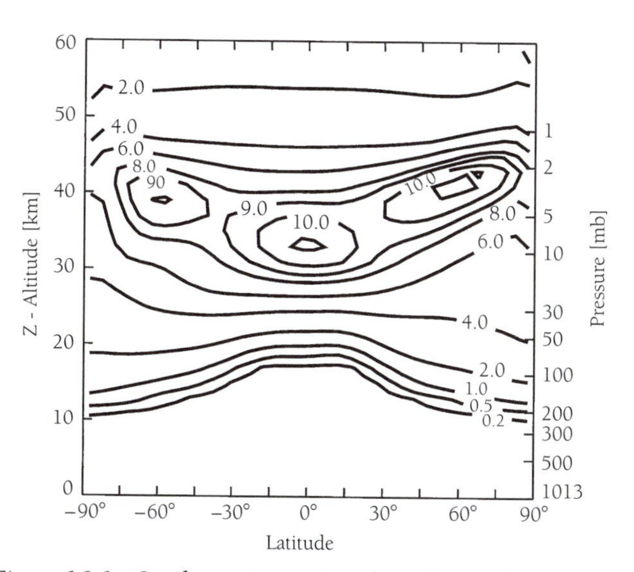

Figure 16.6 Similar to Figure 16.5 but allowing for reactions involving hydrogen and nitrogen radicals.

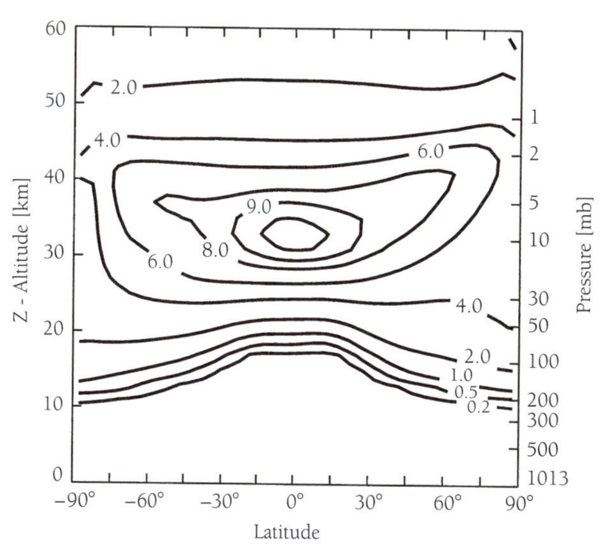

Figure 16.7 Similar to Figure 16.6 but allowing for the influence of reactions involving ClO_x in addition to HO_x and NO_x. Levels of Cl_x refer to conditions in 1985.

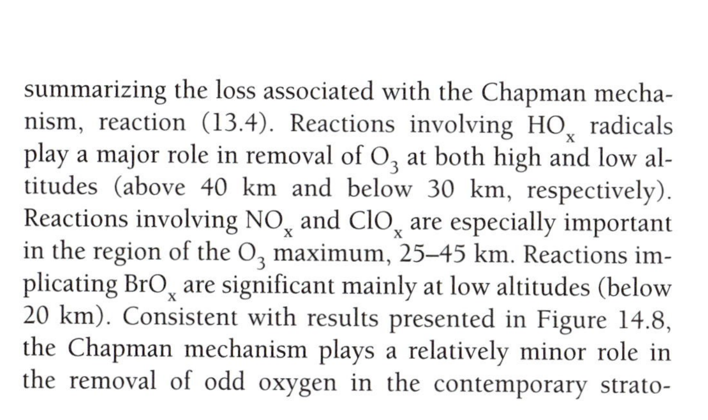
summarizing the loss associated with the Chapman mechanism, reaction (13.4). Reactions involving HO_x radicals play a major role in removal of O_3 at both high and low altitudes (above 40 km and below 30 km, respectively). Reactions involving NO_x and ClO_x are especially important in the region of the O_3 maximum, 25–45 km. Reactions implicating BrO_x are significant mainly at low altitudes (below 20 km). Consistent with results presented in Figure 14.8, the Chapman mechanism plays a relatively minor role in the removal of odd oxygen in the contemporary stratosphere. Loss of O_3 in the present environment is dominated by interconnected catalytic cycles involving NO_x, HO_x, ClO_x, and BrO_x.

Model results for the distribution of the vertical column density of O_3 are displayed as a function of season and latitude in Figures 16.9–16.11. As before, the Figures are intended to illustrate the impact of catalytic chemistry, both omitting (Figure 16.10) and including (Figure 16.11) effects of chlorine. Results obtained using the Chapman scheme are presented for reference in Figure 16.9. Changes

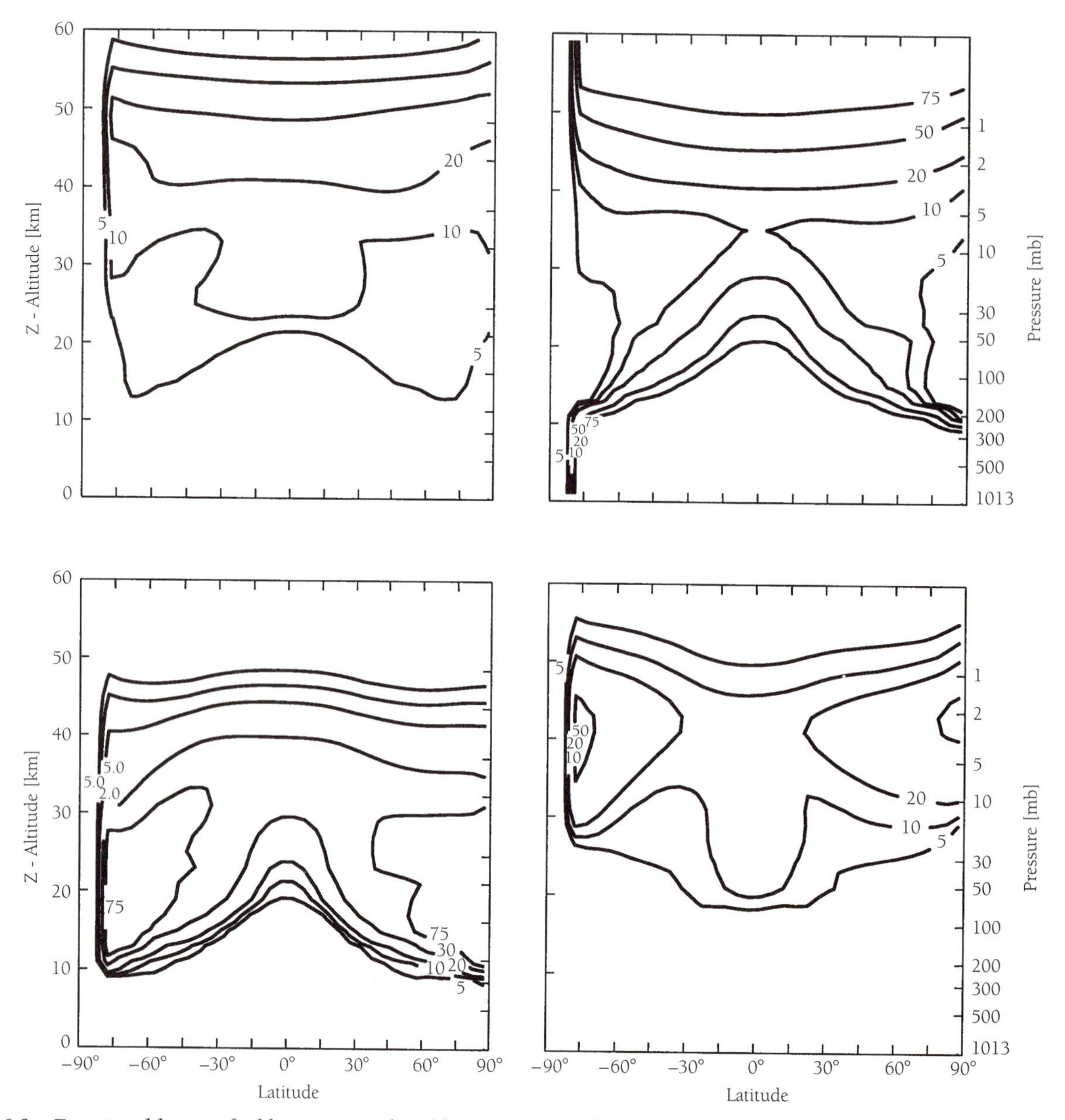

Figure 16.8 Fractional losses of odd oxygen attributable to Chapman chemistry (upper-left panel), HO_x (upper right), NO_2(lower left), and ClO_x (lower right). Results are for equinox. Contours define fractional losses expressed in %.

specifically attributable to chlorine, expressed as percentage changes expected for levels of chlorine applicable in 1990, as compared to 1980, are summarized in Figure 16.12. The reductions in column O_3 depicted in Figure 16.12 are in general agreement with trends inferred from analysis of both ground-based and satellite data reported by the World Meteorological Organization (WMO 1999).[4]

A composite of ground-based and satellite observations of the variation of column O_3 with latitude and season is presented in Plate 5 (bottom) (see color insert). The general pattern of the observational data is reproduced satisfactorily by the model as indicated by a comparison of results in Figure 16.11 with those in Plate 5 (bottom). In particular, the model accounts for the fact that column abundances are lowest in the tropics and highest at high latitudes during late winter–early spring in the Northern Hemisphere. This behavior reflects the fact that the meridional circulation is strongest at this time of year; the model also accounts for the seasonal minimum in column densities observed for northern midlatitudes in late summer–early autumn and for the general asymmetry in column densities observed between the Northern and Southern Hemispheres.[5] Column densities derived using the Chapman formulation are approximately twice as large as values obtained when we allow for the influence of radicals; also, the contrast between tropics and midlatitudes obtained with the Chapman model is much less than that indicated by either the observational data or the results from the more comprehensive models summarized in Figures 16.10 and 16.11. Comparison of observational data for the period 1985–1997 with 1964–1976 indicates a de-

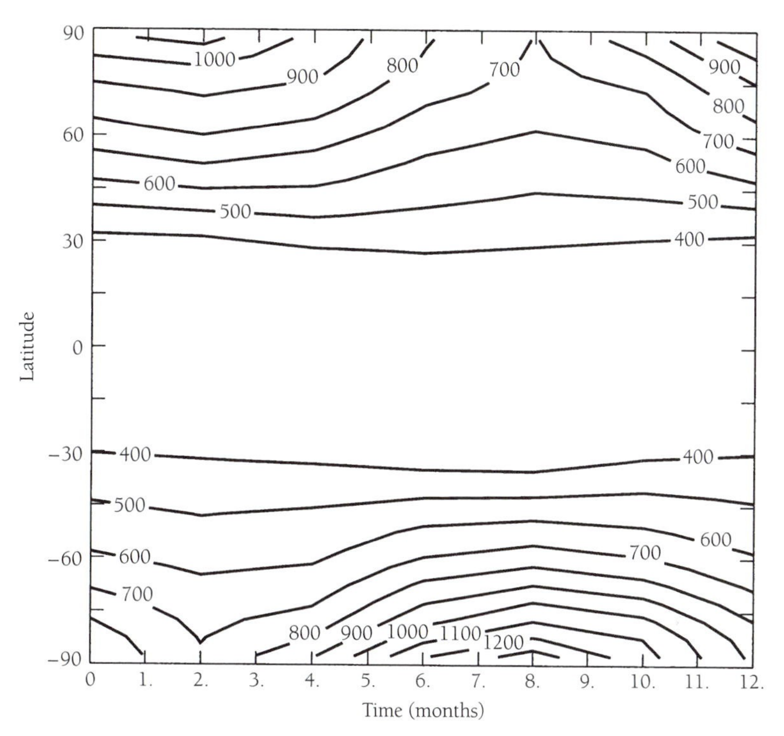

Figure 16.9 Distributions of column O_3 (Dobson units) as a function of time and latitude. Results shown here were calculated using the Chapman scheme. Effects of HO_x, NO_2, and ClO_x were omitted. Months are indicated from January (0) to December (12).

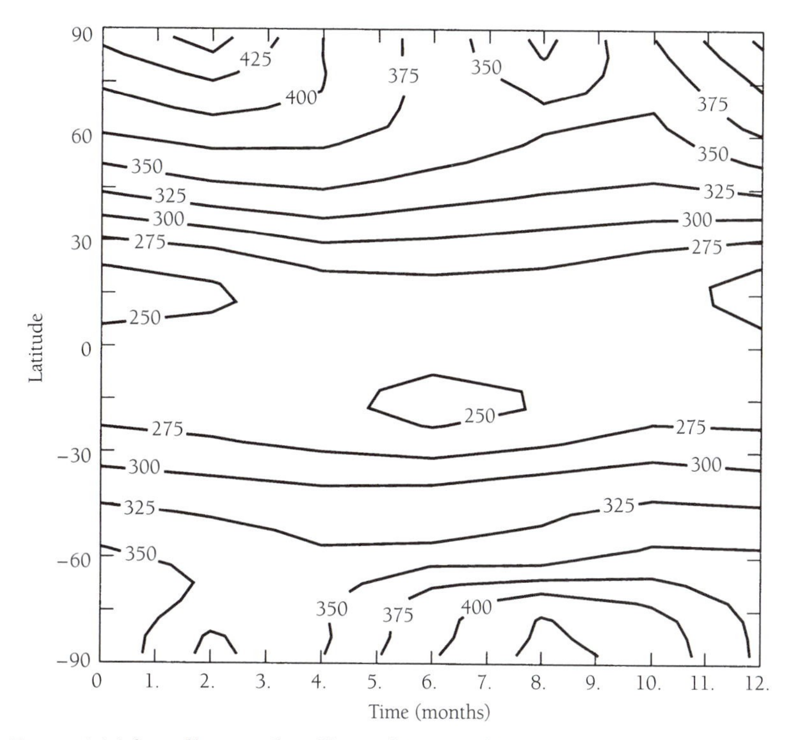

Figure 16.10 Similar to Figure 16.9 but allowing for effects of HO_x and NO_x.

crease in column densities at high northern latitudes during winter by about 12% over this interval. A much larger change is observed for high latitudes of the Southern Hemisphere during local spring (approaching 50%); the change in this case is attributed to the influence of the heterogeneous chemical processes responsible for the ozone hole, as we have discussed in Chapter 15.[6]

The importance of transport in regulating the latitudinal distribution of column O_3 is illustrated in Figures 16.13 and 16.14. Sources and sinks of column O_3 for equinoctial con-

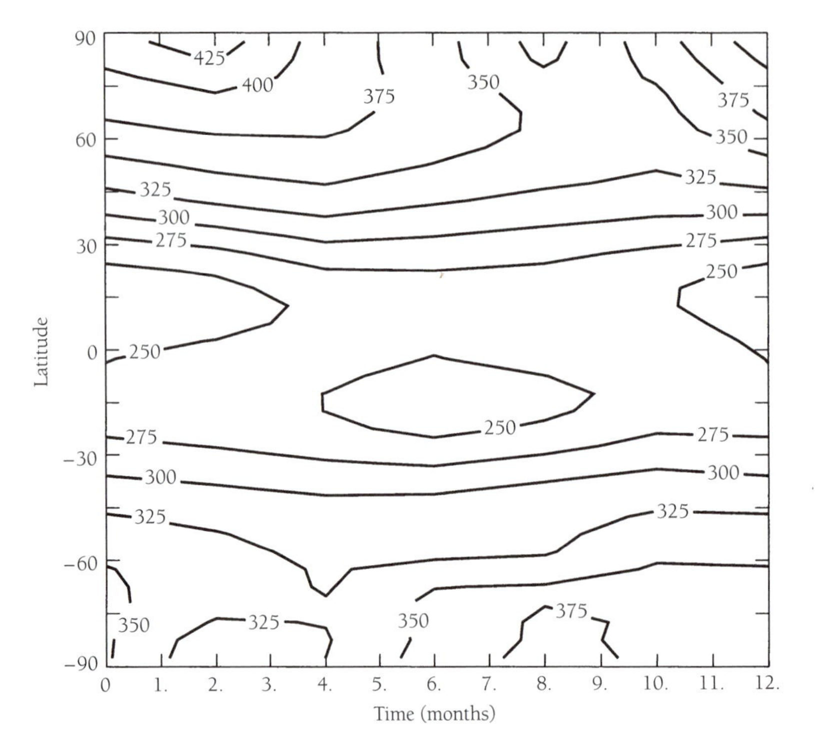

Figure 16.11 *Similar to Figure 16.10 but additionally allowing for effects of ClO_x with Cl_x levels appropriate for 1985.*

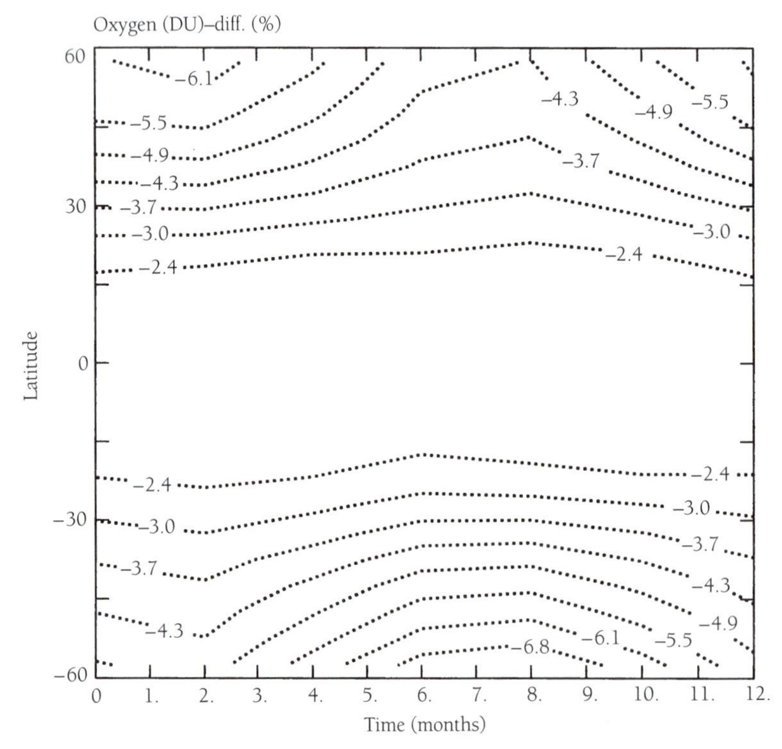

Figure 16.12 *Reductions in column O_3 predicted for the period 1980–1990. The model accounted for effects of BrO_x in addition to HO_x, NO_x, and ClO_x. Results are expressed as percentage changes over the interval.*

ditions, computed using the model accounting for levels of chlorine appropriate for 1990 and expressed in terms of Dobson units per year, are displayed in Figure 16.13. Rates for both production and loss peak in the tropics, dropping by about a factor of two with an increase in latitude to about 60°. The difference between production and loss, displayed in Figure 16.14, accounts for but a small fraction (less than 15%) of the gross source.

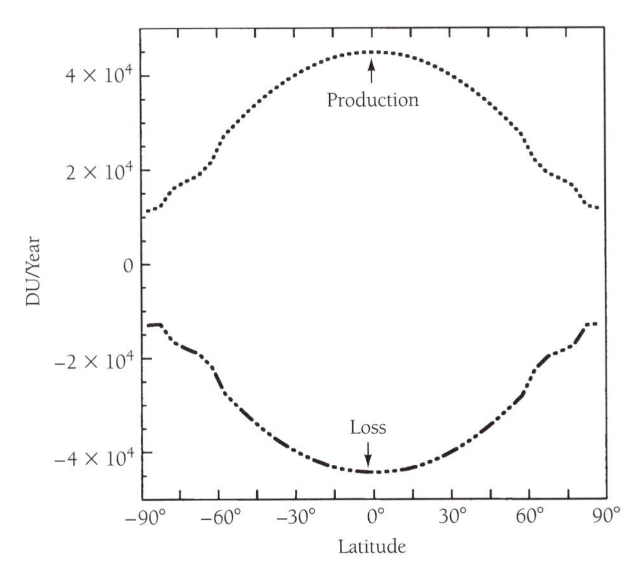

Figure 16.13 Production and loss rates for column O_3 (Dobson units per year) as functions of latitude for equinox. Results refer to the complete model allowing for effects of HO_x, NO_x, ClO_x, and BrO_x.

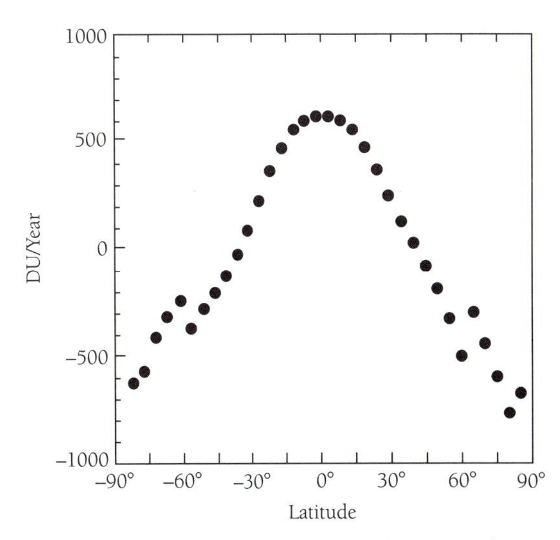

Figure 16.14 The difference between column production and loss rates (Dobson units per year) implied by the data in Figure 16.13. Note the presence of a net source in the tropics (latitudes ±30°) balanced by a net deficit at higher latitudes.

16.3 Summary

We saw that transport plays an important role in determining the temporal and latitudinal distribution of column O_3. The strength of transport depends on forcing associated with upward propagation of dynamical disturbances originating in the troposphere (Rossby and gravity waves). It follows that the abundance and spatial distribution of stratospheric O_3 is sensitive to details of weather in the lower atmosphere.

A change in climate could result in a significant change in stratospheric O_3, but it is difficult to forecast either the magnitude or sign of the impact. The polar regions merit special attention in light of the evidence for major reductions in recent years in the abundance of O_3 over Antarctica and suggestions that a similar phenomenon may be underway in the Arctic. A reduction in wave-forcing could lead to an increase in the temperature of the tropical tropopause with a consequent increase in the abundance of H_2O entering the stratosphere. Such a change might be expected to also result in a temperature decrease for the polar stratosphere. The combination, higher concentrations of H_2O and lower temperatures, would create a more favorable environment for the formation of polar stratospheric clouds.[7] Loss of O_3 would be enhanced accordingly. Alternatively, an increase in wave-forcing could allow for a more rapid recovery of O_3 at high latitudes.

The Chemistry of the Troposphere

17

Chapters 13–16 focused on the processes responsible for production and removal of O_3 in the stratosphere; in a larger sense they concerned the response of the stratosphere to absorption of ultraviolet solar radiation and the sequence of reactions triggered by photodissociation of O_2 at wavelengths less than about 240 nm. We saw that dissociation of O_2 results in the production of O and subsequently of O_3. The presence of O_3 extends the spectral range for absorption of ultraviolet solar radiation by almost 70 nm (to about 310 nm) and plays a critical role in shielding life at Earth's surface from potentially lethal doses of ultraviolet (UV-B) radiation. The abundance of O_3 reflects an overall balance between the rate at which odd oxygen (mainly O + O_3: see definition in Section 13.1) is formed by photodissociation of O_2 and the rate at which it is removed, primarily by reactions catalyzed by trace abundances of nitrogen, hydrogen, chlorine, and bromine radicals. But the balance is not entirely local; odd oxygen formed in one region may be transported over large distances by atmospheric winds and eventually removed far from its point of origin. Transport is particularly important at lower altitudes and at higher latitudes, where the lifetime of odd oxygen is long. A small fraction (about 1%) of odd oxygen formed in the stratosphere makes its way to the troposphere. As we shall see, the supply of O_3 from the stratosphere plays a critical catalytic role in the chemistry of the troposphere.

The troposphere is in a real sense an extension of the biosphere. A large variety of reduced gases is produced and released to the atmosphere as by-products of the metabolism of bacteria, plants, and animals. Methane is formed by bacteria breaking down plant material under anaerobic (low-oxygen) conditions in swamps and rice fields and in the stomachs of cud-chewing ruminants, such as cattle and sheep. Bacteria are also responsible for production of important quantities of nitrogen oxides and hydrogen sulfide, from soils and swamps, respectively. Land-based plants emit a variety of hydrocarbons: notably isoprene and various terpenes. Marine phytoplankton contribute a significant source of dimethylsulfide. Adding to the mix of chemically reduced species is a wide range of gases, including carbon monoxide, methane, and nitric oxide, emitted as by-products of the burning of vegetation (fires naturally initiated by lightning, as well as fires set for various purposes by humans) and the combustion of fossil fuels. Additional production of reduced species is associated with industry and with disposal and decomposition of human and animal wastes. In the absence of mechanisms for removal and recycling of the waste products of life—the waste gases of the biosphere and industry—the composition of the atmosphere would radically change and we would literally choke on our own exhaust; this chapter is concerned with elaboration of the processes regulating the self-cleansing properties of the atmosphere.

In our discussion of tropospheric chemistry, we will be especially interested in processes that contribute to the recycling in the atmosphere of the life-essential elements C, N, and S. Specifically, we are concerned with the fate of these elements as they are emitted in reduced, or partially reduced form (see Section 3.2 for a discussion of the distinction between reduced and oxidized states of elements in specific compounds and for the recipe used to measure the degree of oxidation: the oxidation number) by the biosphere or by industry. The breakdown, transformation, and ultimate removal of reduced forms of C, N, and S involve a series of reactions in which the elements are sequentially converted to higher states of oxidation. As we shall see, depending on circumstances, these reactions can be responsible for significant production or, under some circumstances, important loss of O_3.

Carbon in living tissue (denoted symbolically by CH_2O: see Section 11.1) is characterized by an oxidation number of 0. The oxidation states for carbon in CH_4 and CO are −4 and +2, respectively.[1] Reduced forms of carbon are converted in the atmosphere to the fully oxidized state, CO_2 (carbon oxidation state +4), which is subsequently taken up by plants, completing the cycle summarized by reactions (11.1) and (11.2).

Nitrogen in NO has an oxidation number of +2. Nitrogen in nitric acid (HNO_3), the form in which N is typically removed from the atmosphere, has an oxidation number of +5. Similarly, sulfur in hydrogen sulfide, H_2S, or dimethylsulfide, $(CH_3)_2S$, has an oxidation number of −2. Sulfur is removed from the atmosphere mainly in the oxidation state +6 as sulfuric acid (H_2SO_4). The fully oxidized forms of nitrogen and sulfur are soluble and are readily scavenged from the atmosphere by rain and snow and by deposition on moist surfaces (plant leaves, for example). They are subsequently taken up and reduced by plants, completing the cycle. A steady state is established for N and S, as it is for C: the release of N and S originally contained in plant material into the atmosphere is balanced by oxidation in the atmosphere, followed by either dry or wet deposition (for definition of these terms, see Endnote 7, Chapter 15) and compensatory uptake and reduction by plants.

Oxidation of reduced species is initiated in the troposphere for the most part by reaction with OH. The critical importance of OH in tropospheric chemistry was first recognized in 1971 by Hiram Levy, then a post-doctoral research fellow at Harvard University. Photolysis of O_3 (reaction 14.15) provides a source of $O(^1D)$; subsequent reaction of $O(^1D)$ with H_2O (reaction 14.23) results in the production of OH. As noted above, oxidation of reduced gases in the troposphere can be associated with either production or consumption of O_3. Whether O_3 is produced or consumed in the course of oxidation depends on the supply of NO. In the absence of NO, oxidation is limited by the supply of O_3; in the presence of NO, oxidation may be associated with significant tropospheric production of O_3.

Conversion of CO to CO_2 may be summarized, as we shall see, by either of the bulk or net reactions:[2]

$$CO + 2O_2 \rightarrow CO_2 + O_3 \tag{17.1}$$

or

$$CO + O_3 \rightarrow CO_2 + O_2 \tag{17.2}$$

Similar bifurcation is associated with oxidation of CH_4 and other hydrocarbons. We will be concerned primarily with elaboration of chemical paths for oxidation of reduced gases in the troposphere and with the processes regulating the abundance of O_3. Our focus will be on the mechanisms through which the chemical system in the atmosphere is able to dispose of the rich suite of compounds emitted by industry and by the biosphere, with the self-cleansing feature of the atmosphere noted above. A comprehensive treatment of oxidation presupposes an understanding of sources and sinks for O_3.

In sufficiently high concentration, O_3 can pose a serious problem for human health. It is unclear at what, if any, level O_3 may be considered environmentally benign. Models suggest that in the absence of human activity—specifically, in the absence of industrial sources of NO, CO, and hydrocarbons—ambient levels of O_3 in the troposphere should lie in the range 10–20 ppb. Chronic exposure to O_3 at levels above 50 ppb is known to cause damage to sensitive plants. Levels above 300 ppb are judged serious enough in cities such as Los Angeles to force the issuance of health alerts. Standards for permissible levels of O_3 differ from country to country. The limit is set at 84 ppb in the United States; by comparison, the standard for Switzerland is 60 ppb. Setting standards for O_3 involves necessarily a measure of pragmatism. There is little sense in legislating requirements so strict that they cannot be met; on the other hand, they should be stringent enough to stimulate creative measures to ensure protection of the environment at a reasonable cost.

As we will see, O_3 is formed in urban areas as a product of chemical reactions triggered by local industrial emissions of hydrocarbons and NO_x. Elevated levels of O_3 are also observed in regions that would normally be classified as rural: on occasions, for example, during summer in large areas of the United States east of the Mississippi. Production of O_3 is believed to result in this case from reactions fueled by a combination of natural and anthropogenic inputs; hydrocarbons from trees and NO_x from electric power plants, trucks, and automobiles. As the Governor of California, Ronald Reagan was embroiled in a controversy over the cutting of redwoods and is reputed to have commented that trees cause pollution. He was partially right—and partially wrong. Trees, emitting hydrocarbons, contribute to elevated O_3, but they require a helping hand from sunlight and human agents, specifically a stimulus from industrial sources of NO_x. Setting sensible standards for O_3 and developing realistic strategies for compliance requires a comprehensive appreciation of the complex chemical and physical factors determining the level of O_3 in the lower atmosphere; this chapter is intended to provide a perspective on these issues.

In a sense the atmosphere operates as the immune system of the biosphere. To what extent is this immune system under stress? Is the stress merely local, as indicated by unacceptably high levels of O_3 in cities as diverse as Los Angeles, Mexico City, and São Paulo, or has it expanded to a regional scale as suggested by the data for O_3 in the summertime eastern United States? Abundances of CH_4 have been rising recently at a rate of close to 1% per year. The gas is removed mainly by reaction with OH and has a lifetime in the atmosphere of about 8.5 yrs. Does the rise in CH_4 reflect, in part at least, a decrease in the globally averaged abundance of OH? Is the growth in CH_4 attributable to a decrease in the potential of the atmosphere to globally dispose of biospheric waste, to a decline in the oxidative or self-cleansing potential of the global immune system? These are the questions that motivate the following discussion.

We begin, in Section 17.1, with an introduction to the processes regulating the abundance of OH. The simple case of oxidation of CO is discussed in Section 17.2. The more complex example of CH_4 is treated in Section 17.3. The chemistry of polluted environments is discussed in Section 17.4. The global distribution of OH and the oxidative potential of the atmosphere are treated in Section 17.5. Global and hemispheric budgets of O_3 and CH_4 are discussed in Sections 17.6 and 17.7 with summary remarks in Section 17.8. We postpone a treatment of the oxidation of sulfur compounds in the context of a discussion of the phenomenon of acid rain until Chapter 18.

17.1 The Chemistry of Tropospheric OH

Photolysis of O_3 yielding $O(^1D)$,

$$h\nu + O_3 \rightarrow O(^1D) + O_2 , \tag{17.3}$$

provides the essential first step in the production of OH. The rate constant for a photolytic process such as (17.3) depends on the sum (or integral) over wavelength of the product of the cross section for the reaction and the radiative flux: more specifically, on the product of the cross section and the radiance, the intensity of light integrated over all directions (see Section 13.2, as well as Endnote 3 of this chapter). Absorption by O_3 provides the dominant path for attenuation of solar radiation shortward of about 310 nm. As illustrated in Figure 17.1, the cross section for (17.3) is largest near 250 nm. Radiation shortward of 300 nm is strongly absorbed by O_3 in the stratosphere, and relatively little radiation in this spectral range penetrates to the troposphere, as indicated in Figure 17.2a–b. Production of $O(^1D)$ is energetically possible at wavelengths less than about 320 nm. The cross section for (17.3) is negligible, however, longward of 310 nm. It follows that production of $O(^1D)$ in the troposphere is regulated primarily by radiation transmitted in a narrow band of wavelengths between 300–310 nm, as illustrated in Figure 17.3.

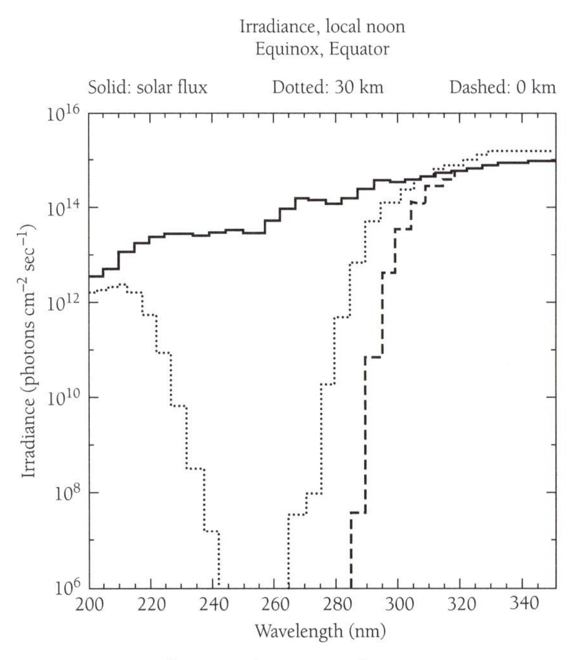

Figure 17.2a Solar irradiance at the equator at equinox, as viewed from outside of the atmosphere (solid line), from 30 km (dotted line), and from the surface (dashed line).

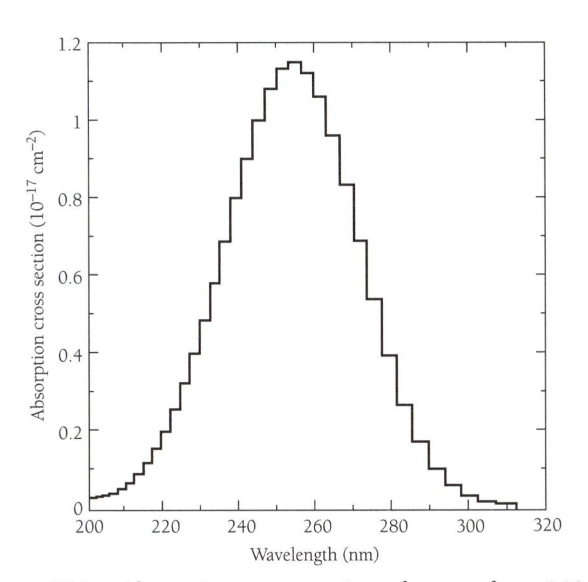

Figure 17.1 Absorption cross section of ozone from 200 to 320 nm.

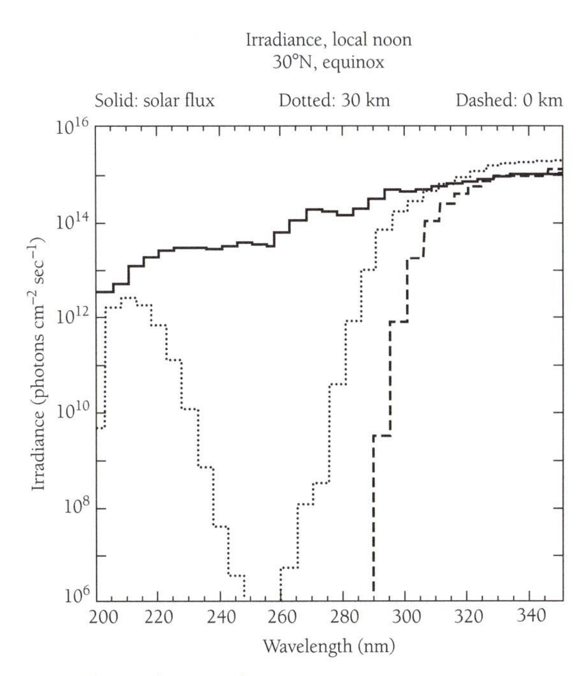

Figure 17.2b Solar irradiance at 30°N at equinox, as viewed from outside the atmosphere (solid line), from 30 km (dotted line), and from the surface (dashed line).

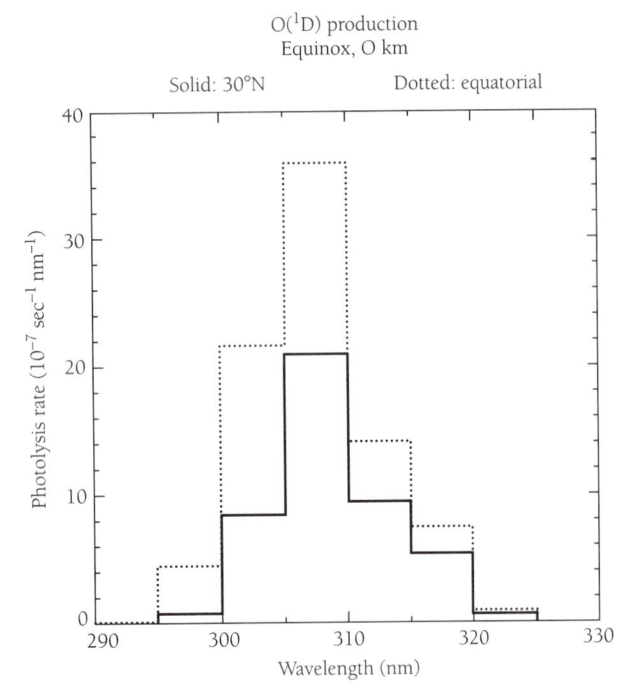

Figure 17.3 Wavelength dependence of ozone photolysis, J_3, for noon under equinoctial conditions near the surface at 30°N (solid line) and at the equator (dotted line).

Example 17.1: Evaluate the contribution of direct solar radiation in the wavelength range 300–310 nm to the rate constant for reaction (17.3) at the equator for conditions corresponding to noon at equinox (solar zenith angle ($\theta = 0$). Assume that the solar flux at the top of the atmosphere in the wavelength bands 300–305 nm and 305–310 nm is given by 3.94×10^{14} and 4.82×10^{14} photons cm^{-2} sec^{-1}, respectively. Radiation is attenuated as a result of absorption by O_3 and Rayleigh scattering by air. The cross sections for absorption by O_3 in the selected wavelength intervals, 300–305 nm and 305–310 nm, are given by 2.47×10^{-19} cm^2 and 1.22×10^{-19} cm^2, respectively. The corresponding cross sections for Rayleigh scattering are 5.47×10^{-26} cm^2 and 5.10×10^{-26} cm^2, respectively. Assume a vertical column density for O_3 of 7.89×10^{18} cm^{-2}, corresponding to an abundance of 294 DU. The column density for air may be taken as 2.15×10^{25} molecules cm^{-2}. Assume that the quantum yield for O(^{1}D), the fraction of photons absorbed by O_3 contributing to production of O(^{1}D), is 1.0 for the interval 300–305 nm and 0.9 for 305–310 nm.

Answer: The first step in this calculation involves evaluation of the optical depth of the atmosphere for the relevant wavelength intervals (see Section 13.1 for a definition of optical depth). For convenience we shall indicate quantities for the 300–305 nm and 305–310 nm intervals by subscripts 1 and 2, respectively:

$$\tau_1 = N_{O_3} Q_{1,O_3} + N_{air} Q_{1,air}$$

$$\tau_2 = N_{O_3} Q_{2,O_3} + N_{air} Q_{2,air},$$

where Q_{i,O_3} denotes the cross sections for absorption of radiation in wavelength interval i by O_3, $Q_{i,air}$ is the corresponding cross section for scattering by air, and N_{O_3} and N_{air} are the vertical column densities for O_3 and air molecules, respectively.

Substituting the appropriate values for cross sections and column densities, we find

$$\tau_1 = \left[(7.89 \times 10^{18}\ \text{cm}^{-2})(2.47 \times 10^{-19}\ \text{cm}^2)\right] + \left[(2.15 \times 10^{25}\ \text{cm}^{-2})(5.47 \times 10^{-26}\ \text{cm}^2)\right]$$

$$= 3.12$$

$$\tau_2 = \left[(7.89 \times 10^{18}\ \text{cm}^{-2})(1.22 \times 10^{-19}\ \text{cm}^2)\right] + \left[(2.15 \times 10^{25}\ \text{cm}^{-2})(5.10 \times 10^{-26}\ \text{cm}^2)\right]$$

$$= 2.06$$

The flux of solar radiation at the surface in the wavelength intervals 1 and 2 is given by

$$F_1 = F_1^{\infty} \exp(-\tau_1)$$

$$= (3.94 \times 10^{14}\ \text{photons cm}^{-2}\ \text{sec}^{-1})(4.37 \times 10^{-2})$$

$$= 1.72 \times 10^{13}\ \text{photons cm}^{-2}\ \text{sec}^{-1}$$

and

$$F_2 = F_2^{\infty} \exp(-\tau_2)$$

$$= (4.82 \times 10^{14}\ \text{photons cm}^{-2}\ \text{sec}^{-1})(1.27 \times 10^{-1})$$

$$= 6.14 \times 10^{13}\ \text{photons cm}^{-2}\ \text{sec}^{-1},$$

where F_i^{∞} defines the appropriate value for the flux at the top of the atmosphere. The contribution, J_3^i, to the rate constant for (17.3) from wavelength interval i is given by

$$J_3^i = E_i^{\infty} Q_{i,O_3} F_i,$$

where E_i is the quantum yield. Substituting appropriate values for E_i, Q_{i,O_3}, and F_i, we find:

$$J_3^1 = (1.0)(2.47 \times 10^{-19}\ \text{cm}^2)(1.72 \times 10^{13}\ \text{photons cm}^{-2}\ \text{sec}^{-1})$$

$$= 4.25 \times 10^{-6}\ \text{sec}^{-1}$$

$$J_3^2 = (0.9)(1.22 \times 10^{-19}\ \text{cm}^2)(6.14 \times 10^{13}\ \text{photons cm}^{-2}\ \text{sec}^{-1})$$

$$= 6.74 \times 10^{-6}\ \text{sec}^{-1}$$

The aggregate value for J_3 is obtained by summing over the contributing wavelength intervals:

$$J_3 = (4.25 \times 10^{-6} + 6.74 \times 10^{-6})\ \text{sec}^{-1}$$

$$= 1.10 \times 10^{-5}\ \text{sec}^{-1}$$

■

Example 17.2: Repeat the calculation described in Example 17.1, ignoring effects of Rayleigh scattering in limiting penetration of solar radiation. The calculation in Example 17.1 implicitly assumes that scattered photons are no longer available to carry out reaction (17.3), meaning, scattering is treated as though it were an absorptive process. The results in Example

(17.1) should thus represent a lower limit for the contribution of radiation in the wavelength interval 300–310 nm to the rate constant for (17.3). The present calculation would be appropriate if scattering were to take place mainly in the forward direction. In practice, the Rayleigh mechanism predicts that scattering should proceed with comparable probability in the forward and backward directions.

Answer: The values of τ_1 and τ_2 are given in this case by the first terms in the expressions derived for these quantities in Example 17.1:

$$\tau_1 = 1.95$$

$$\tau_2 = 0.96$$

It follows that

$$F_1 = (3.94 \times 10^{14} \text{ photons cm}^{-2} \text{ sec}^{-1})(1.42 \times 10^{-1})$$
$$= 5.59 \times 10^{13} \text{ photons cm}^{-2} \text{ sec}^{-1}$$
$$F_2 = (4.82 \times 10^{14} \text{ photons cm}^{-2} \text{ sec}^{-1})(3.83 \times 10^{-1})$$
$$= 1.85 \times 10^{14} \text{ photons cm}^{-2} \text{ sec}^{-1}$$
$$J_3^1 = (1.0)(2.47 \times 10^{-19} \text{ cm}^2) \times (5.59 \times 10^{13} \text{ photons cm}^{-2} \text{ sec}^{-1}) = 1.39 \times 10^{-5}$$
$$J_3^2 = (0.9)(1.22 \times 10^{-19} \text{ cm}^2) \times (1.85 \times 10^{14} \text{ photons cm}^{-2} \text{ sec}^{-1}) = 2.03 \times 10^{-5}$$
$$J_3 = (1.39 \times 10^{-5} + 2.03 \times 10^{-5}) \text{ sec}^{-1}$$
$$= 3.42 \times 10^{-5} \text{ sec}^{-1}$$

■

The probability that scattering should take place into a direction inclined at an angle θ with respect to the direction of incidence of the light is defined by a quantity known as the **phase function**. The phase function for Rayleigh scattering is given by 3/4 $(1+ \cos^2 \theta)$. Forward scattering corresponds to $\theta = 0$, while backward scattering is defined by $\theta = 180°$. According to the Rayleigh phase function, scattering at 90° to the direction of incidence is only half as probable as scattering in either the forward or backward direction. Effects of scattering on photolysis are schematically illustrated in Figure 17.4. Figure 17.4 depicts a situation for which molecules in a representative volume of air near the surface are exposed to a combination of directly transmitted solar radiation, radiation reflected from the surface, radiation scattered once by air molecules (case a), and radiation that is scattered twice (case b).

In practice, the approach adopted in Example 17.2, where effects of scattering are completely ignored, results in a slight overestimate in the values derived for J_3. Also, in the examples above, we omitted contributions to photolysis from shorter and longer wavelengths. This procedure results in an underestimate, by about 30%, in the value derived for J_3. A more detailed calculation, accounting for scattering and allowing for a reasonable reflectivity for the surface (appropriate for an al-

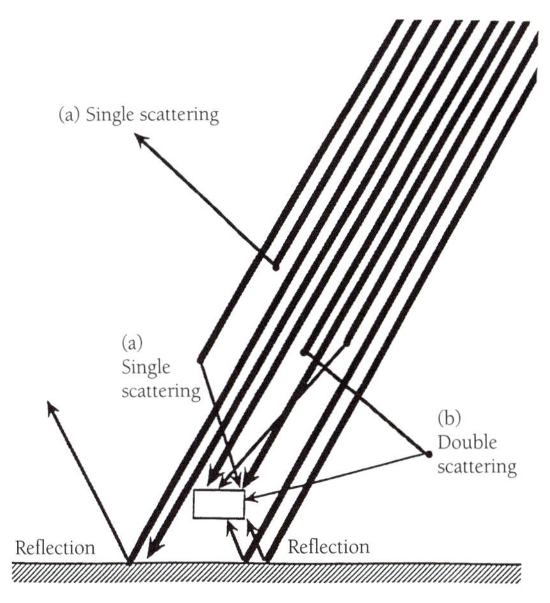

Figure 17.4 Pathways of incoming sunlight.

bedo of 0.1), indicates a value for J_3 corresponding to the overhead Sun at the equator of about 4.2×10^{-5} sec^{-1} for the conditions envisaged in Examples 17.1 and 17.2.

The key reactions regulating the abundance of tropospheric $O(^1D)$ and production of OH, in addition to (17.3), are

$$O(^1D) + N_2 \rightarrow O(^3P) + N_2, \quad (17.4)$$

$$O(^1D) + O_2 \rightarrow O(^3P) + O_2, \quad (17.5)$$

and

$$O(^1D) + H_2O \rightarrow OH + OH \quad (17.6)$$

Adopting the notation used in Chapters 13 through 15, with rate constants for (17.4)–(17.6) denoted by k_4–k_6, the abundance of $O(^1D)$ may be obtained by balancing the rate for production by (17.3) with the rate for removal by (17.4)–(17.6). Thus,

$$J_3[O_3] = [O(^1D)]\ (k_4[N_2] + k_5[O_2] + k_6\ [H_2O]) \quad (17.7)$$

It follows that

$$[O(^1D)] = \frac{J_3[O_3]}{k_4[N_2] + k_5[O_2] + k_6[H_2O]} \quad (17.8)$$

The third term in the denominator of (17.8) typically accounts for 10% or less of the combined contribution to quenching from N_2 and O_2. The relative abundances of N_2 and O_2 are constant, and it is convenient to define an effective rate constant for quenching of $O(^1D)$ by air, according to the relation:

$$k^*[M] = k_4[N_2] + k_5[O_2] \quad (17.9)$$

In this case, (17.8) may be written in the simpler form:

$$[O(^1D)] = \frac{J_3[O_3]}{k^*[M]} \quad (17.10)$$

With typical values for k_4–k_6 (see Table 17.1), the rate for production of OH, p(OH), is given by

Table 17.1 Rate Constants for the Key Reactions Involved in HO_x Chemistry

Reaction	Equatorial		Midlatitudes	
	0 km (T = 300 K)	5 km (T = 272 K)	0 km (T = 292 K)	5 km (T = 266 K)
$O_3 + h\nu \rightarrow O(^1D) + O_2$	4.33×10^{-5} sec^{-1}	6.63×10^{-5} sec^{-1}	2.36×10^{-5} sec^{-1}	3.77×10^{-5} sec^{-1}
$O(^1D) + N_2 \rightarrow O + N_2$	2.60×10^{-11} cm^3 sec^{-1}	2.70×10^{-11} cm^3 sec^{-1}	2.62×10^{-11} cm^3 sec^{-1}	2.72×10^{-11} cm^3 sec^{-1}
$O(^1D) + O_2 \rightarrow O + O_2$	4.04×10^{-11} cm^3 sec^{-1}	4.14×10^{-11} cm^3 sec^{-1}	4.06×10^{-11} cm^3 sec^{-1}	4.16×10^{-11} cm^3 sec^{-1}
$O(^1D) + M \rightarrow O + M$	2.87×10^{-11} cm^3 sec^{-1}	2.97×10^{-11} cm^3 sec^{-1}	2.90×10^{-11} cm^3 sec^{-1}	3.00×10^{-11} cm^3 sec^{-1}
$O(^1D) + H_2O \rightarrow OH + OH$	2.20×10^{-10} cm^3 sec^{-1}	2.20×10^{-10} cm^3 sec-1	2.20×10^{-10} cm^3 sec-1	$2.20 \times 10-10$ cm^3 sec^{-1}
$OH + CO \rightarrow H + CO_2$	2.40×10^{-13} cm^3 sec^{-1}	1.99×10^{-13} cm^3 sec^{-1}	2.40×10^{-13} cm^3 sec^{-1}	1.99×10^{-13} cm^3 sec^{-1}
$OH + CH_4 \rightarrow CH_3 + H_2O$	6.74×10^{-15} cm^3 sec^{-1}	3.62×10^{-15} cm^3 sec^{-1}	5.71×10^{-15} cm^3 sec^{-1}	3.11×10^{-15} cm^3 sec^{-1}
$OH + O_3 \rightarrow HO_2 + O_2$	6.98×10^{-14} cm^3 sec^{-1}	5.07×10^{-14} cm^3 sec^{-1}	6.41×10^{-14} cm^3 sec^{-1}	4.68×10^{-14} cm^3 sec^{-1}
$OH + NO_2 + M \rightarrow HNO_3 + M$	4.63×10^{-31} cm^6 sec^{-1}	7.57×10^{-31} cm^6 sec^{-1}	4.79×10^{-31} cm^6 sec^{-1}	7.87×10^{-31} cm^6 sec^{-1}
$HO_2 + O_3 \rightarrow OH + NO_2$	2.08×10^{-15} cm^3 sec^{-1}	1.75×10^{-15} cm^3 sec^{-1}	1.99×10^{-15} cm^3 sec^{-1}	1.68×10^{-15} cm^3 sec^{-1}
$HO_2 + NO \rightarrow OH + NO_2$	8.51×10^{-12} cm^3 sec^{-1}	9.27×10^{-12} cm^3 sec^{-1}	8.71×10^{12} cm^3 sec^{-1}	9.46×10^{-12} cm^3 sec^{-1}
$NO + O_3 \rightarrow NO_2 + O_2$	1.89×10^{-14} cm^3 sec^{-1}	1.17×10^{-14} cm^3 sec^{-1}	1.66×10^{-14} cm^3 sec^{-1}	1.04×10^{-14} cm^3 sec^{-1}
$NO_2 + h\nu \rightarrow NO + O$	8.67×10^{-3} sec^{-1}	1.14×10^{-2} sec^{-1}	7.92×10^{-3} sec^{-1}	1.09×10^{-2} sec^{-1}
$HO_2 + OH \rightarrow H_2O + O_2$	1.10×10^{-10} cm^3 sec^{-1}	1.20×10^{-10} cm^3 sec^{-1}	1.13×10^{-10} cm^3 sec^{-1}	1.23×10^{-10} cm^3 sec^{-1}
$HO_2 + HO_2 \rightarrow H_2O_2 + O_2$	1.70×10^{-12} cm^3 sec^{-1}	2.08×10^{-12} cm^3 sec^{-1}	1.79×10^{-12} cm^3 sec^{-1}	2.19×10^{-12} cm^3 sec^{-1}
$HO_2 + HO_2 + M \rightarrow H_2O_2 + O_2 + M$	4.75×10^{-32} cm^3 sec^{-1}	6.70×10^{-32} cm^3 sec^{-1}	5.21×10^{-32} cm^3 sec^{-1}	7.29×10^{-32} cm^3 sec^{-1}
$H_2O_2 + h\nu \rightarrow OH + OH$	8.30×10^{-6} sec^{-1}	1.18×10^{-5} sec^{-1}	6.50×10^{-6} sec^{-1}	9.99×10^{-6} sec^{-1}
$H_2O_2 + OH \rightarrow HO_2 + H_2O$	1.70×10^{-12} cm^3 sec^{-1}	1.61×10^{-12} cm^3 sec^{-1}	1.68×10^{-12} cm^3 sec^{-1}	1.59×10^{-12} cm^3 sec^{-1}

$$
\begin{aligned}
p(\mathrm{OH}) &= 2k_6[\mathrm{H_2O}][\mathrm{O(^1D)}] \\
&= \frac{2k_6[\mathrm{H_2O}]J_3[\mathrm{O_3}]}{k^*[\mathrm{M}]} \\
&= 2\frac{k_6}{k^*}f_{\mathrm{H_2O}}J_3[\mathrm{O_3}] \qquad (17.11) \\
&\approx 15f_{\mathrm{H_2O}}J_3[\mathrm{O_3}]
\end{aligned}
$$

Example 17.3: Repeat the calculation outlined in Example 17.2 for noon at 30°N for equinox. Assume a column density for O_3 of 350 DU, corresponding to an abundance of 9.38×10^{18} mol cm^{-2}. Assume, as before, a column density for air of 2.15×10^{25} cm^{-2}.

Answer: The solar zenith angle at noon corresponding to these conditions is 30° ($\cos\theta = \mu = 0.866$: see Example 17.2).

$$\tau_1 = (9.38 \times 10^{18}\ \mathrm{cm^{-2}})\ (2.47 \times 10^{-19}\ \mathrm{cm^2})$$
$$= 2.32$$
$$\tau_2 = (9.38 \times 10^{18}\ \mathrm{cm^{-2}})\ (1.22 \times 10^{-19}\ \mathrm{cm^2})$$
$$= 1.14$$

$$
\begin{aligned}
F_1 &= F_1^{\infty}\exp\left(-\frac{\tau_1}{.866}\right) \\
&= (3.94 \times 10^{14}\mathrm{photons\ cm^{-2}sec^{-1}})\ \exp\left(\frac{-2.32}{.866}\right) \\
&= (3.94 \times 10^{14}\ \mathrm{photons\ cm^{-2}\ sec^{-1}})\ (6.86 \times 10^{-2}) \\
&= 2.70 \times 10^{13}\ \mathrm{photons\ cm^{-2}\ sec^{-1}}
\end{aligned}
$$

$$
\begin{aligned}
F_2 &= (4.82 \times 10^{14}\mathrm{photons\ \ cm^{-2}\ sec^{-1}})\ \exp\left(\frac{-1.14}{.866}\right) \\
&= (4.82 \times 10^{14}\ \mathrm{photons\ cm^{-2}\ sec^{-1}})\ (2.68 \times 10^{-1}) \\
&= 1.29 \times 10^{14}\ \mathrm{photons\ cm^{-2}\ sec^{-1}}
\end{aligned}
$$

$$
\begin{aligned}
J_3^1 &= (1.0)(2.47 \times 10^{-19}\ \mathrm{cm^2}) \times (2.70 \times 10^{13}\ \mathrm{photons\ cm^{-2}\ sec^{-1}}) \\
&= 6.67 \times 10^{-6}\ \mathrm{sec^{-1}}
\end{aligned}
$$

$$
\begin{aligned}
J_3^2 &= (0.9)(1.22 \times 10^{-19}\ \mathrm{cm^2}) \times (1.29 \times 10^{14}\ \mathrm{photons\ cm^{-2}\ sec^{-1}}) \\
&= 1.41 \times 10^{-5}\ \mathrm{sec^{-1}}
\end{aligned}
$$

$$J_3 = 2.08 \times 10^{-5}\ \mathrm{sec^{-1}}$$

A more precise calculation for J_3, allowing for scattering and contributions from other wavelengths, yields a value of 2.36×10^{-5} sec^{-1}. ■

Example 17.4: Estimate the rate for production of OH at the surface at 30°N latitude for noon at equinox. Assume a mixing ratio for H_2O of 1.7×10^{-2} corresponding to a relative humidity of about 80% at a temperature of about 290 K. Assume a mixing ratio for O_3 of 50 ppb, typical for relatively clean rural areas of the United States. Assume a value of 2.4×10^{-5} sec^{-1} for J_3 appropriate for a column density of O_3 of about 350 DU. The total density of air molecules, M, may be taken as 2.5×10^{19} cm^{-3}.

Answer:

$$\begin{aligned} p(\mathrm{OH}) &= 15\mathrm{f}_{\mathrm{H_2O}}J_3[\mathrm{O_3}] \\ &= 15\mathrm{f}_{\mathrm{H_2O}}J_3\mathrm{f}_{\mathrm{O_3}}[\mathrm{M}] \\ &= (1.5 \times 10^1)\,(1.7 \times 10^{-2})\,(2.4 \times 10^{-5}\ \mathrm{sec}^{-1}) \\ &\quad \times (5 \times 10^{-8})\,(2.5 \times 10^{19}\ \mathrm{cm}^{-3}) \\ &= 7.7 \times 10^6\ \mathrm{cm}^{-3}\ \mathrm{sec}^{-1} \end{aligned}$$

■

Example 17.5: Estimate the abundance of O(^{1}D) for the conditions defined in Example 17.4. Assume a value for k^* of 3×10^{-11} cm^3 sec^{-1}.

Answer: The concentration of O(^{1}D) may be evaluated to an adequate approximation using equation (17.10):

$$\begin{aligned} [\mathrm{O^1D}] &= \frac{J_3[\mathrm{O_3}]}{k^*[\mathrm{M}]} \\ &= \frac{J_3}{k^*}\mathrm{f}_{\mathrm{O_3}} \\ &= \frac{(2.4 \times 10^{-5}\mathrm{sec}^{-1})(5 \times 10^{-8})}{3 \times 10^{-11}\ \mathrm{cm}^3\ \mathrm{sec}^{-1}} \\ &= 4 \times 10^{-2}\ \mathrm{cm}^{-3} \end{aligned}$$

■

Example 17.6: Calculate the lifetime of O(^{1}D) at the surface for the conditions defined in Example 17.4.

Answer: The lifetime, $\tau(\mathrm{O(^1D)})$, for O(^{1}D) is determined primarily by the rate at which the species is quenched by air (converted to the ground state form of the atom, O(^{3}P)):

$$\begin{aligned} \tau &= (k^*[\mathrm{M}])^{-1} \\ &= \frac{1}{(3 \times 10^{-11}\mathrm{cm}^3\ \mathrm{sec}^{-1})(2.5 \times 10^{19}\ \mathrm{cm}^{-3})} \\ &= 1.3 \times 10^{-9}\ \mathrm{sec} \end{aligned}$$

■

It is a remarkable fact that the source of OH and, as a consequence, the oxidative potential of the troposphere are controlled in the final analysis by the concentration of a metastable atom, a species whose lifetime in the troposphere is typically less than about 10^{-8} sec (the lifetime is longer in the upper troposphere than at the surface, reflecting the inverse dependence of the lifetime on density, as indicated in Example 17.6). The concentration of tropospheric O(^{1}D) is essentially undetectable: it corresponds to a mixing ratio of less than 10^{-20}.

As indicated by (17.11), production of tropospheric OH depends on the abundance of tropospheric O_3, on the mixing ratio of tropospheric H_2O, and on the value of J_3. The rate constant J_3 is sensitive in turn to the magnitude of the solar zenith angle (time of day) and to the overhead (primarily stratospheric) column abundance of O_3.[4] Values of J_3 are presented in Figure 17.5 for a range of column densities of O_3 at the equator and at 30°N for solar zenith angles appropriate for noon at equinox.[5]

As illustrated in Figure 17.6, the chemistry of OH is inextricably linked to a suite of reactions involving H, HO_2, and H_2O_2 identified as the key components of the tropospheric HO_x family. Reaction with CO provides an important sink for tropospheric OH, with additional contributions from reactions involving CH_4 and H_2O_2:

$$\mathrm{OH + CO \rightarrow H + CO_2}, \tag{17.12}$$

$$\mathrm{OH + CH_4 \rightarrow H_2O + CH_3}, \tag{17.13}$$

and

$$\mathrm{OH + H_2O_2 \rightarrow H_2O + HO_2} \tag{17.14}$$

The H atom produced in (17.12) is immediately converted to HO_2 by

$$\mathrm{H + O_2 + M \rightarrow HO_2 + M} \tag{17.15}$$

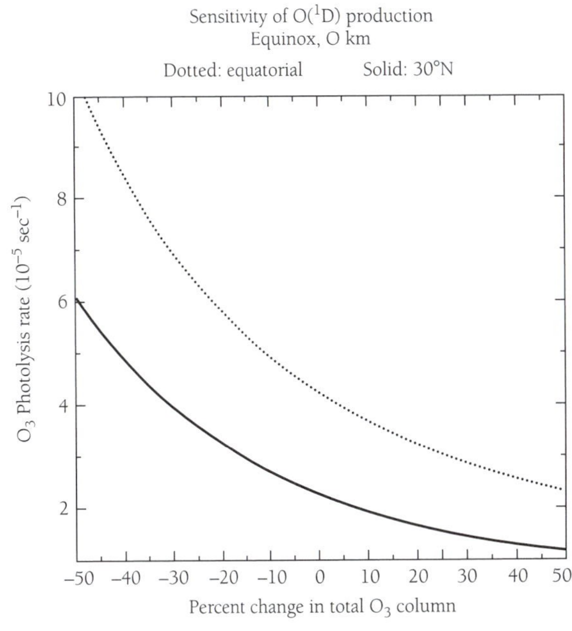

Figure 17.5 Sensitivity of ozone photolysis (J_3) near the surface, at the equator (dotted line), and at 30°N (solid line) to the total column density of ozone. Values are calculated for equinoctial conditions at noon with an equatorial ozone column of 294 DU and a midlatitude ozone column of 343 DU.

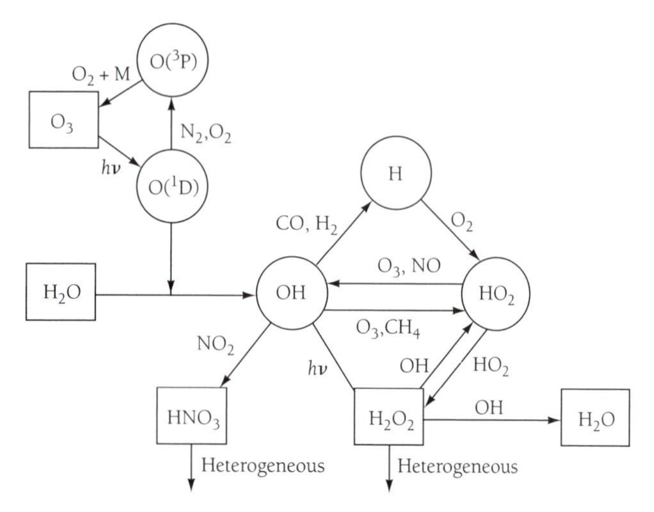

Figure 17.6 Chemical pathways responsible for the production, loss, and partitioning of HO_x.

HO_2 may be cycled back to OH by reactions with NO or O_3,

$$HO_2 + NO \rightarrow OH + NO_2 \qquad (17.16)$$

and

$$HO_2 + O_3 \rightarrow OH + 2O_2, \qquad (17.17)$$

or it may be transformed to H_2O_2 by

$$HO_2 + HO_2 \rightarrow H_2O_2 + O_2 \qquad (17.18)$$

H_2O_2 is removed by photolysis

$$h\nu + H_2O_2 \rightarrow 2OH \qquad (17.19)$$

by reaction with OH, (17.14), and by heterogeneous reactions, either at the surface or on particles (mainly droplets of water) in the atmosphere:

$$H_2O_2 \rightarrow \text{products} \qquad (17.20)$$

Reactions (17.14) and (17.20) provide sinks for two molecules of odd hydrogen (for this purpose H_2O_2 is assumed to represent a pair of HO_x species). Reaction of OH with NO_2 forming HNO_3,

$$OH + NO_2 + M \rightarrow HNO_3 + M, \qquad (17.21)$$

followed by heterogeneous removal of HNO_3 contributes an additional sink for HO_x. One molecule of HO_x is removed in (17.13): on the other hand, (17.13) represents but the first step in oxidation of CH_4. It provides no net sink for radicals; the radical OH is merely replaced by the methyl radical CH_3. Oxidation of CH_4, as we shall see, can represent either a source or a sink for two molecules of HO_x or it may be neutral. The impact of CH_4 oxidation on the abundance of HO_x critically depends on the supply of NO_x.

Example 17.7: Estimate the relative importance of (17.12) and (17.13) as sinks for OH at the surface at 30°N latitude. Assume mixing ratios for CO and CH_4 of 10^{-7} and 1.7×10^{-6}, respectively, with a total air density of 2.5×10^{19} mol cm^{-3}.

Answer: The rates (cm^{-3} sec^{-1}) at which OH is removed by (17.12) and (17.13) are given by

$$R_{12} = k_{12}\,[OH][CO]$$

and

$$R_{13} = k_{13}\,[OH][CH_4]$$

Where k_{12} and k_{13} are the rate constants for (17.12) and (17.13), respectively. The relative importance of (17.12) and (17.13) depends on the values of the corresponding loss frequencies:

$$v_{12} = k_{12}[CO] = k_{12} f_{CO}[M]$$

and

$$v_{13} = k_{13}[CH_4] = k_{13}\, f_{CH_4}\,[M]$$

The ratio v_{12} to v_{13} is given by

$$\frac{v_{12}}{v_{13}} = \frac{k_{12} f_{CO}}{k_{13} f_{CH_4}}$$

Using values presented in *Table 17.1* to substitute for k_{12} and k_{13}, with values for f_{CO} and f_{CH_4} as specified:

$$\frac{v_{12}}{v_{13}} = \frac{\left(2.4\times10^{-13}\ \text{cm}^3\ \text{sec}^{-1}\right)\left(10^{-7}\right)}{\left(5.7\times10^{-15}\ \text{cm}^3\ \text{sec}^{-1}\right)\left(1.7\times10^{-6}\right)}$$

$$= 2.5$$

It follows that (17.12) is 2.5 times more important than (17.13) as a sink for OH. ■

Example 17.8: Estimate the relative importance of (17.16) and (17.17) as sinks for HO_2 at the surface at 30°N equinox. Assume mixing ratios for NO and O_3 of 5×10^{-12} and 5×10^{-8}, respectively, values typical for noon under relatively clean (low NO_x, low O_3) rural conditions.

Answer: Writing rate constants for (17.16) and (17.17) as k_{16} and k_{17}, respectively,

$$v_{16} = k_{16}\,[NO] = k_{16} f_{NO}\,[M]$$

$$v_{17} = k_{17}\,[O_3] = k_{17} f_{O3}\,[M]$$

$$\frac{v_{17}}{v_{16}} = \frac{k_{17} f_{O_3}}{k_{16} f_{NO}}$$

$$= \frac{(2.0 \times 10^{-15}\ \text{cm}^3\ \text{sec}^{-1})(5 \times 10^{-8})}{(8.7 \times 10^{-12}\ \text{cm}^3\ \text{sec}^{-1})(5 \times 10^{-12})}$$

$$= 2.3$$

With our assumptions, removal of HO_2 by O_3 is 2.3 times more efficient than removal by NO. ■

Example 17.9: For conditions defined in Example 17.8, estimate the density of HO_2 for which the rate for removal of HO_2 by (17.18) is comparable to the rate for removal by (17.17).

Answer:

$$R_{17} = k_{17}\,[HO_2][O_3]$$

$$R_{18} = k_{18}\,[HO_2]^2$$

Equal values for R_{17} and R_{18} imply

$$k_{17}\,[HO_2][O_3] = k_{18}\,[HO_2]^2$$

It follows that

$$[HO_2] = k_{17}\,[O_3]\,/\,k_{18}$$

$$= \frac{(2.0 \times 10^{-15}\ \text{cm}^3\ \text{sec}^{-1})(5 \times 10^{-8} \times 2.5 \times 10^{19}\ \text{cm}^{-3})}{(1.8 \times 10^{-12}\ \text{cm}^3\ \text{sec}^{-1} + 1.3 \times 10^{-12}\ \text{cm}^3\ \text{sec}^{-1})}$$

$$= \frac{(2.0 \times 10^{-15}\ \text{cm}^3\ \text{sec}^{-1})(1.25 \times 10^{11}\ \text{cm}^{-3})}{(3.1 \times 10^{-12}\ \text{cm}^3\ \text{sec}^{-1})}$$

$$= 8.1 \times 10^7\ \text{cm}^{-3}$$ ■

The value of k_{18} used here and in subsequent examples accounts for the influence of the three-body reaction, $HO_2 + HO_2 + M \rightarrow H_2O_2 + O_2 + M$, included in Table 17.1. If the concentration of HO_2 is larger than 10^8 cm^{-3}, production of H_2O_2 is more important than cycling back to OH as a sink for HO_2.

Example 17.10: To estimate the abundance of HO_2, assume that HO_x is removed mainly by (17.14) and (17.20). Assume further that photolysis of H_2O_2, reaction (17.19), is less important than (17.14) and (17.20) as a sink for H_2O_2. Under these circumstances, the rate for production of HO_x by (17.6) should equal the rate at which HO_2 is converted to H_2O_2. Estimate the abundance of HO_2 at 30°N equinox, noon, using the HO_x production rate calculated in Example 17.4.

Answer: According to Example 17.4, the rate for production of HO_x is given by

$$p(HO_x) = 7.7 \times 10^6\ \text{cm}^{-3}\ \text{sec}^{-1}$$

and

$$L(HO_x) = 2k_{18}\,[HO_2]^2,$$

where the factor of two accounts for the fact that two molecules of HO_x are removed by (17.18), followed by (17.14) or (17.20). Equating rates for production and loss,

$$7.7 \times 10^6\ \text{cm}^{-3}\ \text{sec}^{-1} = 2k_{18}\,[HO_2]^2$$

$$= 2(3.1 \times 10^{-12}\ \text{cm}^3\ \text{sec}^{-1})[HO_2]^2,$$

it follows that

$$[HO_2]^2 = \frac{7.7 \times 10^6\ \text{cm}^{-3}\ \text{sec}^{-1}}{6.2 \times 10^{-12}\ \text{cm}^{-3}\ \text{sec}^{-1}}$$

$$= 1.24 \times 10^{18}\ \text{cm}^{-6}$$

Thus

$$[HO_2] = 1.1 \times 10^9\ \text{cm}^{-3}$$ ■

The efficiency of reactions such as (17.16), (17.17), and (17.19) in cycling HO_2 and H_2O_2 back to OH may be conveniently measured in terms of a quantity known as the *chain length* for the HO_x cycle. We define the chain length, C, for the HO_x cycle as the ratio of the total rate for production of OH (including contributions from recycling) to the rate for production, solely reflecting the contribution from the primary source (17.6). Allowing for recycling of OH by (17.16), (17.17), and (17.19),

$$C = \big(2k_6[O(^1D)][H_2O] + k_{16}[NO][HO_2] + k_{17}[O_3][HO_2] + 2J_{19}[H_2O_2]\big) / 2k_6[O(^1D)][H_2O], \quad (17.22)$$

where J_{19} defines the rate constant (sec^{-1}) for (17.19). A chain length greater than 1.0 implies that recycling is important. A chain length close to 1.0 implies that recycling is unimportant. The concentration of OH is given to an adequate approximation in this case by equating the rate for production of OH by (17.6) with the rate for removal, mainly by (17.12) and (17.13):

$$2k_6[O(^1D)][H_2O] = k_{12}\,[OH][CO] + k_{13}\,[OH][CH_4] \quad (17.23)$$

It follows that

$$[OH] = \frac{2k_6[O(^1D)][H_2O]}{k_{12}[CO] + k_{13}[CH_4]} \quad (17.24)$$

Allowing for recycling,

$$[OH] = C\left(\frac{2k_6[O(^1D)][H_2O]}{k_{12}[CO] + k_{13}[CH_4]}\right) \quad (17.25)$$

Using (17.10) to substitute for $[O^1D]$, we may rewrite (17.25) in the form

$$[OH] = C\left\{\frac{2k_6 J_3 f_{O_3} f_{H_2O}}{k^*(k_{12}f_{CO} + k_{13}f_{CH_4})}\right\} \quad (17.26)$$

The discussion in Examples (17.8)–(17.10), considering cycling of HO_2 back to OH, suggests a chain length close to 1.0 for conditions as defined there for the surface at equinox at 30°N.

Example 17.11: Estimate the concentration of OH for the surface at 30°N for noon at equinox. Assume the value derived in Example 17.4 for production of OH by (17.6), a chain length of 1.0, and mixing ratios for CO and CH_4 of 10^{-7} and 1.7×10^{-6}, as adopted in Example (17.7).

Answer: The concentration of OH is given by

$$[\text{OH}] = \frac{2k_6[\text{O}(^1\text{D})][\text{H}_2\text{O}]}{k_{12}[\text{CO}] + k_{13}[\text{CH}_4]}$$

$$= \frac{p(\text{OH})}{\nu_{12} + \nu_{13}}$$

$$= \frac{p(\text{OH})}{[k_{12}f_{\text{CO}} + k_{13}f_{\text{CH}_4}][\text{M}]}$$

$$= \frac{7.7 \times 10^6 \text{ cm}^{-3} \text{ sec}^{-1}}{(2.4 \times 10^{-20} \text{ cm}^3 \text{ sec}^{-1} + 9.7 \times 10^{-21} \text{ cm}^3 \text{ sec}^{-1})(2.5 \times 10^{19} \text{ cm}^{-3})}$$

$$= \frac{7.7 \times 10^6}{(3.4 \times 10^{-20})(2.5 \times 10^{19})}$$

$$= 9 \times 10^6 \text{cm}^{-3}$$

■

Example 17.12: For conditions defined in Example 17.11, estimate the lifetime of OH.

Answer: The lifetime $\tau(\text{OH})$ is given by

$$\tau(\text{OH}) = \frac{[\text{OH}]}{p(\text{OH})}$$

$$= \frac{9 \times 10^6 \text{ cm}^{-3}}{7.7 \times 10^6 \text{ cm}^{-3}\text{sec}^{-1}}$$

$$= 1.2 \text{ sec}$$

■

Example 17.13: With the assumptions stated in Example 17.10, estimate the lifetime for HO_2.

Answer: The lifetime $\tau(\text{HO}_2)$ is given by

$$\tau(\text{HO}_2) = \frac{[\text{HO}_2]}{p(\text{HO}_2)}$$

But with our assumptions, $p(\text{OH}) = p(\text{HO}_2)$.

Thus,

$$\tau(\text{HO}_2) = \frac{1.1 \times 10^9 \text{ cm}^{-3}}{7.7 \times 10^6 \text{ cm}^{-3} \text{ sec}^{-1}}$$

$$= 1.4 \times 10^2 \text{ sec}$$

The lifetime for HO_2 is about two minutes. ■

Example 17.14: Estimate the relative importance of (17.14), (17.19), and (17.20) as sinks for H_2O_2. For this purpose, use the density of OH calculated in Example 17.12 and assume a loss frequency of 2.3×10^{-6} sec^{-1} for the heterogeneous process (17.20).

Answer: The relative importance of (17.14), (17.19), and (17.20) are given by the corresponding loss frequencies:

$$\nu_{14} = k_{14}\,[\text{OH}],$$

$$\nu_{19} = J_{19},$$

and

$$\nu_{20} = 2.3 \times 10^{-6} \text{ sec}^{-1},$$

where k_{14} is the rate constant for (17.14).

Using the values for the rate constants given in Table 17.1,

$$\nu_{19} = 6.5 \times 10^{-6} \text{ sec}^{-1}$$

and

$$\nu_{14} = (1.7 \times 10^{-12} \text{ cm}^3 \text{ sec}^{-1})(9 \times 10^6 \text{ cm}^{-3})$$

$$= 1.5 \times 10^{-5} \text{ sec}^{-1}$$

■

It follows that reaction (17.14) is the dominant path for removal of H_2O_2. It is about seven times more efficient than the heterogeneous process (17.20) and twice as important as photolysis, reaction (17.19).

Example 17.15: Calculate the density and lifetime of H_2O_2, assuming that H_2O_2 is removed by (17.14), (17.19), and (17.20), with rates implied by the data in Example (17.14). Estimate the rate for production of H_2O_2, using the concentration of HO_2 obtained in Example 17.10.

Answer: The rate for production of H_2O_2 is given by

$$p(\text{H}_2\text{O}_2) = k_{18}[\text{HO}_2]^2$$

$$= 3.8 \times 10^6 \text{ cm}^{-3} \text{ sec}^{-1}$$

The loss rate is given by

$$L(\text{H}_2\text{O}_2) = [\text{H}_2\text{O}_2]\,(\nu_{14} + \nu_{19} + \nu_{20})$$

It follows that

$$[\text{H}_2\text{O}_2] = \frac{3.8 \times 10^6 \text{ cm}^{-3} \text{ sec}^{-1}}{\nu_{14} + \nu_{19} + \nu_{20}}$$

$$= \frac{3.8 \times 10^6 \text{cm}^{-3} \text{ sec}^{-1}}{2.4 \times 10^{-5} \text{ sec}^{-1}}$$

$$= 1.6 \times 10^{11} \text{ cm}^{-3}$$

$$\tau_{\text{H}_2\text{O}_2} = (\nu_{14} + \nu_{19} + \nu_{20})^{-1}$$

$$= 4.2 \times 10^4 \text{ sec}$$

With our assumptions, the lifetime for H_2O_2 is about12 hours. ■

In calculating the abundance of H_2O_2 in Example 17.15, we assumed a balance between instantaneous (noon time) rates for production and loss; as indicated in Example 17.15; however, the lifetime of H_2O_2 is about 12 hours. It follows that the abundance of H_2O_2 should reflect not the instantaneous rates for production and loss but an average of these quantities over a period comparable to the lifetime of H_2O_2. Our analysis is deficient in another important respect: we allowed for production of HO_x by (17.6), but we did not explicitly account for the requirement that production of HO_x should be balanced by loss. With the simplified chemical scheme introduced here, accepting the indication that (17.14) and (17.20) provide the most important paths for removal of HO_x, it is apparent that the concentration of H_2O_2 must increase to the point where removal of HO_x by (17.14) and (17.20) should be sufficient to balance production by (17.6). To explore implications of these matters further, it is instructive to consider the evolution of HO_x chemistry with time from a hypothetical initial state in which concentrations of individual HO_x species are set equal to zero.

Let the clock begin ticking at midnight. According to our assumption, the abundance of HO_x species is initially zero and remains so until the sun rises and photolysis of O_3 initiates production of $O(^1D)$, OH, and HO_2 (recall that the lifetimes of OH and HO_2 are measured in seconds and minutes, respectively). Concentrations of OH and HO_2 build up during the morning responding to the increase in J_3 related to the decrease in solar zenith angle. They reach a maximum at noon, with values similar to those estimated in Examples 17.9 and 17.11. As rapidly as it is produced, HO_x is converted from OH to HO_2 and hence to H_2O_2. Reflecting cumulative production of HO_x by (17.6), H_2O_2 builds up. Assume a rate for production of OH over the course of the morning equal on average to half the value calculated for noon in Example 17.4. The total number of HO_x molecules formed between 6 A.M. and noon may be obtained by multiplying the average rate for production by the elapsed time (6 hours): $(3.9 \times 10^6$ cm^{-3} sec$^{-1})$ $(2.2 \times 10^4$ sec$)$ = 8.3×10^{10} mol cm^{-3}, equivalent to 4.2×10^{10} mol cm^{-3} of H_2O_2 (recall that a molecule of H_2O_2 contains two molecules of HO_x). The number of H_2O_2 molecules produced over a 24-hour period is equal to approximately twice this value, 8.3×10^{10} cm^{-3} (production of HO_x, is confined to the daylight hours: rates for photolysis of O_3, and consequently for the source of HO_x, should be symmetric with respect to noon). To balance the 24-hour source of HO_x, we require that the concentration of H_2O_2 rise to the point where removal of HO_x by (17.14) and (17.20) is sufficient over a 24-hour period to balance production by (17.6).

Example 17.16: Using the results developed above, calculate the concentration of H_2O_2, such that removal of HO_x by (17.14) and (17.20) should be sufficient to balance production by (17.6) for the surface at 30°N at equinox. As in Example 17.14, assume a rate constant for (17.20) of 2.3×10^{-6} sec^{-1} equivalent to a lifetime for H_2O_2 of about five days. Assume an average loss frequency for (17.14) equal to 1/4 of the value used in Example 17.14, $v_{14} = 3.8 \times 10^{-6}$ sec^{-1}, which is equivalent to assuming that the diurnally averaged concentration of OH is equal to 1/4 of the value calculated for noon.

Answer: According to the analysis above, reaction (17.6) contributes a source, S, of 1.7×10^{11} mol of HO_x over a 24-hour period (production is confined to the hours of daylight).

The instantaneous rate for loss of HO_x, L(cm^{-3} sec^{-1}), is equal to twice the rate for removal of H_2O_2 by (17.14) and (17.20):

$$L = 2\,[H_2O_2]\,(2.3 \times 10^{-6}\ \text{sec}^{-1} + 3.8 \times 10^{-6}\ \text{sec}^{-1})$$

$$= [H_2O_2]\,(1.2 \times 10^{-5}\ \text{sec}^{-1})$$

Thus, the total number (N) of HO_x molecules removed over a 24-hour period (loss can occur at night as well as during the day) is given by

$$N = L \times (8.6 \times 10^4\ \text{sec})$$

$$= [H_2O_2] \times 1.03$$

Requiring that loss balance production implies that

$$S = N$$

or

$$[H_2O_2] = \frac{1.7 \times 10^{11}\ \text{cm}^{-3}}{1.03} = 1.7 \times 10^{11}\ \text{cm}^{-3}$$ ■

Returning now to our time-dependent scenario, the concentration of H_2O_2 will have risen at the end of day one to a level somewhat less than 8.5×10^{10} cm^{-3} (not all of the H_2O_2 molecules produced will survive in the face of reactions 17.14 and 17.20). More than four days (several lifetimes) will be required to establish a steady state such that production of HO_x by (17.6) is balanced on average over a 24-period by removal due to (17.14) and (17.20).[6] The concentration of H_2O_2 will steadily increase over this period, rising towards the steady-state limit. Concentrations of OH and HO_2 will vary diurnally in response to changes in J_3. The amplitude of the diurnal cycle will increase as the system approaches steady state, reflecting production of OH associated with photolysis of H_2O_2.

Example 17.17: Repeat the calculation of [OH] outlined in Example 17.11, accounting for production of OH by photolysis of H_2O_2. Assume a concentration for H_2O_2 equal to the steady-state value derived in Example 17.16. Calculate the chain length for HO_x at noon allowing for the influence of (17.19).

Answer: The concentration of OH is given in this case by

$$[OH] = \frac{2k_6[O^1D][H_2O] + 2J_{19}[H_2O_2]}{k_{12}[CO] + k_{13}[CH_4]}$$

The rate for production by (17.19) is given by

$$2J_{19}\,[H_2O_2] = 2(6.5\times10^{-6}\ \text{sec}^{-1})(1.7\times10^{11}\ \text{cm}^{-3})$$
$$= 2.2\times10^{6}\ \text{cm}^{-3}\ \text{sec}^{-1}$$

Hence,

$$[OH] = \frac{(7.7\times10^{6}+2.2\times10^{6})\ \text{cm}^{-3}\ \text{sec}^{-1}}{8.5\times10^{-1}\ \text{sec}^{-1}}$$
$$= 1.2\times10^{7}\ \text{cm}^{-3}$$

The chain length is

$$C = \frac{2k_6[O^1D][H_2O] + 2J_{19}[H_2O_2]}{2k_6[O^1D][H_2O]}$$
$$= \frac{(7.7\times10^{6}+2.2\times10^{6})\ \text{cm}^{-3}\ \text{sec}^{-1}}{7.7\times10^{6}\ \text{cm}^{-3}\ \text{sec}^{-1}}$$
$$= 1.3$$

■

The discussion of HO_x chemistry presented here is intended to give the reader a feeling for numbers: the importance of the overlying column abundance of O_3 in regulating the primary source of HO_x; the sensitivity of HO_x to ambient levels of tropospheric O_3 and H_2O; the role of NO_x and O_3 in cycling HO_2 back to OH; and the need to allow for a full diurnal cycle in evaluating the overall balance of production and loss of HO_x. The treatment is deficient, however, in its neglect of secondary reactions involved in oxidation of CH_4. These reactions, as we shall see, can provide both sources and sinks for HO_x, the balance depending sensitively on the abundance of NO_x. We postpone a more complete discussion of HO_x chemistry until later, taking up now the chemistry of CO with an emphasis on the role of CO in the budget of tropospheric O_3.

17.2 The Chemistry of CO

Reaction (17.12) provides the primary sink for CO. The impact of CO on the budget of O_3 depends on the fate of HO_2 formed subsequently by (17.15). Allowing for the impact of CO on the overall budget of HO_x, we may identify three distinct sequences of reactions involved in oxidation of CO:

$$\begin{aligned} &CO + OH \rightarrow CO_2 + H \\ &H + O_2 + M \rightarrow HO_2 + M \\ &NO + HO_2 \rightarrow NO_2 + OH \\ &h\nu + NO_2 \rightarrow NO + O \\ &O + O_2 + M \rightarrow O_3 + M, \end{aligned} \qquad (17.27)$$

$$\begin{aligned} &CO + OH \rightarrow CO_2 + H \\ &H + O_2 + M \rightarrow HO_2 + M \\ &HO_2 + O_3 \rightarrow OH + 2O_2, \end{aligned} \qquad (17.28)$$

and

$$\begin{aligned} &h\nu + O_3 \rightarrow O(^1D) + O_2 \\ &O(^1D) + H_2O \rightarrow 2OH \\ &2[CO + OH \rightarrow CO_2 + H] \\ &2[H + O_2 + M \rightarrow HO_2 + M] \\ &2HO_2 \rightarrow H_2O_2 + O_2 \\ &H_2O_2 \rightarrow \text{products} \end{aligned} \qquad (17.29)$$

Sequence (17.27) implies net production of O_3, as summarized by (17.1). Ozone is removed in (17.28) and (17.29). The net reaction for (17.28) is equivalent to (17.2), and the sequence (17.29) is equivalent to:

$$2CO + O_3 + H_2O \rightarrow 2CO_2 + H_2O_2\ \text{products} \qquad (17.30)$$

The critical step in the production of O_3 in (17.27) involves the reaction of NO with HO_2, reaction (17.16). It is convenient here, as it was for the stratosphere, to draw on the concept of odd oxygen, defined for present purposes as the combination of O_3, O, $O(^1D)$, and NO_2.[7] Reaction (17.16) provides the primary source for tropospheric odd oxygen. It is removed by a combination of (17.6), (17.17), (17.21), and

$$OH + O_3 \rightarrow HO_2 + O_2 \qquad (17.31)$$

The rate at which odd oxygen, and consequently O_3, is produced by (17.16) is given by

$$p(O_3) = k_{16}\,[NO][HO_2] \qquad (17.32)$$

If we assume that HO_x is removed by a combination of (17.14) and (17.20), the HO_x balance may be expressed in terms of the source-sink relation

$$2k_6\,[O^1D][H_2O] = 2\epsilon k_{18}\,[HO_2]^2, \qquad (17.33)$$

where ϵ defines the probability that H_2O_2 is lost by the combination of the HO_x sink reactions (17.14) and (17.20), rather than by (17.19).[8] Using (17.10) to substitute for $[O^1D]$,

$$\frac{2k_6 J_3 f_{H_2O}[O_3]}{k^*} = 2\epsilon k_{18}\,[HO_2]^2 \qquad (17.34)$$

It follows that

$$[HO_2] = \left(\frac{k_6 J_3 f_{H_2O}[O_3]}{\epsilon k^* k_{18}}\right)^{1/2}, \qquad (17.35)$$

and the rate for production of O_3, (17.32), is given by

$$p(O_3) = k_{16}\,[NO]\left(\frac{k_6 J_3 f_{H_2O}[O_3]}{\epsilon k^* k_{18}}\right)^{1/2} \qquad (17.36)$$

Assume that (17.6) provides the dominant loss mechanism for odd oxygen. It follows that the rate for production of O_3 will exceed the rate of loss if

$$k_{16}\,[NO][HO_2] > k_6\,[O(^1D)][H_2O] \qquad (17.37)$$

Using (17.35) and (17.10) to substitute for $[HO_2]$ and $[O(^1D)]$, this inequality may be recast in the form

$$[\mathrm{NO}] > \frac{1}{k_{16}}\left(\frac{k_6 J_3 \epsilon k_{18}}{k^*} f_{H_2O}[O_3]\right)^{1/2} \tag{17.38}$$

Reversing the direction of the inequality sign defines the condition under which loss of O_3 dominates production.

Example 17.18: Estimate, for the surface at 30°N, for noon at equinox, the concentration of NO for which production of odd oxygen by (17.16) is comparable to loss by (17.6). As in Example 17.4, assume a value for J_3 of 2.4×10^{-5} sec^{-1}, with mixing ratios of H_2O and O_3 equal to 1.7×10^{-2} and 5×10^{-8}, respectively. Values for other rate constants may be taken from the data included in Table 17.1. The value of ϵ may be taken equal to 1.

Answer: The concentration of NO at which rates of production and loss of odd oxygen are in balance is given by replacing the inequality sign in (17.38) with an equality:

$$[\mathrm{NO}] = \frac{1}{k_{16}}\left\{\frac{k_6 J_3 \epsilon k_{18}}{k^*}\right\}^{1/2} f_{H_2O}^{1/2}[O_3]^{1/2}$$

Assuming a mixing ratio for O_3 of 5×10^{-8} with a total density [M] of 2.5×10^{19} cm^{-3}:

$$[O_3] = 5 \times 10^{-8} \times 2.5 \times 10^{19} \text{ cm}^{-3}$$
$$= 1.25 \times 10^{11} \text{ cm}^{-3}$$

Substituting for the various rate constants and for f_{H_20} and $[O_3]$, omitting units (all quantities are expressed in the *cgs* convention),

$$[\mathrm{NO}] = \frac{1}{8.7 \times 10^{-12}}$$
$$\times \left[\frac{(2.2 \times 10^{-10})(2.4 \times 10^{-5})(3.1 \times 10^{-12})}{3 \times 10^{-11}}\right]^{1/2}$$
$$\times [(1.7 \times 10^{-2})(1.25 \times 10^{12})]^{1/2}$$
$$= 3.9 \times 10^{8} \text{ cm}^{-3}$$

This corresponds to a mixing ratio for NO of

$$f_{NO} = \frac{3.9 \times 10^8}{(2.5 \times 10^{19})} = 1.6 \times 10^{-11}$$

It follows that oxidation of CO may be expected to result in net production of O_3 if the mixing ratio of NO is greater than about 16 ppt. ■

Example 17.19: Assuming that the relative abundances of NO and NO_2 are determined by a balance of

$$NO + O_3 \rightarrow NO_2 + O_2 \tag{17.39}$$

and

$$h\nu + NO_2 \rightarrow NO + O, \tag{17.40}$$

estimate the value of $[NO_2]$, and thus the abundance of NO_x, corresponding to the concentration of NO derived in Example 17.18. Assume rate constants for (17.39) and (17.40) of 1.7×10^{-14} cm^3 sec^{-1} and 7.9×10^{-3} sec^{-1}, respectively.

Answer: Equating production and loss of NO_2:

$$k_{39}\,[NO][O_3] = J_{40}[NO_2],$$

where k_{39} and J_{40} define the rate constants for (17.39) and (17.40), respectively.

Thus,

$$\frac{[NO]}{[NO_2]} = \frac{J_{40}}{k_{39}[O_3]}$$

Omitting units as before,

$$\frac{[NO]}{[NO_2]} = \frac{7.9 \times 10^{-3}}{(1.7 \times 10^{-14})(1.3 \times 10^{12})}$$

Hence,

$$[NO_2] = 1.1 \times 10^9 \text{ cm}^{-3}$$

The abundance of NO_x is given by

$$[NO_x] = [NO] + [NO_2] = 1.5 \times 10^9 \text{ cm}^{-3},$$

corresponding to a mixing ratio, f_{NO_x}, of 60 ppt. ■

Example 17.19 indicates that NO_2 is the dominant form of NO_x, even at noon. The abundance of NO is directly proportional to the abundance of NO_2 (and consequently NO_x) and inversely proportional to the abundance of O_3:

$$[NO] = \frac{J_{40}[NO_2]}{k_{39}\,[O_3]} \tag{17.41}$$

It follows, substituting for NO in (17.36), that production of O_3 is proportional to the abundance of NO_x and inversely proportional to the square root of the concentration of O_3:

$$p(O_3) \sim \frac{f_{NO_x}}{f_{O_3}^{1/2}} \tag{17.42}$$

As indicated by the analysis leading to (17.38), the transition between net loss and net production of O_3 takes place at a concentration of NO_2 (the dominant form of NO_x) given by

$$[NO_2] = \frac{k_{39}}{J_{40}k_{16}}\left\{\frac{k_6 J_3 \epsilon k_{18}}{k^*}\right\}^{1/2}\left[f_{H_2O}\right]^{1/2}[O_3]^{3/2} \tag{17.43}$$

The crossover corresponds to the situation where the rate for reaction (17.16) is comparable to that for (17.18). For concentrations of NO_2 less than the value given by (17.43), cycling of HO_x from HO_2 to OH is unimportant: the chemistry of HO_x involves a unidirectional flow of HO_x from OH to HO_2

to H_2O_2; oxidation of CO is associated with net loss of O_3. For concentrations of NO_2 larger than (17.43), cycling of HO_x between HO_2 and OH (by reactions 17.12 and 17.16) is significant. Odd oxygen is produced each time the cycle is traversed and oxidation of CO results in net production of O_3.

We assumed in the analysis to this point that HO_x is removed mainly as H_2O_2, either by (17.14) or by (17.20): this condition served to fix the concentration of HO_2 as indicated in equation (17.35). For high concentrations of NO_x, we must allow for the removal of HO_x in the form of HNO_3 produced by reaction (17.21). Under conditions where HNO_3 rather than H_2O_2 provides the dominant sink for HO_x, the HO_x balance is given by

$$2k_6\,[O^1D][H_2O] = k_{21}\,[OH][NO_2][M] \qquad (17.44)$$

As we have seen, cycling of HO_x by (17.12) and (17.16) results in production of O_3 and is rapid at high concentrations of NO_x:

$$p(O_3) = k_{12}\,[OH][CO] = k_{16}\,[HO_2][NO] \qquad (17.45)$$

Using (17.44) to substitute for [OH] in (17.45), with $[O^1D]$ given by (17.10), we find

$$p(O_3) = \frac{2k_{12}k_6J_3}{k^*k_{21}} f_{O_3} f_{H_2O} \frac{[CO]}{[NO_2]} \qquad (17.46)$$

It follows, for high concentrations of NO_x, that the rate for production of O_3 is proportional to the abundance of CO and inversely proportional to the abundance of NO_x. The inverse relationship between $p(O_3)$ and $[NO_x]$ reflects the fact that at high NO_x the concentration of OH is fixed by the abundance of NO_x; the larger the abundance of NO_x, the lower the abundance of OH and as a consequence the slower the rate for production of O_3.

Example 17.20: Estimate for 30°N, at equinox and at noon, the concentration of NO_x at which we might expect the switchover to occur from H_2O_2 to HNO_3 control of HO_x. Assume concentrations of O_3, H_2O, and other quantities, as used in the preceding examples.

Answer: The switchover corresponds approximately to the point at which the rate for removal of HO_x in the HO_x–only scheme, mainly by (17.14), is equal to the rate for removal by (17.21):

$$2k_{14}\,[OH][H_2O_2] = k_{21}\,[OH][NO_2][M],$$

where k_{21} denotes the rate constant for (17.21). The factor of two reflects the fact that two molecules of HO_x are removed by (17.14). It follows that

$$[NO_2] = \frac{2k_{14}[H_2O_2]}{k_{21}[M]}$$

Using the concentration of H_2O_2 obtained in Example 17.16,

$$[NO_2] = \frac{2(1.7\times10^{-12}\ \mathrm{cm^3\ sec^{-1}})(1.7\times10^{11}\ \mathrm{cm^{-3}})}{(4.8\times10^{-31}\ \mathrm{cm^6\ sec^{-1}})(2.5\times10^{19}\ \mathrm{cm^{-3}})}$$

$$= 5\times10^{10}\ \mathrm{cm^{-3}}$$ ■

The value of $[NO_2]$ obtained in Example 17.20 corresponds to a NO_x mixing ratio of about 3 ppb. In practice, the concentration of NO_x at the transition point, estimated in this fashion, is a little high: a more complete analysis locates the transition at a NO_x mixing ratio closer to 1 ppb. The discrepancy reflects the assumption that the concentration of H_2O_2 was taken equal to the value derived in Example 17.16. At high concentrations of NO_x, however, the abundance of HO_2 is reduced and production of H_2O_2 is suppressed relative to the value assumed in Example 17.16. It would have been more appropriate here to have used a concentration of H_2O_2 of about 6×10^{10} cm^{-3}.

As indicated in (17.42), the rate for production of O_3 is proportional to the NO_x mixing ratio when the concentration of NO_x is small. For high concentrations of NO_x, it varies inversely as the mixing ratio of NO_x. Detailed calculations, summarized in Figures 17.7 and 17.8, indicate that production of O_3 is a maximum for a concentration of NO_x equal to about 1 ppb.

A slightly different view of O_3 production is presented in Figure 17.9, which shows the response to a combination of emissions of CO and NO_x. We assumed in this analysis that emissions were confined to a thin layer, 500 m thick, near the surface as would arise if the lowest region of the atmosphere was capped by a strong inversion, a common occurrence in cities such as Los Angeles. The results presented here suggest that, if the source of NO_x was less than about 5×10^{10} cm^{-2} sec^{-1}, a reduction in NO_x emissions would lead to a lower rate for production of O_3. If the source of NO_x was larger than 5×10^{10} cm^{-2} sec^{-1}, a drop in NO_x emissions could result in an *increase* in O_3. Production of O_3 is rela-

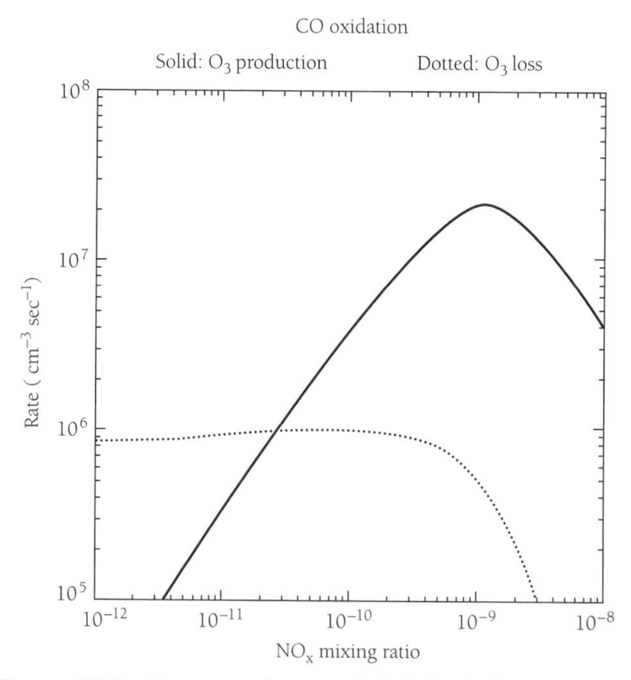

Figure 17.7 Ozone production (solid line) from reaction (17.16) and ozone loss (dotted line) due to reaction (17.17), as a function of NO_x. Values are calculated near the surface at 30°N for equinoctial conditions.

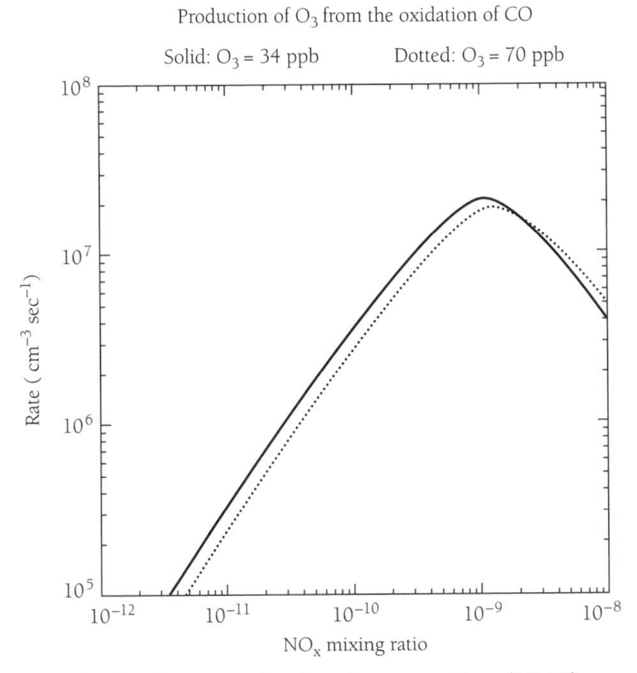

Figure 17.8 Ozone production from reaction (17.16) as a function of NO_x with fixed background mixing ratios of ozone of 34 ppb (solid line) and 70 ppb (dotted line). Values are calculated near the surface at 30°N for equinoctial conditions.

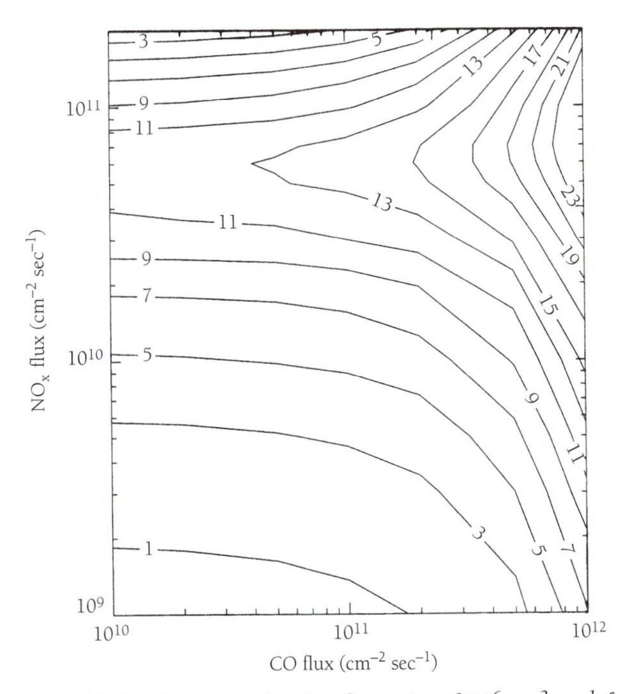

Figure 17.9 Ozone production [in units of $10^6 cm^{-3} sec^{-1}$ from reaction (17.16)] as a function of emission of NO_x and CO. Results are given for 30°N at equinox, with the assumption that the emissions are confined to a 0.5 km layer above the surface.

tively insensitive to emission of CO if the source strength is less than about 10^{11} cm^{-2} sec^{-1} (as indicated by the relatively flat shape of the contours on the left-hand side of Figure 17.9). If emission rates for NO_x and CO are both high (corresponding to the upper right-hand side of Figure 17.9), a reduction in the source of O_3 could be accomplished either by a *decrease* in emissions of CO or by an *increase* in emissions of NO_x.

The behavior illustrated in Figures 17.7–17.9 poses an interesting dilemma for policy makers concerned with regulation of emissions to reduce the buildup of O_3. It suggests that if mixing ratios of NO_x were larger than about 1 ppb (emission rates higher than 5×10^{10} cm^{-2} sec^{-1} in Figure 17.9), a reduction in O_3 could be brought about by an *increase* in emissions of NO_x. This could be interpreted as a license to pollute. The decrease in the rate of production of O_3 at high concentrations of NO_x reflects the role of (17.21) as a sink for OH. The concentration of OH is suppressed, resulting in a decrease in the rate of oxidation of CO.

17.3 The Chemistry of CH_4

The sequence of reactions implicated in oxidation of CH_4 is significantly more complex than that involved in oxidation of CO. As with CO, the sequence is initiated by reaction with OH:

$$OH + CH_4 \rightarrow H_2O + CH_3 \qquad (17.47)$$

The methyl radical (CH_3) formed in (17.47) combines with O_2 to produce the methylperoxy radical (CH_3O_2):

$$CH_3 + O_2 + M \rightarrow CH_3O_2 + M \qquad (17.48)$$

We can think of the methyl radical formed by (17.47) as analogous to the H atom produced when OH reacts with CO, reaction (17.12). The methylperoxy radical formed in (17.48) is chemically similar to the hydroperoxy radical (HO_2) produced in (17.15). As was the case for HO_2, the subsequent chemistry depends on the supply of NO. It is convenient to distinguish from the outset between the low and high NO_x environments.

In the low-NO_x regime, CH_3O_2 is initially converted to methylhydroperoxide (CH_3OOH) by reaction with HO_2:

$$CH_3O_2 + HO_2 \rightarrow CH_3OOH + O_2 \qquad (17.49)$$

Reaction (17.49) is analogous to (17.18), which is responsible for production of H_2O_2. Subsequent reactions, either

$$CH_3OOH + OH \rightarrow CH_2O + OH + H_2O \qquad (17.50)$$

or

$$CH_3OOH + h\nu \rightarrow CH_3O + OH$$

and

$$CH_3O + O_2 \rightarrow CH_2O + HO_2, \qquad (17.51)$$

result in conversion of CH_3OOH to formaldehyde (CH_2O). Sequence (17.47), followed by (17.48), (17.49), and (17.50), is equivalent to

$$OH + HO_2 + CH_4 \rightarrow 2H_2O + CH_2O \qquad (17.52)$$

Reaction (17.47), followed by (17.48), (17.49), and (17.51), is equivalent to

$$CH_4 + O_2 \rightarrow CH_2O + H_2O \qquad (17.53)$$

The radicals OH and HO_2 consumed in (17.52) may be supplied by (17.6), followed by

$$OH + O_3 \rightarrow HO_2 + O_2 \quad (17.54)$$

Accounting for (17.3), (17.6), and (17.54), (17.52) may be written in the summary form

$$CH_4 + 2O_3 \rightarrow CH_2O + H_2O + 2O_2 \quad (17.55)$$

Reactions (17.53) and (17.55) summarize alternate paths in the low-NO_x regime for what may be considered the first stage in the oxidation of CH_4. The oxidation state of carbon is raised by four units in the transition from CH_4 (−4) to CH_2O (0). We shall refer to the sequence of reactions involved in (17.53) and (17.55) as *Ia* and *Ib*, respectively. Molecular oxygen serves as the oxidant for Ia. The oxidant for Ib is O_3: two moles of O_3 are consumed by Ib for every mole of CH_4 converted to CH_2O in Ib. The specific reactions involved in these sequences are summarized in Table 17.2.

Formaldehyde may be removed by photolysis, either by

$$h\nu + CH_2O \rightarrow CO + H_2 \quad (17.56)$$

or by

$$h\nu + CH_2O \rightarrow CHO + H, \quad (17.57)$$

with the H atom in (17.57) converted to HO_2 by (17.15). Reaction with OH provides an additional sink:

$$OH + CH_2O \rightarrow CHO + H_2O \quad (17.58)$$

The CHO radical reacts rapidly with O_2 to form CO and HO_2:

$$CHO + O_2 \rightarrow CO + HO_2 \quad (17.59)$$

Molecular hydrogen formed by (17.56) is removed by reaction with OH, forming H_2O:

$$OH + H_2 \rightarrow H_2O + H, \quad (17.60)$$

with the H atom in (17.60) subsequently converted to HO_2 by (17.15). Sequence (17.3), followed by (17.6), (17.56), (17.60), and (17.15), is equivalent to

$$CH_2O + O_3 \rightarrow CO + OH + HO_2 \quad (17.61)$$

Reaction (17.57), followed by (17.15) and (17.59), is equivalent to

$$CH_2O + 2O_2 \rightarrow CO + 2HO_2, \quad (17.62)$$

while (17.1), followed by (17.6), (17.58), and (17.59), corresponds to

$$CH_2O + O_3 \rightarrow CO + OH + HO_2 \quad (17.63)$$

We shall refer to the sequences of reactions involved in oxidation of CH_2O to CO, (17.61), (17.62), and (17.63), as IIa, IIb, and IIc. The elementary reactions involved in these sequences are summarized in Table 17.3.

Table 17.2 Sequences of Reactions Involved in the Oxidation of CH_4 to CH_2O

Sequence Ia (low NO_x)
$OH + CH_4 \rightarrow H_2O + CH_3$
$CH_3 + O_2 + M \rightarrow CH_3O_2 + M$
$CH_3O_2 + HO_2 \rightarrow CH_3OOH + O_2$
$CH_3OOH + h\nu \rightarrow CH_3O + OH$
$CH_3O + O_2 \rightarrow CH_2O + HO_2$
Net: $CH_4 + O_2 \rightarrow CH_2O + H_2O$
Sequence Ib (low NO_x)
$h\nu + O_3 \rightarrow O(^1D) + O_2$
$O(^1D) + H_2O \rightarrow OH + OH$
$OH + O_3 \rightarrow HO_2 + O_2$
$OH + CH_4 \rightarrow H_2O + CH_3$
$CH_3 + O_2 + M \rightarrow CH_3O_2 + M$
$CH_3O_2 + HO_2 \rightarrow CH_3OOH + O_2$
$CH_3OOH + OH \rightarrow CH_2O + OH + H_2O$
Net: $CH_4 + 2O_3 \rightarrow CH_2O + H_2O + 2O_2$
Sequence Ic (high NO_x)
$OH + CH_4 \rightarrow H_2O + CH_3$
$CH_3 + O_2 + M \rightarrow CH_3O_2 + M$
$CH_3O_2 + NO \rightarrow CH_3O + NO_2$
$CH_3O + O_2 \rightarrow CH_2O + HO_2$
$HO_2 + NO \rightarrow OH + NO_2$
$2\{h\nu + NO_2 \rightarrow NO + O\}$
$2\{O + O_2 + M \rightarrow O_3 + M\}$
Net: $CH_4 + 4O_2 \rightarrow CH_2O + 2O_3 + H_2O$

Table 17.3 Sequences of Reactions Involved in the Oxidation of CH_2O to CO

Sequence IIa
$h\nu + O_3 \rightarrow O(^1D) + O_2$
$O(^1D) + H_2O \rightarrow OH + OH$
$h\nu + CH_2O \rightarrow CO + H_2$
$OH + H_2 \rightarrow H_2O + H$
$H + O_2 + M \rightarrow HO_2 + M$
Net: $CH_2O + O_3 \rightarrow CO + OH + HO_2$
Sequence IIb
$h\nu + CH_2O \rightarrow CHO + H$
$H + O_2 + M \rightarrow HO_2 + M$
$CHO + O_2 \rightarrow CO + HO_2$
Net: $CH_2O + 2O_2 \rightarrow CO + 2HO_2$
Sequence IIc
$h\nu + O_3 \rightarrow O(^1D) + O_2$
$O(^1D) + H_2O \rightarrow OH + OH$
$OH + CH_2O \rightarrow CHO + H_2O$
$CHO + O_2 \rightarrow CO + HO_2$
Net: $CH_2O + O_3 \rightarrow CO + OH + HO_2$

Oxidation of CH_4 is completed by reaction of CO with OH, (17.12). If the abundance of NO_x is low, such that reactions involving NO may be ignored, we see that oxidation of CH_4 is generally responsible for a net loss of O_3. Depending on the details of the reaction path, as many as three or as few as zero molecules of O_3 may be consumed in the oxidation of CH_4 to CO_2 in the absence of NO_x. The sequence Ib followed by IIa, (17.12) and (17.15), for example, is equivalent to

$$CH_4 + 3O_3 \rightarrow CO_2 + H_2O + 2HO_2 + O_2 \quad (17.64)$$

On the other hand, Ia, followed by IIc, (17.12), and (17.15), is equivalent to

$$CH_4 + O_3 + 2O_2 \rightarrow CO_2 + 2HO_2 + H_2O \quad (17.65)$$

The key difference between the path for oxidation of CH_4 in the presence and absence of NO_x concerns the fate of the peroxy radicals HO_2 and CH_3O_2. Given an adequate source of NO_x, these radicals are removed by reaction with NO resulting in a sequence of reactions leading to production of O_3:

$$\begin{aligned} RO_2 + NO &\rightarrow RO + NO_2 \\ h\nu + NO_2 &\rightarrow NO + O \qquad (17.66) \\ O + O_2 + M &\rightarrow O_3 + M \end{aligned}$$

Here R denotes either H or CH_3. Oxidation of CH_4 to CH_2O proceeds in this case by a suite of reactions indicated as sequence Ic in Table 17.2. Sequence Ic leads to production of two molecules of O_3 for each molecule of CH_4 converted to CH_2O through the reaction sequence summarized by

$$CH_4 + 4O_2 \rightarrow CH_2O + 2O_3 + H_2O \quad (17.67)$$

Sequence Ic, followed by IIb and (17.27), is equivalent to

$$CH_4 + 8O_2 \rightarrow CO_2 + 3O_3 + 2HO_2 + H_2O \quad (17.68)$$

If we assume that the HO_2 radicals produced in (17.68) are converted to OH by (17.66), we see that oxidation of a single molecule of CH_4 in the presence of NO_x can account for the production of as many as five molecules of O_3 and two molecules of OH:

$$CH_4 + 10O_2 \rightarrow CO_2 + 5O_3 + 2OH + H_2O \quad (17.69)$$

Reaction (17.69) provides an example of what is known as a **chain mechanism** for production of O_3. The chain is *initiated* by (17.6). It is *propagated* by the complex suite of reactions outlined above. It is *terminated* when radicals are removed by reactions such as (17.20) and (17.21). In the presence of NO_x, oxidation of CH_4 is sustained by a sequence of reactions catalyzed by HO_x and NO_x. In the absence of NO_x, oxidation of CH_4 (and other hydrocarbons) in the troposphere would be limited by supply of O_3 from the stratosphere. A diagrammatic representation of the key steps involved in oxidation of CH_4 is presented in Figure 17.10.

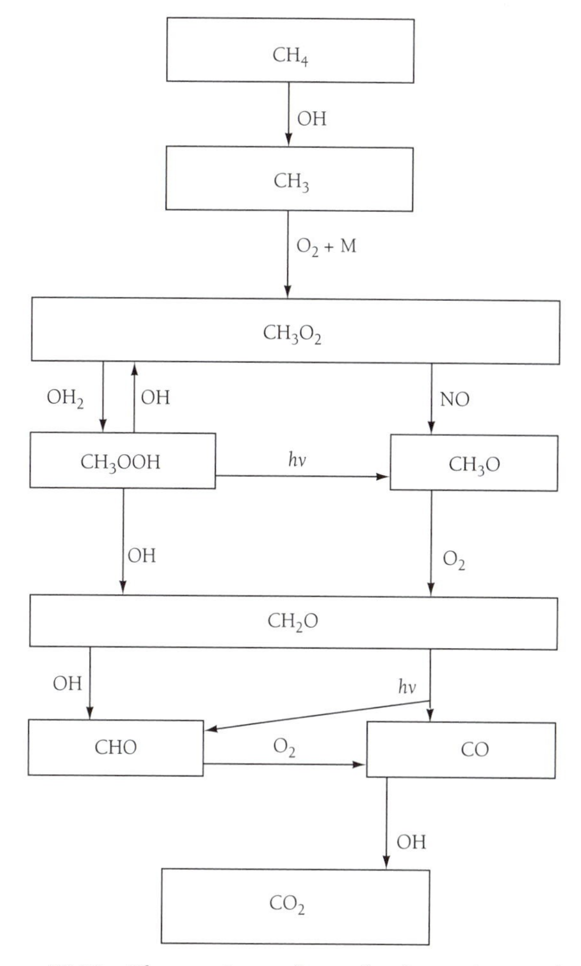

Figure 17.10 The reaction pathway for the oxidation of CH_4 to CO_2. Source: Logan et al. 1981.

17.4 The Chemistry of Smog

We turn our attention now to the chemistry of polluted environments. Specifically, we explore the conditions that lead to production of high levels of O_3 in urban and regional environments subject to high emission rates of anthropogenic hydrocarbons and NO_x.

As was seen for CH_4, oxidation of hydrocarbons is initiated by reaction with OH. Denoting a representative hydrocarbon as *RH*, where *R* denotes an organic group, the first step in oxidation is given by

$$RH + OH \rightarrow R + H_2O \quad (17.70)$$

This is followed by formation of a peroxy radical (analogous to the production of CH_3O_2 in the oxidation of CH_4):

$$R + O_2 + M \rightarrow RO_2 + M \quad (17.71)$$

Ozone is produced subsequently by

$$RO_2 + NO \rightarrow HO_2 + \begin{pmatrix} \text{aldehyde} \\ \text{or} \\ \text{carbonyl} \end{pmatrix} + NO_2, \quad (17.72)$$

$$HO_2 + NO \rightarrow OH + NO_2, \quad (17.73)$$

and

$$2\{NO_2 + h\nu \rightarrow NO + O\}, \quad (17.74)$$

followed by

$$2\{O + O_2 + M \rightarrow O_3 + M\} \quad (17.75)$$

The sequence (17.70)–(17.75), essentially a generalization of Ic presented in Table 17.2 for CH_4, is equivalent to

$$RH + 3O_2 \rightarrow \begin{pmatrix} \text{aldehyde} \\ \text{or} \\ \text{carbonyl} \end{pmatrix} + 2O_3 + H_2O \quad (17.76)$$

Reactions (17.70)–(17.75) provide an example of what we referred to above as a *chain mechanism* for production of O_3. The chain is initiated by production of OH (by reaction 17.6 for example); production of O_3 is catalyzed during chain propagation by HO_x and NO_x; the chain is terminated by reactions resulting in removal of either HO_x or NO_x. The key reactions involved in chain termination are (17.18) and (17.21). Reaction (17.18) dominates when the abundance of NO_x is low; reaction (17.21) is more important when the abundance of NO_x is high.

Noting that the key (or rate-limiting) steps for production of O_3 by (17.70)–(17.75) are represented by (17.72) and (17.73) and that the rates (cm^{-3} sec^{-1}) for these reactions must be equal (they occur as part of a sequence), the rate for production of O_3, P_{O_3}, may be written as either twice the rate for (17.72) or twice the rate for (17.73). Following the earlier discussion of the production of O_3 by oxidation of CO (Section 17.2), we choose to write P_{O_3} in the form

$$P_{O_3} = 2\, k_{73}\, [NO][HO_2] \quad (17.77)$$

In the low-NO_x regime, where HO_x is removed mainly by (17.18), followed by (17.14) or (17.20), it follows (see argument leading up to the derivation of 17.33) that

$$P_{HO_x} = 2\, \epsilon k_{18}\, [HO_2]^2 \quad (17.78)$$

Here P_{HOx} denotes the rate for production of HO_x. Using (17.78) to substitute for $[HO_2]$ in (17.77), the rate for production of O_3 is given by

$$P_{O_3} = 2\, k_{73} \left(\frac{P_{HO_x}}{2\epsilon k_{18}} \right)^{1/2} [NO] \quad (17.79)$$

In the high-NO_x regime, where HO_x is removed mainly by (17.21), the HO_x budget is determined by

$$P_{HO_x} = k_{21}\, [OH][NO_2][M] \quad (17.80)$$

It follows that

$$[OH] = \frac{P_{HO_x}}{k_{21}[NO_2][M]} \quad (17.81)$$

The production rate of O_3 equals twice the rate for reaction (17.70):

$$P_{O_3} = 2\, k_{70}\, [OH][RH] \quad (17.82)$$

Using (17.81) to substitute for OH in (17.82), the rate for production of O_3 in the high-NO_x regime is given by

$$P_{O_3} = \frac{2k_{70} P_{HO_x} [RH]}{k_{21}[NO_2][M]} \quad (17.83)$$

As we have seen, production of O_3 depends on the rate of reaction (17.16). Thus it is proportional to the product of the concentrations of NO and HO_2. The abundance of HO_2 is independent of the abundances of both NO_x and hydrocarbons in the low-NO_x regime. It is determined by a balance between production of HO_x by (17.6) (accounting for additional production as a byproduct of the oxidation chain) and removal by (17.18). At low levels of NO_x, therefore, production of O_3 is proportional to the abundance of NO_x and independent of the abundance of hydrocarbons. Conditions for production of O_3 in this regime are said to be **NO_x limited**. At higher levels of NO_x, production of HO_x is balanced mainly by (17.21). In this case the concentration of OH varies inversely as the concentration of NO_x. The concentration of OH is independent of the concentration of hydrocarbons at high levels of NO_x. An increase in supply of hydrocarbons results in a proportional increase in production of O_3. In this sense, production of O_3 at high levels of NO_x is said to be **hydrocarbon limited.**

The distinction between the high- and low-NO_x regimes is clearly illustrated in Figure 17.11. The Figure indicates concentrations of O_3 expected to develop under summer conditions in the United States as a result of specified rates for emission of NO_x and industrial hydrocarbons. The model used to derive these results allowed for the presence

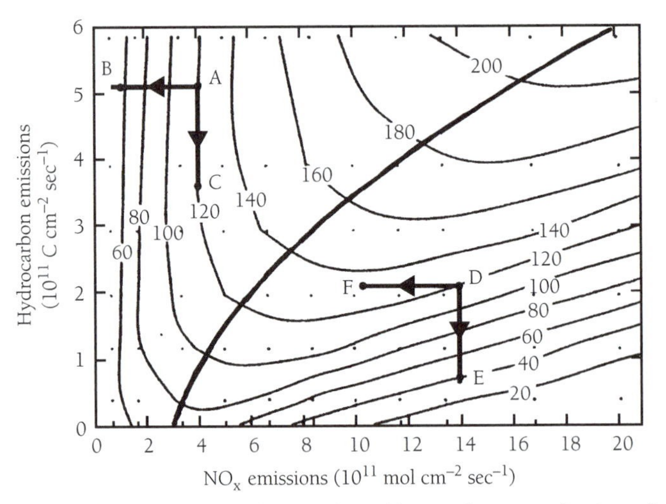

Figure 17.11 Isolines of ozone (in ppb) as a function of NO_x and anthropocentric hydrocarbon emission rates. Results of a two-layer model are shown for 1600 LT on the fourth day. The dots indicate emission rates of NO_x and hydrocarbons for individual simulations. The thick line running from the bottom left to the top right indicates the boundary between NO_x-limited and hydrocarbon-limited regimes. The arrows show successful and futile emissions control choices in each regime. In the NO_x-limited regime, reducing NO_x emissions greatly decreases ambient O_3 concentrations (A→B), while decreasing hydrocarbon emissions have no effect on air quality (A→C). In the hydrocarbon-limited regime, reducing hydrocarbon emissions decreased ambient O_3 concentrations (D→E), while decreasing NO_x emissions actually increases the ambient O_3 concentrations (D→F). Implementing the wrong control strategy might thus have no effect, or even an adverse effect. Source: Sillman, Logan, and Wofsy 1990.

of a boundary layer near the surface in which it was assumed that the atmosphere was efficiently mixed by convective processes during the day. The boundary layer climbed from about 100 m in early morning, extending to about 1800 m during late afternoon, and dropping to the surface at night. It was assumed that these relatively stable conditions persisted for four days. The results in Figure 17.11 refer to 1600 hours local time (LT) on Day 4.

The boundary between NO_x- and hydrocarbon-limiting conditions is approximately indicated by the solid line extending from the bottom left to the upper right-hand of Figure 17.11. Note that above the line, in the NO_x-limited condition, concentrations of O_3 are essentially independent of the magnitude of the source of hydrocarbons. A reduction in input of hydrocarbons, from 5.2×10^{11} to 3.5×10^{11} C atoms cm^{-2} sec^{-1} (as indicated by the segment A→C), leads to no net change in O_3. In contrast, a decrease in emissions of NO_x from 4×10^{11} to 1×10^{11} mol cm^{-2} sec^{-1} (as indicated by A→B), would result in a decrease in O_3 of more than 60 ppb, from 120 to less than 60 ppb. Quite different behavior is observed below the line, in the hydrocarbon-limited regime. A decrease of NO_x from 14×10^{11} to 10×10^{11} mol cm^{-2} sec^{-1} (as indicated by D→F) would result in a small increase in O_3, from 120 ppb to about 130 ppb. Clearly, efforts to improve air quality in the hydrocarbon-limited regime would be better served by working to reduce emissions of hydrocarbons. A reduction in hydrocarbon emissions from 2.2×10^{11} to 0.8×10^{11} C atoms cm^{-2} sec^{-1} (as indicated by the D→E segment) would provide for a reduction in O_3 of about 80 ppb, from 120 to about 40 ppb.

The results in Figure 17.11 considered the effects of emissions of NO_x and *anthropogenic* sources of hydrocarbons; a more realistic simulation should allow additionally for the impact of *natural* sources of hydrocarbons. Of particular importance in this context is emission of isoprene (CH_2=CH-C(CH_3)=CH_2), released as a by-product of photosynthesis. Figure 17.12 summarizes the results of a study designed to simulate production of O_3 over the eastern United States under hot, stagnant conditions characteristic of situations arising several times over the course of a typical summer.

Results in Figure 17.12a refer to an air mass originating on Day 1 in Illinois, moving over Indiana, Ohio, and Pennsylvania and reaching New York four days later. Figure 17.12b refers to a trajectory beginning in Missouri, crossing the Ohio River Valley, and arriving in West Virginia on Day 4. The model considered meteorological conditions typical of environments observed to result in the buildup of unhealthy levels of O_3. It assumed a wind speed of 3 m sec^{-1}, temperatures averaging about 298K, and a boundary layer reaching to a height of about 1500 m in the atmosphere during late afternoon. Emissions of NO_x and anthropogenic hydrocarbons were taken from the National Acid Precipitation Assessment Program (NAPAP) and were selected to represent typical midweek conditions for the region during summer. Emissions of isoprene were calculated using procedures described by Lamb et al. (1987) and accounting for regional

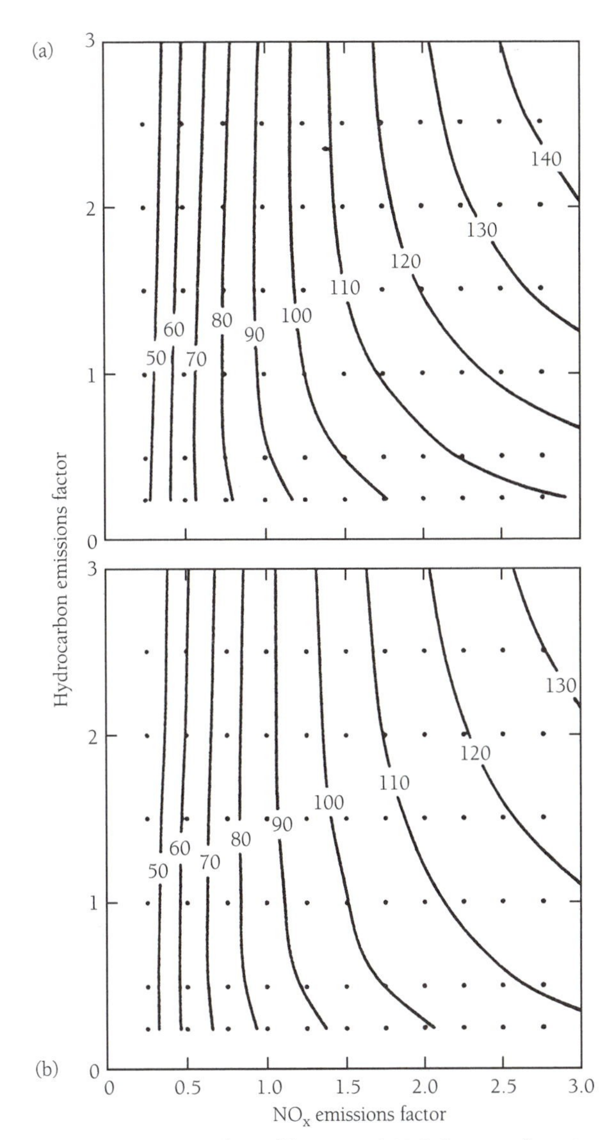

Figure 17.12 Ozone (in ppb) in rural U.S.A., as a function of NO_x and hydrocarbon emissions. The top panel (a) shows results for New York for an air mass that originated four days earlier in Illinois and passed over Indiana, Ohio, and Pennsylvania. The bottom panel (b) shows results for West Virginia for an air mass that originated four days earlier in Missouri and crossed over the Ohio River Valley. Source: Sillman et al. 1990.

differences in vegetative cover (Matthews 1983). Average rates for emission of NO_x were estimated in the range 1.5 – 3.4×10^{11} mol cm^{-2} sec^{-1}. Emissions of anthropogenic hydrocarbons over the region varied between 3.5×10^{11} and 1.5×10^{12} mol cm^{-2} sec^{-1}, as compared with emissions of isoprene in the range $7.5–19 \times 10^{11}$ mol cm^{-2} sec^{-1}.

Emissions in Figure 17.12 are expressed as multiples of emissions rates estimated by NAPAP. Note that, assuming scale factors of unity for both NO_x and anthropogenic hydrocarbons (the most likely range for the associated emissions), predicted levels of O_3 exceed, for both regions, standards mandated by the United States Environmental Protection Agency (EPA). The most recent regulations, issued in 1997, define a maximum 8-hour permissible average concentration for O_3 of 80 ppb, mandating that this standard should not be exceeded on more than three occasions over the course of a year. From the pattern of the results displayed in Figure 17.12, it is clear that conditions for O_3 formation lie in the NO_x-limited regime. Reducing emissions of anthropogenic hydrocarbons would have little effect on O_3; the analysis suggests that if O_3 levels are to be reduced to acceptable levels during smog episodes during summer in the rural eastern United States, it will be necessary to implement steps to reduce emissions of NO_x.

Results for a region subject to the influence of air plumes originating from urban environments are summarized in Figure 17.13. Results are presented in this case as functions of scale factors referenced with respect to the magnitude of emissions of NO_x and anthropogenic hydrocarbons expected for urban regions. Predicted levels of O_3 are significantly higher than values indicated in Figure 17.12, in excess of 140 ppb, assuming the most likely levels of emissions (scale factors of one). Conditions for production of O_3 appear to lie close to the boundary between NO_x and hydrocarbon limitation. A reduction in O_3 would appear to dictate a strategy to reduce emissions of both NO_x and hydrocarbons. Paradoxically, as previously noted, it would be possible to achieve a reduction in O_3 in the vicinity of the urban plume by *increasing* emissions of NO_x. Benefits for the urban hinterland would be achieved, however, at the expense of the down-wind environment. Export of excess NO_x from urban regions would result in enhanced production of O_3 elsewhere.

In summary, it is clear that for most of the United States, while emissions of hydrocarbons are important for cities, the

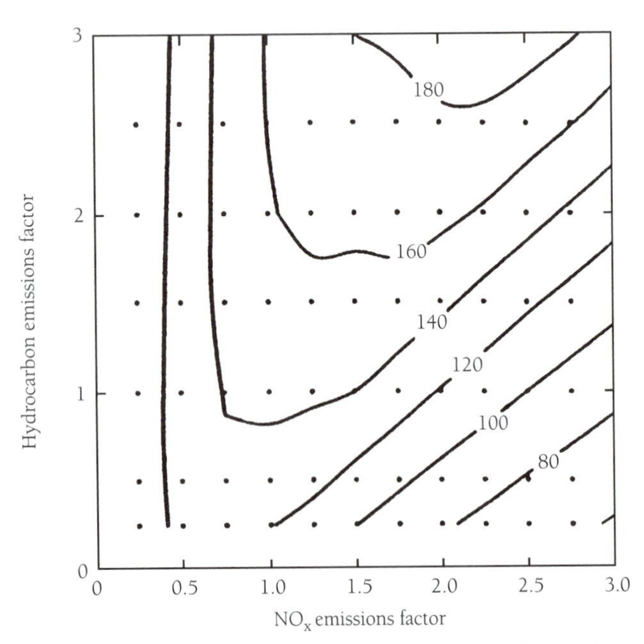

Figure 17.13 Peak ozone (in ppb) in the urban plume of New York, on the fourth day of ozone-episode simulations. Source: Sillman et al. 1990.

buildup of O_3 in the lower atmosphere during summer depends most critically on the supply of NO_x. Aggressive steps have been taken since the Clean Air Act was implemented in 1970 to regulate emissions of both NO_x and hydrocarbons. Emissions of industrial hydrocarbons in the United States declined by about 12 % between 1980 and 1995. It has proved more difficult, however, to reduce emissions of NO_x that remained relatively constant over this period. The transportation sector (cars and trucks) represents a dominant source of NO_x, especially in cities. An additional important contribution is associated with the generation of electric power by the combustion of fossil fuels. The number of vehicle miles driven per year increased by 60% over the past 15 years. It is clear, therefore, that the problem could have been much worse absent the steps taken to reduce emissions. Air quality in heavily polluted cities, such as Los Angeles, has improved significantly, most likely reflecting the reduction in emission of hydrocarbons, but continues to violate national standards. Full compliance with the Clean Air Act will require further, major reductions in emissions of NO_x. It is unlikely that these reductions can be achieved if we continue to rely on gas-guzzling cars for personal transportation. Hybrid vehicles, such as the Toyota Prius, or electrically powered vehicles, or modes of transportation based on new technologies such as the fuel cell, offer hope for the future. In the interim, we may hope for improvements in public transportation systems and for regional plans to reduce the hazards posed by uncontrolled urban sprawl.

17.5 The Global Distribution of Tropospheric OH and the Oxidative Capacity of the Atmosphere

As noted at the outset, reactions with tropospheric OH play a critical role in removal from the atmosphere of a variety of gases ranging from CH_4 to CO, including the suite of industrially produced hydrocarbons discussed above. The abundance of tropospheric OH depends on the flux of ultraviolet solar radiation transmitted through the stratosphere to the lower atmosphere, on the intensity of radiation available to produce the $O(^1D)$ atom implicated in the primary source of OH (see reaction 17.6). It is sensitive, therefore, to the density of the overlying column density of O_3, to the abundance of O_3 present in the stratosphere. It is also sensitive to the abundances of O_3 and H_2O in the troposphere, the latter being, obviously, a function of temperature. We expect concentrations of OH in the troposphere to be highest at low latitudes and to vary significantly with the time of day. A realistic treatment of the role of tropospheric OH as a sink for specific atmospheric gases requires a comprehensive analysis of the factors responsible for its variation in space and time; in short, a global perspective.

Our discussion in this Section draws heavily on results of a recent study of the global, 3-dimensional distribution of tropospheric OH by Spivakovsky et al. (2000). In this study, distributions of O_3, CO, hydrocarbons, water vapor, oxides of nitrogen, temperature, cloud cover, and column O_3 were specified, based on a composite of best available data. Concentrations of OH were calculated, solving the relevant chemical equations defining sources and sinks for OH and allowing appropriately for variations in the intensity of sunlight as a function of time of day. A summary of results, expressed in terms of zonal (longitudinal) averages, is presented for the months of January, April, July, and October in Table 17.3a–d. A graphical representation of these data is displayed in Plate 6 (top) (see color insert). Averaged over space and time, the analysis of Spivakovsky et al. (2000) implies a global average value for the concentration of OH in the troposphere of 1.16×10^6 mol cm^{-3}.

The sensitivity of OH to a number of key variables, notably the concentrations of tropospheric H_2O, O_3, NO_x, CO, and CH_4, and to the abundance of O_3 in the overlying stratospheric column, is summarized in Table 17.4. As expected, an increase (or decrease) in the abundance of either H_2O or O_3 should result in an increase (or decrease) in OH. The change in this case reflects the impact of the postulated change on the rate for production of OH by reaction (17.6). As indicated, an increase of 25% in both H_2O and O_3 should lead to a comparable, though slightly smaller (20%), increase in OH, with the sensitivity to H_2O somewhat greater than that for O_3. An increase (or decrease) in NO_x should result in a modest change of the same sign in OH, reflecting the impact of the change in NO_x on the rate for recycling of HO_x by (17.16). An increase (or decrease) in CO or CH_4 should lead to a small change of the opposite sign in OH, corresponding to the change in the loss frequency for OH (rates for reactions 17.12 and 17.13). Results are most sensitive to an adjustment in the overlying column of O_3. As indicated, a decrease of 25% in column O_3, leading to an increase in transmission of ultraviolet solar radiation, would result in an increase of similar magnitude (26%) in OH. Note that the response of OH to column O_3 is asymmetric: greater when the column density is reduced, reflecting the exponential dependence of the transmission of solar radiation on the column density of O_3.

Observations of methyl chloroform (CH_3CCl_3) provide an important check on the accuracy of the computed distribution of OH. Methyl chloroform is an industrial solvent: chemical industry accounts for the only known source of the gas in the atmosphere. It is removed mainly by reaction with tropospheric OH, with additional loss due to stratospheric oxidation (by photolysis and reaction with OH) and, at a significantly smaller rate, due to uptake by the ocean. It is thought that the source is known to an accuracy of about ±5%. Model results are compared with observations from a number of locations around the world (Prinn et al. 1995) in Figure 17.14. The Figure includes results from model simulations where concentrations of OH were allowed to vary over a range of ±25% (dotted lines). The

Table 17.3a Zonally and Monthly Averaged OH Concentrations for January (10^5 mol cm^{-3}).

	1000 hPa	*900 hPa*	*800 hPa*	*700 hPa*	*500 hPa*	*300 hPa*	*200 hPa*
90°N	--	--	--	--	--	--	--
84°N	--	--	--	--	--	--	--
76°N	--	--	--	--	--	--	--
68°N	0.0	0.0	0.0	0.0	0.0	0.0	0.0
60°N	0.2	0.2	0.2	0.2	0.2	0.2	0.2
52°N	0.4	0.4	0.8	0.6	0.6	0.7	0.8
44°N	0.7	0.9	1.8	1.4	1.6	1.6	1.7
36°N	1.7	2.1	3.6	3.3	3.7	3.5	3.5
28°N	4.3	5.1	6.2	7.1	6.4	4.5	4.6
20°N	7.3	9.2	11.4	12.2	10.0	6.4	5.9
12°N	10.0	13.7	16.3	16.0	14.3	8.5	7.2
4°N	7.0	11.2	15.4	18.5	20.7	11.2	9.2
4°N	7.1	10.9	15.5	20.2	23.0	12.8	9.6
12°N	8.9	13.1	18.4	22.8	26.1	14.2	11.1
20°N	10.2	16.0	24.5	26.6	25.6	14.6	11.2
28°N	10.5	15.8	24.4	26.1	24.3	14.8	11.4
36°N	9.6	13.8	18.4	21.1	21.1	14.7	11.8
44°N	6.2	8.1	11.0	16.2	18.9	14.2	11.4
52°N	4.1	5.1	7.4	12.5	15.5	12.7	10.2
60°N	3.0	3.7	5.2	9.9	12.0	11.0	9.3
68°N	4.5	7.2	7.3	8.6	8.9	9.6	9.0
76°N	4.0	5.0	5.0	6.8	7.3	8.5	8.9
84°N	4.7	5.0	5.0	6.4	6.6	7.9	8.9
90°N	--	--	--	6.5	6.6	7.7	8.6

From Spivakovsky et al. (2000).

Table 17.3b Zonally and Monthly Averaged OH Concentrations for April (10^5mol cm^{-3}).

	1000 hPa	*900 hPa*	*800 hPa*	*700 hPa*	*500 hPa*	*300 hPa*	*200 hPa*
90°N	0.4	0.6	0.8	1.0	0.8	0.9	1.3
84°N	0.3	0.5	0.6	0.8	1.0	1.1	1.5
76°N	0.5	0.7	0.9	1.4	1.4	1.6	2.1
68°N	3.9	3.5	3.4	3.2	2.2	2.3	2.7
60°N	3.8	4.1	5.0	4.4	3.5	3.2	3.3
52°N	4.9	5.3	7.1	6.0	5.1	4.2	4.0
44°N	7.8	8.7	11.0	8.8	7.7	5.6	5.0
36°N	9.0	11.9	13.8	12.7	11.3	7.5	6.4
28°N	12.4	14.8	17.1	18.1	14.8	7.9	6.9
20°N	15.1	18.9	21.4	23.7	17.8	9.6	8.0
12°N	14.3	19.4	22.1	23.5	20.4	10.4	8.4
4°N	7.7	14.1	18.8	22.4	23.5	12.3	10.0
4°N	7.1	13.4	16.0	20.8	22.3	12.7	10.6
12°N	7.3	13.6	15.9	19.1	19.8	12.0	10.7
20°N	8.7	12.8	16.8	18.4	16.7	11.0	9.6
28°N	7.9	10.4	12.9	14.1	13.6	9.7	8.6
36°N	5.8	6.8	8.0	9.6	10.5	8.0	6.9
44°N	2.7	3.1	3.9	5.5	7.5	6.1	5.6
52°N	1.1	1.4	1.7	2.8	4.3	4.0	3.4
60°N	0.4	0.5	0.7	1.3	2.1	2.1	1.8
68°N	0.2	0.3	0.3	0.5	0.7	0.8	0.7
76°N	0.1	0.1	0.1	0.1	0.2	0.1	0.2
84°N	0.0	0.0	0.0	0.1	0.1	0.0	0.0
90°N	--	--	--	--	--	--	--

From Spivakovsky et al. (2000).

model allowed for loss of CH_3CCl_3 in the stratosphere but did not account for removal by the ocean. With the standard OH model (solid line in Figure 17.14), 11% of the global loss of CH_3CCl_3 occurs in the stratosphere with the balance (89%) in the troposphere. Agreement between model and observation is excellent using the computed distribution of OH; it could be improved somewhat if OH concentrations were reduced by about 3% and if we allowed for a small sink due to the ocean. Based on a consideration of all of the available sources of information, Spivakovsky believes that the distributions of OH summarized in Table 17.3 and Plate 6 (top) should be accurate to within about ±10%.

The change with time of the abundance of a gas X, which is well-mixed throughout the atmosphere but removed mainly by reaction with OH in the troposphere, may be expressed in terms of an equation of the form

$$\frac{dN}{dt} = P(t) - (0.85)\, N(t)\, v, \qquad (17.84)$$

where $N(t)$ defines the total number of X molecules present in the atmosphere at time t, $P(t)$ denotes the magnitude of the source (mol sec^{-1}) and v is a frequency (sec^{-1}) associated with the loss of X by reaction with tropospheric OH. The factor of 0.85 accounts for the fact that the loss process is assumed to be confined to the troposphere, and only 85% of the total mass of X in the atmosphere resides in the troposphere.[9] In steady state ($dN/dt = 0$), it follows that

$$N(t) = P(t)\left[(0.85)\, v\right]^{-1} = P(t)\,\tau, \qquad (17.85)$$

where

$$\tau = \left[(0.85)\, v\right]^{-1} \qquad (17.86)$$

may be identified with the lifetime of X with respect to removal by tropospheric OH. Note that τ defines the time required for production to supply a quantity of X sufficient to fill the atmospheric reservoir to content N, starting from zero, if loss were eliminated. Equivalently, if the source were eliminated, τ would define the time required for the concentration of X to decline by a factor of e.[9]

Spivakovsky et al. (2000) showed that the lifetime of a long-lived gas X toward removal by reaction with tropospheric OH may be estimated with adequate precision using the relation

$$\tau_X = (0.85\,[\mathrm{OH}]\, k_X)^{-1}, \qquad (17.87)$$

where [OH] defines an average value for the concentration of OH in the troposphere and k_X denotes an appropriate average value for the rate constant for reaction of OH with X. They showed that little error is introduced by using the spatially and temporally averaged value for the concentration of OH (1.16×10^6 cm^{-3}, as noted above), assuming a value for k_X appropriate for a temperature of 270K. With these assumptions, as indicated by the following Example, the lifetime of CH_3CCl_3 toward removal by OH in the troposphere is estimated at 5.7 yr. The lifetime reflecting removal by the stratosphere and ocean in addition to the troposphere is estimated (Example 17.22) at 4.6 yr. This compares favorably

Table 17.3c Zonally and Monthly Averaged Concentrations of OH for July (10^5mol cm^{-3}).

	1000 hPa	900 hPa	800 hPa	700 hPa	500 hPa	300 hPa	200 hPa
90°N	5.5	7.4	8.3	10.0	9.7	6.9	8.8
84°N	3.8	5.2	6.4	8.6	10.9	7.7	9.5
76°N	3.7	5.1	6.9	9.7	11.5	7.6	8.4
68°N	7.4	11.6	13.9	15.4	13.2	8.2	7.7
60°N	6.2	10.6	16.2	17.0	15.3	9.2	8.0
52°N	6.2	11.8	19.5	19.4	17.5	10.6	8.6
44°N	12.3	23.1	26.9	25.1	22.1	12.6	10.0
36°N	14.9	24.7	27.9	28.0	25.0	14.4	11.3
28°N	12.6	17.9	22.7	24.7	23.3	12.3	8.8
20°N	13.0	17.6	21.9	26.5	24.8	13.7	10.0
12°N	10.9	16.4	20.1	24.4	24.8	13.6	10.0
4°N	10.3	14.6	19.2	22.3	23.2	12.9	10.2
4°N	10.4	15.4	20.4	21.3	20.1	12.4	10.9
12°N	9.9	13.8	17.0	17.8	15.9	10.5	9.9
20°N	7.9	10.9	12.7	13.1	11.3	8.6	7.8
28°N	4.5	5.6	6.2	7.1	7.1	5.8	5.8
36°N	2.6	2.8	3.2	4.0	4.6	3.7	3.3
44°N	0.9	1.0	1.2	1.9	2.6	2.2	1.9
52°N	0.3	0.3	0.4	0.7	1.1	1.0	0.9
60°N	0.1	0.1	0.1	0.2	0.3	0.2	0.2
68°N	0.0	0.0	0.0	0.0	0.0	0.0	0.0
76°N	--	--	--	--	--	--	--
84°N	--	--	--	--	--	--	--
90°N	--	--	--	--	--	--	--

From Spivakovsky et al. (2000).

with the lifetime of 4.8 yr. inferred from a purely empirical analysis of the observational data (Prinn et al. 1995).

Example 17.21: The rate constant for reaction of OH with CH_3CCl_3 has a value of 5.62×10^{-15} cm^{-3} sec^{-1} for a temperature characteristic of the mid troposphere. Assuming an average concentration of tropospheric OH equal to 1.16×10^6 cm^{-3}, estimate the lifetime of CH_3CCl_3 as determined by reaction with tropospheric OH.

Answer: The lifetime is given by equation (17.86). Substituting for [OH] and k_x, we find

$$\begin{aligned}\tau_{CH_3CCl_3} &= \left[(0.85)(1.16\times10^6)(5.62\times10^{-15})\right]^{-1}\text{ sec}\\ &= (5.54\times10^{-9})^{-1}\text{ sec}\\ &= 1.8\times10^8\text{ sec}\\ &= 5.7\text{ years}\end{aligned}$$

■

Example 17.22: Methyl chloroform is also removed in the stratosphere with an effective lifetime estimated at 34 yr. It is removed also by the ocean, with a lifetime estimated at 80 yr. Use the results in Example 17.21 to calculate the composite lifetime for CH_3CCl_3, reflecting removal by the combination of troposphere, stratosphere, and ocean.

Answer: To estimate the composite lifetime, we must combine first the loss frequencies associated with the separate processes. To find the composite lifetime, we must estimate first the net rate for removal. (Note that the composite lifetime *cannot* be derived by adding lifetimes for component loss processes.)

$$\begin{aligned}\tau_{CH_3CCl_3} &= \left[(\tau_{trop})^{-1} + (\tau_{strat})^{-1} + (\tau_{ocean})^{-1}\right]^{-1}\text{ yr}\\ &= \left[(5.7)^{-1} + (34)^{-1} + (80)^{-1}\right]^{-1}\text{ yr}\\ &= [0.175 + 0.029 + 0.013]^{-1}\text{ yr}\\ &= (0.217)^{-1}\text{ yr}\\ &= 4.6\text{ years}\end{aligned}$$

■

Lifetimes set by reaction with tropospheric OH are presented for a number of industrial hydrofluorocarbons (HFCs) and hydrochlorofluorocarbons (HFCs) in Table 17.5. The Table also includes results for CH_4, CH_3Cl, CH_3Br, CH_2CCl_2, and C_2Cl_4. Lifetimes summarized here were calculated in the same manner as for CH_3CCl_3 (also included in the Table).

17.6 The Budget of Tropospheric Ozone

As noted earlier, O_3 enters the troposphere in part by transport downward from the stratosphere, in part by production in situ through reactions (17.72)–(17.75). It is removed in

Table 17.3d Zonally and Monthly Averaged Concentrations of OH for December (10^5mol cm^{-3}).

	1000 hPa	900 hPa	800 hPa	700 hPa	500 hPa	300 hPa	200 hPa
90°N	--	--	--	--	--	--	--
84°N	0.0	0.0	0.0	0.0	0.0	0.0	0.0
76°N	0.1	0.1	0.1	0.1	0.1	0.1	0.1
68°N	0.5	0.5	0.5	0.6	0.5	0.4	0.6
60°N	0.8	0.9	1.2	1.2	1.5	1.3	1.5
52°N	1.7	1.9	2.9	2.6	3.1	2.8	2.9
44°N	4.4	4.7	6.5	5.3	5.7	4.9	5.1
36°N	6.3	8.3	10.1	9.2	9.0	7.3	7.5
28°N	9.4	12.0	13.5	13.2	11.3	7.0	6.5
20°N	10.1	13.9	15.3	16.0	14.6	8.8	8.0
12°N	9.9	15.3	17.4	19.1	20.1	10.6	8.8
4°N	8.4	13.5	19.1	23.7	25.1	12.4	9.9
4°N	11.4	18.5	24.5	28.8	28.0	14.3	11.2
12°N	13.7	19.5	25.5	29.6	26.7	13.9	11.1
20°N	13.1	19.5	23.8	27.1	22.2	13.6	10.4
28°N	9.2	13.0	16.0	19.3	17.2	10.6	9.1
36°N	8.0	9.4	11.0	13.5	13.4	8.9	7.4
44°N	4.9	5.6	7.0	10.2	10.6	7.2	6.1
52°N	3.0	3.5	4.5	6.9	7.5	6.0	5.3
60°N	2.7	2.9	3.2	4.5	5.4	5.6	5.1
68°N	2.7	3.4	3.3	3.7	3.6	5.3	5.2
76°N	1.2	1.6	1.0	1.7	1.9	3.4	3.4
84°N	0.0	0.9	0.9	1.0	1.3	2.3	2.2
90°N	--	--	--	0.8	1.1	1.7	1.6

From Spivakovsky et al. (2000).

Table 17.4 Relative Change (%) in the Global-Mean OH Concentration.

Scaling of Precursors	*−50%*	*−25%*	*25%*	*50%*
H_2O	−24	−11	10	20
O_3	−15	−8	8	17
O_3 and H_2O	−31	−18	20	43
NO_x	−17	−8	8	14
CO	23	10	−8	−14
CH_4	14	6	−5	−10
O_3 column	71	26	−17	−28

From Spivakovsky et al. 2000. Concentrations are in response to a uniform scaling of concentrations of precursors globally by ±25% and ±50%.

the gas phase by (17.6), (17.17), (17.21), and (17.31), with additional loss due to contact with organic matter at the surface (dry deposition).

Production of O_3 requires a supply of reduced species–hydrocarbons and CO–and, most critically, a source of NO_x. It is also sensitive to the flux of ultraviolet solar radiation reaching the troposphere. The lifetime of O_3 in the troposphere ranges from months at high latitudes during winter, decreasing to weeks at high latitude during summer, and declining to days in the tropics. Lifetimes are shortest at low altitude, increasing to hundreds of days in the upper troposphere. The source of O_3 is largest over continents and over industrial regions during summer. Significant production takes place in the tropics, especially in regions subject to seasonal emissions of NO_x and hydrocarbons associated with the burning of biomass. It comes as no great surprise that the abundance of O_3 is variable in space and time; a realistic treatment requires a 3-dimensional perspective. To be useful, a model for tropospheric O_3 must account for spatial and temporal variations in transport, for variations in emissions of hydrocarbons and NO_x, and for changes in the overlying abundance of O_3 in the stratosphere. The analysis presented here is based on results of a 3-dimensional study reported by Wang et al. (1998a–c).

Wang et al. (1998a–c) accounted for both industrial and natural sources of CO, hydrocarbons, and NO_x; they allowed in addition for the influence of biomass burning. A summary of assumptions of their model with respect to sources for NO_x, CO, and various hydrocarbons is given in Table 17.6. Spatial distributions of O_3 calculated for the surface and for the level in the atmosphere corresponding to a pressure of 500 mb are presented for the months of January and July in Plate 6 (bottom) (see color insert). Note the preponderance of elevated levels of O_3 over industrial regions of the Northern

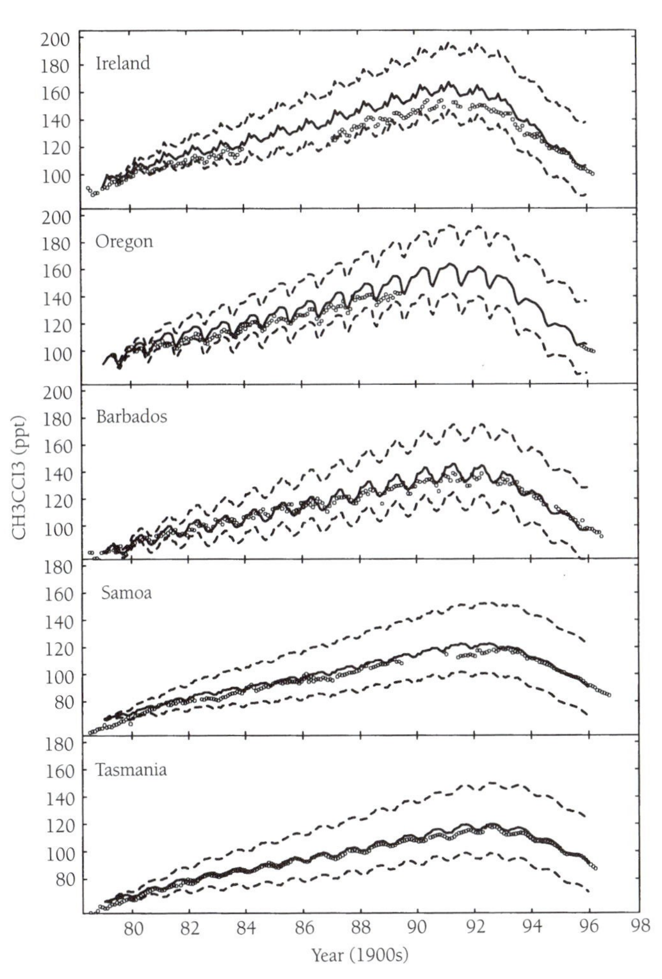

Figure 17.14 Observed long-term trend in CH_3CCl_3 (dots) simulated using standard OH without the ocean sink (solid lines), and with tropospheric loss frequencies reduced and increased by 25% (dashed lines). Source: Prinn et al. 1995.

Table 17.5 Lifetimes for Gases Removed by Reaction with Tropospheric OH.

Species	*Common Name*	*Lifetime (yrs)*
CH_3F	HFC-41	2.8
CH_2F_2	HFC-32	5.3
CHF_3	HFC-23	25.1
CH_2FCl	HCFC-31	1.3
$CHFCl_2$	HCFC-21	2.0
CHF_2Cl	HCFC-22	12.0
CH_3CH_2F	HFC-161	0.3
CH_2FCH_2F	HFC-152	0.5
CH_3CHF_2	HFC-152a	1.5
CH_2FCHF_2	HFC-143	3.6
CH_3CF_3	HFC-143a	53.0
CHF_2CHF_2	HFC-134	10.0
CH_2FCF_3	HFC-134a	13.8
CHF_2CF_3	HFC-125	30.9
CH_3CFCl_2	HCFC-141b	10.2
CH_3CF_2Cl	HCFC-142b	19.1
CH_2ClCF_2Cl	HCFC-132b	3.3
CH_2ClCF_3	HCFC-133a	3.7
$CHCl_2CF_3$	HCFC-123	1.3
$CHFClCF_3$	HCFC-124	6.0
CH_4	--	9.2
CH_3Cl	--	1.4
CH_3Br	--	1.9
CH_3CCl_3	--	5.6
CH_2Cl_2	--	0.4
C_2Cl_4	--	0.3

Hemisphere during summer and over regions of the tropics subject to seasonal inputs due to burning of biomass during the dry season. Zonally averaged values for the chemical lifetime of O_3 as a function of latitude and altitude are given in Figure 17.15. Composite budgets (spatially integrated sources and sinks) for O_3 for the Northern and Southern Hemispheres and for the globe are summarized in Table 17.7.

The results in Table 17.7 suggest that the rate at which O_3 is produced by tropospheric chemical reactions exceeds the rate at which it is supplied by transport from the stratosphere on a global scale by more than a factor of ten. The source of O_3 in the Northern Hemisphere is larger than that in the Southern Hemisphere by more than 70%, reflecting primarily the importance of the higher (a factor of 3.5) source of NO_x in the north. The abundance of O_3 in the Northern Hemisphere is larger than that in the Southern Hemisphere by about 40%. The lifetime of O_3, averaged over the entire global troposphere, is estimated as 25 days, slightly longer for the Southern as compared to the Northern Hemisphere: 28 days as compared to 23 days.

The data in Table 17.6 indicate emissions of NO_x in the Northern Hemisphere of 3.3×10^{13} g N yr^{-1}, resulting in an in situ source of O_3 of 2.62×10^{15} g O_3 yr^{-1} (Table 17.7). This implies a yield for O_3 per molecule of NO_x of 23 molecules per molecule (the factor 14/48 accounts for the difference in masses of O_3 and N): the corresponding yield for the Southern Hemisphere is 46 molecules per molecule. The yield of O_3 per NO_x molecule depends on the relative rates for reactions (17.21) and (17.16), allowing for additional production by (17.72). If we assume that production of O_3 is balanced primarily by in situ chemical loss and that loss proceeds primarily by (17.6), it follows that production of a molecule of O_3 should be associated with production of approximately three molecules of OH.[10] If oxidation of reduced species provides the dominant sink for OH (reactions of the form 17.70), it follows that the integrated rate for production of O_3 should be equal to approximately one-third of the rate at which reduced species are consumed in the troposphere. Assuming that (17.21) provides the dominant sink for NO_x, it follows that the yield for O_3 per molecule of NO_x, ϵ, should be given by

$$\epsilon = \frac{\text{production of } O_3}{\text{loss of } NO_x} = \frac{S_c}{3S_n}, \tag{17.88}$$

where S_c indicates the magnitude of the source for reduced species and S_n defines the corresponding magnitude of the

*Table 17.6 Sources of NO_x, CO, Ethane, Propane, $\geq C_4$ Alkanes, $\geq C_3$ Alkenes, Isoprene, and Acetone.**

	Global	Northern Hemisphere	Southern Hemisphere
NO_x			
Fossil fuel combustion	21	20	1.2
Biomass burning	11.6	6.5	5.1
Soil	6.0	4.2	1.8
Lightning	3.0	1.7	1.3
Aircraft	0.51	0.47	0.04
Stratosphere[a]	0.10	0.006	0.04
Total	42	33	9.4
CO			
Fossil and wood fuel combustion, industry	520	480	40
Biomass burning	520	290	230
CH_4 oxidation[b]	800	460	340
NMHC oxidation[b]	290	170	120
Total	2130	1400	730
Ethane			
Industry	6.3	5.7	0.6
Biomass burning	2.5	1.4	1.1
Total	8.8	7.1	1.7
Propane[c]			
Industry	6.8	6.1	0.7
Biomass burning	1.0	0.92	0.08
Total	7.8	7.0	0.8
$\geq C_4$ alkanes			
Industry	30	27	3
$\geq C_3$ alkenes			
Industry	10.4	9	1.4
Biomass burning[d]	12.6	7	5.6
Total	23	16	7
Isoprene			
Vegetation	597	297	300
Acetone			
Industry	1.0	0.9	0.1
Biomass burning	8.9	5.0	3.9
Vegetation	15	7.5	7.5
Oxidation of propane	6.2	5.3	0.9
Oxidation of higher alkanes[b]	6.2	5.5	0.7
Total	37	24	13

**-As assumed in the 3-dimensional simulation of the chemistry of the troposphere described by Wang et al. 1998a.*

Units are Tg [10^{12} g] N yr^{-1} for NO_x, Tg CO yr^{-1} for CO, and Tg C yr^{-1} for NMHCs.

[a]Downward transport of NO_x across the tropopause. This transport also supplies 0.38 Tg N yr^{-1} of HNO_3 globally.

[b]Computed within the model.

[c]Included in the model as a direct emission of acetone; the yield of acetone from oxidation of propane is specified as 80% (Singh et al. 1994).

[d]Including 6 Tg C yr^{-1} of $\geq C_2$ aldehydes.

source for NO_x.[11] The lower value of the yield for O_3 inferred for the Northern Hemisphere may be attributed to the higher value for the associated source of NO_x.

Equation (17.88) implies that, integrated over a large region (a hemisphere, for example), the source of O_3 (obtained by multiplying the yield per NO_x molecule by the total source of NO_x) should be proportional to S_c: the input of reduced species (hydrocarbons and CO). This analysis suggests that production of O_3 should be independent of S_n, the source of NO_x. It follows with our assumptions, since removal of O_3 is proportional to the abundance of O_3, that the integrated abundance of O_3 should also be proportional to S_c and independent of S_n. This apparently contradicts the result obtained earlier: that production of O_3 in the NO_x-

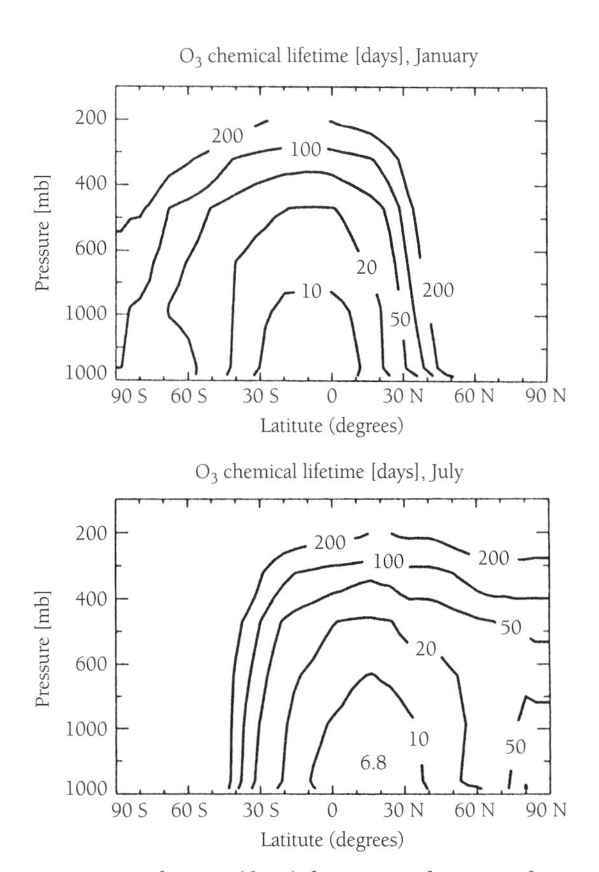

Figure 17.15 Lifetimes (days) for O_3 as a function of pressure (altitude) and latitude, calculated using model values for O_3 concentrations together with model estimates for rates of chemical removal. Source: Wang et al. 1995.

limited regime should be proportional to the abundance of NO_x but independent of the abundance of hydrocarbons. The contradiction arises from our neglect of such loss processes for O_3 as dry deposition, reactions (17.17) and (17.54), which together amount to about one-third of total loss of O_3. In fact, $S_c/3$ provides a lower limit for production of O_3, assuming that the flux of O_3 from the straosphere is small (Wang and Jacob 1998). It follows that strategies designed to address problems of elevated O_3, in not only urban but also remote regions, must provide for reduction of emissions of *all* relevant sources of pollution—hydrocarbons, CO, and NO_x.

While the most serious effects of anthropogenic emissions of hydrocarbons and NO_x on O_3 are as yet confined largely to industrial regions of the Northern Hemisphere and to selected regions of the tropics, there are troubling indications that the impact may be spreading. Wang, Logan, and Jacob (1998b) concluded that Asian pollution is already responsible for a detectable rise in the level and seasonal dependence of O_3 observed at the remote, high-altitude Mauna Loa Observatory in Hawaii. They attribute this effect to long-range transport of O_3 from Asia, noting that the lifetime of O_3 at the altitude of the Observatory (with a pressure level of 650 mb) is about 20 days.

The lifetime of NO_x is relatively brief, about a day. The impact of anthropogenic sources of NO_x may be extended, however, to larger spatial scales, if the compound is converted to longer-lived forms such as peroxyacetylnitrate (PAN, $CH_3C(O)OONO_2$, where (O) indicates that the O atom is double-bonded to the neighboring C atom). PAN is produced by photochemical oxidation of carbonyl species in

*Table 17.7 Contributions to the Budget of Tropospheric Ozone for the Northern Hemisphere, Southern Hemisphere, and Globe.**

	Global	*Northern Hemisphere*	*Southern Hemisphere*
Sources, Tg O_3 yr^{-1}			
In situ chemical production[a]	4100	2620	1480
Transport from stratosphere	400	240	160
Total	4500	2860	1640
Sinks, Tg O_3 yr^{-1}			
In situ chemical loss[b]	3680	2290	1390
Dry deposition	820	530	290
Total	4500	2820	1680
Interhemispheric transport, Tg O_3 yr^{-1}	0	–40	40
Burden, Tg O_3	310	180	130
Residence time, days	25	23	28

**-Due to: (1) in situ chemical production; (2) transport from the stratosphere; (3) in situ chemical loss; (4) loss at the surface, and (5) exchange between the hemispheres. The budget is for the air column extending up to 150 mbar. Annual mean budget terms are given for the odd oxygen family ($O_x = O_3 + O + NO_2 + HNO_4 + 2 \times NO_3 + 3 \times N_2O_5$ + PANs + HNO_3) to account for chemical interconversion between ozone and other components of O_x. Since ozone accounts for over 95% of O_x, the budgets of ozone and O_x can be regarded as equivalent. From Wang et al. 1998b.*

[a]Mainly from the reactions of peroxy radicals with NO.

[b]Mainly from the reactions $O^1D + H_2O$, $O_3 + HO_2$, and $O_3 + OH$.

the presence of NO_x. The key steps in formation of PAN from acetaldehyde (CH_3CHO) may be summarized as follows (Seinfeld and Pandis 1998):

$$CH_3CHO + OH \rightarrow CH_3CO + H_2O \quad (17.89)$$

$$CH_3CO + O_2 \rightarrow CH_3C(O)O_2 \quad (17.90)$$

$$CH_3C(O)O_2 + NO_2 + M \rightarrow CH_3C(O)O_2NO_2 + M \quad (17.91)$$

$$CH_3C(O)O_2NO_2 + M \rightarrow CH_3C(O)O_2 + NO_2 + M \quad (17.92)$$

The lifetime of PAN, determined by reaction (17.92), is a sensitive function of temperature, ranging from hours in warmer regions of the lower atmosphere, to months at colder temperatures characteristic of mid levels of the troposphere. The concern is that PAN, formed in polluted regions of the atmosphere where the abundance of NO_x is high, may be mixed upward into colder regions of the middle troposphere where the compound is comparatively stable. As PAN, NO_x may be transported over large spatial scales. Returning to lower, warmer regions of the atmosphere, it will decompose (by reaction 17.92), providing a distributed source of NO_x, thus extending the influence of local sources of NO_x pollution potentially to global scale.

The Harvard 3-dimensional model has been applied also in an attempt to simulate the abundance of O_3 in the preindustrial troposphere. Wang and Jacob (1998) concluded that the global abundance of O_3 was about 38% lower in the preindustrial environment as compared to today, 4.0×10^{12} moles as compared to 6.5×10^{12} moles. Similar to today, they found that the rate for in situ production of O_3 in the preindustrial troposphere was larger than the rate at which O_3 was supplied from the stratosphere. Similar again to today, they concluded that production of O_3 in the preindustrial environment was limited by NO_x. Globally averaged, they inferred a yield for O_3 of 58 molecules O_3 per molecule NO_x, higher than the contemporary value by about a factor of two, consistent with (17.88), given that the decrease in S_n for the preindustrial environment was larger than that for S_c (by a factor of 4.7, as compared to a factor of 2.6). They concluded that the globally averaged value for the concentration of OH in the preindustrial troposphere was about the same as today, consistent with the expectation that the concentration of OH should vary approximately as $S_n/S_c^{3/2}$, as implied by the analysis in Example 17.23.

Example 17.23: Estimate how the abundances of HO_2 and OH would be expected to vary as functions of S_c and S_n. Assume that production of O_3 is NO_x limited and that HO_2 is removed primarily by reaction with NO rather than O_3. Assume further that the relative abundances of NO and NO_2 are determined by reactions (17.39) and (17.40) and that production of O_3 and the abundance of O_3 are proportional to S_c, as indicated above.

Answer: With our assumptions, $[HO_2]$ is given by (17.35). Hence,

$$[HO_2] \approx [O_3]^{1/2} \approx S_c^{1/2}$$

Loss of NO_x proceeds primarily by (17.21). Thus,

$$[OH] \approx S_n / [NO_2]$$

Since the relative abundances of NO and NO_2 are determined by (17.39) and (17.40), it follows that

$$[NO_2] \approx [NO][O_3]$$

But production of O_3, P_{O_3}, is determined by the rate for reaction (17.16). Thus

$$P_{O_3} \approx [NO][HO_2] \approx [NO]\, S_c^{1/2}$$

Using (17.88), it follows that

$$[NO] \approx P_{O_3} / S_c^{1/2} \approx S_c^{1/2},$$

and hence,

$$[NO_2] \approx S_c^{1/2}\, [O_3] \approx S_c^{3/2}$$

Thus,

$$[OH] \approx \frac{S_n}{[NO_2]} \approx \frac{S_n}{S_c^{3/2}}$$ ■

Wang and Jacob (1998) assumed an abundance of CH_4 for the preindustrial environment of 0.7 ppm, as inferred from studies of gases trapped in polar ice (see Figure 5.5). Given that the abundance of OH for the preindustrial environment is expected to be approximately the same as today, it follows that the increase in the abundance of CH_4 in the atmosphere over the past several hundred years should be attributed primarily to an increase in the magnitude of the source. The increase in concentration from 0.7 ppm to about 1.7 ppm implies an increase in emissions by about a factor of 2.4. We return to this matter below in the context of our discussion of the CH_4 budget.

17.7 The Budget of Atmospheric Methane

As illustrated in Figure 5.5, the mixing ratio of CH_4 was relatively constant, at a little less than 700 ppb, from around A.D. 1000 to about the midpoint of the eighteenth century. It has more than doubled, rising from 650 ppb to close to 1700 ppb over the past 250 years. Accepting the conclusion of Wang and Jacob (1998) that the abundance of OH in the preindustrial environment was similar to today, it follows that production of CH_4 increased by about a factor of 2.4 over the past 250 years, from about 230 Tg yr^{-1} in 1750 (1 Tg = 10^{12} g) to more than 550 Tg yr^{-1} today. The long-term trend in the source, assuming constant OH, is illustrated in Figure 17.16.[12] The growth rate was particularly rapid after World War II, reaching a maximum in the early 1980s (Etheridge et al. 1998) as indicated in Figure 17.17. Observations for the most recent period suggest that production has been relatively constant since about 1984 (Dlugokencky et al. 1998).

Estimates of contributions from individual sources and sinks, reported by a number of authors over the past several years, are presented in Table 17.8. It seems reasonable to assume that production prior to A.D. 1750 was dominated by wetlands (with minor contributions from biomass burning,

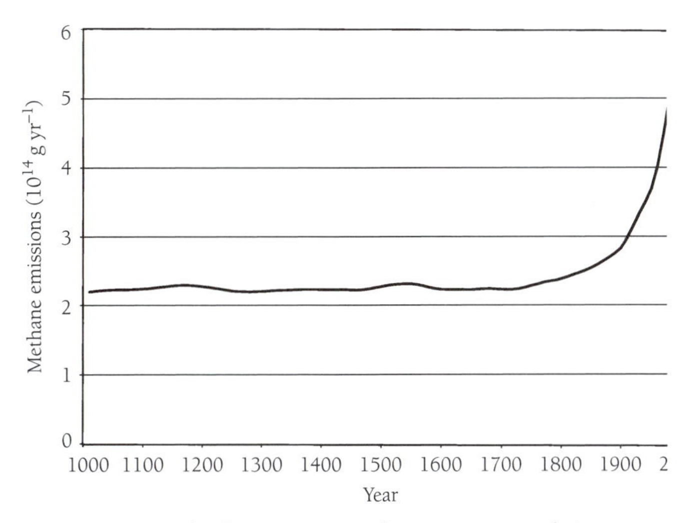

Figure 17.16 The long term trend in emissions of CH_4, assuming constant OH.

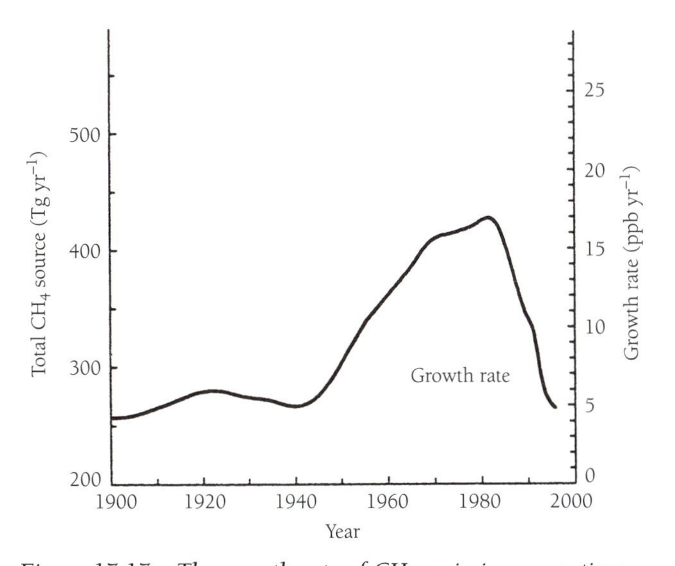

Figure 17.17 The growth rate of CH_4 emissions over time.

termites, and ruminants) and to adopt an estimate for the source strength of about 230 Tg yr^{-1}, as indicated by the argument presented above. We note that this value is consistent with the result for wetlands quoted by Hein, Crutzen, and Heimann (1997). The growth in emission of CH_4 over the past several hundred years must be attributed, it appears, to human activity. The important influences include: agricultural practices favoring cultivation of rice; an expanding population of domestic ruminants, mainly cattle, sheep, and goats; burning of vegetation to clear land, largely in the tropics and primarily for purposes of agriculture; increased use of fossil fuels; and disposal of concentrated organic waste produced by an increasingly urbanized society.

The bulk of the sources indicated in Table 17.8 reflect production by bacteria from the biological kingdom known as the *Archaebacteria*. Commonly referred to as *methanogens*, these organisms are obligatory anaerobes; that is to say, they are capable of growth only under conditions where the abundance of oxygen is extremely low. They proliferate in organic-rich sediments as found, for example, in natural swamps or in flooded, rich paddy fields, where the supply of oxygen to sediments is limited by high rates for biological consumption of oxygen in the overlying water column. They play an important role also in the digestive tracts of grass-feeding (herbivorous) animals, such as cattle, sheep, goats, camels, buffalo, and deer. Plant material eaten by these animals is temporarily stored in the rumen (the first compartment of the stomach, hence the collective reference to these species as *ruminants*), where it is transformed to more digestible organic matter by a complex population of anaerobic bacteria. The relationship between ruminants and methanogenic bacteria is symbiotic: ruminants feed the bacteria; in turn, the bacteria feed the ruminants. In the process, copious quantities of CH_4 are evolved and released to the atmosphere through a combination of belching and flatulence. A typical domestic cow produces as much as 200 liters of CH_4 per day (Wolin 1979). Given that there are more than a billion cattle in the world today (more than 100 million in

Table 17.8 CH_4 Budget (Tg yr^{-1})

	WMO (1995)	*Fung et al.* (1991)	*Hein, Crutzen, and Heimann* (1997)
Wetlands	110	115	232
Termites	20	20	--
Ocean + Freshwater	15	10	--
Hydrates	10	5	--
Energy-related	100	75	103
Sewage Treatment	25	--	--
Landfills	30	40	40
Animal Waste	25	--	--
Ruminants	80	80	90
Biomass Burning	40	55	41
Rice Paddies	60	100	69
Total Sources	515	500	575
Soils	30	10	28
Reaction with OH	445	450	469
Removal in stratosphere	40	--	44
Total sinks	515	460	541

the United States alone), it is not surprising that cattle are thought to provide a significant, contemporary source of CH_4.

Trends in cattle population as a function of time are illustrated in Figure 17.18. The Figure also includes data on the area of land devoted to rice. It is interesting to note the rapid increase in both cattle population and rice acreage that took place after World War II. The data suggest (Khalil and Shearer 1993) that animal husbandry and rice cultivation may have significantly contributed to the rapid growth in emissions of CH_4 observed between 1950–1980. It is interesting to further note that the population of cattle and acreage devoted to rice have been relatively constant over the past several decades: this suggests that the decrease in the growth rate of CH_4 observed in recent years may be attributed at least in part to a slowdown in the growth of the cattle population and in acreage devoted to rice cultivation.

It is clearly important to understand the social and economic factors responsible for the pattern displayed in Figure 17.18. Are the recent trends in cattle population and rice cultivation aberrations resulting in a temporary pause in the growth of emissions of CH_4 or alternately, may they be taken as an indication that sources of CH_4 may have reached an asymptotic limit? The slowdown in the growth of rice acreage may reflect a fundamental limitation imposed by the availability of suitable land; the trend observed in cattle population is more puzzling. A regional analysis could be instructive in elucidating the underlying cause.

Measurements of the ^{14}C content of atmospheric CH_4 suggest that fossil sources account for about 16% of the total contemporary source of atmospheric CH_4.[13] The fossil contribution includes releases from geologic deposits associated with mining of fossil fuels (coal, oil, and natural gas), emissions due to incomplete flaring of gas by the oil industry, and inadvertent losses due to leaks in natural gas pipelines. Contributions from the individual fossil sources are uncertain, and it is difficult to define the temporal trend in associated emissions. It is clear, though, that fossil sources of CH_4 are important and that they may become increasingly significant in the future, reflecting both growth in demand for fossil fuels generally and growth in demand for natural gas specifically. Given the significance of CH_4 as a greenhouse gas, it is important that steps be implemented to eliminate unnecessary, energy-related emissions of CH_4 to the atmosphere.

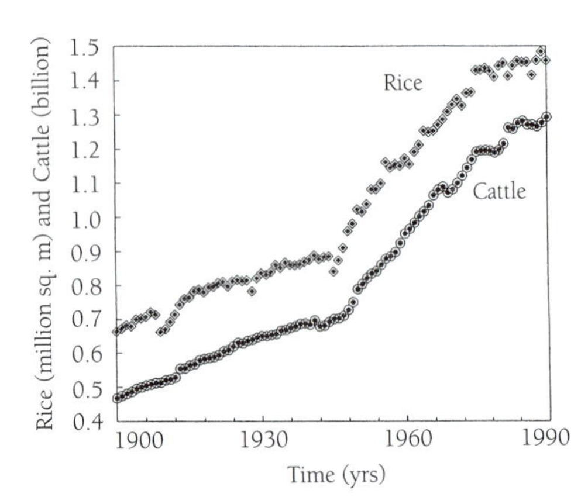

Figure 17.18 Trends in cattle population and rice cultivation over time.

Biomass burning provides an isotopically distinct source of atmospheric CH_4. In contrast to bacterially mediated sources, which tend to favor production of isotopically light carbon (concentrations of $^{12}CH_4$ much larger than $^{13}CH_4$), biomass burning results in the production of CH_4 with isotopic composition comparable to that of the parent fuel. The biomass-burning source has been characterized by a value for $\delta^{13}C$ of about –25‰, typical of C3 plant material (Stevens and Engelkemeir 1988; Wahlen et al. 1989).[14] Bacterial sources, including wetlands, rice cultivation, landfills, and ruminants, exhibit an average value for $\delta^{13}C$ of about –55‰, with a range from about –45 to –70‰ (Quay et al. 1991). The $\delta^{13}C$ of fossil CH_4 is reported as intermediate between that for bacterial production and biomass burning, about –37‰, with a range from as low as –76 to as high as –15‰ (Quay et al. 1991).

Measurements of the isotopic composition of CH_4 trapped in polar ice suggest that $\delta^{13}C$ for atmospheric CH_4 has increased by about 2‰ over the past several decades. The value of $\delta^{13}C$ for CH_4 satisfies an equation of the form

$$\frac{d(N\delta)}{dt} = \sum_i p_i \delta_i^s - N\nu(\delta - \theta), \tag{17.93}$$

where N defines the abundance of CH_4 in the atmosphere (expressed in Tg), δ is the value of $\delta^{13}C$ for atmospheric CH_4 (given in ‰), p_i denotes the rate for production associated with source i (Tg yr^{-1}), δ_i^s (‰) is the value for $\delta^{13}C$ corresponding to source i, ν is the loss frequency (the inverse of the CH_4 lifetime expressed in yr^{-1}), and θ (‰) allows for the selective removal of $^{12}CH_4$ relative to $^{13}CH_4$.

The abundance, N, is given by

$$\frac{dN}{dt} = \sum_i p_i - N\nu \tag{17.94}$$

In steady state, the time derivatives on the left-hand side of (17.93) and (17.94) vanish. In this case,

$$N\nu(\delta - \theta) = \sum_i p_i \delta_i^s \tag{17.95}$$

and

$$N = \nu^{-1} \sum_i p_i = \nu^{-1} p, \tag{17.96}$$

where p denotes the combined source associated with the individual contributions p_i. Combining (17.95) and (17.96), it follows that

$$(\delta - \theta) = \frac{1}{p} \sum_i p_i \delta_i^s \tag{17.97}$$

Under circumstances where either or both of N and δ are observed to vary with time, the derivative on the left-hand side of (17.93) may be expanded and may be rewritten in the form

$$N\frac{d\delta}{dt}+\delta\frac{dN}{dt}=\sum_{i} p_i\delta_i^s - N\nu(\delta-\theta) \quad (17.98)$$

Equation (17.98) may be employed to estimate the production-weighted average value of $\delta^{13}C$ defining the source of CH_4, using data (from observations, for example) for $d\delta/dt$ and dN/dt and assuming appropriate values for N, ν, and θ.

In steady state, the isotopic composition of atmospheric CH_4 is determined by the production-weighted average isotopic composition of the source, corrected to allow for differential removal of $^{12}CH_4$ relative to $^{13}CH_4$.

A budget for atmospheric CH_4, constrained to be consistent not only with information on rates for production by individual sources but also with isotopic data for specific sources, is presented in Table 17.9. The Table distinguishes between sources for the preindustrial environment and more recent contributions associated with the diverse forms of human activity outlined above. The following examples illustrate that the change in $\delta^{13}C$ implied by the data in Table 17.9 is consistent with the trend inferred from analysis of the ice core data. Rows 1–2 refer to the preindustrial environment. Rows 3–7 summarize data attributed to various forms of human activity in the contemporary environment.

Example 17.24: Using the data from Table 17.9, estimate the average value, $\overline{\delta^s}$, for $\delta^{13}C$ associated with preindustrial sources of CH_4.

Answer:

$$\sum_{i=1}^{2} p_i\delta_i^s = -12{,}500$$

$$\sum_{i=1}^{2} p_i = 230$$

$$\overline{\delta^s} = -54.3‰$$ ■

Example 17.25: Using the data in Table 17.9, assuming steady state, estimate the average value, $\overline{\delta^s}$, for $\delta^{13}C$ associated with contemporary sources of CH_4.

Answer:

$$\sum_{i=1}^{7} p_i\delta_i^s = -30{,}820$$

$$\sum_{i=1}^{7} p_i = 590$$

$$\overline{\delta^s} = -52.2‰$$ ■

The last column of Table 17.9 presents values for production attributed to individual sources of CH_4, weighted by the difference between $\delta^{13}C$ for individual sources combined with the average value of $\delta^{13}C$ for the composite source, as estimated in Example 17.25 for the contemporary environment. Positive values of $p_i(\delta_i^s - \overline{\delta^s})$ indicate sources that tend to raise the average value of $\delta^{13}C$ (δ) for CH_4 in the atmosphere. Negative values highlight sources that contribute to a reduction in δ. According to these data, energy-related emissions are primarily responsible for the recent increase in δ for atmospheric CH_4, with a significant additional contribution from biomass burning. The tendency for energy-related emissions and biomass burning to raise the average value of δ for atmospheric CH_4 is partially offset by emissions of isotopically light CH_4 contributed by cattle and rice cultivation.

As indicated above, the rate of increase of atmospheric CH_4 has significantly slowed over the past few decades. The key question is whether this can be taken as a guide for the future: has the concentration of atmospheric CH_4 reached an asymptotic limit or is the rapid rate of increase observed prior to 1980 likely to resume, or even reverse sign, in the future? It seems unlikely that the behavior observed in the most recent period can be taken as a reliable guide for the future.

Table 17.9 Rates for Production and Isotopic Data for Specific Sources of CH_4.

Source	Production, p_i (Tg CH_4 yr^{-1})	δ_i^s (‰)	$p_i\delta_i^s$ (Tg CH_4‰ yr^{-1})	$p_i(\delta_i^s - \overline{\delta^s})$ (Tg CH_4‰ yr^{-1})
1 Wetlands	225	–55	–12,375	–630
2 Biomass burning	5	–25	–125	+136
3 Energy	95	–37	–3515	+1444
4 Landfills	40	–51	–2040	+48
5 Ruminants	80	–62	–4960	–784
6 Biomass burning	35	–25	–875	+952
7 Rice	110	–63	–6930	–1188

Warming at high latitudes could contribute to an increase in the prevalence of stagnant pools of water overlying reservoirs of organic-rich soils. This could give rise to an increase in what might be otherwise considered a natural source of CH_4 (except, of course, that the warming may arise as a consequence of human-induced changes in the concentration of greenhouse gases). Similarly, changes in climate in the tropics favoring warmer and wetter conditions could also result in an increase in emissions. As noted in our discussion of climates of the past, it is clear that natural sources of CH_4 can respond rapidly to regional changes in climate. If steps are to be taken to offset possible climate-related increases in emissions of CH_4, it is essential that we act to reduce, where possible, sources of CH_4 for which we must assume primary responsibility: notably, emissions related to cattle, rice, energy production, biomass burning, and waste disposal. Changes in dietary preferences for meat products, justified on health grounds, could reduce demands for cattle and other ruminants. Selection of strains of rice minimizing emissions of CH_4 to the atmosphere could reduce the rice-related source: CH_4 is transferred from sediments of rice paddy fields mainly through stems of plants, rather than directly through the water column. Steps could be taken to minimize losses of CH_4 associated with unnecessary leaks in pipelines and to capture CH_4 released as a by-product of oil, gas, and coal production. Unnecessary burning of biomass could be eliminated and CH_4 produced by decomposition of organic waste could be captured and used as a substitute for alternative primary fuels.

17.8 Summary

The hydroxyl radical, as we have seen, plays a critical role in tropospheric chemistry: it is responsible for removal of a variety of chemically reduced gases, including CO, CH_4, and a host of hydrocarbons of both natural and anthropogenic origin; it is produced by reaction of $O(^1D)$ with H_2O, with $O(^1D)$ formed by photodissociation of O_3. Production of $O(^1D)$ takes place over a narrow range of wavelengths between about 300 and 310 nm, limited on the long-wavelength limit by the rapid drop-off of the cross section for dissociation of O_3 and limited on the short-wavelength side by the even more rapid decrease in the flux of solar radiation penetrating to the troposphere through the thick protective overlying layer of stratospheric O_3. Not surprisingly, the concentration of OH and the oxidative capacity of the troposphere were found to sensitively depend not only on the abundance of O_3 in the troposphere but also on the height-integrated density of O_3 in the stratosphere.

Ozone is supplied to the troposphere in part by transport from the stratosphere, in part by in situ chemical production. Production in situ occurs as a by-product of the oxidation of reduced gases, such as CO and CH_4. It is catalyzed by reactions involving hydrogen and nitrogen radicals. Three distinct chemical regimes were identified. When the abundance of NO_x is extremely low, oxidation of reduced species results in net *loss* of tropospheric O_3. If the abundance of NO_x is high enough to ensure that peroxy radicals such as HO_2 react with NO, rather than with O_3, oxidation of hydrocarbons and CO is expected to result in net production of O_3. If hydrogen radicals are removed primarily through formation of H_2O_2, the rate for production of O_3 is proportional to the abundance of NO_x. This defines an environment we identified as *NO_x limited*. When concentrations of NO_x are very high, such that hydrogen radicals are removed mainly by formation of HNO_3, the rate for production of O_3 is proportional to the abundance of hydrocarbons. We referred to this regime as *hydrocarbon limited*.

On a global scale, the abundance of O_3 in the troposphere at the present time primarily reflects a balance between in situ chemical production and loss: supply from the stratosphere is small by comparison. On average, the troposphere appears to be NO_x limited, although hydrocarbon limitation may apply in environments where NO_x levels are unusually high (in heavily polluted cities, for example). It is likely that the troposphere was also NO_x limited in the preindustrial environment. The yield of O_3 per NO_x molecule (the number of O_3 molecules produced per molecule of NO_x) varies inversely with respect to the magnitude of the source of NO_x. Production of O_3 is rapid in regions where the abundance of NO_x is high. A large rate for production of O_3 under high-NO_x conditions is compensated by a slower rate for production in regions where concentrations of NO_x are low. Ultimately, globally integrated production of O_3 in the troposphere depends on the rate at which hydrocarbons and CO are made available for oxidation.

The globally averaged abundance of tropospheric OH should be relatively constant in time. This was attributed to the fact that the concentration of OH should vary approximately as $S_n/S_c^{3/2}$ under NO_x-limited conditions, where S_c and S_n define the magnitudes of the sources for hydrocarbons and NO_x, respectively. A change in emission of hydrocarbons may be compensated by a change of similar sign in emissions of NO_x. Adopting the lifetime of CH_4 inferred for the contemporary environment, we used this result in combination with concentrations of CH_4 inferred from studies of gases trapped in ice cores to reconstruct the history of CH_4 emissions over the past 250 years. The rapid growth in emissions in the post–World War II period was attributed primarily to increases in the population of domestic ruminants and in acreage devoted to rice cultivation. The growth in emission of CH_4 has significantly slowed over the past few decades. We suggest, however, that this should not be taken as a reliable guide for the future; a warmer and wetter climate may result in an increase in emissions from natural sources: wetlands, for example. To offset possible increases in emissions from natural sources, steps should be taken to reduce emissions from sources associated more directly with human activity: ruminants, rice cultivation, fossil fuel industry, biomass burning, and waste disposal.

The Chemistry of Precipitation

18

As discussed in Chapter 3, liquid water in its pure form has a pH of 7.0. From the definition of pH, it follows that the concentration of H^+ for pure water is equal to 10^{-7} mol l^{-1}. The abundance of positively charged H^+ in pure water is offset by an identical concentration of negatively charged OH^- as required to ensure charge neutrality. In practice, the concept of **pure water** refers to a state of matter that is rarely, if ever, realized in nature. In contact with the atmosphere, water absorbs a variety of gases at rates determined by the available supply and by relevant Henry's law coefficients (see below). Gases such as CO_2 can react in solution, contributing an additional source of H^+.[1] The acidity of the liquid medium may be directly enhanced by absorption of acidic gases, such as HNO_3 and H_2SO_4. Alternatively, it may be reduced by absorption of a gas such as NH_3, which can combine with H^+ to form NH_4^+. The chemistry of atmospheric water may be further altered by incorporation of aerosols ubiquitously present in the atmosphere. Calcium carbonate, $CaCO_3$, for example, is a common component of wind-blown dust; it decomposes in solution, contributing a source of Ca^{2+} and CO_3^{2-}. Subsequent reaction of CO_3^{2-} with H^+ provides a sink for H^+, resulting in an increase in pH. The pH of atmospheric water depends on a complex interaction of liquid water with both gaseous and condensed materials in the medium with which the water is in contact.

By convention, a medium with a pH less than 7.0 is defined by chemists as **acidic.** A solution with a pH higher than 7.0 is said to be **basic.** Neutrality is identified with a pH of 7.0. By this standard, atmospheric water may be considered naturally acidic. As we shall see, water in equilibrium with CO_2 at present concentrations of the gas in the atmosphere would have a pH of about 5.6. Rain in relatively pristine environments, as indicated in Figure 18.1, exhibits a variety of values of pH, ranging from as low as 3.8 to as high as 6.3. The term **acid rain** is usually associated with environments subject to intense emissions of acidic precursors, such as SO_2 and NO_x, released in conjunction with diverse forms of industrial activity, including the combustion of fossil fuels and the smelting of ore. Rain over industrial regions of the eastern North America and northern Europe is frequently characterized by values of pH as low as 4.0. The pH of precipitation is generally higher than the pH of cloud- or fogwater, reflecting more efficient dilution of acids by higher concentrations of water. Fogwater in southern California, for example, has been observed with a pH as low as 1.7 (Munger et al. 1983), roughly comparable to the pH of limejuice (see Table 3.3).

Acid rain as a phenomenon is not new. The presence of nitrogen- and sulfur-bearing acids in air and rain was reported as early as 1692 in a book titled *A General History of the Air,* published by the British chemist Robert Boyle. Robert Angus Smith is usually credited with coining the term *acid rain*, in a book titled *Air and Rain: The Beginnings of Chemical Climatology,* in 1872. Modern interest in the topic dates mainly to the 1950s and early 1960s. The Canadian ecologist Eville Gorham, working in Canada and

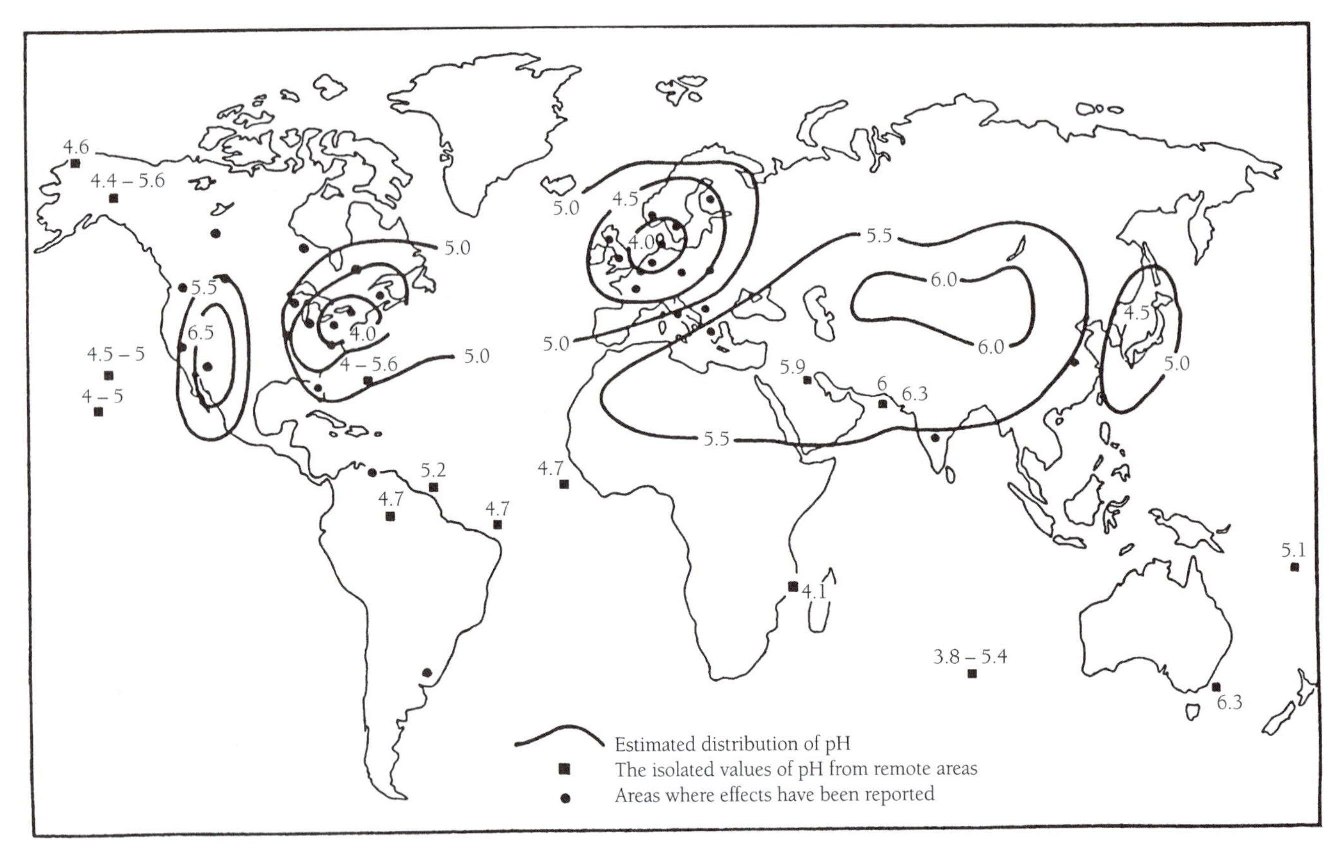

Figure 18.1 The global distribution of the pH of rain as defined in the air pollution monitoring program of the World Meteorological Organization. Source: Seinfeld and Pandis 1997.

England, established the critical connection between industrial emissions of acidic compounds (notably sulfur oxides), the acidity of precipitation, and changes observed in the chemistry of sensitive soils and aquatic systems. The Swedish chemist Svante Odin set up a network of stations in 1961 to monitor changes in the chemistry of lakes and rivers in Scandinavia. He showed that precipitation over Scandinavia and aquatic systems receiving this precipitation were gradually becoming more acidic. He suggested that an increasing concentration of sulfur compounds in polluted air masses originating from industrial regions to the south (specifically the United Kingdom, Germany, and Poland) was primarily responsible for the observed changes. And he speculated that acid rain falling in southern Scandinavia was responsible for a decline in forest growth, for leaching of toxic metals from soils to rivers and lakes, and for an alarming decline in fish populations. Odin's research had a major influence, contributing to a prompt recognition of the need for an internationally coordinated strategy to deal with the problems of long-range transport of air pollutants in Europe.

The impact of acid precipitation on specific ecosystems critically depends on the nature of the underlying soils and bedrock: environments containing high concentrations of carbonate minerals are effectively insulated from the potentially negative effects of acid rain; the acidity of rain water is neutralized as it percolates through these soils. The pH of waters draining these environments is often significantly higher than the pH of water entering in the form of precipitation. Regions underlain by thin soils and granitic bedrock are more vulnerable, however. The acid-neutralizing potential of these environments is low to begin with, and their capacity to neutralize the acidity of rainwater may have been further degraded by exposure to extended episodes of acid precipitation in the past. Regions of North America that are particularly sensitive to effects of acid precipitation are indicated in Figure 18.2.

For healthy growth, plants require a reliable supply of trace cations, such as Ca^{2+} and Mg^{2+}. Prolonged exposure to acid precipitation can result in significant depletion of these elements in sensitive soils. The acid-neutralizing capacity of the soil is accordingly reduced, with additional impacts on the health of overlying vegetation. If the pH of soil water drops below about 5.0, aluminum is transferred from mineral forms to solution and may be subsequently transported to rivers and lakes. At concentrations as low as 0.2 mg l^{-1}, aluminum is toxic to fish. Depletion of fish stocks in rivers and lakes in Scandinavia and eastern North America has been attributed to the combined effects of low pH and the presence of high concentrations of toxic metals, such as aluminum: in both incidences, the problem can be ultimately attributed to deposition of acidic substances from the atmosphere.

High concentrations of aluminum in soilwater can have an effect also on plant growth. As noted, plants require calcium and magnesium for healthy growth. As calcium and magnesium are depleted in soils and aluminum is enriched,

Figure 18.2 Regions of North America that are particularly sensitive to acid rain due to low buffering capacity of soil. Source: Jacob 1999.

plants have a tendency to absorb aluminum in substitution for calcium and magnesium. The fine root systems of plants are particularly sensitive to aluminum, which acts to inhibit cell division. Aluminum has a negative impact also on the ability of plants to absorb nutrients, such as phosphate. Acid precipitation has been invoked to account for the decline of forests in parts of Europe and North America over the past several decades. The details of the impact, however, are unclear. It is difficult to attribute changes observed in specific environments to any single influence. Changes in climate—drought, for example—can affect the health of a forest ecosystem over a particular period of time. Damage may be exasperated by outbreaks of insects or disease or by exposure to air pollutants, such as O_3. It could be argued, indeed, that the impact of acid precipitation may be positive for some systems, under some circumstances. Acid rain can provide an important source of nutrients N and S that might be otherwise in short supply.

This chapter is motivated by the need to understand the factors influencing the chemistry of precipitation, especially the phenomenon of acid rain. We begin, in Section 18.1, with a more basic objective: to outline the approach used to describe solution in the liquid phase of a variety of atmospheric gases, without regard for their subsequent fate in the liquid phase. Specifically, our interest in this case is to identify gases that are especially soluble in water, species for which solution in cloudwater can result in significant depletion of concentrations in the gaseous phase. We discuss the chemistry of CO_2 in Section 18.2, defining a reference value for the pH of water in equilibrium with atmospheric CO_2. We elaborate mechanisms responsible for introduction of nitric and sulfuric acids to the condensed phase in Sections 18.3 and 18.4, respectively. We discuss neutralization of acids in the presence of NH_3 in Section 18.5, whereas we present and discuss observational data on the chemistry of precipitation in Section 18.6. We conclude with summary remarks in Section 18.7.

18.1 The Solubility of Atmospheric Gases in the Liquid Phase

In equilibrium, the concentration in the aqueous phase of a chemical species A is determined by the concentration in the gaseous phase through a relation known as **Henry's law**, first recognized by the British chemist William Henry in 1807:

$$[A(aq)] = H_A P_A \tag{18.1}$$

Here $[A(aq)]$ denotes the concentration of A in the aqueous phase and P_A defines the partial pressure of A in the gaseous phase. The factor H_A in (18.1) is known as the **Henry's law coefficient.** With $[A(aq)]$ expressed in units of mol l^{-1} (the conventional choice of units to describe the concentration of a chemical species in the aqueous phase) and with P_A given in units of atmospheres, H_A has units of mol l^{-1} atm^{-1}, written as M atm^{-1}.

Interpretation of the physical significance of Henry's law is straightforward: an increase in the concentration of A in the atmosphere (or equivalently, the partial pressure) would be expected to result in a proportional increase in the concentration of A in solution. If A is highly soluble in water, the Henry's law coefficient will be relatively large; if A is insoluble, H_A will be small. In general, the solubility of a gas A is greater at low than at high temperatures. We expect, thus, that the magnitude of H_A should decrease as a function of increasing temperature. The solubility of O_2 in liquid water, for example, is about 60% higher at a temperature of 273 K as compared to its value at a temperature of 298 K.

Values of Henry's law coefficients for selected atmospheric gases are presented for a temperature of 298 K in Table 18.1. Note that the magnitude of H_A sensitively depends on the nature of the dissolved chemical species. Large molecules with a tendency to bind to water are generally more soluble than small, chemically simple species. The solubility of SO_2 exceeds that of O_3, for example, by about a factor of one hundred at 298 K. The Henry's law coefficient for SO_2 increases by a factor of 2.7, from 1.23 to 3.28 mol l^{-1} atm^{-1}, as the temperature drops from 298 K to 273 K. The relative increase in the solubility for O_3 over the same temperature range is slighly less than that for SO_2, about a factor of two, corresponding to a change in H_A from about 0.01 to about 0.02 mol l^{-1} atm^{-1}.

The fraction of a species A dissolved in cloudwater at a particular altitude depends not only on the magnitude of the value for H_A but also on the abundance of water in the liquid phase. The liquid content of clouds ranges from about 5×10^{-8} to about 3×10^{-6} g cm^{-3}. The liquid content of fogs is similar, varying from about 2×10^{-8} to about 5×10^{-7} g cm^{-3}.[2] To compare the concentration of A in the liquid phase to the concentration in the gaseous phase, we must ensure that concentrations in both phases are expressed in a common system of units. The perfect gas law in the form favored by chemists, equation (2.17′), may be used to calculate the concentration of A in the gaseous phase, $N(A,g)$. With P_A expressed in units of atmospheres and T in K, $N(A,g)$ is given by

$$N(A,g) = \frac{P_A}{RT} \tag{18.2}$$

With the universal gas constant R taken equal to 0.08206 atm l mol^{-1} K^{-1}, $N(A,g)$, as calculated using (18.2), is expressed in units of moles *per liter of air*. The concentration of A in the aqueous phase, as given by (18.1), is expressed in units of moles per liter *of water*. To convert the concentration of A in the liquid phase to the same system of units adopted for the gaseous phase, the abundance of water must be translated first to units of liters of water per liter of air. Using the facts that a gram of liquid water occupies a volume of 10^{-3} liters (the density of water is equal to 1 g cm^{-3}) and that a liter of air is equivalent to 10^3 cm^3, a concentration of liquid water equal to w g cm^{-3} is precisely the same as a concentration of w expressed in units of liters of water per liter of air. It follows that the abundance of A, $N(A,aq)$, contained in the liquid phase (concentration w g cm^{-3}), expressed in moles per liter of air similar to the units adopted above for the gaseous phase, is given by

$$N(A, aq) = H_A P_A w \tag{18.3}$$

Equal concentrations of A in the liquid and gaseous phases would require

$$H_A P_A w = \frac{P_A}{RT}, \tag{18.4}$$

corresponding to

$$H_A = \frac{1}{wRT} \tag{18.5}$$

Adopting a value for w on the high end of values observed for clouds, 10^{-6} g cm^{-3} and assuming a temperature of 298 K, we conclude that the concentrations of A in the liquid and gaseous phases would be same if the value of H_A were equal to 4.1×10^4 mol l^{-1} atm^{-1}. The concentration in the liquid phase would be equal to at least 10% of the concentration in the gaseous phase if the value of H_A were larger than about 4×10^3 mol l^{-1} atm^{-1}. It follows that the abundance of gases in the atmosphere is unlikely to be affected by solutions in the liquid phase, except for species near the bottom of the list included in Table 18.1. The analysis confirms the conclusions drawn earlier in Chapter 17 that wet deposition should have an important influence on HNO_3 and that it may be marginally important for H_2O_2, but that it should be relatively unimportant for CH_3OOH and CH_2O.

Table 18.1 Henry's Law Coefficients of Some Common Atmospheric Gases

*Species**	*H(M atm^{-1}) at 298 K*
O_2	1.3×10^{-3}
NO	1.9×10^{-3}
C_2H_4	4.8×10^{-3}
NO_2	1.0×10^{-2}
O_3	1.13×10^{-2}
N_2O	2.5×10^{-2}
CO_2	3.4×10^{-2}
H_2S	1.2×10^{-1}
DMS	5.6×10^{-1}
SO_2	1.23
CH_3ONO_2	2.6
CH_3O_2	6.0
OH	2.5×10^{1}
HNO_2	4.9×10^{1}
NH_3	6.2×10^{1}
CH_3OH	2.20×10^{2}
CH_3OOH	2.27×10^{2}
$CH_3C(O)OOH$	4.73×10^{2}
HCl	7.27×10^{2}
HO_2	2.0×10^{3}
HCOOH	3.6×10^{3}
HCHO	6.3×10^{3}**
CH_3COOH	8.8×10^{3}
H_2O_2	7.45×10^{4}
HNO_3	2.1×10^{5}
NO_3	2.1×10^{5}

**The constants given above do not account for dissociation or other aqueous-phase transformations.*

***Allowing for diol formation; otherwise, the Henry's law coefficient equals 2.5.*

18.2 The pH of Atmospheric Water in Equilibrium with CO_2

The concentration of aqueous-phase CO_2 in equilibrium with gaseous CO_2 in the atmosphere may be calculated using the Henry's law expression given by (18.1):

$$[CO_2\ (aq)] = H_{CO_2} P_{CO_2} \tag{18.6}$$

Assuming a mixing ratio for CO_2 of 2.8×10^{-4} as appropriate for the preindustrial environment, equivalent to a value for P_{CO_2} of 2.8×10^{-4} atm, with H_{CO_2} equal to 3.4×10^{-2} mol l^{-1} atm^{-1} as indicated in Table 18.1, we calculate a concentration for dissolved neutral CO_2 equal to 9.52×10^{-6} mol l^{-1}.[3] In solution, CO_2 behaves as a weak acid. As indicated in Chapter 11, it can react with H_2O to form H^+ and HCO_3^-.

Equilibrium between CO_2 (aq) and HCO_3^- may be expressed by the relation

$$CO_2(aq) + H_2O \leftrightarrow HCO_3^- + H^+ \tag{18.7}$$

Assuming an equilibrium constant for this reaction pair of K_1 (see Chapter 11), concentrations of CO_2 (aq), HCO_3^-, and H^+ are related according to the equation

$$K_1 = \frac{[HCO_3^-][H^+]}{[CO_2(aq)]} \tag{18.8}$$

Adopting a value for K_1 at 298 K of 4.3×10^{-7} mol l^{-1} (Seinfeld and Pandis 1997) and using the value for CO_2 (aq) obtained above, it follows that

$$[HCO_3^-][H^+] = 4.09 \times 10^{-12}\ mol^2\ l^{-2} \tag{18.9}$$

If we assume further that charge equilibrium in the liquid reflects primarily a balance between H^+ and HCO_3^-, such that

$$[H^+] = [HCO_3^-] \tag{18.10}$$

it follows that

$$[H^+]^2 = 4.09 \times 10^{-12} \text{ mol}^2 \text{ l}^{-2},$$

$$[H^+] = 2.02 \times 10^{-6} \text{ mol l}^{-1},$$

and

$$pH = -\log_{10} [H^+] = 5.7$$

A somewhat more accurate estimate of pH may be obtained if we were to account for the presence of negatively charged OH^-, in addition to HCO_3^-, in the liquid. However, at a pH of 5.7, the abundance of OH^-, determined by the equilibrium between H_2O, H^+, and OH^- (equation 3.12) is small, 5×10^{-9} mol l^{-1}, and the adjustment to pH is negligible.

Example 18.1: Calculate the pH for atmospheric water in equilibrium with CO_2 at its present-day concentration of 360 ppm.

Answer: Using the results from the analysis above, the concentration of aqueous CO_2 in the present atmosphere should be larger than that in the preindustrial atmosphere by a factor of 360/280.

$$[CO_2(aq)] = (9.55 \times 10^{-6})(360/280)$$

$$= 1.22 \times 10^{-5} \text{ mol l}^{-1}$$

The concentration of H^+ is given by

$$[H^+]^2 = K_1 \, [CO_2(aq)]$$

$$= (4.3 \times 10^{-7})\,(1.22 \times 10^{-5})$$

$$= 5.25 \times 10^{-12}$$

Thus,

$$[H^+] = 2.29 \times 10^{-6}$$

and

$$pH = -\log_{10}[H^+]$$

$$= 6 - 0.36$$

$$= 5.64$$

■

As may be readily shown, an approximate doubling of CO_2 to a concentration of 600 ppm, as may be expected to arise at some point over this century absent vigorous action to reduce the use of fossil fuels, would result in a drop in pH to a value of about 5.5.

At high values of pH and low concentrations of H^+, HCO_3^- can dissociate to form CO_3^{2-} and H^+. The equilibrium in this case is expressed by

$$HCO_3^- \leftrightarrow CO_3^{2-} + H^+ \tag{18.11}$$

It follows for equilibrium conditions that

$$K_2 = \frac{[CO_3^{2-}][H^+]}{[HCO_3^-]}, \tag{18.12}$$

where K_2 defines the value of the associated equilibrium constant. At 298 K, K_2 for atmospheric water has a value of 4.7×10^{-11} mol l^{-1} (Seinfeld and Pandis 1997). Based on equation (18.12) and the analysis above, assuming a concentration of H^+ approximately equal to that of HCO_3^-, it follows that the concentration of CO_3^{2-} should be given by the value for K_2:

$$[CO_3^{2-}] = K_2 = 4.7 \times 10^{-11} \text{ mol l}^{-1} \tag{18.13}$$

At a pH of 5.6, the concentration of CO_3^{2-} is thus much less than that of HCO_3^- (2.29×10^{-6} mol l^{-1} according to the analysis in Example 18.1). This calculation serves to justify the assumption, implicit in the analysis above, that the charge balance for water in equilibrium with atmospheric CO_2 is dominated on the positive side of the ledger by H^+ and that the dominant contributor to the negative complement is HCO_3^-.

We have assumed to this point that the chemistry of atmospheric water is determined exclusively by equilibrium with gaseous atmospheric CO_2. In practice, as noted in the introduction, the pH of atmospheric water may be either higher or lower than the value corresponding to equilibrium with CO_2. As indicated in equation (18.1), the concentration of CO_2 (aq) depends only on P_{CO_2} and HCO_2: it is independent of pH. As may be inferred from (18.8) and (18.12), however, an increase in the concentration of H^+ (a decrease in pH), as could arise as a consequence of the inclusion in the liquid phase of acidic compounds, such as HNO_3 or H_2SO_4, would result in a decrease in the concentrations of both HCO_3^- and CO_3^{2-}. What would happen under these circumstances is that CO_3^{2-} would be converted to HCO_3^-, HCO_3^- would be converted to CO_2 (aq) and excess CO_2 (aq) would be transferred to the gaseous phase. Conversely, a decrease in the concentration of H^+ (an increase in pH) would result in transfer of CO_2 from the atmosphere to the aqueous phase with conversion of excess CO_2 (aq) to HCO_3^- and CO_3^{2-}. The capacity of the liquid phase to absorb carbon rapidly increases as a function of increasing values of pH.

The total concentration of carbon in the liquid phase, $\sum CO_2$, is given by the combination of CO_2 (aq), HCO_3^-, and CO_3^{2-}:

$$\sum CO_2 = [CO_2(aq)] + [HCO_3^-] + [CO_3^{2-}] \tag{18.14}$$

Using (18.8), $[HCO_3^-]$ may be written in the form

$$[HCO_3^-] = \frac{K_1[CO_2(aq)]}{H^+} \tag{18.15}$$

Similarly, combining (18.12) and (18.15), $[CO_3^{2-}]$ may be expressed as

$$[CO_3^{2-}] = \frac{K_1 K_2[CO_2(aq)]}{H^+} \tag{18.16}$$

It follows that

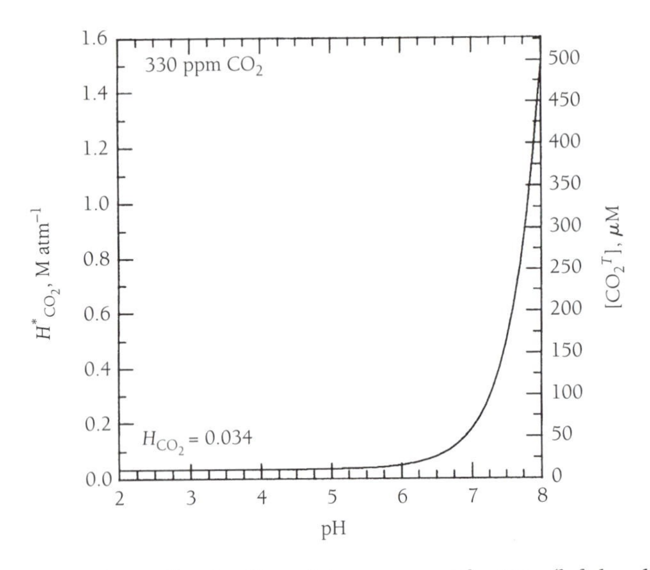

Figure 18.3 Effective Henry's law constant for CO_2 (left-hand vertical scale). Also indicated (right-hand vertical scale) is the concentration of inorganic carbon (CO_2) in solution corresponding to a gas-phase concentration of CO_2 equal to 330 ppm. Source: Seinfeld and Pandis 1997.

$$\sum CO_2 = [CO_2(aq)]\left\{1 + \frac{K_1}{[H^+]} + \frac{K_1K_2}{[H^+]^2}\right\} \quad (18.17)$$

Or, using (18.6) to substitute for $[CO_2\,(aq)]$,

$$\sum CO_2 = H_{CO_2}P_{CO_2}\left\{1 + \frac{K_1}{[H^+]} + \frac{K_1K_2}{[H^+]^2}\right\} \quad (18.18)$$

Equation (18.18) may be used to define an effective Henry's law coefficient, H'_{CO_2}, for incorporation of carbon in all forms in the liquid phase. Defining H'_{CO_2} by the relation

$$H'_{CO_2} = H_{CO_2}\left\{1 + \frac{K_1}{[H^+]} + \frac{K_1K_2}{[H^+]^2}\right\}, \quad (18.19)$$

Henry's law in its modified form is given by

$$\sum CO_2 = H'_{CO_2}P_{CO_2} \quad (18.20)$$

As would be expected, H'_{CO_2} is always larger than H_{CO_2}. Values of H'_{CO_2} are presented as a function of pH in Figure 18.3. The vertical scale on the right-hand side of the Figure indicates values of $\sum CO_2$ that would apply in the presence of a concentration of atmospheric CO_2 equal to 330 ppm.

18.3 The Contribution of Nitrogen Oxides to the Acidity of Atmospheric Water

As indicated in Chapter 17, a combination of wet and dry deposition of HNO_3 provides the dominant sink for nitrogen oxides in the atmosphere. As noted above, with a Henry's law coefficient of 2.1×10^5 mol l^{-1} atm (Table 18.1), we expect significant depletion of gaseous-phase HNO_3 in the presence of either cloud- or fogwater.

Equilibrium between aqueous-phase HNO_3 and the nitrate ion NO_3^- is described by

$$HNO_3(aq) \leftrightarrow H^+ + NO_3^- \quad (18.21)$$

With the equilibrium constant for (18.21) written as K_1, it follows that

$$K_1 = \frac{[H^+][NO_3^-]}{[HNO_3(aq)]} \quad (18.22)$$

The concentration of NO_3^- is equal to that of HNO_3 (aq) when the concentration of H^+ is given by K_1. At a temperature of 298 K (Seinfeld and Pandis 1997), this limit corresponds to

$$[H^+] = 15.4 \text{ mol l}^{-1}, \quad (18.23)$$

equivalent to a pH of 1.19. For concentrations of H^+ less than the limiting value given in (18.23) (pH values higher than 1.19), we expect the abundance of nitric acid in solution to be dominated by the dissociation products H^+ and NO_3^-.

Example 18.2: Suppose that the partial pressure of gaseous HNO_3 at some level in the lower atmosphere initially equals 10^{-9} atm. Assume that a cloud begins to form and that the available source of gaseous HNO_3 is fully incorporated in the liquid phase. Consider a concentration of liquid water equivalent to 10^{-6} g cm^{-3}, on the high end of the range quoted earlier for typical clouds. Assume a temperature of 298 K. Estimate the pH of the resulting solution.

Answer: As noted earlier in Section 18.1, a concentration of liquid water equal to 10^{-6} g cm^{-3} corresponds to an abundance of liquid per liter of air of identical magnitude. A partial pressure of gaseous HNO_3 of 10^{-9} atm corresponds to a molar abundance given by

$$N_{HNO_3}(\text{moles per liter of air}) = \frac{P_{NO_3}}{RT}$$
$$= \frac{10^{-9}}{(8.206 \times 10^{-2})(2.98 \times 10^2)}$$
$$= 4.08 \times 10^{-11}$$

The concentration in the liquid phase, expressed in units of moles per liter of H_2O, is given by

$$N(HNO_3, aq) = (10^6)(4.08 \times 10^{-11})$$
$$= 4.08 \times 10^{-5} \text{ mol l}^{-1}$$

Here, the factor 10^6 reflects the assumption that the abundance of H_2O in the liquid phase is equivalent to 10^{-6} liters of H_2O per liter of air. With our assumptions that gaseous HNO_3 is completely transferred to the liquid phase and that it is fully dissociated, it follows that

$$[H^+] = [NO_3^-] = 4.08 \times 10^{-5} \text{ mol l}^{-1}$$

and that the pH of the resulting solution is given by

$$pH = -\log_{10}[H^+] = 4.39$$ ■

Example 18.3: We expect gaseous and liquid phases to remain in equilibrium as HNO_3 is transferred from the gaseous to the liquid phase. For the conditions implied in Example 18.2, calculate the final equilibrium concentration of gaseous phase HNO_3 evolved in the presence of the cloud. Check the validity of the assumption made in Example 18.2, that transfer from gas to liquid is essentially complete following incorporation of HNO_3 in the liquid.

Answer: The concentration of aqueous-phase HNO_3 may be estimated using (18.22):

$$[HNO_3\ (aq)] = \frac{[H^+][NO_3^-]}{K_1}$$

Assuming $[H^+] = [NO_3^-]$, it follows that

$$[HNO_3\ (aq)] = \frac{[H^+]^2}{K_1}$$
$$= \frac{(4.08 \times 10^{-5})^2}{1.54 \times 10^1}\ \text{mol l}^{-1}$$
$$= 1.08 \times 10^{-10}\ \text{mol l}^{-1}$$

Using the value for the Henry's law coefficient for HNO_3 given in Table 18.1, we find

$$P_{HNO_3} = \frac{[HNO_3(aq)]}{H_{HNO_3}}$$
$$= \frac{1.08 \times 10^{-10}}{2.1 \times 10^5}\ \text{atm}$$
$$= 5.1 \times 10^{-16}\ \text{atm}$$

This implies a reduction in the abundance of gaseous HNO_3 by a factor of 2×10^6, from 10^{-9} to 5.1×10^{-16} atm in the presence of the cloud, verifying that transfer of HNO_3 from gaseous to liquid phase is essentially complete. ■

The lower the abundance of liquid water, the greater the impact of a given concentration of HNO_3 on the resulting acidity of the condensed phase. The following Example illustrates the conditions that may have led to the exceptionally low value of pH observed by Munger et al. (1983) for a dispersed fog in southern California.

Example 18.4: Assume an initial concentration of HNO_3 in the gaseous phase equivalent to 10^{-9} atm. Suppose that a thin, dispersed fog begins to form and that the liquid water content of the fog is initially equal to 2×10^{-9} g cm^{-3}, equivalent to 2×10^{-9} moles liquid H_2O per liter of air. This would be less by a factor of about ten than the lower limit of the range of values quoted for fogwater in Section 18.1. Calculate the pH of the resulting fogwater, assuming a complete transfer of HNO_3 from gas to liquid.

Answer: Using the analysis in Example 18.2, we see that the concentration of aqueous-phase HNO_3 is given in this case by

$$N(HNO_3, aq) = \frac{4.08 \times 10^{-11}}{2 \times 10^{-9}}\ \text{mol l}^{-1}$$
$$= 2 \times 10^{-2}\ \text{mol l}^{-1}$$

The concentration of H^+ is given by

$$[H^+] = 2 \times 10^{-2}\ \text{mol l}^{-1}$$

and

$$\text{pH} = 1.7$$ ■

The following example, adopted from Jacob (1999), illustrates the potential importance of industrial emissions of NO_x to the acidification of rain over the United States.

Example 18.5: NO_x is emitted to the atmosphere over the United States as a by-product of industrial activity at an average rate of about 1.3×10^9 moles per day. Assuming a mean precipitation rate of 2 mm per day over an area of 10^7 km^2 and neglecting the potential effect of additional acidic or basic constituents such as H_2SO_4, NH_3, or $CaCO_3$, calculate the average value for the pH of rain falling on the United States. Assume that NO_x emitted over the United States is removed from the atmosphere by precipitation over the same region.

Answer: A precipitation rate of 1 mm per day (10^{-1} cm per day) corresponds to a mass of H_2O equal to 10^{-1} g or 10^{-4} liters falling on each cm^2. This implies a total rainfall of $10^{17} \times 10^{-4}$ liters per day (1km^2 = 10^{10} cm^2), or 10^{13} liters per day. Assuming that 1 mole of H^+ is produced for each mole of NO_x, it follows that the average H^+ concentration of rain is given by

$$[H^+] = \frac{1.3 \times 10^9}{10^{13}}\ \text{mol l}^{-1}$$
$$= 1.3 \times 10^{-4}\ \text{mol l}^{-1},$$

corresponding to

$$\text{pH} = 3.9$$ ■

Oxides of nitrogen are emitted to the atmosphere primarily as NO. As previously noted, in the presence of O_3, NO is converted to NO_2. The lifetime of NO_x in the lower atmosphere is about a day. Production of HNO_3 provides the dominant sink for NO_x. Emission of NO_x is offset by a combination of dry and wet deposition of HNO_3. The lifetime of HNO_3 toward deposition in the lowest portion of the atmosphere, the boundary layer, is about five days, while the lifetime in the upper troposphere may be as long as two to three weeks (Balkanski et al. 1993). It follows that long-range transport is likely to play a role in determining the spatial pattern for the deposition of nitrate. In practice, not all of the NO_x emitted over the United States will be deposited as acid rain; a portion will return to the surface as a result of dry deposition. Also, a fraction will be exported to down-wind regions of Canada and to the Atlantic Ocean. The Example presented above provides an instructive indication, however, of the potential contribution of industrial NO_x to the acidity of rain over the United States.

18.4 The Contribution of Sulfur to the Acidity of Atmospheric Water

Sulfur dioxide (SO_2) is the most important industrial source of atmospheric sulfur. In combination with combustion-derived sources of NO_x, it is a major contributor to the phenomenon of acid rain over North America, Northern and Central Europe, and large portions of East Asia. Combustion of sulfur-rich fossil fuels and the smelting of sulfur-bearing ores (copper, lead, and zinc, for example) represent the dominant industrial sources of atmospheric SO_2.

Oxidation of SO_2 in the gaseous phase is initiated by reaction with OH:

$$SO_2 + OH + M \rightarrow HSO_3 + M \quad (18.24)$$

Subsequent reactions with O_2 and H_2O,

$$HSO_3 + O_2 \rightarrow SO_3 + HO_2 \quad (18.25)$$

and

$$SO_3 + H_2O + M \rightarrow H_2SO_4 + M, \quad (18.26)$$

result in the production of H_2SO_4. The rate-limiting step for the **gaseous-phase oxidation** of SO_2 is represented by (18.24). The lifetime of SO_2 determined by (18.24) is estimated at between one and two weeks. Observations of high concentrations of sulfate in close spatial proximity to major sources of SO_2 indicate, however, that there must exist alternate, more efficient paths for conversion of SO_2 to H_2SO_4.

It is now thought that oxidation of SO_2 takes place mainly in the aqueous phase. Oxidation in the aqueous phase can proceed either by reaction of HSO_3^- with dissolved H_2O_2 or by reaction of SO_3^{2-} with dissolved O_3. The relevant oxidation reactions are summarized by

$$\begin{aligned} &SO_2(aq) + H_2O \rightarrow HSO_3^- + H^+ \\ &H_2O_2(aq) \rightarrow H^+ + HO_2^- \\ &HSO_3^- + HO_2^- \rightarrow HSO_4^- + OH^- \quad (18.27) \\ &H^+ + OH^- \rightarrow H_2O \\ &HSO_4^- \rightarrow H^+ + SO_4^{2-} \end{aligned}$$

and

$$\begin{aligned} &SO_2(aq) + H_2O \rightarrow HSO_3^- + H^+ \\ &HSO_3^- \rightarrow SO_3^{2-} + H^+ \quad (18.28) \\ &SO_3^{2-} + O_3(aq) \rightarrow SO_4^{2-} + O_2(aq) \end{aligned}$$

The net reaction in the case of (18.27) is given by

$$SO_2(aq) + H_2O_2(aq) \rightarrow SO_4^{2-} + 2\,H^+ \quad (18.29)$$

In the case of (18.28), it is summarized by

$$SO_2(aq) + O_3(aq) + H_2O \rightarrow SO_4^{2-} + 2\,H^+ + O_2(aq) \quad (18.30)$$

Sequence (18.27) is thought to provide the dominant path for oxidation of SO_2 in the present environment. At ambient levels of H_2O_2, in the presence of a cloud, it can result in rates for conversion of SO_2 to sulfate as great as several percent per minute. This accounts for the coincidence noted above, the prevalence of rain with a low pH, and a high concentration of sulfate in regions distinguished by elevated emissions of SO_2.

Oxidation of SO_2 to H_2SO_4 involves a change in the oxidation state of sulfur, from +4 to +6. The key step in (18.27) is represented by the reaction of HSO_3^- with HO_2^-. An important feature of (18.27) is that it can proceed under acidic conditions. The concentration of HSO_3^- is regulated by the equilibrium

$$SO_2(aq) + H_2O \leftrightarrow HSO_3^- + H^+ \quad (18.31)$$

Reaction of HSO_3^- with H_2O_2, mediated by HO_2^-, provides an important sink for HSO_3^-, allowing transfer of SO_2 from the gaseous to the liquid phase to continue, sustaining a continuing high rate for conversion of SO_2 to SO_4^{2-}.

In the absence of an oxidant, the behavior of sulfur in solution would be similar to that of CO_2. Aqueous SO_2 would react with H_2O to form the bisulfide ion (HSO_3^-), maintaining the equilibrium summarized by (18.31). The bisulfide ion would dissociate to form the sulfide ion (SO_3^-), with the equilibrium described in this case by

$$HSO_3^- \leftrightarrow SO_3^{2-} + H^+ \quad (18.32)$$

The equilibrium constants K_1 and K_2 for (18.31) and (18.32) at a temperature of 298 K are equal to 1.3×10^{-2} mol l^{-1} and 6.6×10^{-8} mol l^{-1}, respectively (Seinfeld and Pandis 1997). Abundances of $SO_2(aq)$, HSO_3^- and SO_3^{2-} are presented as functions of pH in Figure 18.4 with the concentration of SO_2 in the gaseous phase taken equal to 1 ppb. Note, as indicated by the following example, that if the pH of atmospheric H_2O were determined by sulfide rather than sulfate, the pH of rain in equilibrium with a gaseous-phase concentration of SO_2 equal to 1 ppb would be equal to about 5.4.

Example 18.6: With the concentration of gaseous phase SO_2 taken equal to 1 ppb, using the Henry's law coefficient for SO_2 given in Table 18.1, calculate the value of pH for which the positive charge on H^+ is balanced primarily by the negative charge supplied by HSO_3^- and SO_3^{2-}. Assume that the concentration of

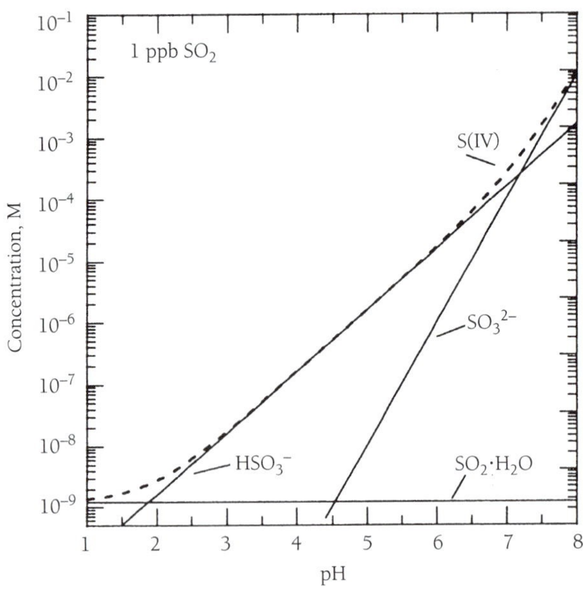

Figure 18.4 Concentrations of aqueous SO_2, HSO_3^-, and SO_3^{2-} in solution as a function of pH corresponding to a gaseous-phase concentration of SO_2 equal to 1 ppb. Source: Seinfeld and Pandis 1997.

HSO_3^- is much larger than SO_3^{2-}. Verify the validity of this assumption.

Answer: The concentration of aqueous-phase SO_2 is given by

$$[SO_2(aq)] = H_{SO_2}[SO_2(g)]$$
$$= 1.23 \times 10^{-9} \text{ mol l}^{-1}$$

The concentrations of $SO_2(aq)$, H^+, and HSO_3^- are related according to

$$K_1 = \frac{[HSO_3^-][H^+]}{[SO_2(aq)]},$$

hence

$$1.3 \times 10^{-2} = \frac{[HSO_3^-][H^+]}{1.23 \times 10^{-9}}$$

Assuming $[HSO_3^-] = [H^+]$, it follows that $[H^+]^2 = 1.6 \times 10^{-11}$ mol^2 l^{-2}. Hence, $[H^+] = 4 \times 10^{-6}$ mol l^{-1} and pH = 5.4. Concentrations of HSO_3^- and SO_3^{2-} satisfy the equilibrium expression

$$K_2 = \frac{[SO_3^{2-}][H^+]}{HSO_3^-},$$

hence

$$6.6 \times 10^{-8} = \frac{[SO_3^{2-}][H^+]}{HSO_3^-}$$

With $[H^+] = [HSO_3^-]$, the concentration of SO_3^{2-} would equal 6.6×10^{-8} mol l^{-1}, small compared to the concentration of HSO_3^- as calculated above. ■

The following Example demonstrates that, if a cloud were to form in the presence of 1 ppb gaseous SO_2, given an adequate source of oxidant, the pH of the resulting solution would be significantly less than the value obtained above.

Example 18.7: Assume that the liquid content of the cloud is equal to 10^{-6} g cm^{-3}. Assume that all of the available SO_2 is converted to SO_4^{2-} in solution. Calculate the resulting value of pH.

Answer: Following the analysis in Example 18.2, the concentration of SO_2 expressed in moles per liter of air is equal to 4.08×10^{-11}. The concentration of S dissolved in the liquid phase expressed in mol l^{-1} of H_2O is equal to 4.08×10^{-5}. Each mole of SO_2 dissolved in the liquid is responsible for production of 2 moles of H^+. Thus,

$$[H^+] = 8.16 \times 10^{-5} \text{ mol l}^{-1},$$

corresponding to

$$\text{pH} = 4.1$$ ■

In practice, we expect the concentration of sulfide resulting from an initial abundance of gaseous-phase SO_2 equal to 1 ppb to be much less than the value derived above in Example 18.6. In the presence of H_2O_2, the bulk of the initial supply of SO_2 would be converted to SO_4^{2-}. The concentration of gaseous-phase SO_2 would be lowered accordingly, with a proportional drop in the abundances of HSO_3^- and SO_3^{2-}.

18.5 Neutralization of the Acidity of Atmospheric Water in the Presence of Ammonia

As indicated in Table 18.1, equilibrium between gaseous-phase and aqueous-phase NH_3 is defined by Henry's law with a Henry's law coefficient equal to 62 mol l^{-1} atm^{-1} at a temperature of 298 K. Ammonia is converted to NH_4^+ in the aqueous phase by reaction with H_2O, establishing an equilibrium represented by

$$NH_3(aq) + H_2O \leftrightarrow NH_4^+ + OH^- \quad (18.33)$$

The equilibrium constant, K, for (18.33) has a value of 1.7×10^{-5} mol l^{-1} at a temperature of 298 K (Seinfeld and Pandis 1997). Thus,

$$K = \frac{[NH_4^+][OH^-]}{[NH_3(aq)]} \quad (18.34)$$

With

$$[OH^-] = \frac{10^{-14}}{[H^+]} \quad (18.35)$$

as given by the equilibrium between H_2O, H^+, and OH^- (equation 3.12), it follows that

$$K = \frac{10^{-14}[NH_4^+]}{[H^+][NH_3(aq)]} \quad (18.36)$$

The concentration of NH_4^+ is comparable to that of $NH_3(aq)$ when the concentration of H^+ is equal to 10^{-14} K^{-1}, meaning, for a concentration of H^+ equal to 5.9×10^{-10} mol l^{-1} corresponding to a pH of 9.2. For the range of pH values encountered in the atmosphere, which are pH values lower than 8, it is safe to assume that the bulk of ammonia included in the aqueous phase is present as NH_4^+ and that the concentration of $NH_3(aq)$ is small by comparison.

The total abundance of NH_3 in solution, $\sum NH_3$, is given by

$$\sum NH_3 = [NH_3(aq)] + [NH_4^+] \quad (18.37)$$

Using (18.36), (18.37) may be rewritten in the form

$$\sum NH_3 = [NH_3(aq)]\,(1 + 10^{14} K[H^+]) \quad (18.38)$$

or, using Henry's law, as

$$\sum NH_3 = H_{NH_3} P_{NH_3}(1 + 10^{14} K[H^+]) \quad (18.39)$$

By analogy with the earlier discussion of the equilibrium between gaseous and dissolved CO_2, summarized by equation (18.20), we can define an effective Henry's law coefficient for NH_3 by

$$H'_{NH_3} = H_{NH_3}(1 + 10^{14} K[H^+]) \quad (18.40)$$

In this case,

$$\sum NH_3 = H'_{NH_3} P_{NH_3} \quad (18.41)$$

Consider now the equilibrium between gaseous-phase and cloudwater NH_3. Suppose that the abundance of gaseous-phase NH_3 is initially equal to P_{NH_3} measured in atmospheres.

Suppose that a cloud forms and that the concentration of liquid water is equal to w g cm^{-3}. Following the approach outlined in Example 18.2, the total initial concentration of NH_3 expressed in units of moles per liter of air for a temperature of 298 K, is given by

$$N(NH_3,g) = \frac{P_{NH_3}}{RT} = \frac{P_{NH_3}}{2.45 \times 10^1} \quad (18.42)$$

The abundance in the liquid phase after equilibration, expressed in the same units (moles per liter of air) is given by

$$N(NH_3,aq) = w\, H'_{NH_3} P_{NH_3} \quad (18.43)$$

The fraction of NH_3 contained in the liquid phase, $f(NH_3,aq)$ may be written, therefore, as

$$f(NH_3,aq) = \frac{N(NH_3,aq)}{N(NH_3,aq) + N(NH_3,g)} \quad (18.44)$$

The following Examples illustrate the importance of dissolution in cloudwater as a sink for gaseous-phase NH_3 for the range of values of pH expected to arise in the atmosphere.

Example 18.8: Calculate the fraction of NH_3 expected to occur in the aqueous phase at a temperature of 298 K, assuming a concentration of cloudwater w equal to 10^{-6} g cm^{-3} with a value for pH of 5.0, in the mid range of results expected for the atmosphere.

Answer:

$$H'_{NH_3} = H_{NH_3}\left(1 + 10^{14}\,K[H^+]\right)$$
$$= 6.2 \times 10^1\,[1+(10^{14})(1.7 \times 10^{-5})\,(10^{-5})]$$
$$= 6.2 \times 10^1\,(1.7 \times 10^4)$$
$$= 1.05 \times 10^6$$
$$f(NH_3,aq) = \frac{2.45 \times 10^1 w\, H'_{NH_3}}{1 + (2.45 \times 10^1 w\, H'_{NH_3})}$$
$$= \frac{2.45 \times 10^1 (10^{-6})(1.05 \times 10^6)}{1 + \left[(2.45 \times 10^1)(10^{-6})(1.05 \times 10^6)\right]}$$
$$= \frac{2.57}{2.67}$$
$$= 0.96$$

■

Example 18.9: Repeat the calculation for the conditions defined in Example 18.8, assuming values for pH equal to 6.0 and 7.0.

Answer: With pH = 6.0,

$$H'_{NH_3} = (6.02 \times 10^1)(1.7 \times 10^3)$$
$$= 1.05 \times 10^5$$
$$f(NH_3,aq) = \frac{2.57}{3.57} = 0.72$$

With pH = 7.0,

$$H'_{NH_3} = 1.05 \times 10^4$$
$$f(NH_3,aq) = \frac{0.257}{1.257} = 0.2$$

■

It follows that cloudwater, and consequently precipitation, may be expected to provide a significant sink for gaseous NH_3. An exception to this conclusion would apply if cloudwater (and subsequent precipitation) included an alkaline material, such as $CaCO_3$, with a concentration sufficient to raise the pH of the liquid phase to a value higher than about 6.5.

Microbial activity in soils results in a significant production of NH_4^+ associated with the decomposition of organic matter; for the most part, however, this NH_4^+ is taken up by plants or it may be converted to nitrite and nitrate by the action of nitrifying bacteria (see Chapter 12). Under alkaline conditions, a portion of the ammonium produced in soils may be converted to NH_3 and released to the atmosphere, providing a regionally significant source of NH_3. Emission of NH_3 over much of the United States is dominated by sources associated with animal waste; the contribution from cattle feedlots in the Midwest is particularly significant. On a global basis, it is estimated that domestic animals are responsible for emission of about 35 million tons per year of nitrogen as NH_3(Mt N yr^{-1}). Additional contributions of about 1.5 Mt N yr^{-1} and 3 Mt N yr^{-1} are associated with humans and wild animals, respectively.[4]

18.6 Observations of the Chemistry of Precipitation

Average values for pH and for concentrations of SO_4^{2-}, NO_3^-, and NH_4^+ in precipitation over the United States are presented in Figure 18.5 (Jacob 1999). Corresponding data for Europe (Sisterson 1990) are displayed in Figure 18.6. Concentrations of SO_4^{2-}, NO_3^-, and NH_4^+ in Figure 18.5 are given in units of μ equivalents l^{-1}.[5] Concentrations of SO_4^{2-} in Figure 18.6 are expressed in units of mg S l^{-1}; concentrations of NO_3^- and NH_4^+ are given in units of mg N l^{-1}.

Note that highest concentrations of SO_4^{2-} in the United States are observed in the industrial heartland: the region of the Ohio River Valley. The highest concentrations of NO_3^- are displaced somewhat to the northeast of the maximum for SO_4^{2-}, consistent with the expectation that the lifetime for HNO_3 is longer than that for SO_2. The lowest values of pH are associated with regions of high SO_4^{2-} and NO_3^-. Concentrations of NH_4^+ are a maximum in the Midwest, consistent with the fact that the source in this case relates largely to agriculture: specifically, to domestic animals (primarily cattle).

Median concentrations of negative and positive charges (anions and cations, respectively) in precipitation falling in rural New York and southwestern Minnesota are presented in Table 18.2. Note the relatively high concentrations of NH_4^+, Na^+, K^+, Ca^{2+}, and Mg^{2+} observed in Minnesota. Agriculture is responsible for the high concentration of NH_4^+: dust from dry alkaline soils to the west is responsible for the high concentrations observed for the other major cations. The pH of rain falling on Minnesota is relatively high (6.31), whereas that over New York is comparatively low (4.34), despite the fact that concentrations of SO_4^{2-} and NO_3^- at the two sites are comparable.

As noted at the outset, the phenomenon of acid rain is not confined simply to regions subject to large emissions of indus-

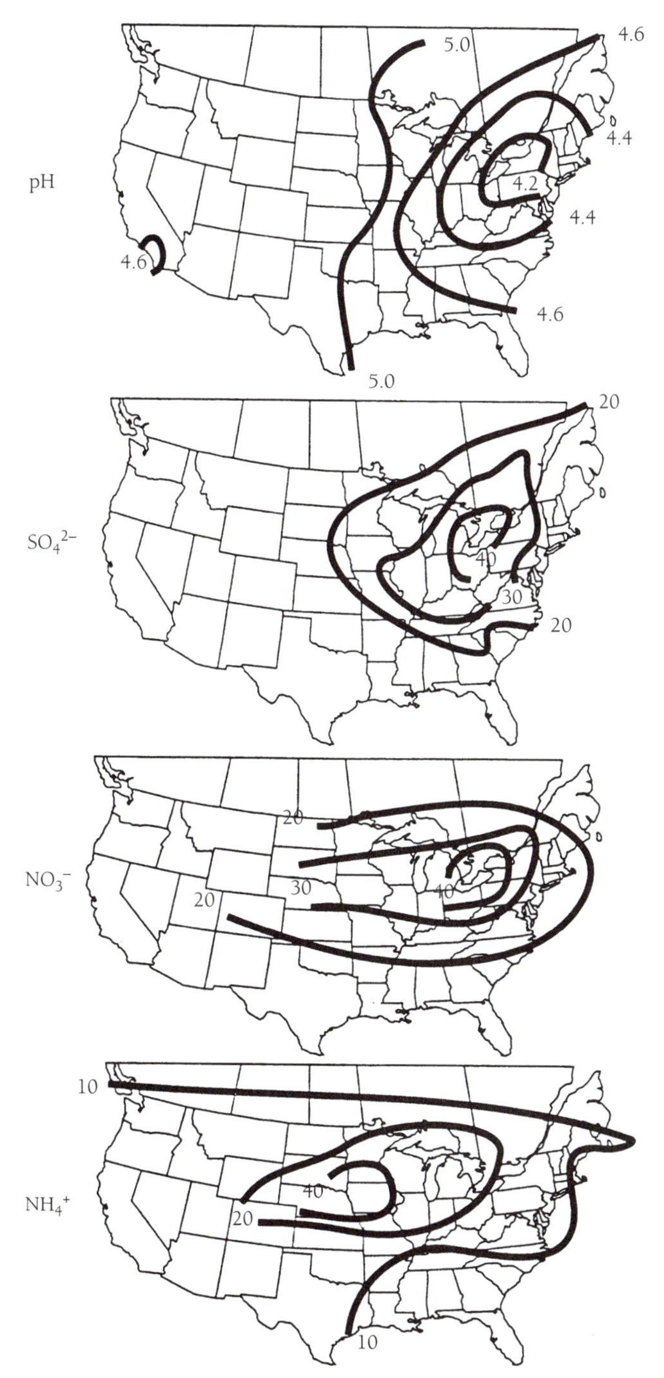

Figure 18.5 Average values for pH and for the concentrations of SO_4^{2-}, NO_3^-, and NH_4^+ (μ eq l^{-1}) as measured in precipitation over North America during the 1970s. Source: Jacob 1999.

trial products such as SO_2 and NO. Values of pH less than 5.0 are observed in environments as diverse as Amsterdam Island in the South Indian Ocean, Poker Flat, Alaska, Katherine, Australia, and Saint Georges, Bermuda, as indicated in Table 18.3. Measurements of the chemical composition of rain provide clues to the source of the excess acidity observed in these regions. Precipitation in all three environments includes high concentrations of sulfate. A portion of this sulfate is derived from sea salt; the contribution of sea salt to the S content of rain may be estimated using measurements of the abundance of other elements present in precipitation for which sea salt is a dominant source (Na and Cl, for example). The additional sulfur, indicated as excess sulfate in Table 18.3, is attributed to oxidation of chemically reduced species emitted by marine organisms. Of particular importance is $(CH_3)_2S$ (dimethyl sulfide, referred to as *DMS*) produced by phytoplankton. The global source of S associated with marine phytoplankton is estimated at about 35 Mt S yr^{-1} (see Chapter 12). Sulfur derived from these organisms is thought to provide a major source of condensation nuclei over the ocean. By influencing the abundance of condensation nuclei and thus cloud cover over the ocean, Charlson et al. (1987) suggest that phytoplankton can have a significant influence on climate; they offer the interaction between phytoplankton and climate exemplified by the capacity of the marine biota to alter the reflectivity of clouds over the ocean as an example of the type of feedback implied in the Gaia hypothesis advanced by James Lovelock (see Chapter 5).

There has been a significant decrease over the past several decades in emissions of acid precursors, such as SO_2 and to a lesser extent NO, over both North America and Europe; this has been achieved mainly as a result of government action responding to the threat of acid rain. Paradoxically, despite the success of efforts to reduce emissions of sulfur and nitrogen oxides, the pH of precipitation has remained relatively constant and uncomfortably low. Hedin and Likens (1996) attribute this bothersome observation to the fact that reductions in emissions of sulfur and nitrogen have been accompanied by simultaneous decreases in emissions of acid-neutralizing cations, such as calcium, magnesium, sodium, and potassium. Trends in concentrations of sulfate and base cations observed over the past several decades in precipitation over Sweden and the northeastern United States are displayed in Figure 18.7. The reduction in cation emissions is attributed at least in part to initiatives taken to regulate emissions of industrial sources of small particles implicated in a variety of negative impacts on public health. Steps taken to simultaneously address two distinct environmental issues (acid rain and atmospheric particulates) may have been to some extent counterproductive. They were successful in achieving a reduction in the emission of particulates and, separately, emissions of acid-forming sulfurous and nitrogenous gases. But the decrease in cations contributed by particles may have cancelled, or even reversed, the impact of the reduction in emission of acidic precursors on the acidity of precipitation. Success in dealing with the problem of acid rain may require further, more aggressive steps to regulate emissions of industrial sources of sulfur and nitrogen oxides.

18.7 Summary

The pH of rain falling in different regions of Earth varies over a wide range, from less than 4.0 to more than 6.0. In equilibrium with CO_2, rain and cloudwater would have a pH of about 5.6. Given a source of SO_2 or HNO_3, the pH of atmospheric water may be lowered considerably with respect to the limit set by CO_2 (recall that a change of pH of one unit corresponds to an adjustment in the concentration of H^+ by a factor of ten). In the presence of NH_3, or of a source of base

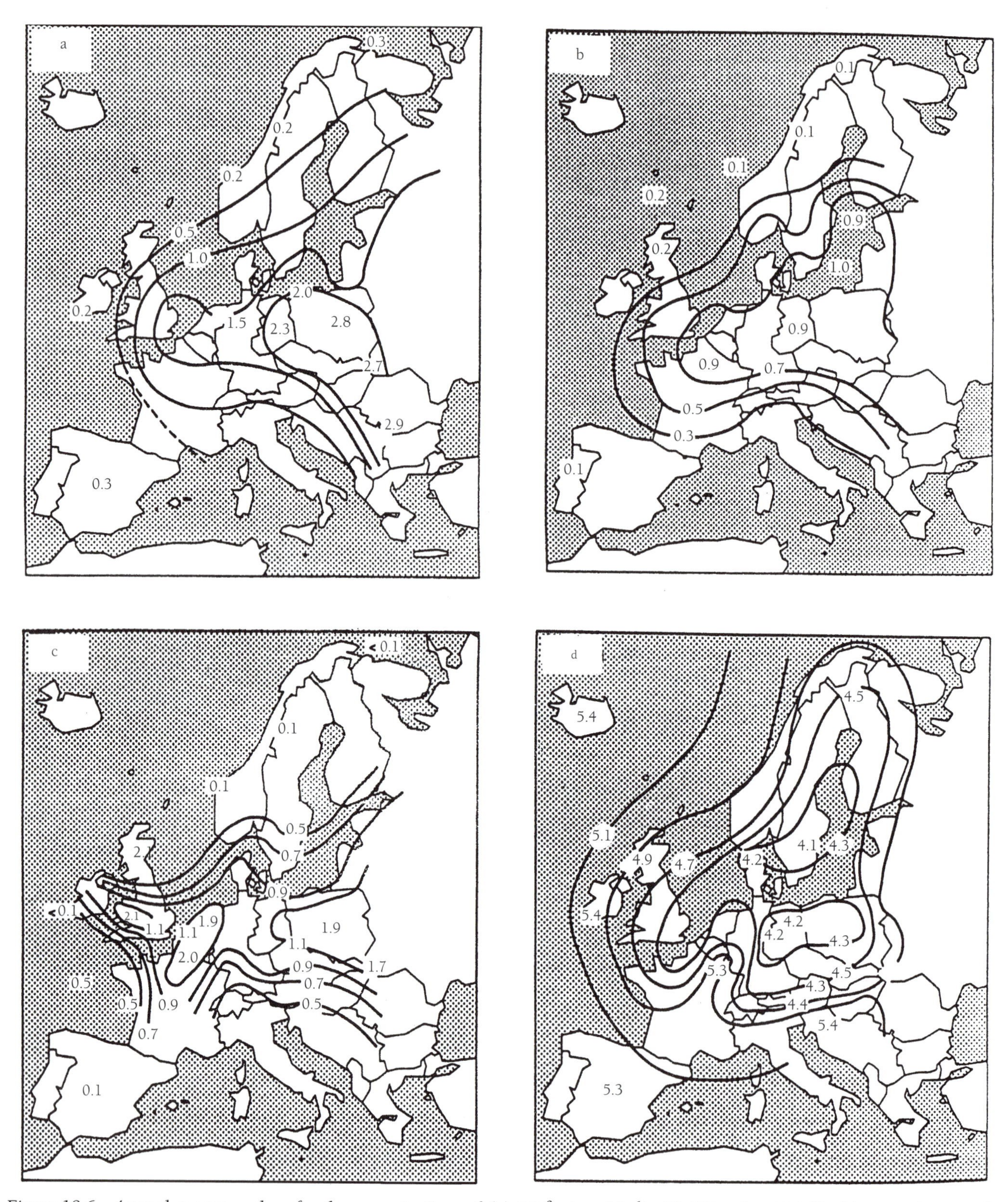

Figure 18.6 Annual average values for the concentrations of (a) SO_4^{2-} (mg S L^{-1}), (b) NO_3^- (mg N L^{-1}), (c) NH_4^+ (mg N L^{-1}), and for (d) pH for precipitation over Europe in 1985. Source: Hales 1995.

cations such as Ca^{2+} or Mg^{2+}, the pH of atmospheric water may be increased above the limit set by CO_2.

Human activity is responsible for much of the sulfate and nitrate observed in precipitation, especially in industrial regions of the Northern Hemisphere. An important source of sulfate in marine environments is supplied by oxidation to SO_2 of reduced forms of sulfur DMS emitted by phytoplankton. Oxidation of SO_2 to H_2SO_4 takes place mainly in the liquid phase. It proceeds rapidly in the presence of H_2O_2 or O_3, ensuring a relatively short lifetime (about a day) for SO_2.

Table 18.2 Median Concentration of Ions (μeq l^{-1}) in Precipitation at Two Typical Sites in the United States

Ion	*Rural New York State*	*Southwest Minnesota*
SO_4^{2-}	45	46
NO_3^-	25	24
Cl^-	4	4
HCO_3^-	0.1	10
Sum Anions	74	84
H^+ (pH)	46 (4.34)	0.5 (6.31)
NH_4^+	8.3	38
Ca^{2+}	7	29
Mg^{2+}	1.9	6
K^+	0.4	2.0
Na^+	5.0	14
Sum Cations	68	89

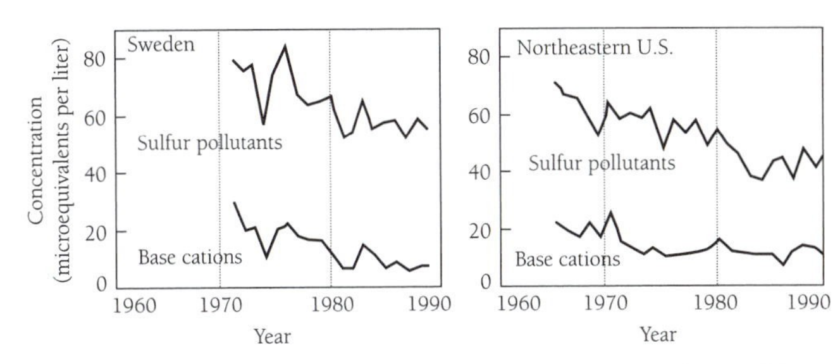

Figure 18.7 Trends in concentrations of acidic sulfur and base cations observed over the past 20 years in Sweden (left-hand panel) and the northeastern United States (right-hand panel). Source: Hedin and Likens 1996.

The lifetime of HNO_3 is longer (about five days), reflecting a slower rate for transfer from gaseous to liquid phases of HNO_3, as compared to SO_2.

We saw that waste from domestic animals and humans makes an important contribution to the budget of atmospheric NH_3. Additional sources of NH_3 are associated with applications of ammonia-based fertilizers and with conversion of NH_4^+ to NH_3 in alkaline soils.

Base cations, such as Ca^{2+} and Mg^{2+}, are supplied to the atmosphere as components of windblown dust and particles emitted as by-products of combustion and other forms of industrial activity; they are removed in the liquid phase as particulate matter is incorporated in atmospheric water. Dust enters the atmosphere both as a result of natural agents (strong winds blowing over land forms devoid of vegetation, in desert regions, for example) and as a consequence of various forms of human activity (construction and vehicles driving on unpaved roads, for instance).

We have presented evidence that the abundance of base cations in precipitation over Sweden and the northeastern United States has declined in recent years, most likely a result of steps taken to reduce emission of industrial sources of particulates. The relative lack of success of policy measures instituted to mitigate the impact of acid rain by suppressing the emission of industrial sources of sulfur and nitrogen oxides may reflect offsetting, counteracting effects associated with reductions in the emission of alkaline particulates.

We have outlined a variety of complex ecological consequences associated with acid rain: fish kills caused by input of acid water to sensitive aquatic systems; indirect effects induced

Table 18.3 Average Values for pH and for the Composition of Precipitation Observed at Selected Locations Based on 1980–1981 Data. (Source: Galloway et al. 1982)

	Location				
Species	*Amsterdam Island (Indian Ocean)*	*Poker Flat (Alaska)*	*Katherine (Australia)*	*San Carlos (Venezuela)*	*St. Georges (Bermuda)*
pH	4.92	4.96	4.78	4.81	4.79
SO_4^{2-}	52.6	10.5	8.3	3.3	48.8
ex–SO_4^{2-}[a]	11.5	10.2	6.9	3.0	21.6
NO_3^-	2.7	2.4	5.9	3.5	7.9
Cl^-	406	4.8	20.6	4.3	264
Mg^{2+}	72.8	0.5	3.6	0.8	49.3
Na^+	334	2.1	11.3	2.7	221
K^+	7.2	1.2	1.2	1.1	6.5
Ca^{2+}	15.5	0.5	5.7	0.5	14.4
NH_4^+	5.1	2.0	2.8	2.3	4.8
H^+	18.2	11.8	19.2	17.0	21.1

All concentration values are in μeqL^{-1}.

[a]*Excess sulfate calculated by subtracting the sea-salt sulfate from the total.*

by mobilization of toxic metals, such as Al; and subtle effects associated with loss from soils of essential nutrients, such as Ca and Mg. Public perception to the contrary, it appears that steps taken to date to deal with the problem of acid rain may be inadequate. There is a need for further, more aggressive initiatives to reduce the emission of acid precursors, such as SO_2 and NO_x, especially in light of the concurrent incentives to maintain, and indeed supplement, controls on emission of particulates.

Prospects for Climate Change

19

Predicting the response of the climate system to anticipated future changes in the concentration of greenhouse gases is a formidable task. The potential of a particular greenhouse gas to alter climate is usefully specified in terms of **radiative forcing**, which identifies the change in the net flux of radiative energy crossing the tropopause in response to a particular change in atmospheric composition: in calculating the magnitude of radiative forcing associated with a specific perturbation, the temperature of the stratosphere is assumed to adjust to the compositional change so as to maintain a balance between local absorption and emissions of radiative energy. The stratosphere is taken to be in what is known as **local radiative equilibrium**. The change in the flux of radiative energy at the tropopause is exactly the same, in this case, as the change at the top of the atmosphere. In calculating the magnitude of radiative forcing, we hold as fixed the temperature at the surface and the profile of temperature and moisture as a function of altitude in the troposphere. The rationale behind the radiative forcing concept is that the change in radiative flux at the tropopause should provide a measure of the energy available to bring about a rearrangement of the climatic system. Selection of the tropopause as the reference level reflects the fact that, although the stratospheric temperature can adjust rapidly (on a time scale of a month or so) to a change in composition, the response of the atmosphere's lower region is more sluggish, limited in part by inertia imposed by the high heat capacity of the ocean.

Consider, for example, the change associated with an instantaneous doubling of the concentration of CO_2. Studies suggest that the upward flux of long-wave radiation crossing the tropopause would be reduced in this case by about 4 watts m^{-2}; in the presence of a higher concentration of CO_2, the upward flux of long-wave radiation traversing the tropopause would originate from a higher, and consequently colder, level of the troposphere. The net flux of short-wave (solar) radiation intercepted by Earth below the tropopause would remain the same (recall that the radiative forcing concept assumes that there is no change in the properties of the surface and lower atmosphere). It follows that Earth below the tropopause would absorb an incremental net flux of energy equivalent to 4 watts m^{-2}. This energy would be available to warm the planet. The average temperature of Earth would increase to restore the balance between inputs and outputs of energy existing prior to the postulated change in composition. We would expect a complex rearrangement of the atmosphere, ocean, biosphere, and cryosphere (the ice world) in response to this additional input of energy.

Estimates of radiative forcing are presented for a number of important greenhouse gases in Table 19.1; results are quoted both in absolute units (watts m^{-2} ppb^{-1}) and also in units that are scaled relative to CO_2 (taken as 1.0). The data included in Table 19.1 are appropriate for changes in concentrations that are small compared to the concentrations of the gases

Table 19.1 Radiative Forcing for a Number of Representative Greenhouse Gases

Species	Chemical formula	Forcing (watts m^{-2} ppb^{-1})	Relative forcing
Carbon dioxide	CO_2	1.8×10^{-5}	1.0
Methane	CH_4	3.7×10^{-4}	2.1×10^{1}
Nitrous oxide	N_2O	3.7×10^{-3}	2.1×10^{2}
Sulfur hexafluoride	SF_6	6.4×10^{-1}	3.6×10^{4}
CFC-11	CCl_3F	2.2×10^{-1}	1.2×10^{4}
CFC-12	CCl_2F_2	2.8×10^{-1}	1.6×10^{4}
Carbon tetrachloride	CCl_4	1.0×10^{-1}	5.6×10^{3}
Methyl chloroform	CH_3CCl_3	5.0×10^{-2}	2.8×10^{3}
HALON 1301	$CBrF_3$	2.8×10^{-1}	1.6×10^{4}
HCFC-22	$CHClF_2$	1.9×10^{-1}	1.1×10^{4}
HCFC-141b	CH_3CFCl_2	1.4×10^{-1}	7.8×10^{3}
HCFC-142b	CH_3CF_2Cl	1.8×10^{-1}	1.0×10^{4}
Perfluoro-methane	CF_4	1.0×10^{-1}	5.6×10^{3}
HFC-23	CHF_3	1.8×10^{-1}	1.0×10^{4}
HFC-32	CH_2F_2	1.1×10^{-1}	6.1×10^{3}

Source: IPCC 1996.

*Table 19.2 Greenhouse Warming Potentials (GWPs) for Gases Included in Table 19.1.**

Species	Chemical formula	GWP (20 yr)	GWP (100 yr)	GWP (500 yr)
Carbon dioxide	CO_2	1.0	1.0	1.0
Methane	CH_4	56	21	6.5
Nitrous oxide	N_2O	280	310	170
Sulfur hexafluoride	SF_6	16,300	23,900	34,900
CFC-11	CCl_3F	4,900	3,800	--
CFC-12	CCl_2F_2	7,800	8,100	--
Carbon tetrachloride	CCl_4	1,900	1,400	--
Methyl chloroform	CH_3CCl_3	300	100	--
HALON 1301	CBrF3	6,100	5,400	--
HCFC-22	$CHClF_2$	4,000	1,500	--
HCFC-141b	CH_3CFCl_2	4,100	1,800	--
HCFC-142b	CH_3CF_2Cl	1,800	600	--
Perfluoro-methane	CF_4	4,400	6,500	10,000
HFC-23	CHF_3	9,100	11,700	9,800
HFC-32	CH_2F_2	2,100	650	200

**Results are presented for time horizons of 20 yrs, 100 yrs, and 500 yrs and are quoted on a scale with CO_2 set equal to 1.*
Source: IPCC 1996.

currently present in the atmosphere; for larger changes, it is necessary to account for nonlinearities in absorption, as discussed by IPPC (1990). Corrections for nonlinearity are particularly important for CO_2. Note the wide range of values for radiative forcing associated with the gases included in the Table. The efficiency of a particular gas as an agent of radiative forcing depends on the strength of the relevant absorption and the spectral region for which the absorption is most important; gases that efficiently absorb in the spectral region 8–14 μ (the window region), where the atmosphere is otherwise relatively transparent (see for example Figure 6.11b), are particularly significant as greenhouse agents. On a molecule-per-molecule basis, SF_6, for example, is 36,000 times more efficient than CO_2.

The impact on climate of the emission of a specific *quantity* of a particular gas depends not only on the magnitude of the associated radiative forcing but also on the lifetime of the gas in the atmosphere. The **global warming potential** (GWP) is defined as the cumulative value of radiative forcing expected over a specified time interval (radiative forcing integrated with respect to time) as a result of a specified emission of the gas. It provides a useful measure of the potential of emission of a particular gas to alter climate. Estimates of GWPs for gases included in Table 19.2 are given in Table 19.2. Results are presented for time horizons of 20, 100, and 500 yrs. The lifetime of SF_6, for example, is exceptionally long (3200 yr: see Chapter 5). Radiative forcing associated with SF_6 is also large. It follows that the impact of a given quantity of emission of SF_6 on the climatic system (the GWP) is especially large; for a time scale of 100 yr, it is 23,900 times greater than the impact associated with a comparable emission of CO_2, according to the data in Table 19.2. It was for this reason that SF_6 was included in the list of gases singled out for attention in the Kyoto Protocol (see Chapter 20).

The Kyoto Protocol seeks to limit emissions of five classes of chemical compounds, identified as CO_2, CH_4, N_2O, hydrofluorocarbons (HFCs), perfluorocarbons (PFCs) and SF_6. It mandates use of the GWP concept to place these gases on a common, carbon-equivalent accounting scale, assigning the charge of defining appropriate values for these coefficients to IPCC. The GWP concept thus assumes not only an important scientific function but also a critical legal role in efforts to limit potential adverse effects of future climatic change. The time scale for GWPs to be adopted in conjunction with the accounting scheme defined by the Kyoto Protocol is yet to be specified.

This chapter intends to provide a perspective on changes in climate expected as a consequence of past, present, and prospective releases of greenhouse gases associated with diverse forms of human activity. The increase in the concentration of greenhouse gases over the past several centuries (since the industrial revolution) is estimated to have contributed a net (positive) increase in radiative forcing equivalent at present to about 2.5 watts m^{-2}, of which about 64% is attributed to past increases in CO_2 (IPCC 1996). As previously discussed (Chapter 10), the globally averaged temperature at Earth's surface has increased by about 0.7°C over the past 150 years; more than half of this increase occurred over the past two decades. The decade of the 1990s was the warmest on record; many scientists believe that the warming trend is likely to continue, or even accelerate, in the future.

Peter D. Ewins, Chief Executive Officer of the U.K. Meteorological Office, and D. James Baker, Undersecretary of the U.S. National Oceanic and Atmospheric Administration, in an unusual joint letter to the British *Independent* newspaper published on 23 December 1999, stated:

> "The rapid rate of warming since 1976, approximately 0.2 degrees C per decade, is consistent with the projected rate of warming based on human-induced effects. In fact, scientists now say that they cannot explain this unusual warmth without including the effects of human-generated greenhouse gases and aerosols."

They concluded:

> "Our agencies are doing their part to provide the best possible data, understanding and forecasts for policy-makers as they deal with these difficult issues. Ignoring climate change will surely be the most costly of all possible choices, for us and our children."

Contrast this view with remarks by John M. Wallace, chair of a panel of the National Research Council of the U.S. National Academy of Sciences that issued a recent report on climatic change (NRC 2000). In a press release dated 12 January 2000, accompanying publication of the report, he commented:

> "The rapid increase in the Earth's surface temperature over the past 20 years is not necessarily representative of how the atmosphere is responding to long-term, human-induced changes, such as increasing amounts of carbon dioxide and other greenhouse gases. The nations of the world should develop an improved climate monitoring system to resolve uncertainties in the data and provide policy-makers with the best available information."

How can we account for the difference between the conclusions of Ewins and Baker and those of Wallace, given their common starting point that the past several decades have been the warmest on record? The reaction of Ewins and Baker may have been influenced by results of coupled atmosphere-ocean models purporting to provide a reasonable simulation of the changes in climate observed over the past 150 years; these models attribute recent warming primarily to the buildup of greenhouse gases, offset to a modest extent by cooling due to reflective sulfate aerosols (see Chapter 10). As discussed below, these models forecast dramatically warmer climates for the decades ahead. Wallace's view may have reflected a more skeptical assessment of the capacity of models to account for the complexity of the climatic system, a more cautionary appraisal of their ability either to simulate the past or predict the future.

This chapter intends to offer a perspective on the complexity of the task associated with climate modeling. To provide a reliable simulation of the past or a credible prediction for the future, models must account not only for changes in the state of the atmosphere but also for changes in the properties of the ocean, biosphere, and cryosphere. We discuss the challenge in Section 19.1. We present and discuss results from one of the more complete climate models, the high-resolution coupled ocean-atmosphere model developed at the Hadley Centre of the U.K. Meteorological Office, in Section 19.2. We conclude with personal views as to the gravity of the threat posed by the potential for future climate change, and with thoughts as to how we might respond in the absence of a detailed, credible, long-range forecast.

19.1 The Challenge for Models

A comprehensive model for our climate must account for vertical, latitudinal, and longitudinal transport of heat by both the atmosphere and ocean; it must allow for an exchange of heat, moisture, and momentum between the atmosphere and the surface over both oceans and continents; it must accommodate the heating and cooling that results from the interaction of the atmosphere and surface with both short- and long-wave radiation; it must account for the absorption and release of energy due to phase changes of water; it must allow for variations of all of these quantities arising as a result of changes in atmospheric composition.

As noted above, an increase in CO_2 will result in an immediate decrease in emission of long-wavelength radiation to space. This may be expected to trigger a complex adjustment in temperatures at the surface and in the lower atmosphere, prompting changes in the circulation of the atmosphere accompanied by changes in patterns of evaporation, precipitation, cloud cover, and the distribution of water vapor. Variations in cloud cover and water vapor will induce feedbacks leading to further adjustments in the interaction of the atmosphere and surface with the radiation field, with implications for the hydrological cycle and for the circulation of both the atmosphere and ocean. Changes in surface temperature, cloudiness, and precipitation may lead to changes in vegetation, soil moisture and water runoff, with implications for exchange of heat, moisture, and momentum. The impact of a postulated change in radiative forcing on climate may be complicated by subsequent changes in surface snow and ice cover: the composite effect will depend on a complex interplay of all of these multifaceted disturbances. Diagnosing the ultimate result for climate requires a realistic simulation of the coupled, interactive dynamics of the atmosphere, ocean, biosphere, soil, hydrosphere, and cryosphere—no small task.

State-of-the-art models of the climatic system typically involve coupled descriptions of the 3-dimensional dynamics (motions) of the atmosphere and ocean, combined with modules simulating land surface processes and changes in sea ice. The atmospheric component of the model seeks to develop numerical solutions to the complex set of differential equations describing conservation of mass, momentum, energy, and the various phases of water. Pressure, density, and temperature are constrained to satisfy the appropriate equation of state. For computational purposes, derivatives must be expressed in algebraic form, as functions of variables evaluated at predetermined spatial-grid points over preselected intervals of time.[1] To evaluate the derivative of a function y with respect to x, for example, we must write

$$\frac{dy}{dx} = \frac{y(x + \Delta x) - y(x)}{\Delta x} \tag{19.1}$$

Recall that this representation is valid strictly only for the limit where $\Delta x \rightarrow 0$.[2] The accuracy of a numerical model and the credibility of a particular simulation of climate thus depends on the resolution of the model, the spacing of the grid points at which the model seeks to define values for the key properties of the climate. The resolution of a suite of 16 coupled atmosphere-ocean models discussed by IPCC (1996) ranged from a high of 2.5 degrees latitude by 3.8 degrees longitude to a low of about 4 degrees latitude by 5 degrees longitude. The results presented below (Section 19.2) were obtained using the most recent version of the Hadley Centre model: this model has a resolution of 1.25 degrees latitude by 1.25 degrees longitude.

A resolution of 1.25 degrees implies a spacing between grid points of about 140 km. It is clear that there are important processes taking place in the atmosphere on spatial scales much smaller than this. With a resolution as course as 140 km (and note that this is a best case: the resolution of most models is much worse), models are incapable of resolving changes in atmospheric circulation associated with the movement of air across major mountain ranges. They are unable to account for rapid (spatial and temporal) variations in horizontal and vertical wind speeds associated with storms, nor the complex physical processes involved in the formation of clouds and precipitation, nor the sharp gradients in the properties of the atmosphere associated with weather fronts and the height distribution of water vapor. Small-scale processes play a critical role in regulating the transport and spatial redistribution of heat, mass, moisture, and momentum in the atmosphere; they must be taken into account if a model is to provide a reasonably accurate representation of climate. They are customarily incorporated in models using a procedure known as **sub-grid scale parameterization**, predicated on an assumption that transport on scales below the resolution of the model may be defined in terms of properties of the atmosphere projected at model grid points.

Different models use different schemes to accomplish this objective. In the absence of a rigorous physical basis for parameterization, the procedure is unavoidably subjective. A necessary condition is that the model should provide a reasonably accurate representation of the present climate: this condition is not in itself sufficient, however, to ensure the validity of the parameterization scheme. It is possible for models to provide a credible account of the gross features of present climate (seasonal variations of surface temperature, for example) and to accomplish this objective with quite different representations of atmospheric circulation, differences in the simulation of the relative importance of transport by eddies and mean motion, for example (Stone and Risbey 1990). These differences may be attributed at least in part to variances in the treatment of sub-grid scale processes. Under such circumstances, it is clear that the predictive capacities of the models must be viewed with caution.

As previously noted, the rise in global average surface temperature predicted by models as a result of the increase in the concentration of greenhouse gases over the past 150 years is significantly larger than the actually observed increase; the discrepancy is usually attributed to the failure of models to allow for the negative radiative forcing (cooling) associated with enhanced reflection of sunlight by sulfate aerosols formed as by-products of industrial emissions of SO_2. As discussed in Chapter 18, oxidation of SO_2 leads to production of H_2SO_4, mainly in the liquid phase. Sulfate aerosols are formed as a result of the subsequent evaporation of cloudwater. Models tend to focus on what is referred to as the *direct* effect of sulfate aerosols, the function of these relatively light-colored (high albedo) particles in reflecting sunlight; it is more difficult to account for their *indirect* impact, their role as condensation nuclei and their consequent effect on precipitation and on the optical properties of clouds. In the presence of a large number of condensation nuclei, it is likely that a given quantity of condensable water will be distributed over a larger number of cloud droplets. This has implications for the average size of resulting cloud particles, their reflective properties, and the probability that individual drops can grow to sizes large enough to precipitate. A more comprehensive analysis of the impact of anthropogenic emissions of sulfur should allow not only for direct but also for indirect effects of sulfate aerosols on the atmosphere's radiation budget. It should account for the impact of other pollutants, the effect of absorptive soot particles, for example. Negative radiative forcing due to sulfur aerosols could be offset, at least partially, by positive forcing associated with soot.

As noted in Chapter 6 of IPCC (1996), "the ability of a model to reproduce the past record does not necessarily imply that its climate sensitivity or the assumed radiative forcing is correct." The focus on global average temperatures as a gauge of model success may obscure difficulties associated with the ability of models to provide a realistic simulation of climate on more regional scales. As suggested in Chapter 10, the Northern Hemispheric warming associated with the increase in greenhouse gases may be offset to some extent by cooling induced by sulfate aerosols. It is doubtful, however, that an allowance for the effect of sulfate aerosols can do much to improve agreement between models and observations for the Southern Hemisphere. Under the circumstances, agreement between the models and the observed values of globally averaged surface temperatures may provide testimony less to the intrinsic credibility of the models than to the importance of the additional flexibility introduced by the challenge to account for the cooling attributed to sulfate aerosols.[3]

Problems are also associated with simulations of oceanic dynamics. The ocean's circulation is driven in part by stresses imposed at the surface by atmospheric motions, in part by exchanges of sensible heat between the ocean and atmosphere, imbalances between evaporation and precipitation affecting the salinity and heat content of surface waters, and changes in density and thermal properties of the ocean associated with the formation and melting of sea ice. A reliable description of forcing imposed at the ocean's surface by the atmosphere is a prerequisite to a satisfactory simulation of oceanic dynamics. For obvious reasons, our understanding of the ocean is significantly less complete than our understanding of the atmosphere. There are a variety of factors contributing to this state of affairs. Observational data are more limited. The treatment of oceanic dynamics is complicated by problems associated with the complexity of the geometry and bathymetry of ocean basins.[4] As for the atmosphere, small-scale motions play a crucial role in regulating the transport of a number of key dynamical properties of the ocean (notably density and heat content). They are critical, for example, in determining the structure and location of the thermocline, the transition region separating cold water in the depths from relatively warm water at the surface (see Chapter 9). As was the case for the atmosphere, a parameterization approach is required to account for the role of subgrid scale transport in ocean models. It is difficult to gauge the success of a particular scheme in accomplishing this objective. It is clear, though, that even the best of ocean models employed in the current generation of climate studies are deficient. They fail to account for a number of important features of the present climate: the El Niño phenomenon, for example. Under the circumstances, their ability to simulate the response of the ocean to changes in forcing at the surface induced by varying levels of greenhouse gases must be viewed with caution.

The land surface module of a climate model is designed to simulate processes involved in the partitioning of precipitation among evaporation, local storage (as either snow or soil moisture), and runoff. It must additionally account for the conversion of net absorbed radiation to either latent or sensible heat. The early generation of surface models adopted what is referred to as a *bucket* approach to the treatment of soil moisture (Manabe 1969). They assumed that soils had a fixed, limited capacity to store water; when input of water from precipitation exceeded this limit, the "bucket" was assumed to be full, and excess water was allowed to runoff (meaning the bucket was permitted to overflow). Recent models allow for increasingly more complex simulations of soils and plants, accounting for variations in vegetative cover and for progressively more realistic treatments of the response of the land surface to inputs of water and energy (Sellers 1992; Dickinson, Henderson-Sellars, and Kennedy 1993; IPCC 1996). It is difficult, however, to assess the ability of these models to simulate the impact of potential future changes in climate on the function of ecosystems in specific regions and on the manner in which these systems might evolve in response to anticipated changes in their environment.[5]

Formation of sea ice, by concentrating water and excluding salt, has an important influence on the density, and thus on the buoyancy, of near-surface ocean water. Ice formed in one region may be transported over large distances in the ocean, melting elsewhere and accounting for important regional scale changes in density. The presence of sea ice acts as an insulating blanket, inhibiting the release of heat and moisture from the ocean. Evaporation and loss of heat from small-scale breaks in sea ice (or *leads*) can result in a rapid cooling of surface waters, with associated increases in salinity and density contributing to intense local convection.[6] As indicated by IPCC (1996), the treatment of sea ice in models can have an important influence on predictions of models for the response of climate to an increase in greenhouse gases, especially for high latitudes over marine environments during winter. Evidence for a decrease in the spatial extent and thickness of ice in the Arctic Ocean in recent years suggests that this is an issue worthy of further attention.

Assuming that all of the aforementioned difficulties are surmounted, there are additional problems to be confronted in developing a credible model for the response of climate to projected increases in the concentration of greenhouse gases; there is a question, for example, as to how the model should be initialized. In principle, it should be possible to begin with a relatively arbitrary configuration of the atmosphere, ocean, and land surface and allow the combined system to evolve as determined by the governing equations. Beginning with a reasonably accurate specification of the initial state, one might hope that the system would settle down over time to a realistic, consistent representation of its interactive components. The model could be allowed to then evolve in response to a specific, imposed change in radiative forcing. The problem with this approach involves the relatively long time required for the ocean module to relax to a consistent steady state, for the ocean to lose track of transients associated with inconsistencies in the imposed initial conditions. Time scales for the atmosphere are on the order of weeks: time scales for the ocean, on the other hand, may vary from decades, to centuries or longer. If we wish to model the atmosphere and ocean as a coupled system, the time scale for integration of the relevant equations will be set by restrictions imposed by the most rapidly changing component of the system: in this case, the atmosphere. Computational times required for the system to relax to a consistent steady state may be prohibitively long.[7]

Faced with these difficulties, most models adopt a pragmatic approach to the definition of the initial state. Boundary conditions (surface temperatures) are selected for the ocean on the basis of observational data. The atmospheric module is run as an independent unit to obtain a seasonally varying description of the atmosphere consistent with the imposed surface boundary conditions. Fluxes of heat, momentum, and water vapor are evaluated at the ocean's surface using output from the atmospheric model. Under ideal circumstances, fluxes derived by this approach may be used to drive the ocean model; in practice, the circulation of the ocean resulting from the use of unadjusted model fluxes often turns out to be unrealistic. The problem is attributed to a combination of factors, in particular to inaccuracies in the choice of surface temperatures, compounded by errors intrinsic to the atmospheric model (associated, for example, with the treatment of clouds). The conventional approach seeks to fix the problem by adjusting values of surface fluxes, to obtain a more realistic representation of the ocean. If the adjustments are small, one might hope that the errors introduced in subsequent simulations of the impact of changing concentrations of greenhouse gases and other agents of radiative forcing should be acceptable: if the adjustments are large, as is often the case (IPCC 1996), it is difficult to assess the significance of uncertainties introduced by the **flux adjustment procedure**. The consistency between atmospheric and oceanic results obtained with the Hadley Centre model (summarized below) was such that it was unnecessary to resort to the problematic option of flux adjustment in this case.

In attempting to forecast the impact of changing concentrations of greenhouse gases and sulfate aerosols on our climate, it is important to allow for the effect of forcing associated with past emissions. The climatic system today, we would expect, is still in the course of adjustment to perturbations in the radiative budget that developed in the past. The surface and lower atmosphere are colder now than would be the case if the system were in equilibrium with the current concentration of greenhouse gases. The flux of infrared radiation to space is thus less than the net flux of short-wave radiation absorbed from the Sun. The departure from equilibrium represented by this imbalance reflects the importance of the relatively large thermal inertia imposed by the ocean—the role of the ocean in slowing the rate of increase in surface temperature associated with past radiative forcing. Models seeking to simulate future climatic changes arising as a result of the buildup of greenhouse gases will tend to underestimate the magnitude of these changes if they fail to account for the imbalance in the contemporary radiative budget; failure to allow for the imbalance is likely to result in a significant underestimation of the rise in globally averaged surface temperatures expected for any particular period in the future, by as much as 0.4°C after 50 yrs, according to Hasselmann et al. (1993). The difficulty can be avoided, or at least minimized, if the model is initialized for a time prior to onset of the recent buildup of greenhouse gases.

19.2 Results from the Hadley Centre Model

A variety of coupled atmosphere-ocean general circulation models have been used to study the response of climate to a continuing buildup in the concentration of greenhouse gases. A summary of results from 16 models (23 model runs) available in 1996 is presented in IPCC (1996). We choose here to highlight results from the most recent version of the Hadley Centre model; they are taken from a report prepared under the auspices of U.K. Department of the Environment, Transport, and the Regions. The report, which includes a discussion of the impacts of climatic change on vegetation, water resources, food supply, coastal communities, and human health, was presented to delegates to the Fourth Conference of the Parties to the U.N. Framework Convention on Climate Change (Buenos Aires, November 1998).

The climate model on which the study is based incorporates a number of refinements over previous versions of the Hadley Centre model. The model was initialized for the period prior to the buildup of greenhouse gases. It accounts for changes in climate since 1850, allowing for effects of both greenhouse gases and sulfate aerosols, and it allows for an improvement in resolution, permitting fluxes computed at the sea surface to be applied without correction. Although problems remain, we expect that the model should provide at least an instructive guide to the type of changes in climate that may develop in the future, offering an opportunity to assess their potential significance.

Plate 7 (top) (see color insert) presents a summary of trends in globally averaged surface temperatures predicted for the period 1860–2100. It includes a comparison with observational data for the period 1860–1998. As previously noted, models accounting solely for the effect of greenhouse gases tend to overestimate the temperature increase observed since 1860. Agreement is improved if they allow for negative forcing associated with sulfate aerosols, as indicated by a comparison of the results summarized by the lower of the two model curves in Plate 7 (top) with the observational data. The model used to obtain these results accounted for both direct and indirect effects of sulfate aerosols. The label on the Plate, however, is misleading: the study did not attempt to provide an inclusive simulation of the impact of all possible anthropogenic influences on climate. It was restricted rather to an assessment of the effects of greenhouse gases and sulfate aerosols. The agreement between the sulfate/greenhouse gas model and observation is not perfect. It would be surprising, indeed, if it were. We would expect the climate system to exhibit a measure of essentially random (stochastic) variability; the model should exhibit similar behavior. The stochastic variability (noise) of the model is illustrated by the results for the control run presented in Plate 7 (bottom) (see color insert). The model fails to account for the increase in global-mean surface temperature observed between 1910–1940. If this rise is attributed to natural variability of the climate, it should be noted that its amplitude, about 0.4°C, is larger than the amplitude of variability associated with the noise exhibited in the control run of the model. It could reflect a response to an increase in output of energy from the Sun. On the other hand, deficiencies in simulation of oceanic circulation could result in a serious underestimate by the model of the stochastic variability of the real climate under unperturbed, natural conditions.[8]

As indicated in Plate 7 (top), the model predicts an increase in global average surface temperature by about 2°C over the next 50 years, climbing to about 4°C by the end of the century. The increase in surface temperatures expected over continental regions is even greater, as illustrated in Plate 7 (bottom): about 4°C by 2050, rising to close to 6°C by 2100. The increase in sea surface temperature is more modest: 2°C by 2050, a little more than 3°C by 2100. The difference in the response of land and ocean environments to forcing imposed by the projected increase in the concentration of greenhouse gases reflects the influence of the greater heat capacity of the ocean. The results presented here refer to the so-called business-as-usual (unconstrained) scenario defined by IPCC (1992). According to this scenario, anthropogenic emissions of CO_2 are projected to grow from their current level of about 7×10^9 tons C yr^{-1} to close to 35×10^9 tons C yr^{-1} by 2100. The concentration of CO_2 would be expected to more than double over this period, rising to more than 800 ppm from the present level of about 360 ppm (IPCC 1995).

The geographic distribution of the increase in surface temperatures projected for Northern Hemispheric winter in 2050 is illustrated in Plate 8 (top) (see color insert). Associated changes in precipitation are presented in Plate 8 (bottom) (see color insert). Increases in temperature are greatest for high latitudes during winter, in excess of 8°C in the Arctic and over portions of northern Canada. Warming over the North Atlantic is more modest: less than 1°C, reflecting a decrease in the strength of the thermohaline circulation attributed to a decline in salinity resulting from an increase in precipitation and runoff.[9] More surprising are the changes predicted for the tropics. The results in Plate 8 imply significant warming (more than 4°C) over portions of Brazil, accompanied by a marked decrease in precipitation (in excess of 2 mm day^{-1}). The deficit in precipitation extends across the tropical Atlantic to Africa, reaching into southern portions of Europe and the Middle East. Decreases in precipitation are projected also for portions of South and South East Asia, offset by significant increases in rainfall over the western tropical Pacific.

Rainfall in the tropics today is concentrated over three regions: Indonesia, Brazil, and Africa; these environments are associated with net upward motion of the atmosphere. This upward motion is supplied by inflow of air at the surface balanced by outflow aloft. Upward motion is restricted to a relatively small region of the tropics, for the most part to the environments mentioned above. Vertical motion in these regions may be locally intense; it is balanced by relatively gentle descent over a much larger area elsewhere. We may think of the convective regions over Indonesia, Brazil, and Africa as a set of interconnected, west–east aligned fountains. The fountains spout air rather than water, and they burp irregularly. By far the most vigorous of the three fountains is the one centered over Indonesia. Exchange of air between the fountains defines a pattern of zonal motion known as the **Walker circulation**. The Walker circulation combines with the Hadley circulation (see Chapter 8) to establish the overall pattern of motion of air in the tropics.

The results in Plate 8 suggest that the function of the Walker circulation may be very different in the future if we fail to arrest current trends in emissions of CO_2 and the other greenhouse gases. The increase in precipitation predicted for the western tropical Pacific combined with the decrease over Brazil and Africa would appear to imply a more dominant role for the fountain near Indonesia. According to the model, upward motion of air in the tropics may be concentrated in the future to a large extent in the region near Indonesia. Upward motion may be replaced by predominantly downward motion over Brazil and Africa. Rainfall over northern Brazil may decline by as much as 500 mm per year. Soils are projected to dry out, temperatures to rise, and rain forests to be replaced by savanna. Elsewhere, in India, Africa, and in portions of North America, tropical grasslands may be transformed to either temperate grasslands or desert. Changes to ecosystems could be extreme with consequences for human societies that are difficult to estimate but potentially very serious, especially for populations lacking the economic resources required for an efficacious response.

The projected changes are not, of course, all negative. Warmer conditions and an extended growing season could lead to an improvement in crop yields for countries at mid and high latitudes. Warmer winters would reduce demands for heating fuel. Benefits would be offset, however, by increased demands for fuel for cooling during summer. A decrease in the areal extent of sea ice in the Arctic could open up a new maritime route for trade between Europe and East Asia, the long-sought Northwest Passage. Decreases in snowfall would ease the difficulties associated with winter travel while reducing, or eliminating, expenses for snow removal.

The consequences of a projected transition from snow to rain as the dominant form of winter precipitation at high latitude are not all positive, however. If precipitation were to fall primarily as rain during winter, we might expect a large fraction of this precipitation to be lost relatively immediately to runoff. If precipitation occurs as snow, as it does in interior regions of continents at high latitudes during winter today, most of the moisture will be retained. A blanket of snow on the ground during winter keeps soils relatively warm, allowing moisture to seep into the ground to feed local aquifers. Melting snow during spring and summer provides a steady, reliable source of water for streams and rivers. A change from snow to rain as the dominant form of winter precipitation could have far reaching consequences for ecosystems, stream flows, and water resources. It is difficult to assess the overall impact, but there is little doubt that it could be serious.

19.3 Summary

The climatic system is extremely complex. We have outlined here the difficulties involved in building a credible model. Even the best of the current generation of models running on the fastest supercomputers are deficient. In the face of uncertainty, how should we respond to the challenge posed by increasing concentrations of greenhouse gases and the threat of potentially serious related changes in climate?

There is no doubt that we are changing the composition of the atmosphere. Concentrations of CO_2 are higher now than at any time over the past 450 thousand years (see

Chapter 5). There is no doubt that consumption of fossil fuels is largely responsible for the contemporary increase. The concentration of CO_2 has risen over the past 200 years, the period of industrial development, from about 280 ppm to close to 360 ppm. Even under the most optimistic scenarios envisaged by IPCC (1992), concentrations are expected to climb to values in excess of 600 ppm, more than twice the level that persisted for more than 10,000 years since the end of the last ice age. Within the lifetime of most of the people alive today, we may expect to see concentrations approaching levels not seen since dinosaurs roamed Earth 60 million years ago. Vestiges of carbon added to the atmosphere by mining and burning fossil fuels will persist for hundreds of thousands of years. Contemporary transfer of carbon from fossil sedimentary reservoirs to the atmosphere constitutes what the late Roger Revelle described as man's first great inadvertent global geophysical experiment.

Human influence on the composition of the atmosphere is not confined to CO_2. As previously seen (Chapter 5), concentrations of CH_4 and N_2O have increased over the past several decades to levels not seen in recent geological history. We are adding to the atmosphere gases such as SF_6 and the industrial halogens, for which there are no natural analogues. If we are unable to accurately forecast the impact of these emissions, should prudence not dictate a cautionary response? While models differ in detail in their projections of the changes in climate expected to arise as a result of a continuing increase in the concentration of greenhouse gases, they agree in one important respect: the changes could be significant. Faced with uncertainty, does it not make sense to buy insurance, to take steps at least to slow the pace of current increases in emissions? Uncertainty is a two-edged sword. We should admit the possibility that disruptions of the climatic system could be more rather than less severe than predicted by the current generation of models. Our survey of past climates (Chapter 10) indicates that global scale changes in climate can develop rapidly. It is worth noting in this context that none of the current generation of models can account for the rapidity of the large-scale rearrangement of the climate system that marked the end of the Younger Dryas cold period 12,000 years ago.

As discussed above, the past decades have been unusually warm. Global average surface temperatures set records for each of sixteen consecutive months between May 1997 and September 1998 (Karl, Knight, and Baker 2000). The increase in temperatures observed over the past 20 years, equivalent to about 2°C per century, is comparable to the increase predicted for the twenty-first century under the IPCC (1992) business-as-usual scenario for greenhouse gases. Skeptics may interpret the recent trend as a statistical aberration. A more responsible reaction might be to consider it a possible harbinger of times to come, to anticipate consequences for the future, and plan accordingly.

Policy Responses to Climate Change

20

We have had occasion frequently in this text to refer to the work of the Intergovernmental Panel on Climate Change (IPCC). The Panel was established in 1988 under the auspices of the World Meteorological Organization (WMO) and the United Nations Environment Program (UNEP). It was charged with assessing the state of climate science, advising on the likelihood that human activities could lead to significant changes in climate, evaluating the impacts of such changes, and identifying options for possible policy responses. The first report of the Panel was published in May 1990. The Executive Summary began with a declarative statement that "there is a natural greenhouse effect which already keeps the Earth warmer than it would otherwise be"; that "emissions resulting from human activities are substantially increasing the atmospheric concentrations of the greenhouse gases carbon dioxide, methane, chlorofluorocarbons and nitrous oxide"; and that "these increases will enhance the greenhouse effect, resulting in additional warming of the Earth's surface." One hundred and seventy scientists from 25 countries contributed to the report, which had an important influence on deliberations at the Second World Climate Conference that met a few months later in Geneva. It prompted a Ministerial Declaration issued at the Conference that underscored the potential significance of the climate issue for policy and was responsible, at least indirectly, for inclusion of the climate issue on the agenda for the United Nations Conference on Environment and Development (UNCED) that met two years later in Rio de Janeiro (in June 1992).

The Rio meeting, known popularly as the Earth Summit, drew an astonishing 25,000 delegates. Never before had a single occasion attracted such a large portion of the world's political leaders (including Presidents and Prime Ministers). It focused unprecedented attention on the climate issue. The important conclusions of the Summit were summarized in a document formally titled "The United Nations Framework Convention on Climate Change." The Convention recognized from the outset "that the largest share of historical and current global emissions of greenhouse gases has originated in developed countries, that per capita emissions in developing countries are still relatively low and that the share of global emissions originating in developing countries will grow to meet their social and development needs." It noted that "low-lying and other small island countries, countries with low-lying coastal, arid and semi-arid areas or areas liable to floods, drought and desertification, and developing countries with fragile mountainous ecosystems are particularly vulnerable to the adverse effects of climate change." It identified as a key objective (Article 2) "to achieve . . . stabilization of greenhouse gas concentrations in the atmosphere at a level that would prevent dangerous anthropogenic interference with the climate system." It went on to elaborate that "such a level should be achieved within a time-frame sufficient to allow ecosystems to adapt naturally to climate change, to ensure that food production is not threatened and to enable economic development to proceed in a sustainable manner." It decreed (in Article 3) that "the Parties should protect the climate system for the benefit of present and future generations of humankind, on the basis of equity and in accordance with their common but differentiated responsibilities and respective capabilities," adding that "developed country Parties should take the lead in combating climate change and the adverse effects thereof." It obligated a group of developed countries and countries from the former Soviet economic zone, identified as "Annex 1 countries," to institute measures "with the aim of returning individually or jointly to their 1990 levels these emissions of carbon dioxide and other greenhouse gases not controlled by the Montreal Protocol." It defined requirements for Parties to report on steps taken to comply with the objectives of the Convention, requiring that such reporting include estimates for "projected anthropogenic emissions of carbon dioxide and other greenhouse gases not controlled by the Montreal Protocol." It decreed that a "Conference of the Parties, as the supreme body of this Convention, shall keep under regular review the implementation of the Convention and any related legal instruments that the Conference of the Parties may adopt, and shall make, within its mandate, the decisions necessary to promote the effective implementation of the Convention." It required that the first session of the Conference of the Parties "take place not later than one year after the date of entry into force of the Convention" and that "thereafter, ordinary sessions of the Conference of the Parties shall be held every year unless otherwise decided by the Conference of the Parties." It invited "any body or agency, whether national or international, governmental or non-governmental, which is qualified in matters covered by the Convention, and which has informed the secretariat of its wish to be represented at a session of the Conference of the Parties as an observer" to be admitted "unless at least one third of the Parties present object." The Convention was to "enter into force on the ninetieth day after the date of deposit of the fiftieth instrument of ratification, acceptance, approval or accession." This milestone was passed on 21 March 1994 when Portugal became the fiftieth country to register ratification on 21 December 1993. As of 10 December 1999, the Convention had been ratified by 181 countries.

The first meeting of the Conference of the Parties (COP-1) took place in Berlin (28 March to 7 April 1995). It agreed in the so-called Berlin Mandate on the need to strengthen actions to be taken by developed countries and to extend the time horizon for action beyond the year-2000 target highlighted by the Convention. It established a number of support organizations to assist the work of the Convention, the Subsidiary Body for Implementation, and the Subsidiary Body for Scientific and Technological Advice. It defined a process, known as *joint implementation* (JI), by which Parties could cooperate on a voluntary basis to develop projects that could contribute to a reduction in the net emission of greenhouse gases and set the stage for the second Conference of the Parties (COP-2) that met in Geneva (8 July to 19 July 1996).

COP-2 endorsed the state of the science as summarized by IPCC and called for legally binding commitments to reduce the growth of greenhouse gases. It established the agenda for the critical third Conference of the Parties (COP-3) in Kyoto (1 December to 10 December 1997). COP-3 concluded with the adoption of the Kyoto Protocol, arguably the most complex initiative ever formulated to address an issue of international concern.

In advance of the Kyoto meeting, the United States was prepared to commit to a Protocol that would require emissions of greenhouse gases over the period 2008–2012 to not exceed levels that applied in 1990. The European Union was ready to argue for a more ambitious target, a reduction of 15% with respect to the 1990 baseline. According to the European strategy, Europe's commitment would be distributed unequally among countries of the Union, with Germany and the United Kingdom assuming the lion's share of the European obligation. Germany would agree to reduce by 25%, the United Kingdom by 20%, with less stringent limitations assumed by other countries—Portugal, for example, would be permitted to grow by 40%. In the opinion of U.S. policy makers, the 1990 baseline artificially favored the European position, in that it allowed Germany to benefit from the elimination of inefficient, energy-intensive industry in East Germany—a one-time boon following integration—and the United Kingdom to take advantage of the demise of the coal industry orchestrated in the late 1980s by Margaret Thatcher. As it turned out, in a compromise worked out in last hours of the Kyoto conference, Europe agreed to reduce by 8% whereas the United States accepted a reduction of 7%. Annex 1 countries as a whole were committed to decrease emissions by 5%.

We discuss key elements of the Kyoto Protocol in Section 20.1. The Protocol is specific in some areas, vague and even ambiguous in others. Section 20.2 offers suggestions as to how the Protocol might be enhanced in the future to improve prospects of meeting the overall objective of the Convention: to stabilize greenhouse gases at a level "that would prevent dangerous anthropogenic interference with the climate system." We present concluding summary remarks in Section 20.3.

20.1 The Kyoto Protocol

The Kyoto Protocol is designed to curb growth in the concentration of greenhouse gases. It targets four gases (CO_2, CH_4, N_2O, and SF_6) and two classes of compounds (hydrofluorocarbons and perfluorocarbons), accepting that a number of additional greenhouse agents were earlier regulated under measures taken to deal with the threat to stratospheric ozone (as addressed in the Montreal Protocol and subsequent amendments). As previously discussed (Chapter 19), it adopts an accounting scheme based on the GWP concept to place all of the targeted gases on a common, carbon-equivalent scale. It recognizes that sources of regulated compounds may be offset by purposeful manipulation of sinks. It imposes requirements on Annex 1

Parties to be met by a "commitment period" identified as 2008–2012. "Assigned amounts" (specific commitments) are defined relative to emissions in 1990. Overall, it seeks to reduce emissions by Annex 1 Parties by at least 5% by 2008–2012 relative to 1990. A list of Annex 1 Parties with a summary of emissions estimated for the individual Parties in 1990 is given in Table 20.1. Emissions are presented in terms of mass of CO_2: to convert to carbon mass, divide by 3.7 (44/12). Note that according to this tabulation, the United States is responsible for 36.1% of all emissions assessed to Annex 1 Parties. The Russian Federation occupies second place (17.4%), followed by Japan (8.5%), Germany (7.4%), and the United Kingdom (4.3%).

Table 20.1 Emission of Greenhouse Gases by Annex 1 Parties in 1990.

Party	*Emissions (Gg)*[a]	*Percentage*[b]
Australia	288,965	2.1
Austria	59,200	0.4
Belgium	113,405	0.8
Bulgaria	82,990	0.6
Canada	457,441	3.3
Czech Republic	169,514	1.2
Denmark	52,100	0.4
Estonia	37,797	0.3
Finland	53,900	0.4
France	366,536	2.7
Germany	1,012,443	7.4
Greece	82,100	7.4
Hungary	71,673	0.5
Iceland	2,172	0.0
Ireland	30,719	0.2
Italy	428,941	3.1
Japan	1,173,360	8.5
Latvia	22,976	0.2
Liechtenstein	208	0.0
Luxembourg	11,343	0.1
Monaco	71	0.0
Netherlands	167,600	1.2
New Zealand	25,530	0.2
Norway	35,533	0.3
Poland	414,930	3.0
Portugal	42,148	0.3
Romania	171,103	1.2
Russian Federation	2,388,720	17.4
Slovakia	58,278	0.4
Spain	260,654	1.9
Sweden	61,256	0.4
Switzerland	43,600	0.3
United Kingdom and Northern Ireland	584,078	4.3
United States of America	4,957,022	36.1
Total	**13,728,306**	**100.0**

[a] *Results are expressed in terms of CO_2 equivalent mass (1 Gg = 10^9 grams = 10^3 tons). To convert to units of carbon, divided by 3.7.*

[b] *Contributions by individual Parties expressed as a percent of the total emission by Annex 1 Parties.*

The Protocol is ambiguous on the matter of sinks. Article 3 states that "net changes in greenhouse gas emissions by sources and sinks resulting from direct human-induced land-use change and forestry activities, limited to afforestation, reforestation and deforestation since 1990, measured as verifiable changes in carbon stocks in each commitment period, shall be used to meet the commitments under this Article of each Party included in Annex 1." It is unclear whether the limitation implied by the phrase following "forestry activities" applies exclusively to activities associated with conversion of land that was never forested to forest (afforestation), return of land that was previously forested to forest (reforestation), and to conversion of land previously forested to other uses (deforestation). Does it account for changes in the carbon content of soils associated with changes in forest cover? Does it permit a more inclusive interpretation of "land-use change"? Would it allow credit, for example, for management practices designed to enhance storage of carbon in agricultural soils? It is important that these ambiguities be resolved.

Annex 1 Parties are required to "provide, for consideration by the Subsidiary Body for Scientific and Technological Advice, data to establish its level of carbon stocks in 1990 and to enable an estimate to be made of its changes in carbon stocks in subsequent years." It is unclear how this task is to be accomplished. Present understanding of quantities of carbon stored in soils and above ground biomass is uncertain to at least a factor of two. Is it realistic to expect to develop a "verifiable" estimate of changes in these stocks at the level of precision required to assess compliance of Parties obligated under the Protocol? Commitments of Annex 1 Parties range from a reduction in net greenhouse gas emissions of 8% for Parties in the European Union to an increase of 10% by Iceland over the period 1990 to 2008–2012. Obligations of individual Parties are summarized in Table 20.2.

The Protocol includes a number of so-called *flexibility mechanisms* by which Parties can satisfy at least part of their obligations by cooperative arrangements with other Parties. Article 6 authorizes Parties in Annex 1 to "transfer to, or acquire from, any other such Party emission reduction units resulting from projects aimed at reducing anthropogenic emissions by sources, or enhancing anthropogenic removals by sinks of greenhouse gases in any sector of the economy, provided that . . . any such project provides a reduction in emissions by sources, or an enhancement of removals by sinks, that is additional to any that would otherwise occur." This provision, referred to, again, as joint implementation, is restricted to transfers between Annex 1 Parties. Article 12 allows Parties in Annex 1 to benefit from projects initiated in countries of Parties not included in Annex 1, so long as these projects result in "certified emission reductions." This second option, known as the *clean development mechanism* (CDM), offers an important opportunity for developing

Table 20.2 Obligations of Individual Annex 1 Parties Expressed as Percent Changes Relative to the Base Year, 1990 in Most Cases.

Party	*Quantified-emission limitation or reduction commitment (percentage of base-year period)*[a]
Australia	108
Austria	92
Belgium	92
Bulgaria*	92
Canada	94
Croatia*	95
Czech Republic*	92
Denmark	92
Estonia*	92
European Community	92
Finland	92
France	92
Germany	92
Greece	92
Hungary*	94
Iceland	110
Ireland	92
Italy	92
Japan	94
Latvia*	92
Liechtenstein	92
Lithuania*	92
Luxembourg	92
Monaco	92
Netherlands	92
New Zealand	100
Norway	101
Poland*	94
Portugal	92
Romania*	92
Russian Federation*	100
Slovakia*	92
Slovenia*	92
Spain	92
Sweden	92
Switzerland	92
Ukraine*	100
United Kingdom and Northern Ireland	92
United States of America	93

[a]*Commitments refer to emissions of greenhouse gases for the period (2008–2012). Sources may be offset by sinks according to procedures defined by the Protocol.*

**Countries undergoing the transition to a market economy.*

countries to acquire investments and technologies that could enhance prospects for environmentally sustainable growth. It specifies that reductions in emissions must be "additional to any that would occur in the absence of the certified project activity." Activities under CDM may involve "private and/or public entities . . . subject to whatever guidance may be provided by the executive board of the clean development mechanism" subject to "the authority and guidance of the Conference of the Parties serving as the meeting of the Parties to this Protocol." Certified emission reductions "obtained during the period from the year 2000 up to the beginning of the first commitment period can be used to assist in achieving compliance in the first commitment period."

Details of how JI and CDM projects could be initiated and certified remain to be worked out. A subsequent meeting of the Conference of the Parties was authorized to "elaborate modalities and procedures with the objective of ensuring transparency, efficiency and accountability through independent auditing and verification of project activities." A critical issue is to determine how projects resulting in reductions in emissions "additional to any that would otherwise occur" should be certified. The problem is particularly serious for CDM, in that countries benefiting from projects are not themselves subject to any defined limits on emissions. If a public or private Party representing interests in an Annex 1 country chooses to invest in, say, a wind-powered power plant in China, how can we be sure that such an investment would not have occurred in the absence of the CDM option? It is easier to see how the JI option can be implemented. A JI project involves an *exchange* of obligations between Parties, both of which are committed to specific *reductions* in emissions. Under no circumstances can it result in a net *increase* in emissions of Annex 1 Parties taken as a whole. In contrast, CDM allows Annex 1 Parties to achieve a net *reduction* in domestic obligations. There is no guarantee that Parties benefiting from CDM transactions will achieve reductions in emissions sufficient to offset the decrease in obligations for domestic action by Annex 1 Parties entering into such arrangements. CDM investments could even be counterproductive: they could provide a stimulus for economic growth in developing countries, thus contributing to a net *increase* in global emissions.

Article 17 envisages an arrangement whereby Annex 1 Parties may buy and sell rights to emissions of greenhouse gases. The assumption is that the existence of a market for trading would allow Parties as a group to pursue an economically more efficient path to reduce emissions compared to what would be the case if they were required to act independently. If the cost to reduce emissions in country A is higher than that for country B, it might make sense for country A to satisfy its obligations by subsidizing a reduction in emissions in country B. The Conference of the Parties is charged with defining the "relevant principles, modalities, rules and guidelines, in particular for verification, reporting and accountability for emissions trading." The Article states that "any such trading shall be supplemental to domestic actions for the purpose of meeting quantified emission limitation and reduction commitments under that Article." Notably, it fails to define what is meant by "supplemental."

The trading option is controversial. If Annex 1 Parties were allocated emission rights consistent with obligations defined by the Protocol, it is clear that there would be a unidirectional flow of economic resources toward countries for which these obligations are least stringent. The Russian Federation and Ukraine can satisfy their commitments under the Protocol if emissions in 2008–2012 are at or below the level that applied in 1990, a challenge that is unlikely to be demanding, given the economic problems that have beset these nations since 1990. If Russia and Ukraine were to sell the emissions for which they have no need to countries whose obligations under the Kyoto Protocol are more restrictive, it is doubtful that this would accomplish the objectives of the Protocol. Countries purchasing the emission rights could emit more. Russia and Ukraine would emit what they would have independent of the arrangement. Financial resources would be transferred with minimum benefit to the environment. The trading option, inserted in the Protocol largely at the initiative of the United States, has been referred to derisively by the European Union as a *license to trade hot air*.

To enter into force, the Kyoto Protocol must be ratified by a minimum of 55 Parties to the Convention. It must be ratified by a sufficient number of Annex 1 Parties so as to obligate at least 55% of emissions assigned to Annex 1 Parties in 1990. It would be difficult for the Protocol to go into effect without the participation of either or both of the United States and the Russian Federation, which together account for 53.5% of 1990 emissions by Annex 1 Parties (see Table 20.2). It would be difficult for it to go into effect without participation by the United States and Japan, which together account for 44.6% of emissions. Prospects for successful ratification of the Protocol clearly depend on the outlook for ratification in the United States.

20.2 Prospects for Enhancement of the Protocol

It is most unlikely that the United States can unilaterally meet the commitment to which it would be obligated under the Kyoto Protocol. Emissions of carbon dioxide in the United States are now almost 10% higher than they were in 1990. A reduction of 7% by 2010 relative to 1990 would require a decrease of close to 2% per year over the next decade. Aggressive public policy initiatives—a major carbon tax or a system of tradable permits, for example—would be required to accomplish this objective, and there is no indication of a political will supporting such bold steps in the United States at this time. Further, it is doubtful that such steps could be instituted at a level required to be effective without doing serious damage to the economy. The United States could meet its obligation by a combination of limited domestic actions encouraging conservation and by taking advantage of the flexibility mechanisms included in the Protocol. It could make a financial arrangement to purchase emission rights not exercised by Russia and the Ukraine. As discussed above, however, such a step would do little to accomplish the objectives of the Protocol. It could take advantage of CDM, but the CDM mechanism has yet to be defined, and as yet there is little enthusiasm for the concept in large developing countries, such as China or India.

It is also unclear how Europe will meet its commitment. There are signs that Germany, the largest European emitter, is beginning to balk. The choice of 1990 as the base year undoubtedly works to Europe's advantage. Had the base year been selected as 1995, Europe's difficulties in meeting its commitment would be just as severe as for the United States.

We need a longer-term strategy to address the climate issue; even if all of the Annex 1 countries were to meet their obligations under Kyoto, emissions of CO_2 will increase to a level in excess of 7×10^9 tons C yr^{-1} by 2010. If emissions were to stabilize at this value for the rest of the century, concentrations of CO_2 would continue to increase, eventually climbing to levels in excess of 750 ppm, more than twice the level applicable today. By 2010, approximately half of global emissions may be attributed to sources in developing countries (more if we allow for emissions associated with deforestation in the tropics: see Chapter 11). China is now the second-largest emitter and certain to be the first before long if its current rate of growth is sustained. If we are to successfully address the climate issue, it is critical that we engage at least the largest developing countries in meaningful participation. In addition to China, India (currently ranked number 5), and Mexico (number 11), we must strive to involve Brazil and Indonesia, the countries most responsible for emissions associated with deforestation. As indicated earlier, deforestation contributes a source of CO_2 today equal to about 30% of the global source contributed by combustion of fossil fuels.

It would be helpful to extend the time horizon envisaged by the Kyoto Protocol while at the same time stiffening requirements. It might be useful to focus on cumulative emissions rather than on emissions for any particular date, recognizing that the level of CO_2 is determined by cumulative emissions over time, rather than by emissions for any particular year. Extending the time horizon to, say, 2030 would permit countries to institute economically more efficient long-term strategies allowing a more orderly transition to less carbon-intensive technologies. It would be useful to encourage cooperation between countries with closely linked economies, with incentives to exploit efficiencies available for joint implementation. This cooperation could be linked to regional systems for trading emission rights. The countries involved in the North American Free Trade agreement (NAFTA)—Canada, the United States, and Mexico—could constitute such a cooperative group and the arrangement could be extended later to include other countries of Central and South America or to join up with similar arrangements elsewhere. It is important to develop equitable formulas to engage developing countries. Commitments for developing countries could be linked to benefits accruing as a result of the expansion in global trade anticipated with the World Trade Organization (WTO). This could help break the current impasse, where developing countries argue for commitments linked to per capita emissions while developed countries argue for commitments based on historical emissions. There would be advantages for countries entering into cooperative agreements to reduce emissions of greenhouse gases. We might expect economies of such countries to lead in development of energy-efficient technologies with ancillary benefits for exploitation of these technologies in countries that failed to take such initiative. The advantages for the United States are clear. We could reduce our dependence on foreign oil with benefits not only for the environment but also for our economy and our national security.

20.3 Concluding Remarks

While uncertainties remain and there is still work to be done, the weight of the evidence suggests that the risk of serious climate change is real. The climate is warmer now than at any time in the past 150 years (Chapter 10). Arguably, it is warmer than at any time in the past 3000 years if data on the isotopic composition of ice at high altitudes in the tropical Andes are taken as a proxy for global climate (Thompson et al. 1998). The industrial revolution was energized by fossil fuels; coal, oil, and more recently, natural gas. People died in Donora, Pennsylvania, and London before we were moved to address the health impacts of carbon monoxide, sulfur oxides, and particulates. Fish died in Scandinavian lakes before we were driven to focus on the problems of acid rain (Chapter 18). People grew sick in Los Angeles before we were motivated to confront the hazards of elevated levels of tropospheric ozone (Chapter 17). Common to all these problems is our reliance on fossil fuels. Expensive technological solutions can be devised to reduce emissions of carbon monoxide, hydrocarbons, nitrogen and sulfur oxides, and particulates. It is more difficult to deal with the end product of fossil fuel use: carbon dioxide.

Levels of carbon dioxide in the atmosphere are higher now than at any time over the past 450,000 years (Chapter 5). Human activity is responsible for globally significant changes in the biogeochemical cycles not only of carbon but also of nitrogen, sulfur, and phosphorus (Chapters 11 and 12). Changes in the composition of the atmosphere attest to the global significance of these changes. Having reaped the benefits of the industrial revolution, the challenge we now face is to adjust to the consequences, to chart an environmentally more conservative path to a sustainable future. Those who benefited most from the industrial revolution, countries in the developed world, have a responsibility to ease the adjustment for those who confronted the problems more recently, the countries of the developing world.

We have made remarkable progress over the past half-century in unraveling the complex physical, chemical, and biological forces that have shaped the history of our planet. Our species is a late arrival on the stage of life. As the only species that has evolved so far with the capacity to think, and to contemplate the wonders of nature, we have a unique responsibility. We have developed the technology to reshape the face of Earth. But we also have evolved the intellectual resources to assess the consequences of our actions, to serve as wise stewards, to safeguard the integrity of the planet's life support system, and to preserve its wonderful diversity. We stand at a critical juncture. Future generations will judge us poorly if we fail to meet the challenge.

Notes

Chapter 5

1. The lifetime of CH_4 is relatively long, about 7 years (see Chapter 17). It follows that changes in the concentration of CH_4 observed for any given location at any given time may be taken as an indication of changes expected to occur at more or less the same time in other environments. Measurements of CH_4 derived from analysis of air trapped in ice cores recovered from different regions, Greenland and Antarctica for example, can be used to establish a common time line or chronology for these cores. Measurements of the isotopic composition of ice in a particular core can be independently employed to obtain a record of changes in local climate. Combining these measurements using the common time line inferred from analyses of CH_4, we can develop a sense of how changes of climate in one region are linked to changes in another (see Chapter 10).

Chapter 6

1. Measurements of ^{14}C concentration in organic material provide a useful means to estimate its age. A plant, growing by photosynthesis, incorporates CO_2 in proportion to its abundance in air. Subsequent decay of ^{14}C is associated with a decrease in the concentration of ^{14}C, and, since the rate of decay is known, this process provides a means to date the age of the sample. The technique was recently applied to date the age of the shroud of Turin. Measurements indicated that the shroud could not have been used to wrap the body of Jesus; it was constructed with linen formed from CO_2 in the atmosphere about 1000 years ago.

Chapter 7

1. The expression for the work done by a gas as it expands can be understood by considering a spherical volume of gas undergoing expansion, as illustrated in Figure 7.7. Suppose that the external pressure is uniform over the sphere, equal to p both before and after expansion. The radius of the sphere increases from R to $R + \Delta R$, where ΔR is assumed to be *small* compared to R. The work done by the expanding gas is equal to the product of the force on the sphere and the distance that the external air is displaced (see Chapter 2):

$$\begin{aligned} \text{Work} &= \text{Force} \times \text{Distance} \\ &= p \times \text{Area} \times \Delta R \\ &= p(4\pi R^2)\,\Delta R \end{aligned} \tag{7n.1}$$

The change in volume of the sphere (ΔV) is given by the difference between the final and initial volumes:

$$\Delta V = \frac{4\pi(R+\Delta R)^3}{3} - \frac{4\pi(R)^3}{3} \tag{7n.2}$$

For small values of ΔR, we get

$$\begin{aligned} (R+\Delta R)^3 &= R^3 + 3R^2\Delta R + 3R(\Delta R)^2 + (\Delta R)^3 \\ &\approx R^3 + 3R^2\Delta R, \end{aligned} \tag{7n.3}$$

and substituting (7n.3) into (7n.2) gives

$$\Delta V \approx 4\pi R^2 \Delta R \qquad (7n.4)$$

The same result may be obtained by differentiating the expression for volume $\left((4/3)\pi R^3\right)$ with respect to R. Substituting (7n.4) into (7n.1) gives the desired relation

$$\text{Work} = p\Delta V \qquad (7n.5)$$

2. Equation (7.27) may be understood by noting that for changes in the pressure and volume of a gas, denoted by Δp and ΔV, respectively, the change in the product pV is given by the difference between the final value of the product $[(p + \Delta p)(V + \Delta V)]$ and the initial value of the product (pV). Mathematically,

$$\Delta(pV) = (p + \Delta p)(V + \Delta V) - pV \qquad (7n.6)$$

The first term on the right-hand side of (7n.6) may be expanded to give

$$(p + \Delta p)(V + \Delta V)$$
$$= pV + (\Delta p)V + p(\Delta V) + (\Delta p)(\Delta V) \qquad (7n.7)$$

For *small* values of Δp and ΔV with respect to the initial magnitude of p and V, respectively, the value of the last term in (7n.7) $[(\Delta p)(\Delta V)]$ will be much less than the other terms, and can be neglected. Ignoring the last term in (7n.7), and substituting (7n.7) into (7n.6), yields equation (7.27). Those familiar with calculus will recognize that equation (7.27) is equivalent to

$$\Delta(pV) = (\Delta p)V + p(\Delta V) \qquad (7n.8)$$

where $\Delta(pV)$, Δp, and ΔV are the differentials of pV, p, and V, respectively.

3. The concept of **buoyancy** can be understood by examining the density of a warm air parcel and that of the surrounding atmosphere. Density (ρ) is given by the perfect gas law ($p = \rho RT$), equation (7.7). Consider two equal volumes of air, the first denoting ambient conditions, with temperature $T_{AMBIENT}$, and the second corresponding to conditions experienced by a rising air parcel having temperature T_{PARCEL} (Figure 7.9). The pressure force exerted on the rising parcel by the external air is the same as that on the representative volume of ambient air. The ambient air is assumed to be at rest; the force of pressure is equal in magnitude, opposite in direction, to the force of gravity. If T_{PARCEL} is warmer than $T_{AMBIENT}$, the density of the rising air parcel will be less than of the ambient air. The downward force of gravity experienced by the rising air parcel is consequently less than that applied to the ambient air, and the imbalance of forces (pressure and gravity) results in a net upward force on the rising air parcel. Under such circumstances, the rising air parcel will continue to rise; it is said to be **buoyant** with respect to its environment. For a rising air parcel whose temperature is less than that of the surrounding air, the imbalance of forces (pressure and gravity) leads to a suppression of the initial upward velocity. For this reason, the atmosphere is very stable with respect to vertical motion when the ambient temperature rises with increasing altitude, a condition known as a **temperature inversion**.

Chapter 8

1. Meteorologists define an easterly wind as one that blows *from* the east. An eastward wind blows *toward* the east.

2. "Climate" refers to the long-term average state of the atmosphere. Prevailing westerly winds at midlatitudes may be considered in this sense as a property of climate. Weather reflects the more transient state of the atmosphere. A low pressure system off the New England coast can result in a strong northeasterly flow for a short period of time and, under some conditions, can be associated with major precipitation. We refer to phenomena such as this as *weather.*

3. Suppose that region B extends over a distance X. The spatial dimension of region A is then $3X$. The average speed, $\bar{v}$, for the combination of A and B is obtained by weighting the speed in the separate regions by their extent and dividing by the combined scale of A plus B:

$$\bar{v} = \frac{(3X)(3) + X(-1)}{4X} = 2\text{m sec}^{-1}$$

4. "Sensible heat" denotes the heat content of air, excluding contributions arising due to phase changes of H_2O. Sensible heat per unit mass is proportional to temperature, given by C_vT. "Potential energy" denotes the energy contained in the air by virtue of its position with respect to the surface. The potential energy of unit mass at z is taken as gz and is identical to the potential energy discussed earlier for a projectile. It is equivalent to the work expended against the force of gravity in moving air from the surface to height z. Rising or sinking motion is associated with a respective decrease or increase in the kinetic energy of the atmosphere.

5. Small variations in the length of a day are indeed observed over the course of a year. The length of the day increases by about 7×10^{-3} sec between July and January, reflecting net super-rotation of the atmosphere during northern winters associated with an observed increase in the strength of the west-east zonal winds. Measurements of the length of the day have been derived by analyzing signals obtained by bouncing laser light off a reflecting surface placed on the moon by Apollo astronauts.

6. The torque associated with mountains depends on the difference in pressure from the west to east sides of the mountain chain. A net torque requires a persistent difference in the west-east pressure gradient. The direction and magnitude of the pressure gradient can change with the season and is normally associated with large-scale features of regional geography, a contrast between land and sea, for example.

Chapter 9

1. Oceanographers measure volume transports of water in units named in honor of the oceanographer Otto Sverdrup (1855–1930). A Sverdrup unit, denoted by 1 Sv, is equal to 10^{12} cm^3 sec^{-1} or 10^6 m^3 sec^{-1}.

2. Surface waters warm up during summer and convection is suppressed in the ocean. If winds are weak, the surface's mixed layer can be very thin. You can experience this effect while swimming in a sheltered lake during summer. Floating at the surface may be very comfortable. Dip a little below, and you may be shocked to find that the temperature has dropped by ten degrees or more.

3. The freezing temperature depends on the salt content of the water. It drops by almost 2°C as salinities increase from near

zero (appropriate for fresh water) to values observed in the ocean, which range from about 34–37 parts ‰ (per thousand).

4. It is thought that heat is transported downward across the thermocline by eddies, by vertical exchange of warm and cold water arising as a consequence of instabilities generated by shears in the flow of water at different depths in the ocean. The downward flux of heat is assumed to occur at a rate proportional to the vertical gradient of temperature, meaning energy is transferred from warm to cold regions of the ocean. Downward diffusive transport of heat must be balanced by a bulk upward motion of cold water. Henry Stommel, author of the first quantitative theory of the thermohaline circulation, and his colleague A. B. Arons estimated values for the magnitude of the upward velocity of water in the thermocline region in the range 2–12 m yr^{-1}. Subsequent analyses of the abundance and distribution of ^{14}C by M. Stuiver and associates indicated velocities of 4, 5, and 12 m yr^{-1} for the Atlantic, Pacific, and Indian Oceans, respectively.

5. Accumulation of sea ice is thought to play an important role in the production of deep water in the Southern Ocean. There are two candidate regions for formation of deep water near Antarctica. Water salinities on the continental shelf, notably in the Weddell and Ross Seas, can be enhanced both by production of sea ice (the ice is formed largely of fresh water leading as a consequence to enhancement of salt in the residual sea water) and additionally by floating ice as it is driven out to sea by off-shore winds. Dense cold waters from these shallow seas, with temperatures close to the freezing point (–1.8°C), spill over the shelf break, plunging deep into the surrounding ocean. They mix with waters in the circumpolar current, providing a source of some of the coldest and densest waters of the world's oceans. Deep water is also formed in the open ocean as a result of production of sea ice during winter. Deep water formed off Antarctica has a salinity of about 34.7 parts per thousand, as compared with salinities of about 34 parts per thousand in surface water.

6. The Younger Dryas cold period persisted for about 1000 years and abruptly ended (over a time interval of only a few decades) 10,750 years before the present (10.75 kyr B.P.). It was preceded by an earlier interval, the Bolling Allerod, lasting about 1500 years, when the climate, at least in the North Atlantic, was anomalously warm.

7. Biological production is limited over much of the ocean by the supply of essential nutrients, such as nitrate and phosphate. These nutrients are taken up by organisms in surface layers of the ocean as fast as they arrive in the photic zone. They are efficiently recycled and returned to the deep as organisms die and as their body parts fall out of the ocean surface's mixed region. Reflecting the efficiency of biological uptake, the concentration of nutrients is generally low in surface waters (high latitudes and winter conditions, where productivity may be limited by light, provide an exception to this general rule). Most of the ocean's store of nutrients resides in the deep sea. The overall biological productivity of the ocean is regulated by the intensity of upwelling.

8. Public awareness of the El Niño phenomenon has markedly increased in recent years, largely as a result of news accounts documenting the spectacular course of the El Niño of 1997–1998. The El Niño of 1997–1998 was associated with unusual weather in many parts of the world: extreme droughts and out-of-control fires in Indonesia; persistent storms with unseasonably heavy rain accompanied by serious beach erosion in southern California; exceptionally mild winter weather in the Northeastern United States; tornadoes in Florida; and a series of major winter storms with hurricane strength winds in Western Europe. The El Niño of 1997–1998 heralded an important success for forecasters who predicted its occurrence almost a year in advance, using computer models for the coupled motions of the atmosphere and ocean and taking advantage of a concentrated array of observation systems in the tropical Pacific. There are intriguing hints that the dynamics of the tropical ocean may have changed significantly over the past several decades. It appears that the warm phase may now be more prevalent than the cold phase, in contrast to earlier years, when the warm and cold phases were equally probable. Whether this signals a long-term shift in climate, or merely a statistical anomaly, is unclear. A transition to a more persistent warm period could be a response to a change in ocean circulation, an indication of a reduction in the flow of cold water into the tropics from midlatitudes or a decrease in the strength of the Conveyor Belt, with an associated buildup of warm water—a deepening of the thermocline—in the tropics. This possibility merits continuing attention. The El Niño of 1997–1998 could be a harbinger of climates to come, an early indication of a subtle but important shift induced by the build up of greenhouse gases.

Chapter 10

1. The point here is the that, if we are trying to describe the climate today or how it may have changed over centuries, or even millennia, in the past, we can take the position of the continents for granted. If we want to investigate what the climate was like 100 million years ago, we need to appreciate that the configuration of continents and ocean basins was very different then. In general, we expect different arrangements of the lithosphere to be associated with different states of the climate system. Changes in climate driven by changes in the lithosphere are likely, however, to evolve slowly over millions of years.

2. This should come as no great surprise. Greenhouse gases act as insulating agents, providing a barrier to the transfer of energy into space. If we add insulation to our homes and burn oil at the same historical rate, we expect the interior to warm up. It would be surprising if it did not. So also with greenhouse gases: if we add CO_2 or other greenhouse gases to the atmosphere, we expect Earth's surface to become warmer. The question is one of *degree*. Is the warming significant or trivial? What change in temperature do we expect for a specific addition of greenhouse agents? Is the associated change in weather detectable? Are the consequences sufficiently serious to require a change in the industrial or agricultural practices responsible for this disturbance in the first place?

3. Low-altitude clouds are particularly important as cooling agents. They have an effect equivalent to that introduced by an increase in surface albedo. A thin, high-altitude cloud may have a minor

effect on incoming solar radiation. It could be important, however, in limiting transmission of radiation at infrared wavelengths, contributing to the greenhouse at higher altitudes, where the effect of the greenhouse on surface temperature is greatest; as previously noted, heat is vertically transferred at low altitudes mainly by atmospheric motions rather than by radiation.

4. The IPCC was established in 1988 by the World Meteorological Organization (WMO) and the United Nations Environmental Program (UNEP) to (i) assess available information on climate change, (ii) assess the environmental and socioeconomic impacts of climate change, and (iii) formulate policy responses.

5. An abrupt increase in CO_2 or other greenhouse gases results in a net instantaneous heating of the planet: the flux of energy emitted into space at infrared wavelengths is less than that absorbed from the Sun. The atmosphere responds quickly to this energy imbalance. But the rise in surface temperature is delayed as heat is shared with an initially colder ocean. The ocean adjusts more slowly than the atmosphere, reflecting its higher heat capacity. Global-mean surface temperature increases in response to net global retention of heat. Eventually the surface-atmosphere-ocean system reaches an equilibrium state where all of the component parts are in balance with respect to inputs and outputs of energy. Models in Figure 10.2 are distinguished by the increase in surface temperature they would predict to occur in equilibrium in response to a doubling of CO_2. Heat transfer from the atmosphere to the ocean is described by using a simple diffusive approximation. The downward flux of heat from the atmosphere to the ocean, and within the ocean, is assumed to depend solely on the vertical temperature gradient (the rate at which temperature changes with elevation or depth). The efficiency of heat transfer is regulated by a constant of proportionality known as the **eddy conductivity coefficient.** The approach to equilibrium (essentially the inertia of the surface thermal regime) is associated with time scales extending into decades or longer. The higher the value of the eddy conductivity coefficient, the larger the volume of ocean water required to share in the excess heating of the atmosphere and thus the slower the rate at which surface temperature approaches equilibrium.

6. The upwelling diffusion–energy balance model considers land and ocean regimes in each hemisphere as separate boxes. Oceanic heat transfer is assumed to take place through a combination of bulk vertical motion (upwelling) and diffusive mixing.

7. The process outlined here is known as **Rayleigh distillation,** named in honor of the English physicist Lord Rayleigh (1842–1919). It is easy to rationalize why light water molecules are able to escape the liquid phase more easily than heavy; at any given temperature they are moving faster. The distinction between light and heavy diminishes with increasing temperature. The isotopic selection in precipitation is the reverse of that for evaporation.

8. Acquisition of the low-latitude ice cores, from which the data in Figure 10.7 were derived, is a tribute to the skills and ingenuity of a remarkable group of scientists led by a husband and wife team, Lonnie and Ellen Mosley-Thompson, at Ohio State University. The logistical problems that had to be overcome to obtain these cores were truly formidable. First, the team had to conduct extensive surveys to find sites where ice of glacial age might be preserved and where the state of preservation would permit a reasonably convincing interpretation of the depth/age relationship. Environments where these conditions are met in the tropics are rare and invariably located at high altitudes in regions remote from roads and other logistical services. The Bolivian site is located near the summit of Sajama, an extinct volcano at an altitude of more than 6500 m above sea level. Imagine the problem of drilling in this environment, through more than 130 m of ice, using a specially designed solar-powered drill, and, even more demanding, the challenge of returning the ice in its pristine frozen state to a laboratory half the world away in Columbus, Ohio. As field general of this incredible operation, Lonnie Thompson ranks, in our opinion, in the company of Amundsen, Byrd, Scott, and the other great explorers of the twentieth century.

9. An elliptical orbit may be specified by giving the magnitudes of the semi-major (a) and semi-minor (b) axes. Alternatively, it may be defined by giving the magnitude of either a or b and the value of the eccentricity, e. Eccentricity is specified in terms of a and b by the expression $e(a, b) = \sqrt{1 - (b^2/a^2)}$. As a approaches b, e approaches 0, and the ellipse evolves to become a circle. When a is much larger than b, e approaches 1 and the ellipse looks as if it has been squished by an angry celestial giant. If the value of the eccentricity is larger than 1, the orbit is parabolic. Comets, for example, are often on highly eccentric orbits. They make brief, often spectacular, appearances in the inner solar system before returning to the frozen wastes of deep space. If the eccentricity of the orbit is less than 1, we may expect the comet to return. If e is greater than 1, the comet must be regarded as a one-time visitor: when it leaves the inner solar system, it is destined never to reappear.

10. It is relatively easy to understand why this should be the case. A change in obliquity essentially corresponds to a change in latitude. An increase in obliquity means, for example, that the Arctic Circle will tend to move southward. On the other hand, a change in obliquity has a minimal effect on the elevation of the Sun at low latitudes. There a change in the flux of sunlight incident on unit area at any given season depends more on whether Earth is close or relatively distant from the Sun.

11. A wonderfully readable historical account of early attempts to explain the origin of ice ages is presented by Brown University geologist John Imbrie and his daughter K. P. Imbrie (1986). In it, they highlight the contributions of Milutin Milankovitch. During World War I, Milankovitch was captured by the Austro-Hungarian army. He was confined to Budapest for the duration of the war. He took advantage of his four years of confinement to perform pain-staking calculations of the effect of planetary perturbations on the time-varying flux of solar radiation received at different locations on Earth.

12. It is possible, of course, that the rise in sea level could reflect release of ice from Antarctica associated with the early indication of warming evidenced by the Antarctic ice cores. We judge this unlikely for two reasons. First, we might expect the

warming in Antarctica to be accompanied (initially at least) by an *increase*, rather than a *decrease*, in precipitation and thus by a small *decrease* in sea level. Second, the basic tenets of the Milankovitch theory are undoubtedly correct: it is reasonable to expect that the increase in summer insolation setting in roughly 20 kyr B.P. in the Northern Hemisphere should result in a decrease in the volume of water stored in the Northern Hemispheric ice sheets.

13. Under natural conditions (i.e., in the absence of an anthropogenic contribution), methane is produced mainly by bacteria processing organic matter in environments where oxygen is in scarce supply. Stagnant water overlying organic rich soils offers an ideal environment. Waterlogged soils, both in the tropics and at high latitudes, are thought to play important roles in methane production today. Given the presence of the ice sheets for the period of interest here, it is reasonable to assume that the rise in methane observed during the deglaciation epoch primarily reflects a pick-up of production in the tropics. Alternatively, the increase in methane could be attributed to a reduction in the efficiency with which the gas is removed by reaction with OH. Based on the discussion of tropospheric chemistry developed later in Chapter 17, it seems unlikely that warming in the tropics should be associated with a reduction in OH. Indeed, it could be argued that the concentration of OH is more likely to *increase* in response to higher abundances of atmospheric water vapor and likely *increases* in the concentration of tropospheric ozone.

14. As previously noted, this is the unit favored by oceanographers to measure the volume of water transported per unit time by ocean currents: $1 \text{ Sv} = 10^6 \text{ m}^3 \text{ sec}^{-1} = 3 \times 10^{19} \text{ cm}^3 \text{ yr}^{-1}$.

15. There is a steady loss of nutrients (nitrate and phosphate) as waters in the near-surface region of the Atlantic flow northward today to feed production of deep water at high northern latitudes. As a consequence, the nutrient content of waters sinking to the depths in the North Atlantic is significantly less than that of deep water formed in the Southern Ocean in the vicinity of Antarctica. Measurements of N or P can be used to distinguish the origin of deep water: waters with high N or P originate in the south; waters with low N or P derive from the north. In general, waters with a high nutrient content are distinguished by relatively low values for the abundance of ^{13}C relative to ^{12}C; production of nutrients is associated with the decomposition of organic matter falling from above; decomposition leads to release of isotopically light carbon. The isotopic composition of carbon in a particular oceanic region is recorded in the isotopic composition of shells growing in this medium. MIT geochemist Ed Boyle has shown that the ratio of cadmium to calcium in shells can also be used as a surrogate for phosphate. The qualitative representation of the circulation of the glacial Atlantic displayed in Figure 10.17 was presented by Boyle and Broecker, based on distributions of nutrients inferred from measurements of carbon and cadmium in glacial-age shells recovered from deep-sea sedimentary cores.

16. An outflow of water from the large neo-glacial Lake Agassiz may have played an important role in triggering the reversal of the Atlantic circulation. Located to the north and west of the Great Lakes, draining an area of several million square kilometers in what is now southern Manitoba, Lake Agassiz provided a large storage reservoir for glacial melt water. In the early stages of deglaciation, it drained mainly southward down the Mississippi. Roughly coincident with the onset of the Younger Dryas, however, a retreat of the ice sheet opened an important new channel for drainage to the east, through the Great Lakes and the Saint Lawrence. The geological evidence suggests that the lake level dropped by as much as 40 m in a relatively short period of time, sending a vast flood of fresh water directly into the North Atlantic.

17. A variety of independent data supports the view that the tropics during the LGM were significantly colder than today. The evidence is particularly compelling for continental environments. Snow lines in regions as diverse as Mount Kilimanjaro and the Colombian Andes were about 1 km lower than they are today (Rind and Peteet 1985; Broecker and Denton 1989). In a particularly clever approach, Martin Stute and colleagues at Columbia University used measurements of noble gases in ground water to conclude that temperatures in Eastern Brazil (7°S) were colder, by 5.4 ± 0.5 °C, during glacial times than today. Their strategy exploited the fact that, as rainwater percolates into the ground, it has time to absorb gases and come to equilibrium with air present in the upper, unsaturated portions of the soil horizon. In particular, it takes up noble gases such as Ne, Ar, Kr, and Xe. Since the solubility of these gases depends on temperature, a measurement of their relative abundance may be used to infer the temperature of the surface layer of the soil in the region where the water was last in contact with air. The age of ground water (the time since it was last in contact with the atmosphere) is obtained by measuring the ^{14}C age of dissolved HCO_3^-. Up to a decade or so ago, our view of the glacial tropics was influenced largely by data obtained from analyses of the types of organisms that lived in the tropical ocean, as inferred from studies of deep-sea cores. Based on these data, John Imbrie at Brown University and his colleagues in the CLIMAP program concluded that temperatures in the glacial tropical ocean differed little from today (CLIMAP 1981). More direct observations of sea surface temperatures, from a study of oxygen isotopes and Sr/Ca ratios in coral drilled off of the coast of Barbados (Guilderson, Fairbanks, and Rubenstone 1994), have called this interpretation of the CLIMAP data into question. The coral measurements imply that temperatures during the LGM, at least near Barbados, were about 5 to 6 K colder than today, precisely as expected based on the snow-line and noble-gas data. It is difficult to reconcile the CLIMAP results with these other, more direct, sources of information. Our conclusion that the glacial tropics were significantly colder than today reflects, we believe, the weight of the evidence: data on snow lines; the isotopic composition of snow in the high Andes (Figure 10.7); measurements of noble gases in ground water; studies of vegetation (Bonnefille, Roeland, and Guiot 1990); and the measurements of coral from Barbados.

18. What we envisage here is essentially two distinct modes for the energy budget of the tropics. Today, the tropics are relatively warm. Heat is efficiently transferred from the oceanic surface to the atmosphere by evaporation and subsequent precipitation (as

previously noted, evaporation cools the ocean; heat removed from the ocean is transferred to the atmosphere when vapor condenses to form precipitation). Trade winds are relatively weak and so, too, are the westerlies at midlatitudes (if the trade winds are weak, the westerlies must also be weak in order to satisfy constraints imposed by the need to conserve angular momentum). Oceanic circulation is relatively sluggish. As previously described in Chapter 7, the rate at which solar energy is absorbed in the tropics exceeds the rate at which energy is lost from the tropics by radiation into space. This net input of radiative energy to the tropics is balanced by the transport of the excess to midlatitudes, with the ocean and atmosphere playing comparable roles in this energy redistribution. In glacial times, wind speeds were significantly higher. The ocean's circulation was correspondingly enhanced. A larger flux of cold water to the surface of the tropical ocean would have resulted in colder temperatures, lower rates of evaporation, and, as a consequence, a reduction in the rate at which heat was transferred from the ocean to the atmosphere. Assuming a reduction in meridional (Hadley) mass transport as a consequence of a lower rate of atmospheric heating (Gates 1976), we speculate that the stronger flow of tropical surface easterlies during the LGM may have reflected a shift in atmospheric circulation to a more zonal (Walker) pattern maintained by a larger east-west gradient in heating rates, caused in part by the higher albedo of the desiccated continents.

19. As will be discussed in more detail in Chapter 11 (see in particular Figure 11.9), a decrease in the temperature of surface water would result in a reduction in the concentration of CO_2 in equilibrium with water containing a fixed quantity of total inorganic carbon, assuming that alkalinity is held constant. The temperature dependence of the relevant equilibrium constants ensures that, with a drop in temperature, there is a net conversion of CO_3^{2-} and CO_2 to HCO_3^-. A shallower circulation of the glacial ocean would ensure that deep water was isolated from the atmosphere to a larger extent than today. This would result in an accumulation in the deep sea of carbon and nutrients released by decomposition of organic matter. According to Hausman and McElroy (1999), a 5°C decrease in the temperature of the low-latitude ocean would result in a decrease in atmospheric CO_2 by about 50 ppmv. The change in circulation postulated for the glacial ocean would result in a further reduction of similar magnitude.

20. To take all of these factors into account would require a detailed model for the time-dependent chemistry and dynamics of the ocean as it undergoes its complex transition from glacial to interglacial conditions. Such a model must also allow for an important transfer of carbon from the ocean to the atmosphere and also to the land, as the terrestrial biosphere is reestablished with the retreat of continental ice sheets. Release of CO_2 from the ocean is accompanied by enhanced deposition of carbonate in sediments modulated by an initial drop in the position of the lysocline. The alkalinity content of the ocean decreases as a result of the additional carbonate sink. The rate at which alkalinity is removed from the ocean is limited, however, by the rate at which carbonate shells are produced as a by-product of biological activity. Given this limitation, an appreciable time (5000 years or so) is required to restore the balance between the input of alkalinity from land and its removal to sediments. Under the circumstances, it is not surprising that a period of close to 8 kyr is required to complete the glacial-to-interglacial transition in atmospheric CO_2.

21. The $\delta^{18}O$ value for shells is higher than that of seawater by an average of about 40‰. The heavier isotope is preferentially incorporated in shells at low temperatures—the $\delta^{18}O$ value of shell carbonate increases by 0.23% for every degree (°C) drop in temperature. As previously noted, isotopically light water evaporates more readily than heavy; the equilibrium vapor pressure for water with two ^{16}O atoms is about 1% higher than that for water when one of the ^{16}O atoms is replaced by ^{18}O. As a result, the isotopic composition of continental ice is significantly lighter than parent ocean water (by about 40‰). Cold temperatures and growth of continental ice both contribute to cause an increase in the $\delta^{18}O$ value for shell carbonate.

22. "Benthos" (adjective: benthic) refers to the plant and animal species living at the bottom of a body of water, in this case the ocean. Foraminifera are unicellular animals that form calcareous shells.

23. The data in Figure 10.22 are plotted on an inverted scale: values of $\delta^{18}O$ decrease with height on the vertical axis. We choose this form of presentation to emphasize the associated changes in climate. Small $\delta^{18}O$ values (peaks in Figure 10.22) reflect interglacial conditions; high $\delta^{18}O$ values (minima in Figure 10.22) indicate periods of maximum glaciation. Major ice ages end abruptly. The transition from glacial to interglacial conditions is referred to as a **termination.** Terminations are sequentially numbered beginning with the end of the last glacial period approximately 20 kyr B.P.

24. By *large-scale variability* we mean the variations that would show up if we were to apply a temporal filter with a width of, say, a few thousand years to average the data for CH_4 presented in Plate 2 (bottom) (see color insert).

25. The presence of these sharp peaks in $\delta^{18}O$ was first detected by a team of scientists led by the Danish glaciologist Willy Dansgaard. Separately, Hans Oeschger and his colleagues from Bern, Switzerland, reported that fluctuations in $\delta^{18}O$ were associated with rapid variations in CO_2. To date, there is no independent evidence to support the reality of the fluctuations reported for CO_2. It is now thought that they may reflect an artifact peculiar to the Dye 3 Greenland core from which the CO_2 data were derived.

26. The idea is that, in the presence of a thick ice sheet, the mantle (the layer of Earth's interior separating the core on the inside from the crust on the outside) is compressed. Material in the upper region of the mantle flows away the region where it is subjected to the additional pressure imposed by a thick overlying ice sheet. As the ice sheet begins to recede, it retreats into a giant hole, an area from which mantle material has been extruded by the ice. Compression of the underlying rock material contributes to the instability of the remaining ice sheet, and it eventually tends to collapse under its own weight. Flow of mantle material in response to the weight imposed by the ice—or the absence of the weight as the load is removed—is regulated by its viscosity. There are regions of North America

where land is still recovering from the last ice age: rising in some cases, sinking in others, in response to imbalances in the local force field. In regions where land is rising (due to an inflow of mantle material), mainly in regions previously covered by ice, the sea level appears to decline. In regions where land is sinking (as mantle material is removed), mainly on the periphery of the earlier ice sheet, the sea level appears to rise.

27. *Regolith* is the geologic term referring to a layer of solid material lying on top of bedrock. It includes soil and alluvium and fragments of rock produced by the weathering of bedrock.

28. North and South America have been separated for most of the past 100 million years. Motion of the Caribbean plate resulted in the formation of the Central American isthmus approximately 3 million years ago, effectively blocking the previously efficient exchange of water between the Pacific and the Atlantic in this region. We would expect the salinity contrast between the Atlantic and the Pacific to have been relatively minor prior to the time of closure of this land bridge. It is likely that an increase in the salinity of surface waters in the Atlantic following closure would have had a significant impact on the circulation of the North Atlantic. In particular, it may have resulted in an invigorated thermohaline circulation (the North Atlantic Conveyor Belt), accompanied by an inevitable increase in poleward transport of warm water. Evaporation from this relatively warm water at high latitudes during winter may have played an essential role in the initial formation of the continental ice sheets distinguishing the Pleistocene.

29. Porphyrins are derivatives of chlorophyll. As such, it is expected that their isotopic composition should be representative of the bulk isotopic composition of organic matter formed by photosynthesis. Organic matter formed by photosynthesis is isotopically light in carbon with respect to the inorganic carbon employed in photosynthesis. The extent of fractionation depends on the abundance of CO_2 in solution as shown by Popp et al. (1989) and Rau, Takahashi, and Des Marais (1989). Freeman and Hayes (1992) used measurements of the $\delta^{13}C$ of porphyrins in sedimentary organic matter to infer concentrations of CO_2 in solution. Concentrations of atmospheric CO_2 were obtained assuming equilibrium between the atmosphere and ocean. A critical uncertainty of the method relates to the accuracy of temperatures used to translate concentrations of CO_2 in solution to concentrations of CO_2 in the atmosphere.

30. The model published by Berner, Lasaga, and Garrels (1983) is referred to in the geochemical literature as the "BLAG" model, an abbreviation based on the last names of the authors of the paper.

Chapter 11

1. Emissions of CO_2 in early years were dominated by the combustion of coal. The fraction of emissions resulting from oil and natural gas has grown rapidly in recent years. Cumulative emissions of CO_2 from 1860 to 1985 amounted to almost 190×10^9 tons C.

2. The addition of 1×10^9 tons (1×10^{15}g) C to the atmosphere would be sufficient to raise the mixing ratio of CO_2 by 0.47 ppm. The total mass of the atmosphere is 5.12×10^{21}g, and the average molecular weight is 28.97g mol^{-1}; the change in mixing ratio is given by:

$$\frac{1 \times 10^{15}\ \text{g}}{5.12 \times 10^{21}\text{g}} \times \frac{28.97\ \text{g mol}^{-1}}{12.01\ \text{g mol}^{-1}} = 0.47\ \text{ppm}$$

The atmosphere today, with a mixing ratio of CO_2 equal to 355 ppm, contains 753.2×10^9 tons C. The calculation is the same, except for the use of the total mass of carbon:

$$\frac{753.2 \times 10^{15}\ \text{g}}{5.12 \times 10^{21}\ \text{g}} \times \frac{28.97\ \text{g mol}^{-1}}{12.01\ \text{g mol}^{-1}} = 355\ \text{ppm}$$

3. Measurements of gases trapped in polar ice suggest that the value adopted here for the CO_2 content of the preindustrial atmosphere may be somewhat high. It might be more appropriate to consider a mixing ratio of 280 ppm, corresponding to an atmospheric abundance of 594×10^9 tons C.

4. The global rate of photosynthesis, associated with the combination of ground vegetation (15×10^9 tons C yr^{-1}), non-woody parts of trees (22×10^9 tons C yr^{-1}), and woody parts of trees (25×10^9 tons C yr^{-1}), is given in Figure 11.5 as 62×10^9 tons C yr^{-1}. This may be compared with the rate implied by the seasonal amplitude in CO_2 observed at Mauna Loa, 6 ppm. If we assume that the seasonal signal recorded at Mauna Loa is representative of the annual balance of photosynthesis with respiration-decay for deciduous plants in the Northern Hemisphere as a whole, we would infer an associated exchange rate for carbon of about 6×10^9 tons C yr^{-1}. Accounting for deciduous plants in the Southern Hemisphere, it might be reasonable to raise this number to about 10×10^9 ton C yr^{-1} on a global scale. Recognizing that tropical vegetation negligibly contributes to the seasonal amplitude of the variation in atmospheric CO_2, it seems reasonable to increase this value by a factor of six in scaling to global dimensions. It must be admitted, though, that the global value is uncertain; it could be as small as 30×10^9 tons C yr^{-1} or as large as about 80×10^9 tons C yr^{-1}.

5. As previously noted in Chapter 9, in order to sink to the bottom, surface water must become denser than the water underneath. The density of seawater is determined by a combination of temperature and salinity. Conditions appropriate for deep-water formation are realized in today's ocean only in limited regions of the North Atlantic and Antarctic Oceans; specifically, they are not achieved in the North Pacific, where surface waters are relatively fresh. The circulation of the deep sea has been described by the Columbia University geochemist Wallace Broecker as a giant conveyor belt. The conveyor begins its trip when water sinks from the surface to the bottom in the North Atlantic. It flows southward in the Atlantic before circling Antarctica, where it receives a boost from water sinking from the Weddell Sea. The deep circum-Antarctic current feeds the deep regions of the Pacific and Indian Oceans. Waters upwells slowly to the surface as the conveyor describes its long (approximately 1000 yr) pass through the abyssal regions of the world's oceans. The flow of water at depth from the Atlantic to the Pacific and Indian Oceans is balanced by a compensating return flow near the surface.

6. Current studies suggest that photosynthesis in the world's oceans is responsible for the fixation of about 60 to 80×10^9 tons C yr^{-1}, as compared with the value of 62×10^9 tons C yr^{-1}

indicated for terrestrial systems in Figure 11.5. Rates of ocean photosynthesis are largest in coastal regions and in zones characterized by especially intense upwelling of nutrient-rich water from the deep (waters off the coast of Peru, renowned for their fish harvest, provide an example of the latter). Despite the significance of locally productive environments, ocean photosynthesis on a global scale is thought to be dominated by the open sea; its areal extent is more than sufficient to offset the comparatively higher productivity observed near the coast.

7. Nitrogen and phosphorus are believed to represent the limiting nutrients for life in the ocean. The word *limiting*, as used in this context, implies that an increase in the abundance of these chemicals would be expected to enhance the rate of biological productivity. Most of the ocean's store of N and P resides in the deep; an increase in the rate at which these elements are supplied to the surface region would be expected to cause a proportional increase in the rate of photosynthesis.

8. Shells of a large number of marine organisms are composed of $CaCO_3$. The reaction responsible for production of $CaCO_3$ may be written as

$$Ca^{2+} + 2HCO_3^- \rightarrow CaCO_3 + CO_2 + H_2O$$

9. A more complete discussion of marine chemistry should allow for contributions to the charge balance due to $B(OH)_2O^-$. Boron is present in the ocean in two forms: the electrically neutral species $B(OH)_3$ and the negatively charged compound $B(OH)_2O^-$. The relative abundance of these compounds is regulated by the local value of pH.

10. Chemists express alkalinity in units of *charge equivalents per liter.* A charge equivalent of 1 is numerically equal to the charge contributed by a mole of singly charged species. A singly charged species, whose concentration is given by [X] expressed in units of mol l^{-1}, contributes an amount [X] charge equivalents liter^{-1} to alkalinity. If the compound is doubly charged, its contribution equals 2[X] charge equivalents liter^{-1}. The charge equivalent unit measures the contribution from a particular species to the local charge balance of the medium.

11. This follows, for example, from (11.14). The fractional change in $[HCO_3^-]$ is small. The change in $[CO_3^{2-}]$ is offset largely by the change in pCO2, such that the product is held constant.

12. The ratios of C:N:P in typical marine organic matter are usually referred to as the **Redfield ratios** (more accurately, perhaps, we should call them the **Redfield proportions**), named in honor of the distinguised marine biologist Alfred C. Redfield.

13. The giant ice sheets covering Greenland and Antarctica include bubbles of air that provide a unique opportunity to sample the composition of the atmosphere in the past. Bubbles form and are isolated from the atmosphere when overlying snow reaches a depth of about 50 meters. The older the ice, the older the air contained in the bubbles. If snow accumulation is rapid, bubbles close quickly and the age of the air in the bubbles is similar to that of the ice in which they are contained. If accumulation is slow, ages of air and ice can differ by as much as thousands of years. The results in Figure 11.13 were obtained by analyzing the composition of air extracted from a core of ice drilled at Siple Station in a coastal region of west Antarctica. Accumulation rates at Siple are relatively rapid (about half-a-meter per year) compared with rates in the continental interior, providing an opportunity to sample the composition of the atmosphere with excellent temporal resolution (every 20 years or so) from the mid eighteenth century up to about 1970. The ice-core data from Siple overlap in time with the modern direct measurements pioneered by C. D. Keeling in Hawaii and at the South Pole and yield compatible results. Ice-core studies of CO_2 using ice cores were pioneered by teams of Swiss and French scientists led by Hans Oeschger and Claude Lorius, respectively, and have provided an invaluable window to the past, an essential context in which to view changes in the composition of the atmosphere taking place today.

14. It appears that as much as 20–30% of the organic carbon and fixed nitrogen (see Chapter 9) included in the rich soils of the Midwest may have been lost in the early years of farming. The carbon and nitrogen contents of these soils are now more or less stable, reflecting inputs of fixed nitrogen in the form of chemical fertilizer and agricultural practices favoring cultivation of nitrogen-fixing crops, such as legumes. Release of carbon associated with conversion of pristine systems to agriculture in the eighteenth and nineteenth centuries could have made a significant contribution to the early rise in the concentration of atmospheric CO_2 indicated in Figure 8.13.

15. A first-principles model would require a comprehensive understanding of how the abundance of carbon in terrestrial systems, both above and below ground, should vary in response to changes in environmental conditions (temperature, rainfall, cloud cover, chemical inputs, etc.). Implementing such a model would presume detailed, spatially and temporally resolved information on carbon stocks and relevant features of the environment in specific systems, including a definition of changes with time. Our understanding of biospheric processes and relevant soil chemistry is not yet sufficient to justify formulation of such a model, nor would our understanding of environmental conditions be complete enough to warrant its implementation. In the absence of a first-principles model, we adopt a more empirical approach, using measurements of changes in the abundances of CO_2 and O_2 in the atmosphere to differentiate between roles of the biosphere-soil and the ocean as sinks for contemporary emissions of CO_2 from the combustion of fossil fuels.

16. The energy liberated by burning coal is derived mainly from oxidation of chemically reduced carbon contained in the coal. A significant fraction of the energy released by burning natural gas is associated with oxidation of hydrogen. Carbon dioxide is the dominant product of coal combustion. In burning coal, CO_2 is produced and O_2 is consumed in molar proportions close to 1:1. Burning natural gas produces H_2O and CO_2 in molar proportions of 2:1. Approximately twice as much O_2 is consumed relative to CO_2 produced in burning natural gas as compared with coal. The reduced hydrogen content of oil is intermediate between coal and natural gas. The quantity of O_2 consumed relative to CO_2 released in burning oil is intermediate between the quantities appropriate for coal and natural gas.

17. It is easier to carry out a measurement of the ratio O_2 concentration of relative to N_2 than to measure the absolute abun-

dance of O_2. Furthermore, a measurement of the change in the ratio O_2/N_2 relative to a standard can be carried out with higher accuracy than a measurement of the ratio itself. In terms of our application, we are concerned with changes in the ratio, not with the value of the ratio itself.

18. The analysis of O_2 and CO_2 data for the 1991–1994 period summarized here is based on the paper by Keeling et al. (1996).

19. An increase of 1 ppmv in CO_2 due to combustion of fossil fuels, according to our assumptions, is associated with a decrease in the mixing ratio of O_2 of 1.39 ppmv, equivalent to $-(1.39 \times 4.77) = -6.63$ per meg change in O_2. The slope of AB is given then by $-(1.0 / 6.63) = -0.15$ ppmv meg. Similarly, an increase of 1 ppmv in CO_2 due to release of carbon from the biosphere or soil would be associated with a reduction in O_2 of 1.1 ppmv, equivalent to a decrease in delta of $(1.1 \times 4.77) = 5.24$ per meg. The slope of the corresponding segment in this case is given by $-(1.0 / 5.24) = -0.19$ ppmv meg.

20. The isotopic composition of carbon in CO_2 is reported in terms of a quantity $\delta^{13}C$, defining the fractional change in the ratio of the concentration of ^{13}C relative to ^{12}C, with respect to the ratio in a standard (a sample of calcium carbonate from the Peedee formation in South Carolina): $\delta^{13}C = \left\{ \left[(^{13}C/^{12}C)_{sample} - (^{13}C/^{12}C)_{ref} \right] / (^{13}C/^{12}C)_{ref} \right\} \times 10^3$. The convention used to define $\delta^{13}C$ is similar to that introduced earlier to measure the ratio of the concentration of O_2 relative to N_2 (delta), except that the units in this case are parts per thousand (or, per mille). In photosynthesis, ^{12}C is preferentially used relative to ^{13}C. Plant material is consequently isotopically lighter in carbon than is the atmosphere, typically by between 6 and 19 per mille, depending on whether photosynthesis proceeds by the path favored by C4 or C3 plants, respectively (maize, sugar cane, and crabgrass provide examples of C4 plants; C3 plants include all trees and agricultural crops, such as wheat, barley and rice). Burning fossil fuels introduces light carbon to the atmosphere; growth of the biosphere or sequestration of organic carbon in soils has the opposite effect. An isotopic anomaly in atmospheric CO_2 is more readily removed by exchange with the ocean than an anomaly in the concentration of CO_2 itself. The difference relates to the fact that the isotopic signal is readily shared among all forms of organic carbon dissolved in surface water, while incorporation of excess CO_2 requires reaction with the relatively minor species CO_3^{2-}. Observations of $\delta^{13}C$, though less definitive than observations of O_2, provide a useful constraint allowing us to distinguish, partially at least, between the roles of the ocean and biosphere-soils in the budget of fossil fuel–derived CO_2.

21. During an El Niño, as warm water spreads across the Pacific temporarily capping the upwelling region in the ocean off Peru, the zone of strong convective activity in the tropical atmosphere shifts from the western Pacific (near Indonesia) to the central Pacific, closer to the date line. Significant changes in rainfall and temperature are observed over a wide range of longitudes and latitudes. El Niño is associated, for example, with droughts and forest fires in Indonesia and Australia and with decreases in rainfall in regions as far flung as India and parts of Africa (but rainfall increases in the central Pacific). Under the circumstances, it is not surprising that the El Niño condition is accompanied by significant changes in rates for exchange of carbon not only between the atmosphere and ocean but also between the atmosphere, soils, and the biosphere.

Chapter 12

1. The term *fixed nitrogen* is used to describe the family of species that can be transformed into forms incorporable in biological tissue with a minimum expenditure of energy. With this definition, the fixed nitrogen family includes the suite of organic species, inorganic compounds such as NH_4^+, NO_2^-, and NO_3^-, and gases such as NH_3, NO, HNO_2, NO_2, HNO_3, NO_3, and N_2O_5. For the most part the definition encompasses species containing single atoms of nitrogen. Inclusion of N_2O_5 reflects the fact that N_2O_5 is readily converted to HNO_3 in the presence of liquid H_2O. Molecular nitrogen and nitrous oxide (N_2O) are explicitly excluded from the suite of compounds identified as *odd nitrogen*.

2. Biologists use the term *fixation* to describe the conversion of N as N_2 to biologically available forms of fixed nitrogen. The opposite reaction, return of N from the fixed N reservoir to biologically unavailable forms, such as N_2 and N_2O, is referred to as *denitrification*.

3. A *heterotrophic* organism is one that relies on other organisms or on their remains for essential food and essential nutrients.

4. Additions of large concentrations of phosphate associated with widespread use of phosphate-based detergents and runoff from agricultural sources resulted in the occurrence of persistent blooms of blue-green algae in a variety of aquatic systems in the United States during the 1970s and 1980s. The algae provided an additional source of fixed nitrogen, leading to enhanced rates of primary productivity. Subsequent decay of organic matter (including blue-green algae) created a demand for O_2 in the water that could not be met by transfer from the atmosphere. As a consequence, many of these systems became anoxic. Fish and other forms of aquatic life died, and the ecology was altered, in some cases irreparably. Publicity associated with these events resulted in a decision by the United States government to ban the use of phosphate-based detergents and to limit the runoff of nutrients contributed by agriculture.

5. A concentration of N_2O equal to 1 ppb would be equivalent to a global abundance of N as N_2O equal to 4.74 Mt N. The abundance of N in the atmosphere as N_2O in 1988 was therefore equal to 1455 Mt.

6. Oxygen is removed from the water column primarily by oxidation of organic matter; a portion of the oxygen loss is due to oxidation of C; the balance is associated with oxidation of organic forms of N. Apparent oxygen utilization is calculated by first estimating what the oxygen concentration would be in the absence of biological activity, that is, the concentration that would apply in equilibrium with the atmosphere at the observed ambient temperature. AOU refers to the difference between the measured concentration of O_2 and the concentration that would have been expected in the absence of biologically mediated consumption.

7. Given a source of NH_4^+, phytoplankton are known to assimilate N as NH_4^+ in preference to NO_2^- or NO_3^-. The oceanic measurements define a relationship between production of N_2O and consumption of O_2. Relating consumption of O_2 to oxidation of C

requires that we make an assumption concerning the contribution of N to the observed loss of O_2. The yield of N_2O per unit of carbon oxidized as quoted here assumes that N is oxidized together with C with relative contributions determined by the ratio of the elements in typical organic material (see Chapter 11). If we assume that nitrogen is taken up in reduced form by the biota, and if we ignore the contribution of N to the observed consumption of O_2, we would be forced to assume that a larger quantity of C was oxidized to accommodate a particular source of N_2O. Adopting the yield of N_2O per unit of carbon quoted here provides an upper limit to the source of N_2O associated with any particular value of marine primary-carbon productivity.

8. This estimate assumes a contemporary human population of 5.6 billion. The NAS study estimates that the N content of human waste for the United States should be equal to about 1.1 Mt N yr^{-1}. Corresponding animal waste for the United States is estimated at 3.7 Mt N yr^{-1}. Given that animals play a larger role in the nutrition of people in the developing world as compared to the developed world, our assumption, that production of N by humans on a world-wide basis should be multiplied by a factor of four to allow for the additional contribution from domestic animals, may be conservative.

9. We should caution that while the magnitude of the *total* anthropogenic source of N_2O inferred here is relatively well constrained, assignment of specific emissions to particular influences is significantly more uncertain. We believe that our general conclusion—that the increase in N_2O is a consequence primarily of perturbations to the nitrogen cycle driven by needs to feed an ever increasing human population and to dispose of associated organic wastes—is reasonably robust. Purely industrial sources of N_2O, however, while relatively small at present, could significantly grow in the future if steps are not taken to constrain future growth. Current industrial sources include emissions associated with new generations of catalytic converters and with production of adipic acid employed in the manufacture of nylon. The IPCC (1990) estimates that the contemporary industrial source could be as large as 1.3 Mt N yr^{-1}, quoting a range of values for emissions from 0.7 to 1.8 Mt N yr^{-1}.

10. The quantity of P applied to agricultural systems in 1980 amounted to 14.5 Mt P yr^{-1}, almost 50% higher than our estimate for mobilization of P due to erosion of continental rocks under natural, preagricultural conditions.

11. A higher input of P into the ocean in recent times should have a negligible impact on contemporary marine productivity. As we shall see, the lifetime of reactive P in the ocean is estimated to be about 10^5 yr. An increase in supply of P due to human activity, even by as much as a factor of three, should have a minimal effect on the budget of ocean P given that it would have been applied for less than 0.1% of the imputed residence time. On a time scale of 10^5 yr, we would expect the source of reactive P contributed by rivers to be balanced by removal to marine sediments. On shorter time scales, it is possible that the rate for supply by rivers could differ from the rate for removal to sediments. Under conditions of high sea level, as applies, for example, today, it is conceivable that a significant fraction of P carried by rivers could be tied up in estuarine and coastal sediments. Under these circumstances, loss of P from the ocean could exceed supply. This condition would be most likely temporary, however. It is probable that it would reverse under glacial conditions, when sea level is lower. Erosion of estuarine and coastal sediments could account in this case for a net source of P, ensuring a long-term (10^5 yr) balance between sources and sinks for marine P.

12. The value quoted here for the lifetime of N reflects an estimate of the time for depletion of the ocean's budget of NO_3^- due to inclusion of fixed N in sediments. The lifetime for oceanic fixed N is actually much shorter than this, about 10^4 yr: regulated on the loss side by the rate for denitrification, on the source side by inputs associated with river runoff and in situ fixation. In contrast to the situation for C and P, the atmosphere represents a major reservoir for N. While nitrogen in the atmosphere (present mainly as N_2) is not readily available for incorporation in biological tissue (it must first be fixed), the atmosphere must be considered an important potential reservoir to replenish N transferred in fixed form from the ocean to sediments. This supply is not, however, unlimited. The nitrogen content of the atmosphere would be depleted as a consequence of the transfer of fixed N from the atmosphere to the ocean, and thence to sediments on a time scale of about 4×10^8 years in the absence of a mechanism to recycle N from sediments back to the atmosphere.

Chapter 13

1. The cross section (measured in units of cm^2) defines the effective area of the target offered by an individual molecule for the process under consideration. Its value depends on both the nature of the process and on the identity of the target. The cross section for absorption of ultraviolet radiation by O_3 lies in the range 10^{-18}–10^{-17} cm^2 for most of the spectral interval important for the stratosphere (approximately 200–300 nm). Considering its interaction with ultraviolet light, we can think of the O_3 molecule as a sphere of radius R with R equal to about 10^{-9} cm. The cross section (the target outlined on a screen by a hypothetical molecule) would have area given by πR^2. With $R = 10^{-9}$ cm, the cross section for O_3 would be equal to a little more than 3×10^{-18} cm^2. The cross section for absorption of ultraviolet radiation in the stratosphere by O_2 is much smaller: it ranges from about 10^{-24}–10^{-23} cm^2 over the important wavelength interval 240–200 nm. The difference in the values of the cross sections for O_3 and O_2 depends on details of the internal structure of the two molecules and could not have been easily predicted from first principles. Cross sections for absorption of radiation by individual molecules are usually specified on the basis of careful measurements in the laboratory.

2. Equations (13.5) and (13.6) can be rearranged to write

$$J_{13.1} = \frac{R_{13.1}}{[O_2]}$$

$$J_{13.3} = \frac{R_{13.3}}{[O_3]}$$

It is clear, in this case, that the dimension of $J_{13.1}$ and $J_{13.3}$ must be given by $\frac{\mathrm{cm}^{-3}\ \mathrm{sec}^{-1}}{\mathrm{cm}^{-3}} = \mathrm{sec}^{-1}$. Equations (13.7) and (13.8) can be rearranged to give

$$k_{13.2} = \frac{R_{13.2}}{[O][O_2][M]}$$

$$k_{13.4} = \frac{R_{13.4}}{[O][O_3]}$$

Dimensional analysis in this case indicates that

$$k_{13.2} = \frac{\text{cm}^{-3}\ \text{sec}^{-1}}{[\text{cm}^{-3}][\text{cm}^{-3}][\text{cm}^{-3}]} = \frac{\text{sec}^{-1}}{\text{cm}^{-6}} = \text{cm}^6\ \text{sec}^{-1},$$

while

$$k_{13.4} = \frac{\text{cm}^{-3}\ \text{sec}^{-1}}{[\text{cm}^{-3}][\text{cm}^{-3}]} = \frac{\text{sec}^{-1}}{\text{cm}^{-3}} = \text{cm}^3\ \text{sec}^{-1}$$

3. The problem of estimating the residence time for a chemical species by dividing its concentration by the rate at which it is produced or lost, is similar to the problem involved in calculating the time required to travel a specified distance given the speed of motion. Suppose an automobile is traveling at 50 miles per hour and we wish to find the time, τ, required to travel 100 miles. The calculation may be summarized as follows:

$$\begin{aligned}\tau &= \text{Distance/Speed}\\ &= 100\ \text{miles}/50\ (\text{miles hr}^{-1})\\ &= 2\ \text{hr}.\end{aligned}$$

Actually, there is a subtle distinction. To the extent that the rate at which a chemical compound is removed is proportional to its concentration, the time derived by dividing its concentration by its rate of removal refers not to the time for its total removal but to the time for its concentration to decrease by a factor of e (exp (l) = 2.718), meaning the time required for it to fall to approximately 40% of its initial value. The appearance of the exponential is a consequence of the dependence of the loss rate on concentration: the loss rate decreases with time as a result of the decrease in concentration. The distinction is unimportant for present purposes; values of τ derived here are intended to provide merely qualitative estimates for the efficiency of particular chemical processes affecting the concentration of specific chemical species.

4. Chemical species move around a circle of latitude relatively rapidly, typically in about a week. Transport in the vertical and north-south directions proceeds more slowly. Our understanding of stratospheric circulation patterns was initially based in large measure on studies of the movement of chemical, primarily radioactive, species added to the stratosphere in the late 1950s and early 1960s associated with the testing of nuclear weapons. Additional insights were obtained from observations of the spread of materials episodically added to the stratosphere by volcanic eruptions. Our choice of a month for the critical time constant is based mainly on these data; it reflects a rough estimate of the time required for chemical species to spread to a significantly different latitude or altitude, to a region where rates for photolysis are distinctly different from values applicable in the region of interest.

Chapter 14

1. As previously discussed in Chapter 3, a *radical* is a species containing an odd number of electrons. Its chemical reactivity is directly related to the presence of unpaired electrons in its valence (or outer) shell.

2. An electronically excited atom would normally be expected, with the emission of radiation, to rapidly decay to a state of lower energy, ultimately to the lowest energy (ground) state of the species. The term **metastable** refers to a state for which this decay path is inhibited, for reasons beyond the scope of the present discussion, governed by the laws of quantum mechanics. The abundance of $O(^1D)$ in the stratosphere is determined by a balance of production by (14.15) with the loss induced by collisions with N_2 and O_2; reaction (14.14) represents a minor path for removal of $O(^1D)$ in the atmosphere.

3. The compounds $CFCl_3$ and CF_2Cl_2 were developed at the General Motors Research Laboratories in 1930 by Thomas Midgley, Jr., one of the great chemical inventors of the twentieth century. Midgley responded to a challenge posed by the chief engineer of the Frigidair Division to find a nontoxic, nonflammable replacement for the hazardous compounds then in common use as refrigerants—ammonia (NH_3) and sulfur dioxide (SO_2). Midgley had a flair for the dramatic: when he appeared to present his new invention, he was armed with a bell jar of the compound and a candle. Taking a breath of the new gas into his lungs, he blew out the candle, demonstrating in this fashion his complete confidence in the safety of his product. He died in 1944, strangled by a harness he had devised to make it easier for him to rise from bed (he had been paralyzed as a young man by polio). At the time of his death, he was totally unaware of the dangers to stratospheric O_3 posed by his brainchild, CFCs.

4. Actually, the mixing ratio of chlorine in all forms may be expected to vary as a function of the age of air in the stratosphere. This effect arises since the abundance of chlorine in the atmosphere as a whole, and in the troposphere in particular, is increasing with time. Until recently, at least, the rate at which chlorine was added to the atmosphere in the form of the industrial halocarbons exceeded the rate at which it was removed (mainly in the troposphere as hydrochloric acid in precipitation). Air remains in the stratosphere for about five years on average. If the stratosphere was sampled at any given time, we would expect to encounter air that had entered recently, in addition to air that had been present for several years. The abundance of chlorine (in all forms) in the older air would reflect the abundance in the troposphere at an earlier time; it would be less, therefore, than that for air of more recent origin. It follows that an accurate measurement of the total abundance of chlorine could be used to obtain information on the residence time of air in the stratosphere. In general, we expect the abundance of total chorine to decrease with altitude, reflecting greater isolation of air at high altitudes from air in the troposphere (it is likely that high-altitude air is older).

5. The more stable forms of chlorine are sometimes referred to as **reservoir species.** The term focuses on their role as temporary sinks, or holding systems, for chlorine radicals. Chlorine radicals are inactivated as they are transferred to holding reservoirs by reactions such as (14.41)–(14.43); they are reconstituted as they are released by reactions such as (14.44)–(14.46). The

computations presented in this section use levels of chlorine appropriate for 1985. The choice is motivated by the fact that extensive measurements of the chemical composition of the stratosphere were performed in this year using an experiment carried by the Space Shuttle. The experiment is known by the acronym **ATMOS**, which stands for Atmospheric Trace Molecule Spectroscopy Experiment.

6. The **Total Ozone Mapping Spectrometer** (TOMS) is an instrument designed to measure the intensity of sunlight reflected by the atmosphere in the ultraviolet region of the spectrum. Data from TOMS can be interpreted to obtain a measure of the total column density of O_3. The operating principle is relatively simple: the intensity of ultraviolet sunlight reflected by the atmosphere is set by a competition between scattering of incident sunlight, mainly by the more abundant molecular constituents of the atmosphere (N_2 and O_2), and absorption by O_3. Clever algorithms (mathematical procedures implemented on a computer with minimal human involvement) are employed to translate the optical signals measured by TOMS, to obtain densities for the underlying column of O_3. Results for O_3 can be displayed in the form of false-color coded maps, or even movies, permitting the analyst to view large quantities of data in a relatively painless manner. The TOMS instrument was launched on the Nimbus 7 satellite in October 1978. Since then, it and its successors have provided essentially continuous coverage of stratospheric O_3.

7. Marine organisms are thought to provide the dominant natural source of CH_3Br. The gas is also produced by chemical industry. Industrial applications involve mainly the use of CH_3Br to kill unwanted microbes, either in soils or on food held in storage—much of the fruit and vegetables entering the United States, for example, is treated with the compound. Our understanding of the origin, distribution, and chemistry of brominated gases in the atmosphere is significantly less developed than that for chlorine. This is understandable for two reasons. First, the abundance of brominated species in the atmosphere is typically much less than that of comparable chlorinated compounds, 600 ppt for CH_3Cl, for example, as compared with about 12 ppt for CH_3Br; measurements of brominated compounds in the atmosphere are, consequently, more difficult than those for the analogous chlorinated species. Second, chlorine has simply received more attention; this may be understood as the element has been perceived as the major threat to stratospheric O_3 since Molina and Rowland's paper appeared in 1974. It is likely that bromine will receive increasing attention in the future. As will be discussed below, there are reasons to believe that bromine, acting in consort with chlorine, may be significantly implicated in the loss of O_3 reported to have taken place in recent years in both the polar regions and at midlatitudes.

Chapter 15

1. The results noted here were abstracted from a major report published in 1985 entitled *Atmospheric Ozone 1985, Assessment of Our Understanding of the Processes Controlling its Present Distribution and Change*. Approximately 150 scientists from more than 11 countries contributed to the preparation of this report: it was sponsored by the National Aeronautics and Space Administration, the Federal Aviation Administration, the National Oceanic and Atmospheric Administration, the United Nations Environment Program, the World Meteorological Organization, the Commission of the European Communities, and the Bundesministerium für Forschung und Technologie. It provided a key input to the policy process underway in the 1980s, including the Convention for the Protection of the Ozone Layer held in Vienna, Austria, in March 1985, and culminating in the Montreal Protocol on Substances that Deplete the Ozone Layer, an influential agreement drawn up at an historic international convention held in September 1987 in Montreal, Canada. The results quoted here are based on a combination of 1- and 2-dimensional models (altitude alone in the first case, altitude and latitude in the second). The 2-dimensional models available in 1985 (simulating effects of gas-phase chemistry only) suggested that, in steady state (a hundred years or more in the future), a constant release of CFC-11 and CFC-12 at 1980 levels might be expected to cause a rise in the abundance of stratospheric chlorine to about 8 ppb and to result in a reduction in the column density of O_3 by about 4% in the tropics, 9% at midlatitudes, and as much as 14% at high latitudes, corresponding to a globally averaged reduction of about 9%. More modest reductions, a factor of two or so less, were anticipated when models allowed for increases in CO_2 (resulting in a decrease in stratospheric temperatures) and CH_4 (resulting in enhanced conversion of radical chlorine to HCl). Reductions in the column density of O_3 expected for the decade of the 1980s on the basis of gas phase models were small, about 1% or so; the total chlorine loading of the stratosphere increased over this interval from about 2.5 to 3.6 ppb.

2. Vertical-column densities of O_3 are usually expressed in terms of a system of units named in honor of the British meteorologist, G. M. B. Dobson, who pioneered early observations of O_3: 1 Dobson unit (DU) corresponds to a vertical column density of 2.7×10^{16} mol cm^{-2}, equivalent to the number of molecules contained in a column of length 10^{-3} cm and unit cross-sectional area (1 cm^2) maintained at STP (standard temperature and pressure for which the number density of molecules is equal to 2.7×10^{19} cm^{-3}). The data in Plate 5 reflect a synopsis of results from the network of ground-based stations, the Dobson network, which maintains a schedule of routine observations of O_3.

3. The vortex results from the large decrease in temperature that occurs in passing from mid to high latitudes, especially during winter. The temperature gradient is associated with a corresponding pressure gradient; pressure at a given altitude at high latitudes is less than at a comparable altitude at midlatitudes (reflecting requirements of the barometric law and the difference in scale heights: see Section 7.1). The rapid drop in pressure from north to south (in the Antarctic environment) results in a high-speed wind blowing from west to east. This is required to satisfy the force balance implied by geostrophy (see Section 8.6). As a consequence of the strong zonal (westerly) flow, air inside the vortex is essentially isolated from outside air. Thus, ozone lost inside the vortex is not readily replaced by transport from the exterior.

4. A small concentration of aerosols composed primarily of sulfuric acid is present throughout the stratosphere. According to present understanding, the sulfur in these particles is derived mainly from SO_2 vented directly into the stratosphere by volcanoes, with a possible background contributed by the oxidation of COS. As the sulfate aerosols cool, they grow by absorbing H_2O. Water accounts for about 50% of the mass of the sulfate particles at a temperature of about 200 K. Nitric acid freezes in combination with H_2O at a temperature of about 196 K to form NAT, the freezing temperature depending on the abundance of both H_2O and HNO_3. We now believe that the clouds appearing in the polar stratosphere in this temperature range, and which are visible from the ground, are composed for the most part of NAT. Interestingly, while PSCs had been observed and described in the scientific literature by the Norwegian physicist Carl Stormer as early as 1929, the role of NAT was discovered only in 1986 (Figure 15.4). Water freezes under stratospheric conditions when temperatures drop below about 188 K. Particle sizes dramatically increase at this point. The clouds resulting from freezing of H_2O in the polar stratosphere are referred to as *Type 2 PSCs*.

5. The ER-2 aircraft is operated as a high-altitude research platform by the National Aeronautics and Space Administration (NASA). A single-piloted aircraft patterned after the U2 spy plane, it can operate to altitudes as high as about 20 km.

6. The instrument used to measure ClO and BrO on the ER-2 was developed by James Anderson and associates at Harvard University over a period of less than a year in 1986–1987, responding to the need for data on the composition of the Antarctic stratosphere. Adapted from an earlier design used to measure ClO from high-altitude balloons, the instrument worked the first time it was deployed on the ER-2, a tribute to the remarkable experimental and engineering skills of Anderson and his colleagues.

7. *Dry deposition* is the term used to describe removal from the atmosphere of gases or aerosols due to their interaction with materials at Earth's surface. Capture by vegetation is particularly significant in this context, although loss due to solution in surface liquids, such as the ocean, is also important, particularly for soluble compounds such as sulfates. The term *wet deposition* refers to removal by precipitation, by either rain or snow. Use of the adjective *dry* to describe loss by transfer to the ocean may seem paradoxical in this context; the epithet as employed here refers to the form of the compound when it arrives at the surface, without regard to possible transformation after it crosses the air-surface boundary.

8. A particle of radius R presents a cross section, or target area, of magnitude πR^2 for collision with an incoming N_2O_5 molecule (the "size" of the molecule is assumed to be small compared with the size of the particle). Using arguments similar to those employed in introducing the perfect gas law earlier in Section 2.9, it should be clear that the number of N_2O_5 molecules colliding with a particular aerosol in unit time (sec^{-1}) may be expressed as a product of the number density of N_2O_5 molecules, their average speed, and the area of the aerosol target (check the dimensions: $[N_2O_5]\, V_{N_2O_5}\, \pi R^2 = (cm^{-3})(cm\ sec^{-1})(cm^2) = sec^{-1}$. Suppose for simplicity that all particles have the same radius R, and let N (cm^{-3}) denote the number of particles present in unit volume. It follows that the number of collisions between N_2O_5 and aerosols taking place in unit volume in unit time is given by $N \times [N_2O_5](V_{N_2O_5})(\pi R^2)$. The surface area of an individual aerosol is given by $4\pi R^2$. The total surface area included in unit volume, *SA*, may be obtained by adding up the areas of all the individual particles, $SA = N(4\pi R^2)$. With this expression for *SA*, the rate ($cm^{-3}\ sec^{-1}$) for N_2O_5–aerosol collisions may be expressed in the form

$$R = \frac{(SA)V_{N_2O_5}}{4}$$

The expression (15.11) allows for a range of particle sizes so long as *SA* is interpreted as the cumulative or integrated area for the surfaces of particles present in the sample (unit) volume.

9. Under these circumstances, we say that the rate-limiting step for transfer of NO_x to HNO_3 is imposed by reaction (14.19) rather than (15.9). Imagine the flow of traffic along a highway that narrows from four lanes in region A to three lanes in region B to only two lanes in region C. Under rush-hour conditions, traffic will back up, and we expect to find a high density of frustrated motorists in region A. The bottleneck, or rate-limiting step, is imposed in this case by region C. The traffic problems can be alleviated to some extent by adding another lane in region C. Adding two lanes, making region C similar to A, has little additional effect; the bottleneck or rate-limiting step switches now to the restrictions represented by region B. Turning on heterogeneous chemistry is equivalent to adding lanes to region C: it results in an increase in the rate at which NO_x in region A is converted to HNO_3 in region C. Eventually, however, the flow of NO_x in A will be limited by the rate at which NO_x can be converted to N_2O_5 in region B.

Chapter 16

1. The expression for θ given by (16.1) follows from an integration of equation (7.29) with $\Delta Q = 0$:

$$\frac{\Delta T}{T} = \left(\frac{R}{c_v + R}\right)\left(\frac{\Delta p}{p}\right)$$

Integrating from the level where $p = 1000$ mb, denoting local values of temperature and pressure by T and p, respectively, it follows that

$$\log\frac{T}{\theta} = \left(\frac{R}{c_v + R}\right)\log\frac{p}{1000}$$

Thus,

$$\frac{T}{\theta} = \left(\frac{p}{1000}\right)^{\frac{R}{c_v + R}}$$

and

$$\theta = T\left(\frac{1000}{p}\right)^{\frac{R}{c_v + R}}$$

Here θ is equivalent to the temperature that would apply if air was adiabatically displaced ($\Delta Q = 0$) to a pressure of 1000 mb. For dry air, $R/(c_v + R) = 0.288$.

2. The dynamical properties of an air mass are conveniently defined in terms of a quantity known as **potential vorticity**, which is related to local angular momentum. Potential vorticity is

conserved to a significant extent as air masses move around the atmosphere. The potential vorticity of air in the stratosphere differs significantly from that of air in the troposphere. Air entering the troposphere from the stratospheric region defined by the shaded area in Figure 16.1 will tend to retain its distinct values of potential vorticity. Under this circumstance it may be expected to return to the stratosphere with minimal mixing between air of stratospheric and tropospheric origins. Occasionally, though, transfer is irreversible. Air enters the stratosphere from the troposphere primarily in the tropics; it returns to the troposphere mainly by irreversible exchange across the boundary of the shaded region in Figure 16.1. Irreversible exchange is usually associated with complex deformations of the local tropopause (extensive equatorward displacements or localized vertical folds, for example). The mechanisms responsible for the return of air from the stratosphere to the troposphere are not well understood; it is clear, though, that the flux of mass entering the troposphere from the stratosphere in the midlatitudinal regime must be approximately equal to the flux of air entering the stratosphere in the tropics. The processes regulating upward transfer in the tropics are elaborated later in this Section.

3. The poleward direction of the flow induced by breaking Rossby waves may be understood as a consequence of the Coriolis force. The breaking waves are responsible for a westward motion. The Coriolis force causes this flow to deviate to the right (north) in the Northern Hemisphere, to the left (south) in the Southern Hemisphere, as discussed in Chapter 8.

4. WMO (1999) reported evidence for a drop in column O_3 by an average of about 2% for the latitude band 60°S to 60°N over the decade of the 1980s. Reductions were smallest in the tropics, greatest at higher latitudes in both hemispheres. The rate of decrease of column O_3 slowed significantly in the 1990s, presumably reflecting (among other factors) a decrease in the growth rate of the stratospheric chlorine concentrations. The mixing ratio of inorganic chlorine in the stratosphere increased by approximately 50% over the decade of the 1980s. It reached a maximum in 1995. It has now begun a slow secular decrease in response to measures implemented to limit emissions of industrial precursors (see data presented in Chapter 5).

5. The relatively high values of the column densities of O_3 at high latitudes reflect the tendency for O_3 to accumulate in this region, especially during winter, when the lifetime of O_3 is comparatively long. A large fraction of the O_3 column at high latitudes is located in the lower stratosphere. Column densities predicted by the model for high latitudes are thus sensitive to assumptions made concerning the location of the tropopause. Column densities could be increased or decreased, by lowering or raising, the height of the tropopause. They may also be sensitive to intrusion of air from midlatitudes associated with dynamical processes operating in a shallow boundary layer in the immediate vicinity of the tropopause; since these processes are not resolved by the model, its results are, consequently, uncertain in detail. The agreement between model and observed values of the O_3 column at high latitudes may be considered fortuitous to some extent. Further work is required to improve the simulation of transport in the lower stratosphere at high latitudes. Despite these limitations, the model provides a reasonable first-order description of the importance of transport in regulating the global distribution of O_3. In particular, it quantitatively accounts for the low abundance of column O_3 observed in the tropics and at least qualitatively for the sense of the gradient between low and high latitudes.

6. The model used here provides a reasonable representation of the chemical reactions (gaseous phase and heterogeneous) involved in removal of O_3 at low and mid latitudes; it is deficient, however, in that it does not allow for the influence of reactions taking place on the surface of polar stratospheric clouds. It is unable therefore to account for the conspicuous loss of O_3 observed in recent years at high latitudes of the Southern Hemisphere during austral spring, as indicated by the data included in the lower panels of Plate 5.

7. The impact could be exacerbated by higher concentrations of CH_4. As previously noted, the oxidation of CH_4 provides a significant source of stratospheric H_2O. This source is especially important at higher latitudes in the descending branches of the stratospheric circulation cells.

Chapter 17

1. Methane is a product of bacterially mediated fermentation. The bulk reaction involved in synthesis of CH_4 may be written in the form

$$CH_2O + CH_2O \rightarrow CH_4 + CO_2$$

The carbon atom in CH_4 is reduced relative to the carbon in the biological source material CH_2O, from 0 to –4, while the carbon in CO_2 is oxidized by a comparable amount, from 0 to +4. Carbon monoxide is produced by the oxidation of biomass material in fires; the oxidation state of carbon is raised in this case from 0 to +2. Combustion involves the release of energy by oxidation of carbon. Emission of CO from fire indicates that the combustion process is incomplete. More efficient combustion, a larger yield of energy from a given supply of fuel, would require more complete oxidation of carbon: a larger yield of CO_2 relative to CO. As will be seen below, CO is also produced as a by-product of the oxidation of CH_4 in the atmosphere. The oxidation state of carbon is raised in this case from –4 to +2.

2. The term **bulk reaction** is intended to indicate the net result of what may involve a complex set of reactions. It is important to distinguish between a reaction equation describing a specific kinetic path and one denoting a bulk process. In the latter case, products may be evolved in the absence of direct interaction between the specified reactants. The rate of a bulk reaction is not necessarily proportional to the product of the concentrations of reactants; to estimate the rate of a bulk reaction, one must describe the individual chemical steps, the so-called **elementary reactions**, contributing to the overall bulk process.

3. Photolysis can occur as a result of the absorption of either direct or scattered solar radiation. Scattering by small particles

and air molecules is particularly important for the short wavelengths implicated in reaction (17.3). The molecular contribution is known as *Rayleigh scattering* in honor of the English physicist Lord Rayleigh. The cross section for Rayleigh scattering varies inversely as the fourth power of the wavelength; short waves are scattered more readily than long, accounting for the blue color of the sky on a clear day.

4. As indicated in Figure 10.34, column abundances of O_3 are largest in the Northern Hemisphere at high latitudes during late winter or early spring, lowest year-round in the tropics. The peak is shifted in the Southern Hemisphere toward lower latitudes, reflecting differences between the hemispheres in the strength and variability of winds in the stratosphere responsible for meridional redistribution of O_3. The value of J_3 is sensitive to the total quantity of O_3 intercepted by sunlight as it passes through the stratosphere. It depends, therefore, on both the column density of O_3 and the obliquity of the path followed by sunlight as it passes through the atmosphere, the latter determined by the solar zenith angle. We expect production of OH to be largest for a given day at a given location at local noon (when the zenith angle is a minimum), to be largest for a given latitude during summer (when solar zenith angles are smallest and when column abundances of O_3 are least), and to be greatest overall in the tropics (where concentrations of tropospheric H_2O are highest and where column abundances of O_3 are least).

5. We included a relatively large range of values for the column density of O_3 in Figure 17.5, in part to provide perspective on how J_3, and consequently tropospheric chemistry, may have varied in the past. Relatively little information is available to constrain the abundance of stratospheric O_3 for climates very different from today—for ice ages, for example, or for the warm conditions that applied in the Cretaceous. One can imagine circumstances that would have resulted in abundances of stratospheric O_3 either higher or lower than today: the concentrations of gases centrally involved in O_3 chemistry (H_2O, CH_4, and NO_x) could have been very different; the dynamical regime could have fluctuated contributing to a change in the balance of chemistry and dynamics in regulating O_3. Changes in stratospheric O_3, by modulating the value of J_3, could have been responsible for significant variations in the abundance of OH in the past with implications for a range of important constituents of the atmosphere, including CH_4 and CO. Reductions in stratospheric O_3 resulting from human activity today could contribute to slightly higher abundances of tropospheric OH, compensating, in part at least, for a possible decrease in OH associated with industrial emissions of CO (higher concentrations of CO would result in accelerated removal of OH).

6. The lifetime of HO_x, set by reactions (17.14) and (17.20), is approximately 12 hours. Only a fraction of the H_2O_2 produced in a 12-hour interval will survive, however, reflecting the influence of the removal processes. The concentration of H_2O_2 will continue to rise until the steady-state limiting condition is attained. Several lifetimes are typically required to achieve this result. We implicitly assumed in our discussion here that removal of H_2O_2 by (17.20) is a continual process. In practice, heterogeneous loss will take place episodically, mainly in conjunction with precipitation (wet deposition is the dominant mechanism involved in 17.20). It follows that we might expect a degree of spatial variability in the concentration of H_2O_2. The concentration would be relatively low in air masses subject recently to precipitation; it would reach its full steady-state limit only under circumstances where the air mass remained dry (rain or snow absent) for a period of several days or longer.

7. In our treatment of the stratosphere in Chapters 13–16, it was instructive, in discussing the chemistry of an oxygen-nitrogen-hydrogen atmosphere, to include both NO_2 and HO_2 in our definition of odd oxygen. We included HO_2 in recognition of the fact that the compound was produced in the stratosphere mainly by reaction of OH with O_3, meaning it was formed by adding an O atom to OH. In the troposphere, however, HO_2 is produced mainly by reaction of H with O_2, reaction (17.15). Including HO_2 in the odd oxygen family is clearly less appropriate in this case. With the definition of odd oxygen adopted for the stratosphere, reaction of OH with O_3 forming HO_2 involves no net change in the abundance of odd oxygen. In contrast, with HO_2 omitted from the definition of the odd oxygen in the troposphere, reaction of OH with O_3 forming HO_2 must be classified as a sink for tropospheric odd oxygen.

8. More specifically, ϵ is given by the rate for removal of HO_x expressed as a ratio with respect to the total rate for removal of H_2O_2:

$$\epsilon = \frac{k_{14}[\mathrm{OH}] + k_{20}}{k_{14}[\mathrm{OH}] + J_{19} + k_{20}}$$

9. If we assume that $P(t) = 0$, equation (17.84) admits a solution of the form:

$$N(t) = N(t_0)\exp\{-(0.85)v(t - t_0)\}$$

It is assumed here that the mixing ratio of gas X is effectively constant throughout the atmosphere. Implicitly, this presumes that the lifetime of the gas is long compared with the time required for mixing over large spatial scales. The time required for a gas to be mixed efficiently between the Northern and Southern Hemispheres is a little longer than one year. The formulation described here should not be applied, therefore, to gases with lifetimes of less than a few years. For such applications, it would be necessary to account for the spatial distribution of X and for the spatial distribution of its source and sink.

10. Reaction (17.6) is associated with production of two molecules of OH. Production of odd oxygen (NO_2) by (17.16) accounts for an additional OH. Note that (17.6) is but one of several possible loss mechanisms for O_3. Reaction (17.31), for example, accounts for a loss of both OH and O_3. To the extent that it is followed by (17.16), however, net loss of both OH and O_3 may be neglected. The analysis pursued here implicitly assumes that the abundance of NO_x is sufficiently high that

reaction (17.16) should provide a significantly larger sink for HO_2 than reaction (17.17).

11. The factor of three assumes that oxidation of a reduced species in the presence of NO_x is associated with production of one molecule of O_3. This conclusion is rigorously valid for CO. As discussed in Sections 17.3 and 17.4, however, oxidation of hydrocarbons may be associated with production of several molecules of O_3. The yield of O_3 should be increased accordingly.

12. The rates for production of CH_4 quoted here were derived assuming an overall lifetime for CH_4 of 8.4 yr. This result accounts for removal by reaction with OH in the troposphere, corresponding to a lifetime of 9.24 yr, as derived by Spivakovsky et al. (2000), and for removal in the stratosphere (mainly by reaction with OH, with an additional contribution due to reaction with Cl), accounting for a lifetime of 93 yr (Volk et al. 1997).

13. The ^{14}C content of fossil fuels is essentially zero, reflecting the fact that these fuels have been isolated from the atmosphere, and consequently from sources of ^{14}C, for a long time, compared with the lifetime of ^{14}C. Wahlen et al. (1989) estimate that fossil sources contribute 21 ±3% to the contemporary budget of atmospheric CH_4. Quay et al. (1991) favor a somewhat smaller contribution, about 16%, but estimate uncertainty in specification of this result at ±12%. The data in Table 17.8 envisage fossil sources ranging from 15–19%.

14. The globally averaged value of $\delta^{13}C$ for CH_4 produced by burning of biomass is most likely somewhat larger than the value quoted here. It should be raised to account for the higher value of $\delta^{13}C$ associated with combustion of C4 plants in savanna regions.

Chapter 18

1. In the introductory discussion of acids and bases presented in Chapter 3, we referred to protons in the liquid phase as either H_3O^+ or H^+_{aq}. The former convention emphasized the fact that protons in the liquid phase are generally attached to water molecules; they are present mainly as hydrates. In the present discussion, aqueous phase protons are denoted simply as H^+.

2. Values for the liquid content of clouds and fogs quoted here were taken from Seinfeld and Pandis (1997). The difference between the mixing ratio of H_2O at the ground and the mixing ratio corresponding to the equilibrium-saturation vapor pressure within the cloud provides a measure of water available to form liquid. Similarly, the difference between the mixing ratio of H_2O prior to fog formation and the mixing ratio corresponding to equilibrium between vapor and liquid at the temperature prevailing when the fog is present allows an estimate to be made of the potential liquid content of air present in the fog.

3. The careful reader will note that the values of the equilibrium constants, including the Henry's law coefficient, quoted here differ significantly from values used in Chapter 11 to describe the equilibrium of carbon species in the ocean. The difference is real, reflecting differences in the chemical properties of salt water in the ocean, as compared to the chemical properties of relatively pure water in the atmosphere.

4. Klassen (1991) estimates emission factors (kg NH_3 per head per year) for various animals as follows: cattle, 23; sheep and goats, 2.2; horses, 12.5; pigs, 5.0; poultry, 0.3. The value for humans is given as 0.3 kg NH_3 person^{-1} yr^{-1}.

5. Each molecule of SO_4^{2-} contributes two negative units of charge. Thus, 1 μ mol l^{-1} of SO_4^{2-} corresponds to 2 μ mol charge equivalent l^{-1} of SO_4^{2-} (1 μ mol = 10^{-6} mol). Each molecule of NO_3^- contributes one negative unit of charge, while a molecule of NH_4^+ is associated with one positive unit of charge. It follows that 1 μ mol l^{-1} of either NO_3^- or NH_4^+ is equivalent to 1 μ mol charge equivalent l^{-1} of either NO_3^- or NH_4^+.

Chapter 19

1. Remember that, while computers are able to carry out large numbers of arithmetic operations with great speed and precision, they are less capable when it comes to confronting the challenge presented by even the most elementary manipulations involved in differential calculus.

2. The accuracy of equation (19.1) as a representation of the derivative depends on the details of the variation of y as a function of x on space scales of order Δx. If variations are large, as measured by the second and higher derivatives, the approximation afforded by (19.1) will be unreliable. The resolution of the model must be increased in this case (meaning the value of Δx must be reduced) to ensure a satisfactory linear approximation to the derivative.

3. The point here is that, when models allow only for the change in radiative forcing associated with greenhouse gases, they predict an increase in surface temperature larger than was observed. Accounting in a simple manner for cooling associated with the direct effect of sulfate aerosols provides an additional degree of freedom and guarantees to improve the agreement between model and observed values of global-mean surface temperature. A more realistic treatment of the impact of sulfur could have resulted in either a better or a worse agreement between model and observations. Also, in the absence of a focused research program, it is difficult to gauge the importance of other pollutants, such as soot. Under the circumstances, we must conclude that the improvement in the agreement with observations reported by the current generation of models could be fortuitous. It cannot be taken as an unqualified proof of success for the ability of the models to provide a realistic simulation of the role of greenhouse gases for either the past or the future.

4. Take a look at a map defining the boundaries of the world's major ocean basins. Consider the problem to be confronted if you are required to represent these boundaries with a set of grid points spaced by intervals of no closer than a few degrees of latitude and longitude. The problem is even more daunting if you need to additionally simulate details of bottom topography.

5. The nature of vegetative cover in a particular region is determined ultimately by climate, by the seasonal pattern of variation in temperature and precipitation. A change in climate can lead to a change in vegetation. This can affect the albedo of the region and thus its capacity to absorb solar radiation. Through changes in the nature of root systems, a change in vegetative cover may result in changes in the physical properties of soils, with implications for the capacity of soils to retain water.

Changes in vegetation can have an influence also on transpiration rates, with additional consequences for the water budget of the region. To properly allow for these feedbacks, a climate model must account not only for equilibrium properties of ecosystems but also for the manner in which these systems evolve in response to external stimuli imposed by environmental change.

6. Recall that temperatures of ocean water are generally much higher than temperatures of the overlying atmosphere during winter at high latitudes. Differences in air and water temperatures could range as high as 40°C or even more.

7. The problem arises as a consequence of dynamical imbalances introduced in the selection of initial conditions. These imbalances trigger motions in both the atmosphere and ocean, for which there are no analogues in nature. In this sense they are nonphysical. One would hope that the imbalances would damp out over time and that the atmosphere and ocean would evolve to a more realistic configuration; the adjustment, however, may require changes in the properties of the deep ocean. The residence time of water in the deep ocean is about 1000 yr in the present environment, indicating that the time scale for relaxation could be extremely long. In fact, there is no guarantee that the system will evolve to an acceptable final state, given the inevitable accumulation of computational errors.

8. The time scale for variability of climate associated with fluctuations in oceanic circulation is likely much longer than the time scale associated with random variations in atmospheric circulation. As previously indicated, models employed in climate studies are unable to account for changes in climate associated with the El Niño and/or La Niña phenomenon. It would not be surprising if they were deficient also in their ability to simulate effects of fluctuations in oceanic circulation on longer time scales. We have previously suggested (Chapter 10) that the changes in climate associated with the Medieval Optimum and the Little Ice Age may be due to century-scale variations in the circulation responsible for the supply of relatively cold water from the subtropics to the tropics in the Pacific. We have speculated that these changes could have arisen as a result of stochastic fluctuations in the salinity of surface waters in the subtropics. A relatively modest temporary change in ocean circulation, a decrease in the strength of the conveyor belt carrying cold water form the subtropics to the tropics, could have been responsible for the warming observed between 1910–1940. This raises an interesting question: if the model is unable to account for the increase in the global-average temperature of 0.4°C observed over a period of 30 years in the early part of the twentieth century, and if the increase is ascribed to a fluctuation associated with natural noise in the climate system, can we be sure that the warming (of comparable magnitude) observed over the last 30 years of the century is not simply the manifestation of a similar, natural oscillation in the complex dynamics of the coupled atmosphere-ocean climate system?

9. As discussed in Chapters 9 and 10, the water that sinks from the surface at high latitudes is supplied ultimately by the northward flow of water from lower latitudes, in particular by a northward extension of the Gulf Stream. A decrease in the strength of the thermohaline circulation results in a decrease in poleward heat transport by the ocean. The Gulf Stream would assume a more southerly track in its passage across the Atlantic. Warming contributed by higher concentrations of greenhouse gases is offset partially by a decrease in heat transport by the ocean.

References

Ahrens, C. D. *Meteorology Today: An Introduction to Weather, Climate and the Environment.* Fifth edition. St. Paul: West Publishing Company, 1994.

Anderson, J. G., W. H. Brune, and M. H. Proffitt. "Ozone Destruction by Chlorine Radicals Within the Antarctic Vortex: The Spatial and Temporal Evolution of $ClO\text{-}O_3$ Anticorrelation Based on In Situ ER-2 Data." *Journal of Geophysical Research* 94 (1989): 11465–79.

Anderson, J. G., D. W. Toohey, and W. H. Brune. "Free Radicals Within the Antarctic Vortex: the Role of CFC's in Antarctic Ozone Loss." *Science* 251 (1991): 39–46.

Balkanski, Y. J., D. J. Jacob, G. M. Gardner, W. M. Graustein, and K. K. Turekian. "Transport and Residence Times of Continental Aerosols Inferred from a Global 3-Dimensional Simulation of ^{210}Pb." *Journal of Geophysical Research* 98 (1993): 20573–86.

Bard, E., M. Arnold, P. Maurice, J. Duprat, J. Moyes, and J.C. Duplessy. "Retreat Velocity of the North Atlantic Polar Front During the Last Deglaciation Determined by ^{14}C Accelerator Mass Spectrometry." *Nature* 328 (1987): 791–94.

Barendregt, R. W., and E. Irving. "Changes in the Extent of North American Ice Sheets During the Late Cenozoic." *Canadian Journal of Earth Sciences* 35 (1998): 504–09.

Barnola, J. M., D. Raynaud, C. Lorius, and N. I. Barkov. 1999. Historical CO_2 record from the Vostok ice core. In *Trends: A Compendium of Data on Global Change.* Oak Ridge, Tenn.: Carbon Dioxide Information Analysis Center, Oak Ridge National Laboratory, U.S. Department of Energy.

Baron, W. R. "The Reconstruction of Eighteenth-Century Temperature Records through the Use of Content Analysis." *Climat. Change* 4 (1982): 385–89.

Battle, M., M. Bender, T. Sowers, P. P. Tans, J. H. Butler, J. W. Elkins, J. T. Ellis, T. Conway, N. Zang, P. Lang, and A. D. Clark. "Atmospheric Gas Concetrations Over the Past Century Measured in Air from Fern at the South Pole." *Nature* 382 (1996): 231–35.

Berger, W. H., C. G. Adelseck Jr., and L. A. Mayer, "Distribution of carbonate in surface sediments of the Pacific Ocean." *Journal of Geophysical Research* 81 (1976): 2617–27.

Berner, R. A. "Phosphate Removed from Sea-Water by Absorption on Volcanogenic Ferric Oxides." *Earth and Planetary Science Letters* 18 (1973): 57–86.

Berner, R. A., A. C. Lasaga, and R. M. Garrels. "The Carbonate-Silicate Geochemical Cycle and Its Effect on Atmospheric Carbon Dioxide Over the Past 100 Million Years." *American Journal of Science* 283 (1983): 641–83.

Birchfield, G. E., J. Weertman, and A. T. Lunde. "A Paleoclimate Model of Northern Hemisphere Ice Sheets." *Quaternary Research* 15 (1981): 126–42.

Birchfield, G. E., and W. S. Broecker. "A Salt Oscillator in the Glacial Atlantic? 2. A 'Scale Analysis' Model." *Paleoceanography* 5 (1990): 835–43.

Biscayne, P. E., V. Kolla, and K. K. Turekian. "Distribution of Calcium Carbonate in Surface Sediments of the Atlantic Ocean." *Journal of Geophysical Research* 81 (1976): 2595–603.

Blunier, T., J. Chappellaz, J. Schwander, A. Dallenbach, B. Stauffer, T. F. Stocker, D. Raynaud, J. Jouzel, H. B. Clausen, C. U. Hammer, and S. J. Johnsen. "Asynchrony of Antarctic and Greenland Climate Change During the Last Glacial Period." *Nature* 394 (1998): 739–43.

Bonnefille, R., J. C. Roeland, and J. Guiot. "Temperature and Rainfall Estimates for the Past 40,000 Years in Equatorial Africa." *Nature* 346 (1990): 347–49.

Broecker, W. S. "The Biggest Chill." *Natural History* 97 (1987): 74–82.

Broecker, W. S. *The Glacial World According to Wally.* Palisades, N.Y.: Eldigio Press, 1995.

Broecker, W. S., G. Bond, and M. Klas. "A Salt Oscillator in the Glacial Atlantic? 1. The Concept." *Paleoceanography* 5 (1990a): 469-77.

Broecker, W. S., and G. H. Denton. "The Role of Ocean-Atmosphere Reorganizations in Glacial Cycles." *Geochimica et Cosmochimica Acta* 53 (1989): 2465–501.

Broecker, W. S., and T. H. Peng. *Tracers in the Sea.* Palisades, N.Y.: Eldigio Press, 1982.

Broecker, W. S., and J. Van Donk. "Insolation Changes, Ice Volumes and the ^{18}O Record in Deep-Sea Cores." *Reviews of Geophysics and Space Physics* 8 (1970): 169–98.

Brook, E. J., T. Sowers, and J. Orchardo. "Rapid Variations in Atmospheric Methane Concentration During the Past 110,000 Years." *Science* 273 (1996): 1087–91.

Butler, J. H., S. A. Montzka, A. D. Clarke, J. M. Lobert, and J. W. Elkins. "Growth and Distribution of Halons in the Atmosphere." *Journal of Geophysical Research* 103 (1998a): 1503–11.

Butler, J. H., J. W. Elkins, S. A. Motzka, T. M. Thompson, T. H. Swanson, A. D. Clarke, F. L. Moore, D. F. Hurst, P. A. Romashkin, S. A. Yvon-Lewis, J. M. Lobert, M. DiCorleto, G. S. Dutton, L. T. Lock, D. B. King, R. E. Dunn, E. A. Ray, M. Pender, P. R. Wamsley, and C. M. Volk. "Nitrous Oxide and Halocompounds." In *Climate Monitoring and Diagnostics Laboratory Summary Report No. 24 1996–1997.* Eds. D. J. Hofmann, J. T. Peterson, and R. M. Rosson. 91–121. Springfield, Va.: National Technical Information Service.

Callis, L. B., and M. Natarajan. "Ozone and Nitrogen Dioxide Changes in the Stratosphere during 1979–84." *Nature* 323 (1986): 772–77.

Carter, L. D., J. Brigham-Grette, L. Marincovich Jr., V. L. Pease, and J. W. Hillhouse. "Late Cenozoic Arctic Ocean Sea Ice and Terrestrial Paleoclimate." *Geology* 14 (1986): 675–78.

Catchpole, A. J. W., and T. F. Ball. "Analysis of Historical Evidence of Climatic Change in Western and Northern Canada." In *Syllogeous 33: Climatic Change in Canada.* 96–148. Ottawa: National Museum of Natural Science, 1981.

Chamberlain, J. W., and D. M. Hunten. *Theory of Planetary Atmospheres: An Introduction to their Physics and Chemistry.* New York: Academic Press, 1987.

Chance, K. V., D. G. Johnson, and W. A. Traub. "Measurement of Stratospheric HOCl–Concentration Profiles, Including Diurnal-Variation." *Journal of Geophysical Research* 94 (1989): 11059–69.

Chapman, S. "A Theory of Upper Atmospheric Ozone." *Memoirs of the Royal Meteorological Society* 3 (1930): 103–23.

Charlson, R. J., J. E. Lovelock, M. O. Andreae, and S. G. Warren. "Oceanic Phytoplankton, Atmospheric Sulphur, Cloud Albedo, and Climate." *Nature* 326 (1987): 655–61.

Clark, P. U., and D. Pollard. "Origin of the Middle Pleistocene Transition by Ice Sheet Erosion of Regolith." *Paleoceanography* 13 (1998): 1–9.

CLIMAP Project Members. "Seasonal Reconstruction of the Earth's Surface at the Last Glacial Maximum." *Geological Society of America* Map and Chart Series, no. 36. (1981).

Cooke, D. W., and J. D. Hays. "Estimates of Antarctic Ocean Seasonal Sea-Ice Cover During Glacial Intervals." In *Antarctic Geoscience.* Ed. C. Cradduck et al. Madison, Wis.: University of Wisconsin Press, 1982.

Crutzen, P. J. "SST's: A Threat to the Earth's Ozone Shield." *Ambio* 1 (1972): 41–51.

Dawson, M. R., R. M. West, W. Langston Jr., and J. H. Hutchinson. "Paleogene Terrestrial Vertebrates: Northernmost Occurrence, Ellesmere Island, Canada." *Science* 192 (1976): 781–82.

Delwiche, C. C. *The Biosphere.* San Francisco: W. H. Freeman, 1970.

Dickinson, R. E., A. Henderson-Sellers, and P. J. Kennedy. *Biosphere-Atmosphere Transfer Scheme (BATS) Version Ie as Coupled to the NCAR Community Climate Model.* Report TN–387+ STR. Boulder, Colo.: National Center for Atmospheric Research, 1993.

Dixon, R. K., S. Brown, R. A. Houghton, A. M. Solomon, M. C. Trexler, and J. Wisnieki. "Carbon Pools and Flux of Global Forest Ecosystems." *Science* 163 (1994): 185.

Dlugokencky, E. J., K. A. Masarie, P. M. Lang, and P. P. Tans. "Continuing Decline in the Growth Rate of the Atmospheric Methane Burden." *Nature* 393 (1998): 447–50.

Douglas, R. G., and F. Woodruff. "Deep Sea Benthic Foraminifera." In *The Sea, Volume 7: The Oceanic Lithosphere.* Ed. C. Emiliani. New York: Wiley-Interscience, 1981.

Duplessy, J. C., A. W. H. Be, and P. L. Blanc. "Oxygen and Carbon Isotopic Composition and Biogeographic Distribution of Planktonic Foraminifera in the Indian Ocean." *Palaeogeography, Palaeoclimatology, Palaeoecology* 33 (1981): 9–46.

Duetsch, H. U. "The Ozone Distribution in the Atmosphere." *Canadian Journal of Chemistry* 52 (1974): 1491–504.

Dye, J. E., D. Baumgardner, B. W. Gandrud, S. R. Kawa, K. K. Kelly, M. Loewenstein, G. V. Ferry, K. R. Chan, and B. L. Gary. "Particle Size Distribution in Arctic Polar Stratospheric Clouds, Growth and Freezing of Sulfuric Acid Droplets, and Implications for Cloud Formation." *Journal of Geophysical Research* 97 (1992): 8015–34.

Eddy, J. A. "The Solar Output and Its Variations." Ed. O. R. Whitz. Boulder, Colo.: Colorado Associated Universities Press, 1977.

Elkins, J. W., S. C. Wofsy, M. B. McElroy, C. E. Kolb, and W. A. Kaplan. "Aquatic Sources and Sinks for Nitrous Oxide." *Nature* 275 (1978): 602–06.

Elkins, J. W., S. C. Wofsy, M. B. McElroy, and W. A. Kaplan. "Nitrification and Production of N_2O in the Potomac: Evidence for Variability." In *Estuaries and Nutrients*. Ed. B. J. Neilson and L. E. Cronin. Totawa, N.J.: Humana Press, 1980.

Elkins, J. W., T. M. Thompson, T. H. Swanson, J. H. Butler, B. D. Hall, S. O. Cummings, D. A. Fisher, and A. G. Raffo. "Decrease in the Growth Rates of Atmospheric Chlorofluorocarbons 11 and 12." *Nature* 364 (1993): 780–83.

Elkins, J. W. Private Communication. 1999.

Emanuel, W. R., G. G. Killough, W. M. Post, and H. H. Shugart. "Modeling Terrestrial Ecosystems in the Global Carbon Cycle with Shifts in Carbon Storage Capacity by Land Use Change." *Ecology* 65 (1984): 970–83.

Emiliani, C., "Pleistocene Temperatures." *Journal of Geology* 63 (1955): 538–78.

Ennever, F. K., and M. B. McElroy. "Changes in Atmospheric CO_2: Factors Regulating the Glacial to Interglacial Transition." In "The Carbon Cycle and Atmospheric CO_2: Natural Variation Archean to Present." *Geophysical Monograph* 32 (1985): 154–62.

Etheridge, D. M., L. P. Steele, R. L. Langenfelds, R. J. Francey, J. M. Barnola, and V. I. Morgan. "Natural and Anthropogenic Changes in Atmospheric CO_2 over the Last 1000 Years from Air in Antarctic Ice and Firn." *Journal of Geophysical Research* 101 (1996): 4115–28.

Fahey, D. W., K. K. Kelly, G. V. Ferry, L. R. Poule, J. C. Wilson, D. M. Murphy, M. Loewenstein, and J. R. Chan. "In Situ Measurements of Total Reactive Nitrogen, Total Water, and Aerosol in a Polar Stratospheric Cloud in the Antarctic." *Journal of Geophysical Research* 94 (1989): 11299–315.

Fahey, D. W., K. K. Kelly, S. R. Kawa, A. F. Tuck, M. Loewenstein, K. R. Chan, and L. E. Heidt. "Observations of Denitrification and Dehydration in the Winter Polar Stratosphere." *Nature* 344 (1990): 321–24.

Fairbanks, R. G. "A 17,000-Year Glacio-Eustatic Sea Level Record: Influence of Glacial Melting Rates on the Younger Dryas Event and Deep-Ocean Circulation." *Nature* 342 (1989): 637–42.

Fairbridge, R. W. "Little Ice Age." In *The Encyclopedia of Climatology*. Ed. J. E. Oliver and R. W. Fairbridge. 547–51. New York: Van Nostrand Reinhold, 1987.

Fan, S., M. G. Gloor, J. Mahlman, S. Pacala, J. Sarmiento, T. Takahashi, and P. Tans. *Science* 282 (1998): 442–46.

Farman, J. C., B. G. Gardiner, and J. D. Shanklin. "Large Losses of Total Ozone in Antarctica Reveal Seasonal ClO_x/NO_x Interaction." *Nature* 315 (1985): 207–10.

Farmer, C. B., O. F. Raber, and F. G. O'Callaghan. *Final Report on the First Flight of the ATMOS Instrument During the Spacelab 3 Mission, April 29 through May 6, 1985*. JPL Publication 87–32. Pasadena, Calif.: Jet Propulsion Lab, 1985.

Farrell, B. F. "Equable Climate Dynamics." *Journal of Atmospheric Sciences* 47 (1990): 2986–95.

Ferrel, W. "An Essay on the Winds and the Currents of the Ocean." *Nashville Journal of Medicine and Surgery* 2 (1856): 287–301.

Freeman, K. H., and J. M. Hayes. "Fractionation of Carbon Isotopes by Phytoplankton and Estimates of Ancient CO_2 Levels." *Global Biogeochemical Cycles* 6 (1992): 185–98.

Froelich, P. N., M. L. Bender, N. A. Luedtke, G. R. Heath, and T. DeVries. "The Marine Phosphorus Cycle." *American Journal of Science* 282 (1982): 474–511.

Funder, S., N. Abrahamsen, O. Bennike, and R. W. Feyling-Hanssen. "Forested Arctic: Evidence from North Greenland." *Geology* 13 (1985): 542–46.

Fung, I., J. John, J. Lerner, E. Matthews, M. Prather, L. P. Steele, and P. J. Fraser. "Three-Dimensional Model Synthesis of the Global Methane Cycle." *Journal of Geophysical Research* 96 (1991): 13033–65.

Galloway, J. N., G. E. Likens, W. C. Keene, and J. M. Miller. "The Composition of Precipitation in Remote Areas of the World." *Journal of Geophysical Research* 89 (1982): 1447–58.

Gates, W. L. "The Numerical Simulation of Ice-Age Climate With a Global General Circulation Model." *Journal of Atmospheric Sciences* 33 (1976): 1844–73.

Goody, R.M., and J.C.G. Walker. *Atmospheres*. Foundations of Earth Series. (Englewood Cliffs, N.J.: Prentice-Hall, 1972).

Goreau, T. J., W. A. Kaplan, S. C. Wofsy, M. B. McElroy, F. W. Valois, and S. W. Watson. "Production of NO_2^- and N_2O by Nitrifying Bacteria at Reduced Concentrations of Oxygen." *Applied Environmental Microbiology* 40 (1980): 532–62.

Gorham, E. "Northern Peatlands: Role in the Carbon Cycle and Probable Responses to Climatic Warming." *Ecological Applications* 1 (1991): 182–95.

Goulden, M. L., S. C. Wofsy, J. W. Harden, S. E. Trumbore, P. M. Crill, S. T. Gower, T. Fries, B. C. Daube, S. M. Fan, D. J. Sutton, A. Bazzaz, and J. W. Munger. "Sensitivity of Boreal Forest Carbon Balance to Soil Thaw." *Science* 279 (1998): 214–17.

Goulden, M. L., J. W. Munger, S. M. Fan, B. C. Daube, and S. C. Wofsy. "Exchange of Carbon Dioxide by a Deciduous Forest: Response to Interannual Climate Variability." *Science* 271 (1996): 1576–78.

Gross, M. G. *Oceanography*. Englewood Cliffs, N.J.: Prentice Hall, 1990.

Gross, M.G., and E. Gross. *Oceanography: A View of the Earth.* Seventh edition. Englewood Cliffs, N.J. Prentice Hall, 1996.

Guilderson, T.P., R.G. Fairbanks, and J.L. Rubenstone. "Tropical Temperature Variations Since 20,000 Years Ago: Modulating Interhemispheric Climate Change." *Science* 263 (1994): 663–65.

Halley, E. "An Historical Account of the Trade-Winds and Monsoons Observable in the Seas Between and Near the Tropicks With an Attempt to Assign the Physical Cause of Said Winds." *Philosophical Transactions* 26 (1686): 153–68.

Hanel, R.A., B. Schlachman, D. Rugers, and D. Vanous. "Nimbus 4 Michelson Interferometer." *Applied Optics* 10 (1971): 1376–82.

Hasselman, K., K.R. Sausen, E. Maier-Reimer, and R. Voss. "On the Cold Start Problem in Transient Simulations With Coupled Ocean-Atmosphere Models." *Climate Dynamics* 9 (1993): 53–61.

Hastenrath, S. *The Glaciation of the Equadorial Andes.* Rotterdam: Balkhema, 1981.

Hausman, E.D. "The Reorganization of the Global Carbon Cycle at the Last Glacial Termination." Ph.D. diss., Earth and Planetary Sciences, Harvard University, 1997.

Hausman, E.D., and M.B. McElroy. "Role of Sea-Surface Temperature and Ocean Circulation Changes in the Reorganization of the Global Carbon Cycle at the Last Glacial Termination." *Global Biogeochemical Cycles* 13 (1999): 371–81.

Hays, J.D., J. Imbrie, and N.J. Shackleton. "Variations in the Earth's Orbit: Pacemaker of the Ice Ages." *Science* 194 (1976): 1121–32.

Hedin, L.O., and G.E. Likens. "Atmospheric Dust and Acid Rain." *Scientific American.* December 1996: 80–92.

Hein, R., P.J. Crutzen, and M. Heimann. "An Inverse Modeling Approach to Investigate the Global Methane Cycle." *Global Biogeochemical Cycles* 11 (1997): 43–76.

Henderson-Sellars, A. "North American Total Cloud Amount Variations This Century." *Palaeogeography, Palaeoclimatology, Palaeoecology* 75 (1989): 175–94.

Heusser, C.J., and J. Rabassa, "Cold Climatic Episode of Younger Dryas Age in Tierra del Fuego." *Nature* 328 (1989): 609–11.

Hofmann, D.J., J.W. Harder, S.R. Rolf, and J.M. Rusen. "Balloon-borne Observations of the Development and Vertical Structure of the Antarctic Ozone Hole in 1986." *Nature* 326 (1987): 59–62.

Hoffman, P.F., A.J. Kaufman, G.P. Halverson, and D.P. Schrag. "A Neoproterozoic Snowball Earth." *Science* 281 (1998): 1342–46.

Holland, H.D. *The Chemistry of the Atmosphere and Oceans.* N.Y.: John Wiley and Sons, 1978.

Holton, J.R., P.H. Haynes, A.R. McIntyre, A.R. Douglas, R.B. Rood, and L. Pfister. "Stratosphere-Troposphere Exchange." *Reviews of Geophysics* 33 (1995): 403–39.

Holton, J.R. "The Role of Gravity Wave Induced Drag and Diffusion in the Momentum Budget of the Mesosphere." *Journal of Atmospheric Sciences* 39 (1982): 791–99.

Hsiung, J. "Mean Surface Energy Fluxes Over the Global Ocean." *Journal of Geophysical Research* 91 (1986): 10585–606.

Hyde, W.T., and W.R. Peltier. "Sensitivity Experiments with a Model of the Ice Age Cycle: The Response to Harmonic Forcing." *Journal of Atmospheric Sciences* 42 (1985): 2170–88.

Hyde, W.T., and W.R. Peltier. "Sensitivity Experiments with a Model of the Ice Age Cycle: The Response to Milankovitch Forcing." *Journal of Atmospheric Sciences* 44 (1987): 1351–74.

Imbrie, J., J.D. Hays, D.G. Martinson, A. McIntyre, A.C. Mix, J.J. Morley, N.G. Pisias, W.L. Prell, and N.J. Shackleton. "The Orbital Theory of Pleistocene Climate: Support from a Revised Chronology of the Marine $\delta^{18}O$ Record." *Milankovitch and Climate, Part 1.* Ed. A.L. Berger et al. Dordrecht, Holland: D. Reidel, 1984. 269–305.

Imbrie, J., and K.P. Imbrie. *Ice Ages: Solving the Mysteries.* Cambridge, Mass.: Harvard University Press, 1986.

Independent, 23 December 1999.

IPCC. *Climate Change: The IPCC Scientific Assessment: Report Prepared by Working Group I.* Cambridge, U.K.: Cambridge University Press, 1990.

IPCC. *Climate Change 1992: The Supplementary Report to the IPCC Scientific Assessment.* Cambridge, U.K.: Cambridge University Press, 1992.

IPCC. *Climate Change 1995: The Science of Climate Change.* Contribution of Working Group I to the Second Assessment Report of the Intergovernmental Panel on Climate Change. New York: Cambridge University Press, 1996.

Jacob, D.J. *Introduction to Atmospheric Chemistry.* Princeton, N.J.: Princeton University Press, 1999.

Johnsen, S.J., H.B. Clausen, W. Dansgaard, K. Fuhrer, N. Gundestrub, C.U. Hammer, P. Iversen, J. Jouzel, B. Stauffer, and J.P. Steffensen. "Irregular Glacial Interstadials Recorded in a New Greenland Ice Core." *Nature* 359 (1992): 311–13.

Johnston, H.F. "Reduction of Stratospheric Ozone by Nitrogen Oxide Catalysts from SST Exhaust." *Science* 173 (1971): 517–22.

Jouzel, J., C. Lorius, J.R. Petit, C. Genthur, N.I. Barkov, V.M. Kotlyakov, and V.M. Petrov. "Vostok Ice Core: A Continuous Isotope Temperature Record Over the Last Climatic Cycle (160,000 Years)." *Nature* 329 (1987): 403–08.

JPL. *Chemical Kinetics and Photochemical Data for Use in Stratospheric Modeling, Evaluation Number II.* JPL Publication 94–26. Pasadena, Calif.: Jet Propulsion Lab, 1994.

Kallel, N., L.D. Labeyrie, M. Arnold, H. Okada, W.C. Dudley, and J.C. Duplessy. "Evidence of Cooling During the Younger Dryas in the Western North Pacific." *Oceanologica Acta* 11 (1988): 369–75.

Karl, T.R., R.W. Knight, and B. Baker. "The Record Breaking Global Temperatures of 1997 and 1998: Evidence for an Increase in the Rate of Global Warming?" *Geophysical Research Letters* 27 (2000): 719–22.

Katz, M. E., D. K. Pak, G. R. Dickens, and K. G. Miller. "The Source and Fate of Massive Carbon Input During the Latest Paleocene Thermal Maximum." *Science* 286 (1999): 1531–33.

Keeling, C. D., R. B. Bacastow, A. F. Carter, S. C. Piper, T. P. Whorf, M. Heimann, W. G. Mook, and H. Roeloffzen. "A Three-Dimensional Model of Atmospheric CO_2 Transport Based on Observed Winds: 1. Analysis of Observational Data." In *Aspects of Climate Variability in the Pacific and the Western Americas.* Ed. D. H. Peterson. Washington: American Geophysical Union, 1989. 165–236.

Keeling, C. D., T. P. Whorf, M. Whaler, and J. van der Plicht. "Interannual Extremes in the Rate of Rise of Atmospheric Carbon Dioxide Since 1980." *Nature* 375 (1995): 666–70.

Keeling, R. F., S. C. Piper, and M. Heimann. "Global and Hemispheric CO_2 Sinks Deduced From Changes in Atmospheric O_2 Concentration." *Nature* 381 (1996): 218–21.

Kennett, J. P., K. Elmstrom, and N. K. Penrose. "The Last Deglaciation in Ocra Basin, Gulf of Mexico: High-Resolution Planktonic Foraminifera Changes." *Palaeogeography, Palaeoclimatology, Palaeoecology* 50 (1985): 189–216.

Khalil, M. A. K., and M. J. Shearer. "Sources of Methane: An Overview." In *Atmospheric Methane: Sources, Sinks, and Role in Global Change.* Ed. M. A. K. Khalil. NATO ASI Series Volume 13. Berlin, N.Y.: Springer, 1993.

Kirschvink, J. L. "Late Proterozoic Low-Latitude Global Glaciation: The Snowball Earth." In *The Proterozoic Biosphere.* Ed. J. W. Schopf and C. Klein. Cambridge, U.K.: Cambridge University Press, 1992.

Klassen, G. *Part I of Report on Project 64.19.23.01.* Laxenburg, Austria: International Institute for Applied Systems Analysis, 1991.

Kolla, V., A. W. H. Be, and P. E. Biscayne. "Calcium Carbonate Distribution in the Surface Sediments of the Indian Ocean." *Journal of Geophysical Research* 81 (1976): 2605–16.

Kudrass, H. R., E. Erlenkeuser, R. Vollbrecht, and W. Weiss. "Global Nature of the Younger Dryas Cooling Event Inferred from Oxygen Isotope Data from Sulu Sea Cores." *Nature* 349 (1991): 406–09.

LaMarche, V. C. Jr. "Potential of Tree Rings for Reconstruction of Past Climatic Variations in the Southern Hemisphere." In *Proceedings of WMO/IMAP Symposium on Long-Term Climatic Fluctuations.* 21–30. Geneva: World Meteorological Organization, 1975.

Lamb, H. H. "Some Studies of the Little Ice Age of Recent Centuries and Its Great Storms." In *Climatic Change on a Yearly and Millenial Basis.* Ed. N. A. Moerner and W. Karlen. Dordrecht, Holland: D. Reidel, 1984. 309–29.

Lamb, B., A. Guenther, D. Gay, and H. Westberg. "A National Inventory of Biogenic Hydrocarbon Emissions." *Atmospheric Environment* 21 (1987): 1695–705.

Leuenberger, M., and U. Siegenthaler. "Nitrous Oxide Contents of the Atmosphere Over the Past 40,000 Years as Reconstructed From Ice Cores." *Nature* 360 (1992): 449–51.

Lindgren, S., and J. Neumann. "The Cold and Wet Year 1695: A Contemporary German Account." *Climatic Change* 3 (1981): 173–87.

Lindzen, R. S. "Some Coolness Concerning Global Warming." *Bulletin of the American Meteorological Society* 71 (1990): 288–99.

Logan, J. A., M. J. Prather, S. C. Wofsy, and M. B. McElroy. "Tropospheric Chemistry: A Global Perspective." *Journal of Geophysical Research* 86 (1981): 7210–54.

Ludlum, D. *Early American Winters: 1604–1820.* Boston: American Meteorological Society, 1966.

Manabe, S., and F. Moller. "On Radiative Equilibrium and Heat Balance of the Atmosphere." *Monthly Weather Review* 89 (1961): 503.

Manabe, S., and R. F. Strickler. "Thermal Equilibrium of the Atmosphere With a Convective Adjustment." *Journal of Atmospheric Sciences* 21 (1964): 361–85.

Manabe, S. "Climate and the Ocean Circulation: I. The Atmospheric Circulation and the Hydrology of the Earth's Surface." *Monthly Weather Reviews* 7 (1969): 739–74.

Masarie, K. A., and P. P. Tans. "Extension and Integration of Atmospheric Carbon Dioxide Data into a Globally Consistent Measurement Record." *Journal of Geophysical Research* 100 (1995): 593–610.

Matthews, E. "Global Vegetation and Land Use: New High-Resolution Data Bases for Climate Studies." *Journal of Climate and Applied Meteorology* 22 (1983): 474–87.

McElroy, M. B., "Chemical Processes in the Solar System: A Kinetic Perspective," In *International Review of Science.* Physical Chemistry Series Two. Consultant editor A. D. Buckingham, volume editor D. R. Herschback (1976): 127–211.

McElroy, M. B. "Climate of the Earth: An Overview." *Environmental Pollution* 83 (1994): 3–21.

McElroy, M. B., and D. B. A. Jones. "Evidence for an Additional Source for Atmospheric N_2O." *Global Biogeochemical Cycles* 10 (1996): 651–59.

McElroy, M. B., R. J. Salawitch, and K. Minschwaner. "The Changing Stratosphere," *Planet. Space Science* 40 (1992): 373–401.

McElroy, M. B., R. J. Salawitch, S. C. Wofsy, and J. A. Logan. "Reductions of Antarctic Ozone Due to Synergistic Interactions of Chlorine and Bromine." *Nature* 321 (1986): 759–62.

McKenna, M. "Eocene Paleolatitude, Climate, and Mammals of Ellesmere Island." *Palaeogeography, Palaeoclimatology, Palaeoecology* 30 (1980): 349–62.

Milankovitch, M. *Canon of Isolation and the Ice Age Problem.* RSA Special Publication 133. Trans. Israel Program for Scientific Translation, Jerusalem, 1969. Belgrade: Royal Serbian Academy, 1941.

Mitchell, J. F. B., T. J. Johns, J. M. Gregory, and S. B. F. Tett. "Climate Response to Increasing Levels of Greenhouse Gases and Aerosols." *Nature* 376 (1995): 501–04.

Molina, L. T., and M. J. Molina. "Production of Cl_2O_2 from the Self-Reaction of the ClO Radical." *Journal of Physical Chemistry* 91 (1987): 433–36.

Molina, M. I., and F. S. Rowland. "Stratospheric Sink for Chlorofluorocarbons: Chlorine Atom Catalyzed Destruction of Ozone." *Nature* 249 (1974): 810–14.

Montzka, S. A., J. H. Butler, J. W. Elkins, T. M. Thompson, A. D. Clarke, and L. T. Lock. "Present and Future Trends in the Atmospheric Burden of Ozone-Depleting Halogens." *Nature* 398 (1999): 690–94.

Munger, J. W., D. J. Jacob, J. M. Waldman, and M. R. Hoffmann. "Fogwater Chemistry in an Urban Atmosphere." *Journal of Geophysical Research* 88 (1983): 5109–21.

National Research Council, Committee on Nitrate Accumulation. *Accumulation of Nitrate.* Washington: National Academy of Sciences, 1972.

Nesje, A., and M. Kvamme. "Holocene Glacier and Climate Variations in Western Norway: Evidence for Early Holocene Glacier Demise and Multiple Neoglacial Events." *Geology* 19 (1991): 610–12.

Nevison, C. D., R. F. Weiss, and D. J. Erickson III. "Global Oceanic Emissions of Nitrous Oxide." *Journal of Geophysical Research* 100 (1995): 15809–20.

Newman, P. A., and M. R. Schoeberl. "Horizontal Mixing Coefficients for Two-Dimensional Models Calculated from National Meteorological Center Data." *Journal of Geophysical Research* 91 (1986): 7919–24.

Newman, P. A., M. R. Schoeberl, R. A. Plumb, and J. E. Rosenfield. "Mixing Rates Calculated from Potential Vorticity." *Journal of Geophysical Research* 93 (1988): 5221–40.

Norris, R. D., and U. Rohl. "Carbon Cycling and Chronology of Climate Warming During the Palaeocene/Eocene Transition." *Nature* 401 (1999): 775–78.

Norton, F. L. "Regulation of Tropical Climate by Radiation, Ocean Circulation and Atmospheric Dynamics." Ph.D. diss., Earth and Planetary Sciences, Harvard University, 1995.

Oerlemans, J. "Model Experiments on the 100,000-yr Glacial Cycle." *Nature* 287 (1980): 430–32.

Pagani, M., K. H. Freeman, and M. A. Arthur. "Late Miocene Atmosphere CO_2 Concentrations and the Expansion of C_4 Grasses." *Science* 285 (1999): 876–79.

Pasachoff, J. A. *Astronomy: From the Earth to the Universe.* Philadelphia: Saunders College Publishing, 1979.

Pastouret, L., H. Chamley, G. Delibrias, J. C. Duplessy, and J. Thiede. "Late Quaternary Climatic Changes in Western Tropical Africa Deduced from Deep-Sea Sedimentation Off the Niger Delta." *Oceanologica Acta* 1 (1978): 217–32.

Peixoto, J. P., and A. H. Oort. *Physics of Climate.* New York: American Institute of Physics, 1992.

Peltier, W. R., and W. T. Hyde. "A Model of the Ice Age Cycle." In *Milankovitch and Climate, Part 2.* Ed. A. L. Berger et al. 565–80. Dordrecht, Holland: D. Reidel, 1984.

Petterson, O. "The Connection Between Hydrographical and Meterological Phenomena." *Royal Meteorological Society Quarterly Journal* 38 (1912): 173–91.

Pickard, G. L., and W. J. Emery. *Descriptive Physical Oceanography: An Introduction: Fifth Enlarged Edition (in SI Units).* Oxford, N.Y.: Pergamon Press, 1990.

Pollard, D. "A Coupled Climate-Ice Sheet Model Applied to the Quaternary Ice Ages." *Journal of Geophysical Research* 88 (1983a): 7705–18.

Pollard, D. "A Simple Ice Sheet Model Yields Realistic 100 kyr Glacial Cycles." *Nature* 296 (1982): 334–38.

Pollard, D. "Ice-Age Simulations with a Calving Ice-Sheet Model." *Quaternary Research* 20 (1983b): 30–48.

Pollard, D. "Some Ice-Age Aspects of a Calving Ice-Sheet Model." In *Milankovitch and Climate, Part 2.* Ed. A. L. Berger et al. 541–62. Dordrecht, Holland: D. Reidel, 1984.

Pollard, D., A. D. Ingersoll, and J. G. Lockwood. "Response of a Zonal Climate-Ice Sheet Model to the Orbital Perturbations During the Quaternary Ice Ages." *Tellus* 32 (1980): 301–19.

Popp, B. N., R. Takigiku, J. M. Hayes, J. W. Louda, and E. W. Baker. "The Post-Paleozoic Chronology and Mechanism of ^{13}C Depletion in Primary Source Organic Matter." *American Journal of Science* 289 (1989): 27–47.

Porter, S. C. "Equilibrium-Line Altitudes of Late Quaternary Glaciers in the Southern Alps, New Zealand." *Quaternary Research* 5 (1975): 27–47.

Porter, S. C. "Glaciological Evidence of Holocene Climatic Change." In *Climate and History: Studies of Past Climates and Their Impact on Man.* Ed. T. M. L. Wigley, M. J. Ingram, and G. Farmer. 82–110. Cambridge, U.K.: Cambridge University Press, 1981.

Prinn, R. G., R. F. Weiss, B. R. Miller, J. Huang, F. N. Alyea, D. M. Cunnold, P. J. Fraser, D. E. Hartley, and P. G. Simmonds. "Atmospheric Trends and Lifetime of CH_3CCl_3 and Global OH Concentrations." *Science* 269 (1995): 187–92.

Quay, P. D., S. L. King, J. Stutsman, D. O. Wilbur, L. P. Steele, I. Fung, R. H. Crammon, T. A. Brown, G. W. Farwell, P. M. Grootes, and F. H. Schmidt. "Carbon Isotopic Composition of Atmospheric CH_4: Fossil and Biomass Burning Source Strengths." *Global Biogeochemical Cycles* 5 (1991): 25–47.

Quay, P. D., M. Stuiver, and W. S. Broecker. "Upwelling Rates for the Equatorial Pacific Ocean Derived from the Bomb ^{14}C Distribution." *Journal of Marine Research* 41 (1983): 769–92.

Rau, G. H., T. Takahashi, and D. J. Des Marais. "Latitudinal Variations in Plankton $\delta^{13}C$: Implications for CO_2 and Productivity in Past Oceans." *Nature* 341 (1989): 516–18.

Raymo, M. E., W. F. Ruddiman, and P. N. Froelich. "Influence of Late Cenozoic Mountain Building on Ocean Geochemical Cycles." *Geology* 16 (1988): 649–53.

Rind, D., and D. Peteet. "Terrestrial Conditions at the Last Glacial Maximum and CLIMAP Sea-Surface Temperature Estimates: Are They Consistent?" *Quaternary Research* 24 (1985): 1–22.

Rosenlof, K. H., and J. R. Holton. "Estimates of the Stratospheric Residual Circulation Using the Downward Control Principle." *Journal of Geophysical Research* 98 (1993): 10465–79.

Salawitch, R. J., S. C. Wofsy, P. O. Wennberg, R. C. Cohen, J. G. Anderson, D. W. Fahey, R. S. Gao, E. R. Keim, E. L. Woodbridge, R. M. Stimpfle, J. P. Koplow, D. W. Kohn, C. R. Webster, R. D. May, L. Pfister, E. W. Gottlieb, H. A. Michelsen, G. K. Yue, J. C. Wilson, C. A. Brock, H. H. Jonsson, J. E. Dye, D. Baumgardner, M. H. Proffitt, M. Loewenstein, J. R. Podolske, J. W. Elkins, G. S. Dutton, E. J. Hintsa, A. E. Dessler, E. M. Weinstock, K. K. Kelly, K. A. Boering, B. C. Daube, K. R. Chan, and S. W. Bowen. "The Distribution of Hydrogen, Nitrogen, and Chlorine Radicals in the Lower Stratosphere: Implications for Changes in O_3 Due to Emission of NO_y from Supersonic Aircraft." *Geophysical Research Letters* 21 (1994): 2547–50.

Schneider, H. R., M. K. W. Ko, R. Shia, and N. Sze. "A Two-Dimensional Model With Coupled Dynamics, Radiative Transfer, and Photochemistry. 2. Assessment of the Response of Stratospheric Ozone to Increased Levels of CO_2, N_2O, CH_4 and CFC." *Journal of Geophysical Research* 98 (1993): 20441–49.

Schreiner, J., C. Voigt, A. Kohlmann, F. Arnold, K. Mauersberger, and N. Larson. "Chemical Analysis of Polar Stratospheric Cloud Particles." *Science* 283 (1999): 968–70.

Schweitzer, H. J. "Environment and Climate in the Early Tertiary of Spitsbergen." *Palaeogeography, Palaeoclimatology, Palaeoecology* 30 (1980): 297–311.

Seinfeld, J. H., and S. N. Pandis. *Atmospheric Chemistry and Physics.* New York: John Wiley and Sons, 1997.

Sellers, P. J. "Biophysical Models of Land Surface Processes." In *Climate System Modeling.* Ed. K. E. Trenberth. Cambridge, U.K.: Cambridge University Press, 1992.

Sellers, W. D. *Physical Climatology.* Chicago: University of Chicago Press, 1965.

Severinghaus, J. P., T. Sowers, E. J. Brook, R. B. Alley, and M. L. Bender. "Timing of Abrupt Climate Change at the End of the Younger Dryas Interval from Thermally Fractionated Gases in Polar Ice." *Nature* 391 (1998): 141–46.

Shackleton, N. J. "Oxygen Isotope Analyses and Pleistocene Temperatures Re-assessed." *Nature* 215 (1967): 15–17.

Shackleton, N. J., and N. D. Opdyke. "Oxygen Isotope and Paleomagnetic Stratigraphy of Equatorial Pacific Core V28–238: Oxygen Isotope Temperatures and Ice Volumes on a 10^5-Year and 10^6-Year Scale." *Quaternary Research* 3 (1973): 39–55.

Shindell, D. T., D. Rind, and P. Lonergan. "Increased Polar Stratospheric Ozone Losses and Delayed Eventual Recovery Due to Increasing Greenhouse Gas Concentrations." *Nature* 392 (1998): 589–92.

Sillman, S., J. A. Logan, and S. C. Wofsy. "The Sensitivity of Ozone to Nitrogen Oxides and Hydrocarbons in Regional Ozone Episodes." *Journal of Geophysical Research* 95 (1990): 1837–51.

Singh, H. B., D. Ohara, D. Herdth, W. Sachse, D. R. Blake, J. D. Bradshaw, M. Kanakidon, and P. J. Crutzen. "Acetone in the Atmosphere—Distribution, Sources, and Sinks." *Journal of Geophysical Research* 99 (1994): 1805–19.

Sisterson, D. L. *NAPAD State of the Science Report 6.* Washington, D.C.: National Acid Precipitation Assessment Program, 1990.

Skinner, B. J., and S. C. Porter. *The Blue Planet: An Introduction to Earth System Science.* New York: John Wiley & Sons, 1995.

Sloan, L. C., J. C. G. Walker, T. C. Moore, Jr., D. R. Rea, and J. C. Zachos. "Possible Methane-Induced Polar Warming in the Early Eocene." *Nature* 357 (1992): 320–22.

Solomon, S., R. R. Garcia, F. S. Rowland, and D. J. Wuebbles. "On the Depletion of Antarctic Ozone." *Nature* 321 (1986): 755–58.

Spivakovsky, C. M., J. A. Logan, S. A. Motzka, Y. J. Balkanski, M. Foreman-Fowler, D. B. A. Jones, L. W. Horowitz, A. C. Fusco, C. A. M. Brenninkmeijer, M. J. Prather, S. C. Wofsy, and M. B. McElroy. "Three-dimensional Climatological Distribution of Tropospheric OH: Update and Evaluation." *Journal of Geophysical Research* 105 (2000): 8391–980.

Stevens, C. M., and A. Engelkemeir. "Stable Carbon Isotopic Composition of Methane from Some Natural and Anthropogenic Sources." *Journal of Geophysical Research* 93 (1988): 725–33.

Stommel, H. "The Abyssal Circulation." *Deep Sea Research* 5 (1958): 80–82.

Stommel, H., and A. B. Arons. "On the Abyssal Circulation of the World Ocean: An Idealized Model of the Circulation Pattern and Amplitude in Oceanic Basins." *Deep Sea Research* 6 (1960): 217–33.

Stone, P. H., and J. S. Risbey. "On the Limitations of General Circulation Models." *Geophysical Research Letters* 17 (1990): 2173–2176.

Stuiver, M., P. D. Quay, and H. G. Ostlund. "Abyssal ^{14}C and the Age of the World Oceans." *Science* 219 (1983): 849-51.

Thompson, L. G., M. E. Davis, E. Mosley-Thompson, T. A. Sowers, K. A. Henderson, V. S. Zagurodnov, P. N. Lin, V. N. Mikhalenko, R. K. Campen, J. F. Bolzan, J. Cole-Dai, and B. Brancou. "A 25,000-Year Tropical Climate History from Bolivian Ice Cores." *Science* 282 (1998): 1858–1864.

Thompson, L. G., E. Mosley-Thompson, and B. M. Arnao. "El Niño-Southern Oscillation Events Recorded in the Stratigraphy of the Tropical Quelccaya Ice Cap, Peru." *Science* 226 (1984): 50–53.

Thompson, L. G., E. Mosley-Thompson, J. F. Bolzan, and B. R. Koci. "A 1500-Year Record of Tropical Precipitation in Ice Cores from the Quelccaya Ice Cap, Peru." *Science* 229 (1985): 971–73.

Tolmazin, D. *Elements of Dynamic Oceanography.* Winchester, Mass.: Allen and Unwin, 1985.

Toon, O. B., P. Hamill, R. P. Turco, and J. Pinto. "Condensation of HNO_3 and HCl in the Winter Polar Stratosphere." *Geophysical Research Letters* 83 (1986): 5501–04.

Tung, K. K., M. K. W. Ko, J. M. Rodriguez, and N. D. Sze. "Are Antarctic Ozone Variations a Manifestation of Dynamics or Chemistry?" *Nature* 322 (1986): 811–14.

Turco, R. P. *Earth Under Siege: From Air Pollution to Global Change.* Oxford: Oxford University Press, 1997.

Vaccaro, R. F. *Chemical Oceanography,* Volume 1. Ed. J. P. Riley and G. Skirrow. New York: Academic Press, 1965.

Vakhrameev, V. A. "Main Features of Phytogeography of the Globe in Jurassic and Early Cretaceous Time." *Paleontological Journal* 2 (1975): 123–33.

Volk, C.M., J.W. Elkins, D.W. Fahey, G.S. Dutton, J.M. Gilligan, M. Loewenstein, J.R. Podolske, K.R. Chan, and M.R. Gunson. "Evaluation of Source Gas Lifetimes from Stratospheric Observations." *Journal of Geophysical Research* 102 (1997): 102, 25, 543–25, 564.

Wahlen, M., N. Tanaka, R. Henry, B. Deck, J. Zegler, J. S. Vogel, J. Southon, A. Shemish, R. Fairbanks, and W. Broecker. "Carbon-14 in Methane Sources and in Atmospheric Methane: The Contribution from Fossil Carbon." *Science* 245 (1989): 286–90.

Waibel, A. E., T. Peter, K. S. Carslaw, H. Oelhaf, G. Wetzel, P. J. Crutzen, U. Poschl, A. Tsias, E. Reimer, and H. Fischer. "Arctic Ozone Loss Due to Denitrification." *Science* 283 (1999): 2064–69.

Wang, S. C., and Z. C. Zhao. "Droughts and Floods in China, 1470–1979." In *Climate and History: Studies of Past Climates and Their Impact on Man.* Ed. T. M. L. Wigley, M. J. Ingram, and G. Farmer. Cambridge, U.K.: Cambridge University Press, 1981. 271–88.

Wang, Y. H., and D. J. Jacob. "Anthropogenic Forcing on Tropospheric Ozone and OH Since Preindustrial Times." *Journal of Geophysical Research* 103 (1998): 31123–35.

Wang, Y. H., D. J. Jacob, and J. A. Logan. "Global Simulation of Tropospheric O_3-NO_x-Hydrocarbon Chemistry. 1. Model Formulation." *Journal of Geophysical Research* 103 (1998a): 10713–26.

Wang, Y. H., J. A. Logan, and D. J. Jacob. "Global Simulation of Tropospheric O_3-NO_x-Hydrocarbon Chemistry. 2. Model Evaluation and Global Ozone Budget." *Journal of Geophysical Research* 103 (1998b): 10727–56.

Wang, Y. H., D. J. Jacob, and J. A. Logan. "Global Simulation of Tropospheric O_3-NO_x-Hydrocarbon Chemistry. 3. Origin of Tropospheric Ozone and Effects of Non-Methane Hydrocarbons." *Journal of Geophysical Research* 103 (1998c): 10757–68.

Waters, J. W., R. A. Stachnik, J. C. Hardy, and R. F. Jarnot. "ClO and O_3 Stratospheric Profiles—Balloon Microwave Measurements." *Geophysical Research Letters* 15 (1988): 780–83.

Weertman, J. "Milankovitch Solar Radiation Variations and Ice-Age Ice-Sheet Sizes." *Nature* 261 (1976): 17–20.

Weikinn, C. "Katastrophale Duerrejahre während des Zeitraums 1500–1850." *Acta Hydrophysica (Berlin)* 10 (1965): 33–54.

Willson, R. C. "Measurement of Solar Irradiance and its Variability." *Space Science Reviews* 38 (1984): 203.

Wofsy, S. C., M. B. McElroy, and Y. L. Yung. "The Chemistry of Atmospheric Bromine." *Geophysical Research Letters* 2 (1975): 215.

Wolin, M. J. "The Rumen Fermentation: A Model for Microbial Interactions in Anaerobic Ecosystems." In *Advances in Microbial Ecology,* vol. 3. Ed. M. Alexander, 49–77. N.Y.: Plenum, 1979.

World Meteorological Organization. *Atmospheric Ozone 1985.* Report No. 16, Chapter 11. Geneva: World Meteorological Organization, 1986.

World Meteorological Organization. *Scientific Assessment of Ozone Depletion, 1994.* Global Ozone Research and Monitoring Project–Report No. 37. Geneva: World Meteorological Organization, 1995.

World Meteorological Organization. Scientific *Assessment of Ozone Depletion, 1998.* World Meteorological Organization, Global Ozone Research and Monitoring Project–Report No. 44. Washington: National Oceanic and Atmospheric Administration, 1999.

Yamamoto, T. "On the Nature of the Climatic Change in Japan Since the 'Little Ice Age' Around 1800 A.D." In *Japanese Progress on Climatology.* (1972): 97–110.

Yoshino, M. M., and S. Xie. "A Preliminary Study on Climatic Anomalies in East Asia and Sea Surface Temperatures in the North Pacific." *Tsukuba University Institute of Geoscience Science Reports* A4 (1983): 1–23.

Yung, Y. L., J. P. Pinto, R. T. Watson, and S. P. Sander. "Atmospheric Bromine and Ozone Perturbations in the Lower Stratosphere." *Journal of Atmospheric Sciences* 37 (1980): 339–53.

Zander, R., M. R. Gunson, C. B. Farmer, C. P. Rinsland, F. W. Irion, and E. Mahieu. "The 1985 Chlorine and Fluorine Inventories in the Stratosphere Based on the ATMOS Observations at 30° North Latitude." *Journal of Atmospheric Chemistry* 15 (1992): 171–86.

Zeilik, M., and E. V. P. Smith. *Introductory Astronomy and Astrophysics.* New York: Saunders College Publishing, 1987.

Appendix

Constants

	MKS Units	*CGS Units*
Stefan-Boltzmann constant(σ)	5.67×10^{-8} W m^{-2} K^{-4}	5.67×10^{-5} erg sec^{-1} cm^{-2} K^{-4}
Earth's radius	6.38×10^{6} m	6.38×10^{8} cm
Acceleration due to gravity	9.81 m sec^{-2}	981 cm sec^{-2}
Solar radiation incident at top of Earth's atmosphere	1.38×10^{3} W m^{-2}	1.38×10^{6} erg sec^{-1} cm^{-2}
Gas constant for dry air (R) ($P = \rho RT$)	287 J kg^{-1} K^{-1}	2.87×10^{6} erg g^{-1} K^{-1}
Pressure at sea level	1.013×10^{5} N m^{-2}	1.013×10^{6} dyn cm^{-2}
Density of air at 0°C, at sea level	1.275 kg m^{-3}	1.275×10^{-3} g cm^{-3}
Density of pure water at 0°C	10^{3} kg m^{-3}	1.0 g cm^{-3}
Angular velocity of Earth (Ω)	7.3×10^{-5} sec^{-1}	7.3×10^{-5} sec^{-1}
Dry adiabatic lapse rate	9.8 K km^{-1}	
Boltzmann's constant (k)	1.381×10^{-23} J K^{-1}	1.381×10^{-16} erg K^{-1}

Conversions

	MKS Units	*CGS Units*
1 m = 100 cm	K = °C + 273.15	°C = (°F – 32)(5/9)
1 kg = 1000 g	1 bar = 1000 millibars (mb)	1 atm = 1.013 bars
1 mb = 100 N m^{-2}	1 mb = 1000 dyn cm^{-2}	1 Pascal (Pa) = 1 N m^{-2}
1 erg = 1 dyn cm	1 Joule = 1 N m	1 Angstrom (Å) = 10^{-10} m
1 Å = 10^{-4} micron (μ)	1 atm = 76 cm Hg	1 erg = 10^{-7} Joules
1 cm^{3} = 10^{-3} liters	Avogadro's number = 6.02×10^{23}	1 Watt = 1 J sec^{-1}

Units

	MKS Units	CGS Units
Basic Units		
Length	meter (m)	centimeter (cm)
Mass	kilogram (kg)	gram (g)
Time	seconds (sec)	seconds (sec)
Temperature	Kelvin (K)	Kelvin (K)
Composite Units		
Velocity	m sec^{-1}	cm sec^{-1}
Acceleration	m sec^{-2}	cm sec^{-2}
Force	Newtons (N)	dynes
	1 N = 1 kg m sec^{-2}	1 dyn = 1 g cm sec^{-2}
Pressure	N m^{-2}	dyn cm^{-2}
Energy and Work	Joules (J)	ergs
	(work = F × distance)	(Kinetic energy = 1/2 mv^2)
	1 J = 1 kg m^2 sec^{-2}	1 erg = 1 g cm^2 sec^{-2}

Index

A

B

C

D

E

F

G

H

I

J

K

L

M

N

O

P

R

V

W

Y